Material-Integrated Intelligent Systems

Material-Integrated Intelligent Systems

Technology and Applications

Edited by Stefan Bosse, Dirk Lehmhus, Walter Lang, and Matthias Busse

Verlag GmbH & Co. KGaA

Editors

Dr. Stefan Bosse
University of Bremen
Dept. Math. & Comp. Science
and ISIS Sensorial Materials
Scientific Centre
Robert-Hooke-Str. 5
28359 Bremen
Germany

Dr.-Ing. Dirk Lehmhus
University of Bremen
ISIS Sensorial Materials
Scientific Centre
Wiener Straße 12
28359 Bremen
Germany

Prof. Walter Lang
University of Bremen
Institute for Microsensors
-actuators and -systems (IMSAS)
and ISIS Sensorial Materials
Scientific Centre
Otto-Hahn-Allee 1
28359 Bremen
Germany

Prof. Matthias Busse
Fraunhofer Institute for
Manufacturing Technology and
Advanced Materials (IFAM) and
University of Bremen ISIS Sensorial
Materials Scientific Centre
Wiener Straße 12
28359 Bremen
Germany

■ All books published by **Wiley-VCH** are carefully produced. Nevertheless, authors, editors, and publisher do not warrant the information contained in these books, including this book, to be free of errors. Readers are advised to keep in mind that statements, data, illustrations, procedural details or other items may inadvertently be inaccurate.

Library of Congress Card No.: applied for

British Library Cataloguing-in-Publication Data
A catalogue record for this book is available from the British Library.

Bibliographic information published by the Deutsche Nationalbibliothek
The Deutsche Nationalbibliothek lists this publication in the Deutsche Nationalbibliografie; detailed bibliographic data are available on the Internet at http://dnb.d-nb.de.

© 2018 Wiley-VCH Verlag GmbH & Co. KGaA, Boschstr. 12, 69469 Weinheim, Germany

All rights reserved (including those of translation into other languages). No part of this book may be reproduced in any form – by photoprinting, microfilm, or any other means – nor transmitted or translated into a machine language without written permission from the publishers. Registered names, trademarks, etc. used in this book, even when not specifically marked as such, are not to be considered unprotected by law.

Cover Design Formgeber, Mannheim, Germany
Typesetting Thomson Digital, Noida, India
Printing and Binding C.O.S. Printers Pte Ltd Singapore

Print ISBN 978-3-527-33606-7
ePDF ISBN 978-3-527-67925-6
ePub ISBN 978-3-527-67926-3
Mobi ISBN 978-3-527-67923-2
oBook ISBN 978-3-527-67924-9

Printed on acid-free paper

10 9 8 7 6 5 4 3 2 1

Contents

Foreword *XV*
Preface *XIX*

Part One Introduction *1*

1 **On Concepts and Challenges of Realizing Material-Integrated Intelligent Systems** *3*
   Stefan Bosse and Dirk Lehmhus
1.1 Introduction *3*
1.2 System Development Methodologies and Tools (Part Two) *7*
1.3 Sensor Technologies and Material Integration (Part Three and Four) *8*
1.4 Signal and Data Processing (Part Five) *15*
1.5 Networking and Communication (Part Six) *17*
1.6 Energy Supply and Management (Part Seven) *21*
1.7 Applications (Part Eight) *21*
   References *24*

Part Two System Development *29*

2 **Design Methodology for Intelligent Technical Systems** *31*
   Mareen Vaßholz, Roman Dumitrescu, and Jürgen Gausemeier
2.1 From Mechatronics to Intelligent Technical Systems *32*
2.2 Self-Optimizing Systems *36*
2.3 Design Methodology for Intelligent Technical Systems *38*
2.3.1 Domain-Spanning Conceptual Design *41*
2.3.2 Domain-Specific Conceptual Design *50*
   References *51*

3 **Smart Systems Design Methodologies and Tools** *55*
   Nicola Bombieri, Franco Fummi, Giuliana Gangemi, Michelangelo Grosso, Enrico Macii, Massimo Poncino, and Salvatore Rinaudo
3.1 Introduction *55*
3.2 Smart Electronic Systems and Their Design Challenges *56*

3.3	The Smart Systems Codesign before SMAC	*57*
3.4	The SMAC Platform	*60*
3.4.1	The Platform Overview	*61*
3.4.1.1	System C–SystemVue Cosimulation	*61*
3.4.1.2	ADS and the Thermal Simulation	*63*
3.4.1.3	EMPro Extension and ADS Integration	*64*
3.4.1.4	Automated EM – Circuit Cosimulation in ADS	*64*
3.4.1.5	HIF Suite Toolsuite	*65*
3.4.1.6	The MEMS+ Platform	*66*
3.4.2	The (Co)Simulation Levels and the Design–Domains Matrix	*67*
3.5	Case Study: A Sensor Node for Drift-Free Limb Tracking	*69*
3.5.1	System Architecture	*71*
3.5.2	Model Development and System-Level Simulation	*71*
3.5.3	Results	*73*
3.6	Conclusions	*76*
	Acknowledgments	*77*
	References	*77*

Part Three Sensor Technologies *81*

4 Microelectromechanical Systems (MEMS) *83*
Li Yunjia

4.1	Introduction	*83*
4.1.1	What Is MEMS	*83*
4.1.2	Why MEMS	*84*
4.1.3	MEMS Sensors	*84*
4.1.4	Goal of This Chapter	*85*
4.2	Materials	*85*
4.2.1	Silicon	*85*
4.2.2	Dielectrics	*86*
4.2.3	Metals	*87*
4.3	Microfabrication Technologies	*87*
4.3.1	Silicon Wafers	*87*
4.3.2	Lithography	*88*
4.3.3	Etching	*91*
4.3.4	Deposition Techniques	*93*
4.3.5	Other Processes	*94*
4.3.6	Surface and Bulk Micromachining	*95*
4.4	MEMS Sensor	*95*
4.4.1	Resistive Sensors	*95*
4.4.2	Capacitive Sensors	*99*
4.5	Sensor Systems	*103*
	References	*104*

5 Fiber-Optic Sensors *107*
Yi Yang, Kevin Chen, and Nikhil Gupta

5.1	Introduction to Fiber-Optic Sensors	*107*

5.1.1	Sensing Principles	*108*
5.1.2	Types of Optical Fibers	*108*
5.2	Trends in Sensor Fabrication and Miniaturization	*110*
5.3	Fiber-Optic Sensors for Structural Health Monitoring	*112*
5.3.1	Sensors for Cure Monitoring of Composites	*114*
5.3.2	Embedded FOS in Composite Materials	*114*
5.3.3	Surface-Mounted FOS in Composite Materials	*115*
5.3.4	FOS for Structural Monitoring	*115*
5.3.4.1	Aerospace Structures	*115*
5.3.4.2	Civil Structures	*116*
5.3.4.3	Marine Structures	*116*
5.4	Frequency Modulation Sensors	*117*
5.4.1	Bragg Grating Sensors	*117*
5.4.2	Fabry–Pérot Interferometer Sensor	*118*
5.4.3	Whispering Gallery Mode Sensors	*119*
5.5	Intensity Modulation Sensors	*122*
5.5.1	Fiber Microbend Sensors	*122*
5.5.2	Fiber-Optic Loop Sensor	*123*
5.6	Some Challenges in SHM of Composite Materials	*128*
5.7	Summary	*128*
	Acknowledgments	*129*
	References	*129*
6	**Electronics Development for Integration**	*137*
	Jan Vanfleteren	
6.1	Introduction	*137*
6.1.1	Standard Flat Rigid Printed Circuits Boards and Components Assembly	*137*
6.1.2	Flexible Circuits	*138*
6.1.3	Need for Alternative Circuit and Packaging Materials	*140*
6.2	Chip Package Miniaturization Technologies	*140*
6.2.1	Ultrathin Chip Package Technology	*140*
6.2.2	UTCP Circuit Integration	*142*
6.2.2.1	UTCP Embedding	*142*
6.2.2.2	UTCP Stacking	*143*
6.2.3	Applications	*143*
6.3	Elastic Circuits	*145*
6.3.1	Printed Circuit Board-Based Elastic Circuits	*145*
6.3.2	Thin Film Metal-Based Elastic Circuits	*148*
6.3.3	Applications	*148*
6.3.3.1	Wearable Light Therapy	*148*
6.3.4	Stretchable Displays	*149*
6.4	2.5D Rigid Thermoplastic Circuits	*152*
6.5	Large Area Textile-Based Circuits	*153*
6.5.1	Electronic Module Integration Technology	*154*
6.5.2	Applications	*155*
6.6	Conclusions and Outlook	*157*
	References	*157*

Part Four Material Integration Solutions *159*

7 Sensor Integration in Fiber-Reinforced Polymers *161*
Maryam Kahali Moghaddam, Mariugenia Salas, Michael Koerdt, Christian Brauner, Martina Hübner, Dirk Lehmhus, and Walter Lang
7.1 Introduction to Fiber-Reinforced Polymers *161*
7.2 Applications of Integrated Systems in Composites *164*
7.2.1 Production Process Monitoring and Quality Control of Composites *164*
7.2.1.1 Monitoring of the Resin Flow *166*
7.2.1.2 Analytical Modeling of Resin Front by Means of Simulation *166*
7.2.1.3 Monitoring the Resin Curing *166*
7.2.2 In-Service Applications of Integrated Systems *167*
7.2.2.1 Use for Structural Health Monitoring (SHM) *167*
7.2.2.2 Use As Support to Nondestructive Evaluation and Testing (NDE/NDT) *170*
7.3 Fiber-Reinforced Polymer Production and Sensor Integration Processes *170*
7.3.1 Overview of Fiber-Reinforced Polymer Production Processes *170*
7.3.2 Sensor Integration in Fiber-Reinforced Polymers: Selected Case Studies *175*
7.4 Electronics Integration and Data Processing *179*
7.4.1 Materials Integration of Electronics *180*
7.4.2 Electronics for Wireless Sensing *181*
7.5 Examples of Sensors Integrated in Fiber-Reinforced Polymer Composites *183*
7.5.1 Ultrasound Reflection Sensing *183*
7.5.2 Pressure Sensors *184*
7.5.3 Thermocouples *186*
7.5.4 Fiber Optic Sensors *187*
7.5.5 Interdigital Planar Capacitive Sensors *188*
7.6 Conclusion *192*
Acknowledgments *193*
References *193*

8 Integration in Sheet Metal Structures *201*
Welf-Guntram Drossel, Roland Müller, Matthias Nestler, and Sebastian Hensel
8.1 Introduction *201*
8.2 Integration Technology *204*
8.3 Forming of Piezometal Compounds *205*
8.4 Characterization of Functionality *208*
8.5 Fields of Application *211*
8.6 Conclusion and Outlook *212*
References *212*

9	**Sensor and Electronics Integration in Additive Manufacturing** *217*
	Dirk Lehmhus and Matthias Busse
9.1	Introduction to Additive Manufacturing *217*
9.2	Overview of AM Processes *224*
9.3	Links between Sensor Integration and Additive Manufacturing *228*
9.4	AM Sensor Integration Case Studies *230*
9.4.1	Cavity-Based Sensor and Electronic System Integration *236*
9.4.2	Multiprocess Hybrid Manufacturing Systems *239*
9.4.3	Toward a Single AM Platform for Structural Electronics Fabrication *243*
9.5	Conclusion and Outlook *245*
	Abbreviations *246*
	References *248*

Part Five   Signal and Data Processing: The Sensor Node Level *257*

10	**Analog Sensor Signal Processing and Analog-to-Digital Conversion** *259*
	John Horstmann, Marco Ramsbeck, and Stefan Bosse
10.1	Operational Amplifiers *260*
10.2	Analog-to-Digital Converter Specifications *262*
10.3	Data Converter Architectures *268*
10.4	Low-Power ADC Designs and Power Classification *276*
10.5	Moving Window ADC Approach *277*
	References *279*

11	**Digital Real-Time Data Processing with Embedded Systems** *281*
	Stefan Bosse and Dirk Lehmhus
11.1	Levels of Information *281*
11.2	Algorithms and Computational Models *283*
11.3	Scientific Data Mining *287*
11.4	Real-Time and Parallel Processing *291*
	References *297*

12	**The Known World: Model-Based Computing and Inverse Numeric** *301*
	Armin Lechleiter and Stefan Bosse
12.1	Physical Models in Parameter Identification *302*
12.2	Noisy Data Due to Sensor and Modeling Errors *304*
12.3	Coping with Noisy Data: Tikhonov Regularization and Parameter Choice Rules *306*
12.4	Tikhonov Regularization *308*
12.5	Rules for the Choice of the Regularization Parameter *309*
12.6	Explicit Minimizers for Linear Models *311*
12.7	The Soft-Shrinkage Iteration *312*

12.8	Iterative Regularization Schemes *313*
12.9	Gradient Descent Schemes *314*
12.10	Newton-Type Regularization Schemes *317*
12.11	Numerical Examples in Load Reconstruction *318*
	References *326*

13 The Unknown World: Model-Free Computing and Machine Learning *329*
Stefan Bosse

13.1	Machine Learning – An Overview *329*
13.2	Learning of Data Streams *331*
13.3	Learning with Noise *333*
13.4	Distributed Event-Based Learning *333*
13.5	ε-Interval and Nearest-Neighborhood Decision Tree Learning *334*
13.6	Machine Learning – A Sensorial Material Demonstrator *336*
	References *340*

14 Robustness and Data Fusion *343*
Stefan Bosse

| 14.1 | Robust System Design on System Level *345* |
| | References *348* |

Part Six Networking and Communication: The Sensor Network Level *349*

15 Communication Hardware *351*
Tim Tiedemann

15.1	Communication Hardware in Their Applications *351*
15.2	Requirements for Embedded Communication Hardware *352*
15.3	Overview of Physical Communication Classes *354*
15.4	Examples of Wired Communication Hardware *356*
15.5	Examples of Wireless Communication Hardware *358*
15.6	Examples of Optical Communication Hardware *360*
15.7	Summary *360*
	References *361*

16 Networks and Communication Protocols *363*
Stefan Bosse

16.1	Network Topologies and Network of Networks *364*
16.2	Redundancy in Networks *365*
16.3	Protocols *366*
16.4	Switched Networks versus Message Passing *368*
16.5	Bus Systems *369*

16.6	Message Passing and Message Formats *370*	
16.7	Routing *370*	
16.8	Failures, Robustness, and Reliability *377*	
16.9	Distributed Sensor Networks *378*	
16.10	Active Messaging and Agents *381*	
	References *382*	

17 Distributed and Cloud Computing: The Big Machine *385*
Stefan Bosse

17.1 Reference *386*

18 The Mobile Agent and Multiagent Systems *387*
Stefan Bosse

18.1 The Agent Computation and Interaction Model *389*
18.2 Dynamic Activity-Transition Graphs *394*
18.3 The Agent Behavior Class *395*
18.4 Communication and Interaction of Agents *396*
18.5 Agent Programming Models *397*
18.6 Agent Processing Platforms and Technologies *404*
18.7 Agent-Based Learning *415*
18.8 Event and Distributed Agent-Based Learning of Noisy Sensor Data *416*
References *420*

Part Seven Energy Supply *423*

19 Energy Management and Distribution *425*
Stefan Bosse

19.1 Design of Low-Power Smart Sensor Systems *426*
19.2 A Toolbox for Energy Analysis and Simulation *430*
19.3 Dynamic Power Management *434*
19.3.1 CPU-Centric DPM *435*
19.3.2 I/O-Centric DPM *437*
19.3.3 EDS Algorithm *438*
19.4 Energy-Aware Communication in Sensor Networks *440*
19.5 Energy Distribution in Sensor Networks *442*
19.5.1 Distributed Energy Management in Sensor Networks Using Agents *443*
References *446*

20 Microenergy Storage *449*
Robert Kun, Chi Chen, and Francesco Ciucci

20.1 Introduction *449*
20.2 Energy Harvesting/Scavenging *451*
20.3 Energy Storage *452*

20.3.1	Capacitors	*452*
20.3.2	Batteries	*458*
20.3.3	Fuel Cells	*467*
20.3.3.1	Low-Temperature Fuel Cells	*469*
20.3.3.2	High-Temperature Fuel Cells	*469*
20.3.4	Other Storage Systems	*469*
20.4	Summary and Perspectives	*470*
	References	*470*

21   Energy Harvesting   *479*
Rolanas Dauksevicius and Danick Briand
- 21.1   Introduction   *479*
- 21.2   Mechanical Energy Harvesters   *480*
- 21.2.1   Piezoelectric Micropower Generators   *482*
- 21.2.2   Micropower Generators Based on Electroactive Polymers   *489*
- 21.2.3   Electrostatic Micropower Generators   *490*
- 21.2.4   Electromagnetic Micropower Generators   *491*
- 21.2.5   Triboelectric Nanogenerators   *492*
- 21.2.6   Hybrid Micropower Generators   *493*
- 21.2.7   Wideband and Nonlinear Micropower Generators   *494*
- 21.2.8   Concluding Remarks   *495*
- 21.3   Thermal Energy Harvesters   *496*
- 21.3.1   Introduction to Thermoelectric Generators   *496*
- 21.3.2   Thermoelectric Materials and Efficiency   *499*
- 21.3.3   Other Thermal-to-Electrical Energy Conversion Techniques   *501*
- 21.4   Radiation Harvesters   *502*
- 21.4.1   Light Energy Harvesters   *502*
- 21.4.2   RF Energy Harvesters   *506*
- 21.5   Summary and Perspectives   *507*
References   *512*

Part Eight   Application Scenarios   *529*

22   Structural Health Monitoring (SHM)   *531*
Dirk Lehmhus and Matthias Busse
- 22.1   Introduction   *531*
- 22.2   Motivations for SHM System Implementation   *536*
- 22.3   SHM System Classification and Main Components   *540*
- 22.3.1   Sensor and Actuator Elements for SHM Systems   *542*
- 22.3.2   Communication in SHM Systems   *550*
- 22.3.3   SHM Data Evaluation Approaches and Principles   *552*
- 22.4   SHM Areas and Application and Case Studies   *555*
- 22.5   Implications of Material Integration for SHM Systems   *561*
- 22.6   Conclusion and Outlook   *562*
References   *564*

23 Achievements and Open Issues Toward Embedding Tactile Sensing and Interpretation into Electronic Skin Systems *571*
Ali Ibrahim, Luigi Pinna, Lucia Seminara, and Maurizio Valle
23.1 Introduction *571*
23.2 The Skin Mechanical Structure *573*
23.2.1 Transducers and Materials *573*
23.2.2 An Example of Skin Integration into an Existing Robotic Platform *575*
23.3 Tactile Information Processing *579*
23.4 Computational Requirements *582*
23.4.1 Electrical Impedance Tomography *582*
23.4.2 Tensorial Kernel *583*
23.5 Conclusions *585*
References *585*

24 Intelligent Materials in Machine Tool Applications: A Review *595*
Hans-Christian Möhring
24.1 Applications of Shape Memory Alloys (SMA) *596*
24.2 Applications of Piezoelectric Ceramics *596*
24.3 Applications of Magnetostrictive Materials *598*
24.4 Applications of Electro- and Magnetorheological Fluids *600*
24.5 Intelligent Structures and Components *601*
24.6 Summary and Conclusion *603*
References *604*

25 New Markets/Opportunities through Availability of Product Life Cycle Data *613*
Thorsten Wuest, Karl Hribernik, and Klaus-Dieter Thoben
25.1 Product Life Cycle Management *613*
25.1.1 Closed-Loop and Item-Level PLM *615*
25.1.2 Data and Information in PLM *615*
25.1.3 Supporting Concepts for Data and Information Integration in PLM *616*
25.2 Case Studies *617*
25.2.1 Case Study 1: Life Cycle of Leisure Boats *617*
25.2.1.1 Sensors Used *618*
25.2.1.2 Potential Application of Sensorial Materials *619*
25.2.1.3 Limitations and Opportunities of Sensorial Materials *619*
25.2.2 Case Study 2: PROMISE – Product Life Cycle Management and Information Using Smart Embedded Systems *620*
25.2.2.1 Sensors Used *620*
25.2.2.2 Potential Application of Sensorial Materials *621*
25.2.2.3 Limitations and Opportunities of Sensorial Materials *621*
25.2.3 Case Study 3: Composite Bridge *622*
25.2.3.1 Sensors Used *623*

25.2.3.2	Potential Application of Sensorial Materials *623*
25.2.3.3	Limitations and Opportunities of Sensorial Materials *623*
25.3	Potential of Sensorial Materials in PLM Application *623*
	Acknowledgment *624*
	References *624*

26 Human–Computer Interaction with Novel and Advanced Materials *629*

Tanja Döring, Robert Porzel, and Rainer Malaka

26.1	Introduction *629*
26.2	New Forms of Human–Computer Interaction *630*
26.3	Applications and Scenarios *633*
26.3.1	Domestic and Personal Devices *633*
26.3.1.1	The Marble Answering Machine *633*
26.3.1.2	Living Wall: An Interactive Wallpaper *634*
26.3.1.3	Sprout I/O and Shutters: Ambient Textile Information Displays *634*
26.3.1.4	FlexCase: A Flexible Sensing and Display Cover *635*
26.3.2	Learning, Collaboration, and Entertainment *635*
26.3.2.1	Tangibles for Learning and Creativity *635*
26.3.2.2	inFORM: Supporting Remote Collaboration through Shape Capture and Actuation *636*
26.3.2.3	The Soap Bubble Interface *637*
26.4	Opportunities and Challenges *637*
26.5	Conclusions *639*
	References *639*

Index *645*

Foreword

The vision of materials with intelligent behavior has been tantalizing material and computer scientists for many decades. The benefits of such materials, which would more resemble living systems than classical engineered structures, would indeed be tremendous: Materials that can sense and change their properties such as shape, appearance, and other physical properties in response to the environment would allow us to create structures, robots, and other autonomous systems that interact with the environment with animal-like agility and with the robustness common to biological, living systems. In the long run, such materials could even self-assemble and self-heal, and fundamentally change the way how things are made. This vision is particularly nagging as Nature vividly demonstrates these possibilities and their feasibility on a daily basis; yet, progress has been slow and tedious. Unlike conventional engineered structures, Nature tightly integrates sensors, muscles, and nerves with structure. Examples range from our own skin that helps us regulate temperature and provides us with tactile sensing at very high dynamic range to the most complex structure in the known universe, our brains, and more exotic functionality such as camouflage of the cuttlefish or the shape-changing abilities of a bird wing.

In computer science, interest into intelligent material goes back to Toffoli's concept of "Programmable Matter" in the 1990s, and was accelerated by the advent of microelectromechanical structures (MEMSs), which has led to the concept of "smart dust" and the field of sensor networks. At the same time, advances in composite manufacturing have led to the field of "multifunctional materials," and it seems the time has finally come to unite these two fields of which the present book is a first attempt.

Stefan Bosse, Manuel Collet, Dirk Lehmhus, Walter Lang, and Matthias Busse present here one of the first attempts to bridge the currently disparate fields of computer science, robotics, and material engineering. Their diverse backgrounds are reflected in the organization of this book, which follows the same layered approach that has become customary to abstract the inner workings of networked communication from their applications to organize the challenges of material-integrated intelligence in both material science and computing. Establishing this common language and hierarchy is an important first step as it allows the different disciplines to understand where they fit in, the scope of their contribution within the bigger picture, and where the open challenges are.

While this view will be very helpful for the two disparate communities to find common ground, this book does not oversimplify the problem. It remains clear throughout that material-integrated intelligence and structural functionality are indeed at odds. Every additional sensor, communication infrastructure, and computation a computer scientist would wish to integrate into a structure, for example, to perform structural health monitoring, jeopardizes the very structural health of the structure. Similarly, adding the capability for structures to morph, for example, to save fuel during different phases of flight, adds weight to an extent that very likely outweighs the very savings any morphological change could possibly provide. While these constraints seem overly limiting, more pedestrian (in the true sense of the word) applications might not have enough value to justify multifunctional composites.

Yet, natural systems impressively show us that trade-offs in multifunctionality with net benefits are indeed possible and often the only way these systems can survive in a changing environment. It might therefore be worthwhile to put immediate applications aside and indulge in the intellectual challenges of design and distributed computation until emerging applications such as robotics, orthotics, and autonomous systems in general – which will strongly benefit from material-integrated computation – become more mainstream. This book provides the red thread for such a pursuit by providing an overview of recent progress in function scale integration of sensors, power and communication infrastructure, as well as the key computational concepts that such integration could enable, bridging the worlds of the continuous, that is material physics, and the discrete algorithmic world.

Here, one problem becomes very clear: It is not possible to design material-integrated intelligence without understanding both the underlying material physics and the algorithms such intelligence requires. Physics determines the bandwidth, dynamic range, and noise characteristics of sensors and actuators, which define the available inputs and outputs to an algorithm designer. Likewise, certain computational problems require a minimum amount of real estate, energy, and communication bandwidth that the material designer needs to foresee. These challenges are already well understood in robotics, where they are reconciled by a probabilistic view on sensing, actuation, and algorithmic planning. Describing the specific problems that material-integrated intelligence poses will help the community to recognize the similarities between such structures and robotics, and possibly help to leverage insights from probabilistic state estimation, planning, and control that this community has produced in the past decade.

Combining a physical and a computational perspective in a single book will not only help this fledgling field to organize key insights but might also serve as a starting point for a new generation of scientists and engineers who will have their feet comfortably in both the computational and the physical – a prospect that offers tremendous opportunities. For example, it is possible to shift computation into the material and vice versa. The cochlea, which effectively performs a discrete Fourier transform, trades computational with spatial and sensing requirements. Similarly, insect eyes organize their lenses in spatial arrangements that simplify the neural circuitry for rectification of the compound signal, an approach that has become known as "morphological computation." Finally, combinations of active materials can be used to create simple feedback controllers and oscillators. Innovating in this space will require scientists and engineers that are equally at ease with materials and computational concepts.

Another area that is currently untapped in material sciences and builds up on the foundations laid out here is to leverage the principles of self-organization and swarm intelligence to equip future materials, possibly consisting of thousands of pin-head sized computing devices, with intelligence. Receiving interest from both material and computer science, for example, in their study of self-assembly or pattern formation, also known as "Turing Patterns" (after the computer science pioneer Alan Turing who first studied such systems), both communities have not yet connected on creating materials that self-organize – a concept that allows realizing almost limitless functionality.

Back to the here and now, however, the immediate impact research in material-integrated intelligence might have is to provide a new platform for material and computer scientists alike to apply and exchange their tools. Solving the problems of integrating sensors/actuators, computation, communication, and power into smart composites and mass producing them will allow us to create sensor networks and distributed computers of unprecedented scales. Such systems pose the opportunity to perform model predictive control or machine learning *inside* the material, providing unprecedented capabilities and challenges. Likewise, applying the full breadth of what is possible computationally will spur the development of novel sensors and actuators with higher bandwidth, smaller footprints, and lower energy requirements, eventually approximating and transcending natural systems. In order to get there, readers of this book will need to set their disciplinary goggles aside and join us in the quest to make materials computers and computers materials.

Nikolaus Correll
Department of Computer Science
Material Science Engineering
University of Colorado at Boulder
Boulder, CO, USA
February 10, 2016

Preface

This book addresses a topic that has, to the editors' knowledge, not been covered as comprehensively as this before: material-integrated intelligent systems.

The topic links up with recent, current, and emerging trends like smart system integration, ambient intelligence, or structural electronics. Its background is the understanding that to obtain truly smart objects, it is simply not sufficient to tag them with a sensor node and some associated electronics for data evaluation – nor is it satisfactory to merely embed sensors in materials. Instead, the ultimate goal is to have materials that actually feel in a manner that can be compared with our own capabilities as human beings. The skin, its sensory equipment, and the further processing of data acquired through it is undoubtedly the most referenced model of such a material: Here we find different types of sensors situated in the most suitable places and at high resolution wherever necessary – thousands of them, actually, on your fingertip or on the palm of your hand, capturing pressure together with its first and second derivatives as well as secondary information like temperature or humidity.

Besides the sensors, you have filtering of signals, sensor fusion, and information preprocessing. You have communication of aggregated information to a hierarchy of higher level control systems – your spinal cord, or ultimately your brain, with the information passed finally reaching your consciousness.

The complexity of this system is such that research on its basic principles and capabilities even in humans is an ongoing effort involving researchers from several disciplines like biology, medicine, or neuroscience, to name but a few.

The aim defines the approaches that lead to its technical realization, and technology does not differ from biology here: The topic is highly interdisciplinary irrespective of the world we look at, be it natural or artificial. It requires contributions, in various fields, of materials science, but also of production engineering, microelectronics, microsystems technology, systems engineering, and computer science. In fact, it needs more than contributions, but rather close cooperation. It thus faces all the challenges transgressing the boundaries of scientific disciplines commonly entails–starting from a quite natural lack of fully grasping the capabilities and limitations of neighboring research fields to the simple problem of scientific languages that just do not match and thus impede creating the necessary mutual understanding.

At the same time, the promise linked to solving these difficulties is just about as large as the challenge. Material-integrated intelligent systems encounter an economic environment that moves more and more toward computational capabilities and communication technologies dispersed and networked throughout our daily environment. The concept of smart dust may be seen as earlier formulation of this issue. Currently, it is moving back into focus as a potential facet of the Internet of Things (IoT), today one of the main stimulators of research in material-integrated intelligent systems.

The associated fundamental technological enablers and their interdependence is what we primarily intend to illustrate in this book.

With this vision in mind, we have conceived this book, to which several authors with a broad scientific background have contributed.

Our ambition with it is to help bridge the gaps between the various disciplines they represent by allowing these eminent scientists to present their own views on their area's part in the embracing context.

So whom do we expect to benefit most from this extensive collation of information?

Naturally, the answer to this question determines the form and content of the fundamental chapters of our work.

As primary readers, we have professionals in product development and engineering design in mind who are tasked with the problem of integrating mechanical and electronic systems in a classic mechatronics approach, but on the new level of material integration, which requires additional, new solutions in materials and production processes or data evaluation.

Thus, the perspective of the assumed reader is that of an expert in one of the technological fields involved who needs to gain insight into the adjacent ones to be able to devise and evaluate integrated solutions involving several areas. The background of such a reader can either be academic or industrial. Among the industrial readership, we particularly hope to gain the interest of potential applicants of material-integrated intelligent systems: Our book should provide the necessary pathways and perspectives to help them understand the possibilities the combination of the current technological state of the art from various disciplines can offer.

Besides, we have graduate and postgraduate students in mind who seek an introduction to the field. The character of the book, and the intention to lead professionals to new realms beyond their usual field of practice, is reflected in the attempt to structure and formulate the individual chapters in a way that will allow people with an engineering or natural sciences background to easily follow the discussion even if it is not their particular area of expertise or application that is covered.

For reasons of simplicity and easy access, the book is organized in parts that reflect major research areas. Prior to this, the topic is outlined in an introductory section (Part One) that explains the term material-integrated intelligent systems in more detail, puts it in perspective with past, present, and future scientific and technological trends, and thus provides the motivation for engaging in research and new technology development.

Part Two, System Development, assumes the product development perspective and describes methodologies for designing smart systems on different levels of abstraction.

Part Three, Sensor Technologies, provides fundamental information about different types of sensors and discusses the need for adaptation they face in view of material integration, as well as technological solutions developed toward this end.

Part Four, Material Integration Solutions, swaps perspectives from the electronics and microsystems technology point of view toward mechanical and materials engineering. In this part, we consider the integration problem based on specific material classes (metals, polymers) and the closely associated manufacturing processes.

Part Five, Signal and Data Processing: The Sensor Node Level, describes the fundamentals of this area of expertise and relates them to specific problems of material-integrated systems. The perspective is that of the individual, smart sensor node.

Part Six, Networking and Communication: The Sensor Network Level, extends the scope toward the combination of several such sensor nodes and thus also covers information exchange between them, as well as data evaluation in sensor networks.

Part Seven, Energy Supply, discusses ways of ensuring the availability of sufficient amounts of energy – and levels of power – for a material-integrated system to operate, touching upon aspects like storage of energy and management of resources as well as generation of energy through harvesting or scavenging approaches.

Part Eight, Application Scenarios, either provides examples of realized material-integrated intelligent systems or explains how different areas of application like Structural Health Monitoring or Human–Machine Interaction and/or cooperation could profit from future availability of them.

Common to all parts is a general concept that provides entry points for readers with diverse backgrounds and thus strongly deviating levels of competence in the areas covered. In this sense, we do not see our work as giving definite answers across the width of its scope, but rather as defining and providing the cross-disciplinary interfaces between the various elements that need to be connected to generate what is the topic and the vision of this book – material-integrated intelligent systems.

Putting together this book has required a considerable amount of time and commitment from all the many people involved. We are thus extremely grateful to those researchers who have volunteered to supply a contribution to this work. Essentially, it is their dedication, their effort, their perseverance, and not the least their patience that have made possible the result you, the reader, can hold in your hands today.

We are indebted to Dr. Martin Preuss of Wiley-VCH, who discussed the topic, content, and organization of the book with us at its very beginning and thus helped create the original framework that has now been realized in the form you, as reader, are holding in your hands. For this realization, in turn, we owe gratitude to Nina Stadthaus and Stefanie Volk (Wiley-VCH) who accompanied us throughout the whole process of gathering and organizing primary input as well as secondary information with immeasurable patience, dedication, and a lot of good advice.

Finally, we acknowledge the financial support of the Federal State of Bremen, which facilitated the formation and the initial research work of the ISIS (Integrated Solutions in Sensorial Structure Engineering) Sensorial Materials Scientific Centre at the University of Bremen, without which the achievement this work represents could not have been accomplished.

Stefan Bosse
Dirk Lehmhus
Walter Lang
Matthias Busse

Part One

Introduction

1

On Concepts and Challenges of Realizing Material-Integrated Intelligent Systems

Stefan Bosse[1,2] and Dirk Lehmhus[2]

[1]*University of Bremen, Department of Mathematics and Computer Science, Robert-Hooke-Str. 5, 28359 Bremen, Germany*
[2]*University of Bremen, ISIS Sensorial Materials Scientific Centre, Wiener Str. 12, 28359 Bremen, Germany*

1.1 Introduction

Material-integrated intelligent systems constitute materials that are able to "feel." This is the shortest possible definition at hand for the subject of the present book. What it implies will be discussed below, while detailed descriptions of individual aspects and application scenarios will follow in its main parts.

As a concept, material-integrated intelligent systems have implicitly been around for quite some time. To a considerable degree, this is because the concept as such is not so much a human invention, but rather something that is deeply rooted in nature: The human skin and the human nervous system are the typical examples cited pertaining to material-integrated intelligent systems, such as sensorial materials [1–3], robotic materials [4], nervous materials [5], or sensor-array materials [6].

These natural models taken together nicely illustrate the differences between materials with integrated sensor(s) and material-integrated intelligent systems: For one thing, the skin contains a multitude of sensors which do not only capture force or pressure, but also additional aspects like the first and second derivative of pressure or temperature. At the same time, the impression we get when we touch an arbitrary surface is not that of a separate awareness of these factors, but a combined one that is derived from fusion of sensory information.

Besides, we do not base the decisions we make in response to a tactile sensation on quantitative values of pressure, temperature, and so on, and on a deterministic model that links these values to an intended action and its potential outcome. Instead, we rely on experience, that is, on a learned relationship between an action and its outcome in relation to the associated sensory information in one way or another. Translated to technical terms, we thus follow a model-free approach.

Material-Integrated Intelligent Systems: Technology and Applications, First Edition.
Edited by Stefan Bosse, Dirk Lehmhus, Walter Lang, and Matthias Busse.
© 2018 Wiley-VCH Verlag GmbH & Co. KGaA. Published 2018 by Wiley-VCH Verlag GmbH & Co. KGaA.

Having said this, we can derive a list of characteristics a material would need for us to concede that it can actually "feel." Such a material must be capable of

- capturing sensory data;
- aggregating data through some local preprocessing, performing data reduction of individual data points;
- further processing this data to derive some higher-level information, gaining knowledge;
- using this knowledge for decision-making, putting it to some internal/local use, or communicating it to higher system levels;
- coping with damage by being dynamic and reconfigurable; and
- achieving a state of awareness of host material and environment, that is, the derivation of a context knowledge.

If the above list represents a functionality-centered perspective, the question that immediately arises is how a technical implementation of this concept could be achieved, and which research domains would need to contribute to it.

On a generic level, material-integrated intelligent systems follow the universal trend in the microelectronics industry, which is typically described as having two orthogonal, primary directions: on the one hand, miniaturization or the "more Moore" development line, and on the other, diversification through the integration of additional, usually analog, functionalities such as sensing, energy supply, and so on – the "more than Moore" approach. In both cases, reference is made to Moore's law, which predicts (from a 1965 point of view) that transistor count in densely packed integrated circuits would double every 2 years, and which has since then approximately been met by actual developments, although with some indications of slowing down since about 2011. Technologically, "more Moore" is usually associated with system on chip (SoC) solutions, whereas "more than Moore" is linked to system in package (SiP) technologies. However, both merge diagonally combining both SoC and SiP approaches to create higher value systems. Clearly, this is the domain into which material-integrated intelligent systems fall. As a consequence, the following research topics need to be addressed in their development:

- miniaturization on component and system level to limit "footprint" within host material;
- system resilience against effects of processing conditions during integration;
- system compliance with host material properties in the embedded state;
- energy supply solutions that support autonomy, like cooperative energy harvesting and storage, and (intelligent) management of resources;
- reliable and robust low power internally and externally directed communication approaches;
- distributed, reliable, and robust low power data evaluation; and
- multiscale design methodologies that span the scope from chip design to smart products and environments.

Mark Weiser, in his landmark 1991 article that predicted many evolutions in computer science we have witnessed since, has set the scene by stating that "in the 21st century, the technology revolution will move into the everyday, the small and the invisible" [7]. Weiser thus anticipated a development that is connected to terms such as ambient intelligence and ubiquitous or pervasive computing.

Material-integrated intelligent systems will both profit from and contribute to the realization of this prediction through their potential of endowing many of the passive materials surrounding us today with perceptive capabilities, and ultimately even adaptive behavior. A large part of the novelty of this approach has its foundations in the notion that miniaturization of systems will allow integration on a level that provides the added functionality without compromising suitability for the primary role to be fulfilled by the material in question. A prominent example in this respect is structural health monitoring (SHM). This application scenario is relevant for safety-critical, load-bearing structures. Safety can be enhanced, or safety factors relaxed, if the exact structural state is known at any moment in time. If material-integrated intelligent systems were selected for this task, a necessary prerequisite would be that the systems themselves do not adversely affect mechanical characteristics of the host material. In other words, the materials designed thus should not afford considering any property degradation caused by the material-integrated systems during the layout of the structure for its primary task. In a further evolution of the concept, the materials themselves could thus be envisaged as semifinished materials in the same way as sheet metal: Their capabilities, including their smartness, would be available as an asset not necessarily targeted at a specific application, but providing for several ones. For production of material-integrated intelligent systems, such a scenario could open up economy of scale effects significantly enhancing their economic viability. At the same time, this would afford production techniques able to cope with the associated large production volume.

It has been suggested that the implementation of material-integrated sensing can either follow a top-down or a bottom-up approach [2]. Focusing specifically on the sensing function, Lang *et al.* [8] propose an even finer distinction, which demarcates a top-down as opposed to a bottom-up approach:

- top-down approach:
  - hybrid integration
  - local additive buildup
- bottom-up approach:
  - generic (intrinsic) sensing properties of materials
  - local growth of sensors using, for example, bioinspired processes

From our current perspective, Lang *et al.*'s proposal excludes the intelligent side of material-integrated intelligent systems and its prerequisites like energy supply by concentrating on the transducer effect and the hardware to implement it. Specifically, the bottom-up approaches still fail to offer solutions that could provide these system components. This is apparent particularly for the generic sensing properties of materials, which remain ineffectual even as sensor until at least some means of detecting (i.e., sensing) the intrinsic effect is added.

The example shows that at least on the level of full intelligent systems, bottom-up approaches do not yet respond satisfactorily to the questions of realization.

An exception, though a theoretical one, is the notion of programmable matter proposed by Toffoli and Margolus. Their original concept assumes spatially distributed computing elements similar to smart sensor nodes capable of nearest neighbor interaction only. Together, they form a material with the inherent capability of information processing. Practically, this concept is reminiscent of physical realizations of cellular or lattice gas automata [9,10].

Later, alternative or extended definitions of programmable matter stress the ability of such materials to alter their physical characteristics in a controlled fashion – controlled either by a user from the outside or autonomously from within the material. In the latter case, the programmable matter makes use of its data evaluation capabilities to respond, for example, to sensor signals. Under this headline, several materials have been understood to represent forms of programmable matter. Material-integrated intelligent systems would fall into this category, too. Since the wider definitions of programmable matter include spatial reconfiguration building on autonomous objects (cells) as building blocks besides information processing, sensing, actuation, and adaptivity, both sensorial materials [1] and robotic materials [4] can be seen as intermediate-level representatives of this overall class of intelligent materials.

Realization of programmable matter thus depends on the scope of properties and the definition adopted. The full spectrum is usually represented by the material being built up of individual, autonomous units of microscopic scale that form the matter itself by docking to each other in different configurations, an ability that requires some relative locomotion, too. "Utility fog" is another designation for a system of microscale autonomous units having such abilities, with the spatial rearrangement not based on the so-called "foglets" as the smart units, but on the flow of the fluid in which they move. Reaching a certain macroscopic shape thus does not have to depend on deliberately moving to a certain location, but can rely on making or refusing connections once the opportunity is there [11]. To date, to the knowledge of the authors, no physical material is available that combines the full set of characteristics envisaged by Toffoli or Hall [10,11].

Obviously, much nearer to implementation is the top-down approach, which essentially coincides with the intermediate, diagonal path in between the pure "more Moore" and the "more than Moore" trend: What the term top-down implies is a hybrid integration approach in which suitable components are adapted to material integration needs and combined to form the required smart sensor network.

First practical developments leading toward sensor nodes combining subsets of the features required by material-integrated intelligent systems – above all, a minute size, a certain level of energy autonomy, and data evaluation as well as communication capabilities – sailed under the "smart dust" flag from the end of the 1990s to the early twenty-first century [12]. Warneke *et al.* concentrated on developing, as they termed it, a "cubic millimeter-sized computer" fully endowed with sensing, energy storage, and data evaluation, plus communication that could create an *ad hoc* network when dispersed, like dust, in a given environment. Clearly, a handful of smart dust sensor nodes embedded in a host material would conform to our own definition of material-integrated intelligent systems and sensorial materials.

Figure 1.1 provides an overview of the main elements of such a system, which mostly form part of the smart dust mote concept, too. The sensorial material as such consists of a material-integrated network of smart sensor nodes that may

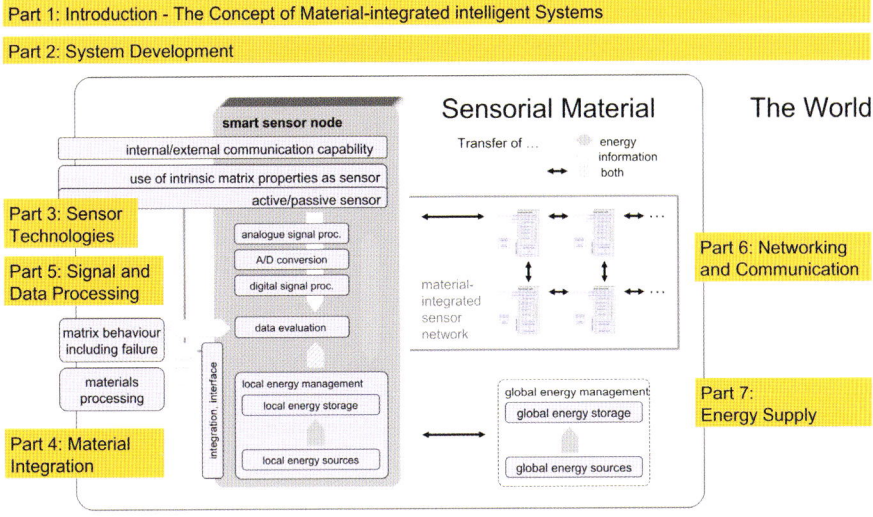

Figure 1.1 The fundamental elements of a material-integrated intelligent system, also called sensorial material, as envisaged by Lehmhus et al. [2] – the aspect of material integration and the resulting interactions between integrated system and host material during production and life cycle are included, as well as the link to the environment in which the system exists: The world may be the source or cause of sensor data, but may also provide communication partners, both as recipients and as suppliers of information, the latter, for example, if the material-integrated system is embedded in a higher-ranking IoT solution. Besides, it may exchange energy with the sensorial material, acting either as source or sink.

incorporate features like data evaluation and energy supply on local as well as global level. Figure 1.1 also links the various technological aspects to the respective parts of the present book.

1.2 System Development Methodologies and Tools (Part Two)

Basing a product on material-integrated intelligent systems leads to increased complexity in product development. Whereas in more conventional solutions, development of smart system and component can follow parallel paths for much of the design process, material integration affords a tighter coupling: Peters et al. [13] have highlighted this problem in an aerospace SHM context.

This calls for adapted design methodologies and tools that reflect the multidisciplinary, multidomain nature of the topic. In order to do so, such tools must incorporate simulation capabilities that address the various scientific disciplines involved and at least allow transferring results between them, if not even integrating them. As has been pointed out, for example, by Lehmhus et al. [14,15], this requirement does incorporate multiscale and multiphysics aspects, but extends beyond these in some respects by linking the process of gathering data and turning it first into information, and then using these to make decisions on physical aspects of the structure: Examples include adaptive structures that allow property change in response to loads or damage reconstructed from sensor signals, and base their decisions on incremental learning approaches [16]. Assessing the safety of such structures will have to include considerations about physical effects within the material potentially causing damage, the probability of detection of damage both of the structure itself and the sensors employed, and the developing nature of the learned model. Besides, energy issues may have to be considered if, for example, energy management adapts the number of sensors interrogated, the communication activity between sensor nodes, or the algorithms used in load or damage assessment in relation to the energy status of sensor nodes and network. If energy harvesting is used to beef up energy resources, probability of detection and so on may even be influenced by the past, present, and predicted (if a forward-looking energy management algorithm is used) service life of the structure, which determines the amount of energy available for scavenging. For such factors to have an impact, strictly speaking not even adaptivity is needed.

Design methodologies will be discussed on a global level and from the point of view of smart system design in part two of the present work.

1.3 Sensor Technologies and Material Integration (Part Three and Four)

The variety of sensors used in material-integrated sensing echoes the variety of application scenarios. Classification of sensors can, for example, be based on the manufacturing process (printed sensors, MEMS sensors, etc.), the material group (organic versus silicon sensors and electronics), the nature of the measurand (mechanical sensors, temperature sensors, chemical sensors, etc.), and the underlying principle of measurement (optical sensors, piezoresistive sensors, piezoelectric sensors, etc.). All the above classes, and more, have relevance for material-integrated intelligent systems and will be discussed in part three of this book.

For the case of sensors, material integration itself is typically described by means of several different levels. Figure 1.2 provides an overview in this respect, starting with an entirely external observation, in this case represented by a camera (level I). Surface application of sensors is the next step: The sensor is applied, usually via its substrate, to the material to be monitored (II). Surface integration in contrast assumes a tighter coupling between sensor and material or part (III). Examples in this respect include sensors directly printed onto a part or material surface, implying in situ production of the device. In materials featuring a layered buildup, as is the case for continuous fiber reinforced polymers or the so-called fiber metal laminates, surface integration realized on an internal surface can turn into volume integration (IV). Also frequently discussed is the use of intrinsic material properties for material-integrated sensing. The drawback of this approach is that it does not replace the smart sensor node, but at best the transducer element that translates the measurand into an electrical signal. From a material-integrated intelligent system point of view, this is a minor contribution that may be bought at the expense of compromising material properties.

1.3 Sensor Technologies and Material Integration (Part Three and Four)

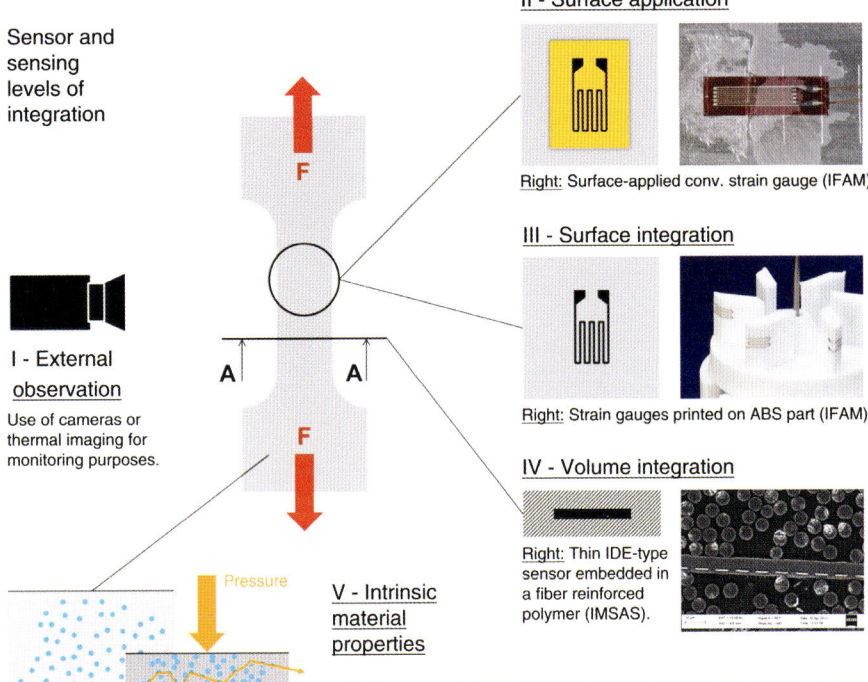

Figure 1.2 Sensor integration as a model for explaining the various levels of material integration, using a tensile test and associated sample as example. (Images courtesy of Fraunhofer IFAM – top and center right – and IMSAS, University of Bremen – bottom right.)

In terms of the challenges of integration, a distinction between life cycle phases is helpful. The beginning-of-life (BoL) phase and especially the production processes typically stress the material beyond the limits it can be expected to see in the middle-of-life (MoL) or application phase. However, unless production process monitoring is among its tasks, the system is only expected to survive in this phase, and not to deliver accurate and reliable information, which simplifies the task to some degree.

The hardships of integration themselves very much depend on the nature of the underlying manufacturing process. Casting or molding processes are typically characterized by higher processing temperature, but generally lower mechanical loads. Forming processes usually involve higher mechanical loads. Processing of polymers naturally occurs at lower temperatures than that of metals. However, in some cases, like the integration of sensors in metal casting, it is not only the absolute temperature that is of interest but also the cooling rate after mold filling: Despite the much higher mechanical loads, it is easier to integrate a sensor in high-pressure die casting than in sand casting because the former reaches cooling rates orders of magnitude higher than the latter. Figure 1.3 illustrates these differences in a qualitative manner. As a consequence, in high-pressure die casting, a thermal protection solution for the sensor system only needs to withstand the elevated temperatures for seconds or minutes at most [17–22].

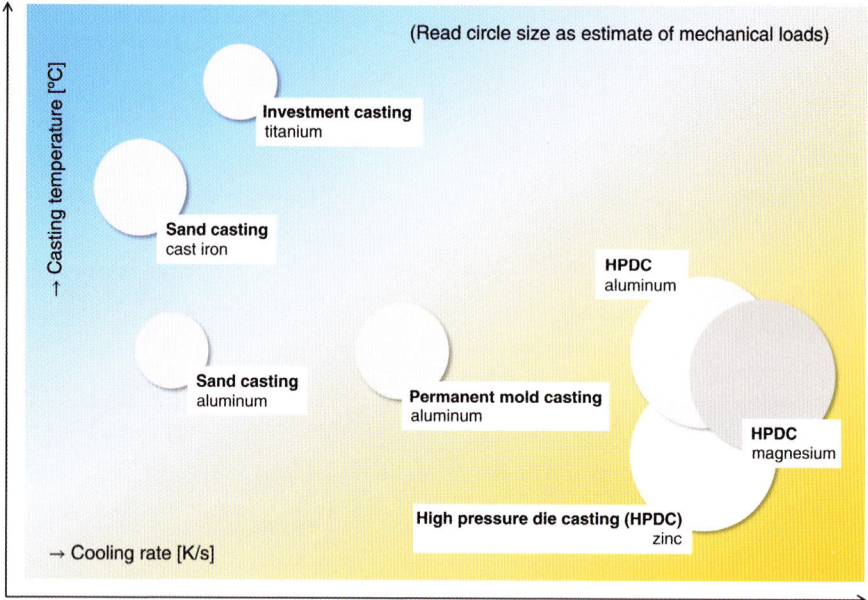

Figure 1.3 Qualitative comparison of production processes (here metal casting processes) in terms of thermal loads exerted on embedded sensors and sensor system components, showing temperature level versus cooling rates reached following mold filling.

Coming to terms with thermal loads can be discussed based on structural state consideration and thus oriented at the melting point of the various materials found in a sensor system for functionality reason. However, in many cases, the maximum sustainable temperature is not defined by this criterion, but by lower boundaries that mark the loss of core functional capabilities. Examples in this respect are semiconductor devices that depend on doping, which may be irreversibly lost if activation energies for dopant diffusion are reached and sufficient time (here the process-dependent cooling rate comes in once again) at elevated temperature is available. Since diffusion paths are short due to the high resolution of current semiconductor device technology, even limited time intervals may already harm the system. Similarly, as ceramics, piezoelectric materials like lead zirconate titanate or PZT can, in principle, withstand high temperatures, but lose their polarization once the Curie temperature (roughly between 230 and 500 °C for PZT, depending on exact composition) is passed. Naturally, polarization of an already integrated PZT sensor is difficult if the encasement is of metallic nature. Both examples underline that material integration may have to rely on specific materials with properties adapted to the integration process. Figure 1.4 enumerates solutions discussed in the field of high-temperature electronics (HTE) and contrasts their application temperatures with typical casting temperatures, illustrating the fact that not only the active components must be considered but also passives and supporting materials like solders.

If thermal stability cannot be guaranteed via an adjusted choice of materials, thermal protection is a second option. A drawback of this solution is that in case of a

1.3 Sensor Technologies and Material Integration (Part Three and Four)

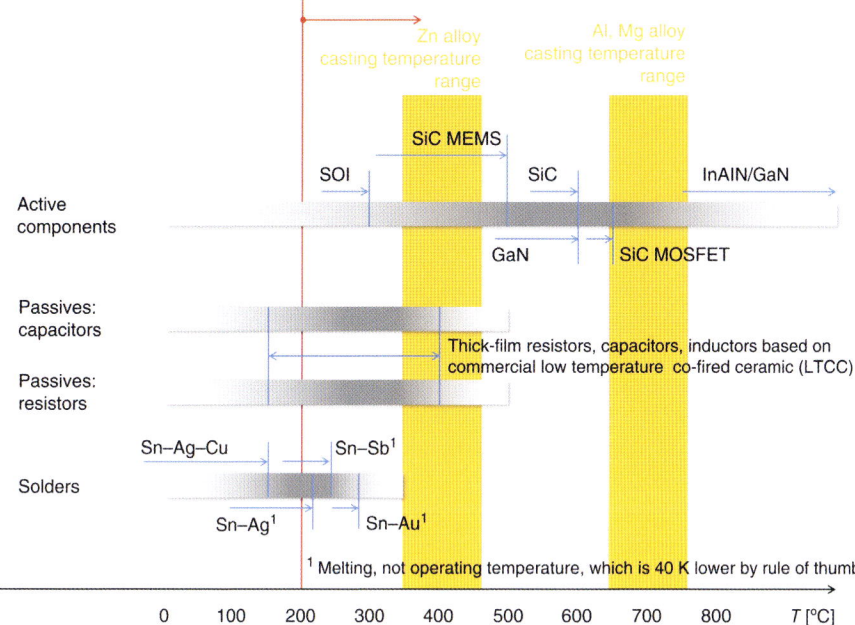

Figure 1.4 Exemplary overview of high temperature electronics material choices for different types of components (active and passive components, solders) contrasted to thermal conditions in metal casting [23–28].

sensor, it means detaching it from the structure to be monitored, which complicates the interpretation of signals gathered. In case of an electronic component, a reversal of the heat flux direction can be observed if usage of the component generates waste heat: In this case, the protective insulation designed against the requirements of the BoL phase will counteract heat dissipation during the MoL phase, that is, the product's service life (see Figure 1.5 for an schematic illustration of the problem).

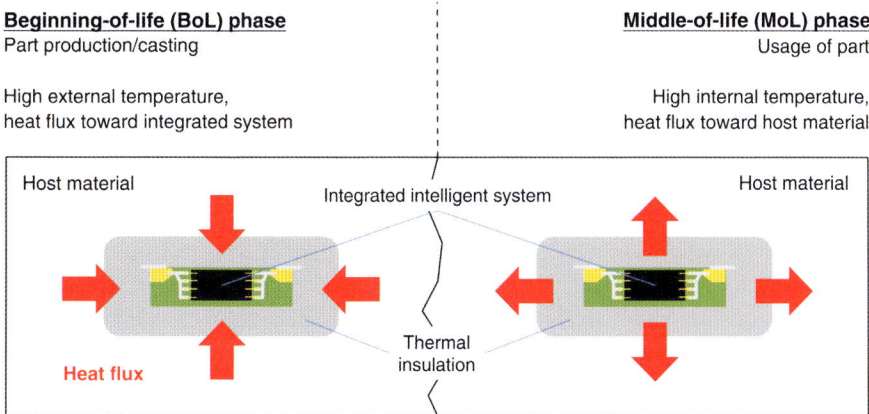

Figure 1.5 Heat flux reversal from beginning-of-life (BoL) to middle-of-life (MoL): Can being too protective do harm?

Thus, once the production process is over, service life emerges as a challenge in its own right. Already in the later phases of thermal production processes, compliance of the typical host and sensor system – functional and packaging – materials become an issue. The main aspects to consider are thermal and mechanical compliance. Differences in the coefficient of thermal expansion (CTE) on the one hand and the Young's modulus on the other affect internal stress distribution in host material and sensor system. If applied deliberately and in a controlled manner, these effects can actually be employed to improve system performance. An elegant example of such a strategy has been presented by Heber *et al.*, who have integrated piezoceramic elements at elevated temperatures in polymers with different CTE values. As a consequence, after cooling, the higher CTE of the enclosing polymers results in compressive residual stresses in the piezoceramic and tensile ones in the polymer. The main benefit is a certain alleviation of the brittle fracture behavior of the piezoceramic material [29]. In a somewhat similar approach, Choi *et al.* have deposited a low-CTE conductive material on a high-CTE substrate at an adjusted temperature to create compressively prestressed interconnects able to better sustain tensile loads and even allow some amount of stretching. A comparable, but mechanical rather than thermal effect has been reported by the same authors relying on deposition of the conductive material on a mechanically prestressed substrate that is released afterward [30].

In contrast to the above examples, deviations in the Young's modulus typically come into effect as soon as mechanical loads are applied. In a fiber-reinforced composite material, the high stiffness fibers rather than the low stiffness matrix see the highest stress levels. The same will happen in a complex material-integrated intelligent system. On a larger length scale, the effect is observable between host material and embedded device. Dumstorff *et al.* [31] have demonstrated this via FEM simulations of an assumed silicon-based device in a "virtual" matrix with varied elastic properties. Similar studies have been performed by Lecavelliers des Etangs-Lavellois *et al.* [32], who looked on a much smaller length scale at the stress distribution caused by external loads like bending within the various materials and subcomponents that together constitute a silicon-based device.

Naturally, selection of materials is neither entirely free on the functional and packaging side nor on that of the host material: The choice is always dominated by the main functional and structural requirements imposed by the product specifications. Product designers must thus attempt to reach a best compromise in terms of the compliance issue, including all system levels, essentially solving a multiobjective optimization problem across several scientific and engineering disciplines. Once again, this stresses the need for dedicated design methodologies and tools. Besides, it motivates further efforts toward limiting, at least, the mechanical loads acting on material-integrated intelligent system in structures subjected to external loads. Naturally, the solution to be adopted in this respect depends on the actual loading conditions. Nevertheless, three main groups of approaches can be distinguished:

- neutral plane engineering
- flexible/bendable/stretchable materials
- rigid islands, flexible interconnects

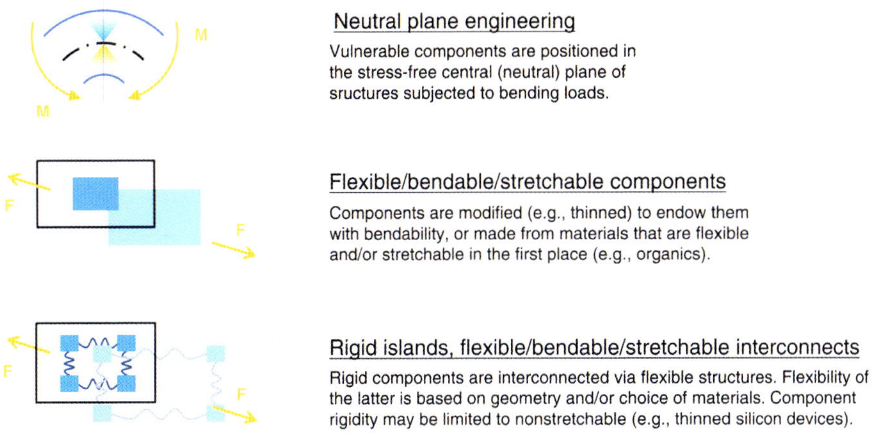

Neutral plane engineering
Vulnerable components are positioned in the stress-free central (neutral) plane of sructures subjected to bending loads.

Flexible/bendable/stretchable components
Components are modified (e.g., thinned) to endow them with bendability, or made from materials that are flexible and/or stretchable in the first place (e.g., organics).

Rigid islands, flexible/bendable/stretchable interconnects
Rigid components are interconnected via flexible structures. Flexibility of the latter is based on geometry and/or choice of materials. Component rigidity may be limited to nonstretchable (e.g., thinned silicon devices).

Figure 1.6 Mechanical compliance: schematic representation of fundamental concepts.

These fundamental concepts, which can partially overlap in practice, are illustrated schematically in Figure 1.6.

Flexibility can stand for several characteristics in this context: In some cases, it is meant to be understood as "bendability," in other cases as "stretchability." Furthermore, it is important to keep in mind that the benchmark for flexibility usually is a silicon device, which is characterized by brittle failure and a high Young's modulus and strength, which together translate into a very limited elongation at break. The level of flexibility that needs to be attained is thus comparatively low for most applications under consideration: Stretchability of a few percent is very often more than satisfactory.

Neutral plane engineering is specifically relevant for flat products subjected to bending loads: In this case, a stress-free (neutral) plane exists in or near which vulnerable components can be positioned. Naturally, if the task of the integrated system encompasses strain measurement, at least the sensors should be placed outside the neutral plane. Neutral plane engineering ideally requires very thin systems because the neutral plane itself, that is, the stress-free zone, has zero thickness. Thus, the thinner the system is, the lower are stress and strain in its top and bottom layer. This effect is, for example, used in ultrathin chip packaging and similar approaches that reduce the substrate thickness of silicon devices, which is not determined by function, but by the need for certain handling characteristics of the silicon wafers, leaving little more than the structured layers. The resulting flexibility in bending has been described by Wagner and Bauer, stating that "like any other stiff material, circuits become flexible and rollable when their thickness is reduced to 1/1000 of the desired radius of curvature" [33]. Since it is possible, as Lecavelliers des Etangs-Lavellois *et al.* [32] have shown, to reduce a complementary metal-oxide-semiconductor (CMOS) microchip to a mere 5.7 μm thickness, bending radii of a few millimeters can be realized this way.

Use of materials that inherently offer flexibility and/or stretchability is an approach that can either be extended to the full system, or limited to the interconnects between components like sensors, microprocessor, or energy

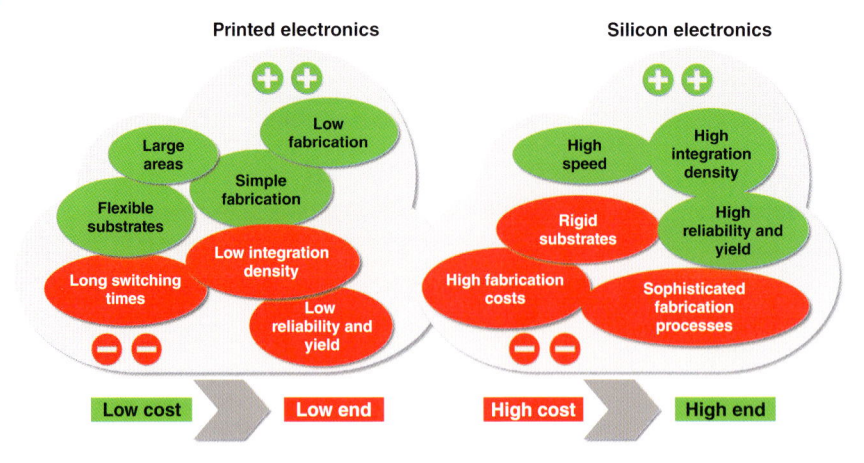

Figure 1.7 Qualitative comparison of organic printed and conventional silicon microelectronics.

harvesting and storage devices. The latter case is essentially the rigid island/flexible interconnect solution. Flexibility of interconnects can be achieved in different ways, of which some, that is, the introduction of residual stresses and/or deformation via mechanical measures or CTE mismatch, have already been described above. Further options have been summarized by Lang *et al.* [8]. In contrast, stretchability of functional devices is usually associated with organic solutions. A secondary advantage of these, besides a certain level of stretchability linked to low values of Young's modulus, is the fact that for these materials, cost-efficient large area manufacturing processes like roll-to-roll printing techniques exist. The price to be paid for these characteristics is related to performance. For one thing, taking microprocessors as an example, the resolution achievable via printed and organic electronics processes falls short of the CMOS process by roughly three orders of magnitude. In other words, on a given unit of area, the CMOS process would deliver 10^6 times the number of, say, transistors a printing process could. Furthermore, signal delays and clock rates associated with organic devices are once again orders of magnitude lower than those common for silicon microelectronics devices. These differences are illustrated in Figure 1.7.

Despite their obvious deficiencies in performance, organic electronics remain very interesting for material-integrated intelligent systems when it comes to simple tasks in signal and data preprocessing or for realizing interconnects on large areas. The printing processes available for this area also favor System-in-Foil packaging approaches, which yield full sensor systems in the form of thin, bendable foils. These can, in principle, be integrated very easily in a broad variety of materials that share a layered buildup, like continuous fiber reinforced composites or fiber metal laminates.

However, even if full compliance was achieved, sensor systems integrated in structural material would still represent a potential failure initiation side: Thus, the systems that are meant to survey structural health may become a danger to it themselves. For this reason, many efforts beyond compliance are being pursued with the aim of reducing the footprint of the sensor system within the host

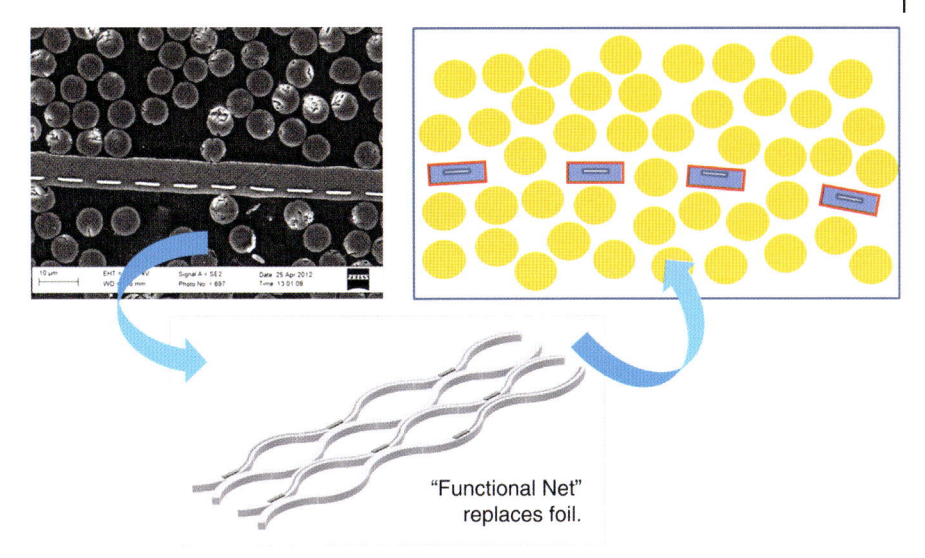

Figure 1.8 A sensor or sensor network realized as functional net, shown here to illustrate this principle. The wavy shape of remaining substrate and interconnects between the actual sensors is meant to provide stretchability. (Image courtesy of Institute of Microsensors, -Actuators and -Systems (IMSAS), University of Bremen.)

material. Elimination of substrate volume, either by continued thinning or by matching substrate and host material to allow the joining of both during integration, is one possibility to approach this aim. One step further in this direction is the layout of sensors as so-called functional nets as exemplified in Figure 1.8. Lang *et al.* [8] have coined the term "function scale integration" for such measures, highlighting that the substrate, which is needed in production only, should ideally not form a major part of the final, integrated system to the performance of which it does not contribute.

Part three of this book includes a section on sensor, electronics, and interconnect adaptation for material integration. Specific solutions for incorporation of sensors in fiber-reinforced polymers, sheet metal, and structures produced via additive manufacturing techniques are discussed in more detail in part four.

1.4 Signal and Data Processing (Part Five)

An active (aka. smart) sensor node should be able to operate autonomously and independently of the environment and other sensor nodes. It consists of different modules and units:

1) Digital signal processing and Data processing units (DSP/DPU: microprocessor, virtual machine, and digital logic and Register-Transfer Level architectures)
2) Communication units (CMU: physical, link, and data level)
3) Digital storage (DS)
4) Energy management units (EM: hardware and software control)

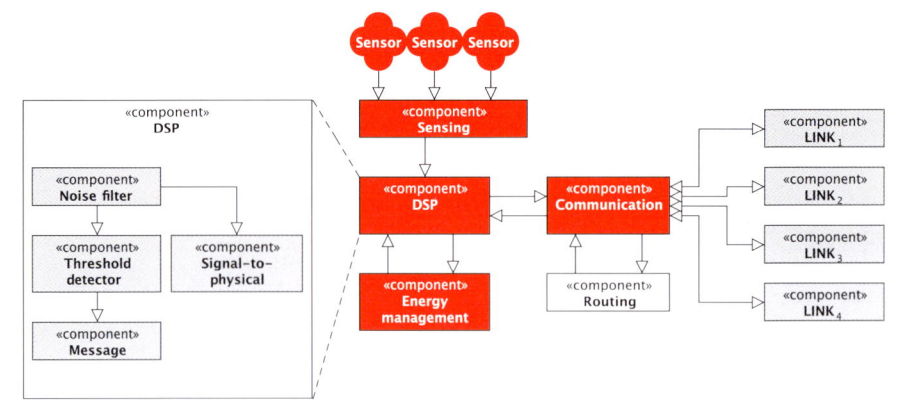

Figure 1.9 The information communication technology (ICT) architecture of a Smart Sensor Node consisting of analog and digital components.

5) Analog signal processing (ASP: analog electronics)
6) Analog-digital conversion (ADC)
7) Energy supply and Energy storage (ES)

The relation of the single units and the data flow is illustrated schematically in Figure 1.9. Typically, a sensor node processes low-level sensor data, improving the quality and accuracy using noise and fusion filter techniques. The sensor data or preprocessed information is passed from this information source to an information sink node by using message passing routed in some kind of communication network.

A smart sensor node integrated in materials must operate under harsh environmental conditions, especially concerning power, size, and technical failure constraints. Smart and distributed sensing systems are technological cornerstones of the Internet-of-Things (IoTs), wearable electronic devices, future transportation, environmental monitoring, and smart cities, shifting toward the Internet-of-Everything.

On the one hand, cloud computing relies on large-scale computers and reliable distributed services, whereas computing in materials focuses on low-scale computers, low power consumption, and technical failures as the usual case, not an exception. But both computing paradigms have to deal with a large amount of data (big data), and efficient algorithms suitable for information and pattern extraction (data mining). The Internet-of-Things is layered between material computing and cloud computing, and can be already considered as a bridge technology, but today still consisting of a diversity of solutions for specific situations and environments using traditional generic computer algorithms and paradigms.

An ongoing trend in microsystem technologies enables the integration of computational units in materials and technical structures, but they are characterized by limited computation power, storage, and reliability. Usually such a computational unit consists of a single microchip integrating the entire information and communication technologies (ICTs), and commonly performing sensing with integrated or attached sensors. The size can be reduced to the mm^3 range, with densities up to 10 million loosely coupled computing units per m^3 (e.g., smart dust [12,34,35]).

Traditional information processing methodologies using generic computers and computer networks cannot be applied and transferred directly to material-integrated embedded system networks due to their insufficient scaling, adaptability, configuration, and reliability. Material computing (not to be confused with computing of materials or materials informatics) facing the special constraints of digital computation in materials is distinct from traditional computing. This concerns programming, communication, and processing architecture models.

The deployment of digital logic enables the optimized application-specific design and implementation of various data processing architectures far beyond traditional generic microprocessor systems, though they are commonly implemented with large-scale digital logic, too. Digital logic itself is inherently parallel. Functional units can be easily parallelized on data path level, requiring only a data dependency analysis. But a parallel system on control level usually requires synchronization for resolving competition (due to concurrent access of shared resources) and providing coordination [36], demanding automated high-level synthesis approaches far beyond traditional software compilers.

Today, the CMOS technology is the prominent transistor architecture used for digital logic designs. A CMOS cell consists of two complementary field-effect transistors (FETs). The IC fabrication process requires clean-room conditions and is divided into multiple cycles of lithography, doping, material deposition and etching. The design of semi- or fully customized ASICs requires expert knowledge.

Due to the expensive and time-consuming ASIC fabrication process, rapid prototyping processes are very attractive, that is, using field programmable gate arrays (FPGA) not only during the development phase of digital logic circuits but also for the deployment in the production phase of small-lot productions.

Low-power design of all modules of a sensor node is mandatory. But low-power design of sensor nodes is not limited to the component view; it is moreover facing the algorithmic view by optimizing the data processing and communication algorithms. Energy-aware Algorithmic Scaling to microchip level can be considered as an important design methodology. At run-time, power management is performed on hardware and software level trying to minimize the energy consumption by guaranteeing the system liveliness.

Reliability necessitates robustness of the entire system in the presence of sensor, entire node, connection, data processing, and communication failures. Interaction between nodes is required to manage and distribute information.

Sensor fusion can improve the overall sensing quality and certainty significantly. It can be performed locally on node level by combining different sensors attached to the node or globally on network level by including neighborhood data, too.

To summarize, this part outlines the principles of different components of a sensor node and the challenges to integrate sensor nodes in materials and mechanical structures.

1.5 Networking and Communication (Part Six)

A smart sensor network is composed of a set of smart sensor nodes, as already discussed in part five, arranged in some kind of a regular communication structure (the network topology). A smart sensor node operates basically on locally acquired sensor data, whereas a smart sensor network can be seen as big virtual machine operating on global sensor data, possibly segmented. The network composition requires different processing and communication layers:

1) Communication hardware – the physical layer
2) Network communication protocols for messaging
3) Message passing and routing in arbitrary network topologies
4) Distributed and cloud computing addressing large-scale networks, distributed storage, and distributed data processing.
5) Agent-based computing addressing autonomous and reliable communication and data processing in large-scale networks with a unified programming and execution model.

Since the advent of networked computing in the 1960s, using simple peer-to-peer networks, there was a rapid advancement of computer and network technologies regarding increasing computational power and decreasing size – initially driven by Moore's law predicting an exponential increase in transistor density and an expected growth in storage and computational power, which is currently nearing saturating. Parallel and distributed computing is considered as solution to overcome well-known limitations of single-threaded computing. Moreover, single-point-of-failure (SPoF) issues must be avoided to ensure reliable operation of the entire system. In the past two decades, there was a shift from powerful centralized to distributed computing consisting of a wide range of different computer and communication technologies making the WEB a big virtual machine, recently driven by the advent of the IoT and cloud-based computing. These large-scale networks consisting of billions of devices demand for new communication and coordination methodologies with loosely coupled distributed information processing and big-data management, shown in Figure 1.10. There is a seamless transition from high-level computing in clouds to data processing performed in sensor networks.

The growing complexity of computer networks and their heterogeneous composition with devices ranging from servers with high computational power and high resource volumes down to low-resource mobile devices with low computational power demands unified and scalable new data processing paradigms and methodologies.

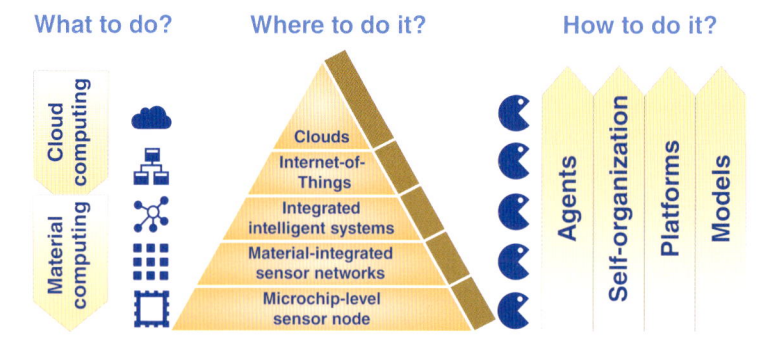

Figure 1.10 The progress: from material-integrated intelligent systems to Internet clouds with one unified information processing and coordination methodology using mobile agents.

1.5 Networking and Communication (Part Six)

The IoT is one major example and use-case emerging in the past decade, strongly correlated with cloud computing and big data concepts, and extending the Internet cloud domain with distributed autonomous sensor networks consisting of miniaturized low-power smart sensors. These smart sensors, for example, embedded in technical structures, are pushed by new trends emerging in engineering and microsystem applications.

Today, sensor nodes equipped with computation and communication capabilities can be scaled down to the cubic millimeter range (e.g., the smart dust mote), enabling the design of large-scale sensor network with loosely coupled nodes and reconfiguration at run-time.

Smart and distributed sensing systems are one of the technological cornerstones of the Internet-of-Things, wearable electronic devices, future transportation, environmental monitoring, and smart cities.

Furthermore, these sensor networks are used for sensorial perception or structural monitoring (load and health monitoring), deployed, for example, in cyber-physical systems (CPS) or Artificial Skin.

In the past decades there was a shift from passive sensors to smart sensor nodes, accompanied by an exponential increase of the sensor node density, shown in Figure 1.11. Distributed material-embedded sensor networks used in technical structures and systems require new data processing and communication paradigms, supporting fundamentally different architectures, establishing core concepts of Informatics in Materials (as with the new emerging Materials Informatics discipline, originally meaning Informatics for the design of materials).

Self-organizing capabilities, robustness, and adaptation to changing environmental conditions, like technical failures of components of the sensor network

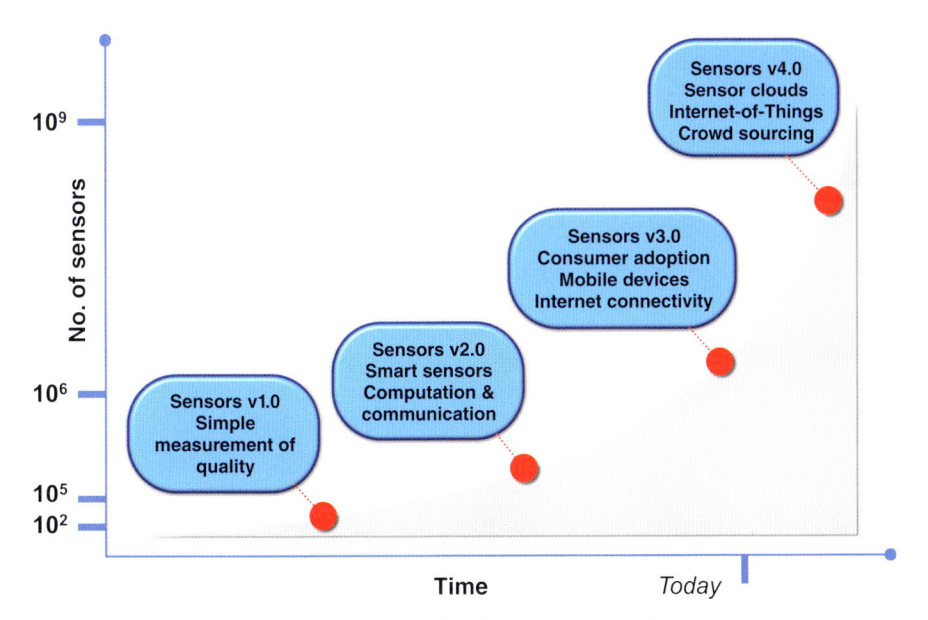

Figure 1.11 From passive sensors to networks of smart sensor nodes.

or a reconfiguration, can be caught by the deployment of mobile multiagent systems (MAS), a well-known concept of artificial intelligence (AI) [37–39].

It can be shown that agent-based computing can be used to partition typical sensing and computation in off-line and online (e.g., in-network and real-time) parts resulting in an increased overall system efficiency (performance and energy demands) and a unified programming interface between off- and online parts [40,41].

The agent model is also capable of providing a programming model for distributed heterogeneous systems crossing different network boundaries. The deployment of MAS in heterogeneous environments is often addressed on the organizational layer.

Multiagent systems are used to enable a paradigm shift from traditionally continuous-data-stream based to event-driven sensor data processing, resulting in increased robustness, performance, and efficiency.

Internet-of-Everything systems [42] require a programming model that is capable of dynamic composition, in conjunction with resource management, which can be provided by MAS.

The design of large-scale sensing networks requires contributions from different disciplines and model levels, shown in Figure 1.12. Beyond the bare sensor processing, self-estimation of the state of the entire system or a device to be monitored using Artificial Intelligence concepts will transform measuring systems in smart perceptive systems. One example is the prediction of the trust in measured

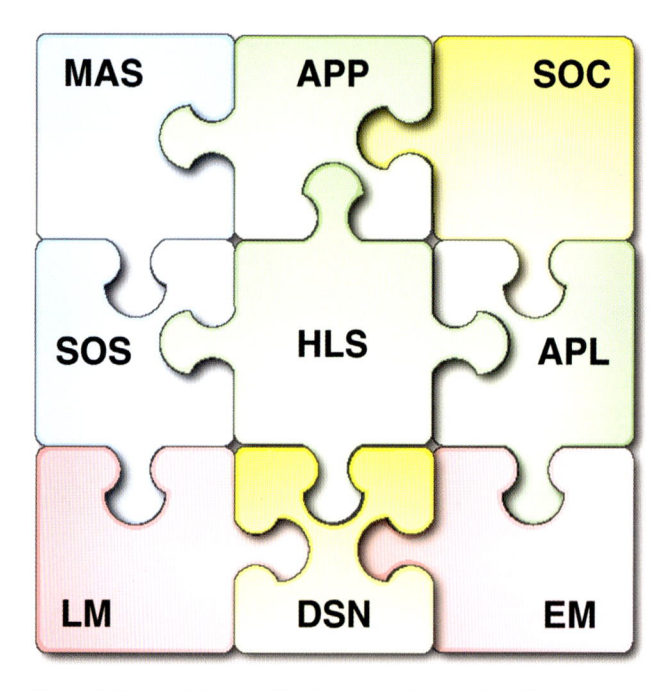

Figure 1.12 Modules contributing to the design of intelligent sensing systems from a computer science viewpoint: multiagent systems (MASs), agent processing platforms (APPs), system-on-chip design (SOC), high-level synthesis (HLS), agent programming model and language (APL), self-organizing systems (SOSs), load monitoring (LM), energy management (EM), distributed sensor networks (DSNs).

and computed results delivered with machine learning (ML). Often, measuring systems deliver ambiguous information that must be further classified. State estimation and ML can require a large amount of computing power. In biological systems like humans a brain with a complex neuronal network is used, not available in sensor networks. But distributed and decentralized state estimation, which can be deployed in large-scale sensor networks, can relax this requirement and can be performed by the low-resource sensor nodes themselves.

1.6 Energy Supply and Management (Part Seven)

Today, one major challenge in deploying smart wireless and material-integrated wired sensors in real use case and environments is related to energy consumption and guaranteeing adequate lifetime. Besides classical external energy supply, self-supply by using energy harvesting technologies gains importance. But energy harvesters still provide only low electrical power, requiring energy storage and repeating active-sleep cycles of the sensor node.

Energy management can be implemented on hardware and software level. In contrast to various commonly deployed energy management approaches, targeting microprocessor and operating system control, *smart algorithmic energy management* (SAEM), can be performed at run-time by applying a dynamic selection from a set of different (implemented) algorithms classified by their demand of computational power, and temporally by varying data processing rates. The smart energy management can be implemented, for example, with decision trees, based on quality-of-service (QoS) and energy constraints. It can be shown that the power and energy consumption of an application-specific SoC design strongly depends on the computational complexity of the used algorithms.

For example, a classical proportional–integral–differential (PID) controller used for the feedback position control of an actuator requires basically only the *P*-part; the *I*- and *D*-parts only increase position accuracy and response dynamics, which are selectable. Depending on the actual state of the system and the actual and estimated future energy deposit, suitable algorithms can be selected and executed optimizing the QoS and the trade-off between accuracy and economy. Energy analysis of data processing systems can be performed at run-time or at design time using simulation techniques, delivering valuable information for the optimization of sensor nodes.

1.7 Applications (Part Eight)

The final part eight of the present work is dedicated to the question of where, in which industries and for which specific purposes, material-integrated intelligent systems can be expected to see commercialization.

SHM has, for a significant time, been considered the killer application for large-scale introduction of material-integrated intelligent systems [43]. The interest in SHM, and thus also in new, more tightly coupled monitoring solutions, has been fueled by independent parallel developments in the receiving industries like the much extended use of fiber-reinforced composites rather than aluminum in the aerospace sector, exemplified by aircraft like the Boeing 787 Dreamliner or the Airbus A350-XWB [44]. Similar interest has been voiced by the wind energy sector, especially since the broader introduction of offshore wind turbines. In both cases, a major motivation is the reduction of maintenance costs that can be achieved through successful implementation of maintenance on demand or predictive maintenance systems supported by a powerful sensor network. Development in these fields is ongoing. One characteristic aspect of contemporary solutions is that in most cases data evaluation is not foreseen as part of the material-integrated system. On the contrary, in many cases it does not even form part of the on-board set of system components. As a consequence, data evaluation tends to be organized in a centralized manner. The vision, however, is clearly oriented toward materials that truly "feel" in our own sense [45], and beyond the structural monitoring aspect to scenarios like "fly-by-feel" [46].

Beyond SHM, production engineering is increasingly becoming aware of the promises of availability of sensorial information at new levels of depths. The term Industry 4.0 is very much en vogue today in Germany and increasingly so in other parts of the world. What it originally implies is that, following the original industrial revolution in the eighteenth century, the advent of mass manufacturing and the emergence of computer numerical control (CNC) and computer-integrated manufacturing (CIM), we are now standing at the verge of a further paradigm shift of comparable importance that is linked to terms like cloud-based design and manufacturing (CBDM) or intelligent production systems. The advent of additive manufacturing with its unparalleled flexibility is an accompanying development, providing a major support to what is at the heart of Industry 4.0: increased autonomy of production systems. In additive manufacturing, geometry representation is entirely digital and not contained in tools as in the case of most conventional manufacturing processes. Needless to say, these systems need information to base their decisions on; they need a connection to the world in which they are meant to function; in other words, they need sensory information and the capability to make sense of it. Ultimately, material-integrated intelligent systems will cover this demand. In conjunction with additive manufacturing, this will facilitate new ways of organizing production systems and product design [47–49].

Similar to the generation-based line of thought behind the term Industry 4.0, Meyendorf *et al.* have suggested a distinction between different ages in the development of a wider industrial world, or maybe even of human society as a whole. Their perspective considers the industrial revolution, set at the turn of the eighteenth century and based, for example, on the availability of iron and steel in large amounts, the steam engine and early complex though still mechanically controlled manufacturing systems, as the first step: Machines replace human muscular power. Electricity, communication and first sensors, and finally the development of computers in the twentieth century herald the next phase, the age of information, in which electronic machines replace human memory. Following this rationale, the twenty-first century becomes the age of machine decision: Smart systems assume the role of humans in decision-making in more and more natural environments. Figure 1.13 graphically depicts some of the major

>2009 Material-integrated Intelligent Systems

e.g. **Sensorial Materials**: ISIS, University of Bremen, Germany
Robotic Materials: University of Colorado, Boulder, USA
Nervous/Sensor Array Materials: King Fahd University, Dhahran, Saudi Arabia

Technology Area

- Computer Science: Artificial Intelligence etc.
- Computer Science: Foundations
- Production Engineering
- Industrial Revolution: Steam Power
- Industrial Revolution: Iron and Steel

Age of Decision — Smart Systems replace human decision

- 2016 Tesla's AutoPilot offering level 2-3 autonomous driving
- 2012 Cloud Based Design and Manufacturing (CBDM)
- 1996 Cloud Computing term coined in the modern sense
- 1990s Intelligent agents
- 1989 First commercial software package using machine learning
- 1988 Weiser coins the term Ubiquitous Computing
- 1987 Tang and Van Slyke's first organic diode
- 1985 Internet of Things (IoT) term coined
- 1980s Embodied cognitive science
- 1971 Texas Instruments TMS 1000 microprocessor
- 1971 First ARPANET email sent
- 1963 Wanlass' patent of the CMOS process
- 1960s Network/distributed computing
- 1960s Computer Numeric Control (CNC) of machine tools
- 1956 Artificial Intelligence coined as a term
- 1947 Bardeen/Brattain/Shockley's first transistor
- 1946 von Neumann architecture
- 1946 Zuse's Plankalkül first high-level programming language
- 1941 Zuse's Z3 programmable computer
- 1936 Universal Turing machine
- 1926 Lilienfeld's patent of a field-effect transistor

Age of Information — Electronic machines replace human memory

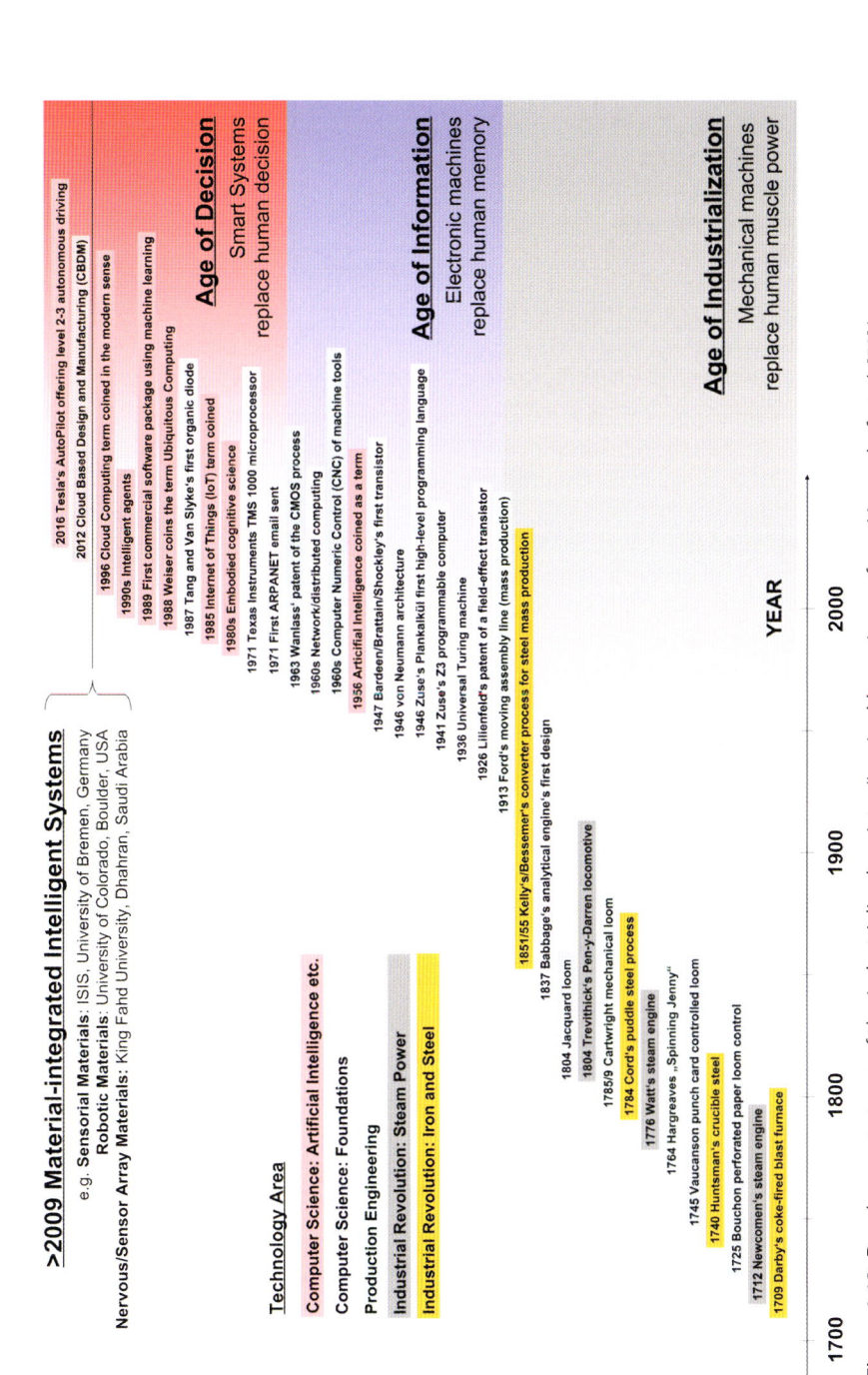

- 1913 Ford's moving assembly line (mass production)
- 1851/55 Kelly's/Bessemer's converter process for steel mass production
- 1837 Babbage's analytical engine's first design
- 1804 Jacquard loom
- 1804 Trevithick's Peny-Darren locomotive
- 1785/9 Cartwright mechanical loom
- 1784 Cord's puddle steel process
- 1776 Watt's steam engine
- 1764 Hargreaves „Spinning Jenny"
- 1745 Vaucanson punch card controlled loom
- 1740 Huntsman's crucible steel
- 1725 Bouchon perforated paper loom control
- 1712 Newcomen's steam engine
- 1709 Darby's coke-fired blast furnace

Age of Industrialization — Mechanical machines replace human muscle power

YEAR: 1700 — 1800 — 1900 — 2000

Figure 1.13 Development stages of the industrialized society. (Inspired by an image from Meyendorf et al. [50].)

developments that have become the hallmarks of these distinct ages, partially taking over Meyendorf *et al.*'s original thoughts [50].

The concept of an age of machine decision leads back to the introductory statement by Weiser that the technology revolution would move into the everyday, the small and the invisible [7]. In terms of application scenarios, this notion evokes the IoT, meant to provide the products surrounding us with the ability to interact and exchange information via the Internet or a similar structure. Such capabilities will open up new ways of using and interacting with objects. Tactile skins as currently used primarily in the field of robotics may have an important part in this, but human–machine interaction may profit on more generic levels, too [51]. Besides, the wide availability of product life cycle data will facilitate new business models, provided that regulatory issues like privacy and ownership of data can be solved. Such questions are treated in part eight of this book.

Looking forward, what will the future hold in store, assuming that material-integrated intelligent systems will prevail?

The vision of the ISIS Sensorial Materials Scientific Centre reads as follows:

"Five senses support us in exploring our world. Engineering components lack such natural abilities. Sensorial materials will help to bridge this gap".

Further to sensing, viable solutions for adaptivity may emerge. Locally initiated and controlled property or shape change will certainly open up additional fields of application [52,53].

What we foresee as a long-term objective linked to the introduction of material-integrated intelligent systems is nothing less than the step from the Internet-of-Things to the Internet-of-Everything – and all that this entails.

To conclude, maybe the single most important lesson learned from the past years of research on material-integrated intelligent systems is the need to establish an interdisciplinary design approach, bringing together five major scientific communities:

- Materials science
- Production engineering
- Electrical and microsystems engineering
- Computer science and artificial intelligence
- Mathematics

This is what the present book is about.

References

1 Lang, W. *et al.* (2011) Sensorial materials – a vision about where progress in sensor integration may lead to. *Sensors and Actuators A: Physical*, **171**, 1–2.
2 Lehmhus, D., Bosse, S., and Busse, M. (2013) Sensorial materials, in *Structural Materials and Processes in Transportation* (eds D. Lehmhus, M. Busse, A.S. Herrmann, and K. Kayvantash), Wiley-VCH, Weinheim.
3 Lehmhus, D. (2015) Material-integrated intelligent systems – notes on state of the art and current trends. Proceedings of the 5th Scientific Symposium of the CRC/TR 39 PT-PIESA, Dresden, Germany, September 14–16.

4 McEvoy, M.A. and Correll, N. (2015) Materials that combine sensing, actuation, computation and communication. *Science*, **347**, 1261689-1–1261689-8.
5 Mekid, S. and Kwon, O.J. (2009) Nervous materials: a new approach for better control, reliability and safety of structures. *Science of Advanced Materials*, **1**, 276–285.
6 Mekid, S., Saheb, N., Khan, S.M.A., and Qureshi, K.K. (2015) Towards sensor array materials: can failure be delayed? *Science and Technology of Advanced Materials*, **16**, 034607 (15 pp.).
7 Weiser, M. (1991) The computer for the 21st Century. *Scientific American*, **265**, 94.
8 Lang, W., Jakobs, F., Tolstosheeva, E., Sturm, H., Ibragimov, A., Kesel, A., Lehmhus, D., and Dicke, U. (2011) From embedded sensors to sensorial materials – the road to function scale integration. *Sensors and Actuators A: Physical*, **171**, 3–11.
9 Toffoli, T. and Margolus, N. (1987) *Cellular Automata Machines: A New Environment for Modeling*, MIT Press, Cambridge, MA.
10 Toffoli, T. and Margolus, N. (1991) Programmable matter: Concepts and realization. *Physica D*, **47**, 263–272.
11 Hall, J.S. (1993) Utility fog: a universal physical substance. NASA. Lewis Research Center, Vision 21: Interdisciplinary Science and Engineering in the Era of Cyberspace, pp. 115–126 (SEE N94-27358 07–12).
12 Warneke, M., Last, M., Liebowitz, B., and Pister, K.S.J. (2001) Smart dust: communicating with a cubic-millimeter computer. *Computer*, **34**, 44–51.
13 Peters, C., Zahlen, P., Bockenheimer, C., and Herrmann, A.S. (2011) Structural health monitoring (SHM) needs S^3(sensor-structure-system) logic for efficient product development. Proceedings of the SPIE 7981, Sensors and Smart Structures Technologies for Civil, Mechanical, and Aerospace Systems 2011, San Diego, CA, March 7–10.
14 Lehmhus, D. *et al.* (2009) Simulation techniques for the description of smart structures and sensorial materials. *Journal of Biological Physics and Chemistry*, **9**, 143–148.
15 Lehmhus, D. *et al.* (2013) When nothing is constant but change: adaptive and sensorial materials and their impact on product design. *Journal of Intelligent Material Systems and Structures*, **24**, 2172–2182.
16 Lehmhus, D. and Bosse, S. (2017) Self-adaptive smart materials: a new agent-based approach. *Proceedings*, **1**, 35, doi: 10.3390/ecsa-3-S2005
17 Mayer, D., Melz, T., Pille, C., and Woestmann, F.-J. (2008) CASTRONICS – direct integration of piezo ceramic materials in high pressure die casting parts for vibration control. Proceedings of Actuator 2008, 11th International Conference on New Actuators & 5th International Exhibition on Smart Actuators and Drive Systems, Bremen, Germany, June 9–11.
18 Pille, C. (2009) Produktidentifikation, Intralogistik und Plagiatschutz - RFID-Integration in Gussbauteile, in (2009), S. V/1-V/4″, presented at the BDG-Fachtagung Gussteilkennzeichnung - Methoden und Datenmanagement - Praxisberichte, S. V/1–V/4.
19 Pille, C. (2010) In-process embedding of piezo sensors and RFID transponders into cast parts for autonomous manufacturing logistics. Proceedings of the Smart Systems Integration (SSI) 2010, Como, Italy, March 23–24.

20 Pille, C., Biehl, S., and Busse, M. (2012) Encapsulating piezoresistive thin film sensors based on amorphous diamond-like carbon in aluminium castings. Proceedings of the 1st Joint International Symposium on System-Integrated Intelligence (SysInt 2012), Hanover, Germany, June 27–29.

21 Schwankl, M., Rübner, M., Singer, R.F., and Körner, C. (2013) Integration of PZT-ceramic modules using hybrid structures in high pressure die casting. *Procedia Materials Science*, **2**, 166–172.

22 Schwankl, M., Rübner, M., Flössel, M., Gebhardt, S., Michaelis, A., Singer, R.F., and Koerner, C. (2014) Active functionality of piezoceramic modules integrated in aluminum high pressure die castings. *Sensors and Actuators A: Physical*, **207**, 84–90.

23 Amalu, E.H., Ekere, N.N., and Bhatti, R. (2009) High temperature electronics: R&D challenges and trends in materials, packaging and interconnection technology. Proceedings of the 2nd International Conference on Adaptive Science & Technology (ICAST 2009), Accra, Ghana, January 14–16. doi: 10.1109/ICASTECH.2009.5409731.

24 Dziedzic, A. and Nowak, D. (2013) Thick-film and LTCC passive components for high-temperature electronics. *Radioengineering*, **22**, 218–226.

25 Herfurth, P., Maier, D., Men, Y., Rötsch, R., Lugani, L., Carlin, J.-F., Grandjean, N., and Kohn, E. (2013) GaN-on-insulator technology for high-temperature electronics beyond 400°C. *Semiconductor Science and Technology*, **28**, 074026.

26 Maier, D., Alumari, M., Grandjean, N., Carlin, J.-F., di Forte-Poisson, M.-A., Dua, C., Chuvilin, A., Troadec, D., Gaquiere, C., Kaiser, U., Delage, S.L., and Kohn, E. (2010) Testing the temperature limits of GaN-based HEMT devices. *IEEE Transactions on Device and Materials Reliability*, **10**, 427–436.

27 Palmour, J.W., Kong, H.S., and Davis, R.F. (1988) Characterization of device parameters in high-temperature metal-oxide-semiconductor field-effect transistors in β–SiC thin films. *Journal of Applied Physics*, **64**, 2168–2177.

28 Watson, J. and Castro, G. (2015) A review of high-temperature electronics technology and applications. *Journal of Materials Science: Materials in Electronics*, **26**, 9226–9235.

29 Heber, T. *et al.* (2014) Production process adapted design of thermoplastic-compatible piezoceramic modules. *Composites Part A*, **59**, 70–77.

30 Choi, W.M., Song, J., Khang, D.-Y., Jiang, H., Huang, Y.Y., and Rogers, J.A. (2007) Biaxially stretchable "wavy" silicon nanomenbranes. *Nano Letters*, **7**, 1655–1663.

31 Dumstorff, G., Paul, S., and Lang, W. (2014) Integration without disruption: the basic challenge of sensor integration. *IEEE Sensors Journal*, **14**, 2102–2111.

32 Lecavelliers des Etangs-Lavellois, A. *et al.* (2013) A converging route towards very high frequency, mechanically flexible, and performance stable integrated electronics. *Journal of Applied Physics*, **113**, 153701.

33 Wagner, S. and Bauer, S. (2012) Materials for stretchable electronics. *MRS Bulletin*, **37**, 207–213.

34 Cook, B.W., Lanzisera, S., and Pister, K.S.J. (2006) SoC issues for RF Smart Dust. *Proceedings of the IEEE*, **94**, 1177–1196.

35 Warneke, M. and Pister, K.S.J. (2002) Exploring the Limits of System Integration with Smart Dust. Proceedings of IMECE'02 2002 ASME International

Mechanical Engineering Congress & Exposition, New Orleans, LA, November 17–22.

36 Bosse, S. and Lehmhus, D. (2010) Smart communication in a wired sensor- and actuator-network of a modular robot actuator system using a hop-protocol with delta-routing. Proceedings of the Smart Systems Integration Conference, Como, Italy, March 23–24. ISBN: 978-3-8007-3208-1.

37 Bosse, S. and Pantke, F. (2013) Distributed computing and reliable communication in sensor networks using multi-agent systems. *Production Engineering, Research and Development*, **7**, 43–51.

38 Bosse, S. (2014) Distributed agent-based computing in material-embedded sensor network systems with the agent-on-chip architecture. *IEEE Sensors Journal*, **14**, 2159–2170.

39 Bosse, S. (2015) Unified distributed computing and co-ordination in pervasive/ubiquitous networks with mobile multi-agent systems using a modular and portable agent code processing platform. *Procedia Computer Science*, **40**. doi: 10.1016/j.procs.2015.08.312

40 Bosse, S. and Lechleiter, A. (2014) Structural health and load monitoring with material-embedded sensor networks and self-organizing multi-agent systems. *Procedia Technology*, **15**, 668–690.

41 Bosse, S., Lechleiter, A., and Lehmhus, D. (2017) Data evaluation in smart sensor networks using inverse methods and artificial intelligence (AI): towards real-time capability and enhanced flexibility. *Advances in Science and Technology*, **101**, 55–61.

42 Batalla, J.M., Mastorakis, G., Mavromoustakis, C.X., and Pallis, E. (eds) (2017) *Beyond the Internet of Things – Everything Interconnected*, Springer.

43 Worden, K., Farrar, C.R., Manson, G., and Park, G. (2007) The fundamental axioms of structural health monitoring. *Proceedings of the Royal Society A*, **463**, 1639–1664.

44 Hermann, A.S. (2013) Polymer matrix composites, in *Structural Materials and Processes in Transportation* (eds D. Lehmhus, M. Busse, A.S. Herrmann, and K. Kayvantash), Wiley-VCH, Weinheim.

45 Renton, W.J. (2001) Aerospace and structures: where are we headed ? *International Journal of Solids and Structures*, **38**, 3309–3319.

46 Salowitz, N. and Chang, F.-K. (2012) Bio-inspired intelligent sensing materials for fly-by-feel autonomous vehicles. *IEEE Explore*. doi: 10.1109/ICSENS.2012.6411534

47 Lehmhus, D., Wuest, T., Wellsandt, S., Bosse, S., Kaihara, T., Thoben, K.-D., and Busse, M. (2015) Cloud-based additive manufacturing and automated design: a PLM-enabled paradigm shift. *Sensors*, **15**, 32079–32122.

48 Lehmhus, D., Aumund-Kopp, C., Petzoldt, F., Godlinski, D., Haberkorn, A., Zöllmer, V., and Busse, M. (2016) Customized smartness: a survey on links between additive manufacturing and sensor integration. *Procedia Technology*, **26**, 284–301.

49 Wu, D., Rosen, D.W., Wang, L., and Schaefer, D. (2015) Cloud-based design and manufacturing: A new paradigm in digital manufacturing and design innovation. *Comput Aided Design*, **59**, 1–14.

50 Meyendorf, N., Frankenstein, B., Hentschel, D., and Schubert, L. (2007) Acoustic techniques for structural health monitoring. IV Conferencia Panamericana de END, Buenos Aires, Brazil, October.
51 Dahiya, R.S. *et al.* (2013) Directions toward effective utilization of tactile skin: a review. *IEEE Sensors Journal*, **13**, 4121–4138.
52 Lehmhus, D. and Bosse, S. (2016) Self-Adaptive Smart Materials: A new agent-based approach. Proceedings of the 3rd Electronic Conference on Sensors and Applications (ECSA-3), November 15–30.
53 Bosse, S. and Lehmhus, D. (2017) Towards Large-Scale Material-Integrated Computing: Self-Adaptive Materials and Agents. Proceedings of the 2017 IEEE 2nd International Workshops on Foundations and Applications of Self* Systems (FAS*W), September 18th-22nd, University of Arizona, Tucson, AZ, USA, DOI: 10.1109/FAS-W.2017.123.

Part Two

System Development

2

Design Methodology for Intelligent Technical Systems

Mareen Vaßholz, Roman Dumitrescu, and Jürgen Gausemeier

University of Paderborn, Heinz Nixdorf Institute, Strategic Product Planning and Systems Engineering, Fürstenallee 11, 33102 Paderborn, Germany

Machines are ubiquitous. Their purpose is to make life easier. The increasing integration of information and communication technology in the field of conventional mechanical engineering implies considerable potential for innovation. Most modern products created in the field of mechanical engineering and related areas, such as automobile technology, already rely on the close symbiotic interaction between mechanics, electrics/electronics, control engineering, and software technology. This is expressed by the term "mechatronics." This term refers to the symbiotic cooperation of mechanics, electronics, control engineering, and software technology in order to improve the behavior of a technical system, which in turn can significantly improve the cost-benefit ratio of familiar products and also stimulate the innovation of new products [1,2]. German industry has a leading position within global competition. The reason is that the progressing development of information and communication technology builds the basis for over 80% of innovations, enabling intelligent technical systems with reduced costs and higher benefits and economic efficiency [3,4]. In the following text we will outline the way from mechatronic systems toward intelligent technical systems (cf. Section 2.1). In Section 2.2, we will present the structure and characteristics of self-optimizing systems as an example of intelligent technical systems.

The design of such systems is an interdisciplinary and complex task. Therefore, effective and continuous cooperation and communication between developers from different domains during the whole development process are required. In Section 2.2, we will present a design methodology for intelligent technical systems to support the developer appropriately with specific expertise like self-optimization. It consists of a reference process, methods, and tools that are based on our experiences within the Collaborative Research Centre (CRC) 614.

Innovations often fail in the market, because they either do not provide enough benefit for the customer or are too costly. Sixty percent of the costs and benefits of a system are defined within the early development. In the conceptual design, developers' often lack the understanding of the origin of costs and the customers' needs. Therefore, the presented approach will also focus on the economic efficiency of intelligent technical systems during the development process [4].

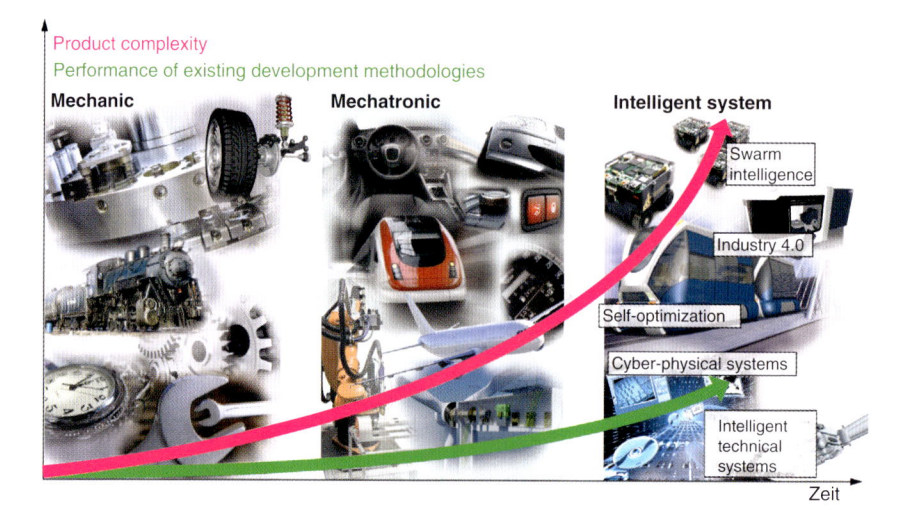

2.1 From Mechatronics to Intelligent Technical Systems

The technical systems of tomorrow will go beyond current mechatronics by incorporating inherent intelligence. The route to intelligent technical systems is determined by three general technology trends:

1) *Miniaturization of electronics*: This development provides numerous advantages by combining the parallelization of information processing by multicore processors, an increase of storage capacity, and a reduction of energy demands. This, in turn, enables the development of suitable hardware for intelligent technical systems [5].
2) *Software technology as driver for innovations*: The increasing degree of software in modern engineering products enables new functionalities. At the same time, the complexity of such systems is increasing rapidly, especially for systems with embedded software. Modern, model-based methods, notations, and tools enable us to cope with the complexity, and nonetheless help to create software of high quality (cf. Figure 2.1) [6].
3) *Networking of information systems*: The "Internet of Things," "ubiquitous computing," "pervasive computing," or " ambient intelligence" are current research areas that deal with electronic, mostly wireless, networking of information processing systems. This technological trend is the basis for intelligent networking systems [6].

Self-optimization refers to the endogenous adjustment of the objectives of a system with regard to changing influences and the resulting, objective-compliant, autonomous adaptation of the system's behavior (cf. Section 2.2) [7].

Figure 2.1 From mechanics to intelligent technical systems.

Cyber-physical systems are systems of the real world with embedded software that are additionally interconnected via a digital communication system. They are connected via both local networks and global networks such as the World Wide Web. Using such connections they are able to solve an underlying task together [8,9].

The described intelligent technical systems have the following four key characteristics in common [6,10]:

Adaptive: They interact with the environment and adapt their operation modes autonomously. In this manner, they can evolve during the runtime within the framework set by the designer and ensure their existence in the long term.

Robust: They are able to operate flexibly and autonomously in a dynamic environment, even in situations that are unexpected or were not foreseen by the developer. Uncertainties or the lack of information can be handled, at least to a certain degree.

Anticipative: Using empirical knowledge as a base, these systems anticipate future impacts and possible states. In this manner, dangers can be identified earlier and appropriate strategies to resolve the problems can be selected and executed. Thus, objectives can be achieved more efficiently.

User-friendly: These systems adapt to user-specific behavior and interact sensibly with the user. Its behavior is comprehensible for the user at all times.

Primarily – but not exclusively – the course of information processing is the driving force for the change from mechatronics to intelligent technical systems. Hence, our technology concept is established on the basic structure of mechatronic systems that is shown in Figure 2.2.

Mechatronic systems are distinguished by the functional and/or spatial integration of sensors, actuators, information processing, and a basic system. Also, of significance is the relationship to humans and the environment in which the mechatronic system operates. In general, mechatronic systems can also be composed of subsystems which themselves are mechatronic systems [11].

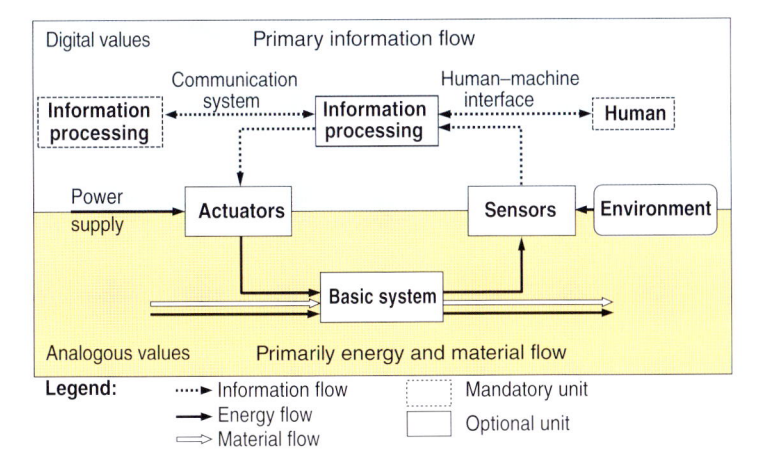

Figure 2.2 Basic structure of mechatronic systems [11].

The basic system is commonly a mechanical structure. Generally, any desired physical system is conceivable as a basic system. The relevant physical (continuous) values of the basic system or its environment are measured using sensors in order to improve the system behavior. The sensors supply the input variables for information processing, which, in most cases, takes place digitally, that is, discretely in terms of value and time. The information processing unit determines the necessary changes of the basic system using the measurement data as well as the user specifications (human–machine interface) and also available information from other processing units (communication system). The information processing unit often consists of control functions. The behavior adaptation of the basic system is caused by actuators [11].

The relationships between the basic system, sensors, information processing, and actuators are represented as flows. In principle, three types of flows can be distinguished: information flows, energy flows, and material flows [12]. As already mentioned, within the field of intelligent technical systems the focus lies on the information flow [11].

A mechatronic system provides a reactive and fixed coupling between sensors and actuators. Similar to cognitive biological creatures, intelligent technical systems must be able to modify these couplings. The cognitive processing must not replace the direct and reactive coupling, but has to coexist with it (cf. Figure 2.3) [11].

The cognitive science refers to cognitive creatures and thus intelligent technical systems that have the following three special characteristics [10]:

1) Active embedding into the environment and the ability to exchange information with it.
2) Flexible and environment-adaptive action control via internal representation of the system-relevant information about the environment.
3) Ability of learning and anticipating of the integrated information processing.

In order to represent these characteristics, Strube developed a three-layer model for a cognitive information processing [13].

The lowest layer includes the noncognitive regulation. That means continuous controlling and fixed reflexes, what we call motoric skills [13].

The midlayer represents the associative regulation. The process of learning is founded by conditioning. Probably the most famous example for this type of learning is Pavlov's dogs [13].

The cognitive regulation takes place in the highest layer. Cognition refers to all types of events to detect information, to process information, and to save it into memory. In addition, the information is used to adapt the creature's behavior. This means, processes like target management, planning, or controlling activities [13].

According to the three-layer model of Strube, the information processing is expanded to create intelligent mechatronic systems [10].

Taking increasing networking into account, we have to go a step further. As already explained, the third trend in technology is the networking of information systems. We, therefore, have expanded the reference structure in order to describe intelligent networked systems. It consists of several intelligent

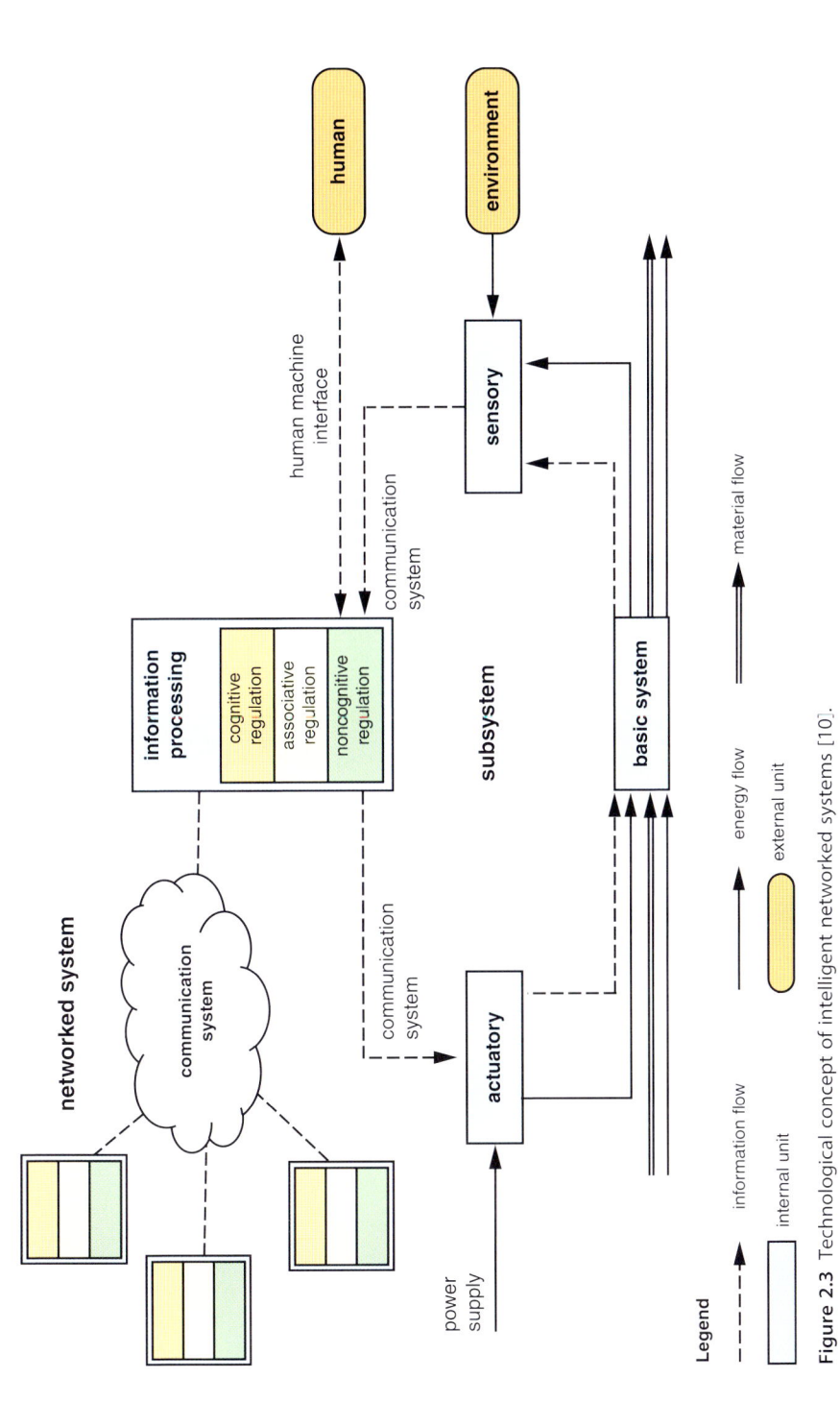

Figure 2.3 Technological concept of intelligent networked systems [10].

mechatronic systems that communicate and cooperate with each other. One example for this kind of systems is described in the following section.

2.2 Self-Optimizing Systems

Self-optimizing systems are systems with inherent partial intelligence. They are bases on mechatronic systems. The integration of cognitive functions (e.g., "share knowledge" or "coordinate behavior") into mechatronic systems enables self-optimizing systems. Self-optimization describes the endogenous adaptation of the system's objectives due to changing operation conditions and the resulting autonomous adjustment of system's parameters or system structure and consequently of the system's behavior [2,7]. The key aspects and the mode of operation of a self-optimizing system are illustrated in Figure 2.4a.

Influences on the technical system originate in its surrounding (environment, users, etc.) or from the system itself. They can support the system's objectives or hinder them. Influences from the environment, for example, strong winds or icy conditions, are unstructured and often unpredictable. They are called disturbance variables, if they hinder the system fulfilling its pursued objectives [2,7].

The user can influence the system, for instance, by choosing preferred objectives. It is also possible that the system itself or other technical systems will influence the system's objectives, for example, if mechanical components are damaged, the objective "max. safety" has to be prioritized [2,14].

The self-optimizing system determines its current pursued objectives (System of Objectives) on the basis of the encountered influences on the system, for example, the environment, the user, or the system itself. New objectives can be added, existing objectives can either be rejected or their priority can be modified during system operations. Therefore, the system of objectives and its autonomous change is the core of self-optimization [2,6,7]. We distinguish between external and inherent objectives. External objectives are set from the outside of the self-optimizing system, by other systems or by the user (e.g., for a driving module this could be "max. comfort"). Inherent objectives reflect the design purpose of the self-optimizing system. An inherent objective of a driving module can be, for example, "max. energy efficiency." Objectives build a hierarchy and each objective can thus be refined by subobjectives (e.g., "min. energy consumption" is a possible subobjective of "max. energy efficiency"). Inherent and external objectives that are pursued by the system at a given moment during its operation are called internal objectives [15].

Adapting the objectives leads to a continuous adjustment of the system behavior to the occurring situation. This is achieved by adapting parameters or reconfiguring the structure (e.g., adapting control strategies) [6]. The self-optimization process consists of the following three actions (cf. Figure 2.4b) [16]:

1) *Analyzing the current situation*: The current situation includes the current state of the system as well as all observations of the environment that have been carried out. Observations can also be made indirectly by communicating with other systems. Furthermore, a system's state contains previous

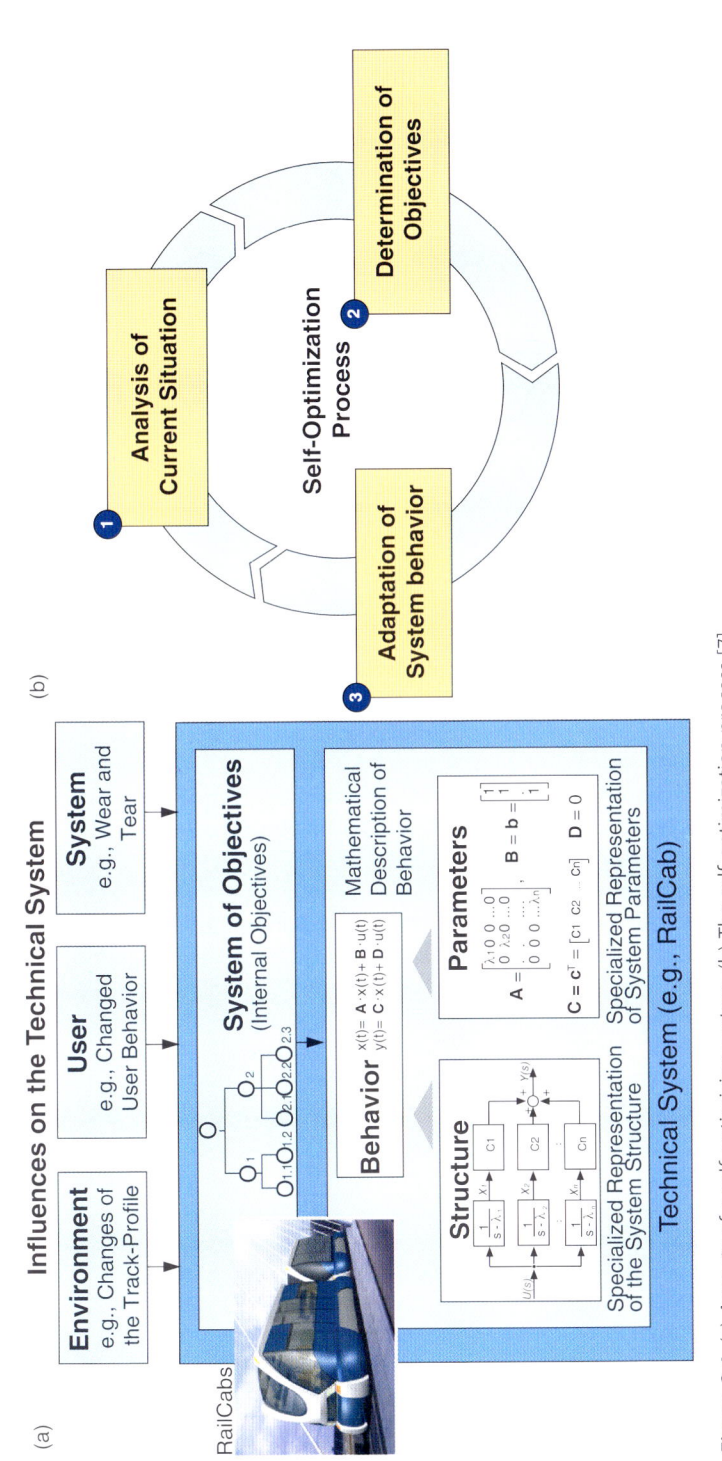

Figure 2.4 (a) Aspects of a self-optimizing system. (b) The self-optimization process [7].

observations that were saved. One basic aspect of this first step is the analysis of the fulfillment of the objectives [7].

2) *Determining the system's objectives*: The system's objectives can be extracted by choice, adjustment, and generation. By choice means the selection of one alternative output of a predetermined quantity of possible objectives. The adjustment of objectives means the gradual modification of existing objectives respective to their relative weighting. Generation means new objectives are being created that are independent from the existing ones [7].

3) *Adapting the system behavior*: The changed system of objectives demands an adaptation of the behavior of the system and its components. As mentioned before, this can be realized by adapting the parameters and, if required, by adapting the structure of the system. There are three different types of behavior adaptation strategies. Parameter adaptation means, for example, changing a control parameter. Structure adaptations affect the arrangement of the system elements and their relationships. Here, we distinguish between reconfiguration, which changes the relationships between a fixed set of available elements and compositional adaptation, in which new elements are integrated into the existing structure or existing elements are removed from it [14]. The self-optimization process leads, according to changing influences, to a new system state. Thus, a state transition takes place. The behavior adaptation finally concludes the self-optimization process [7].

The self-optimization process takes place if the three actions are performed repeatedly by the system while the sequence of the three actions can vary. For example, within the scope of planning, different situations are considered and according to the situations, the objectives are adapted. This results in repeated situation analysis, based on the determination of objectives. Thus, the self-optimization process is executed, if the situation of the system is changed or a planning for possible system scenarios is performed [7].

Thus, self-optimization can be considered as an extension of classical and advanced control engineering [17]. In order to provide an optimal conformity to the environment of the system at any time, self-optimizing systems utilize implemented adaptation strategies, instead.

An example for self-optimizing systems is the innovative railway system RailCab.

2.3 Design Methodology for Intelligent Technical Systems

The development of intelligent technical systems is challenging due to the involvement of different domains, such as mechanical, electrical/electronic, control and software engineering as well as experts from higher mathematics and artificial intelligence. This requires a fundamental understanding of the entire system as well as of its development. Therefore, an effective communication and cooperation between the developers is needed [18].

The design methodology for intelligent technical systems developed by the Collaborative Research Centre (CRC) 614 extends existing methodologies for mechatronic systems, such as the VDI guideline 2206 [19], the approach by

Isermann (2008) [20] or Ehrlenspiel (2007) [21], the iPeM-Modell [22], or the V-Model by Bender (2005) [23], to support the developer appropriately with specific expertise like self-optimization. It consists of a reference process, methods, and tools that are based on the experiences within the CRC 614 [18].

The reference process represents an ideal approach for the development of such complex systems. It needs to be customized according to the development project, the system type, and the environment of the development project in an implementation model. This model provides a detailed process sequence consisting of process steps from the reference process and builds the starting point for the project management. Furthermore, the planned process sequence of the implementation model can vary due to changing development objectives, such as time and costs, during the project execution [18].

In accordance with the existing development methodologies, it can basically be structured into two main phases: the "domain-spanning conceptual design" and the "domain-specific design and development" as shown in Figure 2.5.

Within the *conceptual design* (cf. Section 2.3.1), the basic structure and the operation mode of the system are defined and the system is also structured into subsystems. This division into subsystems has to result in a development-oriented product structure, which integrates the two basic and mostly contradictory views of shape- and function-oriented structure. This is a recursive process, which means that subsystems can also be assembled by further subsystems. All the results of the conceptual design are specified in the so-called principle solution. The principle solution based on the specification technique CONSENS (CONceptual design Specification technique for ENgineering of complex Systems) is built up to create common understanding of the system to be developed. The description of the principle solution is structured into the aspects environment, application scenarios, requirements, functions, active structure, behavior, and system of objectives and shape (cf. Figure 2.6). The aspects are computer-internally represented as partial models. To secure the overall consistency of the principle solution and to manage its complexity, a software support is necessary. The aspects relate to each other and ought to form a coherent system [14,18]. We will describe the aspects in detail within Section 2.3.1.

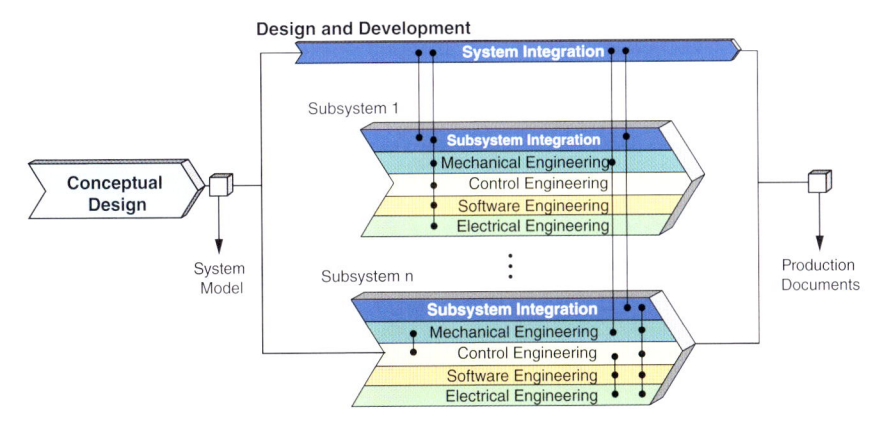

Figure 2.5 Macrocycle for the development of intelligent technical systems [18].

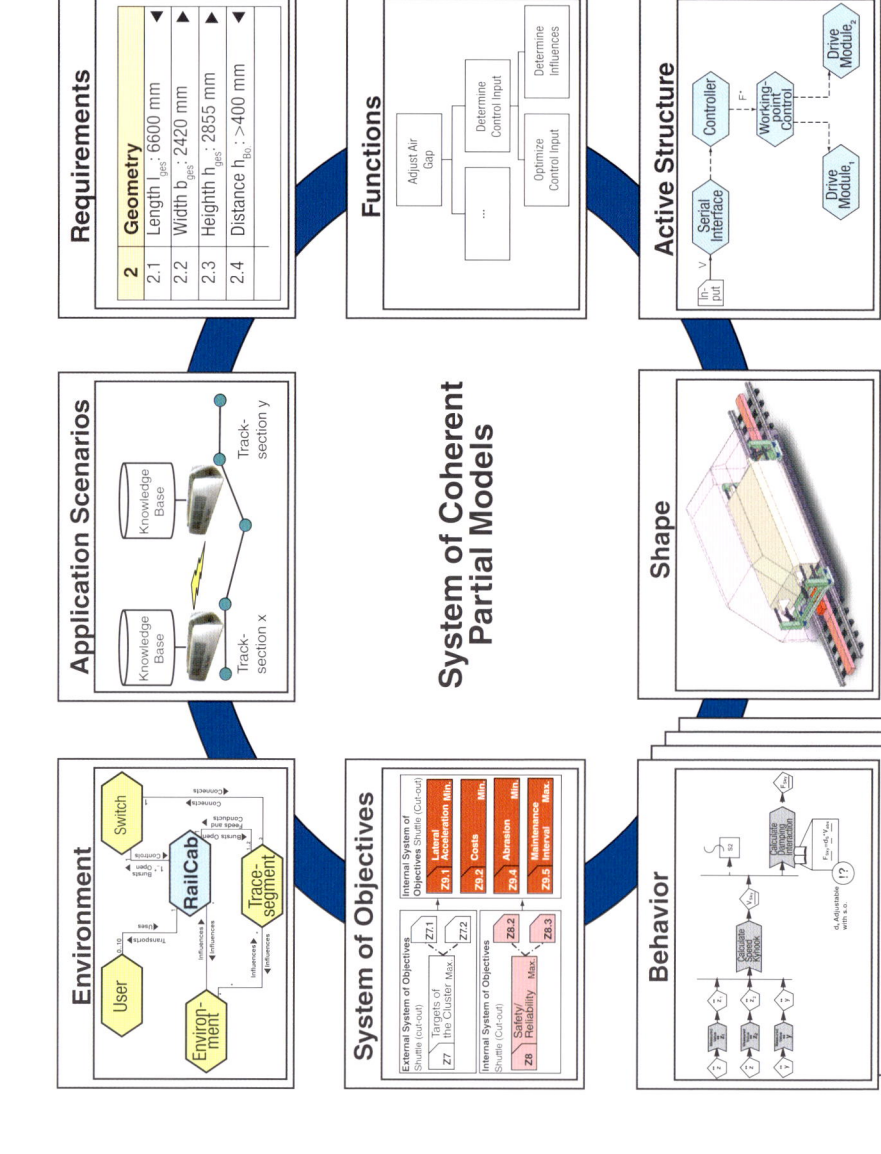

Figure 2.6 Partial models for the domain-spanning description of the principle solution [14].

Based upon the principle solution, the subsequent domain-specific design and development is planned and realized (cf. Section 2.3.1). All defined subsystems are developed in parallel to each other and each subsystem is developed in parallel within the participating domains [24]. During this domain-specific phase, the domains involved, work in parallel with their domain-specific methods and tools, for example, MCAD, ECAD or MATLAB [18].

2.3.1 Domain-Spanning Conceptual Design

The aim of the domain-spanning conceptual design is to develop the principle solution of the system. It determines the framework for the discipline-specific design and development as well as 60% of the systems lifecycle costs and the majority of its future benefit [18,25,26]. Thus, the economic efficiency of the system needs to be considered during this phase. In this early stage of the development, costs and benefits can only be estimated. Nevertheless this degree of precision is sufficient enough to make essential design decisions based on the economic efficiency over the systems life cycle. Additionally, a common understanding about sources of costs and benefits can be created [27,28].

The reference process for the domain-spanning conceptual design consists of four main phases *planning and clarifying the task, conceptual design on the system level, conceptual design on the subsystem level,* and *concept integration* (cf. Figure 2.7).

In *planning and clarifying the task*, the design task of the system and the requirements are identified and evaluated (cf. Section "Planning and Clarifying The Task"). The results are the list of requirements, the environment model, the recommended product structure type, and its design rules application scenarios

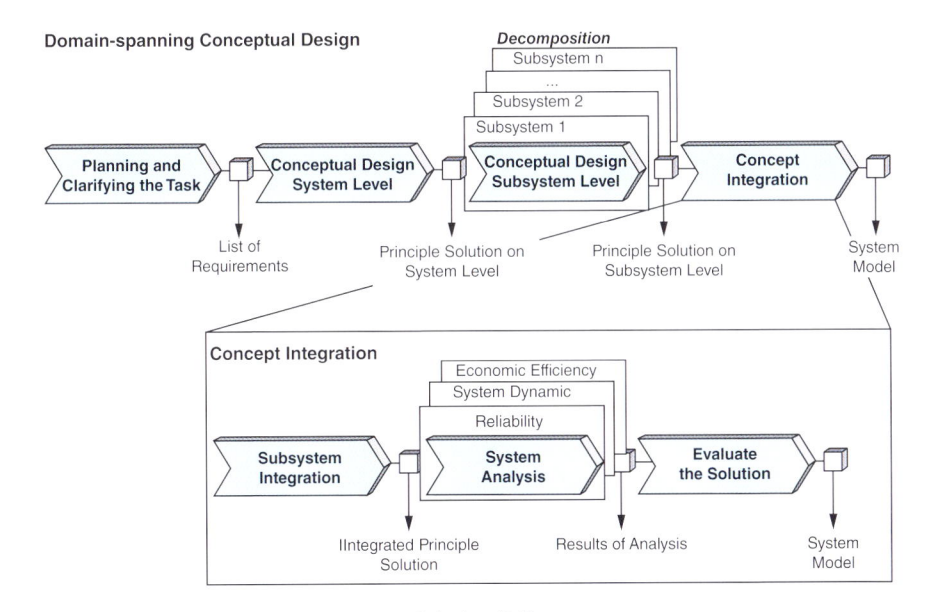

Figure 2.7 Domain-spanning conceptual design [29].

as well as an initial economic efficiency model. Based on the previously determined requirements on the system, solution variants are developed in the *conceptual design on the system level* (cf. Section "Conceptual Design on (Sub-)System Level"). These solution variants are evaluated based on their economic efficiency within this phase. Based on the results, the best one will be chosen and consolidated into the principle solution on the system level. This includes the identification of potential for the use of self-optimization, based on contradictions within the principle solution for the system that can be solved either by compromise or by self-optimization. The principle solution for the self-optimizing system on the system level is the result of this phase. Based on this, the system is modularized and a principle solution for each single subsystem is developed in the phase *conceptual design on the subsystem level* (cf. Section "Conceptual Design on (Sub-)System Level"). This procedure corresponds to the conceptual design on the system level, starting with planning and clarifying the task. The result of this phase is presented by the principle solutions on the subsystem level. This process can be conducted recursively, because the subsystem itself can be a system with subsystems and so forth. Within the phase *concept integration*, the principle solutions for the subsystems are integrated into one principle solution, which represents the complete system, (cf. Section "Concept Integration"). Afterward, the principle solution is analyzed regarding its economic efficiency, dynamical behavior, and dependability. In this analysis, phase contradictions between the principle solutions on the subsystem level are identified. Again it will be checked, if these contradictions can be solved by self-optimization. The result of this phase is an economic efficient principle solution for the complete system that serves as the starting point for the subsequent domain-specific design and development [18,29,30].

Planning and Clarifying the Task Starting point for the conceptual design is the development order that provides a short description of the intelligent technical system with all relevant information about future potential for success, stakeholders, customer requirements, target costs, and dates. In the phase *planning and clarifying the task*, the development task is abstracted in the first step and its core is identified. This includes identifying the relevant stakeholders for the proposed system as well as the potential for benefit [4,29].

In the next step, the most important boundary conditions and influences in the environment are investigated to ensure that the final system will work properly, without any restrictions caused by not considered interactions. Therefore, the aspect *environment* of the principle solution describes the embedding of the system within its environment; the system itself is treated as a "black box". In particular, other elements of the environment (e.g., the user, other technical systems, or the underground) and their interrelation with the system are described [14,29].

The aspect system of objectives describes external, inherent, and internal objectives of the system and their interrelations. Objectives build a hierarchy and each objective can thus be refined by subobjectives (e.g., "max. safety" is a possible subobjective of the objective "max. dependability"; "max. comfort" and "max. driving speed" are possible subobjectives of the objective "max. user satisfaction"). Inherent and external objectives that are pursued by the system at a given moment during its operation are called internal objectives. The selection of internal objectives and their prioritization occurs continuously during the operation of the system. The emerging external objectives (e.g., "max. comfort") as well as disturbances are identified during this phase [14,29].

Beyond that consistent combinations of influences that are called situations are formed. By the combination of characteristic situations with system states, application scenarios occur that describe the most common operation modes of the system and the corresponding behavior in a rough manner. For this step we developed a guideline to define application scenarios over the life cycle systematically [4]. Every application scenario describes a specific situation (e.g., start-up, failure of the system, or an interaction with the user), and the required behavior of the system for this situation. Thus, application scenarios characterize a problem and the possible solution for it. By modeling the application scenarios requirements and potential operational modes for the system can be identified [14,29].

The existing results of the first steps are documented by demands and requests within the list of requirements [30]. It presents an organized collection of requirements that need to be fulfilled by the system under development (e.g., features of the system, overall size, performance, quality, target costs). Requirements allow the development team to expose what is expected from the future system [14,29].

Based on the aspects environment, system of objectives, application scenarios, and requirements, the cost and benefits over the system's life cycle can be identified and structured in the economic efficiency model. The concrete syntax of the economic efficiency model is defined by the graphical notation of the elements and their relations (cf. Figure 2.8). To ensure that the model is plausible

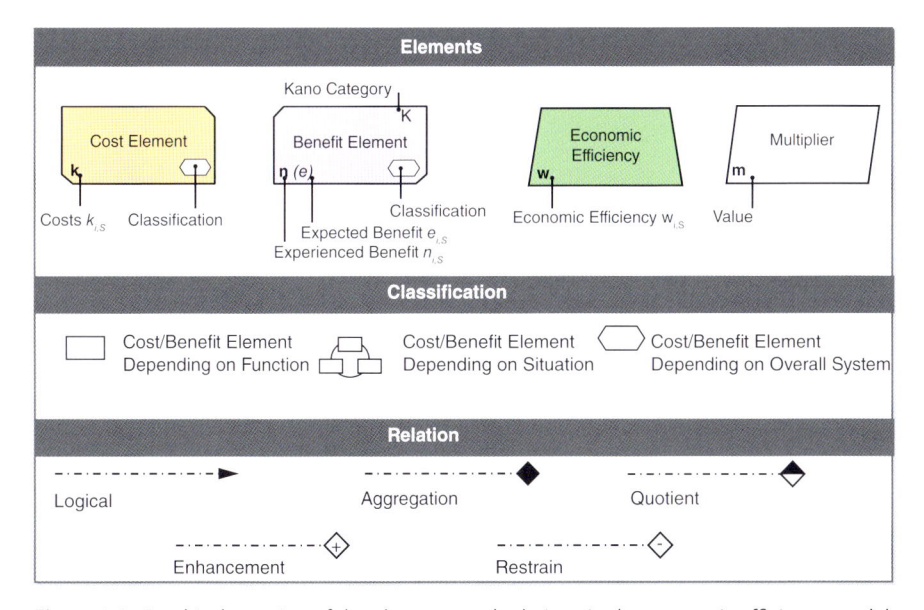

Figure 2.8 Graphical notation of the elements and relations in the economic efficiency model (concrete syntax [4]).

that means comparable, complete, and correct, we developed optional guidelines and mandatory requirements for the modeling process that need to be followed by the engineers [4].

Costs caused by the intelligent technical system are represented by *cost elements* and can be assigned to a life cycle phase and a basic structure following the cost breakdown structure consisting of life cycle costs on the top level can be build. Life cycle costs consist of purchase, commissioning, operation, maintenance, and disposal costs. The purchase costs can be decomposed into the target margin and the cost price and so forth [4].

The benefits over the life cycle are represented by *benefit elements*. The basic structure for the benefits is following Pümpin [31] as well as Keller and Kottler [32]. On the top level, the overall benefit is divided into the benefit for the customer and the benefit for the enterprise that develops and/or operates the system. The customer benefit is decomposed into objective and subjective benefit. The objective benefit is directly resulting from the economic efficiency of the system at the time of acquisition. The subjective benefit results from the behavior of the system, the customers' state of mind as well as synergy effects. The benefit for the enterprise results from the external benefit (enterprise value plus target margin) and the internal benefit like cost cutting, know-how, and synergy potential. The basic structure for costs and benefits over the life cycle can be used as a template and is the starting point for the development of the economic efficiency model [4].

In the first step, the identified cost and benefit elements are decomposed and structured according to the templates. The impact of the elements on a decomposition level to the upper level element can be different. Therefore, the elements are weighted using a weighting matrix (cf. Figure 2.9) [4].

Cost and benefit elements can either be dependent on a single function of the system, the situation during operation, or the overall intelligent technical system. The identified elements are classified accordingly. The quotient of costs and benefit element represents the economic efficiency. Cost, benefit, and economic efficiency elements can be assigned to a stakeholder element via logical relation [4].

Based on the expectations of the stakeholders and a benchmark with competitive systems on the market, the benefit elements are categorized according to Kano *et al.* in basic attributes (B), performance attributes (P), and delighters (D) [33]. Benefit elements with a high expectation that are already fulfilled to a high degree by existing systems are classified as basic attributes. Delighters are benefits that existing systems do not fulfill and that the customer does not expects already. Here, the system has the possibility for differentiation on the market. Benefits with a moderate evaluation of both criteria are classified as performance attributes [4].

The elements within the economic efficiency model have an influence on each other. For example, customer satisfaction has direct impact on the enterprise value. These interdependencies are represented by multiplier elements that are connected by enhancing or restraining relations. Multiplier elements are added in the last step. The result is an initial economic efficiency model (cf. Figure 2.9) [4].

Economic Efficency Model

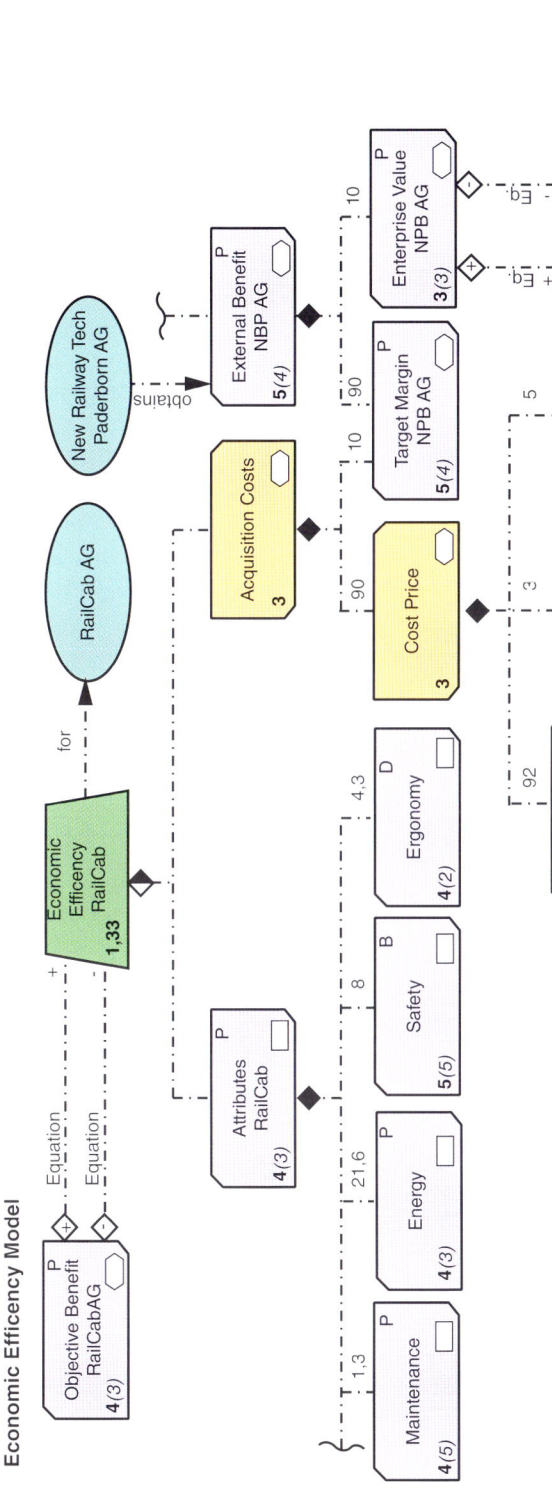

Figure 2.9 Economic efficiency model for the RailCab (extract) [4].

Conceptual Design on (Sub-)System Level Based on the list of requirements, the main functions of the system are identified and set into a function hierarchy in the *conceptual design on the (sub-)system level*. The aspect functions describe the hierarchical subdivision of the desired functionality of the system. A function is the general and required relationship between input and output parameters, with the aim to fulfill a task. For the specification of function hierarchies, we use a catalog with functions that is based on the works of Birkhofer [34] and Langlotz [35]. This catalog has been extended by functions, which describe self-optimizing functionality [10]. Each function has to be fulfilled to satisfy the requirements. Therefore, solution patterns are sought, which can execute the desired functions. Solution patterns represent reusable expertise for problem-specific solutions in a generic mode. Within a morphologic box the solution patterns are combined to consistent solution. A consistency analysis is used, in order to determine useful combinations of solution patterns of the morphologic box [29,36].

In the next step, the solution variants are evaluated if they satisfy the requirements. Therefore, a requirement evaluation matrix can be used that compares the solution variants and the requirements based on a scoring system. The solution variant with the highest score is chosen. In this step more than one solution variant can be promising, the developers have to decide individually, whether they concretize one or more promising solution variant. A reduction of the selected solution variants can be done based on more detailed information at a future point in the development process. The resulting solution must not be self-optimizing at this stage. This is only the case, if self-optimizing functions are explicitly requested in the list of requirements. Otherwise, a mechatronic solution variant results. The potential for use of self-optimization is identified later in this phase [4,29].

The chosen solution variant is then analyzed if it meets the target costs. According to Tanaka [37] and Brink [38], each combination of a function and a solution pattern is analyzed regarding its contribution to the functional benefit for the stakeholder in comparison to its functional cost contribution. In a target cost control diagram, the abscissa represents the functional benefit and the ordinate the functional costs. The objective is to have all functions within the target cost or cost reduction corridor. Functions that exceed the target costs need to be improved, for example, by choosing a different solution pattern [4].

For the relative economic efficient solution variant, the principle solution is completed. The consistent bundle of solution patterns form the basis for the active structure. In contrast to the aspect environment (black box view on the system and its context), the active structure concretize the system (white box view). The active structure defines the internal structure and the operational mode of the system. It describes system elements (e.g., chosen solutions), their attributes as well as the relationships between system elements (material, energy, and information flows as well as logical relationships). Depending on the level of concretization, system elements may be described abstract (e.g., temperature sensor) or specific (e.g., resistance thermometer). If necessary, it is also possible to model elements of the environment (e.g., user) and their interaction with elements of the system (e.g., interaction of the user with the human–machine interface of the system) [14,29].

Based on the active structure, an initial construction structure can be developed, because there are primal details on the shape within the system elements. This aspect describes the first definition of the shape within the conceptual design. In particular, the working surfaces, working places, and frames of the system are described in a rough manner. In mechanical engineering, the aspects shape and active structure form the core of the principle solution [14,29].

In addition, the systems behavior is modeled. In this step, the application scenarios are formalized and the respective behavior is analyzed regarding its consistency. Basically, this concerns the activities, states, and state transitions of the system as well as the communication and cooperation with other systems and subsystems. Subsequently, the principle solution is analyzed regarding conflicting objectives. A potential for self-optimization is given, if the changing influences on the system requires modifications of the pursued objectives. The system needs to adjust its behavior. In this case, the system of objectives is developed and the list of requirements extended. Based on the new requirements, cognitive functions are chosen and the function hierarchy complemented. For these functions, solution patterns for self-optimization are identified to enable self-optimizing behavior continuously. Generally, a pattern describes a recurring problem in our environment and the core of the solution to that [39]. Especially in the context of self-optimizing systems, the pattern approach is an established instrument to externalize and store the knowledge of experts. The overall objective is to integrate specialized knowledge for self-optimizing algorithm during the conceptual design [11,29]. The resulting changes and extensions of the system structure and the system behavior need to be included appropriately for the selected solution pattern. In preparing the self-optimization concept, the self-optimization processes will eventually be defined, the absence of conflicts of the self-optimization process will be analyzed, and the conditions, in which the self-optimization has to be working in, will be defined as well. This iteration runs as long as all conflicting objectives are taken into account. If all conflicting objectives of the system are considered, the system is modularized into subsystems [29].

In the next phase *conceptual design on the subsystem level,* the principle solutions for the subsystems are developed. This is done in the same way as in the conceptual design on the system level. The conceptual design is an iterative process. Subsystems are systems as well and can, therefore, be modularized to subsystems, too [29].

Concept Integration To make sure that the specified subsystems are not in conflict, the phase *concept integration* is performed on each system hierarchy level. In the first step, the principle solutions on the subsystem level are merged to a detailed principle solution on the system level. The final concept is analyzed regarding the dynamical behavior, dependability, and economic efficiency before the design and development is initiated.

Analysis of the Economic Efficiency Based on the values for the cost and benefit elements, the economic efficiency model can be calculated. The multipliers are integrated based on the experience of the development team. Therefore, they need support from experts, for example, from the sales and accounting department. An important multiplier is the customer satisfaction. It results from the comparison of the expected and experienced value. The impact of this comparison on the customer satisfaction varies due to the Kano category of the benefit element (cf. Section "Planning and Clarifying the Task"). In case of basic attributes, the expected value needs to be exactly achieved. A lower experienced value will result in a very unsatisfied customer. But exceeding the expected value will not lead to a higher satisfaction. For performance attributes, the correlation between degree of fulfillment and the resulting customer satisfaction is linear. Delighters are not explicitly demanded by the stakeholders. Therefore, the stakeholder will be indifferent if this requirement is not fulfilled. On the other hand, by fulfilling the requirement the experienced value will lead to a strictly increasing customer satisfaction [33]. These interdependencies can be calculated mathematically according to Buhl *et al.* [40]. The relation between customer satisfaction and enterprise value is a saddle-shaped curve according to Matzler and Stahl [41]. After the calculation of the elements of the economic efficiency model the dependability and the dynamical behavior analysis are performed and their results are integrated into the model [4].

Analysis of the Dependability Due to the autonomous adaptation of the system behavior of an intelligent technical system during operation, the dependability of the system is of high priority. It has to be secured over the entire development process. In the conceptual design specially developed methods can be used, such as the combined failure mode and effect analysis (FMEA) and the fault tree analysis (FTA), based on the principle solution [42]. To be able to evaluate the influence of the systems dependability on costs and benefits, we supplemented the FMEA by approaches of the life cost-based FMEA by Rhee/Ishii [43] and the failure mode and criticality analysis (FMCA) [44]. The resulting failure mode, effect, and failure cost analysis (FMEFA) are conducted based on the existing and on the improved principle solution. The resulting failure and development costs as well as the impact on the customer satisfaction are recorded in the economic efficiency model [4].

To support the developers with search, selection, and planning of dependability engineering methods, which are suitable for their particular development task, a methodology has been developed. Further information about the methodology and some dependability methods are given in Ref. [45].

Analysis of the Dynamical Behavior For the analysis of the dynamical behavior of the system, an idealized simulation model is built based on the relevant aspects of the intelligent technical system. The effort to build the simulation model and the simulation itself increases due to the chosen aspects. The objective is to create a basic understanding of the system's behavior and the resulting situation-dependent costs and benefits, not to model the smallest detail of the system. Therefore, the detail is chosen by the development team based on the resources available [4].

In the first step, the external influences on the system, user demands, and system states are identified that can cause a change of the systems objectives. For the identified influences, the development team specifies different characteristics and their values that can emerge during the systems operation. In this early stage, a qualitative characteristic is acceptable. Afterward, the systems objectives that

2.3 Design Methodology for Intelligent Technical Systems

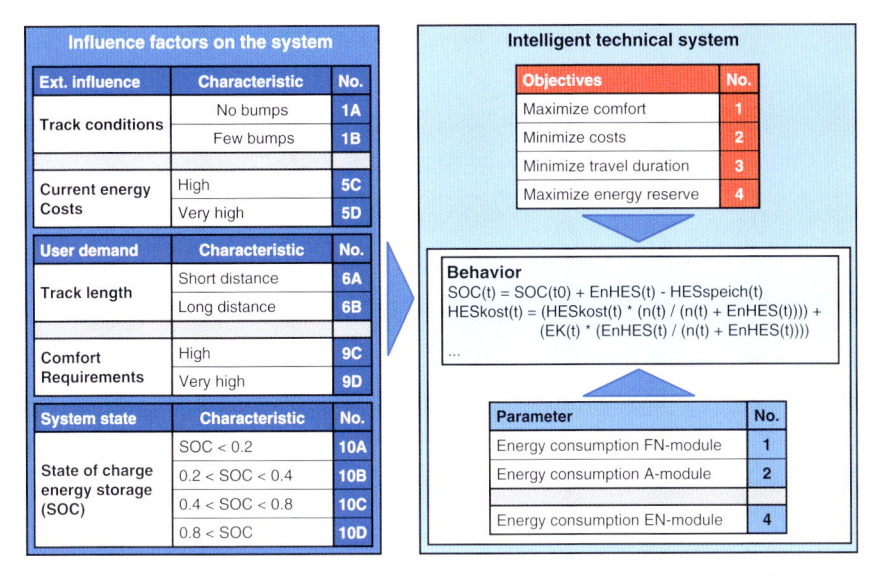

Figure 2.10 Identification of aspects of the intelligent technical system (extract) [4].

Influence factors on the system		
Ext. influence	**Characteristic**	**No.**
Track conditions	No bumps	1A
	Few bumps	1B
Current energy Costs	High	5C
	Very high	5D
User demand	**Characteristic**	**No.**
Track length	Short distance	6A
	Long distance	6B
Comfort Requirements	High	9C
	Very high	9D
System state	**Characteristic**	**No.**
State of charge energy storage (SOC)	SOC < 0.2	10A
	0.2 < SOC < 0.4	10B
	0.4 < SOC < 0.8	10C
	0.8 < SOC	10D

Intelligent technical system

Objectives	No.
Maximize comfort	1
Minimize costs	2
Minimize travel duration	3
Maximize energy reserve	4

Behavior
SOC(t) = SOC(t0) + EnHES(t) - HESspeich(t)
HESkost(t) = (HESkost(t) * (n(t) / (n(t) + EnHES(t)))) + (EK(t) * (EnHES(t) / (n(t) + EnHES(t))))
...

Parameter	No.
Energy consumption FN-module	1
Energy consumption A-module	2
Energy consumption EN-module	4

are relevant for the evaluation of costs and benefits are chosen from the aspect system of objectives and parameters identified that describe the operation mode. Furthermore, the systems behavior and the variable parameters are identified. Reference points are the aspects active structure and behavior. The mathematical description of the system's behavior is performed by the development team based on their experience. For example, the energy storage of the RailCab can be charged or discharged. Figure 2.10 shows an extract of the aspects of the RailCab used for the simulation model [4].

The simulation of the system operation is performed according to the self-optimization process in the three-steps analysis of the current situation, determination of the systems objectives, and adaptation of the systems behavior [4].

Analysis of the current situation: The intelligent technical system is monitoring its environment continuously during operation. In the simulation model, the current state of the environment is described by operation modes. Operation modes consist of a consistent bundle of concrete characteristics of the influence factors. Furthermore, operation modes can be combined to operation use cases. A duration is assigned to each operation mode within the use cases that are simulated [4].

Adaptation of the systems objectives: The adaptation of the system objectives results from changing environmental influences during the simulated operation modes. Based on the identified influences, the system of objectives is adapted. The prioritization of the objectives is conducted by means of the analytic hierarchy process (AHP) by Saaty [46]. A manual evaluation of the objective priority matrix is too expensive, therefore, we automated this process. The development team evaluates an initial objective ranking matrix that compares influence factors with the characteristics of the objectives regarding the ranking of priority. Based on defined transformation rules, the objective priority matrix is completed during the simulation. Based on the results, the system of objective is adapted [4].

Adaptation of the systems behavior: Based on the weighting of the objectives, a new operating point is chosen. In the reality, this will be conducted by look-up tables resulting from dynamic system models and multiobjective optimization. In this early development state the only model we have is the principle solution. Therefore, a simplified look-up table is used. The system parameter matrix is evaluated by the development team based on their expertise. For each influence factor the parameter value is determined. During the simulation the resulting parameter considering restrictions is chosen from the look-up table [4].

These three steps are performed until all possible combinations of the operation modes are simulated and then the situation-dependent elements of the economic efficiency model can be evaluated [4].

Evaluate the Solution In the last phase "evaluate the solution," the solution is evaluated based on its economic efficiency. First, the plausibility check for the economic efficiency model is performed. Based on a checklist, the development team analyzes if the model fulfills the requirements and characteristics. In the next step, all cost elements are verified based on the target cost control diagram. For each benefit element the expected and experienced value is compared. In case that the benefit requirements by the stakeholders are fulfilled, each economic efficiency element is reviewed if its value exceeds 1. Thus, the developed solution is economic efficient for all relevant stakeholders. Afterward, a sensitivity analysis of the model is performed. By changing the value of the multiplier the influence on the economic efficiency is analyzed. In case of a satisfactory result, the monetary value for each cost element is calculated based on the target costs provided by the development order. The result of the economic, efficient, oriented conceptual design is the framework for the following domain-specific design and development [4].

2.3.2 Domain-Specific Conceptual Design

The principle solution is the starting point for the domain-specific design and development. It contains the information that forms the basis for domain-specific development tasks. In classical mechatronic development processes, the four domains mechanical, control, software, and electrical/electronic engineering are involved. For intelligent technical systems, the optimization during operation needs to be additionally taken into account within the domains. The domains involved use their specific methods, tools, and modeling languages to concretize the system. This phase is characterized by a high coordination effort; to ensure the consistency of the system, the results of the domains are continuously integrated. For this purpose, model transformation and synchronization techniques are used. The integrated system is tested as a virtual prototype to identify faults. This allows a short response time concerning design failures and, therefore, reduces time and cost-intensive iterations [47].

To reduce development time and therefore costs, the domains work in parallel where possible. Within the process, important synchronization points are depicted, where the domains exchange their results and get information needed for further development. Process iterations can emerge, in particular, at these synchronization points, although they are possible at every stage of the process [47].

The approach is recursive and conducted on different hierarchy levels of the system. The system itself consists of subsystems, whereby the subsystems are also systems themselves that consist of subsystems, and so forth. When all subsystems are fully implemented into the virtual prototype and the test results are failure-free, the real prototype can be built and evaluated [47].

References

1 Harashima, F., Tomizuka, M., and Fukuda, T. (1996) Mechatronics – What is it?, Why and how? An Editorial. *IEEE/ASME Transactions on Mechatronics.* **1** (1), 1–4.
2 Dellnitz, M., Dumitrescu, R., Flaßkamp, K., Gausemeier, J., Hartmann, P., Iwanek, P., Korf, S., Krüger, M., Ober-Blöbaum, S., Porrmann, M., Priesterjahn, C., Stahl, K., Trächtler, A., and Vaßholz, M. (2014) The paradigm of self-optimization, in *Design Methodology for Intelligent Technical Systems – Develop Intelligent Technical Systems of the Future* (eds J. Gausemeier, F.J. Rammig, and W. Schäfer), Springer, Heidelberg, pp. 1–25.
3 Siemens AG (2012) Digitale Zukunft, Industry Journal 02/2012, Erlangen, pp. 10–18.
4 Vaßholz, M. (2015) Systematik zur wirtschaftlichkeitsorientierten Konzipierung Intelligenter Technischer Systeme, Ph.D. thesis, Fakultät für Maschinenbau, Universität Paderborn, Paderborn.
5 Gausemeier, J. and Feldmann, K. (2006) *Integrierte Entwicklung räumlicher elektronischer Baugruppen*, Carl Hanser Verlag, München.
6 Dumitrescu, R., Anacker, H., and Gausemeier, J. (2013) Design framework for the integration of cognitive functions into intelligent technical systems. *Production Engineering – Research and Developement*, 7 (1), 111–121.
7 Adelt, P., Donoth, J., Gausemeier, J., Geisler, J., Henkler, S., Kahl, S., Klöpper, B., Krupp, A., Münch, E., Oberthür, S., Paiz, C., Porrmann, M., Radkowski, R., Romaus, C., Schmidt, A., Schulz, B., Vöcking, H., Witkowski, U., Witting, K., and Znamenshchykov, O. (2009) *Selbstoptimierende Systeme des Maschinenbaus*, vol. 234, HNI-Verlagsschriftenreihe, Paderborn.
8 ACATECH (2011) Cyber-physical systems. *Innovationsmotor für Mobilität, Gesundheit, Energie und Produktion (acatech POSITION)*, Springer, Berlin, Heidelberg.
9 Broy, M. (2010) Cyber-pysical systems – wissenschaftliche herausforderungen bei der entwicklung, in *Cyber-Physical Systems – Innovation durch Softwareintensive eingebettete Systeme* (ed. M. Broy), Acatech DISKUTIERT, Springer-Verlag, Berlin Heidelberg, pp. 17–31.

10 Dumitrescu, R. (2011) Entwicklungssystematik zur Integration kognitiver Funktionen in fortgeschrittene mechatronische Systeme, Ph.D. thesis, Fakultät für Maschinenbau, Universität Paderborn, Paderborn.
11 VDI 2206 (2004) *Entwicklungsmethodik für mechatronische Systeme*, Beuth Verlag, Berlin.
12 Pahl, G., Beitz, W., Feldhusen, J., and Grote, K.H. (2007) *Engineering Design – A Systematic Approach*, 3rd edn, Springer, Heidelberg.
13 Strube, G. (1998) Modelling Motivation and Action Control in Cognitive Systems. *Mind Modelling*, Pabst, Berlin, pp. 89–108.
14 Frank, U. (2006) Spezifikationstechnik zur Beschreibung der Prinziplösung selbstoptimierender Systeme. Ph.D. thesis, Fakultät für Maschinenbau, Universität Paderborn, Paderborn.
15 Dorociak, R., Gaukstern, T., Gausemeier, J., Iwanek, P., and Vaßholz, M. (2013) A methodology for the improvement of dependability of self-optimizing systems. *Production Engineering- Research and Developement*, **7** (1), 53–67.
16 Gausemeier, J., Frank, U., Donoth, J., and Kahl, S. (2009) Specification technique for the description of self-optimizing mechatronic systems. *Research in Engineering Design*, **20** (4), 201–223.
17 Böcker, J., Schulz, B., Knoke, T., and Fröhleke, N. (2006) Self-optimization as a framework for advanced control systems. Proceedings of the 32nd Annual Conference on IEEE Industrial Electronics, Paris.
18 Gausemeier, J. and Vaßholz, M. (2014) Design methodology for self-optimizing systems, in *Design Methodology for Intelligent Technical Systems – Develop Intelligent Technical Systems of the Future* (eds J. Gausemeier, F.J. Rammig, and W. Schäfer), Springer, Heidelberg, pp. 66–69.
19 VDI 2206 (2004) *Entwicklungsmethodik für mechatronische Systeme*, Beuth Verlag, Berlin.
20 Isermann, R. (2008) *Mechatronische Systeme - Grundlagen*, Springer, Heidelberg.
21 Ehrlenspiel, K. (2007) *Integrierte Produktentwicklung*, 2nd edn, Carl Hanser Verlag, München.
22 Albers, A. (2010) Five hypotheses about engineering processes and their consequences. Proceedings of the 8th International Symposium on Tools and Methods of Competitive, Ancona.
23 Bender, K. (2005) *Embedded Systems – Qualitätsorientierte Entwicklung*, Springer, Heidelberg.
24 Kahl, S., Gausemeier, J., and Dumitrescu, R. (2010) Interactive visualization of development processes. Proceedings of the 1st International Conference on Modelling and Management of Engineering Processes.
25 Koller, R. (1998) *Konstruktionslehre für den Maschinenbau – Grundlagen zur Neu- und Weiterentwicklung technischer Produkte mit Beispielen*, Springer, Heidelberg.
26 Buede, D.M. (2009) *The Engineering Design of Systems – Models and Methods*, 2nd edn, John Wiley & Sons, Inc., Hoboken.
27 Silva, A.P. and Fernandes, A.A. (2006) Integrating life cycle cost analysis into the decision making process in new product development. Proceedings of the International Design Conference, DESIGN 2006, May 15th–18th, 2006, Dubrovnik, pp. 1419–1426.

28 Verein Deutscher Ingenieure (VDI) (2005) *Beschaffung, Betrieb und Instandhaltung von Produktionsmitteln unter Anwendung von Life Cycle Costing*, VDI-Handbuch Betriebstechnik, Teil 4, VDI-Richtlinie 2884, Beuth-Verlag, Berlin.

29 Gausemeier, J. and Vaßholz, M. (2014) Domain-spanning conceptual design, in *Design Methodology for Intelligent Technical Systems – Develop Intelligent Technical Systems of the Future* (eds J. Gausemeier, F.J. Rammig, and W. Schäfer), Springer, Heidelberg, pp. 69–74.

30 Gausemeier, J., Frank, U., Donoth, J., and Kahl, S. (2009) Specification technique for the description of self-optimizing mechatronic systems. *Research in Engineering Design*, **20** (4), 201–223.

31 Pümpin, C. (1980) *Das Dynamik-Prinzip – Zukunftsorientierungen für Unternehmen und Manager*, ECON Taschenbuch Verlag, Düsseldorf, Wien.

32 Keller, K.L. (1993) Conceptualizing, measuring, and managing customer-based brand equity. *Journal of Marketing*, **57**, 1–22.

33 Kano, N., Seraku, N., Takahashi, F., and Fsuji, S. (1984) Attractive quality and must be quality. *Qualiy Journal*, **14** (2), 39–48.

34 Birkhofer, H. (1980) *Analyse und Synthese der Funktionen technischer Produkte*, VDI-Verlag, Düsseldorf.

35 Langlotz, G. (2000) Ein Beitrag zur Funktionsstrukturentwicklung innovativer Produkte. Ph.D. thesis, Institut für Rechneranwendung in Planung und Konstruktion, Universität Karlsruhe, Aachen.

36 Lindemann, U. and Maurer, M. (2006) *Individualisierte Produkte - Komplexität beherrschen in Entwicklung und Produktion*, Springer, Heidelberg.

37 Tanaka, M. (1989) Cost planning and control systems in the design phase of a new product, in *Japanese Management Accounting: A World Class Approach to Profit Management* (eds Y. Monden and M. Sakurai), Productivity Press, Cambridge, pp. 49–71.

38 Brink, V. (2010) Verfahren zur Entwicklung konsistenter Produkt- und Technologiestrategien. Ph.D. thesis, Fakultät für Maschinenbau, Universität Paderborn, Paderborn.

39 Birolini, A. (2007) *Reliability Engineering – Theory and Practice*, 5th edn, Springer, Heidelberg.

40 Buhl, H.U., Kundisch, D., Renz, A., and Sackmann, N. (2007) Spezifizierung des Kano-Modells zur Messung von Kundenzufriedenheit, in *Organisation: Service, Prozess-, Market-Engineering* (edsl. A. Oberweis), 8. Internationale Tagung Wirtschaftsinformatik, Universitätsverlag Karlsruhe, Karlsruhe, pp. 879–896.

41 Matzler, K. and Stahl, H. (2000) Kundenzufriedenheit und Unternehmenswert, DBW 60 5, pp. 626–639.

42 Dorociak, R. (2012) Early probabilistic reliability analysis of mechatronic systems. Proceedings of the Reliability and Maintainability Symposium RAMS, January 23–26, 2012, pp. 1–6.

43 Rhee, S.J.. and Ishii, K. (2002) Life cost-based FMEA incorporating data uncertainty. Proceedings of Design Engineering Technical Conerence DETC 2002, September 29–October 2, 2002, Montreal, pp. 1–10.

44 Deutsches Institut für Normung e.V. (DIN) (2006) *Analysetechniken für die Funktionsfähigkeit von Systemen – Verfahren für die Fehlerzustandsart und -*

auswirkungsanalyse (FMEA) (IEC 60812:2006); Deutsche Fassung EN 60812:2006. DIN EN 60812, Beuth Verlag, Berlin.

45 Gausemeier, J., Rammig, F.J., Schäfer, W., and Sextro, W. (2014) *Dependability of Self-optimizing Mechatronic Systems*, Springer, Heidelberg.

46 Saaty, T.L. (1980) *The Analytic Hierarchy Process – Planning, Priority Setting, Resource Allocation*, McGraw-Hill, Inc., London.

47 Gausemeier, J. and Vaßholz, M. (2014) Development of self-optimizing systems, in *Dependability of Self-Optimizing Mechatronic Systems* (eds J. Gausemeier, F.J. Rammig, W. Schäfer, and W. Sextro), Springer, Heidelberg, pp. 25–36.

3

Smart Systems Design Methodologies and Tools

Nicola Bombieri,[1] Franco Fummi,[1] Giuliana Gangemi,[2] Michelangelo Grosso,[3] Enrico Macii,[4] Massimo Poncino,[4] and Salvatore Rinaudo[2]

[1]Università di Verona, Dipartimento di Informatica, Strada Le Grazie 15, 37134, Verona, Italy
[2]STMicroelectronics, Stradale Primosole 50, 95121 Catania, Italy
[3]STMicroelectronics, Corso Duca degli Abruzzi 24, 10129 Torino, Italy
[4]Politecnico di Torino, Dipartimento di Automatica e Informatica, Corso Duca degli Abruzzi 24, 10129 Torino, Italy

3.1 Introduction

The "smartness" in (smart) electronic systems is typically associated with their capability of sensing environmental conditions of various types, processing these sensed information, and possibly actuating some actions in response to these conditions. While the computational part is intrinsically electronic, sensing and actuation do not just involve the electronic domain but also encompass the interaction between different physical domains – sensors and actuators are nothing but transducers having the electrical domain at one of their sides.

The trend toward the miniaturization of smart systems through the integration into a single silicon chip naturally transforms them into material-integrated systems.

As a matter of fact, integrating a sensor implemented as a microelectrical–mechanical system (MEMS) implementing a transducer onto the same silicon substrate of digital electronics implies solving the same issues of more complex and larger scale material-integrated systems.

The similarities, however, are not limited to the manufacturing-related issues; there are also functionality-related aspects that are typical of material-integrated systems.

An important requirement related to the modeling and simulation of smart systems is the possibility of not just simulating the *functionality* but also some *extra-functional properties* (e.g., power consumption, temperature, reliability). These are of course affected by functionality (what the system does), but depend also on physical parameters of various types. Thermal simulation is a good example of this – analyzing temperature in a smart system requires information about the geometry and the materials (besides activity and power consumption, of course). It is therefore reasonable to consider smart electronic systems as a particular class of material-integrated systems, where the material integration issues are somehow ancillary, but already present both structurally and functionally.

This chapter presents SMAC, a structured approach for the modeling, simulation and validation of such smart electronic systems, focusing mainly on the functional aspects.

The SMAC platform is the result of a European Project (FP7 – SMAC: SMArt systems Co-design) – Integrated Project (IP) of the 7th ICT Call under the Objective 3.2 "Smart components and Smart Systems integration." The project required the joint cooperation of 17 distinct partners, including technology and semiconductor companies, system house, EDA vendors, and research institutions to ensure the platform usability in realistic, industry-strength design flows and environments, with a direct impact on the industrial exploitation.

3.2 Smart Electronic Systems and Their Design Challenges

The challenge in the implementation of Smart Systems lies in the coexistence of a multitude of functionalities, technologies, and materials. As a matter of fact, *integration* is the widely acknowledged keyword in Smart Systems design. Two are the dimensions of integration that represent the main obstacle toward mainstream design of Smart Systems: technological and methodological. As already experienced in specific domains (e.g., in digital and analog design), a solution has been found first for the technological issues. Advanced packaging technologies such as System-in-Package (SiP) and chip stacking (3D IC) with through-silicon vias (TSVs) allow today manufacturers to package all this functionality more densely, combining the various technological domains in a single package. SiP technology works nicely because it allows merging components and subsystems with different processes, and mixed technologies using the state-of-the-art advanced IC packaging technologies with minor impact on the design flow. Therefore, to some extent, technological solutions aimed toward integration are available.

Nevertheless, actual design methodologies are still missing: Current design approaches for Smart Systems use separate design tools and *ad hoc* methods for transferring the nondigital domain to that of IC design and verification tools, which are more consolidated and fully automated. This solution is clearly suboptimal and cannot respond to challenges such as time-to-market and request of advanced sensing functionalities. A big step toward effective large-scale design of Smart Systems is that of changing their design process from an expert methodology to a mainstream (i.e., automated, integrated, reliable, and repeatable) design methodology, with the aim at reducing design costs, shortening time-to-market, and providing design of the various domains to be no longer confined to teams of specialists inside IDMs, thus allowing system miniaturization to be achieved with limited risks.

This chapter presents the SMAC platform, a structured design approach that aims at reaching the goals mentioned above, by explicitly accounting for integration as a specific constraint. The SMAC platform includes methodologies and EDA tools enabling multidisciplinary and multiscale modeling and design, simulation of multidomain systems, subsystems, and components at all levels of abstraction, system integration, and exploration for optimization of functional and nonfunctional metrics.

The SMAC platform relies on two strategic elements. First, it consists of a cosimulation and codesign environment that account for the peculiarities of the basic subsystems and components to be integrated. Then, it relies on modeling and design techniques, methods, and tools that enable multidomain simulation and optimization at various levels of abstraction and across different technological domains. Such a holistic codesign framework aims at closing several technical and cultural gaps by means of a multidisciplinary approach. To do this, the platform development has required the joint cooperation of research and industry partners, including EDA vendors to ensure the platform usability in realistic, industry-strength design flows and environments, with a direct impact on the industrial exploitation.

Behind the growing interest in Smart Systems, there is a potentially huge and quickly growing market, which is expected to grow on the order of $200 billion in 2020, inducing an even larger market of nonhardware services involving all the various devices envisioned in the Internet of Things. Such a market is much larger than those of smart- or feature-phones in terms of number of devices. Over 50 billion devices will be connected to the Internet according to Cisco forecasts, and most of these devices will be Smart Systems. Miniaturized Smart Systems find applications in a broader range of key strategic sectors, including automotive, healthcare, ICT, safety and security, and aerospace.

Also, efficient energy management and environment protection are business sectors in which the utilization of miniaturized Smart Systems may make a difference. The worldwide market for "Monitoring and Control" products and solutions, one of the most important fields of Smart Systems applications, containing solutions for environment, critical infrastructures, manufacturing and process industry, buildings and homes, household appliances, vehicles, logistics and transport or power grids, is around 188 billion euros. This value represents 8% of the total ICT expenditures worldwide, and it is identical to the whole semiconductor industry world revenues and approximately twice that of the world mobile phone manufacturers revenues [1].

3.3 The Smart Systems Codesign before SMAC

The state-of-the-art, specialized design houses, and silicon makers in various forms have been the designers reference point for modeling heterogeneous components and subsystems. The *nonelectrical parts* (including micromechanical structures, electromagnetic fields, thermal phenomena, wave propagation, etc.) are generally designed using partial differential equations (PDE) solvers, like the finite element method (FEM), or, alternatively, schematic-based behavioral libraries. The *analog and RF parts* are designed through the reuse of existing macros, by experienced engineers, and by following a template-based approach. Highly automated synthesis tools (from high-level synthesis to physical synthesis) allow for the design of the *digital parts* by following a top-down paradigm. *System design* is supported by block diagram simulation (e.g., MATLAB-SIMULINK) or newer approaches, which allow obtaining a comprehensive view of the entire system, yet through simple models of the subsystems and of the components. Finally, the amount of *software*, implemented in microcontrollers and DSPs, is significantly increasing, in line with the trend that can be observed in embedded system design.

A clear framework to categorize methodologies and tools, which are adopted in Smart Systems design, should distinguish between the different levels of design abstraction: *system level, device level*, and *physical level*. The interactions of the heterogeneous components and subsystems such as MEMS, RF, analog parts, power sources, digital macrocells, and software with their environment and control electronics can be modeled and simulated at the system level of abstraction. In this context, SystemC [1] is *de facto* reference standard modeling language for system-level design of digital systems, while transaction-level modeling (TLM) [2] is the key paradigm for modeling such digital systems at a high level of abstraction. Because more and more integrated systems have analog components that are tightly coupled with the digital hardware and software, SystemC-AMS can be adopted. SystemC/SystemC-AMS introduces numerous methodologies for efficient modeling and simulation of different domains.

On the other hand, circuit simulation programs, such as SPICE and derivatives, are used for device or circuit-level design. A netlist describes the circuit elements (e.g., transistors, resistors, capacitors) and their connections, and translates such description into nonlinear differential equations to be solved using implicit integration methods, Newton's method, and sparse matrix techniques. MEMS libraries for circuit simulators based on parameterized behavioral models and consisting of electromechanical building blocks, such as beams, plates, comb structures, and electrode models [3–6], are commercially available and are now well established [7]. Recently, 3D visualization of the complex models and simulation results has been added [8]. Latest advancement allows designers to compose MEMS behavioral model-based designs in 3D and pass the model as netlist to an EDA platform for IC design [9].

At physical level, power sources, sensors, MEMS, and other discrete devices have traditionally been modeled in 3D CAD environments and simulated using 3D field solvers based on the finite element method and boundary element method (BEM). Today, a variety of single- and multi-physics field solvers are available, ranging from general-purpose tools to those that address MEMS-specific physics such as electrostatic sensing and actuation, piezoelectric effects, and gas damping [9–11]. In the recent past, substantial improvements have been made in user-friendliness, automatic meshing algorithms, computational efficiency, and results database management to link physical models with system-level models [12]. Model order reduction (MOR), also called reduced order modeling (ROM) [13,14], is an evolving approach for a fast generation of behavioral models based on existing FEM component models, for example, generated in commercial FEM solvers. There are different basic approaches for MOR, such as projection-based methods and simulation-based order reduction.

Recently, some attempts have been made to include different domains into a single cosimulation environment. For example, in Ref. [15], the authors address the problem of conflicting constraints in embedded systems, such as energy and time. The work proposes a mechanism for supporting design decisions on energy consumption and performance of embedded system applications. The authors show that the estimates obtained through the conceived model are 93% close to the respective measures obtained from the real hardware platform.

A different approach has been presented in Ref. [16], in which a HDL design methodology for multistandard RF SoCs covers all the design layers from system design to automatic extraction of the models from circuits and a systematic top-level verification. System- or block-level verification is obtained with models automatically by overnight runs, without the need for extra test benches or designer interaction. This enables short-term detection of functional errors or performance losses. The accuracy of the system-level simulations show a very good match with the measurement results after fabrication.

In Ref. [16], the authors present a multidomain cosimulation approach in the context of smart grids. The data acquisition, protection, and control of smart grid highly depend on advanced information infrastructures, which makes smart grid a coupled system of power and information networks. In order to assess the influence of the uncertainties within information networks on the performance of the real-time controls in smart grid, the work in Ref. [16] presents a cosimulation method based on OpenDSS and OPNET simulators.

Concurrent/simultaneous analysis of passive and active analog/digital parts has been finally investigated in Ref. [17], which proposes a global cosimulation methodology for such systems. An original power-signature concept is introduced to model high-speed digital modules temporal and spatial distribution of their power switching activity through specified chip partitions. Dedicated real-world NXP-Philips-Semiconductors active modules mounted on testboard have been designed and measured for validation of the proposed cosimulation methodology. In the paper, full-wave electromagnetic modeling, broadband SPICE compact model extractions, and measurement results are successfully compared.

Even though all these approaches are optimized in the specific contexts, they suffer from multiple and different types of limitations and bottlenecks when applied to Smart Systems. The design of a Smart System that deals with embedded systems generated from bare die requires a strong link between different worlds of design tools, such as the board/module level, the electronic circuit design level, and the physical level. The tight integration of such differing systems sets high requirements on the system design flow. While FEM simulation allows analyzing detailed problems, models for higher abstraction levels are needed to exceed the complexity barrier imposed by the need of dealing with electrical, electromechanical, and thermal interfaces. Also, the reuse of available components usually requires the creation of *ad hoc* functional models based on datasheets or device characterization.

In conclusion, existing design and simulation frameworks for component/subsystem integration created by system integrators are nonstandard and show several limitations. First, they are greatly fragmented, being a stitching of bits and pieces of existing EDA tools and flows from different vendors. They are very fragile in the scripting, tool interfacing, and data formatting. They are also expensive, since they are licensed from different vendors. They are not flexible, since changes in system architecture, building blocks, models, and workloads often imply significant revision of the framework. Finally, they are customized to specific applications.

The main issues and research objectives for a multilevel design methodology for Smart Systems include several aspects. Some examples are coupling of physical effects toward multidomain modeling approaches, consideration of nonlinear effects and structural discontinuities supported by efficient modeling and simulation algorithms, codesign and cosimulation of electronics (e.g., analog, digital) with nonelectronics (e.g., MEMS, power sources) over multiple scales, and so on. In particular, cosimulation is required to verify the IC design and to predict yield sensitivity to manufacturing variations. One obvious path is to do the cosimulation in the environment used by the IC designers. This requires that the MEMS designers deliver a behavioral model of the MEMS devices expressed in a suitable hardware description language (HDL) such as Verilog-A or VHDL-AMS. Today, MEMS engineers have very limited ability to deliver behavioral models in these formats. Very often they build the model manually, usually in the form of a lookup table, or generate a reduced-order model (ROM) from finite element analysis. The lack of IC-compatible parametric MEMS behavioral models is an impediment to design reuse and to the licensing of MEMS IP. The availability of a library of validated MEMS IP design in the environment that an IC designer or system architect is used to work in would revolutionize the way smart systems are developed today.

There is very limited ability to deliver MEMS component models for architecture-level design. System engineers are usually required to handcraft non-parametric models or lookup tables from publicly available datasheets. System-level tools such as Simulink are often used for functional modeling and simulation. They may also capture continuous-time behavior, but do not target the design of embedded-AMS (E-AMS) systems at an architecture level. HDLs target the design of mixed-signal subsystems close to implementation level, but these languages have limited capabilities to provide efficient HW/SW codesign at high level of abstraction. Existing cosimulation solutions mixing SystemC and Verilog/VHDL-AMS do not provide high-enough simulation performances and lack offering a seamless design refinement flow for modeling mixed discrete-event/continuous-time systems and HW/SW systems at architecture level. In conclusion, component models are not yet integration aware.

3.4 The SMAC Platform

The SMAC platform aims at enabling the design of complex Smart Systems using a top-down approach, from system architecture down to implementation and integration of heterogeneous functions and technologies. It allows enabling multiphysics, multilayer, multiscale, and multidomain full Smart Systems simulation and optimization. The platform consists of modeling and design techniques, methods, and tools that allow simulation and optimization at various levels of abstraction and across different technology domains.

3.4.1 The Platform Overview

The SMAC platform has been developed to allow different tools from different vendors acting together to simulate a system, whose components come from different modeling domains as well as from different abstraction levels. The platform relies on a model generation and conversion flow to bring all the models of the system in an appropriate format, by which they can be simulated either through a set of simulation engines (using a cosimulation flow) or through a single simulation engine at the higher possible abstraction level (i.e., system-level).

The platform supports a set of different modeling formats, including those that are the *de facto* reference standard at the state of the art (e.g., VHDL and Verilog for the Digital Domain). The platform includes different simulators (MEMS, FEM, method of moments, continuous time, synchronous data flow, timed synchronous data flow, circuit, discrete events, instruction set simulators, finite element analysis) for simulating the system components in each different domain.

Several existing tools and their interfaces that are mostly used in the state-of-the-art Smart Systems codesign have been analyzed. Such an analysis has been conducted with the aim of understanding the main lacks of the tools that prevent a complete multidomain and multilevel codesign environment. First, how cosimulation environments can be set up between models belonging to two different domains have been considered. Examples are digital Hardware and Analog-RF cosimulation through the native SystemVue tool,[1] digital Hardware and embedded Software cosimulation through QeMU[2] and SystemC,[3] Analog-RF and embedded Software through SystemVue and MATLAB, and Analog-RF and Discrete and Power Devices through SystemVue and ADS.[4] Even though almost useful to link two domains, all these cosimulation environments do not allow a comprehensive cosimulation between more than two domains. The gaps in the multidomain design have been formulated as requirements and goals to the SMAC project, and have implied extensions and enhancements to the identified modeling tools.

Figure 3.1 shows the main results obtained by the SMAC project regarding the tool collection and interface extension. The bold lines underline the contributions of SMAC to fill the gaps, which consist of new cosimulation environments as well as conversion tools. The conversion tools allow model descriptions to be translated through abstraction levels and through description formats, with the aim of *simulating* (rather than cosimulating) the whole smart system description. All these contributions are summarized in the next sections.

3.4.1.1 System C–SystemVue Cosimulation

The design of analog and digital electronic integrated circuit requires quite different approaches. The design of analog and RF circuitry usually starts with the

---

1 SystemVue ESL Software – Keysight Technologies. http://www.keysight.com/en/pc-1297131/systemvue-electronic-system-level-esl-design-software?nid=-34264.0&cc=GB&lc=eng.
2 QeMU – Open-source processor emulator. www.qemu.org.
3 Accellera Systems Initiative: SystemC. http://www.accellera.org/downloads/standards/systemc.
4 Advanced Design System (ADS) – Keysight Technologies. http://www.keysight.com/en/pc-1297113/advanced-design-system-ads?cc=US&lc=eng.

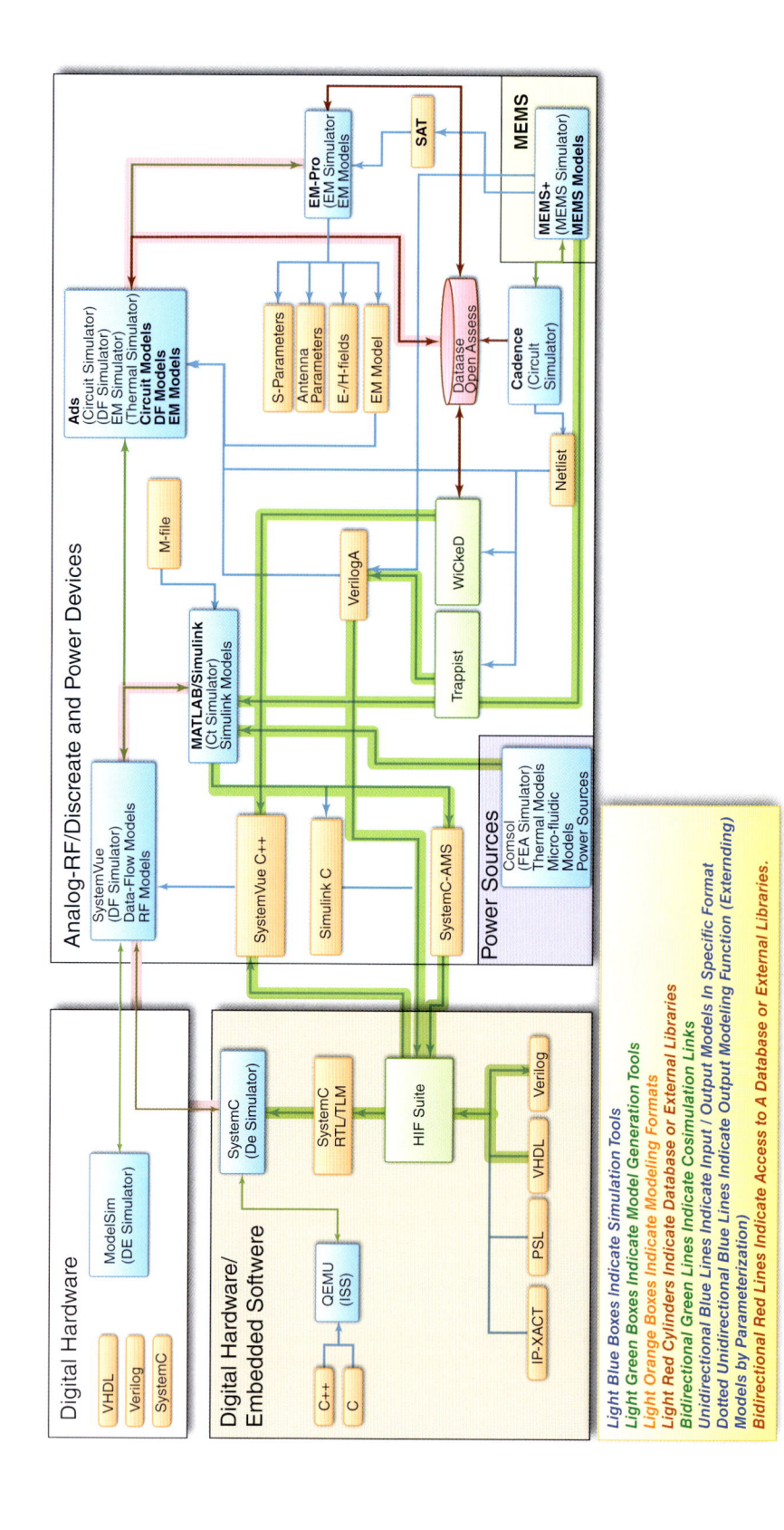

Figure 3.1 The SMAC platform overview.

definition of the circuit schematic by means of Spice networks or *ad hoc* tools such as SystemVue, which can be functionally simulated and verified. Then, the design goes through several physical implementation and refinement steps, where functional and nonfunctional properties (e.g., electromagnetic compatibility) have to be validated against specifications, for obtaining the production masks. Conversely, digital modules are designed starting from behavioral or data-flow (Register Transfer Level) descriptions, and then automated synthesis processes enable the realization of today's highly complex and dense silicon devices.

To fill the gap between the digital/embedded software domain modeled and simulated by SystemC and the analog/RF/MEMS domains modeled and simulated through the SystemVue native data flow simulator, we extended the Keysight Technologies SystemVue ESL platform.[1] The design entry of SystemVue supports model-based design flow (GUI blocks, language-based C++ or math, VHDL). Models are represented as block entities and are connected with each other, forming a block diagram design. The data flow simulation engine executes each of these models based on a schedule computed at the presimulation phase. By exploiting this feature, a new block model, named SystemCCosim, has been developed inside SystemVue to allow cosimulation with SystemC designs. When executed, the block establishes an Interprocess Communication (IPC) link with the SystemC process, which runs externally. Through this link, which is based on shared memory, data are passed back and forth between SystemVue and SystemC processes.[5]

3.4.1.2 ADS and the Thermal Simulation

All devices dissipate electrical power as heat. Heat flow causes devices to operate at different temperatures. Since device electrical properties change with temperature, they need self-consistent powers and temperatures. Traditional methodologies for the design of power modules are based on purely functional modeling, simulation and verification of the circuit at fixed temperature conditions in the expected operating range. The designers have to deal with heat flows by extending design margins and by considering suitable heath sinks.

With the SMAC Platform, Advanced System Design (ADS), which is a native circuit system simulator,[4] has been enhanced with a new electrothermal simulator based on a full 3D thermal solver natively integrated. It incorporates dynamic temperature effects to improve accuracy in "thermally aware" circuit simulation results. The electrothermal analysis provides coupled simulation between a circuit simulator that computes power dissipation and a thermal simulator that computes temperatures. The simulation platform automatically exchanges data between the two engines and iterates to a self-consistent electrothermal solution. The thermal simulator solves the full physical three-dimensional problem, incorporating all of the geometries in the layout and their thermal properties. This integrated methodology allows evaluating the behavior of the device before manufacturing in a more realistic way, thus enabling an increased architectural optimization.

---

5 SystemC Cosimulation, Keysight Technologies. http://edadocs.software.keysight.com/display/ads201101/SystemC+Cosimulation.

3.4.1.3 EMPro Extension and ADS Integration

In the development of electronic smart systems, the electromagnetic properties of the heterogeneous components need to be carefully evaluated during the design since they may have tremendous consequences on the product behavior, especially when the devices have to be integrated in very small volumes. For easing this process, ElectroMagnetic Professional (EMPro) has been extended to be integrated with ADS. EMPro[6] is an EM simulation software for analyzing the 3D EM effects of components such as high-speed and RF IC packages, bondwires, antennas, on-chip and off-chip embedded passives, and PCB interconnects. EMPro designs are stored in a cell, which contains multiple views. These views can be used in ADS layout and schematic. A common database allows 3D components built in EMPro to be placed on ADS schematics and layouts directly. EMPro's increased level of integration with ADS is based on a shared database approach. Three-dimensional objects in EMPro can be saved as ADS design database "cells" for use directly in ADS. Direct integration streamlines the design process and gives circuit designers easier access to full 3D EM modeling capabilities. Furthermore, the use of OpenAccess standard in ADS and EMPro enhances the effort to provide true interoperability, not just data exchange, among IC design tools through an open standard data API.

ADS RF designers can import EMPro parameterized 3D components for EM simulation and optimization in ADS. Because the parameterized 3D component is created just once and integrates with both EMPro and ADS, error prone links between today's standalone 3D EM point tools and ADS are eliminated, resulting in faster design and assured accuracy.

Additionally, ADS circuit designers can transfer their layout from ADS into EMPro to accurately determine effects such as conformal bending of a planar antenna around a curved product package or effectiveness of metal shielding.

EMpro 3D parameterized components such as connectors, solder ball arrays, packages, or any custom components can be transferred into ADS as a library for use in 3D EM cosimulation and co-optimization with circuit and system components, enabling evaluation of complete family of structures in one step.

3.4.1.4 Automated EM – Circuit Cosimulation in ADS

Verifying correct operation at high frequencies often requires EM simulation of complex electronic design. EM simulators can model the behavior of the passive parts of the design, such as interconnects. Active components, such as transistors, cannot be modeled in the same way. A layout design contains both active and passive pieces. Setting up a layout for simulation requires separating both pieces. EM simulation of the passive piece is performed first; then, a subsequent circuit simulation combines the resulting EM model with circuit models for the active piece. The final result is an accurate EM/circuit cosimulation of the whole system.

For complex layouts, manually separating the active and passive pieces and stitching both pieces together in a circuit simulation is a tedious and error-prone job. One of the SMAC contributions is automating this job. First, it allows the user

---

6 EMPro 3D EM simulation software – Keysight Technologies. http://www.keysight.com/en/pc-1297143/empro-3d-em-simulation-software.

to choose whether a piece of the design needs to be EM or circuit simulated. Then, it uses this information to automatically generate the passive piece of layout, including pins and EM ports on all the positions where active components need to be hooked up. The third step is automatically hooking up these active components to the EM model of the passive piece in a circuit simulation. As a result, setting up a proper EM/circuit cosimulation of a complex system becomes a matter of minutes rather than hours or days.

3.4.1.5 HIF Suite Toolsuite

As aforementioned, the development of an integrated circuit requires several steps and the arrangement of different models, usually from a relatively abstract behavioral or schematic description (e.g., VHDL or Verilog) toward more and more refined (and complex) representations (back-annotated gate-level netlists and layouts). During system integration, the analysis of the interactions between different modules ideally requires homogeneous and accurate models (close to the implementation), but dealing with the resulting complexity can be an issue even for today's workstations.

HIFSuite[7] is a toolsuite for the automatic translation, elaboration, and optimization of digital models, able to provide simplified but accurate descriptions by means of abstraction (i.e., in a bottom-up approach). The main HIFSuite goal is the creation of a manipulation and verification environment, independent of any specific HDL. During the project, the framework has been extended to be suitable for smart system codesign in many directions.

To improve the quality of front/backend tools and precisely define the semantics of HIFSuite descriptions, a HIF normal form has been defined. Such a normal form represents HIF descriptions generated after the parsing phase. Each further manipulation performed by manipulation tools (e.g., A2T abstraction tool and HR hierarchy removal) is checked to be compliant with the normal form. *VHDL* and *Verilog* parsers have been redesigned to generate HIF descriptions compliant to the new HIF normal form. The development of a tool (psl2hif) to convert PSL properties into HIF descriptions has been developed. A tool to convert IP-XACT descriptions into HIF code (and vice versa) has been implemented. The HIFSuite A2T abstraction tool, used to generate both SystemC-TLM and C++ descriptions starting from RTL designs, has been improved in several ways. The "dynamic scheduler," which is the default simulation algorithm embedded into the abstracted designs has been optimized to improve the simulation performance. In addition, a static scheduler, which is alternative to the dynamic one, has been developed from scratch. The combination of dynamic and static scheduling leads to faster simulations in designs that have lightweight processes. A2T has been extended to support designs featuring multiple and derived clocks, that is, clocks whose period is a multiple of the period of a main clock. In this case, the main clock is abstracted, and the other derived clocks are treated as signals, thus ensuring the correct elaboration of the design functionality. The C++ code generated by A2T consists of procedures, which performs simulation through steps. This format has been

---

7 HIFSuite – EDALab s.rl.l. hifsuit.com.

developed to be compliant to SystemVue's requirements and thus can be easily integrated for complex cosimulations.

3.4.1.6 The MEMS+ Platform

Advanced design methodologies and a variety of software tools are necessary to analyze the complex geometrical structures that are typical of MEMS devices. This allows accounting for interactions among different physical domains and describing the cooperative play of microarchitectures and connected electronic circuitry or signal processing units. FEM techniques are used for structural, thermal, electromagnetic, and fluid analyses of micrometric elements for sensor and actuator applications. Individual effects and interactions among different physical domains are covered by direct coupling algorithms or analyzed through parameter extraction methods, referred to as reduced order modeling.

MEMS+[8] is an integrated platform for MEMS design and simulation. It has been developed to interface the MEMS design with EDA tools and, at the same time, to allow the MEMS device to be simulated at a high degree of accuracy. MEMS+ enables MEMS designers to assemble a behavioral model of a MEMS device in a 3D graphical user interface based on a library of building blocks mainly including models for mechanics, electrostatics, and piezoelectrics. It has been extended with several new model capabilities to allow packaging–die cosimulation and improved modeling of fluidics damping effects.

In the SMAC platform, the existing system-level simulation possibilities in MEMS+ were not flexible and sufficient enough. The extensions to MEMS+ developed in the SMAC project are described in the following. The simulator allows the user to conduct basic MEMS device analysis to verify the correctness of a device before exporting it to MATLAB, Simulink, or Cadence Virtuoso for system simulation. The simulator can be used to run several analyses, such as DC analysis (steady-state equilibrium with mechanical and electrostatic forcing), DC Sweep (electrostatic pull-in and lift-off, and contact), mechanical modal analysis (includes electrostatic spring softening effects), and small-signal AC (linear frequency response to mechanical and/or electrostatic forcing with Reynolds gas damping effects).

The user can also run a vary analysis, alter operating points, alter variables, and alter sources. The development of this simulator was required in order to allow the majority of the developments needed to interface with the SMAC platform. In addition, it adds a high level of user-friendliness for the MEMS designer, who can now run most the MEMS-only relevant simulation in a 3D environment. Once the 3D model is created in MEMS+ Innovator, it is ready for simulation in MATLAB. MEMS+-for-MATLAB provides a variety of MATLAB scripts that allow model import, manipulation, and simulation directly from the MATLAB command line interface. The scripts provided by MEMS+ support DC, DC transfer, Modal, AC, and transient analyses without requiring any additional toolbox. All MATLAB simulation results can be loaded back into the MEMS+ Scene3D module for 2D and 3D viewing. Users can also create a material database, process, and schematic or modify existing ones from a MATLAB command prompt.

---

8 MEMS+ – Conventor. http://www.coventor.com/mems-solutions/products/mems.

The functionality of exporting Verilog-A models from MEMS+ has been developed to integrate MEMS models easily in most circuit simulators accepting this format, especially Keysight ADS and EDALab HIFSuite. The resulting Verilog-A models are currently linear and nonparametric. Parameterization and a certain degree of nonlinearity are under investigation.

In order to obtain a Verilog-A model from MEMS+, the user runs a DC (or DCSweep) in the MEMS+ simulator plug-in. After the DC has run, a new entry in the context menu appears called "Export Linear Verilog-A." Within one click, a file browser pops-up to choose a file where to save the Verilog-A model. It is a fully automatic way of generating a Verilog-A model. The model can now be used in the majority of circuit simulators and without any license. It includes the system matrix of the MEMS device but no physical information on the design.

Several improvements have been incorporated in the MEMS+ platform related to cosimulation aspects between the MEMS+ library components and Cadence Spectre.[9] Most of the improvements are related to gaining further speed. Of particular interest is the cosimulation for MEMS microphones, where noise simulation is a true challenge and requires full cosimulation between mechanics, electrostatics, fluidics (damping), and electronics – considering noise sources of different physical origin.

3.4.2 The (Co)Simulation Levels and the Design–Domains Matrix

All the simulation tools that are part of the SMAC platform are organized as in the matrix of simulation levels and design domains shown in Figure 3.2. Rows represents the abstraction levels, while columns discriminate application domains. Not every combination is meaningful: This is represented as a white box. A simulation tool can occur in more than one box, and a box can contain more than one tool. The SMAC platform exploits abstraction/refinement techniques and tools to move descriptions across rows. Automatic interface generators enable (co)simulation of tools across columns.

Figure 3.3 also underlines the simulation scenarios through box clusters. Interconnected boxes participate in the (co)simulation. At most one box is selected for a component from each application domain. Not every domain is required. In case there is more than one component belonging to the same application domain, models at different abstraction levels may be used.

In particular, the *functional level* describes, in C/C++, the functionality of a component, and it can be used for simulating a component in isolation, since its execution after compilation represents the module behavior. It can also be used to simulate a set of components of the same domain, since all such components share the same computational model; thus their composite behavior can be obtained by executing the functionality of all modules.

In contrast, the *transactional level* adds a protocol to the functionality of each module to allow the composition of modules of different domains. The protocol allows designers to correctly combine heterogeneous modules to simulate the

---

9 Spectre Circuit Simulator – Cadence. http://www.cadence.com/products/cic/spectre_circuit/pages/default.aspx.

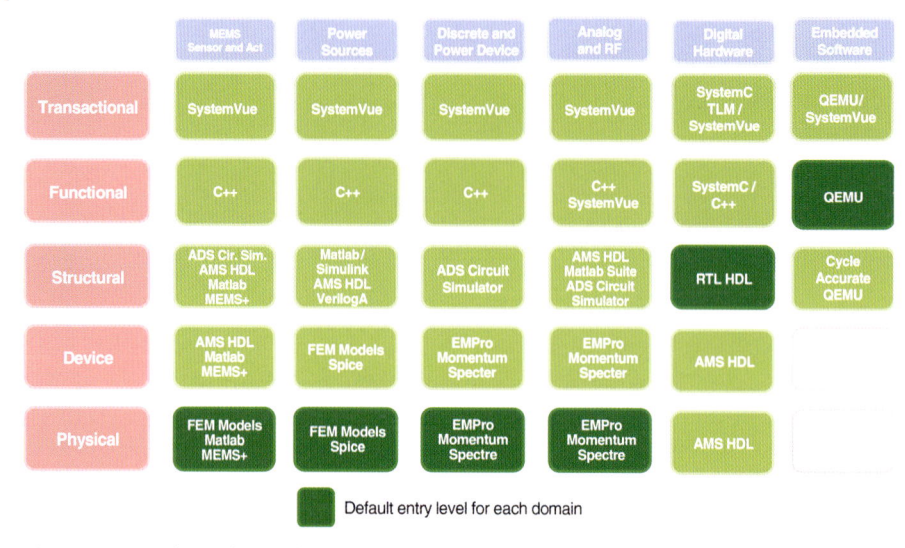

Figure 3.2 Simulation levels/design domains matrix.

composite behavior. A reference example of transactional description is the one standardized by OSCI (now Accellera) for digital hardware and software and implemented in the SystemC TLM library. In SMAC, we propose to use the protocols provided by SystemVue to integrate components of different domains. Each module is thus first abstracted up to the functional level and, then, the protocol is added to allow the integration of heterogeneous modules for a simulation in SystemVue.

All design, simulation scenarios, and models supported by the SMAC platform are represented in Figure 3.2. Some examples of mapping between abstraction levels and design models are the *physical level* (e.g., a TCAD structure/model of a transistor, a PCB geometry in the EM simulator), the *device level* (e.g., the gate-level

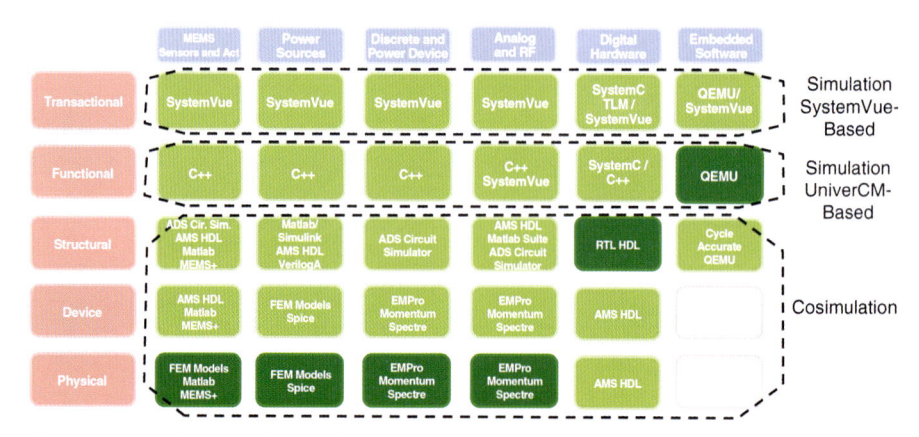

Figure 3.3 Simulation and cosimulation in the SMAC platform.

netlist for a digital block, the SPICE netlist for an analog block, and a SPICE transistor model), the *structural level* (e.g., the RTL description of a digital block, the TRAPPIST VerilogA model of an analog block), the *functional level* (e.g., the abstracted description of a RTL digital block), and the *transactional level* (e.g., the functional description of an analog module completed with the protocol for the simulation in SystemVue).

The SMAC platform allows both cosimulation and simulation, which are selectively applied depending on the involved domains and abstraction levels. To correctly identify which, between cosimulation or simulation approaches, has to be applied, we defined a design domain/simulation-level matrix as shown in Figure 3.3. Models of the lowest abstraction levels are represented by different design languages; therefore, they must be simulated by using their own simulator (e.g., MATLAB, Modelsim, EMPro, etc.). For this reason, a simulation covering more than one domain can be implemented only by using cosimulation techniques that connect different tools by exchanging simulation data from one tool to another.

Moving to the functional level, there is a convergence in the modeling language, that is, all models of the different domains are represented in C++. This would in principle allow simulation to be performed among different domains. However, the computational model implemented into each C++ model may differ from domain to domain. Thus, a simulation cannot be simply obtained by linking functional C++ models of different domains, but such models must be coherent with respect to the same computation model, which must be able to cover all domains. A universal computational model (*univerCM* [18]) has been introduced in the SMAC project to describe models of all design domains. UniverCM is the core of the language abstraction and transformation tools composing the SMAC platform. If functional C++ models are generated by starting from UniverCM, they can be simply linked together to simulate a design covering more than one domain.

Not all functional models can be generated, or are not generated, by using the automatic abstraction tools of SMAC based on U*niverCM*: in this case, we can move to the transactional level where SystemVue is used as general integration methodology to have a simulation covering more than one design domain. In fact, functional C++ models are extended with the protocol at the transactional level and they can thus be connected in SystemVue by using a standard interface defined in the SMAC project. This allows the coherent simulation of functional models based on different computation models, reaching the result of producing the SMAC simulation platform.

3.5 Case Study: A Sensor Node for Drift-Free Limb Tracking

In order to validate the methods developed under the SMAC umbrella and to demonstrate the capabilities of the developed tools that are offered to designers, a number of case studies were selected to build models and perform simulations. Such case studies include *components* belonging to different design domains, such as digital and analog electronics, MEMS, discrete and power devices, as well as some complete (smart) *systems*.

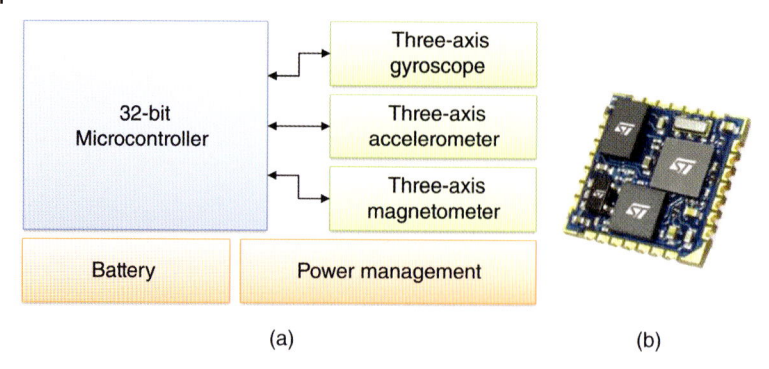

Figure 3.4 System partitioning of a sensor node (a) and an iNEMO M1 System-on-Board (b).

Traditional approaches for the development and optimization of such systems involve several prototype iterations, with a trial-and-error approach. More structured model- and simulation-based methodologies usually require the creation of *ad hoc* descriptions (e.g., C/C++ or MATLAB/Simulink). These derive from generic models that have to be customized in order to correctly represent the specific target design by means of a nontrivial effort, and no information is passed from the component-level design environment to the system-level one.

Conversely, the proposed approach maximizes the reuse of available device descriptions across development phases and abstraction levels to reduce time-to-market and design costs.

In this section, we present some of the results obtained by using the tools and methods of the SMAC platform to obtain a simulatable system-level model of a wearable device for body motion reconstruction based on inertial sensors.

Limb tracking equipment is employed in different fields, from motion capture for computer games and animation movies to sport, healthcare, and fitness. The recent advances in inertial sensors allow a high level of integration of solutions based on inertial measurement units (IMUs) with very low costs and quite good accuracy with respect to vision-based methodologies.

The target system is based on the motion capture system presented in Ref. [19] and uses a set of MEMS sensors to track the relative displacement of limbs. Each node includes three-axis MEMS gyroscopes, accelerometers, and magnetometers and a 32-bit microcontroller (see Figure 3.4a).

The usual laboratory setup for performance characterization of inertial sensor nodes includes expensive vision-based systems employing cameras, markers on the body, and complex motion tracking algorithms.[10] The SMAC precise system-level simulation, taking into account the characteristics of the real components deriving from the designers' lower level representations, is aimed at evaluating the limb tracking accuracy in mission-like conditions in a virtual environment. This simulation is based on models abstracted in a homogeneous language, that is, SystemC/AMS, able to express the complex relationships between the components with a moderate computational effort.

---

10 Vicon Motion Systems Ltd. www.vicon.com/.

3.5.1 System Architecture

The sensor node is based on the iNEMO M1 System-on-Board (SoB) manufactured by STMicroelectronics (Figure 3.4b). The SoB hosts a three-axis digital gyroscope (L3GD20) and a geomagnetic module (LSM303DLHC), a 32-bit ARM CORTEX™-M3 (STM32F103), and an ultra-low dropout, low-noise voltage regulator (LDS3985M33R).

The sensors acquire data with programmable full-scale and sampling rate, while the microcontroller runs an extended Kalman filtering algorithm [20] to perform the *sensor fusion*, that is, the integration of data generated from different sensors, in order to provide precise angular position estimation. The algorithm integrates the gyroscope reading (angular velocity) to compute the angular position in dynamic conditions and corrects the computed values (subject to drift due to the numeric integration) leveraging information about gravity from accelerometers and magnetic field readings from magnetometers. Thus, an absolute angular position is obtained with respect to the gravity vector and to the magnetic North. A set of sensor nodes can be used to track the mutual position of the body parts with a distributed and scalable approach.

The SMAC approach has been used to simulate the limb tracking equipment by means of a system-level model, taking into account multiphysical properties, both functional and extrafunctional. In particular, the goal of the system simulation was to tune the system components and firmware (i.e., the Kalman filtering algorithm) to make the tracking more precise and to render the system immune and capable to compensate environmental effects such as temperature and local magnetic fields.

In detail, each sensor node was required to estimate orientation angles in static conditions with a maximum error of $2°$ to compensate temperature effects between 0 and $50\,°C$ and to be immune to magnetic disturbance caused by vicinity to ferromagnetic elements.

3.5.2 Model Development and System-Level Simulation

The system-level model of each sensor node includes the MEMS accelerometer/magnetometer module, the MEMS gyroscope, and a microcontroller running the sensor fusion algorithms. The analysis and optimization of such a system require modeling techniques merging multidomain models and constraints.

Each of the sensor modules is a heterogeneous subsystem in itself, integrating electromechanical (MEMS) parts, analog, and digital electronics. As an example, Figure 3.5 shows the block diagram of the employed accelerometer model for one axis. Its main composing elements are the mechanical second-order Laplace transfer function, which expresses the relationship between applied acceleration (and consequently force F) and the displacement d of the oscillating mass, a nonlinear module expressing the capacitance of the plates C as a function of their displacement, the analog front-end (AFE) (comprising amplifiers and filters), the analog-to-digital converter (ADC), and the digital front-end (DFE). Mechanical, analog, and digital electronic domains are therefore represented.

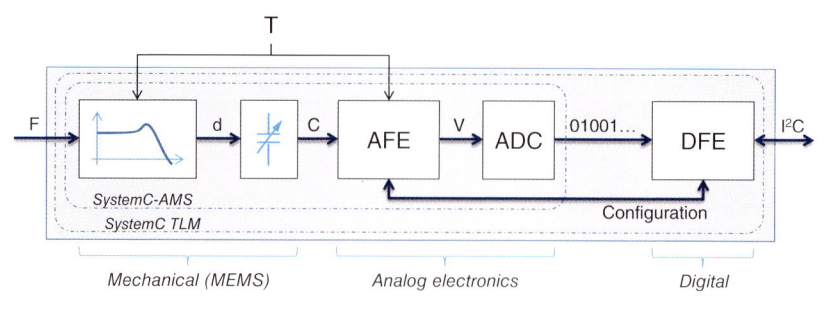

Figure 3.5 Accelerometer model for one axis [21].

The main building blocks have been translated across different abstraction levels. MEMS parts were originally modeled in CAD environments such as ANSYS, COMSOL Multiphysics, and Coventor MEMS+, which support layout definition, process evaluation, mesh generation, and device FEM simulation. From there, partly automated approaches, such as the ones supported by MEMS+, enabled the translation into VerilogA and, then, SystemC-AMS using the Timed Data Flow (TDF) model of computation.

Analog electronic devices, represented as Spice netlists and VerilogA descriptions, were transformed relying on MUNEDA WiCkeD RSM model generation HIFSuite tools into SystemC-AMS modules.

Also the digital electronic parts, whose original descriptions were in VHDL and Verilog, were translated, thanks to HIFSuite (i.e., the A2T abstraction tool) into SystemC TLM models. SystemC TLM has been used to describe the behavior of the sensor digital front-end, including the state machines that manage the device readout configuration and the programmable registers (full-scale selection, output data rate, and digital filters), as well as for the microcontroller interface.

The abstracted microcontroller and the other block models consisting of the limb tracking system have been integrated and cosimulated as a whole at system level in the Keysight SystemVue environment, with digital communications at the system level handled as TLM transactions. This has required the definition of SystemC wrappers that operate the required SystemC to/from SystemC AMS conversions needed for the instantiation of the sensor modules in the SystemVue environment (SystemC–SystemVue cosimulation).

The sensor fusion library, which implements an extended Kalman filtering algorithm, is designed as a C language code that runs on the microcontroller. Such a library is derived from the iNEMO Engine Lite developed and distributed as freeware by STMicroelectronics.[11] The algorithm works on quaternions to avoid singularities introduced by Euler angles.

Finally, in order to test the performance of the sensor node and to develop and evaluate enhancements and customization of the sensor fusion and limb position

---

11 X-CUBE-MEMS1 Motion MEMS and environmental sensor software expansions for STM32Cube - STMicroelectronics s.r.l., http://www.st.com/content/st_com/en/products/embedded-software/mcus-embedded-software/stm32-embedded-software/stm32cube-embedded-software-expansion/x-cube-mems1.html.

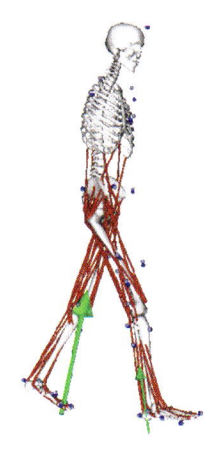

Figure 3.6 Human body representation in the OpenSim environment, with markers placed on the virtual sensor positions.

estimation algorithms, input stimuli have been synthesized, by reflecting real human movements. To obtain such data, the OpenSim open-source platform was selected, for modeling, simulating, and analyzing the neuromusculoskeletal system [22]. The limb movement recordings collected during experiments and simulations and available in the OpenSim online database have been employed: The positions in which the sensor nodes have to be located are defined on an OpenSim virtual human body representation (Figure 3.6). Thanks to the analysis tools, the kinematic properties of the sensor positions were tracked (position and Euler angles) during motion simulation and saved. For each of the tracked points, an *ad hoc* tool generates ideal sensor stimuli for the accelerometer (kinematic acceleration and projections of the gravity vector), magnetic field for the magnetometer (with the possibility of taking into account the Earth's and local magnetic source fields), and angular velocity for the gyroscope.

The same kinematic properties are used as reference to evaluate the error in the angular position estimation obtained by the sensor node model.

The final model thus enables the simulation of multidomain interactions, including mechanical (acceleration and angular velocity), electromagnetic (Earth's magnetic field and disturbance fields), and thermal effects (on the sensor transfer functions), by synthesizing suitable inputs for the sensor models while injecting environmental perturbations representing both EM and temperature noises. In addition, it is possible to evaluate acquisition, elaboration, and data transmission latency.

Figure 3.7 presents a comprehensive view of the simulation environment.

3.5.3 Results

The system-level simulation model obtained by means of the abstraction of lower level component descriptions has been validated by comparison with the outputs of a prototype. Then, it has been employed to evaluate system performance in terms of position reconstruction accuracy and latency. Initial requirements and

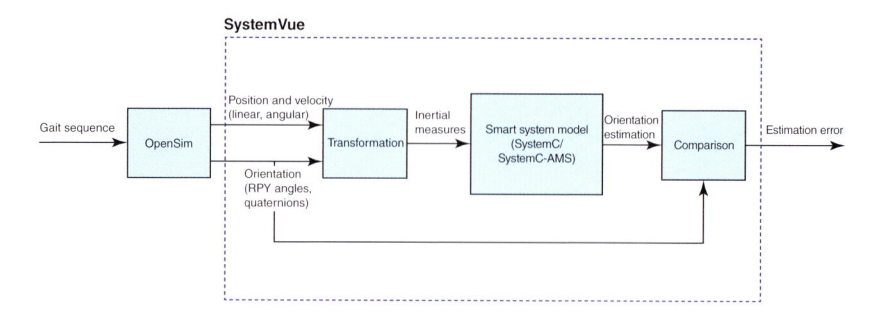

Figure 3.7 Comprehensive view of the simulation environment.

constraints have also been refined and applied to subsystem boundaries as in the top-down approach described in Ref. [23].

The system integrator manages cross-domain and cross-layer constraints by suitably interconnecting functional and nonfunctional signals between the models. As an example, the environmental temperature, and the possible heating of the electronic parts, affects the transfer functions of the MEMS transducers. This effect is visible in system-level simulation and has been corrected by tuning the compensation algorithm running on the microcontroller.

The model has first been employed to improve the performance of position evaluation in static and random positions, as well as to optimize the algorithm accuracy. Noise and uncalibrated sensors are among the main sources of drift, but the final system performance is affected by other causes as well, such as poor magnetic environment modeling, magnetic field disturbances, and temperature changes. Taking into account such effects allowed reducing drift to a negligible value for the targeted applications, that is, reducing the convergence time of angular position estimation while maintaining a low absolute tracking position error.

Figure 3.8a and b shows the results of the simulation of a sensor node in static condition at a nominal temperature of 25 °C: Part (a) shows the estimation of Euler angles (roll, pitch, and yaw) generated by the virtual model running the original sensor fusion algorithm and the reference values, while part (b) highlights the absolute error values for the three axes on a logarithmic scale. After an initial transient, in about 3 s a steady condition is reached for the three angles, with an error of about 6°.

By relying on the SMAC platform models, the sensor fusion algorithm has been customized and improved in different ways. First, sensor calibration has been taken into account to minimize the effect of MEMS and electronic component manufacturing variability, by computing gain and offset values for each sensor axis. Then, the model of the Earth's magnetic field used by the Kalman update step has been improved by considering local declination and inclination. Temperature compensation has been then introduced, by studying the simulated behavior of each sensor and deriving the best fitting parameters. Finally, the parameters of the Kalman noise covariance matrices have been optimized to further reduce the estimation errors.

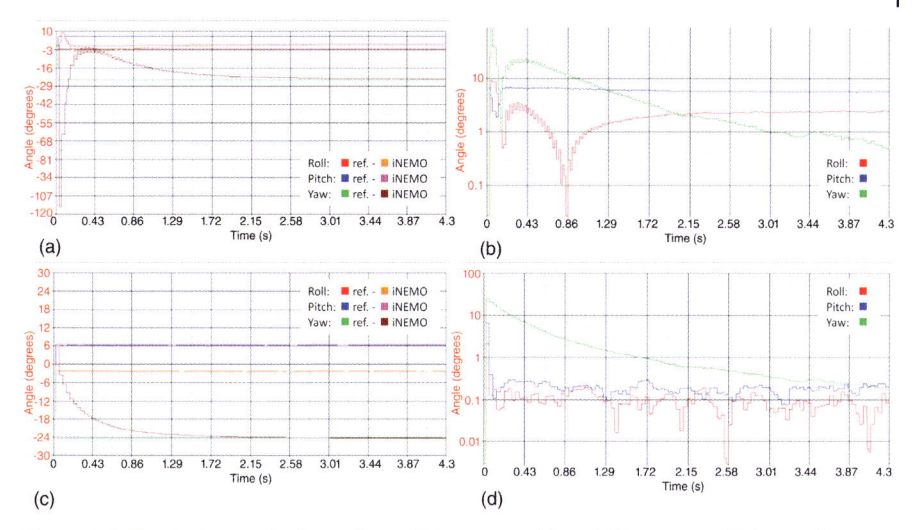

Figure 3.8 Simulation results in static conditions: parts (a) and (b) represent Euler angle estimation and absolute error with the original algorithm, while parts (c) and (d) with the optimized one.

Figure 3.8c and d shows an experiment in static condition after optimization of the sensor fusion algorithm, which reduce the error to less than 0.5° and shorten convergence time.

To study the effects of magnetic field perturbations, deriving, for example, from magnets or electric currents, multiphysics models have been developed and introduced in simulation. To minimize the error introduced by such perturbations in the orientation estimation, we introduced dynamically weighted sensor covariance matrices in the Kalman filter: The weights have been tuned proportionally to the error between the expected field modules and the measured values. The optimization and customization of the software application running in each sensor node allowed satisfying each of the requirements initially set.

The simulation platform also enables studying the system behavior in dynamic conditions, which is usually quite a complex operation to be performed in a laboratory and requires dedicated equipment. Figure 3.9 shows the estimation of Euler angles generated by the virtual model running the optimized sensor fusion algorithm, obtained during a simulated walk at constant temperature and elaborated at a 50 Hz frequency. The sensor is placed on the right femur. The figure shows the comparison between such data with the reference data generated by OpenSim (roll, pitch, and yaw). After an initial setup, the maximum error in dynamic condition is below 5° on each axis.

Each second of simulation of a single sensor node takes about 30 s of CPU time, on a quad-core Intel Core i7 – 2670QM running at 2.20 GHz with 4 GB of RAM. It has to be noted that each SystemC module instantiated within SystemVue runs on a separate thread, thus making the system scalable when more than one sensor node needs to be simulated, directly exploiting multicore parallelism.

The results from the analysis of the case study show that the SMAC platform can effectively support design, validation, and optimization of complex systems

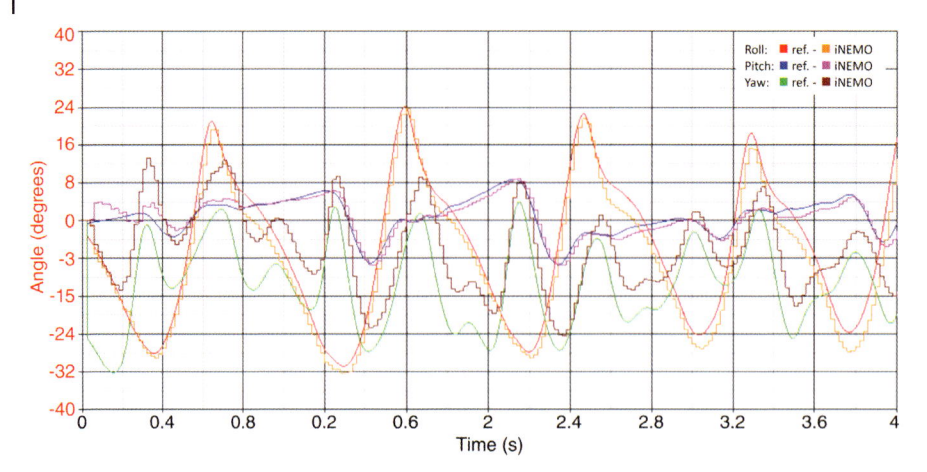

Figure 3.9 Euler angle estimation in dynamic condition (improved algorithm).

taking into account not only the electronic functionality but also the behavior and performance of physical structures that are inherent parts of today's applications. This multiphysical approach captures well the properties of architectures and materials and paves the way for the development of new more and more heterogeneous and integrated devices.

3.6 Conclusions

This chapter presented SMAC, a smart systems codesign platform. The usual design and optimization methods for heterogeneous smart systems involve separate flows: Each smart system component is designed with specific tools and methods, while the system integrator needs to develop from scratch a high-level uniform system representation (e.g., in SystemC or MATLAB). The optimization of embedded algorithms requires measuring system performance and possibly costly prototype iterations. The multidomain and multi-layer modeling and simulation methods developed in SMAC improve the state-of-the-art smart system design techniques, defining faster, more accurate, integrated, and less expensive design flows. The chapter summarized the contributions of SMAC in terms of tools, enhancement of existing tools, conversion flows, and methodologies, to allow a complete multidomain and multilevel codesign environment. The chapter presented preliminary results obtained by applying the SMAC platform to design and simulate the system-level model of a limb tracking smart system to test the efficiency of the platform to a real and complex case of study.

Needless to say, the SMAC platform does not solve all the issues involved in designing smart systems. The consolidation of the platform will consist of addressing the issue of nonfunctional properties – how to model and provide appropriate simulation facilities to assess power consumption and temperature distribution in such systems.

Simulation of power and temperature, as quantities with a concrete physical meaning, imply solving issues typical of material-integrated systems, as already mentioned in Section 3.1. How and where power is lost in a MEMS subsystem is an interesting and partially open research problem and is definitely not just an "electrical" problem. Similarly, a global thermal analysis of a smart system integrated into a System-on-Package (SiP) is a complex problem for which traditional solutions (e.g., FEM analysis tools) are not suitable in terms of computational complexity.

Progress in these aspects of the analysis of smart electronic systems will likely require a closer interaction between the electronic designers and material engineers, as these systems will progressively tend to become *de facto* material-integrated systems.

Acknowledgments

For obvious space reasons, we had to limit the authors' list to the project work package leaders and to the partners with the largest involvement in the project. Needless to say, the achievements of the project would not have been possible without all the other partners and the many people (more than 100) involved in SMAC.

References

1 Pereira, J. (ed.) (2009) *Monitoring and Control: Today's Market and Its Evolution till 2020*, Office for Official Publications of the European Communities, Luxembourg.
2 Cai, L. and Gajski, D. (2003) Transaction level modelling: an overview, in Proceedings of the International Conference on HW/SW Codesign and System Synthesis (CODES-ISSS), pp. 19–24.
3 Romanowicz, B., Ansel, Y., Laudon, M., Amacker, Ch., Renaud, P., Vachoux, A., and Schröpfer, G. (1997) VHDL-1076.1 modelling examples for microsystem simulation, in *Analog and Mixed-Signal Hardware Description Languages* (ed. A. Vachoux), Kluwer Academic Publishers.
4 Zhou, N., Clark, J.V., and Pister, K.S.J. (1998) Nodal analysis for MEMS design using SUGAR v0.5, in Proceedings of the International Conference on Modelling and Simulation of Microsystems, Semiconductors, Sensors and Actuators (MSM), Santa Clara, CA,, pp. 308–313.
5 Vandemeer, J.E., Kranz, M.S., and Fedder, G.K. (1998) Hierarchical representation and simulation of micromachined inertial sensors, in Proceedings of the International Conference on Modelling and Simulation of Microsystems, Semiconductors, Sensors and Actuators (MSM), Santa Clara, CA, pp. 540–545.
6 Lorenz, G. and Neul, R. (1998) Network-type modelling of micromachined sensor systems, in Proceedings of the International Conference on Modelling and Simulation of Microsystems, Semiconductors, Sensors and Actuators (MSM'98), Santa Clara, CA,, pp. 233–238.

7 Lorenz, G. and Repke, J. (2003) Design tools for MEMS, *Sensors Applications, Volume 4, Sensors for Automotive Technology*, Jossey-Bass Publishers, pp. 58–72. ISBN: 3527295534.

8 Lorenz, G. and Kamon, M. (2007) A system-model-based design environment for 3D simulation and animation of micro-electro-mechanical systems (MEMS). Proceedings of APCOM'07 in Conjunction with EPMESC XI, Kyoto, Japan, December 3–6, 2007.

9 Schröpfer, G., Lorenz, G., and Breit, S. (2010) Novel 3D modelling methods for virtual fabrication and EDA compatible design of MEMS via parametric libraries. *Journal of Micromechanics and Microengineering*, **20**, 064003.

10 Frangi, G.S. and Vigna, B. (2006) On the evaluation of damping in MEMS in the slip–flow regime. *International Journal for Numerical Methods in Engineering*, **68**, 1031–1051.

11 Frangi, A. (2005) A fast multipole implementation of the qualocation mixed-velocity–traction approach for exterior Stokes flows. *Journal Engineering Analysis with Boundary Elements*, **29**, 1039–1046.

12 Lorenz, G., Greiner, K., and Breit, S. (2006) A hybrid modelling approach for multi-physics systems. IMSTW 2006, June 2006. Edinburgh, UK.

13 Schneider, P., Reitz, S., Bastian, J., and Schwarz, P. (2005) Combination of analytical models and order reduction methods for system level modelling of gyroscopes. NSTI Nanotechnology Conference and Trade Show 2005, Vol. 3, CCN, MSM, May 8–12, 2005, Anaheim Marriott & Convention Center, Anaheim, CA.

14 Bechtold, T. (2005) Model order reduction of electro-thermal MEMS. Ph.D. thesis, University of Freiburg, Germany.

15 Callou, G., MacIel, P., Tavares, E., Andrade, E., Nogueira, B., Araujo, C., and Cunha, P. (2011) Energy consumption and execution time estimation of embedded system applications. *Microprocessors and Microsystems*, **35** (4), 426–440.

16 Sun, X., Chen, Y., Liu, J., and Huang, S. (2014) A co-simulation platform for smart grid considering interaction between information and power systems. 2014 IEEE PES Innovative Smart Grid Technologies Conference (ISGT'14), Article No. 6816423.

17 Wane, S. and Boguszewski, G. (2008) Global digital-analog cosimulation methodology for power and signal integrity aware design and analysis. Proceedings of the 38th European Microwave Conference (EuMC'08), Article No. 4751746, pp. 1477–1480.

18 Di Guglielmo, L., Fummi, F., Pravadelli, G., Stefanni, F., and Vinco, S. (2013) UNIVERCM: the UNIversal VERsatile computational model for heterogeneous system integration. *IEEE Transactions on Computer*, **62** (1), 225–241.

19 Brigante, C.M.N., Abbate, N., Basile, A., Faulisi, A.C., and Sessa, S. (2011) Towards miniaturization of a MEMS-based wearable motion capture system. *IEEE Transactions on Computers*, **58** (8), 3234–3241.

20 Sabatini, A.M. (2006) Quaternion-based extended Kalman filter for determining orientation by inertial and magnetic sensing. *IEEE Transactions on Biomedical Engineering*, **53** (7), 1346–1356.

21 Bombieri, N., Drogoudis, D., Gangemi, G., Gillon, R., Grosso, M., Macii, E., Poncino, M., and Rinaudo, S. (2015) Addressing the smart systems design challenge: the SMAC platform. *Microprocessors and Microsystems*, **39** (8), 1158–1173.

22 Delp, S.L., Anderson, F.C., Arnold, A.S., Loan, P., Habib, A., John, C.T., Guendelman, E., and Thelen, D.G. (2007) OpenSim: open-source software to create and analyze dynamic simulations of movement. *IEEE Transactions on Biomedical Engineering*, **54** (11), 1940–1950.

23 Crepaldi, M., Grosso, M., Sassone, A., Gallinaro, S., Rinaudo, S., Poncino, M., Macii, E., and Demarchi, D. (2014) A top-down constraint-driven methodology for smart system design. *IEEE Circuits and Systems Magazine*, **14** (1), 37–57.

Part Three

Sensor Technologies

4

Microelectromechanical Systems (MEMS)

Li Yunjia

Xi'an Jiaotong University, School of Electrical Engineering, 710049 Xi'an, China

4.1 Introduction

4.1.1 What Is MEMS

The term "MEMS" might sound very unfamiliar to most of us, but they are actually active in all aspects of our daily life. Take a smart phone for example; there are at least nine different types of MEMS devices inside, including:

- a distance sensor to sense the approach of human face to the screen (and turn off the screen during phone conversation);
- touch sensors to recognize the finger touch on the touch screen;
- image sensors to acquire pictures;
- MEMS microphones to transform sound into electronic signal;
- a pressure sensor to measure the altitude by measuring the atmospheric pressure;
- a MEMS gyroscope to measure the angular speed of the cell phone;
- a MEMS accelerometer to measure the acceleration of the cell phone;
- a magnetometer to measure the orientation of the cell phone by measuring the terrestrial magnetic field;
- temperature and humidity sensors.

MEMS is an acronym for microelectromechanical systems. They are micrometer-scale systems capable of accomplishing tasks that are traditionally carried out by macroscopic systems. The size of MEMS devices ranges from several micrometers to several millimeters. Being small in size, they are mostly produced using micromachining technologies originating from the integrated circuit (IC) industry. The first wave of commercialization of MEMS devices started in the early 1980s, when the MEMS pressure sensors and accelerometer first found broad applications in automotive industry. Since then, various MEMS devices have been successfully commercialized, such as inkjet nozzles, digital micromirror device (for example in projection systems), and radiofrequency (RF) components. From the early 2010s on, MEMS industry has experienced exponential growth, thanks to the rapid development of smart phones.

Material-Integrated Intelligent Systems: Technology and Applications, First Edition.
Edited by Stefan Bosse, Dirk Lehmhus, Walter Lang, and Matthias Busse.
© 2018 Wiley-VCH Verlag GmbH & Co. KGaA. Published 2018 by Wiley-VCH Verlag GmbH & Co. KGaA.

MEMS is a highly interdisciplinary subject that includes a wide range of devices and systems, such as microstructures, transducers, and energy harvesters. Nevertheless, the most widely used MEMS devices up to now are transducers. MEMS transducers convert one type of signal into the signal of another type. More specifically, the MEMS transducer can be further divided into two main subsets, that is, sensors and actuators. Sensors transform external information into a signal (most commonly electrical) that can be processed by the system, and actuators generate force or motion by the electronic signal of the system.

4.1.2 Why MEMS

The first and foremost advantage of MEMS devices is small size. Small size makes MEMS devices superior to their macroscopic counterparts, in terms of ease of integration, lightweight, low power consumption, and high resonance frequency (thus fast response time).

When it comes to the production of submillimeter components, traditional methods such as milling and turning are limited by the high cost as well as insufficient accuracy and repeatability. MEMS devices are usually fabricated from a silicon wafer, by microfabrication (or micromachining) technologies originating from the IC industry. Due to the small size of MEMS devices, hundreds or thousands of devices can be produced in one single run on the same wafer. Such parallel production methods make MEMS devices lower in cost, more accurate in performance, and more repeatable in geometrical parameters.

Another advantage of MEMS devices is the possibility to integrate with other MEMS devices or with their signal processing circuit. Integrating several MEMS devices together can expand the functionality of the chip, while integrating MEMS device with the circuits further decreases the system size and increases the performance (thanks to the lowered noise). For example, Sensirion's SHTW1 humidity and temperature sensor has a size of only $1.3 \times 0.7 \times 0.5$ mm. With such a small (most commonly electrical) footprint, it has included a temperature sensor, a humidity sensor, analog and digital signal processing circuits, A/D converter, calibration data memory, and a digital communication interface. Another example is Bosch Sensortec's integrated environmental sensing unit BME680. With a size of $3 \times 3 \times 0.95$ mm^3, it includes sensors for measuring gas, pressure, humidity, and temperature.

4.1.3 MEMS Sensors

Among various MEMS devices, MEMS sensors are the most widely used and studied devices. Sensors detect or sense information (most commonly physical, chemical, and biological) in their own environment and transform them into a signal that can be processed by the system. There are large numbers of different sensors, and can be categorized in many different ways. However, the following are the three most common ways of categorizing them:

1) *Working principle:* resistive, capacitive, inductive, optoelectronic, Hall, thermoelectric, piezoelectric.

2) *Sensing quantity:* gas sensor, pressure sensor, force sensor, displacement sensor, inertial sensor, temperature sensor, humidity sensor.
3) *Output:* digital sensor and analog sensor.

4.1.4 Goal of This Chapter

This chapter aims at giving the readers, especially those with no background in MEMS, a brief overview on the basic concept of MEMS sensors. For details about MEMS and microfabrication technologies, see books on MEMS devices (Refs [1,2]).

As MEMS devices are inherently material-integrated intelligent systems, (i.e., the MEMS devices are produced by integrating various materials by a series of different process technologies), this chapter is divided into the following sections:

In Section 4.2, the basic materials for MEMS production, including metal, semiconductors, and dielectrics, are presented.
In Section 4.3, the microfabrication technologies, by which the materials are integrated into MEMS devices, are presented.
In Section 4.4, an overview of different sensor devices are presented.
In Section 4.5, the sensor systems and their integration (packaging) technologies are briefly discussed.

4.2 Materials

4.2.1 Silicon

Silicon is the most commonly used substrate material for producing MEMS devices and integrated circuit. The reasons for this dominance are as follows:

- Silicon is the second most abundant material in the earth's crust and very cheap.
- Silicon has excellent processing properties and well-established processing techniques.
- Silicon has outstanding physical properties and very stable chemical properties.
- Silicon MEMS and silicon IC can be integrated to form an entire system.
- A high-quality and stable insulating oxide can be easily grown on Si by oxidation.

Silicon materials for the production of MEMS and IC are mostly in the form of either monocrystalline or polycrystalline silicon. Monocrystalline silicon (mono-Si or single-crystal silicon) is often used as the substrate where MEMS or IC devices are built on (or the structural layer in bulk-micromachined MEMS devices), whereas polycrystalline silicon (polysilicon or poly-Si) is often deposited as functional layers of the devices.

As the name indicates, monocrystalline silicon is a single continuous crystal of silicon, free of any grain boundaries. It has a face-centered cubic lattice structure (diamond structure). Mechanically, it is a brittle material with a Young's modulus of ~190 GPa (close to steel) and widely used as the structural material for MEMS devices. Electrically, pure monocrystalline silicon is an intrinsic semiconductor with a resistivity of $2.3 \times 10^5\,\Omega$ cm. Such high resistivity is not desirable for building either IC or MEMS devices. However, its electrical conductivity can be adjusted by doping with small fractions of impurity materials (e.g., boron or phosphorus). The conductivity of doped silicon can be controlled in the range of 10^{-3}–$10^4\,\Omega$ cm.

Polycrystalline silicon is a solid consisting of small crystals with grain boundaries in between. It is mostly used as gate materials in semiconductor devices and photovoltaic devices, as well as the structural layer in surface-micromachined devices.

4.2.2 Dielectrics

The most commonly used dielectric materials in MEMS production are quartz, glass/SiO_2, silicon nitride, ceramics, and various polymers.

Quartz in MEMS devices is often chemically synthesized monocrystalline silicon dioxide. It is widely used to produce resonators and inertial sensors due to its piezoelectric property. Glass is the amorphous state of quartz. It can be used as an alternative substrate (instead of silicon) of the MEMS devices, especially where the substrates need to be optically transparent or electrically insulated. Thin film glass (SiO_2) can be deposited or grown onto the substrate, used as mask layer, insulation layer, or sacrificial layer.

Silicon nitride (Si_xN_y) usually plays similar role in MEMS devices as SiO_2, that is, mask layer, insulation layer, or sacrificial layer. However, there are also substantial differences between silicon nitride and silicon dioxide: Silicon dioxide can be either chemically/physically deposited or thermally grown, with a layer thickness between few nanometers and 1–2 µm. Silicon nitride can only be deposited with a maximum thickness of few hundred nanometers. Silicon dioxide can be etched by hydrofluoric acid, and silicon nitride can be etched by phosphoric acid. Therefore, whether to use silicon dioxide and silicon nitride has to be carefully chosen according the flow and chemistry of the entire fabrication processes. In some cases, silicon dioxide and silicon nitride can also be deposited one upon the other either to increase the dielectric strength of the dielectric layer or to realize a multistep silicon etching mask. Silicon nitride can be deposited stress free (LPCVD) by choosing the proper deposition parameters, whereas deposited silicon dioxide commonly has compressive stress.

Polymers used in MEMS production include mainly three types: photoresists, adhesive polymers, and structural polymers. Photoresists are light-sensitive polymers coated on the substrates to transfer the MEMS design onto the substrates. More details about photoresists will be discussed in Section 4.3.2. Adhesive polymers, most commonly Parylene and benzocyclobutene (BCB), are used to bond two substrates together (e.g., silicon to silicon or silicon to glass) to form a wafer stack. An emerging new application of polymers in recent years is the structural material of MEMS devices. Polymer-based MEMS devices do not only possess advantages of low cost, easy processing, and potential biocompatibility but also outperforms silicon in specific applications due to polymer's high flexibility and low stiffness. For example, photoresists-based cantilever sensor can be directly fabricated by simple lithographic process [3], and actuators with polymer springs need much lower actuation voltage than traditional actuators with silicon springs [4]. Nevertheless, polymer-based MEMS devices may have their limitations as well, for example, behavior drift induced by viscoelasticity or environmental temperature/humidity change. In practice, these issues are addressed by closed-loop operation or compensation circuit.

4.2.3 Metals

Most frequently, metals are used in MEMS as electrodes and wires. For this purpose, the most used metals are Al and Au. Al is known to form Ohmic contact with silicon and has very good adhesion with Si or SiO_2, whereas Au has a higher conductivity and more stable chemical properties than Al. However, Au is not compatible with many of the microfabrication process (such as dry etching) and bad adhesion to silicon. Therefore, a layer of Cr or Ti is usually added between Au and Si to promote adhesion. Besides basic electrical applications, metals can also be used to form MEMS devices directly. For example, Au can be used as structural layer for RF MEMS [5]; Cu can be used to fabricate high aspect ratio structures such as microheat sinks [6] or through-silicon vias [7]; and Al can be used as reflective layer on optical systems [8]. For these applications, most of the metals can be deposited by physical vapor deposition (PVD) or chemical vapor deposition (CVD) techniques, and have thicknesses between few nanometer and few micrometers. High aspect ratio metal structures up to few hundred micrometers can be fabricated by electroplating process. Details of the fabrication processes will be discussed in Section 4.3.

4.3 Microfabrication Technologies

MEMS devices are produced by micromachining technologies originated from the IC industry. The basic idea of micromachining technologies is to create micrometer-scale three-dimensional (3D) structures, by removing part of the substrates or adding new materials onto the substrates. The technological processes of micromachining include patterning (lithography), etching, and deposition processes. Nevertheless, in addition to these three processes, many other techniques have also been developed and are still being developed to meet the ever-growing complexity of the new MEMS devices. In this section, the micromachining technologies are briefly presented.

4.3.1 Silicon Wafers

Silicon wafers are thin slices of monocrystalline silicon. They are the bases and substrates on which the MEMS devices are constructed. To produce MEMS devices, the silicon wafers usually have to experience a series of fabrication processes, such as lithography, etching, and deposition. At the end, large numbers of individual MEMS devices are separated by dicing the silicon wafer.

Silicon wafers are made of monocrystalline silicon material with nearly no defect and very high purity (99.9999999%). They can be produced using either the

Czochralski or the floating zone process. In a Czochralski process, raw silicon material is first melted in a crucible, in which a seed silicon crystal is immersed and pulled to form a high-purity cylindrical ingot of monocrystalline silicon. Impurities, such as boron or phosphorus, can be added to silicon melt to dope the silicon. Noteworthy is that as the silicon ingot grows, the dopant concentration in the melt changes, so it must be carefully controlled to have a uniform doping profile along the ingot. The ingot is then sliced with a wafer saw, grinded, and polished to form the silicon wafers. In a float zone process, a silicon ingot (e.g., a polysilicon ingot) is fixed with a monocrystalline silicon seed placed below. A heating coil around the ingot moves along the ingot axis to melt the ingot, starting from the region between the polysilicon and monocrystalline silicon seed. A monocrystalline ingot is formed after the process. Compared with the Czochralski process, the float zone process produces silicon wafers with much lower impurities level (e.g., carbon and oxygen). However, it is possible to produce wafers with various diameters (up to 450 mm, 18 in.) by Czochralski process, whereas the maximum wafer diameter can be produced by float zone process is only 200 mm (8 inches). The most commonly used silicon wafer has a (100) or (111) crystal orientation. For wafers with size smaller than 200 mm, one or more flats on the side of silicon wafer are cut to indicate the crystallographic planes (usually (110)) of the wafer.

Another type of silicon wafer being increasingly used recently is the silicon-on-insulator (SOI) wafer. The SOI wafers are usually a stack of three layers: (i) on the bottom, a thick silicon layer (usually with a thickness of 400–500 μm) that provides mechanical stability to the whole stack; (ii) in the middle, a buried oxide layer with thickness ranges from hundreds of nanometers to few micrometers; and (iii) on the top, a thin silicon layer with thickness ranges from few micrometers to a hundred micrometer thick. There are two advantages of using SOI wafers over single silicon wafers, namely, decreasing the complexity of the fabrication process flow and reducing the parasitic capacitance of the fabricated device. However, the price of a SOI wafer is also significantly higher, for example, few hundreds of US dollars for a SOI wafer compared with less than 10 US dollars for a normal silicon wafer. For different needs, SOI wafers can be fabricated with various techniques, including Separation by IMplantation of Oxygen (SIMOX) [9] and wafer bonding techniques, such as Smart Cut [9], NanoCleave [10], and ELTRAN [11].

4.3.2 Lithography

Lithography is a term originating from ancient Greek words "lithos" and "graphein" that means a method of printing. As for MEMS production, lithography is the most crucial process step that transfers the designed micro- or nanoscale pattern to the silicon wafer. Despite the diversity of different lithographic processes, basic steps of lithography include coating the silicon wafer with a thin light-sensitive polymer (i.e., the resist), irradiating certain area of the resist to change its chemical property, and wash away the part of resist that has been irradiated or unirradiated by a liquid solution named developer. The resist remaining on the wafer forms the designed pattern, and the wafer can be

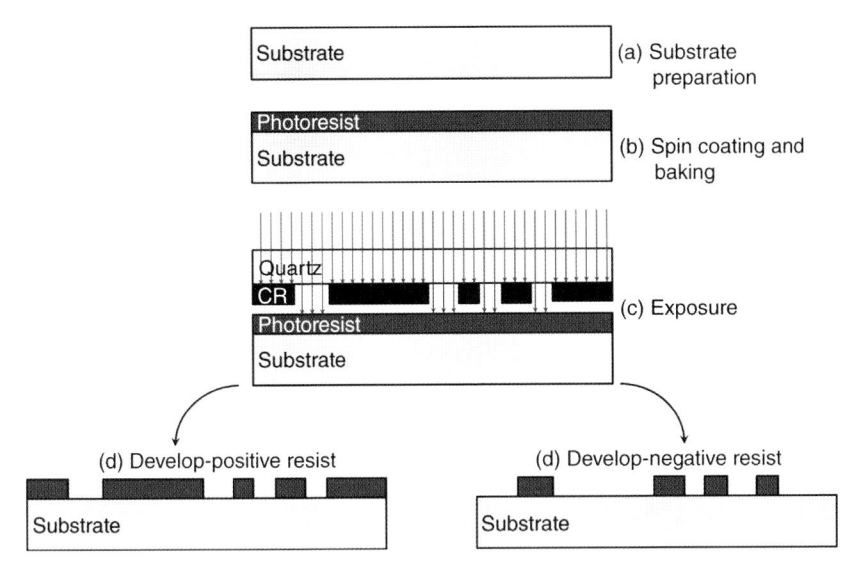

Figure 4.1 Process flow of the photolithography technique.

processed with subsequent steps such as etching or deposition. Depending on the radiation type, there are many different lithography techniques, such as ebeam lithography, scanning probe lithography, X-ray lithography, ion beam lithography, and photolithography. Among these techniques, photolithography is the most widely used process in the MEMS fabrication, for patterns with resolution down to few hundred nanometers.

Photolithography, also known as optical lithography or light lithography, uses light to pattern the resist. The light sources used in photolithography are usually ultraviolet (UV) light with different wavelengths, for example, UV, deep UV, and extreme UV. The general process steps of photolithography include substrate preparation, photoresist coating and baking, exposure, and developing, as shown in Figure 4.1.

- **Substrate preparation**
  Before the coating of photoresists, organic and inorganic contaminants on the wafer surface must be removed. The surface cleaning can usually be carried out by using organic solvent such as acetone, Piranha solution (a mixture of sulfuric acid and hydrogen peroxide), or oxygen plasma ashing process. After the wafer cleaning procedure, an adhesion promoter (such as hexamethyldisilazane, HMDS) is often applied in either liquid or gaseous phase to increase the adhesion of the photoresist to the wafer.
- **Photoresist coating and baking**
  Photoresist is a viscous liquid solution with polymers and photosensitive components dissolved in an organic solvent. The photoresist can be applied onto wafer surface by either spin coating or spray coating. Spin coating is most widely adopted in the standard MEMS fabrication processes, whereas spray coating is usually used to cover wafers with deep structures on the surface.

To spin coat, the photoresist is dispensed onto the wafer surface and subjected to a two-speed spinning process. The wafer first spins at a slow speed (usually around 500 rpm) to distribute the photoresist on the whole wafer surface to form a thick layer, and then the spin speed increases rapidly to 1000–5000 rpm to form a homogeneous thin layer of photoresist. The thickness of the final photoresist layer is determined by the spin-coating speed and usually ranges from 0.5 to 10 μm. The wafer is then baked on a hot-plate or in an oven to drive off excess solvent and solidify the photoresist.

- **Exposure**

    The exposure of photoresist is carried out on a tool called the mask-aligner. It is done by irradiating UV light onto the photoresist-coated wafer, through a photomask (hereinafter, the mask) parallelly placed above the wafer. The mask is usually a quartz plates (around 1.5–3 mm thick) with a thin chromium layer (around 100 nm thick) of the designed pattern on the surface. During the exposure, the UV light is partially radiated through the transparent quartz and partially blocked by the chromium layer. The UV light induces chemical reactions in the irradiated photoresist that allows some of the photoresist to be removed by a certain solution (i.e., the developer). There are two types of photoresists, positive resist and negative resist. For positive resist, the radiated part become soluble in developer due to the breakage of polymer chains by the UV light; whereas for the negative resist, the radiated part become insoluble in the developer due to cross-linking or polymerization induced by the UV light. Sometimes, a postexposure bake (PEB) is performed after the exposure to reduce standing wave phenomena caused by the interference of the incident light. In some other cases where chemically amplified resist (CAR) is used, PEB is also required because the chemical reaction in the resist is only initiated by the exposure, but mostly completed during the PEB.

    Another type of tool for performing the exposure process is the projection exposure systems (steppers or scanners), where a photomask that contains only the pattern of only one or several dies are projected onto the wafer a number of times to finish a complete exposure. The projection lithography will not be detailed in this book. A good overview of lithography processes can be found in Ref. [12].

- **Development**

    After exposure, the wafer is developed in a developer solution, removing partially the photoresist to finally realize the pattern transfer from the photomask to the wafer. For positive resists, the most commonly used developer is sodium hydroxide (NaOH). However, sodium ion degrades the insulating properties of gate oxides in MOSFET, and is thus considered as a contaminant in the IC industry. A commonly used alternative metal-ion-free developer is tetramethylammonium hydroxide (TMAH). For negative resists, as the development is mainly used to wash away unirradiated resist, the commonly used developers are simply organic solvents, such as dimethylbenzene and isopropanol. After development, the wafer is sometimes hard-baked to enhance the durability of the remaining photoresist for the subsequent process steps.

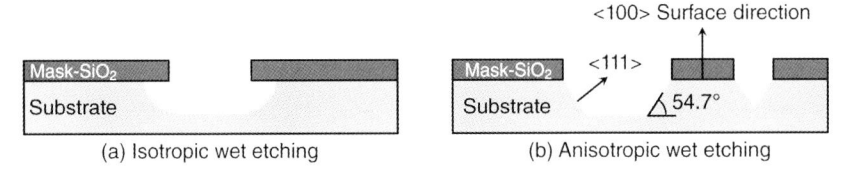

Figure 4.2 Schematic illustration of the isotropic and anisotropic wet etching processes.

4.3.3 Etching

The photolithography step transfers the designed pattern from the mask to the photoresist. The Etching techniques are used to transfer the pattern from the resist to the wafer. The etching processes remove layers of silicon wafer at the surface area where no resist is covered. The patterned photoresist thus serves as a mask material to protect the wafer area not to be etched. However, photoresists are not always durable enough to protect the wafer. Some etching techniques require a more durable mask (namely, the "hard mask") such as silicon dioxide or silicon nitride. Nevertheless, silicon dioxide or silicon nitride must first be etched before they can be used as etching mask. Classic references for the topic of etching are Refs [13,14]. Two major methods of etching in microfabrication processes are wet etching and dry etching. Two major criteria to evaluate the etching processes are the etching selectivity and etching isotropy. Etching selectivity is the etch rate ratio between the target material and the mask material. Therefore, for mask material with a given thickness, a higher selectivity means it can be used to etch the target material for a longer period. The etching isotropy represents the directionality of the etching process. Isotropic etching removes material in all directions with the same speed whereas anisotropic etching indicates a direction-dependent etch rate.

- **Wet etching**

    As the name suggests, wet etching processes are based on liquid phase etchant. The wafer is usually immersed in a batch of etchant, which is usually agitated to ensure a good contact between the etched material and the fresh etchant. Wet etching processes are widely used to etch different materials, such as silicon, dielectrics (e.g., silicon dioxide and silicon nitride), and metals. Depending on the isotropy, the wet etching can be further classified into isotropic an anisotropic wet etching.

    As shown in Figure 4.2a, the isotropic etching process removes material on all directions with the same speed. This induces large undercut during the etching, especially when etching deep cavities. Therefore, the isotropic etching process is more often used to etch thin layers of materials, such as silicon dioxide, silicon nitride, and metals. Sometimes, part of the MEMS device may become completely free to move after the removal of the layer below (i.e., the sacrificial layer): vapor phase wet enchant can be used to avoid the potential stiction of the movable parts induced by the capillary force of liquid etchant. For example, using vapor HF (hydrofluoric) acid to etch silicon dioxide is one of the most commonly used release techniques for movable part in MEMS devices. However, the nature of the vapor HF etching is more a "dry" etching rather than a "wet" etching.

Anisotropic wet etching is a highly directional etching method. Such directionality is a result of different etching rates along different crystalline planes. Therefore, anisotropic etching is mostly used to etch monocrystalline materials, such as silicon and quartz. Most commonly used anisotropic wet etchants for silicon is potassium hydroxide (KOH) whose etch rate is 400 times higher in <100> crystal direction than in <111> direction. However, the potassium ions are considered as a contaminant that degrades the insulation property of gate oxide in integrated circuits. As an alternative, TMAH is often used in the IC industry, with a 37 times higher etch rate in <100> crystal direction than in <111> direction. A common example of anisotropic wet etching of silicon is shown in Figure 4.2b, in which a (100) silicon surface is etched under the protection of SiO_2 masking material with a rectangular opening. The etching process generates a cavity with smooth but sloped sidewall (<111> oriented) and a flat bottom (<100> oriented). The angle between the <111>-oriented sidewalls and the wafer surface has a fixed value of 54.7° (it is the angle between the two crystal directions). With the etching process continues, the area of the flat bottom will decrease and eventually disappear, leaving the cavity with a V-shaped cross section. If the mask opening is in a square shape, the cavity will have a pyramid shape when the etching is completed. In practice, with a proper mask design, anisotropic wet etching can be used to realize quite complicated 3D structures. Anisotropic wet etching can also be used to etch wafers (111) and (110) surfaces, but will not be detailed in this book.

- **Dry etching**

  Dry etching processes mostly utilize plasma to etch materials. The two most commonly used etching techniques are reactive ion etching (RIE) and deep reactive ion etching (DRIE).

  In a RIE system, the wafer to be etched is usually placed at the bottom of a vacuum chamber (a platter electrically isolated from the rest of the chamber), and a mixture of gases is introduced from the top. The chamber pressure and gases flow are strictly controlled during the process. A radiofrequency electromagnetic

(a) 1st Etching step — SF_6 RF plasma

(b) 1st Passivation step — C_4F_8 Deposition

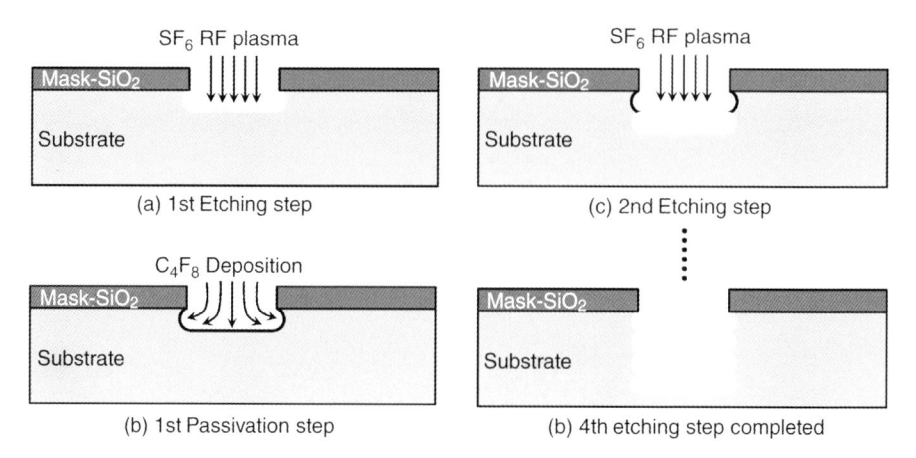

(c) 2nd Etching step — SF_6 RF plasma

(b) 4th etching step completed

Figure 4.3 Schematic illustration of the Bosch processes.

wave (usually with a frequency of 13.56 MHz) is applied to ionize the gas mixture, by oscillating the gas molecules at very high frequency to break down gas molecules into ions and electrons. The ionized gas mixture (reactive ions) is referred to as the plasma. The RIE process utilizes a combination of both physical ion bombardment and chemical reactions to etch materials. As the physical ion bombardment etching is anisotropic and chemical reactions etching is isotropic, the etch directionality and etch rate of RIE etching can be tuned by adjusting the balance between the two etching mechanisms. There are many parameters to be tuned to determine the etching results, for example, plasma power, platter temperature, chamber pressure, etchant gas chemistry, and gas flows. RIE can be used to etch silicon, silicon dioxide, and silicon nitride with gases such as fluorocarbons, chlorocarbons, sulfur hexafluoride, usually in combinations with of oxygen, nitrogen, argon, and helium. Pure oxygen plasma can be used to etch polymers, especially photoresist residues (plasma ashing).

DRIE is a special type of RIE that can be used to etch deep trenches with almost vertical sidewalls. There are two major DRIE technologies: Bosch process and cryogenic process. The Bosch process is named after the German company Robert Bosch GmbH that has originally patented the technology. The Bosch process alternates repeatedly between two independent steps to etch silicon, as shown in Figure 4.3: (i) the etching step: silicon is isotropically etched by SF_6-based plasma from almost a vertical direction; (ii) the passivation step: a chemically inert passivation layer (e.g., Teflon-like C_4F_8) is deposited over the etched silicon. Each of these steps lasts for several seconds. In the passivation step, the passivation layer is deposited homogeneously on the sidewall and bottom of the etched silicon cavity. In the next etching step, the passivation layer on the bottom surface is sputtered away by the physical ion bombardment, whereas the passivation layer on the sidewalls remains. Consequently, the bottom surface is etched and the sidewall is protected by the passivation layer. Repeating these two steps, high aspect ratio trenches with nearly vertical walls can be achieved. The cryogenic DRIE uses low temperature ($-110\,°C$) to slow down the isotropic chemical etching while maintaining the physical ion bombardment etching. In such a way, high aspect ratio trenches with highly vertical sidewalls can be obtained as well. However, in practice, the usage of cryogenic DRIE process is limited due to mainly two issues: (i) many standard etching masks cannot withstand the extreme low temperature; and (ii) by-products of the etching process usually deposit on the substrate, leading to a contamination.

4.3.4 Deposition Techniques

Deposition techniques in micromachining technologies are mainly used to deposit or grow thin film materials on the substrate. Deposited film thickness can range from few nanometers to several micrometers. Most commonly used deposition techniques include Chemical Vapor Deposition (CVD) and Physical Vapor Deposition (PVD).

CVD deposits thin film materials on the wafer surface by the chemical reaction of precursors. Usually, wafer is placed in a chamber, in which vapor-phase precursors are introduced and react on the wafer surface to carry out the material deposition. CVD can be used to produce many different materials, such as silicon, metals, dielectrics (silicon dioxide/nitride/oxynitride/carbide, etc.), and different form of carbon (fiber, nanotube, diamond, and graphene). The form of these deposited materials can also vary among monocrystalline, polycrystalline, amorphous, and epitaxial. To evaluate the CVD deposition processes, following parameters have to be considered: deposition rate, thickness uniformity of the deposited layer over the entire wafer, conformity (step coverage of the deposited layer), as well as deposited layer's chemical composition and crystal structure. To achieve different deposition goals, various CVD techniques have been developed, categorized by reaction types and conditions, such as atmospheric pressure CVD (APCVD), low-pressure CVD (LPCVD), plasma-enhanced CVD (PECVD), and metalorganic chemical vapor deposition (MOCVD), and so on.

Instead of depositing via chemical interactions, PVD uses physical processes to vaporize target materials (usually in vacuum) and deposit them onto wafers. PVD is widely used in different industries to produce thin film coating on surfaces, whose electrical, mechanical, or optical properties need to be modified. In the MEMS fabrication process, PVD is mostly used to produce thin film materials, including semiconductors, metals and metal alloys, as well as organic and inorganic dielectrics. Most commonly used PVD techniques are sputtering and evaporation (such as ebeam evaporation and thermal evaporation). Compared to most of the CVD technologies, the PVD processes are usually faster in deposition rate and more environment-friendly. However, as PVD is usually conducted in a low-pressure environment, the uniformity and conformality of PVD are not as good as CVD techniques.

In addition to CVD and PVD, there are many other deposition/growth techniques, such as thermal oxidation process for growing thick silicon dioxide on silicon wafer and electroplating process for depositing thick metal layers.

4.3.5 Other Processes

- **Lift-off process**

    The lift-off process is used to pattern metal and dielectrics on the wafer surface. It is especially useful for patterning the materials that are hard to dissolve, such as Pt, Au, and silicon-based dielectrics. An example of a lift-off process for patterning Au is shown in Figure 4.4. A photolithography step is

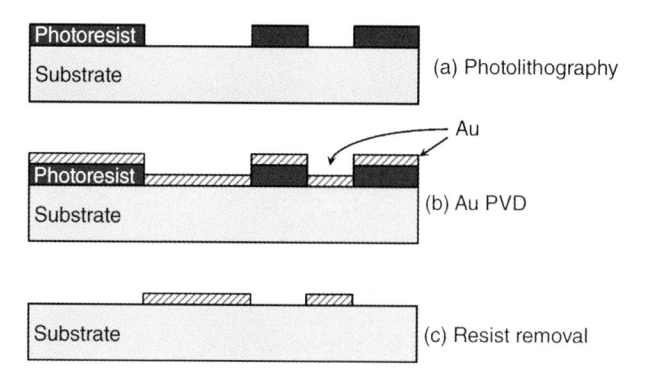

Figure 4.4 Schematic illustration of the lift-off process.

first performed to pattern the resist. Then Au is deposited on the wafer, part on the resist, and part directly on the wafer surface (in the cavities of the resist). Subsequently, the entire resist layer is removed in the solvent, together with the Au covering it. Finally, the Au pattern left on the wafer is the part that was directly deposited in the cavities of the resist.

- **Wafer bonding**

  Wafer bonding is the process of bonding two wafers together. Wafer bonding techniques can be used to create complicated 3D MEMS structures, produce SOI wafers, or package MEMS devices. There are many different bonding techniques. For example, without intermediate medium, two Si wafers can be bonded directly by fusion bonding (around 1000 °C) or plasma-activated direct bonding (below 400 °C), a Si wafer and a glass wafer can be bonded directly by anodic bonding. With intermediate medium, two wafers with same or different materials can be bonded with glass frit bonding, eutectic bonding, adhesive bonding, and thermal compression bonding techniques. A good overview of wafer bonding techniques can be found in Ref. [15].

4.3.6 Surface and Bulk Micromachining

To fabricate MEMS device, usually many steps of different micromachining techniques have to be combined. According to the type of technologies used in the fabrication process, the micromachining techniques can be divided into two types: bulk micromachining and surface micromachining. Bulk micromachining uses etching (dry or wet etching) and bonding techniques to produce MEMS structures. Silicon wafers are selectively etched and the MEMS device is built within the silicon wafer. Surface micromachining uses deposition and layer removal (etching) processes to produce MEMS devices on the silicon wafer. All the layers are usually within the thickness of a few micrometers. Most commonly used materials are polysilicon as structural layer, silicon dioxide as sacrificial layers, and metal as electrodes. Sacrificial layer is usually deposited below the movable part of MEMS device, and removed later to release free the movable part. Figure 4.5 shows a typical process of building a silicon membrane over an Au electrode, realized by both micromachining processes.

4.4 MEMS Sensor

4.4.1 Resistive Sensors

Resistive sensors usually consist of one or several resistors. They detect or sense property variations in their environments by measuring the resistance variations. Resistive sensors can be used to measure a variety of physical and chemical properties that can influence resistance of certain materials, for instance, thermoresistive sensors for measuring temperature, piezoresistive for measuring strain, magnetoresistive sensor for measuring magnetic field, photoresistive sensor for measuring light intensity, chemoresistive sensors for measuring gas concentrations, and resistive humidity sensor for measuring relative humidity.

(a) Wafer 1 lithography and wet etching	(a) Au PVD and lift-off
(b) Wafer 2 lithography and wet etching	(b) Sacrificial layer SiO₂ PVD and etching
(c) Wafer 2 Au PVD and lift-off	(c) Polysilicon CVD and dry etching
(d) Wafer bonding and oxide removal	(d) Sacrificial layer removal
Bulk micromachining process	Surface micromachining process

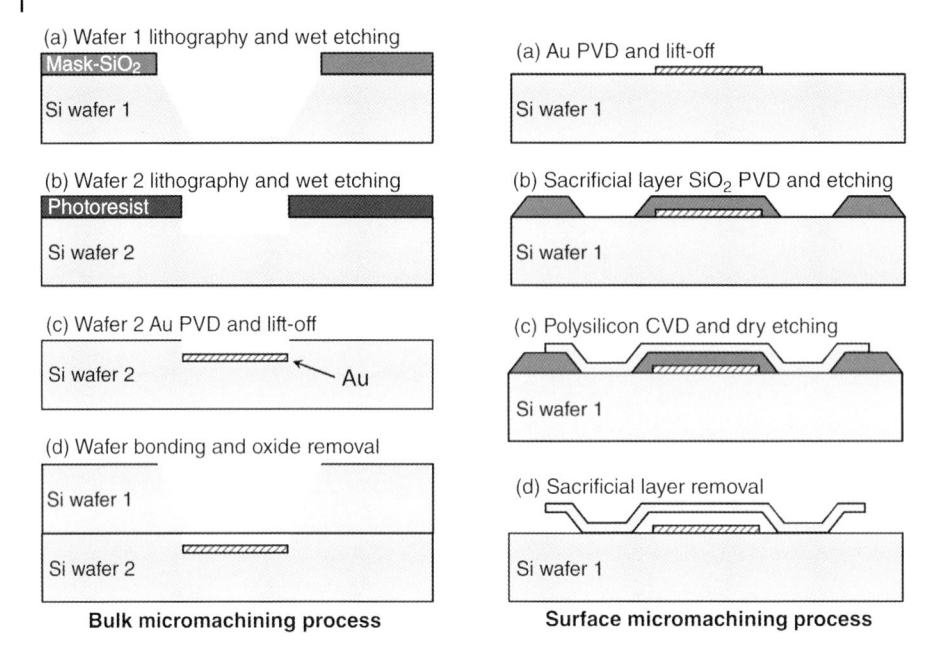

Figure 4.5 Comparison between a bulk micromachining process and a surface micromachining process.

In this section, the mostly commonly used thermoresistive and piezoresistive sensors will be described.

- **Thermoresistive sensor**

  A thermoresistive sensor, also known as resistance thermometers, measures the temperature by measuring the temperature-dependent resistance of materials. A variety of metallic materials can be used to produce the sensor, such as Al, Ti, Pt, Au, and Cu. Pt is the most widely adopted material due to its high accuracy, low drift, and wide range of operation.

  Traditional thermoresistive sensors are fabricated by either wrapping a long thin coiled wire around a ceramic or glass core, or depositing a thin serpentine planar wires on to a ceramic plate. The resistor is always in a long and thin shape in order to achieve a higher resistance and thus higher sensor accuracy (derived from Ohm's law: the resistance of a resistor is in direct proportional to its length and reverse proportional to its cross-sectional area). Nevertheless, micromachining technologies offer the opportunities to significantly decrease the size and cost of temperature sensors while maintaining the performance and production repeatability of the sensors. By micromachining technologies, the temperature sensor can be fabricated by depositing a very thin and narrow serpentine wire, as shown in Figure 4.6. Such a micro temperature sensor can be fabricated either by a simple three-step process (including photolithography, PVD, and lift-off, on a glass substrate), or a single-step process (PVD over a shadow mask). In addition, micro temperature sensor can be directly integrated with signal processing circuit and AD converter to further increase its performance and functionality.

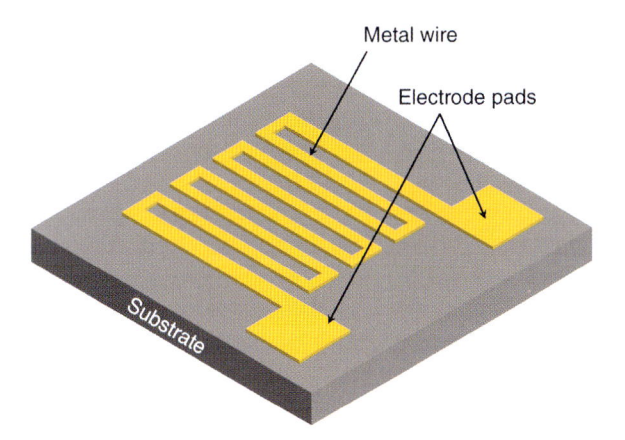

Figure 4.6 Schematic illustration of a MEMS thermoresistive sensor.

Thanks to the flexible combination of different micromachining techniques, the MEMS temperature sensor can be tailored into other applications. For example, if the substrate below the temperature sensor is removed by etching, the metal wire can be suspended by a dielectric film and will be thermally isolated, as shown in Figure 4.7a. If current is applied to the resistor, it can generate heat and thus be used as a microheater. At the same time, the actual temperature of the heater can be monitored by measuring its resistance. Furthermore, a combination of a microheater and two temperature sensors can be used as a mass flow sensor, as shown in Figure 4.7b. During the flow measurement, the microheater generates heat, which is dissipated by the gas flow to be measured. The mass flow of the gas can be obtained by measuring the temperature change across two temperature sensors.

- **Piezoresistive sensor**

Piezoresistive materials, including a variety of semiconductors and metals, change their resistivity under mechanical strain. Piezoresistive sensors utilize this material property to measure the mechanical strain, or some other physical quantities whose variation will induce a strain variation.

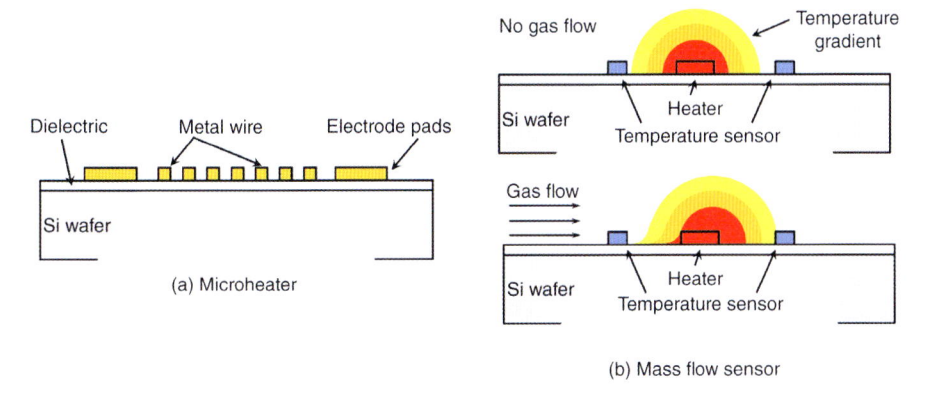

Figure 4.7 Schematic illustrations of (a) a microheater, and (b) a mass flow sensor.

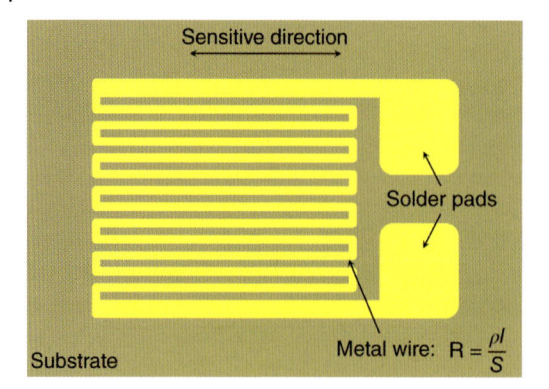

Figure 4.8 Schematic illustrations of a piezoresistive strain gauge.

A simple example of piezoresistive sensor is a metal strain gauge. As shown in Figure 4.8, the sensor consists of serpentine metal wire deposited on a flexible substrate. When mechanical strain is applied on the horizontal direction, the metal wire deforms and changes its resistance. However, such change in resistance is most usually a result of the geometrical change of the metal wire (dominated by the Ohm's law). Only in few metallic materials, the piezoresistive effect can be larger than the geometrical effect, such as nickel.

Semiconductor piezoresistive materials, such as silicon and germanium, change their resistance mainly due to strain-induced variations of their charge carrier mobility. Consequently, semiconductor materials have usually orders of magnitudes larger piezoresistive effect than the geometrical effect. Therefore, semiconductor materials can be used to build highly sensitive strain gauges. Figure 4.9 shows a piezoresistive pressure sensor fabricated with silicon. The pressure sensor is based on a square silicon diaphragm, which can be fabricated by anisotropic wet etching. On the diaphragm, four piezoresistors are formed by silicon doping techniques (diffusion or ion implantation). These four piezoresistors are connected in a Wheatstone bridge configuration to maximize output and to minimize sensitivity to errors. When there is no pressure applied

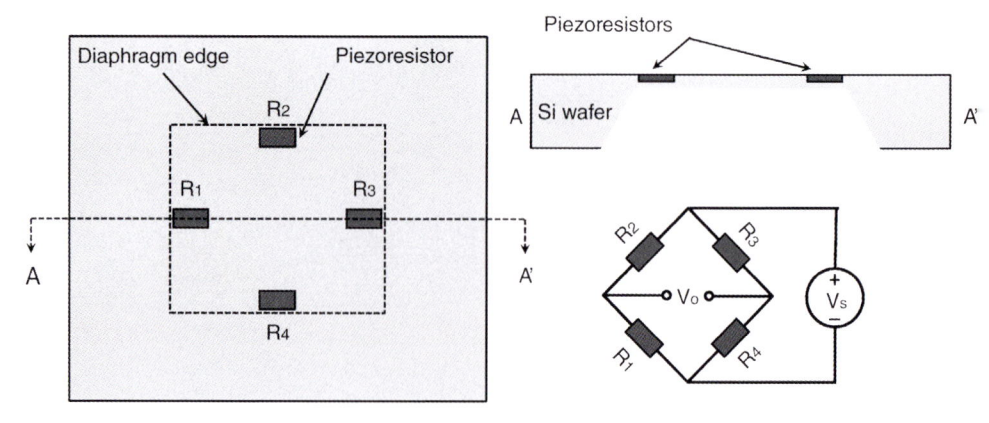

Figure 4.9 Schematic illustrations of a piezoresistive pressure sensor and the Wheatstone bridge circuit.

on the diaphragm, the piezoresistors are not strained and the output of the Wheatstone bridge is zero. When a pressure is applied on the diaphragm, the diaphragm deflects and induces strain in the four piezoresistors. Among the four piezoresistors, R_1 and R_3 are subjected to longitudinal strain, while R_2 and R_4 are subjected to transverse strain. Different strain levels result in different resistance changes between R_1 (R_3) and R_2 (R_4) and thus an output voltage of the Wheatstone bridge. By measuring the output voltage, the pressure can be measured. The piezoresistive pressure sensor is one of the first commercialized MEMS devices, and possesses many advantages such as high sensitivity and accuracy. However, as the charge transfer properties in the piezoresistive material are strongly influenced by the temperature, such sensors are usually characterized by temperature-induced signal drift. Therefore, in practice, piezoresistive sensors are often used with temperature compensation techniques.

4.4.2 Capacitive Sensors

Capacitive sensors sense quantity variations in their environments by measuring capacitance changes. The simplest form of a capacitor, a parallel-plate capacitor, has a capacitance determined by the distance, overlapping area, and dielectric between the two plates. Therefore, capacitive sensing can usually be carried out in three different configurations: (i) variable distance, (ii) variable area, and (iii) variable dielectric. It is worth noting that "variable dielectric" can be realized by either a movable dielectric or a dielectric that changes its electric property (e.g., dielectric constant). Capacitive sensing is one of the most widely adopted sensing techniques due to its high accuracy and simplicity. In this section, different capacitive sensors will be discussed, including humidity sensor, cantilever-based chemical sensor, accelerometer, and MEMS gyroscope.

- **Humidity sensor**

  Capacitive humidity sensor is one of the simplest forms of capacitive sensor. Traditional capacitive humidity sensors usually consist of a humidity-sensitive dielectric layer sandwiched between two electrodes. Humidity variations induce change in the dielectric constant of the dielectric layer that can be measured capacitively. MEMS humidity sensor enabled by micromachining technologies consists of a pair of an interdigitated electrodes covered by a layer of humidity-sensitive dielectrics (most often, polyimide), as shown in Figure 4.10. The micromachining technologies not only decrease significantly the size of the sensor but also offer the possibility to integrate the humidity sensor with a temperature sensor, signal processing circuit, memories, and AD converter. As a result, modern MEMS humidity sensor is capable of outputting temperature-compensated, calibrated, accurate digital data.

- **Cantilever-based chemical and biosensor**

  Cantilever is a beam structure clamped at one end with the other end free to deflect, as shown i Figure 4.11. Micro cantilever is one of the most studied structures among different MEMS devices due to its simplicity and versatility. Micromachining technologies offer a variety of possibilities to fabricate

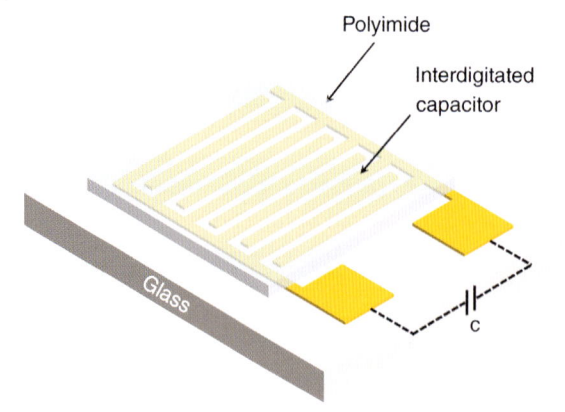

Figure 4.10 Schematic illustration of a MEMS capacitive humidity sensor.

microcantilever structures, by both surface micromachining techniques and bulking micromachining techniques. The cantilever structures have broad applications, for example, gas sensors, chemical sensors, biosensors, inertial sensors, and AFM tips. In this section, a simple chemical sensor enabled by MEMS technology will be discussed.

Gas Chemical sensors can be used to detect or measure the concentration of specific chemical substances. The chemical detection or measurement can be realized by many different devices, for example, surface acoustic wave devices, quartz crystal microbalance, and acoustic plate mode sensors. To use the microcantilever as a chemical sensor, a functional layer sensitive to the target chemical is coated on the surface of the cantilever. When the target chemical substance presents in the surrounding of the cantilever, a chemical interaction or absorption will occur between the target chemical and the functional layer. Such chemical interaction or absorption will then induce stress or mass change

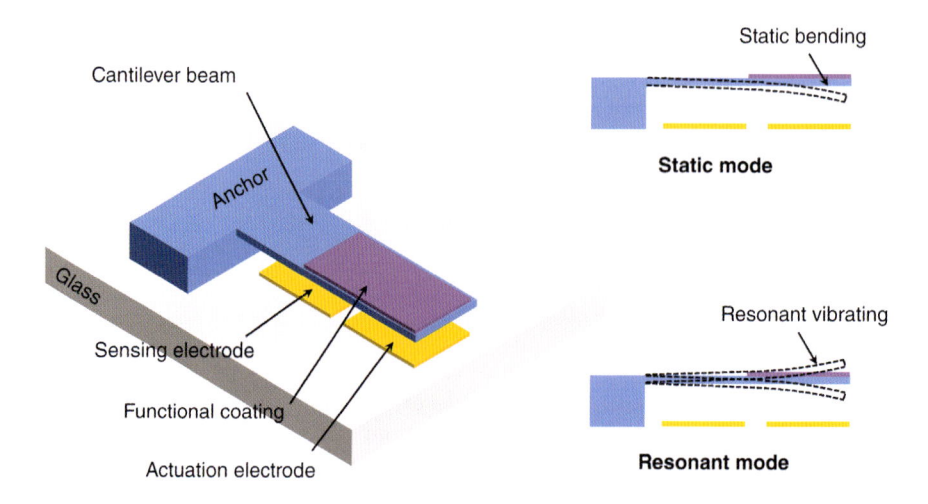

Figure 4.11 Schematic illustration of a cantilever-based gas sensor.

of the cantilever structure, which can be detected capacitively. The capacitive detection can be carried out in either a static mode or a resonant mode. In a static mode, the stress change of the cantilever will result a bending, changing the capacitance between the cantilever and the electrode the substrate. In a resonant mode, the cantilever is actuated by an electronic signal and vibrating constantly at its mechanical resonant frequency. When a mass or stress is introduced onto the cantilever, a shift in the resonant frequency can be detected. The resonant capacitive detection can be carried out by two pairs of electrodes, as shown in Figure 4.11, where one electrode is used to actuate the cantilever and the other electrode is used to measure the resonant frequency of the cantilever capacitively. Such resonant detection is widely used in different resonator-based sensors and can be carried out by different technologies. For example, if a piezoresistor is formed on the surface of the cantilever, the resonant frequency can also be measured piezoresistively.

- **MEMS accelerometer**

  Accelerometers are sensors that measure the acceleration. Traditionally, they are mainly used in the airbag deployment systems for automobiles, the inertial navigation systems for aircraft and missiles, and vibration measurements in the building industry. With the rapid development of micromachining technologies, low-cost high-performance MEMS accelerometers start to emerge in the twenty-first century and being widely used in the portable electronic devices. For instance, they can be used in laptops to shut down the hard drive during a free fall, or in tablet and smart phones to measure the phone position in order to display the screen always in an upright position (the "autorotation" function). Accelerometer combined with gyroscope and magnetometer can be used as the navigation system of the cell phone.

  The working principle of MEMS accelerometers can vary, such as piezoresistive, capacitive, inductive, and optical. Nevertheless, the basic working principle of the accelerometers is usually based on a mass–spring system. Under acceleration, the mass is displaced to a point until stopped by the spring. The value of the acceleration can be calculated from the displacement of the mass. Based on this principle, the simplest form of accelerometer can be realized by a cantilever beam with a proof mass at the free end.

  A more complex three-axis capacitive accelerometer is shown in Figure 4.12. The X, Y direction acceleration is measured by comb-drive finger capacitors ($C1$ and $C3$ for X, $C2$ and $C4$ for Y, all with capacitance C), and Z direction acceleration is measured by the parallel plate capacitor formed by the proof mass and the electrode below ($C5$). The comb-drive capacitor is named after its geometrical similarity to a comb. It is composed of an array of capacitor electrodes arranged in parallel, which effectively increases the overlapping area between the two electrodes. The increased area of the capacitor enhances the capacitance and thus the resolution of the sensor. To eliminate the direction cross-sensitivity and parasitic capacitance, differential output of the capacitors that are in the same direction is used. Such differential sensing scheme uses the capacitance difference between the $C1$ and $C3$ to calculate the X-direction acceleration, and the capacitance difference between the $C2$ and $C4$ to calculate the X-direction acceleration. Considering the sensor is subjected to

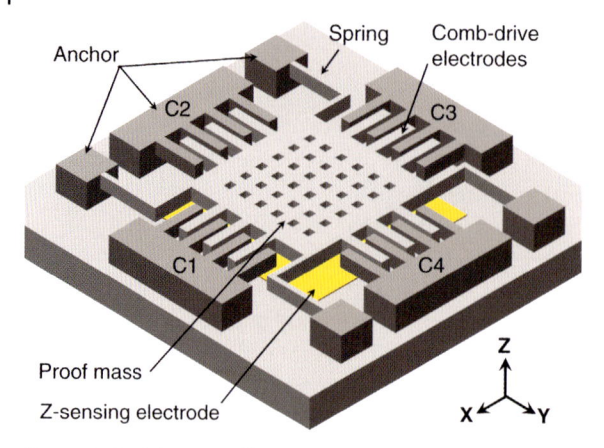

Figure 4.12 Schematic illustration of a three-axis MEMS accelerometer.

acceleration on the X-direction, the displacement of the proof mass induces capacitance changes in both $C1$ and $C3$: $C1$ has become $C+\Delta C13$ and $C3$ has become $C-\Delta C13$. Therefore, the difference between $C1$ and $C3$ is $2\Delta C13$, and it can be used to calculate the X-direction acceleration. At the same time, the X-direction acceleration also changes $C2$ and $C4$, but both to the same value $C+\Delta C24$ (the common mode signal). The difference of $C2$ and $C4$ is still zero. Same principle also works for $C2$ and $C4$ under Y-direction acceleration. Therefore, the X-direction acceleration only changes the differential output of $C1$ and $C3$, and Y-direction acceleration only changes the differential output of $C2$ and $C4$.

- **MEMS gyroscope**

A spinning top is a toy capable of maintaining its balance on one point when spun at a high speed. A gyroscope is a scientific instrument for measuring angular displacement, based on a principle (Coriolis force) similar to that of the spinning top. A traditional mechanical gyroscope uses a rapidly spinning disc mounted on gimbals to measure orientation of objects. It has been widely used in the navigation system of aircrafts and ships. Despite its high accuracy and stability, traditional gyroscopes are too bulky and expensive to be used in the commercial applications. Nevertheless, the market potential for low-cost miniaturized gyroscopes is huge, especially in the consumer electronics and automobiles industry. Therefore, intensive development activities have been conducted on the MEMS gyroscopes in the past decades. Classic reviews on inertial sensor can be found in Refs. [16, 17].

MEMS gyroscopes are all based on the vibratory systems (resonators) due to the absence of microscale bearing and the consequent difficulty to fabricate microscale rotational systems. However, the working principle of the MEMS gyroscopes is also based on the Coriolis force similar to traditional gyroscopes. Figure 4.13 shows a MEMS resonator gyroscope based on comb-drive electrodes. The inner frame is actuated by the comb-drive electrode in the center and vibrates at its resonance frequency. The outer frames and the outer comb-drive electrodes are used to detect angular movements. When an

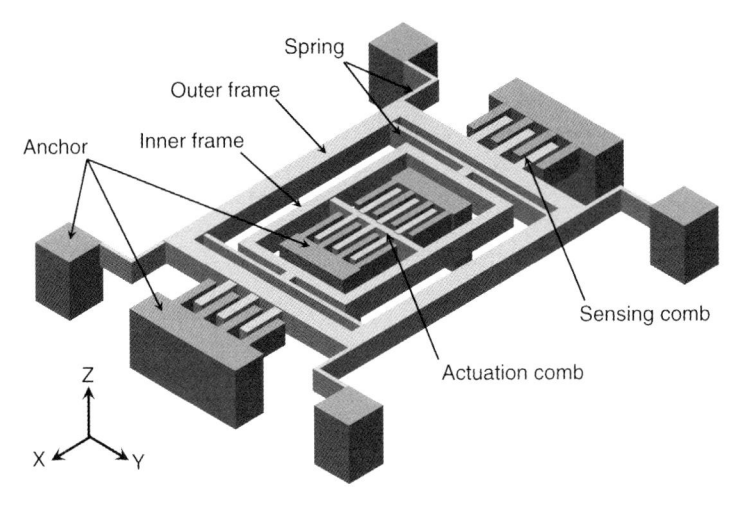

Figure 4.13 Schematic illustration of a one-axis MEMS gyroscope.

external rotation (around axis Z) is applied to the entire system, the anchors (for both springs and fixed comb-drives) will follow the external rotation and rotate to a new location, while the vibratory inner frame tends to still vibrate on its original vibration plane. Such a mismatch will generate a displacement between the outer frame and comb-drive electrode, which can be detected capacitively and used to calculate the angular movement of the gyroscope. Nevertheless, this type of gyroscope is highly sensitive to environmental noise, especially external vibration close to the resonant frequency of the system. Therefore, in practice, two devices of this type are always used in pair to form a differential sensing scheme. In such a way, the environmental noise becomes common-mode signal to the systems and can be canceled.

The two-resonator gyroscope described above is actually a modified version of traditional tuning fork resonator. MEMS gyroscope can actually be realized in many different ways, such as wine-glass resonator, cylindrical resonator gyroscope, vibrating wheel gyroscope, and bulk-acoustic wave [18] gyroscope. Since the twenty-first century, the functionality of MEMS gyroscopes has also been significantly enhanced. Multiple-axis (two-axis and three-axis) gyroscopes based on a single MEMS chip have been developed. For example, STMicroelectronics's L3G4200D uses four vibratory mass resonating in a "beating heart" fashion to sense angular displacement on all three axes. Angular displacements around the three axes of the device change the basic drive resonant mode into three new different corresponding resonant modes [19]. The angular displacement is then calculated from the new resonant modes.

4.5 Sensor Systems

The sensor systems are generally composed of sensor device and circuits [20]. The individual sensor device transforms environmental information into electrical signal, which are often changes in capacitance, resistance, or signal frequency, as mentioned in previous section. These changes are then processed by circuits to be converted into a voltage or current signal. Traditionally, the circuits are usually analog circuits that transform the continuous measurand, such as temperature, pressure, humidity, and distance, into a proportional continuous electrical signal (i.e., the analog signal). However, the analog circuits are usually easily influenced by noise and hard to interface other devices. Therefore, modern sensor systems usually consist of a combination of both analog and digital circuits (mixed-signal circuits). In a mixed-signal sensor circuit, the measurand is first transformed into an analog signal, which is then quantized by an analog–digital converter (ADC) to form a digital signal. The digital signal can be further processed by other digital circuits (e.g., a microcontroller) or the output interface.

Modern IC technologies offer the possibility to include an entire sensor system in a single package. Two popular methods to implement the combination of sensor and circuit are system on a chip (SoC) and system in a package (SiP) technologies [21].

SoC technology uses CMOS fabrication process to produce MEMS devices (the "integrated MEMS") together with analog and digital circuits on the same chip. The advantages of a SoC technology include small footprint, low power, high system complexity and reliability, and low cost.

SiP technology combines several chips (dies) with different functionality in the same package. For example, different MEMS sensors and their circuits can be assembled into the same package and interconnected with wire bonding. Compared to the SoC technology, SiP is capable of heterointegrate chips with different technology node into the same package, and thus has higher degree of freedom and potentially higher performance. The SiP technology has generally a higher manufacturing cost but a lower designing cost than the SoC technology. In practice, SoC and SiP technologies must be carefully selected for different sensor systems depending on the anticipated application, volume, cost, and performance.

References

1 Kovacs, G.T.A. (1998) *Micromachined Transducers-Sourcebook*, McGraw-Hill, New York, NY.
2 Gad-el-Hak, M. (ed.) (2001) *The MEMS Handbook*, CRC Press.
3 Boisen, A. et al. (2011) Cantilever-like micromechanical sensors. *Reports on Progress in Physics*, **74** (3), 036101.
4 Li, Y. et al. (2013) Large stroke staggered vertical comb-drive actuator for the application of a millimeter-wave tunable phase shifter. *Journal of Microelectromechanical Systems*, **22** (4), 962–975.
5 Rebeiz, G.M. (2004) *RF MEMS: Theory, Design, and Technology*, John Wiley & Sons.
6 Wojtas, N.Z. (2014) Microfluidic Thermoelectric Heat Exchangers for Low-Temperature Waste Heat Recovery, Diss., ETH Zürich, Nr., 21783.
7 Kühne, S. and Hierold, C. (2011) Wafer-level packaging and direct interconnection technology based on hybrid bonding and through silicon vias. *Journal of Micromechanics and Microengineering*, **21** (8), 085032.

8 Sandner, T. *et al.* (2005) Highly reflective coatings for micromechanical mirror arrays operating in the DUV and VUV spectral range. MOEMS-MEMS Micro & Nanofabrication, International Society for Optics and Photonics.
9 Maleville, C. and Mazuré, C. (2004) Smart-Cut® technology: from 300mm ultrathin SOI production to advanced engineered substrates. *Solid-State Electronics*, **48** (6), 1055–1063.
10 Current, M.I. (2002) Nanocleaving: an enabling technology for ultra-thin SOI. Extended Abstracts of the Third International Workshop on Junction Technology: IWJT, December 2–3, IEEE, Tokyo, Japan.
11 Yonehara, T. and Sakaguchi, K. (2001) Eltran®, novel SOI wafer technology. *JSAP International*, **4**, 10–16.
12 Lin, B. J. (2009) *Optical Lithography*, SPIE Press, Bellingham, WA, p. 136.
13 Williams, K.R. and Muller, R.S. (1996) Etch rates for micromachining processing. *Journal of Microelectromechanical Systems*, **5** (4), 256–269.
14 Williams, K.R., Gupta, K., and Wasilik, M. (2003) Etch rates for micromachining processing-Part II. *Journal of Microelectromechanical Systems*, **12** (6), 761–778.
15 Najafi, Kh. *et al.* (2008) Wafer bonding, in *Comprehensive Microsystems*, Elsevier, Oxford, pp. 235–270.
16 Kempe, V. (2011) *Inertial MEMS: Principles and Practice*, Cambridge University Press.
17 Yazdi, N., Farrokh A., and Khalil N. (1998) Micromachined inertial sensors. *Proceedings of the IEEE*, **86.8**, 1640–1659.
18 Ayazi, F. and Johari, H. (2009) Capacitive bulk acoustic wave disk gyroscopes. U.S. Patent 7,543,496.
19 *STMicroelectronics* (2011) Everything about STMicroelectronics' three-axis digital MEMS gyroscopes. *Technical article TA0343* (July 2011).
20 Yallup, K. and Krzysztof I. (eds) (2014) *Technologies for smart sensors and sensor fusion*. CRC Press.
21 Hsu, T.-R. (2004) MEMS Packaging, No. 3. IET.

5

Fiber-Optic Sensors

Yi Yang, Kevin Chen, and Nikhil Gupta

New York University Tandon School of Engineering, Department of Mechanical and Aerospace Engineering, Composite Materials and Mechanics Laboratory, Brooklyn, NY 11201, USA

5.1 Introduction to Fiber-Optic Sensors

Fiber-optic sensors (FOSs) have been considered promising in structural health monitoring (SHM) applications for many years. The field has advanced substantially in the past two decades with the invention of new sensors, upgrades in electronics, and miniaturization of systems. Both hardware and software fields have advanced to enable new SHM methodologies. In addition, mechanics of composite materials has advanced to better interpret the sensor response for complex stimuli using FOS. SHM of composite materials using strain gauges, piezoelectric patches, and accelerometers has been reported. However, there are many applications where electrical sensors are not desirable and optical sensors are the next frontier, for example, hydrogen storage tanks made of composite materials would prefer optical sensors over electrical sensors. It is emphasized that an appropriate sensor or combination of sensors will have to be selected for a given component. FOSs are not expected to replace traditional sensors in all SHM applications, but appropriate situations where these sensors present advantages need to be identified for their adoption.

Extensive use of composite materials in aerospace structures, as in the Boeing Dreamliner and Airbus A380 aircraft, and associated requirements for fatigue and vibration monitoring are also stimulating developments in FOS technology for SHM applications. Civil infrastructure is another area where SHM strategies are being implemented for continuous monitoring of load, displacement, and vibration. In civil infrastructure, monitoring of crack growth rate is also a major field of application for sensors. Automotive applications of sensors are rapidly growing in the past 5 years due to the introduction of features related to crash avoidance, lane departure, and parking distance control. The number of sensors used in automobiles is growing rapidly as these systems are tested and introduced in mainstream production models of cars. This growing trend also presents new opportunities for FOS.

Material-Integrated Intelligent Systems: Technology and Applications, First Edition.
Edited by Stefan Bosse, Dirk Lehmhus, Walter Lang, and Matthias Busse.
© 2018 Wiley-VCH Verlag GmbH & Co. KGaA. Published 2018 by Wiley-VCH Verlag GmbH & Co. KGaA.

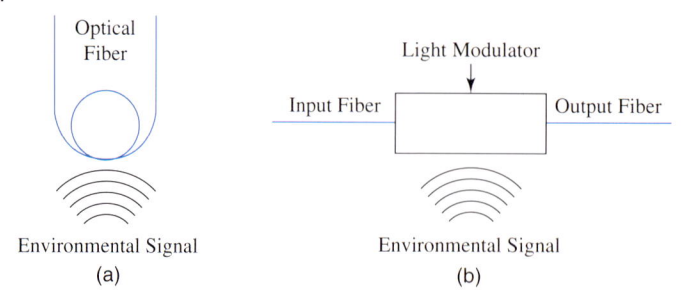

Figure 5.1 Schematic diagram representing (a) intrinsic and (b) extrinsic fiber optic sensor principles [1].

Many text and reference books are now available on FOS technology and SHM. Therefore, discussion in this chapter provides a brief overview of sensing principles, applications of sensors in SHM, and discusses in detail two relatively new sensors, whispering gallery mode sensor and fiber-optic loop sensor, that have potential for use in SHM applications.

5.1.1 Sensing Principles

FOSs are designed based on two sensing strategies, which are called intrinsic or extrinsic. Schematic representation of these sensing strategies is presented in Figure 5.1 [1]. In intrinsic sensors, the light is modulated inside the fiber to change intensity, frequency, polarization, or phase due to the applied stimulus. The change in the property of light can be calibrated with respect to the stimulus to conduct the measurement. In contrast, the modulation of light is conducted outside of the fiber in extrinsic sensors. The modulated light is collected either through the same fiber or from a different fiber. A sensor is designed based on the property that needs to be measured, the type of environment where measurement is conducted, and several other considerations such as size of the specimen, accessibility or location, frequency of measurement, rate of change, and ambient temperature variations, to name but a few.

Among the commonly used sensors, the fiber Bragg grating (FBG) sensor is an example of an intrinsic sensor, while the Fabry–Pérot interferometric (EFPI) sensor is an example of an extrinsic fiber-optic sensor. Numerous other types of sensors are available, some of which will be discussed in detail in this chapter.

5.1.2 Types of Optical Fibers

Optical fibers are used to transmit coherent light (laser light) over long distances without any intensity losses. This property has been used extensively in the telecommunication field. An optical fiber consists of three concentric layers, from innermost to outermost: the core, cladding, and coating, as shown in Figure 5.2. The explanations given in this section are simplified. More technical descriptions of the terms introduced in this section can be obtained in optics textbooks.

The operating principle of the optical fiber is total internal reflection at the core–cladding interface, which keeps the light in the core. According to Snell's

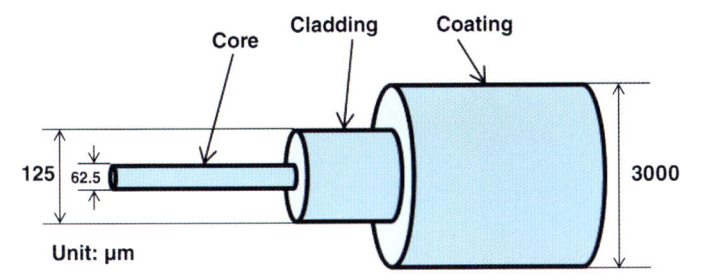

Figure 5.2 Schematic diagram of multimode optical fiber with 62.5 μm core and 3 mm jacket.

law, the refractive index (RI) of core is maintained slightly higher than that of the cladding material. In silica optical fibers, both core and cladding are made of silica but the RI can be changed by doping. In order to obtain total internal reflection, light must enter the fiber core at an angle such that when it hits the core–cladding interface, its incidence angle exceeds the critical angle that allows it to reflect back in the core rather than to refract into the cladding. This keeps the light in the core and allows it to propagate through the cable without loss in the intensity. The purpose of coating is to protect the fiber from damage and for ease of handling. Sometimes multiple optical fibers are bundled together and a plastic jacket is used around them to create a cable. Such cables are commercially used in telecommunication application. Most FOSs rely on a single optical fiber to conduct the sensing operation.

There are two categories of optical fibers, stepped-index and graded-index fibers. For stepped-index fibers, the RI between the three layers is distinctively different and constant within each layer. However, for graded-index fibers, the RI changes gradually throughout the core and cladding layers of the fiber.

Within these two categories of fibers, there are two distinctive types: multimode and single-mode fibers, which are schematically represented in Figure 5.3. A multimode fiber can carry several modes of light within the core. This means that the core diameter is a large multiple of the propagating wavelength so that

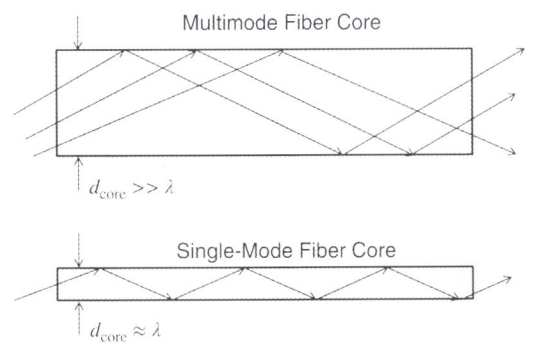

Figure 5.3 Schematic representation of a multimode and a single-mode fiber. The multimode fiber has a core diameter of 50–100 μm for wavelengths of 850–1300 nm. In the single-mode fiber, the diameter of the core is between 8 and 10 μm for wavelengths ranging from 320 to 2100 nm.

light can enter at many different angles creating different modes in the fibers. When the core diameter approaches the operating light wavelength, only one mode of light can propagate in the core of the fiber. Such fibers are called single-mode fibers. Single-mode fibers are useful for fiber-optic sensors because noise is reduced since there is no interference from the other propagating modes that are present in a multimode fiber. However, sensors based on both single- and multimode fibers are available.

Several types of light modulations can occur in optical fibers due to the applied stimulus, which include change in amplitude, phase, polarization, and wavelength. Of these, the latter three are spectral modulations, while the first is a power or amplitude modulation. This splits the realm of optical sensors into spectral modulation and intensity modulation.

Intensity modulation sensors are simpler in construction and operation in the sense that fewer components are needed in their instrumentation and the data analysis is also less intensive. Technically, all that is needed is a light source and a photodetector such as a photodiode connected to the optical fiber. However, in reality, there are many more components attached to any FOS operating under real application conditions. A spectral modulation sensor requires a tunable laser and a spectrum analyzer such as a spectrometer, which are significantly more expensive and require a higher skill level to operate compared to the intensity modulation sensing approach.

5.2 Trends in Sensor Fabrication and Miniaturization

Polymer-based sensor foils with purely optical measurement principles were introduced as a way of realizing optical sensors for material integration because limits of miniaturization has been reached for off-the-shelf component assemblies. When considering systems containing multiple sensors to be attached to the same material for measuring of different parameters such as strain and temperature or for measuring the same parameter at different locations, an alternative way would be to develop foil-based systems containing optical waveguides and sensors that can be easily attached to the surface.

The concept of planar optronic systems is to integrate an optical sensing network with light source, waveguides, sensing elements, and detectors completely into a planar polymer foil [2]. Such self-contained foil-based systems hold great promise for a wide range of applications [3], such as temperature, strain, or chemical concentration sensing [4,5]. For example, strain sensing through a foil-based sensor can be achieved by fabricating a polymer optical waveguide, where mechanical deformation of the waveguide or the sensing element will modulate the optical output and can be calibrated to measure the strain [5,6].

FBG strain sensors have several advantages: They are insensitive to electromagnetic noise and can be embedded in concrete [7], carbon, or glass fiber composite structures [8] for structural health monitoring. To unscramble the spectrum from multiple sensors, an interrogator or an optical spectrum analyzer has to be used, which makes it difficult to embed the entire optoelectronic system using microelectromechanical (MEMS) technology.

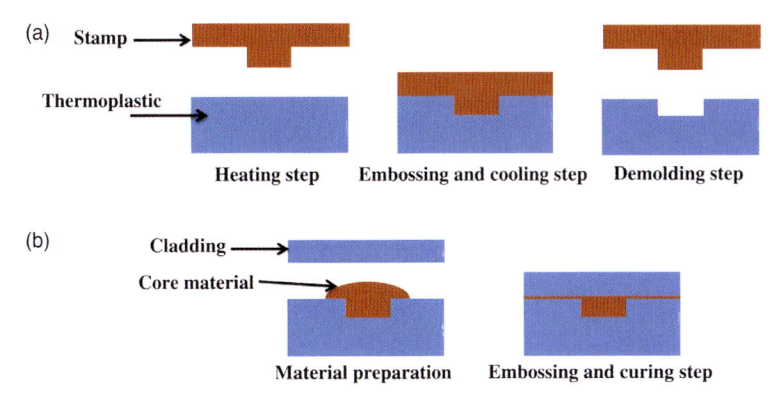

Figure 5.4 (a) The process steps of hot embossing include heating, embossing, cooling, and demolding [3]. (b) Heating and embossing steps are implemented in hot embossing of optical waveguides [3].

An important part of these sensors is the optical waveguide. Various techniques have been implemented such as photolithography [9], flexographic and inkjet printing [10], nanoimprint lithography [11], processing with femtosecond lasers [12], and hot embossing. New opportunities are also arising as the 3D printing field is progressing. Printing components with integrated sensors and sending instrumentation are one of the goals in using 3D printers for developing smart materials. The process of hot embossing of thermoplastic polymers is shown in Figure 5.4a, which transfers a stamp structure onto a substrate [3]. The process is able to replicate structures with dimensions from millimeter scale to nanometer scale. Due to large-scale and inexpensive fabrication, hot embossing is attractive for optical applications [13]. Moreover, different fabrication techniques and various polymer materials are investigated into the process of hot embossing [14–18].

The hot embossing process of optical waveguides is similar to hot embossing of thermoplastics. Figure 5.4b shows the summary of the process [3]. A force is applied on the cladding layers after depositing the liquid core material and then heated to the curing temperature of core material. The force is released after curing is completed. Different materials of cladding and core result in different refractive index [3].

Miniaturization of optical fiber sensors for microscale applications is an important task for structural health monitoring applications. Sensors that are small, highly sensitive, and have fast response time within a large dynamic range are useful. Thus, traditional optical components are being replaced by small-sized devices that enable the sensors to operate on the fiber scales. Structures that have two optical paths in one physical line are considered as an ideal candidate to implement miniaturized fiber-optic interferometric sensors. The inline structure offers several advantages such as easy alignment, high coupling efficiency, and high stability [19,20]. The Fabry–Pérot interferometer has been an attractive sensor in this respect because it provides a miniature size, inline structure, linear calibration response, and high sensitivity.

There are several issues with miniaturization. Nanoscale sensors are more complicated to fabricate and the manufacturing yield is lower. Sensors of this sort tend to require more advanced and expensive equipment to operate, especially for MEMS systems [21–23]. However, the most concerning problem with miniaturization is the decrease of measurement precision, resulting in a poor detection limit. Short-term noise and long-term stability issues are important in this context. It is documented in Refs [24–26] that a part of current, which is converted from photon flux by the photodetector, can be denoted by I and the output of the sensor due to intrinsic noise in the instrumentation generating a current can be called I_{noise}. In an optical sensor, the initial current is measured to change in experiment with a certain percentage. Thus, the current changes from I to αI can be understood, where α is a number with an exact value that depends on the given change in the mass per unit area. The signal-to-noise ratio of the measured current [24,27] is given as

$$\frac{\text{signal}}{\text{noise}} = \frac{\alpha I - I - (I - I_{\text{noise}})}{\sqrt{(\sqrt{I})^2 + \text{var}(I_{\text{noise}})}} = \frac{I(\alpha - 1)}{\sqrt{I + \text{var}(I_{\text{noise}})}}. \tag{5.1}$$

Equation (5.1) represents that signal-to-noise ratio increases with $I \gg I_{\text{noise}}$. However, the reason why small sensors have lower resolution is that it is very hard for a small sensor to get high I. The situation is the same for the case of optical readout since if a larger area is illuminated, a higher photon flux would be analyzed. Therefore, it is noted that miniaturization often leads to higher noise in the measurement [27]. Studies have shown that mechanical stability is also a problem for small-size sensors. For instance, the resolution in single-nanoparticle spectroscopy heavily suffers from small changes in the position of the particle with respect to the surrounding optical components [28]. A miniaturized sensor can be stable but would likely require high-precision placement control and a more complicated experimental setup. A means of countering this effect is to work with microscale sensors. Microscale sensors offer a good compromise between utilizing the advantages of miniaturization while avoiding some of the complications that appear at the finer size.

5.3 Fiber-Optic Sensors for Structural Health Monitoring

Composite materials are now widely used in structural applications in aerospace, marine, ground transport, sports, and civil engineering industries [29]. Several initial applications of composite materials in the transportation industry came in performance cars. However, many composite components are now found in mainstream production cars, further accelerating the use of composite materials. Initial concerns about fiber-reinforced composites such as joining of composite parts, joining of composite parts with metal parts, drilling of holes, and constructing parts with complex shapes have been answered by research and development efforts over past decade. Two examples of large-scale use of composites in modern automobiles are shown in Figure 5.5.

Figure 5.5 Examples of large-scale composite parts in modern automobiles. (a) Ford GT rear diffuser and body panels. (b) Seat frame in the Chevrolet Corvette.

Demand for lighter cars with improved fuel economy is growing and composite materials are among the candidate materials to replace steel in some body panel applications. Electric cars are under immense pressure to increase their driving range, which can be achieved by using lightweight composite materials. In this regard, BMW has been using an entirely carbon composite chassis for the i3 and i8 production models. Similar advancements have been taking place in aircraft structures, where increase in flying range, increase in payload capacity, and reduction in fuel consumption are achieved by using composite materials. Both automobile and aircraft applications desire to have sensors in the structures to monitor structural loading and failure initiation and propagation. A variety of sensors, including piezoelectric patches, strain gauges, and accelerometers, are used in composite material structures for health monitoring. FOSs provide the advantage of being free from electromagnetic interference over electrical sensors and can be used to measure multiple properties using the same instrumentation.

Composite materials have a complex construction. Most of the composites of interest for this chapter are fiber-reinforced laminates. In these composites, the number of layers, layup sequence, and fiber volume fraction are important parameters. Some thick composites may contain hundreds of plies in several different orientations. In such cases there is a possibility of having defects such as voids, dry spots, resin pockets, and ply stack misalignment [30], which can cause internal stresses, warpage, and stress concentration locations, among other possibilities. Repeated severe loading or fatigue loading can cause some of these defects to grow and induce cracks in the material. Crack initiation, crack growth rate, delamination extent, and location of delamination are among the parameters that are of interest in such cases.

5.3.1 Sensors for Cure Monitoring of Composites

Many of the defects in composites are introduced during their fabrication process. Therefore, *in situ* monitoring of composite materials is vital [30–32]. In one study, both FBGs and EFPIs were used and their responses were compared with one another during the curing process of carbon fiber-reinforced composites [31]. Two EFPI sensors, one at $0°$ and another at $90°$, and one FBG sensor at $0°$ were placed on a $0_4/90_4/90_4/0_4$ carbon fiber-reinforced laminate before the cure process. Cure strains were compared between a pristine laminate and a damaged laminate in which a delamination was induced. It was found that the damaged laminate resulted in higher cure strains toward the end of the curing process. Monitoring of the hydrostatic pressure of the resin curing is also useful during fabrication of the laminated composites. An FBG hydrostatic pressure sensor has been developed to measure the differential pressure of fluids [33]. In the sensor, the external fluid pressurizes a silicone oil that is used to mechanically strain an FBG thereby causing a spectral shift. Results showed a linear relationship between wavelength shift and water pressure with a sensitivity of $1.636 \times 10^{-2}\,\text{MPa}^{-1}$. Several other studies on cure monitoring of polymer resins and composites are available [34,35] that focus on measuring cure shrinkage, viscosity of the resin, and residual stresses and strains. The stress and strain results from these studies are difficult to generalize because of their dependence on ply orientation sequence in the composite. However, studies on neat resin to determine the pot life and gel time and temperature are very useful in planning the composite fabrication schedule.

5.3.2 Embedded FOS in Composite Materials

The convenience of using FOS to monitor the curing of composites is that the same sensors can be later used in the cured composites for SHM under applied load. This is important in the construction of smart structures where "nervous systems" are built directly into the structure to monitor its condition. In several existing works [31,32], three- and four-point bend tests have been performed on composites and the results show a linear correlation between wavelength shift and the applied load. Also, the slopes of these curves were different between pristine samples and samples with induced defects, which provided the possibility of defect detection. Different loading conditions correlated to different spectral shifts as well.

Wind turbine blades are made of composite materials in order to obtain lightweight and high bending stiffness in such long aspect ratio structures. Load histories of wind turbine rotors have been continuously monitored using FBGs. In Ref. [36], composite substrate sensor pads were adhered to the surface of a rotor blade. FBG thermal sensors were also installed for a 2-year field test. Sinusoidal load profiles were found and changes in this load profile were correlated back to different acceleration periods of the wind turbine.

5.3.3 Surface-Mounted FOS in Composite Materials

The intrinsic fragility of the fiber sometimes does not allow for embedding the FOS inside a composite material. This is especially the case when trying to embed optical fibers in concrete materials. Since concrete is high in alkalinity, unprotected FOSs are unfavorable in these conditions. In one application [37], FBG FOSs were surface mounted near existing strain gauges on bridge tendons embedded within concrete. Steel strand, carbon fiber composite cable, and Leadline rod tendons were monitored using FOS. The FOS was able to withstand the demands of cyclic loading on a simulated bridge model. Cyclic strains on the FBG sensor were between 0 and $-2000\,\mu\varepsilon$ for 320 000 cycles. However, long-term monitoring of such concrete or cement-based structures with embedded FOS may be a problem due to degradation of fiber. An overview of various types of sensors including FOS in monitoring of bridges is presented in Ref. [38]. Apart from FOS, a variety of sensors have been used for this purpose, including anemometers, strain gauges, accelerometers, displacement transducers, global positioning systems, tiltmeters, seismometers, and video cameras.

In a report on the SHM of bridges in Canada, it was reported that several FOSs were both surface mounted and embedded in several bridges across the country [39]. The following sensing methods were used: (i) FBG sensors embedded into concrete girders for inspection of the carbon fiber cables. (ii) FBG sensors embedded into fiber-reinforced plastic in polypropylene fiber-reinforced concrete. (iii) Fabry–Pérot FOS used on glass fiber reinforcements on girders. The report shows that the sensors were effective in providing data on load and displacement.

FOSs have also been surface mounted on composite material foils. FBGs have been bonded onto a marine hydrokinetic turbine foil [40]. The composite material under observation was a carbon fiber-reinforced laminate. Using the FOS, percent mass gain of the marine turbine foil in a seawater environment and strains under mechanical testing were obtained.

The fragility of optical fibers may be a concern in surface-mounted sensor applications. However, a protective sheath can be used for such conditions. Polymer jackets have been used for corrosive environments, and ceramic sheaths have been used for high-temperature environments to protect the fiber from damage during measurement.

5.3.4 FOS for Structural Monitoring

5.3.4.1 Aerospace Structures

The early failure indications of conventional materials such as metal and wood are well known and have been documented extensively. On composite materials, however, these indications are sometimes difficult to determine as the damage such as delamination or internal ply failure may not be visible from outside and visually the composite may appear to be intact. With the surge in using composite materials in aerospace structures, the SHM of these materials is important for the safety of the operators and the ease of diagnosis by the maintainers.

Acoustic emission, ultrasonic imaging, and dye penetration are among the methods used for detecting defects and damage in composite aircraft structures. Conventional use of acoustic emission sensors has been known to be complex in setup and signal processing. In addition, acoustic emission can only detect active cracks that are growing. This method cannot be used after the aircraft flight to detect the damage that is now passive. Use of FBGs has been demonstrated to replace conventional electric sensors with comparable results [41].

5.3.4.2 Civil Structures

In terms of safety, SHM is of great importance to ensure the integrity of civil infrastructure because they are designed to use for long periods and encounter vastly varying loading conditions due to traffic, wind, seasonal thermal expansion, and other factors. Polymer optical fibers (POFs) have been embedded into composite 3D orthogonal woven preforms for civil structures [42]. A straight POF embedded within the composite material along with an optical time domain reflectometer (OTDR) allowed for detection of attenuation losses due to three-point bend and impact tests.

On the other end of the spectrum, old historical structures have also been reinforced or retrofitted with composite materials. In one study, carbon fiber-reinforced sheets were used to reinforce wooden floor beams that were damaged during an earthquake [43]. External FBGs were placed on the beam and the composite sheets. The FBGs show that there was a 27% reduction in beam strain and an increase in damping coefficient due to the composite reinforcement. Results were also in agreement with Doppler laser vibrometer measurements.

5.3.4.3 Marine Structures

Strains on ship hulls have been monitored using FOS. SHM of ship structures is important throughout the lifetime of the vessel. In the early stages such as during ship design and construction, sensors help validate and verify static and dynamic finite element analysis (FEA) results. During ship operations, strain data could help determine fatigue lives for the ship structure and components. Aside from the usual benefit of determining damage in the structure, sensors can also help the operator in determining operating condition limits when the structure is overloaded in conditions such as high sea states.

Data from FBGs are seen to be comparable to that obtained from conventional sensors such as strain gauges. Since FOSs are much less susceptible to electromagnetic interference (EMI), noise levels are below $0.16\,\mu\varepsilon\cdot\text{Hz}^{-1/2}$ [44]. FBGs have demonstrated that they can detect the dynamic response of composite sandwich panels under slamming loads such as those seen on the hull of a ship [45].

In the shipbuilding industry, the use of glass fiber-reinforced plastics is now common. Oftentimes, adhesives are used to reinforce the composite material joints. Li et al. studied the sensitivity of embedded FBGs to detect damage to adhesive composite joints [46]. Sensitivity to intentionally placed disbonds in the bond-line was assessed via FEA and compared with results obtained from embedded FBGs. It was found that the FBGs were effective at detecting large strain concentrations at defect tips and were able to detect the strain variation as a function of load. On the other hand, larger Bragg gratings were found to be less effective in detecting strain variation in high strain gradient areas.

5.4 Frequency Modulation Sensors

5.4.1 Bragg Grating Sensors

Fiber Bragg grating sensors are an example of intrinsic FOS that modulate propagating light via gratings that are inscribed in the fiber core. A laser source is used to create permanent, periodic, RI changes in the core of a single-mode optical fiber to create the grating. These grating planes act to scatter the light in the guided mode into other modes such as cladding, radiation, or another core mode. In the case when scattered light at each grating plane constructively interferes in phase with each other and when the conditions are such that it can propagate back through the fiber core, a backward propagation mode is observed. The peak light intensity of this mode occurs at the Bragg wavelength described by

$$\lambda_B = 2n_{eff}\Lambda, \tag{5.2}$$

where λ_B is the Bragg wavelength, n_{eff} is the effective RI of the grating, and Λ is the period, or distance between each grating plane in the set.

Since some of the input light is reflected backward at a particular wavelength, the interaction of the light with the Bragg gratings can be seen through spectral analysis of the transmitted and reflected light, as seen in Figure 5.6 [47]. When an environmental signal in the form of strain or temperature interacts with the gratings and causes a change in effective RI or the distance between Bragg gratings, a spectral shift of the reflected and transmitted light can be observed and

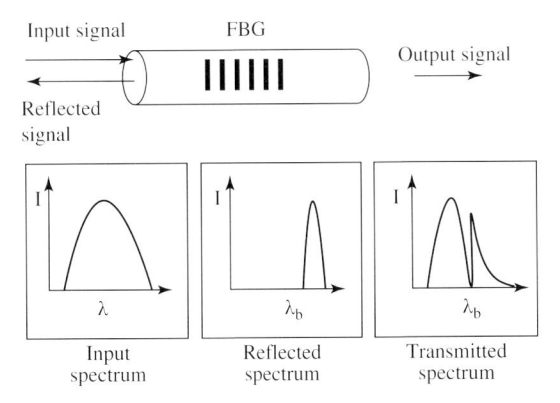

Figure 5.6 Illustration of the sensing principle of a fiber Bragg grating optical sensor [47].

correlated back to the change in strain or temperature. Changes in temperature and strain are direct measurands from the FBGs.

The benefits of the FBG are that it can be multiplexed, that is, multiple gratings can be placed along the length of a single fiber line. Multiplexing allows conducting measurements in different places using the same fiber. This also reduces the cost of instrumentation because the same laser source and photodiode can be used to operate the multiple sets of gratings placed on the same optical fiber. Another benefit is that they have a low sensitivity to humidity that makes them suitable for long-term monitoring of structures. However, the FBG sensors are intrinsically sensitive toward both strain and temperature, which would lead to a cross-sensitivity issue while using these sensors at high temperatures. Temperature compensation schemes are required where such an issue can arise.

5.4.2 Fabry–Pérot Interferometer Sensor

Another FOS that has been used for SHM is the extrinsic Fabry–Pérot interferometer (EFPI) sensor. A schematic of the EFPI sensing principle is shown in Figure 5.7 [31]. This sensor is constructed by cleaving the ends of an optical fiber, polishing them to mirror finish, and inserting them into a capillary tube. This creates two parallel reflective mirrors inside the capillary. The light is reflected back and forth between the two fibers and Fabry–Pérot interference occurs. When a mechanical or thermal strain is applied to the capillary hosing, the distance between the gaps causes a spectral shift in the reflected light that can be correlated back to the measurand. Fabry–Pérot sensors can be either intrinsic or extrinsic based on how the gap between the two reflective surfaces is set up.

The distance between the two ends of the fibers in a Fabry–Pérot sensor is given as

$$d = \frac{m\lambda_1\lambda_2}{2(\lambda_2 - \lambda_1)}, \qquad (5.3)$$

where d is the gap distance, m is an integer, and λ_1 and λ_2 are the specified wavelength range [30,48]. The change of cavity length Δd is expressed as

$$\Delta d = L\epsilon + A\Delta T, \qquad (5.4)$$

where ϵ is the strain of the sensor, L is the gauge length, ΔT is the change of environment temperature, and A is the length change of the sensor that can be

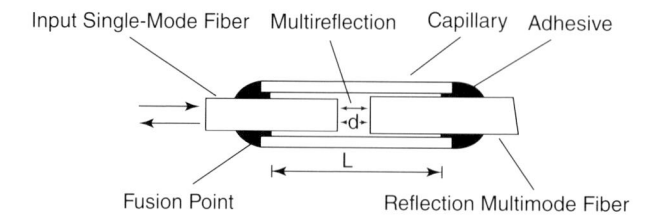

Figure 5.7 Schematic diagram of the operation EFPI [31].

Table 5.1 Sensitivities of FBG and EFPI sensors to strain and temperature.

Sensor	Strain sensitivity	Temperature sensitivity
FBG	1.2 pm/με (at 1550 nm) [47]	13 pm/°C [47]
EFPI	1.5 pm/με [49]	0.59 pm/°C [49]

expressed by $A = L(\alpha_q - \alpha_f)$. α_q and α_f are thermal expansion coefficients of the quartz capillary and optical fiber, which are very close to each other. The EFPI sensor is relatively insensitive to the temperature. The experimental results on EFPI sensor showing a shift in the reflection spectrum after the EFPI sensor was subjected to strain are available in the literature [49].

Studies have compared the response of EFPI and FBG sensors attached to aluminum and composite specimens subjected to three-point bend test [50,51]. Results show that these sensors provide comparable results for flexural loading and can be used to inspect for structural damage such as cracks in an aluminum plate as well as delamination in composite laminates. The flexural strain of damaged structures is much higher compared with corresponding undamaged structures under the same bending load.

The comparison presented in Table 5.1 shows that the EFPI sensors are better suited for strain monitoring applications with high ambient temperatures compared to the FBG sensors. Temperature insensitivity also enables their use in high-temperature environment [52].

5.4.3 Whispering Gallery Mode Sensors

The whispering gallery mode (WGM) is a wave phenomenon that was first explained by Rayleigh [53]. In this effect, a sound wave can travel around a curved surface through reflection without significant loss. Similarly, under appropriate conditions, total internal reflections along a curved surface is observed in optics [54]. In 1939, Richtmyer was the first to analyze the electromagnetic wave WGM dielectric resonator [55]. Theory predicts that the optical WGM sensors have extremely high quality factor (Q-factor). With the development of laser research facilities, researchers at Bell Laboratory used cylindrical CaF_2 media in 1961 to produce a microwave cavity to experimentally realize the WGM [56].

A typical WGM sensor instrumentation is schematically shown in Figure 5.8 [57]. The WGMs of dielectric resonators can be excited by coupling light from a tunable diode laser through an optical fiber. Light coupling occurs

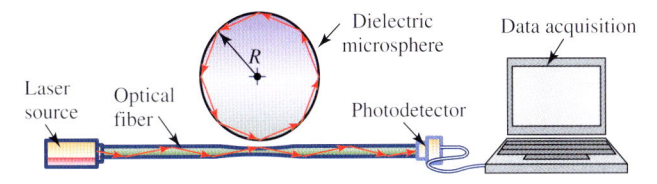

Figure 5.8 Representation of WGM sensor and the sensing instrumentation [57].

from the exposed section of an optical fiber to a particle resonator that is placed in the evanescent field. If the RI of the surrounding medium is lower than that of the particle, laser light coupled into the particle travels around the inner surface of the particle through total internal reflection. An optical resonance (WGM) will occur when the optical path length of the light traveling inside the particle is an integer multiple of the light wavelength. With the laser diode tuned, WGMs can be observed as sharp dips in the transmission spectrum. Even a very small physical or optical change around the microsphere will perturb the morphology (shape, size, or RI) of the particle and lead to a shift in the observed WGMs.

In the case when the radius of the microsphere (R) is much larger than the laser wavelength ($R \gg \lambda$), the condition of resonance can be expressed as

$$2\pi R n = l\lambda, \tag{5.5}$$

where n is the RI of the particle, λ is the vacuum wavelength of the laser light, and l is an integer number. Hence, any changes in the RI and radius of the particle, Δn and ΔR respectively, will lead to a WGM shift of $\Delta \lambda$ as per

$$\frac{\Delta R}{R} + \frac{\Delta n}{n} = \frac{\Delta \lambda}{\lambda}. \tag{5.6}$$

The value of $\Delta\lambda/\lambda$ is measured directly based on the applied force in experimental studies and is calibrated with respect to the quantity that is measured using the sensor. The value of $\Delta R/R$ and $\Delta n/n$ are not obtained separately.

It can be envisioned that the applied force changes the radius of the microparticle resonator. It is also known that the stresses induced in the particle would cause a change in RI. The correlation between change in RI and stress is defined by elasto-optic constants. Using these constants, the relation between change in radius and RI can be computed in order to obtain the total WGMs shift. These effects are computed through analytical and FEA methods in a model set of calculations for silica and polymethyl methacrylate (PMMA). Hertz contact theory [58–60] and finite element analysis [61–64] values for $(d\lambda/\lambda)/F$ are compared in Figure 5.9 with respect to the resonator sphere diameter [65]. In the analysis geometry, the resonator is compressed by two platens to obtain

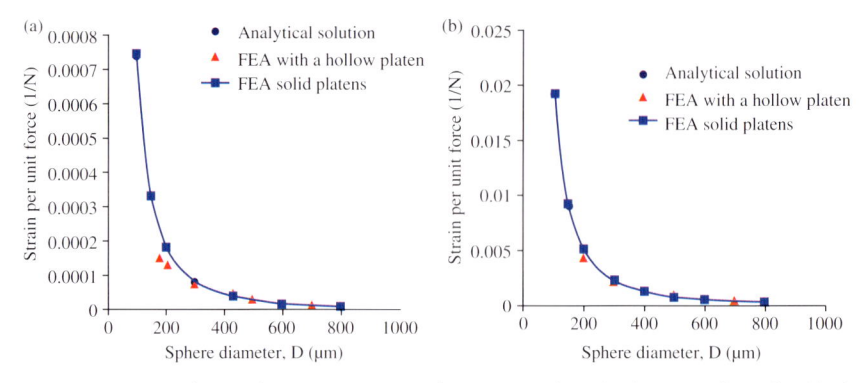

Figure 5.9 Circumferential strain comparison between analytical solution and FEA for (a) silica and (b) PMMA microspheres [65].

deformation and WGM shift. In one case, both platens are solid; in the second case, the platens are hollow (like the experimental case) to allow the optical fiber to pass through. The FEA results for solid platen case overlap with the analytical solution for both material types. It is also observed that the FEA solution for the hollow platen case deviates for spheres that are smaller than about 300 μm, which is likely due to the nonuniform deformation around the hole that is present in the platen. PMMA spheres show significantly greater sensitivity because of low stiffness of PMMA material.

Previous discussion was mainly focused on spherical optical resonators. However, a variety of resonator geometries exist, from toroidal to tubular to microbubble, each with their own sets of merits [66–71]. With regard to toroidal resonators, the process of reflow smoothing, which is applied in the fabrication of these resonators, generates a great deal of difficulties when fabricating larger resonators. To solve this problem, wedge geometries were designed, which push the mode away from the scattering surface in shallow wedge angle applications. Other processes are now available, which allows creation of larger resonators with larger wedge angles and Q-factors of 10^9 [72]. WGM characteristics in conical [73] and polygonal [74] polymer resonators are of interest and advancements have been made to achieve larger Q-factors using such geometries [75].

Studies have shown that the choice of material and geometry of the resonator can help in changing the sensitivity of the sensor. In engineering applications such as SHM of composite materials, where the noise level may be high, having ultrahigh sensitivity like WGM sensors may not be very useful. WGM sensors have shown the possibility of directly measuring force of 1 nN and displacement on the order of 1pm [76,77]. In such cases, appropriate resonators of high stiffness can be selected for lower sensitivity. However, in cases where such sensitivity is required, resonators such as hollow particles can be useful. The effective modulus of a hollow particle is lower than that of the solid particle of the same material. Given the same amount of force, a hollow particle would deform more and provide greater shift in WGMs. Many composite materials already have particles and fibers in their microstructure as shown in Figure 5.10 for a fiber and hollow particle reinforced syntactic foam [65]. The size of the particles used in composite

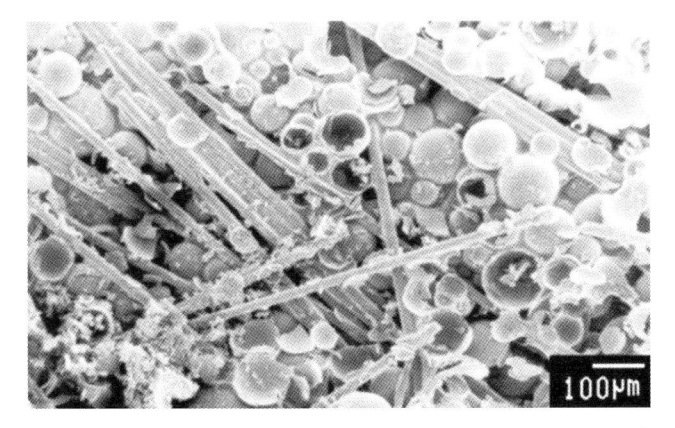

Figure 5.10 Micrograph of a syntactic foam reinforced with glass fibers [65].

materials is similar to that used in WGM sensors. Such composite microstructures can be embedded with WGM sensors without the questions of structural integrity loss due to an embedded sensor [65]. Due to low density, high specific strength, and high damage tolerance [78–80], syntactic foams are extensively used in aerospace and marine structural application. Smart syntactic foams can provide enhanced benefits in these applications.

5.5 Intensity Modulation Sensors

5.5.1 Fiber Microbend Sensors

Fiber microbend sensors are designed for detecting and measuring temperature, chemical species, force, pressure, displacement, strain, and acceleration, among others [81–88]. First demonstrated around 1980 [89,90], it was shown that the losses and impulse response of multimode fibers can be affected by creating microbends [91,92]. This kind of sensor is created by bending an optical fiber in a set of micrometer size kinks in a localized region [87,93]. The intensity loss through the microbend section is calibrated with respect to the strain or temperature as the sensing principle [87,94,95]. Schematic representation of the microbend sensing concept and instrumentation is shown in Figure 5.11 (adapted from Ref. [96]). Two common microbend deformer geometries are shown in Figure 5.12 (adapted from Ref. [96]). Other types of bender geometries

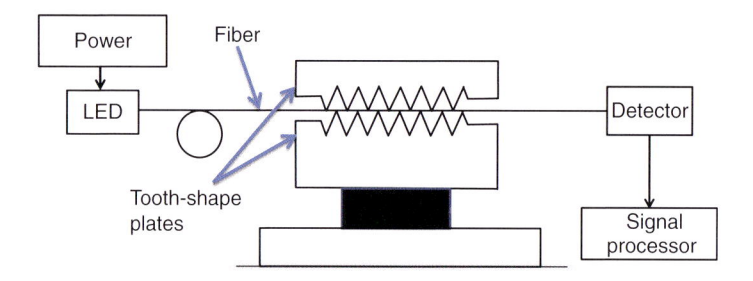

Figure 5.11 Schematic of experimental apparatus of optical fiber microbend sensor [96].

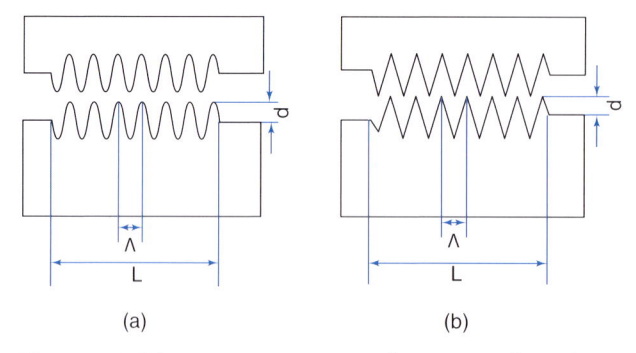

Figure 5.12 Schematic representation of two types of microbend deformer geometries [96].

such as corrugated plates [97], chain rollers [98], or wire winders [99] have also been used.

In the microbend sensor instrumentation, the optical fiber passes through the bending zone, for example, corrugated plates, which keeps the fiber under bending conditions. Applied force or strain changes the distance between the plates, which result in increase or decrease in the losses in the transmitted power. The more acute the bending, the higher the intensity losses. The results show that in a certain pressure range, microbending optical fiber sensors have strong linearity in the optical power loss with respect to the displacement. Such results are found to be reversible and repeatable, which allows creating a sensor that can last many loading–unloading cycles.

Microbends lead to redistribution of light power among the many modes in the multimode optical fiber. Coupling from the guided core mode to cladding mode or radiation mode causes the loss of transmitted power. The part of power from guided core mode coupling with the cladding mode is dissipated rapidly due to the high losses of the cladding mode [100,101]. Theoretical modeling of the microbend sensor concept has been presented in Refs [87,102]. These sensors are also shown to work in low temperature ranges and across a wide frequency spectrum [103]. Approximately linear response is obtained for transmitted power loss with respect to microbend radius (or displacement) for these sensors [84], which simplifies their calibration and use.

The microbend sensor can be used for measuring temperature, strain, force, and pressure. Other parameters that can be transformed to the change in distance between the bending plates can be measured through this kind of sensing scheme [101]. In composite materials, such sensors can potentially be useful as they can be braided into the fabric with microbends created during the braiding process. Cured resin would keep the optical fiber in place, but applied strains would change the characteristics and enable sensing. Patches of microbend sensors can be applied to the surface of laminates or composite components. The size of the bending plates is sometimes a limiting factor for their application in composites.

5.5.2 Fiber-Optic Loop Sensor

The microbend sensors described in the previous section use multimode fibers. The ease of calibration due to the linear profile of power loss with respect to strain is an asset for these sensors. However, one of the major limitations of microbend sensors is the small measurement range for strain. Use of single-mode fibers based on the same principle of optical power loss leads to significant differences in the physics of intensity losses compared to multimode fibers [104,105]. Single-mode optical fibers are used to create fiber-optic loop sensors (FOLS), which overcome the limitations of microbend sensors and provide several other advantages. In single-mode optical fibers, resonances are found at certain bend radius values. These resonances are repeatable in compression and decompression of the fiber loop. The general trend of intensity loss with respect to loop radius reduction and the presence of resonances are used as the sensing principles of FOLS.

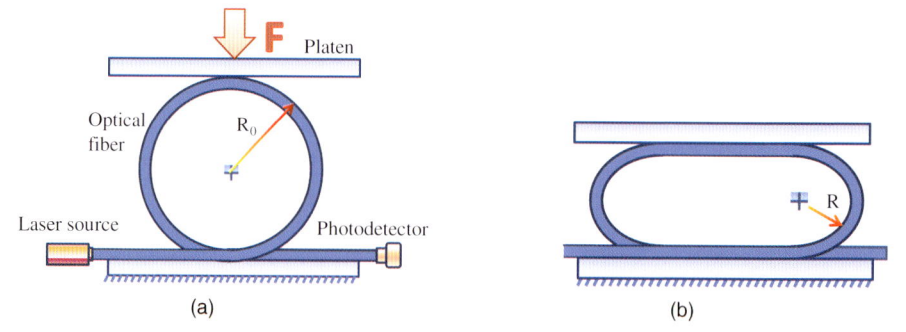

Figure 5.13 Deformation of a FOLS. (a) Original loop sensor between two compression platens. (b) Compressed loop showing smaller radius of curvature [106].

By taking a straight, single-mode fiber and bending it to form a loop of radius smaller than a critical radius, a sensing element is formed in FOLS. Previous studies have shown that the FOLS can be used to detect and analyze displacements, forces [106], vibrations [107], temperature [108], and humidity [109]. FOLS based on intensity and wavelength modulation have been used in these studies. However, the intensity-based FOLS have the benefit of being simple, robust, and relatively inexpensive compared to other types of optical sensors.

When the single-mode optical fiber is curved in the form of a loop of smaller than a critical radius, measurable losses occur from the curved section as the guided modes leak out through the fiber cladding. The intensity loss can be calibrated with respect to the displacement to obtain calibration curve. Figure 5.13 shows the structure and operating principle of FOLS [106]. This sensing principle has been used for the detection of displacements and vibrations of syntactic foams. The quality factor of a FOLS made of very small diameter fibers is demonstrated to be as high as 10^{10} [110].

The power transmission through an ideal step-index single-mode optical fiber having an infinitely thick cladding is expressed by the Marcuse model as [111]

$$\frac{P_{out}}{P_{in}} = \exp\left(-2\alpha_B l_B^e\right), \tag{5.7}$$

where $l_B^e = 2\pi R^e$ and R^e is the effective bend radius, while the effect of bending stress on the index of refraction of the fiber core material is considered. R^e differs from the actual bend radius R by an elasto-optic correction factor (1.28 for SMF28e fiber). The geometrical parameters and RI of Corning SMF28e fiber are shown in Table 5.2 [106,112,113]. In addition, the bend loss coefficient of a fiber with infinite coating ($2\alpha_B$) is defined by

$$2\alpha_B = \frac{1}{2}\left(\frac{\pi}{\gamma^3 R^e}\right)^{1/2} \frac{\kappa^2}{V^2 K_1^2(\gamma a)} \exp\left(-\frac{2\gamma^3 R^e}{3\beta_0^2}\right), \tag{5.8}$$

Table 5.2 Geometrical parameters and index of refraction of Corning SMF28e fiber.

Fiber layer	Radius (μm)	Refractive index
Core	4.1	1.4517
Cladding	62.5	1.447
Coating	125	1.4786

where V, κ, and γ are defined as follows:

$$V = ak\left(n_{co}^2 - n_{cl}^2\right)^{1/2}. \tag{5.9}$$

$$\kappa = \left(k^2 n_{co}^2 - \beta_0^2\right)^{1/2}. \tag{5.10}$$

$$\gamma = \left(\beta_0^2 - k^2 n_{cl}^2\right)^{1/2}. \tag{5.11}$$

Here k is the wave number, $k = 2\pi/\lambda$, n_{co} and n_{cl} are RI of the core and cladding, respectively, and a is the core radius. K_1 is the modified Bessel function of the first order of the second kind and β_0 is the coefficient of the leaky fundamental modes in the straight fiber.

Equation (5.8) defines behavior that resembles that of multimode fibers and does not predict the intermediate resonance that appears in the single-mode fiber. A modified model developed by Renner takes into account finite cladding thickness for single-mode optical fibers to calculate the bend loss coefficient of finite coating thickness ($2\alpha_{BC}$) as [104]

$$2\alpha_{BC} = 2\alpha_B \frac{2(Z_{ct} Z_{cl})^{1/2}}{(Z_{ct} + Z_{cl}) - (Z_{ct} - Z_{cl})\cos(2\theta_0)}, \tag{5.12}$$

where subscripts ct and cl are used for coating and cladding, respectively, and parameters Z_{cl}, Z_{ct}, and θ_0 are defined by

$$Z_{cl} = k^2 n_{cl}^2 (1 + 2b/R^e) - \beta_0^2, \tag{5.13}$$

$$Z_{ct} = k^2 n_{ct}^2 (1 + 2b/R^e) - \beta_0^2, \tag{5.14}$$

$$\theta_0 = \frac{\gamma^3 R^e}{3k^2 n_{cl}^2} \left(\frac{R_c}{R^e} - 1\right)^{3/2}. \tag{5.15}$$

Here, b is the cladding radius and R_c is defined as the critical radius that will generate total internal reflection and is defined as

$$R_c = \frac{2k^2 n_{cl}^2 b}{\gamma^2}. \tag{5.16}$$

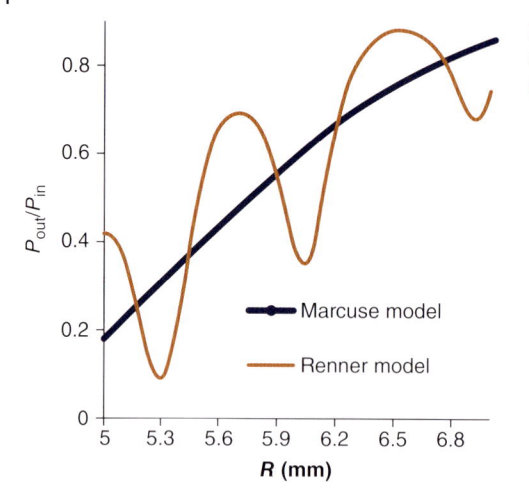

Figure 5.14 Comparison of predictions obtained from Renner and Marcuse models [106].

The appearance of extrema in optical power can be predicted as [104]

$$\frac{4b\gamma R}{3\pi R_c}\left(\frac{R_c}{R^e}-1\right)^{3/2} = \begin{cases} 2m - 1/2 \text{ max} \\ 2m - 3/2 \text{ min} \end{cases}, \quad (5.17)$$

where m is a positive integer. Figure 5.14 shows predictions from the Marcuse and Renner models [106]. The Renner model is successful in predicting the resonances that appear in the actual transmitted power profile from a single-mode optical fiber. The Marcuse model predictions present an overall trend of the data without catching the localized details. This figure shows that the sensor can be operated at very high sensitivity within the peak for a single resonance. For example, catching the Renner model trend between loop radius $R = 5.35$–5.60 mm would change the power output by fivefold. However, within the same radius range, Marcuse model predictions show change in intensity of about 30%. Calibration studies show that the major advantage of the FOLS is that it can be used with fine resolution over a small displacement range; at the same time, the average trend can help in using this sensor over a much larger displacement range of several millimeter.

Calibration results on FOLS of 5, 6, 7, and 8 mm radius are shown in Figure 5.15 [106]. Smaller radius FOLS show higher effective loop stiffness. Also, the measurement range of smaller radius loop is lower but the sensitivity is higher. The relative transmitted power-displacement slopes are 6×10^{-4} and 3.1×10^{-3} µm^{-1} for the 5 and 7 mm loops, respectively. Through the linear approximation in the transmitted power curve over the entire length of the displacement, the sensitivity of 6 mm radius loop is approximated to be 10^{-2} N for force and 100 µm for displacement.

The sensor was also tested for fatigue performance by conducting a cyclic loading test. The sensor showed no change in the response for over 10^4 loading and unloading cycles with a displacement rate of 0.4 mm/s for 6 mm of total displacement on a 5 mm radius sensing element. The effect of loading rate on the sensor response is also explored and the results showed that the relative

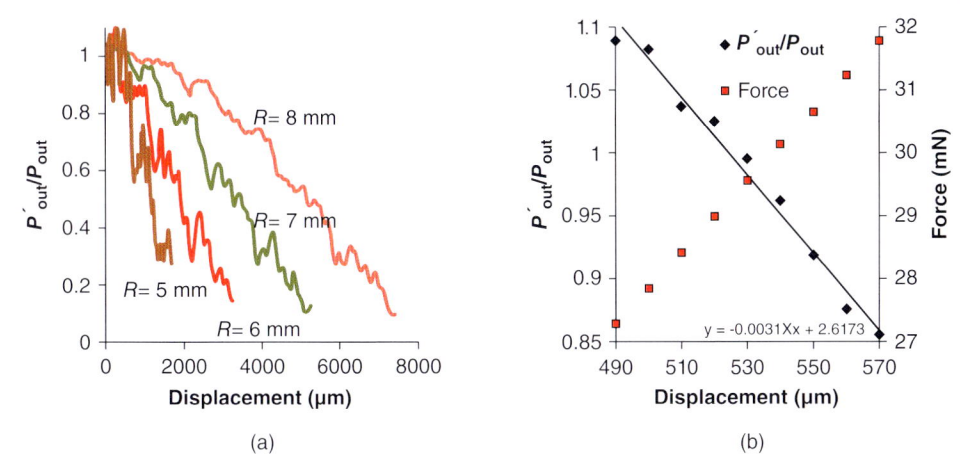

Figure 5.15 (a) Optical power–displacement relationships for loop sensors of four different radii. (b) High sensitivity range of loop sensors having $R = 5$ mm [106].

transmitted power–time and the force–time graphs show correspondence with each other for all loading rates tested [81].

Vibration tests were also performed on FOLS to assess their capability of vibration measurement [107]. The experiments were conducted on a vibration-isolated optical table and the FOLS was excited by 1310 nm single-mode laser. Data were acquired at 5 kHz sampling rate. The measurements were performed at frequency values of 10 and 100 Hz using a minishaker, which was directly connected to the FOLS. The frequency domain response of the FOLS is presented in Figure 5.16, which shows that FOLS captures peaks for both 10 and 100 Hz frequencies [107]. Further experiments were conducted on composite materials to capture the vibration response of lightweight syntactic foam composites. With composites such as syntactic foams, the lightweight of FOLS is a major advantage. The experimental results showed that even with these very lightweight materials, bonding of sensor on the specimen surface did not cause change in the vibration frequency [107].

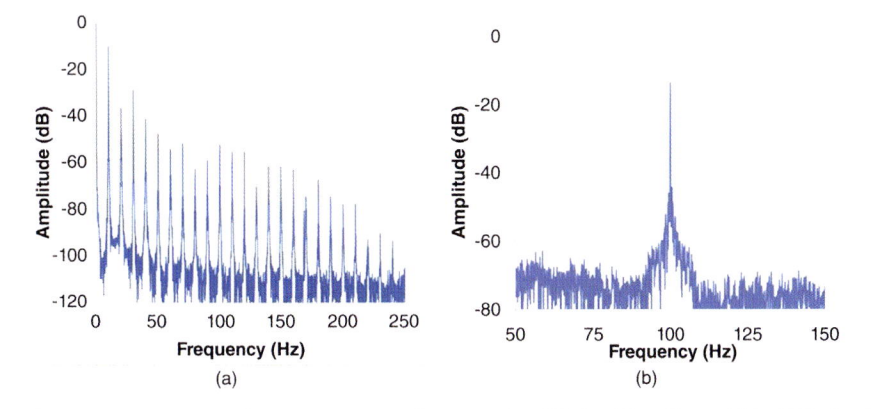

Figure 5.16 Frequency domain response of 8 mm radius FOLS at (a) 10 Hz and (b) 100 Hz [107].

5.6 Some Challenges in SHM of Composite Materials

Past research has studied the possibility of embedding optical sensors in composite materials [114–116]. However, there are several persistent questions and challenges in this field. It was always a concern that optical sensors may act as defects or inclusions in the composite specimens and create stress concentration locations [117,118]. There are also concerns because the optical fibers with coating have diameter of 250–3000 μm, compared to the diameter of reinforcing fibers of about 10 μm. One optical fiber would replace an entire tow of reinforcing carbon or glass fibers and may cause a long defect line. These concerns are weighed against benefits of the capabilities of SHM in these composites.

There are several manufacturing challenges with embedding sensors in laminates and other composite materials, some of which are as follows:

- Optical fibers need to be connected to laser and photodiodes. If the optical fiber breaks during the manufacturing process, the sensing scheme would not work. This concern usually requires embedding several optical fibers in the component.
- A laminate containing optical fiber cannot be cut, trimmed, or shaped because these operations would damage the optical fiber. Automatic fiber placement machines are useful in this regard as they can provide net-shaped components with high fiber placement accuracy.
- Entry and exit points need to be designed for the optical fiber in the laminate. Reinforcing fibers would be displaced in that region.
- Most sensors such as EFPI and microbend sensors have components that are very large to be embedded inside the sensor.
- Large size of the sensing element also restricts their use on the surface of composites because they can disrupt the fluid flow around the component or affect the natural vibration profile.

Development of new sensors is an ongoing process. Lightweight and small size sensors are of great interest for composite materials for SHM purposes.

5.7 Summary

Several fiber-optic sensors relevant to structural health monitoring of composite materials are discussed. Fiber-optic sensors are designed based on frequency or intensity modulation approaches. Usually the instrumentation is less expensive and simpler in the intensity modulation approach. Examples of both types of sensors are presented. Fiber Bragg's grating sensor, extrinsic Fabry–Pérot interferometric sensor, whispering gallery mode sensor, microbend sensor, and fiber-optic loop sensor are discussed for their operating principles. The use of these sensors in SHM applications is also discussed. There has been significant progress in the past two decades in understanding the effect of embedded sensors on the structural integrity of composite materials. However, benefits of the possibility of health monitoring need to be weighed against the possibility of compromise in the structural integrity at known locations. Some of the most widely used fiber-optic sensors are large and bulky and are not the right option for embedding inside composite materials. Continuous development of optical sensors is important for this field to find new sensors that can conduct health monitoring without causing any perturbations.

Acknowledgments

We acknowledge funding from NYU Technology Acceleration and Commercialization Grant. Environmental Protection Agency fellowship to Kevin Chen is also acknowledged. We thank Steven Zeltmann for his help in preparation of the text.

References

1 Udd, E. and Spillman, W.B., Jr. (2011) *Fiber Optic Sensors: An Introduction for Engineers and Scientists*, John Wiley & Sons, Inc., Hoboken, NJ.
2 Körner, M., Prucker, O., and Rühe, J. (2015) Polymer hybrid materials for planar optronic systems. Proceedings of the SPIE 9626, Optical Systems Design 2015: Optical Design and Engineering VI, paper ID: 96262O.
3 Thoben, K.-D., Busse, M., Denkena, B., Gausemeier, J., Rezem, M., Günther, A., Rahlves, M., Roth, B., and Reithmeier, E. (2014) Challenges for product and production engineering hot embossing of polymer optical waveguides for sensing applications. *Procedia Technology*, **15**, 514–520.
4 Baaske, M. and Vollmer, F. (2012) Optical resonator biosensors: molecular diagnostic and nanoparticle detection on an integrated platform. *ChemPhysChem*, **13** (2), 427–436.
5 Kelb, C., Reithmeier, E., and Roth, B. (2014) Planar integrated polymer-based optical strain sensor. SPIE MOEMS-MEMS, article ID: 89770Y.
6 Missinne, J., Kalathimekkad, S., Van Hoe, B., Bosman, E., Vanfleteren, J., and Van Steenberge, G. (2014) Stretchable optical waveguides. *Optics Express*, **22** (4), 4168–4179.
7 Soh, C.-K., Yang, Y., and Bhalla, S. (2012) *Smart Materials in Structural Health Monitoring, Control and Biomechanics*, Springer Science+Business Media.
8 Arsenault, T.J., Achuthan, A., Marzocca, P., Grappasonni, C., and Coppotelli, G. (2013) Development of a FBG based distributed strain sensor system for wind turbine structural health monitoring. *Smart Materials and Structures*, **22** (7), 075027.
9 Ma, H., Jen, A., and Dalton, L.R. (2002) Polymer-based optical waveguides: materials, processing, and devices. *Advanced Materials*, **14** (19), 1339–1365.
10 Wolfer, T., Bollgruen, P., Mager, D., Overmeyer, L., and Korvink, J.G. (2014) Flexographic and inkjet printing of polymer optical waveguides for fully integrated sensor systems. *Procedia Technology*, **15**, 521–529.
11 Han, T., Madden, S., Zhang, M., Charters, R., and Luther-Davies, B. (2009) Low loss high index contrast nanoimprinted polysiloxane waveguides. *Optics Express*, **17** (4), 2623–2630.

12 Wang, S., Vaidyanathan, V., and Borden, B. (2009) Polymer optical channel waveguide components fabricated by using a laser direct writing system. *Journal of Applied Science & Engineering Technology*, **3**, 47–52.

13 Worgull, M. (2009) *Hot Embossing: Theory and Technology of Microreplication*, William Andrew.

14 Becker, H. and Heim, U. (2000) Hot embossing as a method for the fabrication of polymer high aspect ratio structures. *Sensors and Actuators A: Physical*, **83** (1), 130–135.

15 Heckele, M. and Schomburg, W. (2003) Review on micro molding of thermoplastic polymers. *Journal of Micromechanics and Microengineering*, **14** (3), R1.

16 Charest, J.L., Bryant, L.E., Garcia, A.J., and King, W.P. (2004) Hot embossing for micropatterned cell substrates. *Biomaterials*, **25** (19), 4767–4775.

17 Reyes, D.R., Iossifidis, D., Auroux, P.-A., and Manz, A. (2002) Micro total analysis systems: 1. Introduction, theory, and technology. *Analytical Chemistry*, **74** (12), 2623–2636.

18 Eldada, L. and Shacklette, L.W. (2000) Advances in polymer integrated optics. *IEEE Journal of Selected Topics in Quantum Electronics*, **6** (1), 54–68.

19 Wei, T., Han, Y., Li, Y., Tsai, H.-L., and Xiao, H. (2008) Temperature-insensitive miniaturized fiber inline Fabry–Pérot interferometer for highly sensitive refractive index measurement. *Optics Express*, **16** (8), 5764–5769.

20 Liao, C.R., Hu, T.Y., and Wang, D.N. (2012) Optical fiber Fabry–Pérot interferometer cavity fabricated by femtosecond laser micromachining and fusion splicing for refractive index sensing. *Optics Express*, **20** (20), 22813–22818.

21 Hwang, K.S., Lee, S.-M., Kim, S.K., Lee, J.H., and Kim, T.S. (2009) Micro-and nanocantilever devices and systems for biomolecule detection. *Annual Review of Analytical Chemistry*, **2**, 77–98.

22 Nirschl, M., Rantala, A., Tukkiniemi, K., Auer, S., Hellgren, A.-C., Pitzer, D., Schreiter, M., and Vikholm-Lundin, I. (2010) CMOS-integrated film bulk acoustic resonators for label-free biosensing. *Sensors*, **10** (5), 4180–4193.

23 Makowski, M.S. and Ivanisevic, A. (2011) Molecular analysis of blood with micro-/nanoscale field-effect-transistor biosensors. *Small*, **7** (14), 1863–1875.

24 Dahlin, A.B., Chen, S., Jonsson, M.P., Gunnarsson, L., Kall, M., and Höök, F. (2009) High-resolution microspectroscopy of plasmonic nanostructures for miniaturized biosensing. *Analytical Chemistry*, **81** (16), 6572–6580.

25 Dahlin, A.B., Zahn, R., and Vörös, J. (2012) Nanoplasmonic sensing of metal–halide complex formation and the electric double layer capacitor. *Nanoscale*, **4** (7), 2339–2351.

26 Dahlin, A.B., Dielacher, B., Rajendran, P., Sugihara, K., Sannomiya, T., Zenobi-Wong, M., and Vörös, J. (2012) Electrochemical plasmonic sensors. *Analytical and Bioanalytical Chemistry*, **402** (5), 1773–1784.

27 Dahlin, A.B. (2012) Size matters: problems and advantages associated with highly miniaturized sensors. *Sensors*, **12** (3), 3018–3036.

28 Curry, A., Nusz, G., Chilkoti, A., and Wax, A. (2007) Analysis of total uncertainty in spectral peak measurements for plasmonic nanoparticle-based biosensors. *Applied Optics*, **46** (10), 1931–1939.

29 Devendra, P.G., Mohammed, A.Z., and Gary, L.A. (2001) Current and potential future research activities in adaptive structures: an ARO perspective. *Smart Materials and Structures*, **10** (4), 610.

30 Nielsen, M.W., Schmidt, J.W., Høgh, J.H., Waldbjørn, J.P., Hattel, J.H., Andersen, T.L., and Markussen, C.M. (2014) Life cycle strain monitoring in glass fibre reinforced polymer laminates using embedded fibre Bragg grating sensors from manufacturing to failure. *Journal of Composite Materials*, **48** (3), 365–381.

31 Leng, J. and Asundi, A. (2003) Structural health monitoring of smart composite materials by using EFPI and FBG sensors. *Sensors and Actuators A: Physical*, **103** (3), 330–340.

32 Murukeshan, V., Chan, P., Ong, L., and Seah, L. (2000) Cure monitoring of smart composites using fiber Bragg grating based embedded sensors. *Sensors and Actuators A: Physical*, **79** (2), 153–161.

33 Ganapathi, A., Maheshwari, M., Joshi, S.C., Chen, Z., Asundi, A., and Tjin, S.C. (2015) Fibre Bragg grating sensors for *in-situ* measurement of resin pressure in curing composites, in *International Conference on Experimental Mechanics, November 15–17*, International Society for Optics and Photonics, Singapore.

34 Lekakou, C., Cook, S., Deng, Y., Ang, T.W., and Reed, G.T. (2006) Optical fibre sensor for monitoring flow and resin curing in composites manufacturing. *Composites Part A: Applied Science and Manufacturing*, **37** (6), 934–938.

35 Kim, H.-S., Yoo, S.-H., and Chang, S.-H. (2013) *In situ* monitoring of the strain evolution and curing reaction of composite laminates to reduce the thermal residual stress using FBG sensor and dielectrometry. *Composites Part B: Engineering*, **44** (1), 446–452.

36 Schroeder, K., Ecke, W., Apitz, J., Lembke, E., and Lenschow, G. (2006) A fibre Bragg grating sensor system monitors operational load in a wind turbine rotor blade. *Measurement Science and Technology*, **17** (5), 1167.

37 Maaskant, R., Alavie, T., Measures, R., Tadros, G., Rizkalla, S., and Guha-Thakurta, A. (1997) Fiber-optic Bragg grating sensors for bridge monitoring. *Cement and Concrete Composites*, **19** (1), 21–33.

38 Ko, J.M. and Ni, Y.Q. (2005) Technology developments in structural health monitoring of large-scale bridges. *Engineering Structures*, **27** (12), 1715–1725.

39 Tennyson, R., Mufti, A., Rizkalla, S., Tadros, G., and Benmokrane, B. (2001) Structural health monitoring of innovative bridges in Canada with fiber optic sensors. *Smart Materials and Structures*, **10** (3), 560.

40 Schuster, M., Fritz, N., McEntee, J., Graver, T., Rumsey, M., Hernandez-Sanchez, B., Miller, D., and Johnson, E. (2014) Externally bonded FBG strain sensors structural health monitoring of marine hydrokinetic structures. Proceedings of the 2nd Marine Energy Technology Symposium (METS'14), April 15–18, 2014, Seattle, WA.

41 Mendoza, E., Prohaska, J., Kempen, C., Esterkin, Y., and Sun, S. (2013) In-flight fiber optic acoustic emission sensor (FAESense) system for the real time detection, localization, and classification of damage in composite aircraft structures. in Proceedings of the SPIE 8720, Photonic Applications for Aerospace, Commercial, and Harsh Environments IV, 87200K, May 31, 2013, Baltimore, MD.

42 Hamouda, T., Seyam, A.-F.M., and Peters, K. (2015) Polymer optical fibers integrated directly into 3D orthogonal woven composites for sensing. *Smart Materials and Structures*, **24** (2), 025027.

43 Rossi, G. and Speranzini, E. (2008) Fiber Bragg grating strain sensors for *in situ* analysis and monitoring of fiber-reinforced historical civil structures. *Eighth International Conference on Vibration Measurements by Laser Techniques: Advances and Applications*, International Society for Optics and Photonics, Ancona, Italy.

44 Wang, G., Pran, K., Sagvolden, G., Havsgård, G., Jensen, A., Johnson, G., and Vohra, S. (2001) Ship hull structure monitoring using fibre optic sensors. *Smart Materials and Structures*, **10** (3), 472.

45 Jensen, A., Havsgård, G., Pran, K., Wang, G., Vohra, S., Davis, M., and Dandridge, A. (2000) Wet deck slamming experiments with a FRP sandwich panel using a network of 16 fibre optic Bragg grating strain sensors. *Composites Part B: Engineering*, **31** (3), 187–198.

46 Li, H., Herszberg, I., Mouritz, A., Davis, C., and Galea, S. (2004) Sensitivity of embedded fibre optic Bragg grating sensors to disbonds in bonded composite ship joints. *Composite Structures*, **66** (1), 239–248.

47 Majumder, M., Gangopadhyay, T.K., Chakraborty, A.K., Dasgupta, K., and Bhattacharya, D.K. (2008) Fibre Bragg gratings in structural health monitoring: present status and applications. *Sensors and Actuators A: Physical*, **147** (1), 150–164.

48 Liu, T., Brooks, D., Martin, A., Badcock, R., Ralph, B., and Fernando, G.F. (1997) A multi-mode extrinsic Fabry–Pérot interferometric strain sensor. *Smart Materials and Structures*, **6** (4), 464.

49 Kaur, A. (2014) Femtosecond laser micro-machined optical fiber based embeddable strain and temperature sensors for structural monitoring. *Electrical and Computer Engineering*, Missouri University of Science and Technology.

50 Leng, J., Barnes, R., Hameed, A., Winter, D., Tetlow, J., Mays, G., and Fernando, G. (2006) Structural NDE of concrete structures using protected EFPI and FBG sensors. *Sensors and Actuators A: Physical*, **126** (2), 340–347.

51 Leng, J.S. and Asundi, A. (2002) Non-destructive evaluation of smart materials by using extrinsic Fabry–Pérot interferometric and fiber Bragg grating sensors. *NDT & E International*, **35** (4), 273–276.

52 Aref, S.H., Zibaii, M.I., and Latifi, H. (2009) An improved fiber optic pressure and temperature sensor for downhole application. *Measurement Science and Technology*, **20** (3), 034009.

53 Rayleigh, L. (1910) The problem of the whispering gallery. Philosophical Magazine Series 6, vol. 20, Taylor & Francis.

54 Born, M. and Wolf, E. (1999) *Principles of Optics: Electromagnetic Theory of Propagation, Interference and Diffraction of Light*, 7th edn, Cambridge University Press, Cambridge, UK.

55 Richtmyer, R.D. (1939) Dielectric resonators. *Journal of Applied Physics*, **10** (6), 391–398.

56 Garrett, C.G.B., Kaiser, W., and Bond, W.L. (1961) Stimulated emission into optical whispering modes of spheres. *Physical Review*, **124** (6), 1807–1809.

57 Nguyen, N.Q. and Gupta, N. (2010) Whispering gallery mode sensor for phase transformation and solidification studies. *Philosophical Magazine Letters*, **90** (1), 61–67.

58 Chau, K., Wei, X., Wong, R., and Yu, T. (2000) Fragmentation of brittle spheres under static and dynamic compressions: experiments and analyses. *Mechanics of Materials*, **32** (9), 543–554.

59 Timoshenko, S., Goodier, J., and Abramson, H.N. (1970) Theory of elasticity. *Journal of Applied Mechanics*, **37**, 888.

60 Wu, S. and Chau, K. (2006) Dynamic response of an elastic sphere under diametral impacts. *Mechanics of Materials*, **38** (11), 1039–1060.

61 Bellouard, Y., Colomb, T., Depeursinge, C., Dugan, M., Said, A.A., and Bado, P. (2006) Nanoindentation and birefringence measurements on fused silica specimen exposed to low-energy femtosecond pulses. *Optics Express*, **14** (18), 8360–8366.

62 Rudd, J. and Gurnee, E. (1957) Photoelastic properties of polystyrene in the glassy state: II. Effect of temperature. *Journal of Applied Physics*, **28** (10), 1096–1100.

63 Ay, F., Kocabas, A., Kocabas, C., Aydinli, A., and Agan, S. (2004) Prism coupling technique investigation of elasto-optical properties of thin polymer films. *Journal of Applied Physics*, **96** (12), 7147–7153.

64 Primak, W. and Post, D. (1959) Photoelastic constants of vitreous silica and its elastic coefficient of refractive index. *Journal of Applied Physics*, **30** (5), 779–788.

65 Nguyen, N.Q., Gupta, N., Ioppolo, T., and Ötügen, M.V. (2009) Whispering gallery mode-based micro-optical sensors for structural health monitoring of composite materials. *Journal of Materials Science*, **44** (6), 1560–1571.

66 Vollmer, F. and Yang, L. (2012) Review label-free detection with high-Q microcavities: a review of biosensing mechanisms for integrated devices. *Nanophotonics*, **1** (3–4), 267–291.

67 Ward, J.M., Dhasmana, N., and Chormaic, S.N. (2014) Hollow core, whispering gallery resonator sensors. *The European Physical Journal: Special Topics*, **223** (10), 1917–1935.

68 Wang, J., Zhan, T., Huang, G., Chu, P.K., and Mei, Y. (2014) Optical microcavities with tubular geometry: properties and applications. *Laser & Photonics Reviews*, **8** (4), 521–547.

69 Pöllinger, M., O'Shea, D., Warken, F., and Rauschenbeutel, A. (2009) Ultrahigh-Q tunable whispering-gallery-mode microresonator. *Physical Review Letters*, **103** (5), 053901.

70 Sumetsky, M. (2004) Whispering-gallery-bottle microcavities: the three-dimensional etalon. *Optics Letters*, **29** (1), 8–10.

71 Sumetsky, M. (2010) Mode localization and the Q-factor of a cylindrical microresonator. *Optics Letters*, **35** (14), 2385–2387.

72 Lee, H., Chen, T., Li, J., Yang, K.Y., Jeon, S., Painter, O., and Vahala, K.J. (2012) Chemically etched ultrahigh-Q wedge-resonator on a silicon chip. *Nature Photonics*, **6** (6), 369–373.

73 Grossmann, T., Hauser, M., Beck, T., Gohn-Kreuz, C., Karl, M., Kalt, H., Vannahme, C., and Mappes, T. (2010) High-Q conical polymeric microcavities. *Applied Physics Letters*, **96** (1), 013303.

74 Kudo, H., Ogawa, Y., Kato, T., Yokoo, A., and Tanabe, T. (2013) Fabrication of whispering gallery mode cavity using crystal growth. *Applied Physics Letters*, **102** (21), 211105.

75 Foreman, M.R., Swaim, J.D., and Vollmer, F. (2015) Whispering gallery mode sensors. *Advances in Optics and Photonics*, **7** (2), 168–240.

76 Kozhevnikov, M., Ioppolo, T., Stepaniuk, V., Sheverev, V., and Otugen, V. (2006) Optical force sensor based on whispering galley mode resonators. 44th AIAA Aerospace Science Meeting and Exhibit, January 9–12, 2006, Reno, Nev.

77 Ioppolo, T., Kozhevnikov, M., Stepaniuk, V., Ötügen, M.V., and Sheverev, V. (2008) Micro-optical force sensor concept based on whispering gallery mode resonators. *Applied Optics*, **47** (16), 3009–3014.

78 Gupta, N., Woldesenbet, E., and Sankaran, S. (2001) Studies on compressive failure features in syntactic foam material. *Journal of Materials Science*, **36** (18), 4485–4491.

79 Gupta, N., Woldesenbet, E., and Mensah, P. (2004) Compression properties of syntactic foams: effect of cenosphere radius ratio and specimen aspect ratio. *Composites Part A: Applied Science and Manufacturing*, **35** (1), 103–111.

80 Gupta, N., Zeltmann, S.E., Shunmugasamy, V.C., and Pinisetty, D. (2014) Applications of polymer matrix syntactic foams. *JOM*, **66** (2), 245–254.

81 Remouche, M., Mokdad, R., Chakari, A., and Meyrueis, P. (2007) Intrinsic integrated optical temperature sensor based on waveguide bend loss. *Optics & Laser Technology*, **39** (7), 1454–1460.

82 Remouche, M., Mokdad, R., Lahrashe, M., Chakari, A., and Meyrueis, P. (2007) Intrinsic optical fiber temperature sensor operating by modulation of the local numerical aperture. *Optical Engineering*, **46** (2), 024401–024401.

83 Otsuki, S., Adachi, K., and Taguchi, T. (1998) A novel fiber-optic gas-sensing configuration using extremely curved optical fibers and an attempt for optical humidity detection. *Sensors and Actuators B: Chemical*, **53** (1–2), 91–96.

84 Asawa, C., Yao, S., Stearns, R., Mota, N., and Downs, J. (1982) High-sensitivity fibre-optic strain sensors for measuring structural distortion. *Electronics Letters*, **18** (9), 362–364.

85 Lee, S.T., George, N.A., Sureshkumar, P., Radhakrishnan, P., Vallabhan, C.P.G., and Nampoori, V.P.N. (2001) Chemical sensing with microbent optical fiber. *Optics Letters*, **26** (20), 1541–1543.

86 Pierce, S.G., MacLean, A., and Culshaw, B. (2000) Optical frequency-domain reflectometry for microbend sensor demodulation. *Applied Optics*, **39** (25), 4569–4581.

87 Lagakos, N., Cole, J.H., and Bucaro, J.A. (1987) Microbend fiber-optic sensor. *Applied Optics*, **26** (11), 2171–2180.

88 Ilev, I.K. and Waynant, R.W. (1999) All-fiber-optic sensor for liquid level measurement. *Review of Scientific Instruments*, **70** (5), 2551–2554.

89 Fields, J., Asawa, C., Ramer, O., and Barnoski, M. (1980) Fiber optic pressure sensor. *Journal of the Acoustical Society of America*, **67** (3), 816–818.

90 Fields, J. (1980) Attenuation of a parabolic-index fiber with periodic bends. *Applied Physics Letters*, **36** (10), 799–801.

91 Personick, S. (1971) Time dispersion in dielectric waveguides. *Bell System Technical Journal*, **50** (3), 843–859.

92 Marcuse, D. (1972) Pulse propagation in multimode dielectric waveguides. *Bell System Technical Journal*, **51** (6), 1199–1232.

93 Lagakos, N., Litovitz, T., Macedo, P., Mohr, R., and Meister, R. (1981) Multimode optical fiber displacement sensor. *Applied Optics*, **20** (2), 167–168.

94 Jeunhomme, L. and Pocholle, J.P. (1975) Mode coupling in a multimode optical fiber with microbends. *Applied Optics*, **14** (10), 2400–2405.

95 Roriz, P., Ramos, A., Santos, J.L., and Simões, J.A. (2012) Fiber optic intensity-modulated sensors: a review in biomechanics. *Photonic Sensors*, **2** (4), 315–330.

96 Liu, Y., Liu, J.P., Zhu, Z., Liu, Y., and Wang, A. (2008) Experimental research on microbend fiber optic pressure sensor. *Instrument Technique & Sensor*, **37** (1), 4–7.

97 Ivanov, O.V. (2004) Wavelength shift and split of cladding mode resonances in microbend long-period fiber gratings under torsion. *Optics Communications*, **232** (1–6), 159–166.

98 Marvin, D.C. and Ives, N.A. (1984) Wide-range fiber-optic strain sensor. *Applied Optics*, **23** (23), 4212–4217.

99 Jay, J.A. (2010) An Overview of Macrobending and Microbending of Optical Fibers. White paper of Corning, pp 1–21. Available at http://www.corning.com/media/worldwide/coc/documents/Fiber/RC-%20White%20Papers/WP-General/WP1212_12-10.pdf.

100 Berthold, J., III (1998) Microbend fiber optic sensors, in *Optical Fiber Sensor Technology* (eds K.T.V. Grattan and B.T. Meggitt), Kluwer Academic Publishers, London, UK, pp. 225–240.

101 Xuejin, L., Yuanlong, D., Yongqin, Y., Xinyi, W., and Jingxian, L. (2008) Microbending optical fiber sensors and their applications, in *Proceedings of the 2008 International Conference on Advanced Infocomm Technology*, ACM, Shenzhen, China, pp. 1–4.

102 Marcuse, D. (28 1991) Chapter 6 – Theory of the directional coupler, in *Theory of Dielectric Optical Waveguides*, 2nd edn, Academic Press, London, UK, pp. 251–279.

103 Dahl-Petersen, S., Larsen, C.C., Povlsen, J.H., Lumholt, O., Bjarklev, A., Rasmussen, T., and Rottwitt, K. (1991) Simple fiber-optic low-temperature sensor that uses microbending loss. *Optics Letters*, **16** (17), 1355–1357.

104 Renner, H. (1992) Bending losses of coated single-mode fibers: a simple approach. *Journal of Lightwave Technology*, **10** (5), 544–551.

105 Faustini, L. and Martini, G. (1997) Bend loss in single-mode fibers. *Journal of Lightwave Technology*, **15** (4), 671–679.

106 Nguyen, N.Q. and Gupta, N. (2009) Power modulation based fiber-optic loop-sensor having a dual measurement range. *Journal of Applied Physics*, **106** (3), 033502.

107 Nishino, Z.T., Chen, K., and Gupta, N. (2014) Power modulation-based optical sensor for high-sensitivity vibration measurements. *IEEE Sensors Journal*, **14** (7), 2153–2158.

108 Nam, S.H. and Yin, S. (2005) High-temperature sensing using whispering gallery mode resonance in bent optical fibers. *IEEE Photonics Technology Letters*, **17** (11), 2391–2393.

109 Mathew, J., Semenova, Y., Rajan, G., Wang, P., and Farrell, G. (2011) Improving the sensitivity of a humidity sensor based on fiber bend coated with a hygroscopic coating. *Optics & Laser Technology*, **43** (7), 1301–1305.

110 Sumetsky, M., Dulashko, Y., Fini, J.M., Hale, A., and DiGiovanni, D.J. (2006) The microfiber loop resonator: theory, experiment, and application. *Journal of Lightwave Technology*, **24** (1), 242–250.

111 Marcuse, D. (1976) Curvature loss formula for optical fibers. *Journal of the Optical Society of America*, **66** (3), 216–220.

112 Wang, Q., Farrell, G., and Freir, T. (2005) Theoretical and experimental investigations of macro-bend losses for standard single mode fibers. *Optics Express*, **13** (12), 4476–4484.

113 Schermer, R.T. and Cole, J.H. (2007) Improved bend loss formula verified for optical fiber by simulation and experiment. *IEEE Journal of Quantum Electronics*, **43** (10), 899–909.

114 Delobelle, B., Perreux, D., and Delobelle, P. (2014) Failure of nano-structured optical fibers by femtosecond laser procedure as a strain safety-fuse sensor for composite material applications. *Sensors and Actuators A: Physical*, **210**, 67–76.

115 Lammens, N., Luyckx, G., W., Van Paepegem., and Degrieck, J. (2016) Finite element prediction of resin pocket geometries around arbitrary inclusions in composites: case study for an embedded optical fiber interrogator. *Composite Structures*, **146**, 95–107.

116 Lesiak, P., Szeląg, M., Budaszewski, D., Plaga, R., Mileńko, K., Rajan, G., Semenova, Y., Farrell, G., Boczkowska, A., Domański, A., and Woliński, T. (2012) Influence of lamination process on optical fiber sensors embedded in composite material. *Measurement*, **45** (9), 2275–2280.

117 Bhargava, A., Shivakumar, K., and Emmanwori, L. (2003) Stress concentration and failure around embedded fiber optic sensor in composite laminates. 44th AIAA/ASME/ASCE/AHS/ASC Structures, Structural Dynamics, and Materials Conference, Norfolk, Virginia, April 7–10, 2003, American Institute of Aeronautics and Astronautics.

118 Huang, Y. and Nemat-Nasser, S. (2007) Structural integrity of composite laminates with embedded micro-sensors, *Proceedings of SPIE*, **6530 65300W-1**, 277–786.

6

Electronics Development for Integration

Jan Vanfleteren

Ghent University – IMEC, Centre for Microsystems Technology (CMST),
Technologiepark 15 - iGent Building, 9052 Gent-Zwijnaarde, Belgium

6.1 Introduction

6.1.1 Standard Flat Rigid Printed Circuits Boards and Components Assembly

In order to produce an electronic system, the different electronic components (chips, passives, sensors, etc.) must be fixed on a carrier and electrically interconnected. Since tens of years this has been done making use of printed circuit board (PCB) technology. The board contains a flat, rigid electrically insulating material, carrying conductors, which, similar as wiring, provide the interconnection between two nodes of the electronic circuit; these nodes belong to two different components or the same component (e.g., two pins of the same semiconductor chip package). The most commonly used board material today is Flame Retardant 4 (FR4), which has an epoxy resin, a woven glass fabric reinforcement, and a brominated flame retardant as the main constituents. On this carrier material, plain Cu sheets with thicknesses in the range normally between 10 and 100 μm (mainly used standard thickness are 17 and 35 μm) are laminated. The plain Cu sheets are patterned using photolithography and wet etching, thus forming the interconnection tracks. PCBs with multiple conductor layer can be produced using stacking (by lamination) of single- or double-sided single PCBs. Electrical connections between different layers are realized by drilling holes (mechanically or using a laser) through the different Cu layers and metallizing the drilled holes by electroless deposition and electroplating of Cu. Top and/or bottom of the PCB might finally be coated with a solder mask, which determines the flow of solder, used during the assembly of the components on the board. To prevent oxidation of the copper and facilitate component assembly by soldering, an additional finishing of the contact pads is applied, which consists of electroless Ni/Au, Ag, Sn, or an organic surface protection (OSP) coating. Figure 6.1 shows a finished printed circuit board before component assembly.

Material-Integrated Intelligent Systems: Technology and Applications, First Edition.
Edited by Stefan Bosse, Dirk Lehmhus, Walter Lang, and Matthias Busse.
© 2018 Wiley-VCH Verlag GmbH & Co. KGaA. Published 2018 by Wiley-VCH Verlag GmbH & Co. KGaA.

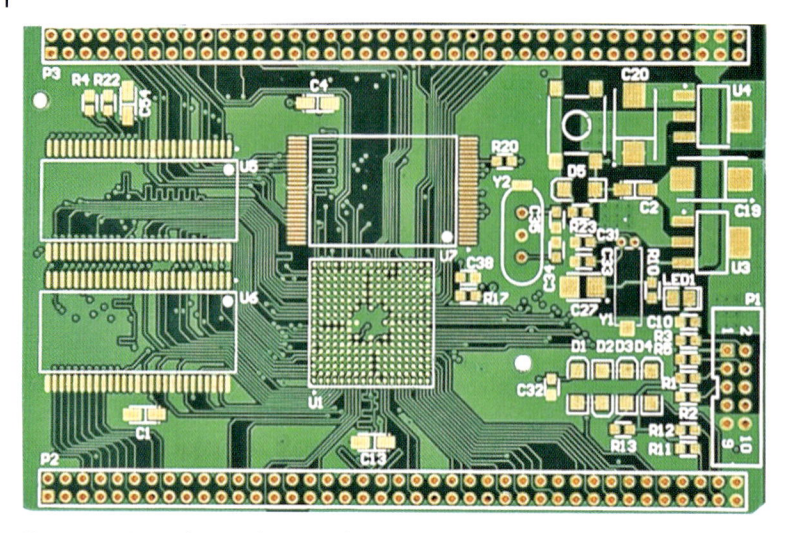

Figure 6.1 Printed circuit board before component assembly. The component contact pads have a Ni/Au finish. The applied solder mask has a green color.

After manufacturing of the PCB, components are assembled. On an industrial scale this is normally done in a company different from the PCB manufacturer. Recently, components are mainly assembled using surface mount technology (SMT) of which the principle is shown in Figure 6.2a: on the board carrier material 6 with patterned conductors 5 solder paste 3 is applied by screen printing or stencil printing technology. The solder paste consists of a metal powder (currently usually lead-free SnAgCu (SAC) alloys), dispersed in a flux, which temporarily acts as an adhesive. Subsequently, the electronic component 1 with its solderable contacts 2 is accurately placed on the board and the solder is heated until it melts. The solder wets both the contact pads 5 and the component contacts 2, and thus, after cooling down and solidification, realizes a proper electrical and mechanical connection. Sometimes an underfill material 4 is applied to improve the mechanical robustness of the assembly. A commonly used solder alloy is Sn96.5Ag3.0Cu0.5 with a melting temperature of 217–220 °C, and a corresponding peak temperature of the reflow solder oven of 245–250 °C. In an industrial electronic assembly process, the solder paste is applied on the entire PCB, subsequently all components are placed using pick-and-place machines, which can place up to 20 components per second, and finally the entire PCB is placed in a reflow oven, thus soldering all components at once. Figure 6.2b shows a fragment of an assembled PCB. For more details on PCB fabrication and assembly, we refer to Ref. [1].

6.1.2 Flexible Circuits

The production flow, described above, using flat rigid substrates, provides an easy way for electronic circuit production. In a growing number of applications, the flat format is, however, not desirable, and the circuit should, for example, take the shape of the object onto or into which it is integrated. In a first step, the rigid

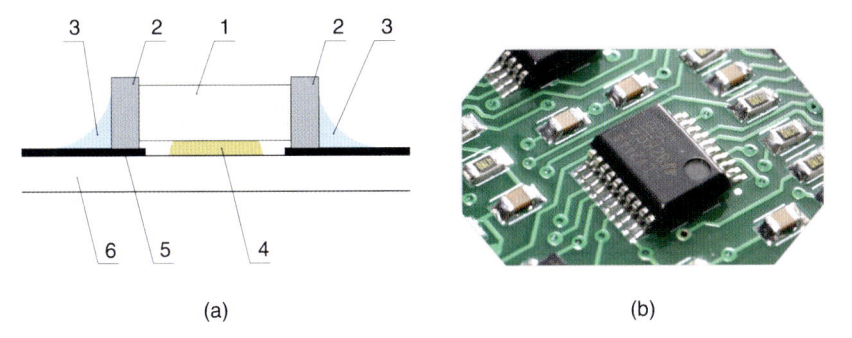

Figure 6.2 Component assembly using SMT technology. (a) Principle; (b) fragment of a PCB with assembled passives and semiconductor components.

epoxy carrier is replaced by a mechanically flexible alternative. In this case flexible foils are used; the rigid epoxy carrier is replaced by bendable thin (typically 25 or 50 µm thick) sheets of materials like polyimide (PI), or the cheaper but chemically and mechanically less stable polyethylene naphthalate (PEN) or polyethylene terephthalate (PET) materials. As conductors, still Cu can be used as in rigid boards; alternatively also printed conductors (often Ag) are applied. Components are assembled using soldering (in combination with Cu conductors) or using conductive adhesives (also mainly based on Ag flakes) for assembly on printed conductors. The operations of circuit production and component assembly are done with the flex substrate in a flat state. Bending to the final shape is done after circuit production, during the final assembly step. As an example Figure 6.3 shows the inside of an Olympus Stylus camera, where the electronics are assembled on a

Figure 6.3 Inside of a Olympus Stylus camera, showing flexible circuit assembly. A polyimide flex with Cu conductors and solder-assembled components is used. (*Photo Credit:* Steve Jurvetson, Creative Commons.)

polyimide flex, which is subsequently folded during the final assembly and thus fits inside the camera.

For more details on flexible circuit fabrication and assembly, we refer to Ref. [2].

6.1.3 Need for Alternative Circuit and Packaging Materials

Figure 6.3 illustrates the current trend toward a so-called Internet-of-Things (IoTs) society where we are surrounded by smart objects, which are fixed (smart lighting), mobile (smart car interiors), portable (all kinds of mobile computing and communication tools), wearable (smart textiles or patches), and even inside our bodies (smart implants). Common features and requirements for the sensor and electronics hardware of all these systems are that it should be unobtrusive, almost nonnoticeable. Therefore, the electronics should be low-volume, lightweight, and preferably completely integrated in the smart object. One way to achieve this is to miniaturize the electronics by, for example, miniaturizing the components or the component packages. When the circuit still has a considerable size (due to its complexity or due to its function, for example, requiring a large area (e.g., a display or a distributed sensor network)), then another strategy is to be chosen, consisting of using technologies that allow the circuit to follow the shape of the smart object and that allows for complete integration/embedding in the object. In the remainder of this chapter, we present a number of novel technologies that were developed for the future fabrication of smart objects. Section 6.2 of this chapter describes the development and application of an ultrathin chip package (UTCP), allowing for extreme miniaturization of electronic assemblies. Section 6.3 deals with the technology for dynamically deformable, that is, elastic circuits. These circuits go beyond what can be obtained by flexible circuits, as they allow the deformation from a flat circuit to a randomly shaped one. In this case, elastomeric materials have to be used as circuit carriers, instead of rigid epoxies or flexible foils. When the elastomers are, furthermore, replaced by thermoplastic materials, then shape retaining one-time deformable circuits can be obtained. This technology is described in Section 6.4. All these technologies are based on the printed circuit board technology, described above, or on thin film technology, where the conductors are sputter-deposited metals with a thickness below 1 μm. However, these technologies are not cost-effective for long interconnection distances (>1 m). In this case, it can be favorable to use individual conductive wires in the form of textile substrates with embedded woven conductive yarns. The technology for using such large area textile-based interconnection substrates, combining them with small, more conventional functional electronic nodes, is presented in Section 6.5.

6.2 Chip Package Miniaturization Technologies

6.2.1 Ultrathin Chip Package Technology

The UTCP acronym stands for ultrathin chip packaging. In this patented technology, a thinned chip is packaged in a polyimide membrane, onto which

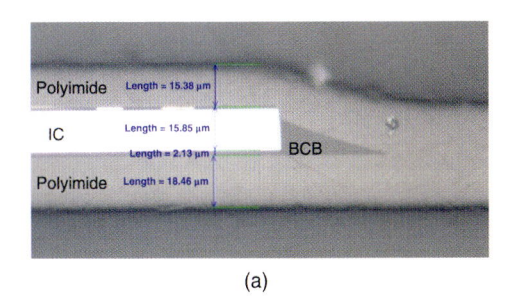

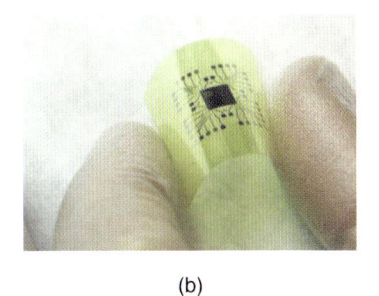

Figure 6.4 The cross section of an ultrathin chip package (a) illustrates the embedded IC, packaged between two polyimide layers. The result is a flexible chip package with a metal fan-out, providing the connectivity to the chip (b).

a metal fan-out pattern is deposited and patterned using thin film techniques. In this way, a flexible ultrathin polymer interposer is created, which allows the packaging of individual custom-off-the-shelf ICs (Figure 6.4).

The UTCP technology differs from other approaches for thin chip embedding by its choice for combining the base material of the package (polyimide) with thin film Cu deposition techniques for electrical interconnectivity. Polyimide is interesting as a base material since it is inert to most chemicals and has a thermal budget up to 350 °C; hence, it does not limit the later back end processes that are required for embedding the packaged chip in a system, such as chip embedding in a circuit board (see also Section 6.2.2.1).

The technology of embedding thinned dies in a polyimide foil was developed and presented by Christiaens *et al.* in 2006 [3]. The process was based on nonphotosensitive polyimide PI2611 (HD microsystems), which is spin-coated and patterned using laser ablation. This packaging process was later used to demonstrate 3D integration by embedding a UTCP-packaged microcontroller MSP430F149 in a flexible circuit board (FCB) [4]. To increase the production capacity and improve the yield of the UTCP packaging technology, a number of changes with respect to the initial process flow were made:

- The introduction of a photosensitive type of polyimide allows increasing the production capacity by exposing multiple vias, and eventually multiple dies at once, as opposed to sequentially opening the vias using laser ablation.
- The use of a new release strategy based on KCl.
- Embedding the chip in a cavity improves the step coverage of the polyimide and avoids the risk of shorts in the fan-out.

The UTCP is realized on a glass substrate, where a polyimide membrane (HD4110) is spun and cured. As this flavor of polyimide is self-priming, a KCl release layer is evaporated prior to spinning. The thinned dies are placed on this base membrane using BCB. Next, the die is covered with two layers of photo-patternable polyimide: a first layer is structured using the thinned die as a self-aligning mask, thus leveling the topography of the embedded die. The top layer of polyimide is patterned for opening vias to the necessary contact pads. Finally, a copper fan-out is applied by sputtering a seed layer and subsequently electroplating of the layer using a photoresist mask. Once the seed layer is removed, the packaged die can be tested and released from its carrier. The presented process flow has been tested and optimized on different types of ICs, proving the feasibility of realizing yield figures of 87% in a laboratory environment.

Several competing technologies with similar objectives as the ultrathin chip package are being developed. Some of these show a strong resemblance to the UTCP technology, as, for example, described in Refs. [5,6]. While the first technology is almost identical, the second approach uses a two-polymer method for encapsulating the chip. The actual encapsulation is done using BCB, but due to its brittle nature, an additional polyimide layer is needed to increase the mechanical robustness of the package. This double layer makes the processing more complex as both layers need to be crossed to realize the electrical contact to the chip.

6.2.2 UTCP Circuit Integration

6.2.2.1 UTCP Embedding

The embedding of a UTCP into an FCB is based on the standard process flow for realizing multilayer FCBs. In this process, multiple copper-clad polyimide sheets are laminated using an acrylic sheet adhesive (Dupont LF0100). An example construction of a four-layer circuit board is illustrated in Figure 6.5. In this buildup a double-clad polyimide sheet with a thickness of 50 µm is laminated to two single-sided sheets with the UTCP located between the first and the second conductor layer. The cross-sectional image shows the UTCP is completely encapsulated by the acrylic adhesive, resulting in a local deformation of the board. This can be avoided by adding extra planarization layers surrounding the UTCP.

The UTCP is (manually) aligned to the metal pattern on the inner layers of the middle flex with an alignment error smaller than 25 µm. The UTCP sample is fixed to the LF0100 adhesive by locally heating the adhesive. The complete stack is then laminated by applying pressure and heat in a vacuum environment. The drilling and electroplating of the vias and the optional application of solder mask

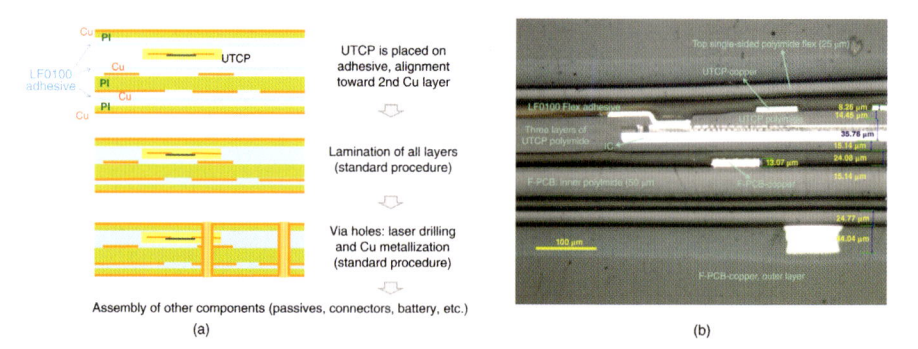

Figure 6.5 Summarizing schematic overview of the embedding process (a) and a cross section of the realized flexible circuit board (b).

or silkscreen were executed in the conventional way by a commercial PCB manufacturer (ACB NV, Dendermonde, Belgium).

6.2.2.2 UTCP Stacking

An alternative approach toward extreme miniaturization of systems with multiple ICs in a package is to reduce the area that is occupied. This can be done by stacking multiple UTCP-packaged dies on top of each other using lamination technology, combined with the fabrication of the interconnections between the layers using via-drilling and via-filling by electroplating of the metallization. Such a stacked UTCP is not flexible anymore, but an extreme degree of miniaturization is realized: in a package of about 300 µm thick, four dies can be embedded and interconnected [7]. The final result is shown in Figure 6.6: a carrier substrate with multiple flat UTCPs, and the finally obtained stacked EEPROM system, consisting of four stacked UTCPs and being an example of extreme miniaturization.

6.2.3 Applications

Consider a microcontroller packaged in a conventional plastic quad flatpack (QFP), and placed on a reference FCB. The total volume occupied by the package is about 240 mm^3, while it occupies about 150 mm^2 of the FCB. The thickness of the FCB is 300 µm (four-layered FCB). The UTCP-packaged microcontroller occupies an area of 1.1025 mm^2, while the FCB thickness increased to 350 µm. The use of embedded UTCP packages allows recycling about 50% of the occupied area for placing passives and routing. Note that the vias to the UTCP also occupy a considerable amount of area. Limiting the amount of I/O to the bare necessity will further enhance the miniaturizing effect of embedded ultrathin chip packages. Figure 6.7 illustrates in one image the power of UTCP for minimizing small electronic systems. The total volume of the UTCP-based module is <60% of the original module.

This level of miniaturization is of strong interest to medical and wearable applications. Thin and light systems can be worn close to the body without

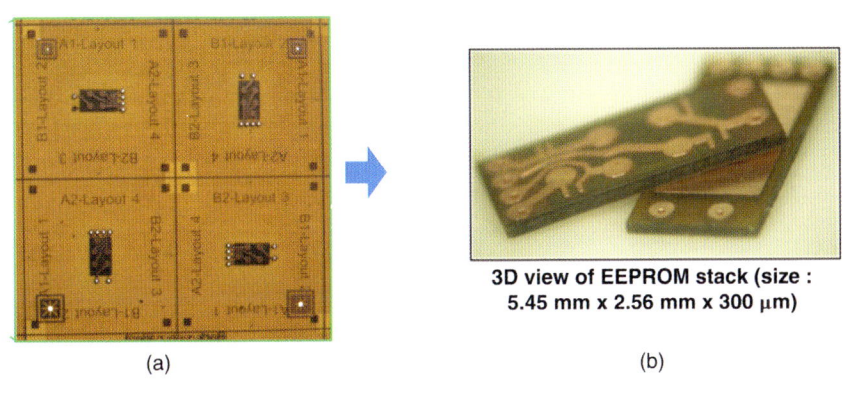

Figure 6.6 A UTCP carrier substrate with four UTCPs, ready for stacking (a); the EEPROM memory device consisting of four stacked dies, all in a package of only 300 µm thick (b).

Figure 6.7 On the left side a system based on a CSP-packaged ZL70102 is compared with its replica based on an embedded UTCP.

interfering with the comfort of the user. Heart rate monitoring outside through a wearable ECG patch is made possible, thanks to the UTCP technology. Figure 6.8 shows a functional demonstrator before final encapsulation of the device, indicating the locations of the three embedded UTCPs: a microcontroller, a radio communication chip, and an analog ASIC.

Hearing aids are at the forefront of miniaturization, relying on origami-like construction to fit all the electronics inside the device. The rigid-flex PCBs required for this are complex to produce and costly. Thanks to the UTCP stacking technology, four EEPROM chips can be packaged in a thickness and size comparable to that of a single bare EEPROM die. Next to the increased functional density, this technology allowed to avoid the use of the costly rigid-flex PCBs (Figure 6.9).

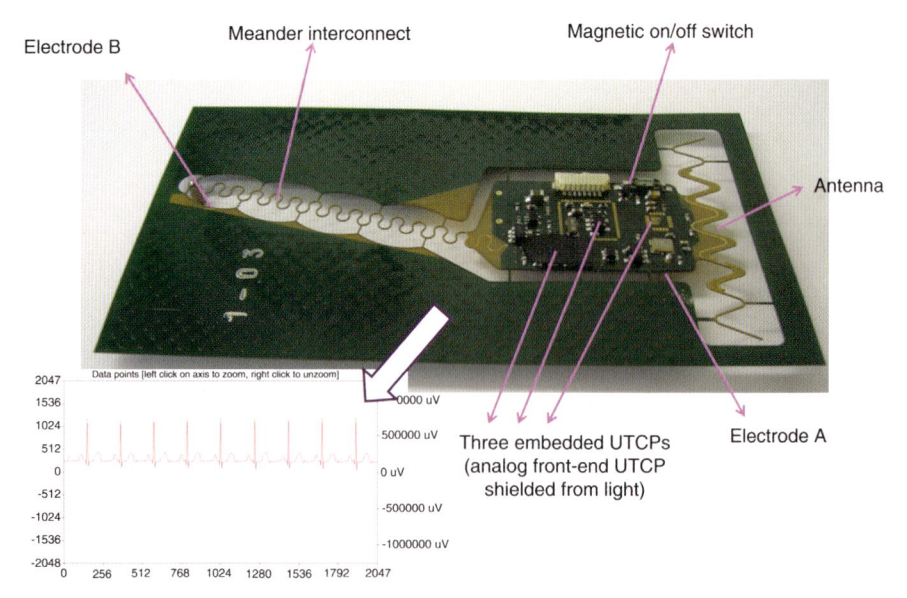

Figure 6.8 Wearable ECG patch containing three embedded off-the-shelf ICs in UTCP format: a microcontroller, a radio communication chip, and an analog ASIC.

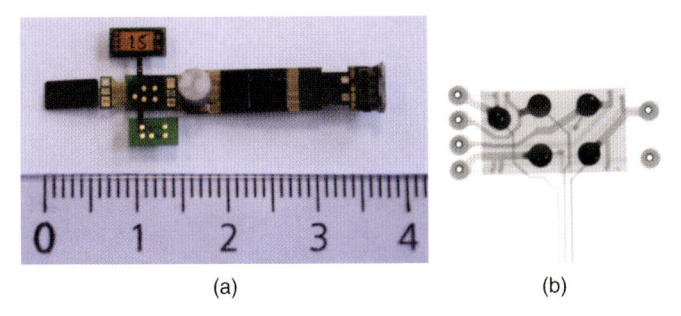

Figure 6.9 Four stacked EEPROM chips in a single UTCP, located in the top flap of the existing rigid-flex PCB (a). (b) The X-ray picture shows the chips inside the package (light gray), the routing (dark gray), as well as the solder balls underneath the package (black).

6.3 Elastic Circuits

6.3.1 Printed Circuit Board-Based Elastic Circuits

In the past years, a technology has been developed that enables the realization of stretchable systems from flexible foils [8]. The basic principle is shown in Figure 6.10.

The electronic functionality is distributed onto islands that are interconnected to each other by meander-shaped interconnects. The whole system is embedded into a stretchable rubber to keep everything together and to protect the device. An advantage of this technology over competing stretchable electronics technologies [9] is that it uses standard flex foil manufacturing technologies.

The process makes use of PCB techniques in order to produce an electronic circuit attached with a temporary adhesive to a rigid carrier (Figure 6.11). The electronic circuit is designed in a way that it contains parts that will become stretchable, this by defining in-plane meander-shaped interconnects, and parts that contain rigid electronic components that remain rigid/flexible in the final device. Thus, flexible/rigid islands containing surface mount devices (SMD),

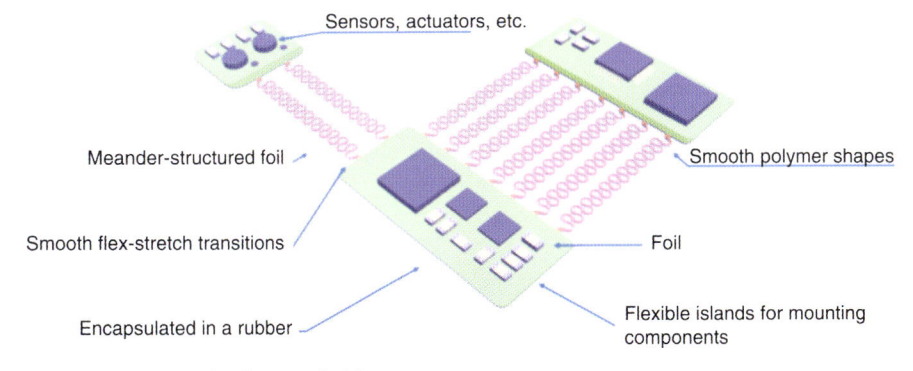

Figure 6.10 Principle of a stretchable system.

6 Electronics Development for Integration

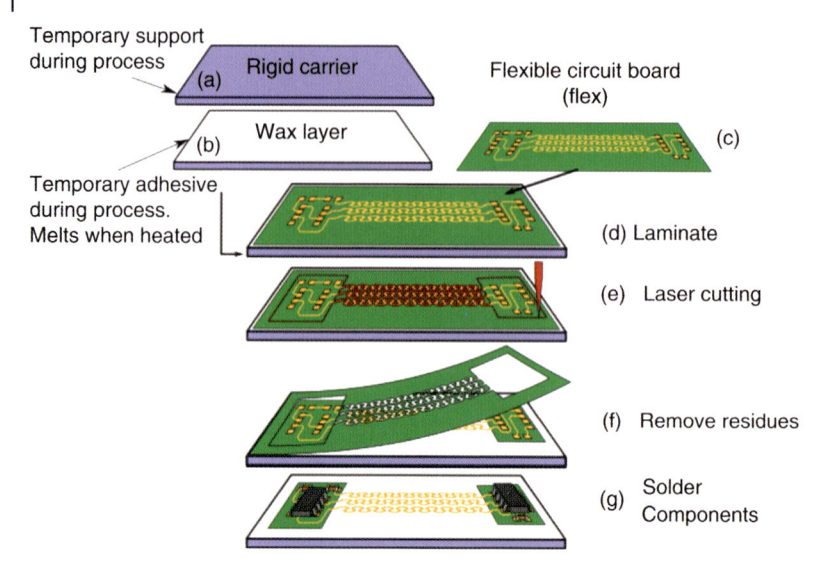

Figure 6.11 Stretchable substrate fabrication on rigid carrier.

Labels in figure:
- (a) Rigid carrier — Temporary support during process
- (b) Wax layer — Temporary adhesive during process. Melts when heated
- (c) Flexible circuit board (flex)
- (d) Laminate
- (e) Laser cutting
- (f) Remove residues
- (g) Solder Components

including resistors, capacitors, microcontrollers, and so on, are interconnected with each other by the stretchable interconnects. This electronic circuit with dedicated design is transferred from the temporary carrier into a stretchable polymer (Figure 6.12), for example, polydimethyl siloxane (PDMS), resulting in a conformable, stretchable electronic device. The encapsulant provides elasticity and conformability for the stretchable system. Owing to the fact that the polymer matrix comes into play at the end of the process, the type of the encapsulant can be chosen without having major impact on the preceding processes. Depending on the application, encapsulants varying in softness, UV stability, and moisture absorbance can be selected. The circuitry can be transferred into any other thermoplastic or thermosetting polymer besides PDMS.

The two main parts of the process are demonstrated in Figures 6.11 and 6.12 and are described in more detail below.

The process starts from a normal full area flex foil (polyester, polyimide, or other plastic foil materials). Circuitry is made with standard printed circuit board technologies (e.g., printing or lithography). Components are also assembled using traditional interconnect technologies (adhesives, soldering). Once finished, the meanders are structured into the foil using, for example, a laser or die cutting. Finally, the structured foil is embedded inside a rubber material by lamination of a polyurethane or overmolding with silicone (Figure 6.12). Overall good stretchability and reliability can be achieved with the technology.

Figure 6.13 shows an overview of typical stretchabilities for stretchable interconnects made on different supporting foil types embedded in silicone (Dow Corning, Sylgard 186). Unsupported meanders can be stretched almost a thousand times at 10% elongation. The use of a support material like polyimide will increase the lifetime to more than 100,000 times.

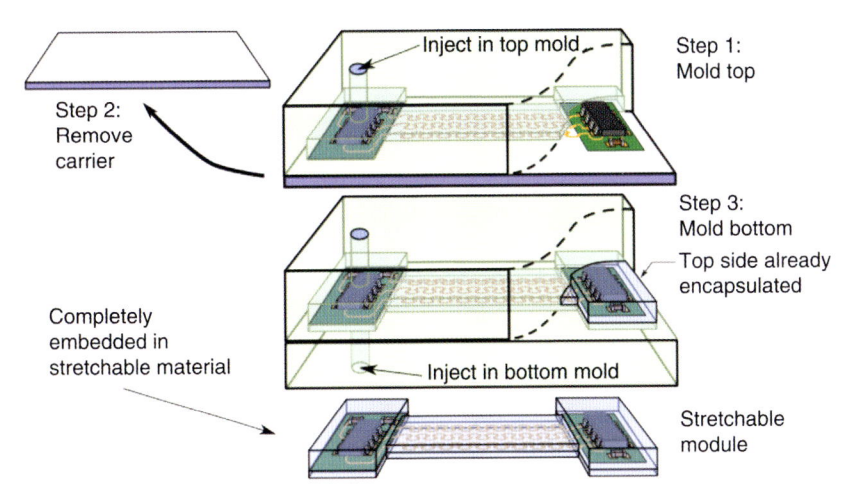

Figure 6.12 Two-step molding process for stretchable circuits.

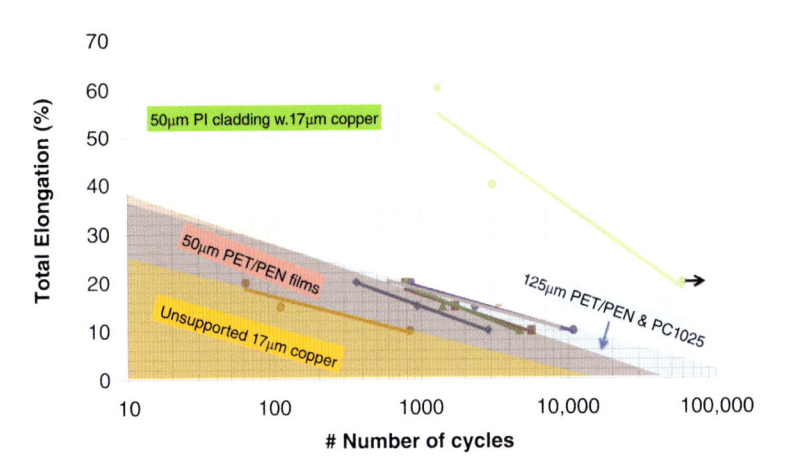

Figure 6.13 Number of stretching cycles that can be achieved in function of elongation, for different technologies.

An aim of the design and modeling activities was to minimize stress and strain concentrations in the meander under deformation. Therefore, the stress has to be distributed as much as possible along the meander. The current meander shape of choice is the horseshoe shape. For minimal stress, the width (W) of the track should be as small as possible. This minimum W is determined mainly by technological constraints. In case of conventional PCB manufacturing, where the conductors are patterned by lithography and wet etching of a 17 or 35 µm thick Cu sheet, conductor tracks with $W = 100$ µm provide circuits with sufficient process yield. The meander radius (R) and angle (θ) are design parameters. The values chosen for these parameters will depend on the application requirements (e.g., the maximum desired stretchability of the circuit). A further improvement of the mechanical reliability of the meanders is achieved by supplying the meanders with a flexible support (polyimide, PEN, PET, etc.). Numerical modeling showed that the width is the main parameter, influencing (in fact reducing, compared with nonsupported meanders) the maximum plastic strain in the Cu meander [10,11]. This can be explained by the redistribution of the plastic strain in the metal. Figure 6.13 gives an overview of the typical stretch reliability that can be achieved.

6.3.2 Thin Film Metal-Based Elastic Circuits

As mentioned in the previous section, the ratio between the radius of the meander and the width of the track should be as large as possible to minimize the stress and strain concentrations in the meander under deformation. Moving from conventional PCB structuring to thin film-based processing allows the realization of finer lines, resulting in smaller meanders and thus strongly reducing the area needed for the stretchable interconnections.

In addition, the use of spin-on polyimide allows more freedom in the design of the polymer support, further reducing the plastic strain in the metallic conductors. Since the processing is done on a temporary carrier, transfer to an elastomeric substrate is performed in a final stage of the fabrication process. This implies a degree of freedom in the choice of the elastomeric substrate material, as its characteristics (properties, thickness, shape, etc.) do not affect the fabrication process up to the point of embedding.

A schematic representation of the fabrication process for the stretchable electrical interconnections is shown in Figure 6.14. The fabrication is started by cleaning a glass substrate (a) and depositing a release stack (b). Next, polyimide is spin-coated (c) and metallization is sputter-deposited (d). The metallization is patterned using photolithography and wet etching (e), and covered by depositing another polyimide film (f). A hard mask is then applied (g and h) and both supporting polyimide films are patterned in a single dry etch step to form polyimide-covered meandering conductors (i and j). At a final stage of the fabrication process, the circuit is transferred to PDMS by a two-step embedding procedure (k and l).

6.3.3 Applications

As an illustration of the advancements in the technology, some application example prototypes are described below.

6.3.3.1 Wearable Light Therapy

Well-being applications demand unobtrusive treatment methods in order to reach user acceptance. In the field of light therapy, this needs to be carefully addressed because, in most cases, light treatment system size has to be significant with respect to human body scale. Once scaled up, standard flexible electronics (FCB) fail to conform to body curvatures leading to decrease in comfort. A solution to this problem demands a highly conformable technology based on our PCB-based stretchable circuit technology. In Figure 6.15, a demonstrator device for blue light therapy of wrist RSI syndrome is shown [12].

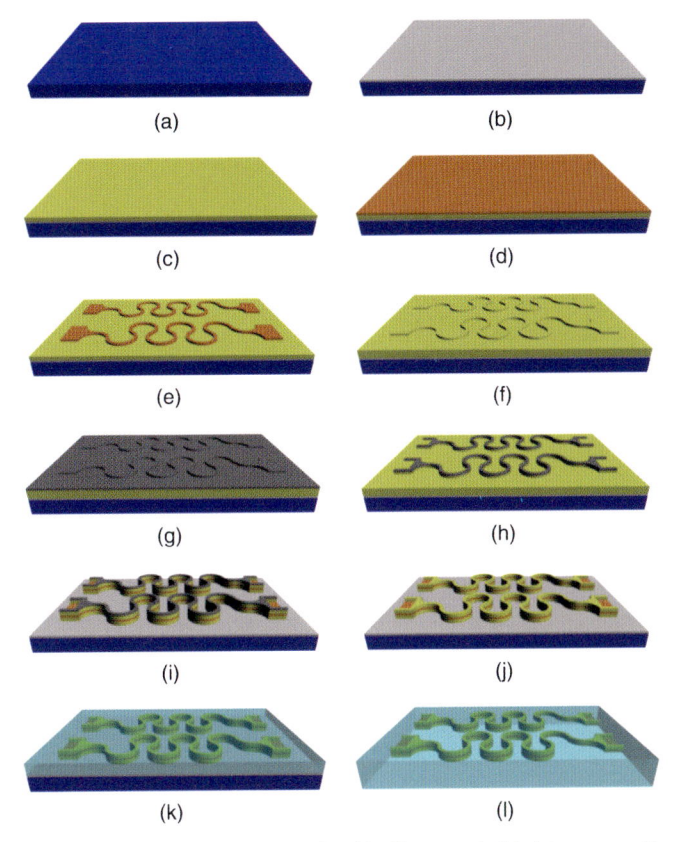

Figure 6.14 Fabrication process for thin film stretchable interconnections (not drawn to scale).

We have demonstrated that this wearable LED engine delivers breathability comparable with highly permeable textiles as well as mechanical compliance along with reliability under stretching deformations (120 000 cycles under 9% strain led to 1% efficiency loss). We have also shown that in this form factor the device allows for up to 15 min exposure with high energy visible wavelengths of 450 nm, with average irradiance of 12.8 mW/cm^2 (120 mW/cm^2 peak irradiances) over an area of 158 cm^2 in DC driving mode without exceeding skin thermal safety of 42 °C.

6.3.4 Stretchable Displays

Stretchable and conformable displays that can be stretched and bent into any direction provide new opportunities since they can conform to complex surfaces and locally stretch where needed. The fabrication of stretchable and conformable display technologies is a missing link toward realizing the next generation applications such as wearable displays in/on textiles and digital signage currently achieved by projector mapping. It could as well provide new opportunities to the

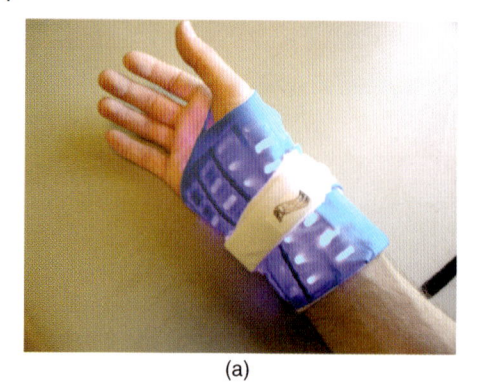

Figure 6.15 RSI wrap tear-down. (a) LED engine inserted into a textile wrap and worn on the wrist. (b) LED engine taken out of the textile wrap. (c) LED engine alone.

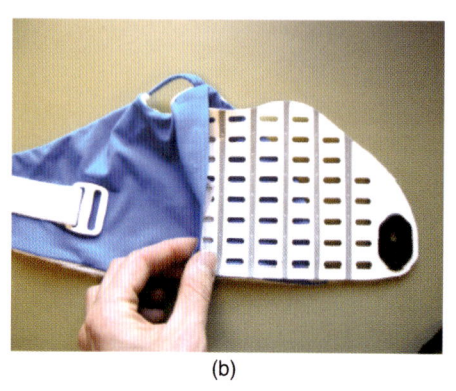

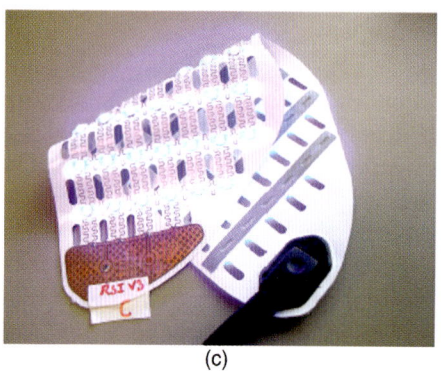

automotive industry where displays could be integrated on the A pillars, the steering wheel, dashboard, and even in the car seats.

A stretchable and foldable passive matrix-driven display has been realized using 45 by 80 RGB LEDs mounted on a four-layer stretchable PCB circuit embedded in a 200 μm thin polyurethane film [13]. The meander interconnections have been optimized with respect to their electrical and mechanical properties to provide a display with a 3 mm pitch between the pixels and a stretchability of up to 10%. At an operating supply voltage of 5 V, the brightness of the display exceeds 30 cd/m^2 (Figure 6.16).

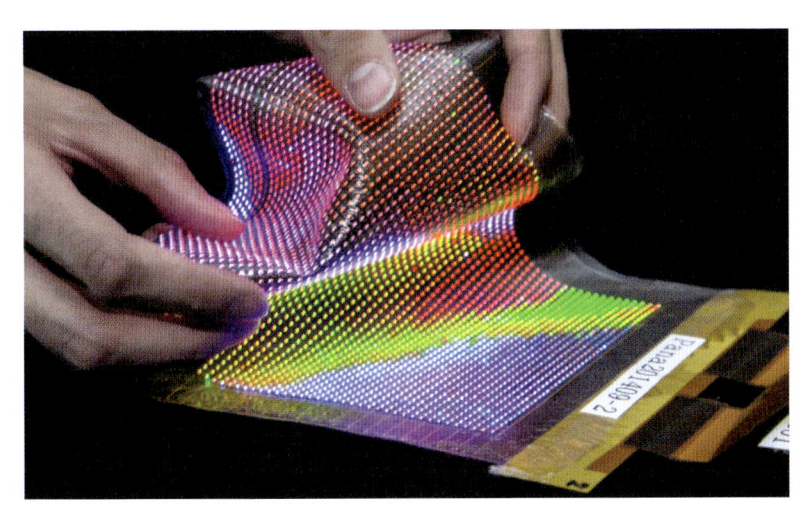

Figure 6.16 Stretchable display including 80 × 45 RGB LEDs with a pitch of 3 mm.

As a final example, Figure 6.17 shows a first proof-of-concept of a stretchable LED-based microdisplay using thin film stretchable interconnections. The base substrate consists of 0.5 mm sized "LED islands" that are interconnected to each other by 0.5 mm sized meanders for making the display stretchable. A multitude of bare die LEDs (CREE, size $200 \times 200 \times 50\,\mu m^3$) were flip-chip assembled onto the base foil using conductive adhesives. Thin sheets of thermoplastic polyurethane (50 µm) were laminated on both sides to encapsulate and protect the device. Initial stretchability tests show that the device can be stretched up to 10% without damage.

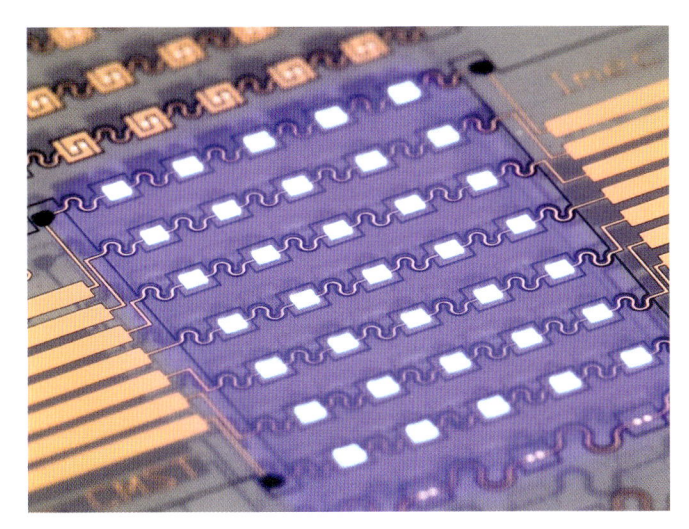

Figure 6.17 Stretchable LED microdisplay consisting of bare die LEDs assembled on a thin film stretchable interconnect substrate.

6.4 2.5D Rigid Thermoplastic Circuits

An alternative application for the circuit board-based stretchable circuits introduced in Section 6.3 is the integration of electronic circuits in thermoplastics, resulting in free-form rigid devices that can be manufactured in a flat state. This offers advantages such as late-stage configuration and usage of default PCB equipment [14]. These thermoplastic polymers can be processed by heating them past their glass transition temperature, at which point they are easily deformed using a wide range of thermoforming processes that operate in a temperature range of 120–200 °C [15]. The common variant called vacuum forming is illustrated in Figure 6.18; there a vacuum is used to draw the thermoplastic sheet toward the mold. After cooling the plastic retains its shape until heated again [16]. Combined with the low setup and tooling cost this makes these processes an attractive option for many applications (e.g., white goods and lighting). The key to integrating the existing elastic circuit technologies is replacing the silicone rubber with thermoplastic elastomers (TPEs) such as thermoplastic polyurethane (TPU). These materials show excellent adhesion to most rigid thermoplastics and circuit board materials and are available in film or pellet form.

The production of these circuits starts using the same method as described in Section 6.3. The key differences are the carrier board and the encapsulation process; the wax layer is replaced by a reusable high-temperature pressure-sensitive silicone adhesive (PSA) that does not stick to TPEs. In the meanwhile,

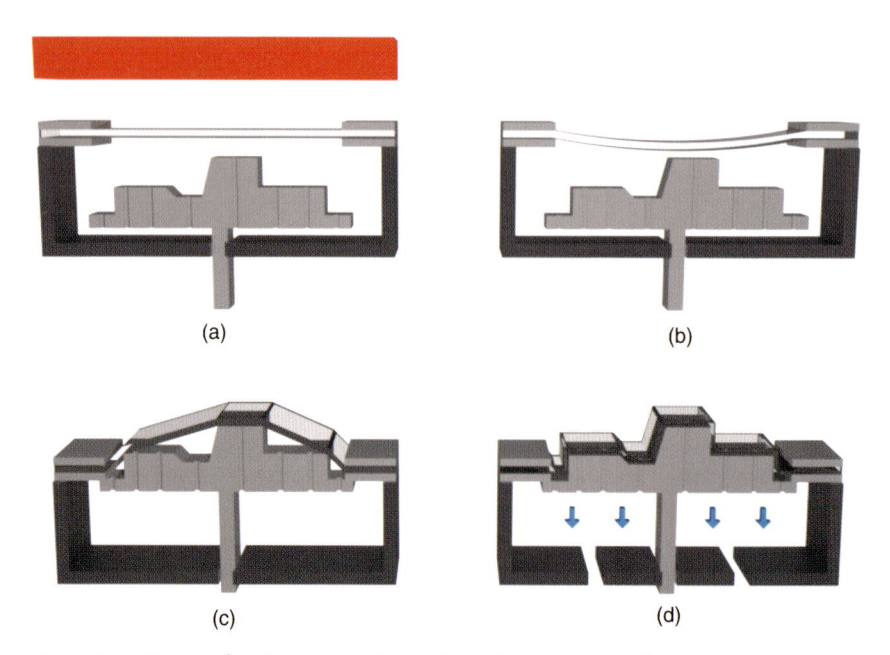

Figure 6.18 Vacuum forming process illustration. (a) Sheet is heated using an infrared heater past its glass transition temperature. (b) Heater is removed as the sheet starts to sag. (c) Mold is pushed up against the sheet. (d) Sheet is drawn against the mold using a vacuum.

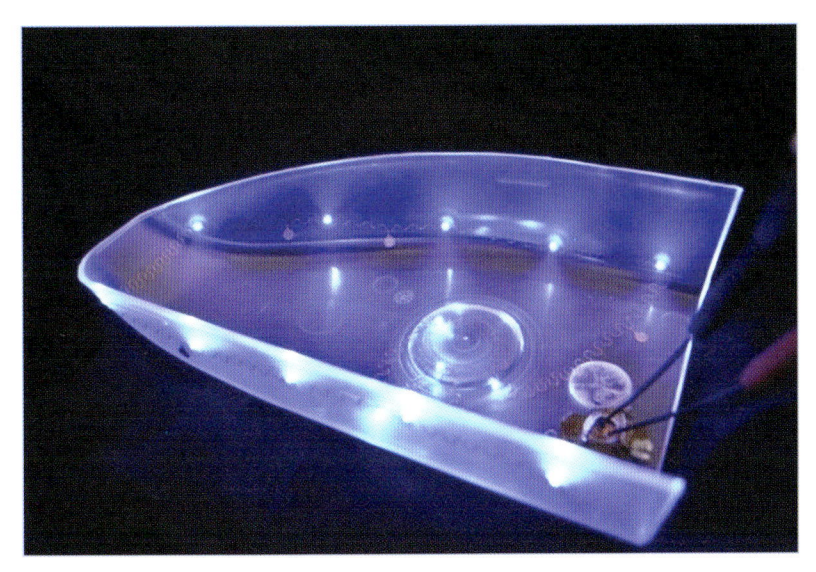

Figure 6.19 Thermoformed lighting module made by IMEC (Ghent, Belgium) in cooperation with IPC (Bellignat, France) for the EC FP-7 TERASEL project. In addition, this module was overmolded to demonstrate the further integration potential of this technology.

the liquid molding encapsulation step is replaced by vacuum lamination. For this a TPE film is placed on top of the circuit with a release foil on top; to prevent damage to the electronics components, a silicone foam rubber press pad is placed on top of the circuit to evenly distribute the pressure. This press book is then inserted in the vacuum press for up to half an hour with temperatures ranging from 90 °C up to 200 °C and pressures up to 20 N/cm^2. Because the TPE does not stick to the PSA, it is possible to easily peel the circuit from the carrier board. This stack is then laminated further using rigid thermoplastics such as polycarbonate (PC) or polystyrene (PS) to build a symmetric laminate that completely embeds the device.

Typical applications for this technology are readily found in the consumer market. Possible examples are intricately shaped control panels, 3D light sources, car interiors, and white goods. An example of such a light source can be found in Figure 6.19.

6.5 Large Area Textile-Based Circuits

Stretchable modules, as described in the previous sections, can be used to realize e-textiles. For an e-textile, it is important that the introduction of electronics has minimal impact on the properties of the textile that are important for the application. The use of stretchable modules in combination with integrated conductive yarns can be a solution for this.

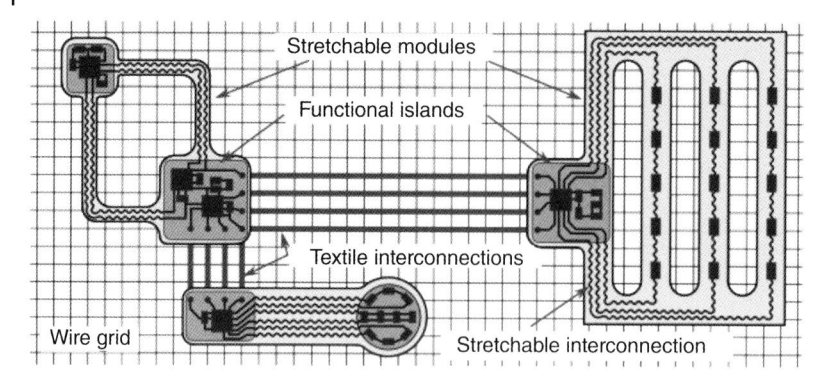

Figure 6.20 Stretchable modules interconnected via the wire grid in the fabric (concept drawing).

A concept drawing of an e-textile making use of stretchable modules is given in Figure 6.20. The different stretchable modules are made up of several functional islands, which are interconnected with each other by stretchable interconnections as was described before. Contact pads at the back of the module, which have the same spacing as the conductors in the textile wire grid, are then used for the textile-based interconnection. As most of the signal routing is kept inside the stretchable modules (where there is also more routing freedom), the number of textile-based interconnections can be limited. As a side note, note that the electronic components in the stretchable circuit technology are already protected and insulated by the used encapsulant, whereas the direct mounting of electronic components onto textile may induce reliability issues if no additional mechanical protection is provided.

6.5.1 Electronic Module Integration Technology

All electronic components must be interconnected and assembled to form a functional and operating system. The electrical interconnections in a system can be assigned to different levels of interconnection. At the lowest level, the interconnections are found inside the plastic packages of active and passive components. One level higher, these components are interconnected with each other at board level by the copper tracks on the stretchable modules. It is possible that a certain application requires multiple stretchable modules, distributed at different positions on a textile; for example, different modules including movement sensors for body posture tracking. This ensures that interconnection at module level may be necessary.

For our technologies, we start with fabrics that include conductive yarns. These yarns can be woven in a repetitive manner inside the fabric during the production process or embroidered on the textile after the fabric production process. Both approaches have their (dis)advantages, depending on the application. The conductive yarns are used to distribute electrical current to the different electronic modules. They are also used to allow communication between the different

6.5 Large Area Textile-Based Circuits

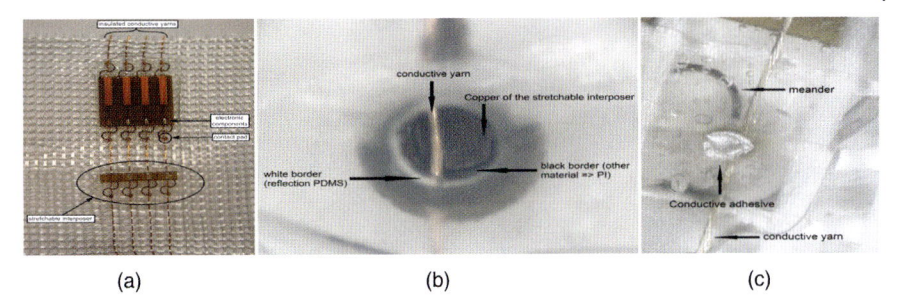

(a) (b) (c)

Figure 6.21 Interconnection method for connecting conductive yarns with electronic modules. (a) Electronic modules are glued on the fabric. (b) By laserablation, insulation is removed. (c) By dispensing conductive adhesive, an electrical connection is realized.

electronic modules. The yarns are selected in a way that they have the required conductivity for the power needs of the application.

Once the conductive yarns are in position, a fully encapsulated stretchable interposer is glued on the fabric in a way that contact pads on the stretchable interposers are aligned to the conductive yarns inside the fabric (Figure 6.21a). After attaching the stretchable interposer on the textile fabric, an electrical connection between the stretchable interposer's contact pads and the conductive yarns has to be made. This is done by a laserablation process, which removes the insulation from the yarn and the insulation from the stretchable interposer (Figure 6.21b). Once the contact pad is accessible and the yarn coating is removed, a conductive adhesive is used to realize the electrical interconnection between both (Figure 6.21c).

6.5.2 Applications

A smart bed linen has been developed that is able to measure moisture, for example, urine and sweat. This intelligent textile will improve the care work in hospitals, elderly homes, and home care. Geriatric nurses and hospital nurses will be informed when bedwetting occurred. Furthermore, bed linen hygiene can be automatically monitored, improving the patient's comfort. To be able for this detection, sensor yarns are woven into a high-quality cotton fabric.

The sensor yarns are used for moisture detection. The placement of the integrated sensor yarns is shown in Figure 6.22. The sensor yarns are non-insulated yarns that are put in groups of three. When moisture is applied on the yarns, electrical current can flow between the yarns, which is detected by the sensor nodes at the side of the bed linen. In Figure 6.23, the captured signal is shown when moisture is applied to the bed linen.

The stretchable interposers are fixed on the crossing points between the sensor yarns (weft) and signal communication yarns (warp) (Figure 6.24). A central communication unit is connected with the hospital environment. The target is to connect the bed linen with the normal used equipment in hospitals and elderly homes.

Figure 6.22 Smart bed linen capable of detecting moisture.

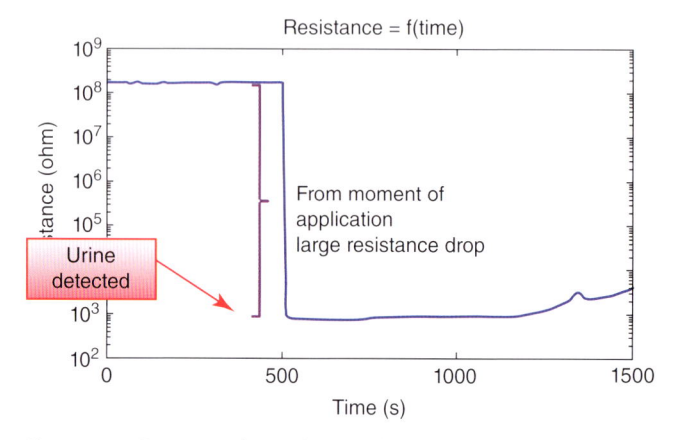

Figure 6.23 Resistivity drop indicating the presence of moisture between the conductive yarns.

The backside of the fabric is coated to avoid that moisture is running into the mattress. To have the coating on the backside, the textile character will not be disturbed on the topside where the patient lies. The bed linen will have the same lifecycle compared with normal bedsheets in hospitals. The bed linen has been evaluated on home laundry washing and is able to withstand at least 25 washing cycles at temperatures up to 90°.

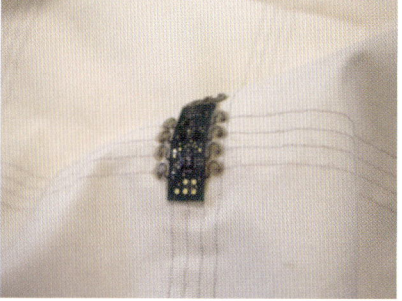

Figure 6.24 Stretchable interposers fixed at the crossing points between sensing yarns (weft) and communication yarns (warp).

6.6 Conclusions and Outlook

In this chapter we have shown how materials can be favorably used to embed functionality and intelligence under the form of electronic and sensor circuits. Longstanding circuit production technologies like rigid epoxy-based PCBs or polyimide-based flexible circuits will in future be replaced by alternatives that are more suitable to help in realizing the Internet-of-Things vision. At the same time also component packaging is evolving, resulting in ultracompact and low weight formats. Technologies are under development that allow for the integration of the electronics in the plastic shell or textile surface of smart objects. Development on an industrial scale of technologies for smart textiles or smart plastics requires the intensive cooperation of the electronics circuit production and assembly industry with polymer processing or textile manufacturing industries. This eventually will result in new value chains, where the competences of these different industries will be combined.

References

1 Coombs, C.F., Jr. (2008) *Printed Circuits Handbook*, 6th edn (1st edn, 1967), The McGraw-Hill Companies.
2 Fjelstad, J. (2011) *Flexible Circuit Technology*, 4th edn (1st edn, 1994), Br Publishing.
3 Christiaens, W., Bosman, E., and Vanfleteren, J. (2010) UTCP: a novel polyimide-based ultra-thin chip packaging technology. *IEEE Transactions on Components and Packaging Technologies*, **33** (4), 754–760.
4 Christiaens, W., Torfs, T., Huwel, W., Van Hoof, C., and Vanfleteren, J. (2009) 3D integration of ultra-thin functional devices inside standard multilayer flex laminates. European Microelectronics and Packaging Conference (EMPC).
5 Kuo, T.-Y., Shih, Y.-C., Lee, Y.-C., Chang, H.-H., Hsiao, Z.-C., Chiang, C.-W., Li, S.-M., Hwang, Y.-J., Ko, C.-T., and Chen, Y.-H. (2009) Flexible and ultra-thin embedded chip package. 59th Electronic Components and Technology Conference, pp. 1749–1753.
6 Hassan, M.-U., Schomburg, C., Penteker, E., Harendt, C., Hoang, T., and Burghartz, J.N. (2011) Imbedding ultrathin chips in polymers. STW ICT.OPEN 2011, Veldhoven, The Netherlands.
7 Priyabadini, S., Sterken, T., Cauwe, M., Van Hoorebeke, L., and Vanfleteren, J. (2014) High-yield fabrication process for 3D-stacked ultrathin chip packages using photo-definable polyimide and symmetry in packages. *IEEE Transactions on Components, Packaging and Manufacturing Technology*, **4** (1), 158–167.
8 Bossuyt, F., Vervust, T., and Vanfleteren, J. (2013) Stretchable electronics technology for large area applications: fabrication and mechanical characterization. *IEEE Transactions on Components and Packaging Technologies*, **3**, 229–235.
9 Rogers, J.A, Someya, T., and Huang, Y. (2010) Materials and mechanics for stretchable electronics. *Science*, **327**, 1603–1607.

10 Hsu, Y.Y., Gonzalez, M., Bossuyt, F., Vanfleteren, J., and De Wolf, I. (2011) Polyimide-enhanced stretchable interconnects: design, fabrication, and characterization. *IEEE Transactions on Electron Devices*, **58**, 2680–2688.

11 Hsu, Y.Y., Gonzalez, M., Bossuyt, F., Axisa, F., Vanfleteren, J., and De Wolf, I. (2011) The effects of encapsulation on deformation behavior and failure mechanisms of stretchable interconnects. *Thin Solid Films*, **519**, 2225–2234.

12 Jablonski, M. (2014) Conformable light emitting modules. Dissertation. Ghent University, Faculty of Engineering and Architecture, Ghent, Belgium.

13 Ohmea, H., Tomita, Y., Kashara, M., Schram, J., Smits, E.C.P., Brand, J., van den Bossuyt, F., Vanfleteren, J., and de Baets, J. (2015) Stretchable 45x80 RGB LED display using meander wiring technology. *SID Symposium Digest of Technical Papers*, **1** (46), 102–105.

14 Plovie, B., Dunphy, S., Dhaenens, K., Van Put, S., Vandecasteele, B., Bossuyt, F. *et al.* (2015) 2.5D smart objects using thermoplastic stretchable interconnects. *International Symposium on Microelectronics*, **2015**, 868–873.

15 Vanfleteren, J., Chtioui, I., Plovie, B., Yang, Y., Bossuyt, F., Vervust, T. *et al.* (2014) Arbitrarily shaped 2.5D circuits using stretchable interconnections and embedding in thermoplastic polymers. *Procedia Technology*, **15**, 208–215.

16 Throne, J.L. (2008) *Understanding Thermoforming*, 2nd edn, Hanser, Munich.

Part Four

Material Integration Solutions

7

Sensor Integration in Fiber-Reinforced Polymers

Maryam Kahali Moghaddam,[1] Mariugenia Salas,[2] Michael Koerdt,[3] Christian Brauner,[3] Martina Hübner,[1] Dirk Lehmhus,[4] and Walter Lang[1,4]

[1]University of Bremen, Institute for Microsensors, -Actuators and -Systems (IMSAS), Otto-Hahn-Allee NW1, 28359, Bremen, Germany
[2]Friedrich-Wilhelm-Bessel-Institut (FWBI) mbH, Otto-Hahn-Allee NW1, 28359, Bremen, Germany
[3]Faserinstitut Bremen e.V., Department of Composite Structures and Processes, Am Biologischen Garten 2, 28359 Bremen, Germany
[4]University of Bremen, ISIS Sensorial Materials Scientific Centre, Wiener Str. 12, 28359 Bremen, Germany

7.1 Introduction to Fiber-Reinforced Polymers

Nowadays, composite materials are well known as the material of the future in sectors, such as transportation, where reduction of weight is required for lighter designs. Previously, polymer composites used in military, energy and aerospace applications were generally used for a broad range of applications such as blades for wind turbines, in the building sector for bridges, for sports application or for special purpose manufacturing machinery like robot arms. The distinctive feature of this material is that it allows reaching tailored designs that cover the aspects of material efficiency, lightweight, high strength, high stiffness, and durability. The use of composite materials for intelligent systems could be achieved with the additional benefit of weight savings and in some cases reduction of manufacturing costs. Besides of all the mentioned advantages, on the other hand, production of composite materials can be more expensive compared to conventional materials like metals. Especially in mass production, these high material and manufacturing costs are an important factor that limits its industrial use. Therefore, adding an additional function like sensing during production or lifetime could be an enabling technology to reach higher efficiency. Sensor strategies can be resourcefully used; for example, during manufacturing of an aircraft wing made of composite material, information about the curing process could be available using the *online process monitoring*. Additionally, later on, these sensors could also be used for structural health monitoring (SHM) purposes to detect defects related to environmental loads.

Fiber composites can be basically described as reinforcing fibers embedded in a rigid polymer matrix. The main advantage of composite materials is that two or more materials are combined to reach resulting properties that are better than the properties of each single materials. The components are sharing their tasks; here the fiber takes mechanical loads and the matrix polymer fixes and supports the position of the fiber and transfers the stress to the fibers. The fibers used in composite materials can be small particles, whiskers, or continuous filaments. In almost all applications requiring high stiffness, strength, and fatigue resistance, composites are reinforced with continuous fibers rather than small particles or whiskers [1]. There are many different possibilities of combining fiber. Roving is the form in which fibers are supplied; it consists of filament strands wound into a package that can be woven into different fabric constructions, such as plain weave, twills, and satin weave.

The fiber materials are classified as natural fiber such as cotton and flax; organic fibers such as polyethylene (PE), polypropylene (PP), polyamide (PA), aramid, and carbon; inorganic fibers such as glass, and quartz basalt; and metallic fibers such as steel, aluminum, and copper. Most common fibers used in industrial applications are glass, carbon, and aramid fibers.

Moreover, the matrix system can be selected from different polymer systems classified into thermoset, thermoplastic, and elastomer. The most common types of matrix are thermoset systems with epoxy or polyester formulation. Typically, aerospace epoxy resins cure at 120–180 °C in an autoclave or closed cavity tool at high pressures, occasionally with a postcure at a higher temperature. Nevertheless, in the last years, a trend is observed toward thermoplastic systems such as polyetheretherketone (PEEK), polyphenylene sulfide (PPS) for aerospace applications or polypropylene (PP), and polyamide (PA) for automotive applications.

Most of the time composite materials are used in layered structures; where plies are combined to a so-called laminate. These laminates are draped over a complex structure in a way based on the mechanical loads affecting the part. As a consequence, some aspects need consideration in a multiscale approach. First, attention is directed to the micro level with fiber and matrix on the scale of micrometer. Concerns are later made in a mesoscale defined by the fabric textile structure, which builds the ply at a level of 0.1–1 mm. The consideration jumps to the laminate level (1–30 mm) and finally to the structural level depending on the size of the part (1–100 m). Therefore, the distinctive feature of composite materials is that the resulting properties depend not only on the selection of these different materials but also on the selection of the manufacturing process itself. The manufacturing process defines the quality and cost requirements leading to the desired fiber volume content; which is the ratio of fiber to matrix and defines the homogenized properties of a ply.

The largest proportion of carbon fiber composites used in primary class-one structures is fabricated by placing layer upon layer of unidirectional (UD) material to the designer's requirement in terms of ply profile and fiber orientation [2]. A scanning electron microscope (SEM) image of a UD glass fiber composite is shown in Figure 7.1a. For aerospace applications, ply stacking sequences typically involve a combination of three-ply angles: $\pm 0°$, $\pm 45°$, and $\pm 90°$ called quasi-isotropic (QI) [3].

7.1 Introduction to Fiber-Reinforced Polymers

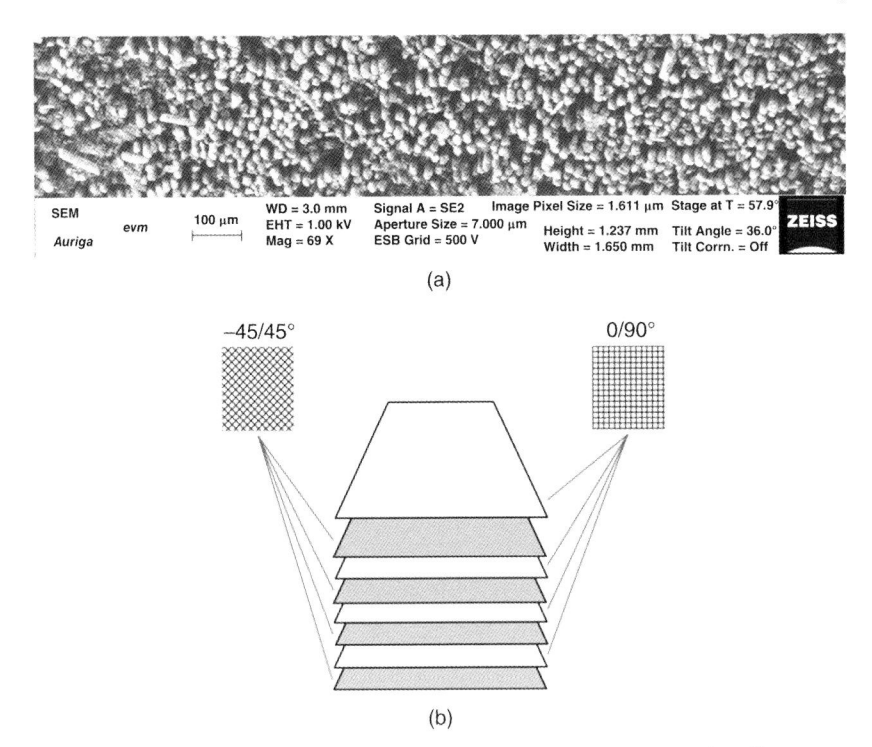

Figure 7.1 (a) Scanning electron microscope (SEM) image of an isotropic glass fiber composite. (b) Layup of quasi-isotropic fiber composite.

The angles are relative to the main loading direction of the applied load. Such a QI laminate is schematically shown in Figure 7.1b.

A variety of different processes are available to manufacture thermoset fiber-reinforced composite parts. In general, there are process methods using prepeg materials or dry textile materials. Prepregs are the textiles in which the fibers are preimpregnated with high-viscosity matrix systems. The dry textile is impregnated with low-viscous resin by infusion (vacuum driven) or injection (pressure driven). Dry textile-based processes are called liquid composite manufacturing (LCM) and can be classified into open mold concepts like resin transfer infusion (RTI) and closed mold concepts like resin transfer molding (RTM). To reach high fiber volume contents and a minimum amount of voids, accelerating resin impregnation times is usually done by additional pressure. This additional pressure can easily be performed in the case of a closed mold. In the case of an open mold, an autoclave is needed to add pressure on the vacuum bag. On the other hand, the RTM process has the advantage that a heating system can be applied by a heated press that has lower energy consumptions than an autoclave. A further development of the RTM process is to integrate the heating system into the mold itself using, for example, a fluid heating system. This is the so-called Fast RTM process. Using this technology, the uniform temperature can be applied with accelerated heat or cooling rates up to 30 °C/min [4,5]. Figure 7.2 shows the scheme of RTM and RTI in an autoclave.

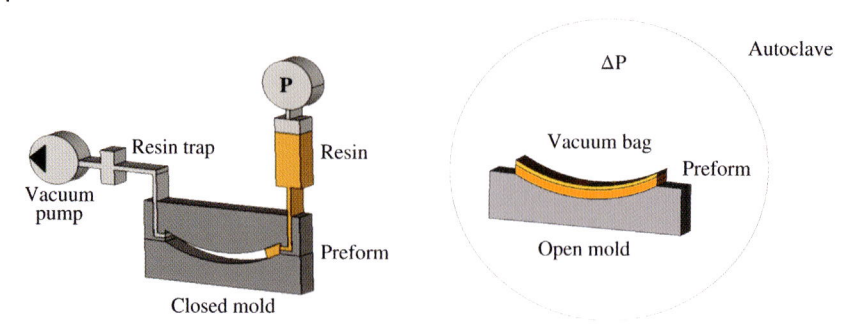

Figure 7.2 RTM process – RTI process [5].

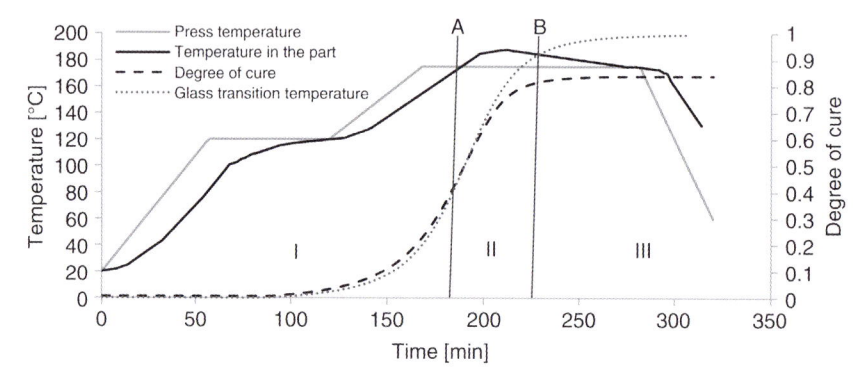

Figure 7.3 Curing process [6].

For example, in a classical RTM process, the following process steps have to be fulfilled. Dry textile preforms are placed into a mold, heated up to 100–120 °C prior to resin injection. Then the resin is injected into the mold. The mold will be heated to the curing temperature, which is 180 °C for RTM6 resin. The mold will keep this temperature for 2 h to cure the resin and then the mold will be cooled down to room temperature.

During the curing, the thermoset resin passes through three different morphologic states. The resin converts from a liquid (part I in Figure 7.3) to a rubbery state (part II in Figure 7.3) and finally into a solid state (part III in Figure 7.3). These morphologic changes are gradual, but from an engineering point of view, the transition can be defined as the gel point (part A in Figure 7.3) and the vitrification point (part B in Figure 7.3).

7.2 Applications of Integrated Systems in Composites

7.2.1 Production Process Monitoring and Quality Control of Composites

The quality of the final fiber-reinforced polymer (FRP) composite product is directly influenced by the manufacturing process. Mainly, the quality of a FRP composite is affected by the existence of dry spots, incomplete saturation of fiber with resin, porosity, voids and microvoids in fiber tows as well as under-curing of the resin. By monitoring the manufacturing process, from the flow of the resin in LCM processes to the curing process of the resin, the cost-effectiveness and optimum properties will be achieved. The most critical elements of the manufacturing process are information about resin flow, the hydrostatic pressure of the resin, curing/exothermal overheating, effects that change the fiber volume content, and process-induced deformations.

In vacuum-assisted resin transfer molding (VARTM), the resin infusion scenario depends on different factors such as number of resin inlets to the laminate, infusion or injection, and ventilation gates, having a point or line distribution of inlet and ventilation and triggering methods of the infusion or injection gates. These parameters determine the impregnation of the dry textile. Tracking of the resin flow can be foreseen by use of different simulation methods. To get the advantage of the simulation methods, the viscosity of the resin and porosity and permeability of the fiber should be known. The viscosity of the resin changes with the temperature and curing time. Finding out the permeability of the dry textile is also a hard task. Plentiful of researchers have studied various methods to measure and to analytically model the permeability [7–12].

The objective of an *online process monitoring* is to gain knowledge about what is happening in the process during composite manufacturing and to use this knowledge for process optimization.

Online process monitoring and direct methods to control the curing process of polymer composite structures are not commonly used in the industry. This missing monitoring methods lead to conservative manufacturing management and longer cycle times. The quality assurance has to be completely performed after the manufacturing process; which is of course time-consuming, expensive, and only a scalar quantity. The process history can only be reconstructed by the temperature. Using a robust cure monitoring system, it is possible to react on process by a control system and improve the productivity. Especially for aerospace structures, which have to fulfill a high-quality standard, online process monitoring is required to achieve the full potential of composite materials. Therefore, *in situ* cure monitoring methods are one step further to decrease manufacturing costs and simultaneously increase the quality.

In the last three decades, different curing monitoring methods like dielectric, resistance, piezoelectric, and optical measurements have been investigated by different researchers. The correlation of the electrical properties of the resin with the curing progress is required for the monitoring of composites manufacturing. The most common methods are dielectric analysis or measuring the electrical resistivity. The change of the resin properties from arrival and during curing is important.

- Dielectric-based systems have already been successfully used to follow the curing process using commercially available sensors like from Netzsch or Lambient Technologies LLC. Using theses dielectric sensors resin arrival during the infusion process and developing the degree of cure can be measured by the change in the ion conductivity.
- Resistance-based systems are similar to the previous dielectric systems but in general simpler to use because only resistance is directly measured [13].

- Piezoelectric methods apply an ultrasonic excitation from an emitter that is transferred to a receiver. A material during curing will change the speed of sound and the damping. Thus, the time of one period and the amplitude can be measured and analyzed. Piezoelectric sensors have the advantage that they offer results for many different materials compared to dielectric sensors. The disadvantage will be that mounted sensors have to be used and a tool sensor interaction has to be specified [14].
- Fiber optic sensors can be used to analyze the fabrication strains and, therefore, resin arrival, gel point, vitrification point, and change due to thermal expansion can be measured. A fiber optical sensor (FOS) can be placed between two plies and measure deformations by a change of a reflected wavelength shift. The advantages are that this type of sensors are robust and can be afterward used for measuring the impact of external loading. Different studies have been performed in Refs [15–17].

In summary, all these four types of methods are measuring different physical quantities and have their own unique advantages.

7.2.1.1 Monitoring of the Resin Flow

To monitor the resin flow, *nonembedded* and *embedded* sensing elements are used. Using a camera and ultrasound methods for monitoring the flow of the resin belongs to the *nonembedded* sensing processes. Using a camera is a common visualization technique used to detect the resin flow only in transparent molds [18–20]. Ultrasound transducers, consisting of the sender and receiver, can be placed in the mold for both resin flow and resin cure monitoring [14]. The direct reflection and direct transmission are two commonly used methods in RTM processing.

There are some *embedded* sensing methods to observe the resin flow. *Embedded* sensing comprises pressure sensors [21–24], optical fiber [25], and thermocouples [25,26]. To get a product with uniform specification, tracking the flow of the resin and the impregnation of the fibers is necessary.

7.2.1.2 Analytical Modeling of Resin Front by Means of Simulation

There are numerous studies on analytical modeling of fiber permeability that directly influences the resin flow. Some researchers tried to model the in-plane permeability of fabrics to find a predictive model [7–9]. Besides, the trans-plate permeability is studied [27,28] theoretically and practically to explain 3D permeability of fiber laminates. Another approach is finding the resin flow using equivalent permeability; for example, increasing the thickness of the preform to get equivalent permeability for different fabrics [10–12]. Darcy's law and Kozeny–Carman equation are widely used to demonstrate the impregnations of the fiber in porous material such as fabrics in FRP composite production [29,30].

7.2.1.3 Monitoring the Resin Curing

After full impregnation of the fibers, it is necessary to monitor the curing of the resin to guarantee a high-quality product, which can afford the expected mechanical properties. To measure the degree of resin cure various techniques are used such as measuring the *electromagnetic properties* by dielectric analysis (DEA) [31–36], electrical time domain reflectometry [37], *optical properties* (optical fiber refractometers [38], spectrometers [39]). Between these approaches for cure monitoring, DEA shows the biggest potential in online cure monitoring. Some advantages of DEA are one-sided measurement, high sensitivity, and the availability of cheap disposable sensors or reusable sensors, with the possibility to integrate a thermocouple inside the sensors. Both ultrasonic transducers and optical fiber methods have some disadvantages. The ultrasonic technique uses a sender and a receiver for detecting the state of cure in FRP manufacturing molds; there is a big influence of the fiber volume content on the ultrasound signal. The fiber optic can be broken easily due to mechanical treatment.

7.2.2 In-Service Applications of Integrated Systems

7.2.2.1 Use for Structural Health Monitoring (SHM)

Structural health monitoring is defined by the relevant regulatory document, the SAE's Aerospace Recommended Practice (ARP) 6461, "Guidelines for the Implementation of Structural Health Monitoring on Fixed Wing Aircraft," as "the process of acquiring and analyzing data from on-board sensors to determine the health of a structure" [40]. This definition does not explicitly ask for material-integrated sensing, but includes it as a possibility. From a capability point of view, SHM system performance has been described as rising in complexity from load monitoring to damage detection, damage localization, damage extent identification up to remaining lifetime prediction, self-analysis and, ultimately self-healing [41].

Based on their high performance-to-weight ratio, fiber-reinforced polymers are increasingly used in applications with a need for high levels of lightweight design. Prominent examples in this respect are aerospace applications. Here, the possibility of switching from an aluminum-centered design to composite materials has been demonstrated for large, twin aisle commercial aircraft by the Boeing 787 and the Airbus A350 alike. One significant side aspect of switching the primary structural material is the fact that new failure mechanisms irrelevant for metals need to be accounted for. An important example in this respect is the susceptibility of fiber-reinforced polymers to impact damage, which can, for example, be expressed in the form of delamination. Damage of this kind may occur without being visually detectable from the outside, for example, during the ramp check of a plane. This observation has led to the definition of the so-called barely visible impact damage (BVID) [42,43]. The concept implies that composite (used here as synonym for fiber-reinforced polymers, which in the aerospace industry typically translates to carbon fiber-reinforced polymers or CFRP) aerostructures need to be dimensioned in such a way that even under BVID conditions, the static ultimate design strength must still be borne at the usual safety margins.

Effectively, this means that the baseline for design is not defined by the healthy structure. The rationale behind this approach is that if such damage existed, and went undetected, the aircraft would still be as safe to fly as a healthy metal-based design. In other words, if means existed that guaranteed that if such damage was present – which is rarely the case – it would immediately be detected and the appropriate countermeasures initiated, allowing the safety margins for the structures in question to be significantly relaxed. Naturally, this would directly translate into weight savings, which would in turn lead to reduced fuel consumption and thus lower direct costs as well as environmental benefits. Thus, one of the main arguments in favor of developing SHM solutions for composite aerostructures is to achieve just that aim. A secondary motivation, demonstrated in Ref. [44] that can be sufficient to reach economic viability, is the control and scheduling of maintenance operations based on the availability of detailed information on structural state. A general overview of SHM solutions for composite materials has recently been published by Jian et al. [45].

A more detailed discussion of structural health monitoring as application concept for material-integrated intelligent systems is provided in Chapter 22. Classification of SHM systems distinguishes between operation and damage monitoring. The latter employs sensors to either directly detect damage (e.g., rip wire or electrical crack gauge approaches [46] comparative vacuum monitoring/CVM [47]) or through its effects on aspects like ultrasonic-guided wave propagation (e.g., Lamb wave methods) or global behavior of the structure (e.g., modal analysis [48]). Operation monitoring in contrast is aimed at directly or indirectly capturing the loads acting on the structure, derive a load history through incorporating their accumulation, and establish damage within the structure based on models that describe introduction and evolution of damage in relation to this load history. In either case, depending on whether effects are evaluated on local or global scale, a further distinction between local (e.g., CVM) and global monitoring is common practice. The actual generation of data provides a further criterion for classification of SHM systems: Passive monitoring employs sensors that merely listen, while in active monitoring, either separate actuators or sensors usable both in a sensor and actuator mode probe the structure by emitting a signal. This signal is either detected by further sensors after having based through part of the structure (pitch-catch mode), or its reflection at interfaces, surfaces, or damage sites is again recorded at its source (pulse-echo mode). These approaches are typically based on piezoelectric sensor arrays or piezoelectric wafer active sensors (PWAS), which introduce the ultrasonic waves via the piezoelectric effect that converts electrical charge change in mechanical vibration [49]. Coordination of several such sensors in a phased array allow directed probing of the structure. Among the types of ultrasonic waves that can be generated are so-called Lamb waves, that is, guided-plate waves. These waves propagate in two dimensions allowing for sparse sensor arrays, are easily excited and measured, and most importantly highly sensitive to most types of damage occurring in composites.

Table 7.1 provides an overview of sensor concepts that have been studied for SHM of composite materials. The list is not conclusive, but provides an idea of the variety of approaches studied and can be matched with the previous comments of sensors for production process monitoring, and with the integration examples provided later in this text.

As Table 7.1 shows, in their practical implementation, structural health monitoring concepts can be based on several different types of sensors and sensing principles. Many of these can also be employed in production process

Table 7.1 Selected examples of sensors and sensing principles employed in SHM of composite materials.

Type of sensor	Scenario description	Monitoring/sensing approach	Reference
Comparative vacuum monitoring	Detection of crack propagation below sensor patch through detection of loss of vacuum in evacuated channels (galleries) of the CVM sensor patch.	Damage monitoring, local monitoring, passive monitoring	[47]
Acoustic emission	Detection of acoustic signals related to the initiation or growth of damage within a composite wind turbine rotor blade.	Damage monitoring, local monitoring, passive monitoring	[50]
Fiber-optic sensors (FOS)	Review of FOS application in composites, scenarios discussed incl. strain measurements using fiber Bragg gratings (FBG) for local strain monitoring	Operation monitoring, local monitoring, passive monitoring (FBG scenario)	[51]
PWAS piezoelectric sensor arrays	Monitoring of a door surround structure (application example)	Damage monitoring, local monitoring, active monitoring	[52]
Interdigital transducer (IDT)	Introduction of surface acoustic waves through IDTs capturing of same in pitch-catch or pulse-echo mode	Damage monitoring, local monitoring, active monitoring	[53]

monitoring. This allows exploiting synergetic effects, as a single-sensor concept can be designed for both tasks.

Besides, process monitoring and quality control data can be employed to provide each individual component manufactured with an individualized baseline on which in-service monitoring can build. This way, the accuracy of information gained from the SHM system can be increased considerably.

An aspect of sensor integration that is of significant importance for composite materials, and even more so in the case of a monitored component, is the question of the sensor footprint. As already discussed, SHM implementation indicates that the structure equipped thus faces higher level mechanical performance requirements. In composites, an integrated sensor or electronic system constitutes for one thing, a possible failure initiation site. Besides, the sensor may locally reduce structural characteristics, for example, via influencing the matrix curing process, or by causing fiber undulations. For this reason, limiting the size of the sensor in all possible dimensions bears a high importance for sensor integration in composites. This issue has been discussed both by Lang *et al.* and by Dumstorff *et al.*, who introduced the concept of function scale integration: The notion is to drastically reduce the volume of sensors to what is absolutely indispensable for maintaining functionality, up to the level of so-called functional nets [54,55].

7.2.2.2 Use As Support to Nondestructive Evaluation and Testing (NDE/NDT)

The damage detection and structural performance evaluation approaches described in the previous section have the potential, once real-time, on-board data or even fully material-integrated data evaluation is available, to continuously survey composite structures, to (re-)set the timing for maintenance, and in case of damage detected during operation, to redefine the allowable service envelope in order to maintain safety levels. Naturally, such application requires extremely high levels of confidence. Besides, if operational safety of the monitored structure is directly based on the SHM system, it affords measures and procedures that can cope with partial system failure, which makes system implementation an even more demanding task. For this reason, it is not unlikely that in-service application of material-integrated structural sensing for composites may first be practically realized as support measure to nondestructive testing. On a lower level, such application will save a considerable amount of cost, if, for example, sensor placement in hard-to-access locations will save disassembly time and effort during regular maintenance. From a structural design point of view, a higher level advantage would be that provision of material-integrated sensing as an NDE/NDT tool could allow designing structures not acceptable today for lack of suitable testing approaches.

Finally, conceiving the SHM system for use during maintenance only would eliminate problems caused by the superposition, in service, of damage-related signals with those originating from typical service situations (e.g., engine vibrations) and environmental influence (e.g., temperature, humidity). The dependency of composite material properties on environmental influences is well known, and the added effect of temperature and humidity variation on SHM sensing principles like Lamb wave propagation have recently been discussed in detail by Schubert *et al.* [56,57]. A maintenance environment with greatly reduced noise and controlled environmental conditions would thus greatly improve the performance of the SHM system in terms of characteristics like probability of detection (PoD).

7.3 Fiber-Reinforced Polymer Production and Sensor Integration Processes

7.3.1 Overview of Fiber-Reinforced Polymer Production Processes

There is a multitude of techniques to fabricate fiber-reinforced composites. Selected examples are described further, accompanied by case studies highlighting sensor integration techniques associated with them.

The simplest disambiguation of processes distinguishes between short, long, and continuous fiber-reinforced composites. Typically, the following rough measures are used to define these classes:

- Short fiber-reinforced polymers (typical fiber lengths <1 mm, processing similar to nonreinforced polymers)
- Long fiber-reinforced polymers (typical fiber length 1–50 mm)
- Continuous fiber-reinforced polymers (typical fiber length >50 mm, highest performance materials, dedicated specialized manufacturing processes)

The mechanical characteristics of a fiber-reinforced polymer are controlled by the materials combined, but also by factors like fiber volume content, fiber orientation, or the type of fabric of the reinforcement. An overview of fiber and matrix properties is given in Tables 7.2 and 7.3, respectively. In the present context, continuous fiber-reinforced materials are of greatest importance, since their use implies an application demanding the highest performance levels. Naturally, these are also the applications that profit most from the detailed monitoring in production and service offered by material-integrated systems. For this reason, the following brief overview of fiber-reinforced polymer production processes is focused on this group of materials.

A comprehensive overview of composite material manufacturing processes has been provided by Herrmann *et al.* [59] as part of the work by Lehmhus *et al.* [60]. Since a composite is by definition made up of the two components matrix and reinforcement, a fundamental distinction between production processes can be based on the question at which point in the process both components meet. Effectively, this leads to a distinction between dry fiber and prepreg processes. During processes of the former class, fibers are typically positioned first, and the matrix is added – usually in a liquid state – in a subsequent step. In contrast, in so-called prepreg processes, the fibers are already impregnated with the matrix material prior to the part shaping process. An overview of types of fibers and matrices is provided in Tables 7.2 and 7.3. Naturally, in terms of the matrix, the type once again influences production processes, as, for example, thermoplastic matrices allow forming operations.

Separated by the matrix being either thermoset or thermoplastic, and by the positioning of the process in the manufacturing chain, the main fiber-reinforced polymer production processes are as follows[59]:

- Preforming processes (dry fibers)
  - Tailored fiber placement
  - Continuous preforming
    - UD braiding
    - winding–draping process
- Composite manufacturing processes
  - Thermoset matrix
    - Dry fiber processes
      - Hand/manual layup process
      - Liquid resin infusion process
        -- (Modified) vacuum infusion
        -- Seeman composites resin infusion molding process (SCRIMP)
        -- Vacuum-assisted processes (VAP)
      - Resin transfer molding processes
        -- Fast RTM
        -- High-pressure RTM
      - Pultrusion processes
        -- Pultrusion
        -- ACROSOMA process
      - Filament winding process

Table 7.2 Some types of reinforcing fibers used in fiber-reinforced polymers, data based on Ref. [59].

Fiber Type	Glass			Carbon[a]			Aramid (here: Kevlar)		
	E	S/R	M	HT	HST	UHM/HMS	29	49	149
Density (g/cm^3)	2.55–2.6	2.49–2.55	2.89	1.74–1.8	1.78–1.81	1.8–2.18	1.44	1.44	1.47
Young's modulus (GPa)	73	87	125	228–240	235–250	440–830	67–83	124–140	185
Tensile strength (MPa)	2500–3000	4050	—	3400–3600	3850–5000	2100–2200	3620	3620	3440
Breaking strain (%)	3.5–4.5	5.4	≈5.5	1.0–1.6	1.65–2.1	0.38	3.3–4	2.3–2.9	2

a) Precursor: PAN.

Table 7.3 Some types of matrix materials used in fiber-reinforced polymers [58].

Class of matrix material	Thermoset			Thermoplastic		
Type	Polyester	Epoxy	Polyimide	PP	PC	PEEK
Density (g/cm^3)	1.1–1.5	1.1–1.4	1.43–1.89	0.9	1.06–1.2	1.32
Young's modulus (GPa)	1.2–4.5	2–6	3.1–4.9	1–1.4	2.2–2.4	3.6
Tensile strength (MPa)	40–90	35–135	70–120	25–38	45–70	92–100
Elongation (%)	2–5	1–8.5	1.5–3	300	50–100	150
Glass transition temperature (°C)	50–110	50–250	280–320	−20 to −5	133	143

- Prepreg processes
    - Automated prepreg layup processes
        -- Automated tape laying (ATL)
        -- Automated fiber placement
- Thermoplastic matrix
    - Automated stamping of TP blanks
    - Thermoplastic laminates made of hybrid textiles

Wet/hand layup is widely used for many years for composite fabrication. Dry reinforcement textiles are laid up into a mold and manually impregnated with resin usually using brushes and rollers. With the hand layup method endless reinforcement fibers can be handled and higher fiber content than with spray layup, a long fiber process, is achieved. Vacuum bagging is a further advanced technique to get higher fiber-to-resin ratios than with hand layup. In this method, a vacuum bag is used in order to draw air and excess resin out of the stacked resin-soaked plies.

Vacuum infusion processes may serve as example for techniques of this kind. Figure 7.4 illustrates the process in which dry fiber layers are positioned on a tool

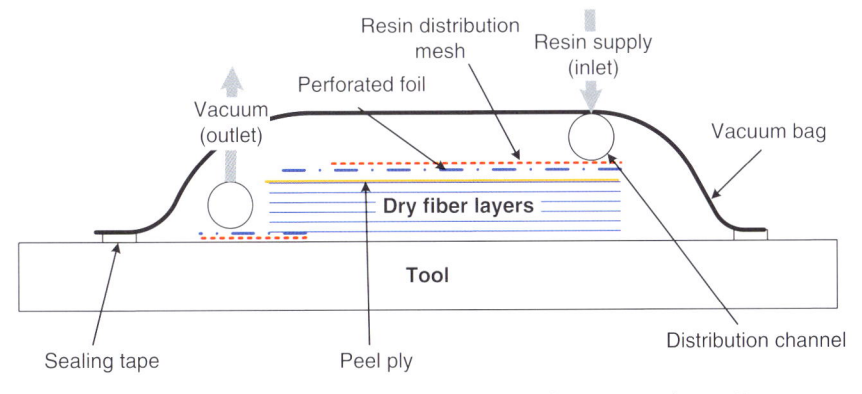

Figure 7.4 Modified vacuum infusion process. Schematic layout according to Herrmann et al. [59].

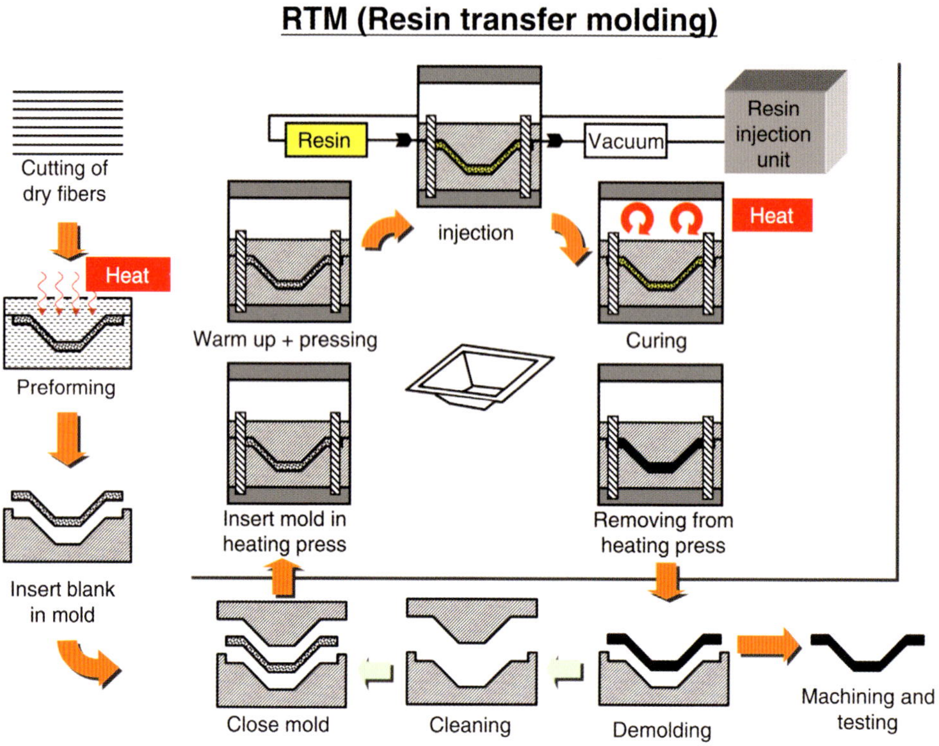

Figure 7.5 Basic principle of the resin transfer molding (RTM process) according to Ref. [59].

that represents a negative of the part's surface. The whole buildup, including, for example, the resin distribution mesh and a peel-ply for easy removal of the latter and any other sacrificial layers is enclosed by a vacuum bag with a resin inlet through which the matrix material is sucked into the vacuum bag. The resin distribution mesh creates a resin flow across the top surface of the layup, with parallel infiltration in thickness direction. Once infiltration is complete, the resin flow is stopped and temperature applied to initiate curing of the matrix. Figure 7.4 schematically illustrates both layup and tools [59].

Resin transfer molding processes are characterized by a closed mold in which the dry fiber preform (which may in turn be generated using, for example, *tailored fiber placement* processes) is placed. The liquid resin is injected into the mold, where it infiltrates and saturates the dry fibers. Curing of the resin is also initiated within the mold to consolidate the part and thus allow ejection. Vacuum assisted processes (VARTM) create a vacuum that supports, for example, a plunger in forcing the resin into the mold cavity. Figure 7.5 depicts the various steps in producing an RTM part, in this case following a VARTM sequence [59].

In *pultrusion* processes, preforming and resin infusion are done in a setup adapted to continuously produce profiles with different cross sections. The process can be fully automated. The production line itself has a fiber storage, for example, equipped with creels, on one side, and alternating grippers which pull the profile and thus eventually the fiber rovings. In between, the rovings are preoriented toward the intended cross section, then pulled through a liquid resin bath and thus saturated with the later matrix material. The forming tool or die is next in line and finally shapes the cross section of the impregnated fibers. Besides, through introduction of heat, it also initiates the curing of the resin. Thus, the profile is consolidated when it reaches the aforementioned grippers, which exert the necessary longitudinal force to move the material through the individual processing stages. Variants of the process use resin injection in the die or additional, movable furnaces or heaters to finalize the curing before the profile reaches the grippers. Cross sections of the profiles may either be open or closed.

Filament winding uses dry fibers that are passed through a resin supply before being led through an orifice or delivery eye from which they are transferred tangentially and under strain onto a rotating mandrel on which the resin-impregnated fibers are thus wound up. As the delivery eye is movable parallel to the axis of rotation, the direction of winding can be adjusted relative to the latter. This allows tailoring the orientation of fibers in accordance with the demands of the application. Once winding is completed, the matrix resin is cured either via furnace or radiation heating. If the part geometry features reentrant angle and thus forbids simple removal of the mandrel, the latter is made of water-soluble materials.

Prepreg processes are based on semifinished fabrics that have been impregnated with the matrix resin and can then be laid up on a single-sided tool. Fiber placement in manual or automated form can be used to define the build-up and thus the properties of the composite. The layup is then enclosed by a vacuum bag, which is evacuated prior to the prepreg being placed inside an autoclave, where temperature and pressure serve to consolidate the part and cure the resin. Ramakrishnan *et al.* named temperatures between 120 and 200 °C and pressure levels of roughly 100 psi, which translates to 6.89 kPa, as typical processing conditions [51].

Automated stamping of TP blanks uses preconsolidated sheets of continuous fiber-reinforced thermoplastic polymers, so called "organic blanks," and provides them with their final shape through the combined action of temperature and mechanical forces. Effectively, the blanks are heated toward the softening temperature of the matrix and shaped using a forming press. Typical values of temperature and exerted pressure for high-performance thermoplastic matrices like PEI, PPS, or PEEK are 300–420 °C and 4.5 MPa, respectively. Homogeneous heating can, for example, be achieved using infrared radiation [59].

7.3.2 Sensor Integration in Fiber-Reinforced Polymers: Selected Case Studies

Integration of sensors in fiber-reinforced composites can rely on a large variety of techniques that are naturally influenced by the composite manufacturing process itself. In dry fiber processes, for example, sensor integration is usually linked to the preforming process, that is, the positioning and lay up of the reinforcement, to which the sensor can then be added. Since the basis of the layup process is 2D fabrics, flat sensor patches can easily be positioned between the layers and thus be integrated in the final material. Similarly, techniques, such as embroidering, can be used to attach nonplanar sensors to individual layers. Direct deposition of sensors on the fabric has also been evaluated. The following section contains examples of such sensor integration techniques.

The integration of fiber optical sensors in fiber-reinforced polymers must seem an almost natural approach, since sensor and reinforcement share the same geometry. Nevertheless, the apparent ease is alleviated to some degree due to the fact that at least in carbon fiber-reinforced polymers, the typical size of the sensor exceeds that of the matrix by at least one order of magnitude. Ramakrishnan *et al.* discuss several ways of tightly integrating a fiber optical sensor within a polymer matrix composite [51]. The examples they cite range from the manual and prepreg layup processes to filament winding and complex 3D part production.

Manual and prepreg layup both allow simple placement of sensor fiber on the lower of the two fabric layers between which the sensor is meant to be integrated. Depending on the load case to be monitored, positioning in the fabric stack's center plane is either beneficial (in case no bending loads are meant to be recorded) or detrimental (in case bending load is of primary interest). In addition, Ramakrishnan *et al.* observe that typically some prestraining of the sensor fibers is applied to avoid bending as well as deviation from the intended position [51]. This procedure has some practical implications, since layup for shaped parts is done on a mold half representing a negative of the final part geometry, which means that prestraining may be difficult if the mold has a concave shape. During consolidation of the part, for example, by vacuum bagging or in an autoclave, the liquid matrix resin will enclose the optical fiber and thus tightly embed it. It may be noted that in carbon fiber-reinforced composites, the size (diameter) of the reinforcement tends to be significantly smaller than that of the sensor (typically 125 µm), which thus represents a potential weakness within the material due to effects like locally lowered volume content levels of reinforcement fibers and reinforcement fiber undulations.

Koerdt *et al.* report about the production of polymer fiber-based fiber-Bragg grating-type sensors, and the inscription of the sensor in the fiber, which is made of CYTOP™, an amorphous fluoropolymer, using UV light of 248 nm wavelength. The advantage of the suggested sensors in contrast to conventional glass fiber sensors is the greatly increased failure strain, which reaches 10% for the perfluorinated polymer optical fibers under scrutiny and is assumed to allow strain measurements up to a level of 6%, which even exceeds typical strain gauges. Thus, these materials may alleviate problems observed in production and use of conventional optical glass fibers due to their brittleness. The embedding process has been evaluated by Koerdt *et al.* in a vacuum infusion process: The polymer optical fibers were placed between uniweave glass fabric layers and oriented parallel to the reinforcement fibers. A standard, transparent room temperature hardening resin was used as matrix (CYTOP glass transition temperature is 108 °C, limiting the application temperature of the smart composite to roughly 70 °C), and integrity successfully tested after curing by coupling red laser light into one of the free ends of the polymer optical fiber [61].

Degrieck *et al.* have investigated fiber optical sensor integration in pressure vessels produced by filament winding. The sensors used were of fiber Bragg grating (FBG)-type and embedded within outer of the final three hoop windings (winding angle 90°) covering the initial polar windings with winding angles of 12.7 and 16.5°, respectively. The composite itself was based on glass fibers prestressed during winding. Curing of the matrix, Araldite LY 5052 epoxy with hardener HY 5052, was done for 15 h at 50 °C. The mandrel was made from water-soluble materials to allow facile removal following the winding process [62]. Kang et al. have studied a similar approach, also looking at pressure vessels as application scenario. However, their application concept includes production monitoring and multiple sensors, stressing the multiplexing capabilities of FBG sensors. Furthermore, the positioning of the sensors is within the inner layers of the pressure vessel. The pretensioning of the fibers is seen as one specific problem in the study, because it cannot be transferred directly to the sensor fibers for fear of causing excessive fiber breakage among them. In fact, after initial tests that saw only 1 of 4 sensor arrays survive production and handling, Kang et al. concentrated their research efforts on better mechanical performance of the optical fibers and protective actions effective during production. Measures taken include a change of the optical fiber type to eliminate the need for the mechanically degrading hydrogen loading, the elimination of splices in the sensor fiber, and a strengthening of the optical using acrylate coating and adhesive films. Altogether, these procedures helped to greatly improve survivability of the sensors during embedding. In a final evaluation of the embedded sensors, measurement results agreed well with externally applied electrical strain gauges used as reference [63].

Hufenbach et al. developed a process that initially creates piezoelectric element with interdigitated input electrodes encapsulated in a thermoplastic polymer foil. The material of the electrode is either Cu in the case of a PEEK carrier film or Ag for the PA variant. The production process is conceived to allow simple upscaling to large-scale series production [64]. When integrating these so-called TPM-modules in fiber-reinforced polymers with matrices matching the TPM substrate material, a homogeneous integration is achieved that eliminates a possible weakness in the buildup originating from the dissimilar material interface between matrix and substrate that is common in conventional solutions for piezoelectric patch integration like the SmartLayer™ approach, which builds on a polyimide substrate [65,66]. A suitable processing route for this type of sensor integration would be the automated stamping of TP blanks, with the TPM module secured between blanks prior to the combined forming and consolidation step.

Boll et al. studied cure monitoring of fiber-reinforced polymer matrix resins using an interdigitated electrode (IDE) sensor. The proposed sensor follows the principle of function scale integration as formulated by Lang et al. [55] in being tailored to a thickness of no more than 6 µm, which is in the same order of magnitude as the carbon fiber filaments of the carbon fiber-reinforced polymer studied. Furthermore, it features cut outs that allow interconnection of the matrix across the sensor plane. In view of processing conditions, including a pressure of 6 bar and temperatures of up to 200 °C, polyimide with a tensile strength of 390 MPa and heat resistance up to 500 °C was selected as substrate material. The sensor itself uses MEMS techniques. The actual sensor area is 18.4 mm^2 and contains 1024 electrodes with equal width and spacing of 3 µm. For integration, the sensor was placed between four layers of unidirectional dry carbon fiber fabric. As part production process, resin transfer molding has been selected. In the process, the fabrics were infiltrated with standard RTM6 epoxy resin at a temperature of 85 °C. For curing, the closed mold was held at a temperature of 180 °C. No data about the actual pressure during processing was reported besides the aforementioned limit of 6 bar. The study confirmed the suitability of the sensors for cure monitoring for a sensor miniaturized to avoid detrimental effects on macroscopic mechanical properties of the composite. The outlook section proposes a further reduction of the sensor footprint achieved via cut outs in between the individual electrodes, allowing matrix resin to permeate the sensor to an even higher degree [36].

Dumstorff et al. have studied the integration by printing of a strain gauge sensor in a fiber-reinforced thermoset. The process studied relies on dry fiber processing in its initial steps and follows the sequence described further: First, an epoxy insulation layer is printed onto the woven dry fiber ribbon. This base layer is then cured, during which process it partly infiltrates the fiber bed, sinking into it. A second epoxy layer is then deposited on top of the first, cured again, and a strain gauge-type sensor plus the necessary interconnects and contact pads applied to it using screen printing. The structure is then completed by a final epoxy insulation layer and can then be integrated in a CFRP structure in such a way that the contact pads are still accessible. In the original study by Dumstorff and Lang, the process chosen for material integration is prepreg based, that is, the sensor plus fiber-reinforced substrate is positioned between prepregs that are then cured, forming a beam-like structure supported to further mechanical testing, for example, under bending loads. Positioning of the sensor in thickness direction is outside the central neutral plane, making sure that bending loads, too, cause a measurable change in resistance. The material used for sensor and interconnects is a commercial silver conductor paste. The process allows producing the strain gauge conductors with a width of 200 μm and interconnects with a width of 1 mm. Their thickness typically reaches 5–10 μm [67]. Further studies on integration of printed sensors in glass, carbon, and hybrid fiber-reinforced polymer layups have been conducted at the Fraunhofer Institute for Manufacturing technology and Advanced Materials, Fraunhofer IFAM. In this case, too, processing of prepregs provides the basis for printing insulation layers and sensors. A significant deviation of this approach from the aforementioned is the use of maskless printing techniques like Aerosol Jet™ or inkjet printing. The respective techniques typically reach higher resolutions and can thus help to further reduce the "wound" in the material that an embedded structure may represent [68].

Braided composites containing fiber optical sensors have been scrutinized by Yuan et al. In their study, the sensor fibers have been cobraided with the reinforcement fibers and are thus aligned with them, causing limited perturbation. A secondary effect of cobraiding is that the direction of measurement of the sensors thus coincides with the expected primary direction of internal loads, according to which the reinforcement is oriented [69]. Also, in view of the production of 3D composite parts, Antonucci et al. have examined integration of fiber optical sensors during stitching of fiber preforms. The motivation of fiber sensor integration is to evaluate the permeability of different types of fabrics for a better understanding of the boundary conditions of resin infusion- and injection-based processes. The material preform used in the present study consists of eight stacked unidirectional plies stitched together in thickness direction. The plies themselves consist of dry carbon tows with matching orientation. These are stitched together in a zigzag pattern with a pitch of 5 mm. Fiber optic sensors functioning according to the Fresnel reflection principle are stitched to some of the preforms and thus integrated between plies 1 and 2, 4 and 5, and 7 and 8. Their orientation is perpendicular to the reinforcement fiber and thus also to the main resin flow direction, which makes sense because the main objective is monitoring the progress of resin flow and explicitly not in-service monitoring of the composite under mechanical loads. The actual embedding of the sensors occurs during resin injection and curing. The resin used was a standard RTM6 material, which was injected at 80 °C with the semimold heated to 100 °C. Curing was done in two steps, starting with 75 min at 120 °C and a subsequent holding at 180 °C for 2 h. Size of the final part is 150 mm × 150 mm. The effect of the transversal fiber orientation on imperfections and defects in the composite has not been recorded [70].

7.4 Electronics Integration and Data Processing

The electronics essential for each application will vary according to the data being processed. For instance, higher frequencies or low signal amplitudes can reduce the possibility to implement some analog-to-digital converters. Depending on what is going to be measured during manufacturing the electronics can be implemented within the sensing node, in a cluster of several sensors, or where possible the electronics might be left entirely outside. The actual limitations for implementing certain electronic components are mostly physical, like those high temperatures and forces that occur during manufacturing. Other special considerations are to be taken in structural health monitoring (SHM) regarding size, weight, and data volume. For example a piezo disc element used for Lamb wave generation and detection is less than 1 mg and 1 cm in weight and diameter, respectively, and they are usually surface bonded every 15 to 10 cm; however, adding electronics, power sources, and wires necessary for communication makes the system undesirably large and heavy. This is why one of the main issues is how energy comes in and signals go out of the composite material. Not only are smaller sizes desired in order to have the least influence on the mechanical behavior of the structure, but also for applications, such as aerospace and wind energy, it is highly recommended to maintain the lightweight of the structure, so dense wiring presence is unwanted. Therefore, wireless sensor networks are of high interest nowadays, yet existing electronics for wireless transfer are still much larger than the sensors [54]. Radio frequency identification (RFID) is one of the available technologies that could be easily adapted for material integration. The market ready solutions can usually result in complex systems that by no means can be integrated within any material.

An approach being evaluated is the use of the inductive coupling similar to RFID technology, but keeping the data transmission analog [71]. This is a simple solution realized by using planar coils, but that can only be surface applied when integrated in CFRP due to the electrical conductivity of the material that causes shielding. Depending on the application the data rates may vary. A problem is that at low frequency the coils get larger; nevertheless they actually work better for CFRP structures. An example of such technology achieved at Fraunhofer IIS [72] is the embedding of high-frequency RFID coils and even Antennas within GFRP.

Furthermore, new technologies such as flexible electronics are becoming more attractive to be integrated within the FRP material. Examples of such technologies are flexible printed circuits, stretchable flexible networks, foil technology, and polymer waveguides. Future trends point toward thin ASICs and system on chip (SoC). These approaches, however, will only be economically viable in large-scale production.

7.4.1 Materials Integration of Electronics

Embedding electronics in composites is of major interest in future products such as smart structures. As mentioned before, some of already embedded devices are strain and acoustic wave sensing sensors; however, there is a need to add computing power to such systems. Processing the sensor signal without long interconnects between sensor and electronics reduces signal interference and the need for shielding. Furthermore, low-level data processing and filtering can reduce the amounts to be transmitted from the sensor [54]. For self-sufficient operation of the sensor system, the power source should be integrated with the circuit where the available energy has to support the lifetime of the device. Another approach is energy harvesting. To handle this power issue, wireless power transfer can be implemented, as explained in Section 7.4.2, but also ultralow power circuit techniques can be used for circuit design.

During the embedding process the electronics have to withstand thermal and mechanical stress. The electronic systems need to endure temperatures up to 180 °C and high pressures during the manufacturing processes of FRP. Higher operating temperatures and extreme environmental conditions may corrupt the designs based on CMOS technology, for example, the shifting of operating points in analog circuits. Therefore, new design techniques and progress in semiconductor technology are needed.

Moreover, the influence on the mechanical strength due to the presence of foreign objects needs to be considered, simulated, and tested. Embedded electronics produces material discontinuity around the insertions. This in turn produces stress concentrations. The influence of piezoelectric and optical sensors on the structure has been estimated by many authors who have simulated the influence of embedded sensors as well as performed mechanical tests [54,73–75]. For a reduction of the vertical dimension, thinning of integrated circuits (ICs) or manufacturing of ultrathin chips are of great interest.

Stepping out of the wireless technologies, a minimal invasive integration can be achieved by electric interconnects without acting as a wound. Thin wires as well as foil technology can be applied. With foil technology, the sensor or actuator device, the electronics, and interconnects are built together by fabricating all within the same technology at the wafer level. The design of such a system has to set the ease of use for the embedding process as the highest priority.

7.4.2 Electronics for Wireless Sensing

Using state-of-the-art RFID technology, sensor nodes can be designed according to specific requirements. Systems can be, therefore, passive, semipassive, or active. A passive system will use no additional sources of energy other than the one provided by the wireless link. Meanwhile, semipassive and active systems rely on other sources of energy. In a semipassive system, it is very common to use energy harvesters, while in active systems a battery is implemented. Undeniably, passive and semipassive systems get usually more attention not only due to their simplicity, but also because the absence of batteries allows the system to have a longer life span.

Passive wireless power transfer occurs by implementing the principle of inductive coupled coils. Two industrial, scientific, and medical (ISM) bands are commonly used in the international community. The first one is high frequency (HF) working at 13.56 MHz and the second one is low frequency (LF) located from 125 to 134.2 kHz [76]. RFID coils that work reliably even when embedded in GFRP parts have been developed recently by Fraunhofer IIS [72]. An application of these technologies is found in SHM by means of piezo wafer active sensors (PWAS) implementing low-frequency signals with significantly higher power consumption.

A fully passive and inductively powered system for easier integration in the fiber composites is the main target. Focusing on the coil antenna systems, it can be evaluated how the material properties have a direct impact on the strength of the electromagnetic field [77]. CFRP is known to behave as a conductor, where electrical characteristics are highly dependent on the production techniques, that is, anisotropic and quasi-isotropic laminates. The conductivity of CFRP laminates is frequency dependent and increases with frequency when the incident electric field is perpendicular to the fiber direction as explained in Ref. [78], where models of the electrical anisotropy of CFRP were presented. Depending on their frequency, radio waves can be shielded by such materials due to eddy currents. These effects constitute a problem for the implementation in real SHM environments. In the high-frequency range, the influence is higher compared to the low frequency range. This phenomenon was evaluated in Ref. [77] both at HF and LF range. At the 13.56 MHz range, the efficiency decreases to about one third of its value once the CFRP was introduced to the antenna system. Meanwhile, at the 125 kHz range, efficiency decreases to half of the reference value. The coil input reflection coefficient (S_{11}) measured using a vector network analyzer is the indicator of radiation capabilities of the receiver coils that can be studied under the presence of CFRP. In Figure 7.6, the reflection coefficient S_{11} of the antenna is attenuated from −50.4 to −14.4 dB, implying that the quality of the antenna coil degrades when placed at a distance of 1 cm from the CFRP material. The same effect is worse at the HF range, where the reflection is not only attenuated from −43.9 to −9.9 dB, but also the resonance frequency shifted to 14.52 MHz. This topic is presented in Ref. [79] also, where the challenges of HF passive system integration were introduced.

Receiver coils for high frequency have been embedded in the CFRP in order to determine the changes in impedance and quality factor. Here, unidirectional

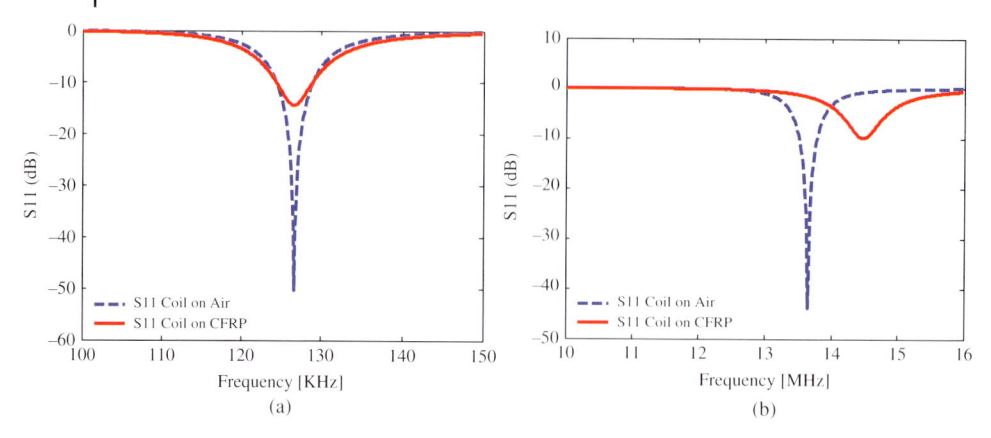

Figure 7.6 Influence of CFRP to the resonance frequency. (a) LF Band. (b) HF Band [77].

woven carbon fiber was chosen with a design of 0°/90°, which was processed with resin RTM6 at 180 °C for 2 h. Two versions of a coil were produced. The first one has an FR4 substrate with a self-resonance at 32 MHz. The same coil was made of flexible material to reduce the thickness of the antenna and insulated with DuPont™ Kapton; it presented a self-resonance at 38 MHz. Finally, this flexible coil was placed in between four unidirectional CFRP layers. After resin infusion and curing the resonance shifted to 15.5 MHz due to parasitic capacitances, likewise the quality factor diminished.

Before a structure with a fully functional wireless sensor network might be achieved a first step needs to happen, where the embedding of a single sensor or actuator along with simple circuitry for signal conditioning and amplification is done. Incorporating LF coils into FRP enables the opportunity to couple enough energy for such simple systems. In Ref. [80], the integration of LF coils to power and communicate with sensors and circuits is achieved by tailored fiber placement (TFP). Here a coil is stitched on glass fiber fabric by a modified embroidery machine (Tajima TMHL-G108). The passive wireless actuator, which is connected to the sewed coils, is currently being developed [71,80]. The node consists of a piezo driver for lamb wave generation that requires about 4 W of wireless power (Figure 7.7).

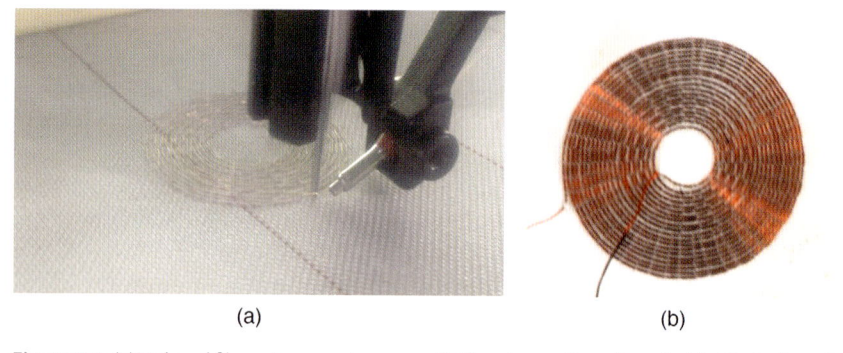

Figure 7.7 (a) Tailored fiber placement process. (b) Receiver coil produced with 25 windings [80].

7.5 Examples of Sensors Integrated in Fiber-Reinforced Polymer Composites

In general, sensors in composites can be used for different purposes of monitoring and controlling the manufacturing of a material or structure and for monitoring its service life. The production of the composite material can be divided into the two phases of impregnation of the fibers and curing of the resin. When the mold is set up or the preform is ready, the resin gets infused or injected through the inlet line/point. The fully impregnation of the fiber at the infusion phase minimizes the voids in the final product, as well as reducing the waste regarding dry spots. There are different sensors or techniques that can be applied to monitor the resin flow with the aim of guaranteeing full fiber saturation before curing of the resin. After the fiber laminate is fully saturated with resin, the resin has to be cured. Monitoring of resin cure is as important as tracking the fabric impregnation. There are different methods of real-time resin cure monitoring in composite material production. Those listed here are the most common ones applied today. Some of the sensors discussed here can be utilized for individual production-related criteria only, while a few span both the component's beginning of life phase in full, as well as the middle of life phase. The latter covers service life applications, including SHM and support to nondestructive testing in general. The text below contains an overview of available types of sensors for these different application scenarios.

7.5.1 Ultrasound Reflection Sensing

There are two different ultrasound reflection methods, which are commonly used in RTM productions; the direct reflection and the direct transmission techniques. In the direct reflection, only one ultrasonic transducer is used. This transducer sends and receives the acoustic signal. The recorded data are reflections of the sent signal at the boundary surface of two different materials, for example, steel (mold material) and fabrics (laminate material). The direct transmission comprises of one transmitter and one receiver. This method is widely used to investigate resin flow in RTM process [14,81,82].

The ultrasonic transducer produces an acoustic signal by means of a signal generator. This signal is detected by the sensor, which is a receiver. The received signal will be transformed into digital form to be stored and displayed to the user. Both signals, which are sent and received, have the sound velocity. The transmission time and the amplitude of the wave signal are useful to detect the flow front during injection and monitoring the resin cure afterward.

Besides detecting the resin front in the laminate, the ultrasonic technique can be used for measuring the degree of resin cure. The principle of measurement is schematically shown in Figure 7.6. When a low-intensity ultrasound wave propagates in a curing resin, it excites the resin with a high-frequency oscillation. The attenuation and the velocity of the sound are highly sensitive to the viscosity of the resin. Therefore, these two parameters are correlated to the state of the resin cure. The velocity of the sound relates to the density and modulus of the resin, while the attenuation is affected by energy dissipation of the resin. There

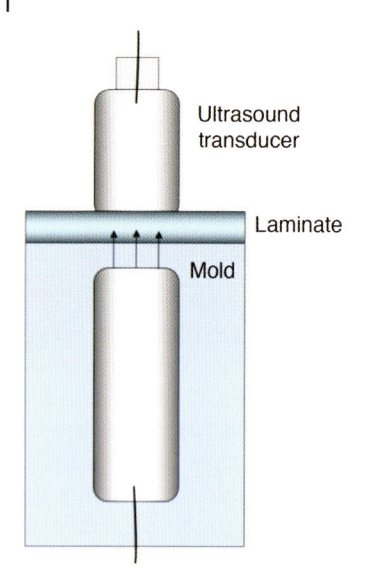

Figure 7.8 Scheme of resin cure measurement using ultrasound transducer.

are two main methods of ultrasonic measurements: *contact* and *noncontact* measurement. In the *contact* method, there is a fluid medium between the transducer and sample. In the *noncontact method*, which is also called air-coupled, there is not any contact between the transducer and the sample (Figure 7.8).

Some advantages of using ultrasound techniques for cure monitoring of the resin are low cost, potential of online cure monitoring without even being in contact with the material under test, and a simple geometry. The main disadvantage is a big influence of the fiber volume content on the ultrasound signal [83,84].

7.5.2 Pressure Sensors

Pressure sensors are also used to measure the pressure in vacuum-assisted resin infusion (VARI) composite manufacturing [21,23]. In laboratory trials and in production of the wind turbine blades, VARI processes using a vacuum bag technique is common. The inner pressure of the bag is usually a few millibar by means of a vacuum pump connected to the venting point. At the evacuation time, the pressure sensors are useful to detect the air leakages to the sealed vacuum bag. After checking the airtightness of the vacuum bag, the inlet of the resin to the laminate will be opened. Due to the pressure gradient between the resin inlet and the venting point, the resin will be infused through the laminate and impregnate the fibers. When the resin impregnates the fibers and reaches the embedded pressure sensors, which were already in a vacuum, the sensor's membrane will be deflected. The sensor will measure the pressure above the vacuum, which in this case is the hydrostatic pressure of the resin. The absolute types of piezoresistive pressure sensors for barometric application are suitable for such an application. The pressure sensors can be placed at the bottom of the mold [23], which results in a much localized pressure measurement. Another way is to place the pressure

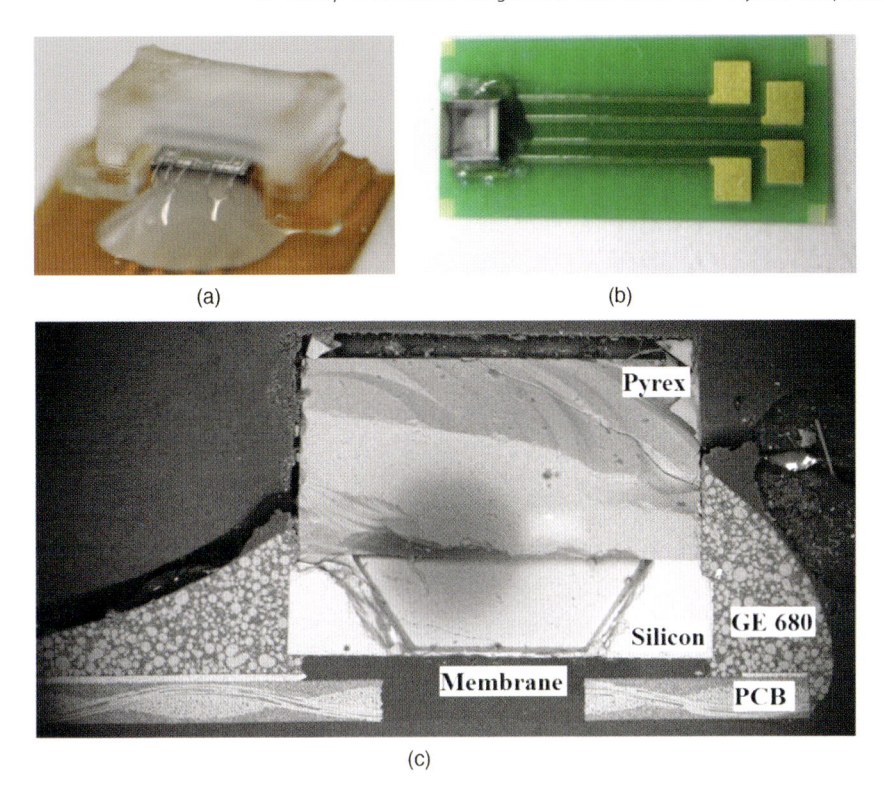

Figure 7.9 (a) A wire-bonded piezoresistive pressure sensor chip on Kapton PCB. (b) A flip-chipped piezoresistive pressure sensor chip on the thin flexible PCB. (c) Microscope image of the longitudinal cross section of a flip-chipped and embedded in FRP composite.

sensors through the thickness of the laminate [21] and monitor the pressure values at different points during the resin infusion.

The pressure sensors can be wire bonded for electrical connection and data acquisition. In this way, the chip is glued to the PCB and usually aluminum wire with the dimension of 20–80 μm connects the sensor to the PCB electrically. Since the sensors are embedded between fabrics in the laminate and as a result of evacuation the whole laminate will be compacted, the membrane of the sensors needs to be protected by a cap to prevent stimulation of the sensor by either weight of fabrics on the top or fabric's compaction after vacuum starts. The wire-bonded chip on the Kapton PCB is shown in Figure 7.9a.

To reduce the size of the embedded sensor in FRP composites, the pressure sensors can be flip-chipped on the PCB. The PCB needs to have a hole in front of the sensor's membrane. The hole lets the chip tracking the pressure changes inside the laminate. A flip chip-bonded sensor does not need a protection cap, since the sensor is flipped on the PCB and the membrane is faced to the PCB. This sensors and PCB are of a smaller size. Figure 7.9b and c shows a top view and the cross section of the embedded pressure sensor, respectively.

The development of the pressure gradient inside a vacuum-assisted FRP composite is shown in Figure 7.10. Figure 7.10a shows the measurement of

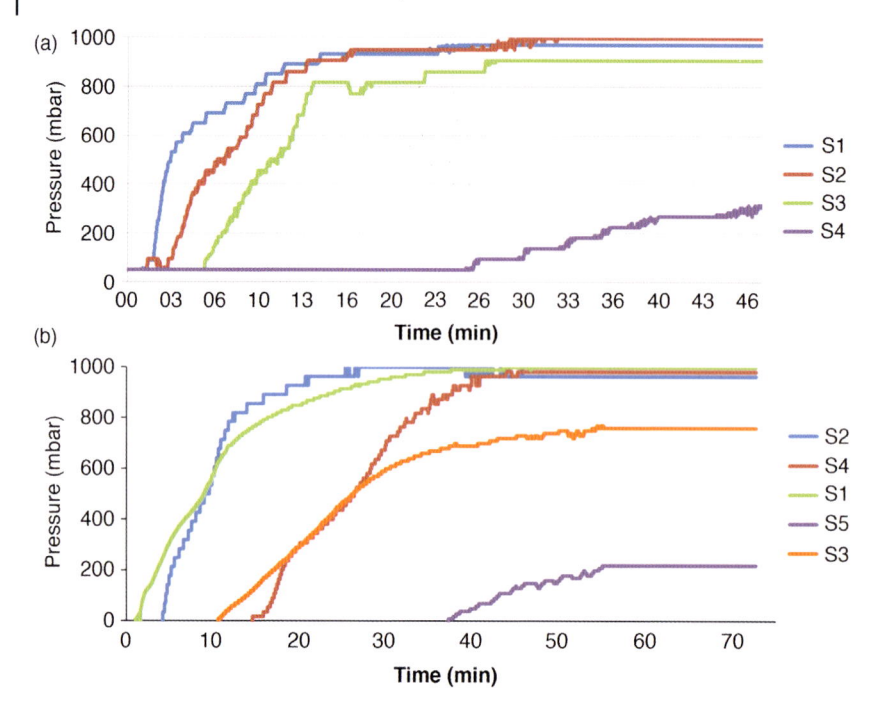

Figure 7.10 Online measured pressure values in CFRP (a) wire-bonded sensors [21] (b) flip-chipped sensors.

the pressure changes in CFRP by means of the wire-bonded sensors as shown in Figure 7.9a. In Figure 7.10b, the changes of pressure at different locations in CFRP are measured by means of the flip-chipped sensors.

The piezoresistive pressure sensors are very cheap (less than 2 Euro per sensor die). The main disadvantage of using such a sensor is the size of the chip (around 4 mm^2) and the rigidity of that. Another disadvantage of the piezoresistive pressure sensors is a dependency of the output voltage on the operating temperature. Using a constant current circuit will eliminate the reduction of the sensor's sensitivity with increase in temperature. For high-temperature applications, the sensors have to be calibrated at two different pressure and temperature points.

7.5.3 Thermocouples

If the temperature of the resin and preform are different, thermocouples can be used for detection of the resin flow [25,26]. The local temperature changes as the resin arrives. The thermocouples can be distributed through the thickness of the preform and in different planes. The thermocouples of type-K with the resolution of 0.1 °C are available in market to track the temperature changes in an acceptable accuracy.

Using thermocouples for resin flow monitoring is an indirect detection of the resin arrival. In case of using carbon fiber, the measurement may get disturbed by high conductivity of carbon fibers. Moreover, if the preform is not

7.5.4 Fiber Optic Sensors

The resin flow and curing degree can be monitored by measuring the refractive index of the optical fiber during filling of the mold with the resin [25,37–39]. In principle, between two media with different refractive indices of n_1 and n_2, there is a reflection coefficient of R that can be calculated based on Fresnel laws

$$R = \left(\frac{n_1 - n_2}{n_1 + n_2}\right)^2. \tag{7.1}$$

In fact, the refractive index is sensitive to the changes of medium density. Therefore, Lorentz–Lorentz law demonstrates precisely the relation between refractive index and the density of a liquid medium

$$\frac{n^2 - 1}{n^2 + 2} = \frac{R_M}{M}\rho, \tag{7.2}$$

where R_M is the molar refractivity, M is the molar mass, and ρ is the density.

The external reflection influences the reflected intensity at the end of the optical fiber. Thus, the resin front and the degree of cure can be obtained by observing the intensity of the reflected indices. Figure 7.11 shows the simplified principle of the measurement using optical fiber.

One of the advantages of using the optical fiber is its small size, which means the optical fiber can stay in the laminate without disrupting the mechanical properties of that. Using fiber Bragg grating sensor, temperature-induced cross-effect from the measured data can be separated. The most important disadvantages are a very expensive measurement tool and highly brittleness of the fiber. Another disadvantage of using this type of sensor is distortion in measured signal when the sensor is under stress. This stress is unavoidable in multilayers composite laminates. Therefore, there are always substantial measurement errors due to the difficulty of finding the center of Bragg wavelength [39].

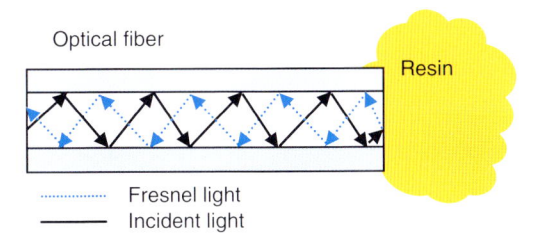

Figure 7.11 Fresnel reflection at fiber end–resin interface, n_f and n_m are the refractive indices of fiber and resin.

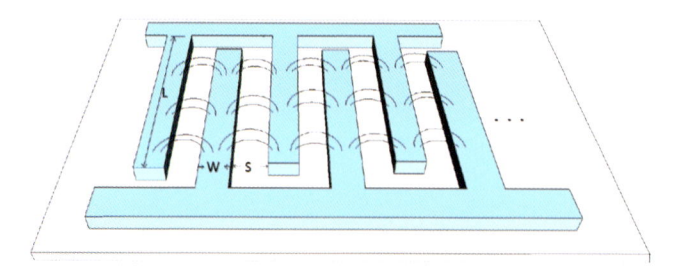

Figure 7.12 An interdigitated array and the electric field between electrodes. S is the spacing, W is the width of electrodes, and L is the mutual length of each of the two neighbor electrodes [85].

7.5.5 Interdigital Planar Capacitive Sensors

The most common method to ensure the completeness of resin curing is dielectric analysis of the resin during composite manufacturing. Planar interdigitated sensors for DEA analysis have been widely used for *in situ* cure monitoring of resins [31–36]. DEA is in accordance with a thermoanalytical method standard of ASTM 2038. An AC excitation voltage generates an electric field between interdigitated electrodes [85]. Figure 7.12 shows the simplified scheme of the electric field between neighboring electrodes in an interdigital planar capacitive sensor. In this electric field, dipole polarization and ion migration occur in the material under test. The ion migration can only be measured by DEA. Polarization of the dipole and ion migration due to the external field frequency (ω) causes energy loss (resistive) and energy storage (capacitive) behavior.

When the curing of the resin starts, the state of the resin changes from low-viscosity liquid to gel, and later on by increasing the number of cross-linking in the resin structure, from gel to solid [15]. The solidification of the resin reduces the movements of the charges and the alignments of the dipoles to nearly zero. This reduction in movement results in decrement of the loss factor. The loss factor of the material is the imaginary part of the relative permittivity of the material. If the permittivity remains at a constant value the resin is fully cured. The complex relative permittivity of the resin can be described as

$$\varepsilon_r^* = \varepsilon_r' - i\varepsilon_r''. \tag{7.3}$$

ε_r^* has a phase lag, δ, compared to the applied voltage. In ε_r^*, the real part (ε_r') is the relative permittivity and the imaginary part (ε_r^*) is the loss factor and i is the square root of -1

$$\tan(\delta) = \frac{\varepsilon_r''}{\varepsilon_r'}. \tag{7.4}$$

The relative permittivity and the loss factor for a capacitor with parallel plate are driven as

$$\varepsilon_r' = \frac{C}{\varepsilon_0 A/d}, \tag{7.5}$$

$$\varepsilon_r'' = \frac{G}{\omega\varepsilon_0 A/d}. \tag{7.6}$$

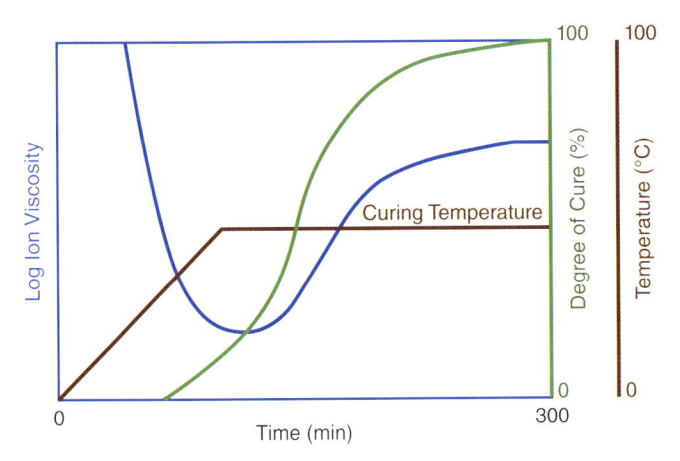

Figure 7.13 Ion viscosity versus degree of resin cure [85].

The loss factor (resistive behavior) is in correlation to the degree of cure in DEA measurement and to the inverse of frequency. For a parallel plate capacitor, C is the capacitance (F), G is the conductance (S), and A/d is the ratio of the electrode's area to the electrode's distance d (m), ϵ_0 is the vacuum permittivity (8.85×10^{-14} F/cm) and ω is the angular frequency (Hz). The resistivity can be measured

$$\rho = \frac{A}{Gd} = \frac{1}{\omega \epsilon_0 \epsilon''_r}. \tag{7.7}$$

The relationship between the ion viscosity (resistivity, ρ) and the degree of curing is shown schematically in Figure 7.13. When the ion viscosity reaches its maximum limit and stays constant, the resin is cured.

There are types of commercial detachable interdigital sensors (Netzsch-Gerätebau GmbH and Lambient Technologies LLC) that stay in the mold. Figure 7.14 shows photographs of these two commercially available sensors. The sensors made by Lambient Technologies LLC have the line spacing of 0.02″ and substrate

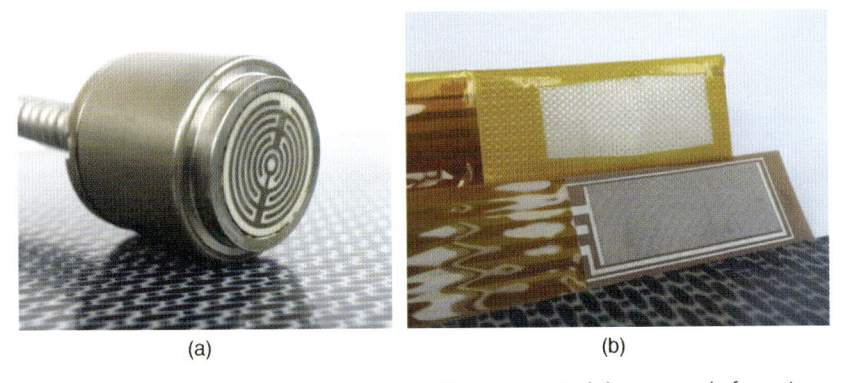

Figure 7.14 (a) Reusable tool mount sensor with an integrated thermocouple for resin cure monitoring made by Netzsch. (b) Planar interdigital capacitive sensor for resin cure monitoring made by Netzsch.

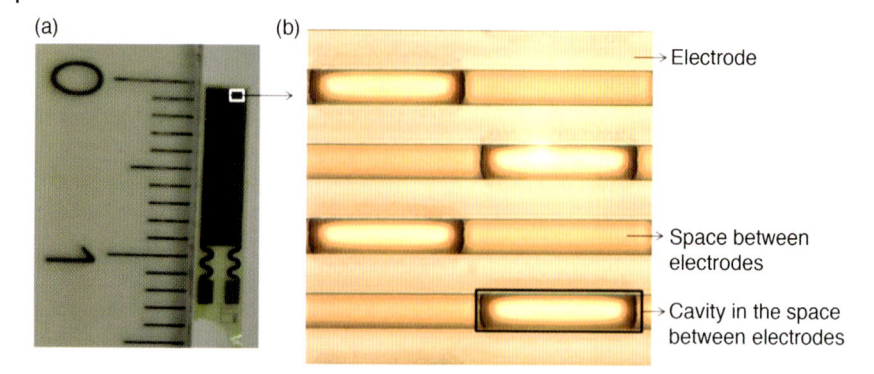

Figure 7.15 (a) A photograph of the microscale planar capacitive sensor made at IMSAS (scale in mm). (b) Photo showing the porous substrate of the sensor using optical microscope [85].

of alumina. These sensors are integrated in the mold and will be in touch with the surface of the laminate. These sensors can be used over and over, but they just provide much localized information about the state of resin at the surface on which they are attached and not about the internal state of the laminate. This caused a problem when the laminate is thick (e.g., blades of wind turbines) and there is no information about the state of resin far above where the sensor is placed.

Compared to the widely used commercial type of the interdigital sensor, Institute for Microsensors, -Actuators and -Systems (IMSAS) designed and fabricated microscale planar interdigital capacitive sensors. Figure 7.15 shows a photograph of microscale interdigital sensor of IMSAS. These sensors are thin, flexible and having a small porous sensing area to a thickness of about 5 μm and a sensing area of 18 mm^2. The sensor cannot be made infinitely small, since the sensor sensitivity depends on the number and length of electrodes.

The thickness of the novel IMSAS sensor is about 5 μm, which is comparable to the diameter of glass/carbon filaments. In Refs [54,74,75] the effect of the embedded microsensors on the mechanical properties of the fiber laminate composite is investigated, showing that the impact of embedding a *thin, flexible,* and *perforated* inlay on the composite properties can be neglected in comparison to an embedded rigid inlay. There is a high demand for miniaturized sensors that can *in situ* monitor the state of resin cure at different points in the composite without downgrading properties of the finished product. This new microscale interdigital capacitive sensor determines the cure degree of the resin based on the most common and reliable method of DEA. The porous substrate of the

Figure 7.16 Longitudinal cross section of an embedded microscale IMSAS capacitive sensor, the resin filled the cavities of the perforated substrate.

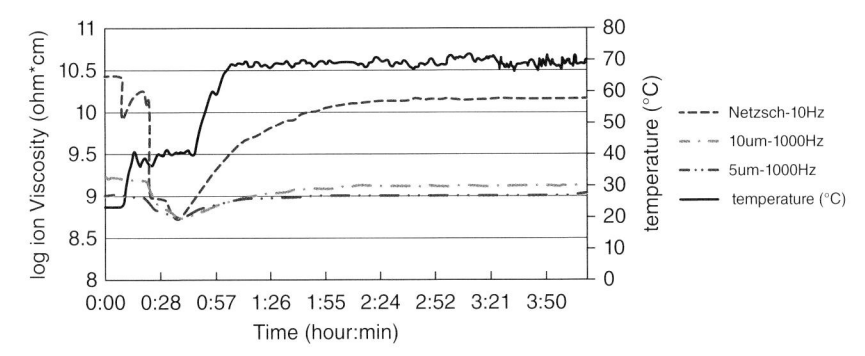

Figure 7.17 Real-time monitoring of the resin cure using microscale interdigital sensors with electrode's width and spacing of 5 and 10 μm, respectively, in comparison to the commercial Netzsch sensor with width and spacing of 115 μm [85].

sensor lets the resin go through the sensor and reinforce the sensor in a laminate. Figure 7.16 shows a longitudinal cross section of the embedded microscale interdigital sensor in composite. The resin filled the cavities on the sensor's substrate. These are promising points showing that the newly developed microscale sensors can be used in a real composite production to ensure the complete curing of the resin and remain in the laminate without degrading its mechanical properties.

Figure 7.17 shows *in situ* measuring of the resistivity ρ of resin by microscale and Netzsch planar capacitive interdigital sensors. The resistivity ρ is correlated to the ion viscosity and the degree of the resin cure (Eq. (7.5)). The difference between measured ion viscosity at different frequencies of 10 Hz and 1 kHz is due to inverse relation to the excitation frequency, while the difference of measured ion viscosity between 5 and 10 μm sensors is due to different geometry of the sensors and to the one-point calibration method.

The major disadvantage of interdigital sensors is getting a short cut by embedding in conductive fabrics, for example, carbon. This problem can be solved by using a nonconductive coating as an insulator on top of the sensing electrodes of an interdigital sensor. In some sensors, a layer of glass fiber fabric is used as an insulator. This thick layer of glass fabric reduces the responsivity of the sensor and there is a possibility to have trapped air between the sensor and the fabric. In the recent and novel approach by IMSAS, a thin metal oxide insulation layer is grown up from the metallization as an insulator. This thin oxide layer is in a range of nanometer. Therefore, it does not increase the thickness of the sensor.

When the curing of the resin is completed, the embedded sensor will remain inside the composite part. The commercial planar interdigital capacitive sensors for DEA measurements from Netzsch or Lambient Technologies LLC are usually a printed pattern on a polycarbonate or polyimide film with thickness of 150 μm. The dimension and the thickness of the both commercial sensors are majorly harming the uniformity of the composite. Therefore, they can be used only in laboratory experiments. This big integrated sensor will degrade the mechanical properties of the composite part drastically and behaves like a wound in the host material. The composites having large embedded sensor will fail under the load during operation. The small sensors, however, can measure the curing degree of resin during manufacturing and remain in the final part for SHM during the lifetime of the composite.

7.6 Conclusion

Fiber-reinforced polymers have been in the very focus of studies on material integration of sensor and intelligent systems for several reasons. These include the following:

- The comparatively moderate processing conditions, which – depending on the exact manufacturing approach chosen – limit both thermal and mechanical loads exerted on the integrated system during manufacturing of a component.
- The possibility of monitoring and documenting the production process through material-integrated sensing, which can be employed for product optimization and/or a more detailed characterization of the component made.
- The general interest in in-service monitoring of components with the aim of
  - realizing further weight reductions at highest safety levels and
  - the reduction of costs via need-based planning and execution of maintenance tasks.
- In aerospace structures, the implementation of material-integrated sensing systems as the exclusive NDE/NDT approach for advanced structural design solutions for which other no economical NDE/NDT solution would have been available.

For optimizing time and costs of the manufacturing process of fiber reinforced polymer composites, different types of the sensors can be embedded in the laminate or in the mold. Such sensors can monitor different parameters like resin flow, hydrostatic resin pressure, fiber compaction, curing degree of the resin, and so on. This *in situ* monitoring of different parameters helps optimization of the process. However, the embedded sensors behave as a wound in the host material and may cause delamination of the fiber layers, reduction of local properties, and thus potentially the premature failure of the loaded part. Therefore, the size of the sensors has to be minimized or ideally, the sensors have to be produced from the same fibers or resins as reinforcement or matrix.

The combined implementation of a production process and service life monitoring system bears great promise for fiber-reinforced polymers, as it allows enhancing predictive capabilities of the material-integrated SHM system through availability of individual baselines for each component manufactured.

Possible effects of material-integrated sensors on mechanical performance have been analyzed by several authors. Based on their typical buildup and associated failure mechanisms, composite materials have raised specific concerns in this respect. Among the potentially detrimental effects discussed in the literature are delaminations, for which the integrated sensor may act as starting point, resin pockets formed in the vicinity of the sensor, or fiber undulations. A detailed overview of such phenomena plus an evaluation of their effects is found in the work by Shivakumar and Emmanwori [86]. Luyckx *et al.* have provided another survey of the discussion, which is more recent and stresses the fact that not only geometry and size of the sensor itself are of importance in this respect, but also the type of fabric and the orientation and positioning of the sensor relative to the reinforcing fibers – the latter is specifically true for fiber optic sensors [87]. Finally, most studies come to the conclusion that for most types of sensors, solutions can be found that alleviate any negative impact on material mechanics. For example, Luyckx *et al.* note that even standard optical fibers of 125 µm diameter cause minimum levels of perturbation as long as they are embedded in parallel to the reinforcement, and for small diameter fibers, not even resin pockets may be identified adjacent to the embedded fiber [87]. For nonfiber, planar types of sensors, Lang *et al.*, Dumstorff *et al.*, and Boll *et al.* have extensively discussed and demonstrated the possibility of reducing the footprint of microsystem technology sensors [36,54,55].

Acknowledgments

German Research Foundation, DFG numbers LA 1471/17-1 and HE 2574/31-1, funds part of this work.

References

1 Tong, L., Mouritz, A.P., and Bann, M.K. (2002) Introduction, in *3D Fibre Reinforced Polymer Composites*, Elsevier Science, Oxford, pp. 1–12.
2 Soutis, S. (2015) Introduction: engineering requirements for aerospace composeite materials, in *Polymer Composites in the Aerospace Industry*, Woodhead Publishing, pp. 1–18.
3 Galehdar, A., Rowe, W., Ghorbani, K., Callus, P.J., John, S., and Wang, C.H. (2011) The effect of ply orientation on the performance of antennas in or on carbon fiber composites. *Progress In Electromagnetics Research*, **116**, 123–136.
4 Kleineberg, M., Liebers, N., and Kühn, M. (2013) Interactive manufacturing process parameter control, in *Adaptive, Tolerant and Efficient Composite Structures*, (eds W. Martin and S. Michael), Springer, Berlin Heidelberg, pp. 363–372.
5 Brauner, C. (2013) *Analysis of Process Induced Distortions and Residual Stresses in Composite Structures*, Logos Verlag, Berlin.
6 Johnston, A.A. (1996) An integrate model of the development of process-induced deformation in autoclave processing of composites structures, Ph.D. thesis, The University of British Columbia.
7 Wang, T.J., Wu, C.H., and Lee, L.J. (1994) In-plane permeability measurement and analysis in liquid composite molding. *Polymer Composites*, **15**, 278–288.
8 Demaria, C., Ruiz, E., and Trochu, F. (2007) In-plane anisotropic permeability characterization of deformed woven fabrics by unidirectional injection. Part I: experimental results. *Polymer Composites*, **28** (6), 797–811.

9 Demaria, C., Ruiz, E., and Trochu, F. (2007) In-plane anisotropic permeability characterization of deformed woven fabrics by unidirectional injection. Part II: prediction model and numerical simulations. *Polymer Composites*, **28** (6), 812–827.
10 Advani, G.S., and Calado, V.M.A. (1996) Effective average permeability of multi-layer preforms in resin transfer molding. *Composites Science and Technology*, **56** (5), 519–531.
11 Chen, R., Dong, C., Liang, Z., Zhang, C., and Wang, B. (2004) Flow modeling and simulation for vacuum assisted resin transfer molding process with the equivalent permeability method. *Polymer Composites*, **25** (2), 146–164.
12 Dong, C. (2006) An equivalent medium method for the vacuum assisted resin transfer molding process simulation. *Journal of Composite Materials*, **40** (13), 1193–1213.
13 Pantelelis, N.G. and Bistekos, E. (2010) Monitoring and control for the production of CRFP components. Proceedings of SAMPE, Seattle.
14 Schmachtenberg, E., Schulte zur Heide, J., and Töpke, J. (2005) Application of ultrasonics for the process control of resin transfer moulding (RTM). *Polymer Testing*, **24**, 330–338.
15 Lawrence, C.M., Nelson, D.W., Bennett, T.E., and Spingarn, J.R. (1998) An embedded fiber optic sensor method for determining residual stresses in fiber-reinforced composite. *Journal of Intelligent Material Systems and Structures*, **9**, 788–799.
16 Kang, H., Kang, D., Bang, H., Hong, C.-S., and Kim, C. (2002) Cure monitoring of composite laminates using fiber optic sensors. *Smart Materials and Structures*, **11**, 279.
17 O'Dwyer, M.J., Maistros, G.M., James, S.W., Tatam, R.P., and Partridge, I.K. (1998) Relating the state of cure to the real-time internal strain development in a curing composite using in-fibre Bragg gratings and dielectric sensors. *Measurement Science and Technology*, **9**, 1153.
18 Lee, Y., Jhan, Y., Chung, C., and Hsu, Y. (2011) A prediction method for in-plane permeability and manufacturing applications in the VARTM process. *Engineering*, **3**, 691–699.
19 Sun, X., Li, S., and Lee, L.J. (1998) Molding filling analysis in vacuum-assisted resin transfer molding, part I: scrimp based on a high-permeable medium. *Polymer Composites*, **19**, 807–817.
20 Ni, J., Li, S.J., Sun, X.D., and Lee, L.J. (1998) Mold filling analysis in vacuum-assisted resin transfer molding. Part II: SCRIMP based on grooves. *Polymer Composites*, **19**, 818–829.
21 Kahali Moghaddam, M., Breede, A., Brauner, C., and Lang, W. (2015) Embedding piezoresistive pressure sensors to obtain online pressure profiles inside fiber composite laminates. *Sensors*, **15**, 7499–7511.
22 Xin, C., Gu, Y., Li, M., Li, Y., and Zhang, Z. (2011) Online monitoring and analysis of resin pressure inside composite laminate during zero-bleeding autoclave process. *Polymer Composites*, **32**, 314–323.
23 Amico, S. and Lekakou, C. (2001) An experimental study of the permeability and capillary pressure in resin-transfer moulding. *Composites Science and Technology*, **61**, 1945–1959.

24 Lynch, K., Hubert, R., and Poursartip, A. (1999) Use of a simple, inexpensive pressure sensor to measure hydrostatic resin pressure during processing of composite laminates. *Polymer Composites*, **20** (4), 581–593.

25 Wang, P., Molimard, J., Drapier, S., and Vautrin, A. (2012) Monitoring the liquid resin infusion (LRI) manufacturing process under industrial environment using distributed sensors. ICCM17 proceedings.

26 Wang, P., Demirel, M., Drapier, I.S., and Molimard, J. (2008) In-plane and transverse detection of the fluid flow front during the LRI manufacturing process. The 9th International Conference on Flow Processes in Composite Materials, Montréal (Québec), Canada.

27 Wu, C.H., Wang, T.J., and Lee, L.J. (1994) Trans-plane permeability measurement and its application in liquid composite molding. *Polymer Composites*, **4** (289–298), 15.

28 Nedanov, P.B. and Advani, S.G. (2002) A method to determine 3D permeability of fibrous reinforcements. *Journal of Composite Materials*, **36** (2), 241–254.

29 Gebart, B.R. (1992) Permeability of unidirectional reinforcements for RTM. *Journal of Composite Materials*, **26** (8), 1100–1133.

30 Simacek, P., Neacsu, V., and Advani, S.G. (2010) A phenomenological model for fiber tow saturation of dual scale fabrics in liquid composite molding. *Polymer Composites*, **31** (11), 1881–1889.

31 Vassilikou-Dova, I.K.A. (2009) Dielectric Analysis (DEA), in *Thermal Analysis of Polymers, Fundamentals and Applications*, John Wiley & Sons, Inc., Hoboken, pp. 497–613.

32 Breede, A., Kahali Moghaddam, M., Brauner, C., and Lang, W. (2015) Online process monitoring and control by dielectric sensors for a composite main spar for wind turbine blades. 20th International Conference on Composite Materials, Copenhagen, Denmark.

33 Lee, H.L. (2014) *The Handbook of Dielectric Analysis and Cure Monitoring*, Lambient Technologies, LLC.

34 Kim, J.S. and Lee, D.G. (1996) Measurement of the degree of cure of carbon fiber epoxy composite materials. *Journal of Composite Materials*, **30** (13), 1436–1457.

35 Dave, R.S. and Loos, A.C. (2000) *Processing of Composites*, Hanser/Gardner Publications.

36 Boll, D., Schubert, K., Brauner, C., and Lang, W. (2014) Miniaturized flexible interdigital sensor for *in situ* dielectric cure monitoring of composite materials. *IEEE Sensors Journal*, **14** (7), 2193–2197.

37 Pandey, G., Deffor, H., Thostenson, E.T., and Heider, D. (2013) Smart tooling with integrated time domain reflectometry sensing line for non-invasive flow and cure monitoring during composites manufacturing. *Composites Part A: Applied Science and Manufacturing*, **47**, 102–108.

38 Buggy, S.J., Chehura, E., James, S.W., and Tatam, R.P. (2007) Optical fibre grating refractometers for resin cure monitoring. *Journal of Optics A: Pure and Applied Optics*, **9**, 60–65.

39 Sampath, U., Kim, H., Kim, D., Kim, Y., and Song, M. (2015) *In-situ* cure monitoring of wind turbine blades by using fiber brag grating sensors and frenchel reflection measurement. *Sensors*, **15**, 18229–18238.

40 Society of Automotive Engineers SAE (2013) Aerospace recommended practice (ARP) 6414: Guidelines for Implementation of Structural Health Monitoring on Fixed Wing Aircraft.

41 Lehmhus, D., Brugger, J., Muralt, P., Pané, S., Ergeneman, O., Dubois, M.-A., Gupta, N., and Busse, M. (2013) When nothing is constant but change: adaptive and sensorial materials and their impact on product design. *Journal of Intelligent Material Systems and Structures*, **24**, 2172–2182.

42 Sproat, W. and Lewis. W. (1986) Darely visible impact damage (BVID) detection in aircraft composites. Proceedings of the AIAA/SOLE 2nd Aerospace Maintenance Conference, San Antonio, Texas.

43 Heida, J.H. and Platenkamp, D.J. (2012) In-service inspection guidelines for composite aerostructures. Proceedings of the 18th World Conference on Nondestructive Testing, Durban, South Africa.

44 Pattabhiraman, S., Gogu, C., Kim, N.H., Haftka, R.T., and Bes, C. (2012) Skipping unnecessary structural airframe maintenance using an onboard structural health monitoring system. *Proceedings of the Institution of Mechanical Engineers Part O: Journal of Risk and Reliability*, **226**, 549–560.

45 Jian, C., Lei, Q., Shenfang, Y., Lihua, S., PeiPei, L., and Dong, L. (2012) *Structural Health Monitoring for Composite Materials, Composites and Their Applications* (ed. N. Hu), InTech.

46 Sbarufatti, C., Manes, A., and Giglio, M. (2014) Application of sensor technologies for local and distributed structural health monitoring. *Structural Control and Health Monitoring*, **21**, 1057–1083.

47 Roach, D. (2009) Real time crack detection using mountable comparative vacuum monitoring sensors. *Smart Structures and Systems*, **5** (4), 317–328.

48 Farrar, C.R. and Doebling, S.W. (1997) An overview of modal-based damage identification methods. Proceedings of the DAMAS 97 International Workshop, Sheffield, UK.

49 Giurgiutiu, V. (2008) *Structural Health Monitoring With Piezoelectric Wafer Active Sensors*, Academic Press, Burlington.

50 Kirikera, G., Shinde, V., Schulz, M., Sundaresan, M., Hughes, S., van Dam, J., Nkrumah, F., Grandhi, G., and Ghoshal, A. (2008) Monitoring multi-site damage growth during quasi-static testing of a wind turbine blade using a structural neural system. *Structural Health Monitoring*, **7**, 65–83.

51 Ramakrishnan, M., Rajan, G., Semenova, Y., and Farrell, G. (2016) Overview of fiber optic sensor technologies for strain/temperature sensing applications in composite materials. *Sensors*, **16** (1), 99.

52 Schmidt, D., Kolbe, A., Kaps, R., Wierach, P., Linke, S., Steeger, S., von Dungern, F., Tauchner, J., Breu, C., and Newman, B. (2015) Development of a door surround structure with integrated structural health monitoring system. Smart Intelligent Aircraft Structures (SARISTU): Proceedings of the Final Project Conference, Heidelberg, Springer Verlag, pp. 935–945.

53 Kirikera, G.R. and Krishnaswam, S. (2009) Series-connected interdigitated surface acoustic wave sensors for structural health monitoring. *Structural Longevity*, **2**, 11–24.

54 Dumstorff, G., Paul, S., and Lang, W. (2014) Integration without disruption: the basic challenge of sensor integration. *IEEE Sensors*, **14** (7), 2102–2111.

55 Lang, W., Jakobsa, F., Tolstosheevaa, E., Sturma, H., Ibragimov, A., Kesel, A., Lehmhus, D., and Dicke, U. (2011) From embedded sensors to sensorial materials – the road to function scale integration. *Sensors and Actuators A: Physical*, **171**, 3–11.

56 Schubert, K.J. and Herrmann, A.S. (2012) On the influence of moisture absorption on Lamb wave propagation and measurements in viscoelastic CFRP using surface applied piezoelectric sensors. *Composite Structures*, **94**, 3635–3643.

57 Schubert, K.J., Brauner, C., and Herrmann, A.S. (2014) Non-damage-related influences on Lamb wave-based structural health monitoring of carbon fiber-reinforced plastic structures. *Structural Health Monitoring – an International Journal*, **13**, 158–176.

58 Herrmann, A.S., Stieglitz, A., Brauner, C., Peters, C., and Schiebel, P. (2013) *Polymer Matrix Composites, Structural Materials and Processes in Transportation*, Wiley-VCH Verlag GmbH, Weinheim, Germany.

59 Lehmhus, D., Busse, M., Herrmann, A.S., and Kayvantash, K. (2013) *Structural Materials and Processes in Transportation*, Wiley-VCH Verlag GmbH, Weinheim, Germany.

60 Hyer, M.W. (1998) *Stress Analysis of Fiber Reinforced Composite Materials*, WCB/McGraw Hill, New York, USA.

61 Koerdt, M., Kibben, S., Bendig, O., Chandrashekhar, S., Hesselbach, J., Brauner, C., Herrmann, A.S., Vollertsen, F., and Kroll, L. (2016) Fabrication and characterization of Bragg gratings in perfluorinated polymer optical fibers and their embedding in composites. *Mechatronics*, **34**, 137–146.

62 Degrieck, J., De Waele, W., and Verleysen, P. (2001) Monitoring of fibre reinforced composites with embedded optical fibre Bragg sensors, with application to filament wound pressure vessels. *NDT & E International*, **34** (4), 289–296.

63 Kang, D.H., Kim, C.U., and Kim, C. (2006) The embedment of fiber Bragg grating sensors into filament wound pressure tanks considering multiplexing. *NDT&E International*, **39** (2), 109–116.

64 Hufenbach, W., Gude, M., and Heber, T. (2009) Development of novel piezoceramic modules for adaptive thermoplastic composite structures capable for series production. *Sensors and Actuators A*, **156** (1), 22–27.

65 Hufenbach, W., Gude, M., and Heber, T. (2011) Embedding versus adhesive bonding of adapted piezoceramic modules for function-integrative thermoplastic composite structures. *Composites Science and Technology*, **71**, 1132–2113.

66 Lin, M., Kumar, A., Beard, S.J., and Xinlin, Q. (2001) Built-in structural diagnostic with the SMART LayerTM and SMART SuitcaseTM. *Smart Materials Bulletin*, 7–11, doi: 10.1016/S1471-3918(01)80123-4

67 Dumstorff, G. and Lang, W. (2016) Strain gauge printed on carbon weave for sensing in carbon fiber reinforced plastics. Proceedings of the 2016 IEEE SENSORS Conference, Orlando, FL.

68 Fraunhofer Institute for Manufacturing Technology and Advanced Materials (2017) Part monitoring with printed sensors. Fraunhofer IFAM. Availableat www.ifam.fraunhofer.de/en.html (accessed February 01, 2017).

69 Yuan, S., Huang, R., and Rao, Y. (2004) Internal strain measurement in 3D braided composites using co-braided optical fiber sensors. *Journal of Materials Science and Technology.*, **20**, 199–202.

70 Antonucci, V., Esposito, M., Ricciardi, M., Raffone, M., Zarrelli, M., and Giordano, M. (2011) Permeability characterization of stitched carbon fiber preforms by fiber optic sensors. *Express Polymer Letters*, **5**, 1075–1084.

71 Salas, M., Focke, O., Stoltenberg, G., Herrmann, A.S., and Lang, W. (2015) Wireless sensor network for structural health monitoring by means of Lamb-waves. Proceeding of the 10th International Workshop on Structural Health Monitoring, Stanford University.

72 Bernhard, J., Dräger, T., Grabowski, C., Sotriffer, I., and Philipp, T. (2011) Integrating RFID in fibre-reinforced plastics. ITG-Fachbericht 229, Dresden, Germany.

73 Prabhugoud, M. and Peters, K. (2006) Finite element model for embedded fiber Bragg grating sensor. *Smart Materials and Structures*, **15** (2), 550–562.

74 Kahali Moghaddam, M., Lang, W., and Boll, D. (2014) Embedding rigid and flexible inlays in carbon fiber reinforced plastics. IEEE/ASME International Conference on Advanced Intelligent Mechatronics.

75 Lang, W., Tolstosheeva, E., Pistor, J., Hoeffmann, D., Rotermund, T., Schellenberg, T., Boll, D., Hertzberg, T., Gordillo-Gonzales, V., Mandon, S., and Peters-Drolsha, D. (2013) Neural implants: a challenge for microtechnology and microelectronics. 1st Bernstein Sparks Workshop on Cortical Neurointerfaces.

76 Finkenzeller, K. and Müller, D. (2010) *RFID Handbook*, 3rd edn, John Wiley & Sons, Ltd, Chichester.

77 Salas, M., Focke, O., Lang, W., and Herrmann, A. (2014) Wireless power transmission for structural health monitoring of fiber-reinforced-composite materials. *IEEE Sensors Journal Special Issue on Material-integrated Sensing*, **17**, 2171–2176.

78 Galehadar, A., Nicholson, K., Rowe, K.S.T., and Ghorbani, K. (2010) The conductivity of unidirectional and quasi isotropic carbon fiber composites. Preceedings of the 40th European Microwave Conference, 882–885.

79 Mayordomo, I., Dräger, T., and Bernhard, J. (2011) Technical challenges for the integration of passive HF RFID. IEEE International Conference on RFID-Technologies and Applications.

80 Salas, M., Focke, O., Lang, W., and Herrmann, A. (2014) Low-frequency inductive power transmission for piezo-wafer-active-sensors in the structural health monitoring of carbon-fiber-reinforced-polymer. 2nd International Conference on System-Integrated Intelligence, Bremen, Germany.

81 Döring, J., Stark, W., and Splitt, G. (1998) On-line process monitoring of thermosets by ultrasonic methods. *NDT.net*, **3** (11), 1–4.

82 Shepard, D.D. and Smith, K.R. (1997) Ultrasonic Cure monitoring of 42nd international SAMPE symposium.

83 Lionetto, F. and Maffezzoli, F. (2013) Monitoring the cure state of thermosetting resins by ultrasound. *Materials*, **6**, 3783–3804.

84 Liebers, N., Raddatz, F., and Schadow, F. (2012) Effective and Flexible Ultrasound Sensors for Cure Monitoring for Industrial Composite Production. Deutsche Luft - und Raumfahrt Kongress.

85 Kahali Moghaddam, M., Breede, A., Chaloupka, A., Bödecker, A., Habben, C., Meyer, E.M., Brauner, C., and Lang, W. (2016) Design, fabrication and embedding of microscale interdigital sensors for real-time cure monitoring during composite manufacturing. *Sensors and Actuators A: Physical*, **243**, 123–133.

86 Shivakumar, K. and Emmanwori, L. (2004) Mechanics of failure of composite laminates with an embedded fiber optic sensor. *Journal of Composite Materials*, **38**, 669–680.

87 Luyckx, G., Voet, E., Lammens, N., and Degrieck, J. (2011) Strain measurements of composite laminates with embedded fibre bragg gratings: criticism and opportunities for research. *Sensors*, **11**, 384–408.

8

Integration in Sheet Metal Structures

Welf-Guntram Drossel,[1] Roland Müller,[2] Matthias Nestler,[2] and Sebastian Hensel[2]

[1]*Fraunhofer IWU, Fraunhofer Institute for Machine Tools and Forming Technology IWU, Scientific Field Mechatronics and Lightweight Structures, Reichenhainer Str. 88, 09126 Chemnitz, Germany*
[2]*Fraunhofer IWU, Fraunhofer Institute for Machine Tools and Forming Technology IWU, Scientific Field Forming Technology and Joining, Reichenhainer Str. 88, 09126 Chemnitz, Germany*

8.1 Introduction

The manufacturing process for fiber-reinforced composites allows for an excellent direct integration of piezoceramic sensors and actuators into such structures. Prepackaged piezoceramic fiber foils are placed between fiber layers and are laminated together with the surrounding fiber-reinforced plastic within the production process. Rodgers and Hagood presented in Ref. [1] an active rotor concept that includes a composite spar with laminated active fiber composite (AFC) foils. The actuation of the blade leads to a rotor twist rate of $1.26°/m$. Paradies and Ciresa [2] manufactured a wing profile with six macrofiber composite (MFC) foils, applied on a foam core and covered by carbon- and glass-reinforced plastic layers (Figure 8.1). With the optimized prototypic design, they demonstrated the flight control ability of the rudderless wing. Tien and Goo investigated a piezocomposite generating element (PCGE) in Ref. [3], which can be used in energy harvesting applications. A lead zirconate titanate (PZT) ceramic layer is embedded in a composite layup and laminated to produce the glass/carbon/epoxy element. The thermal residual stresses caused by the production process were evaluated within a numerical simulation. Furthermore, the device was tested with an experimental setup to assess its electricity generating performance.

At least on a theoretical level, Ref. [4,5] discuss the integration of piezoelectric transducer elements into fiber–metal laminates (FML) such as GLARE® or ARALL®. The piezoelectric layer substitutes one ply locally. Hence, it is integrated as part of the stack sequence. The authors of the first study provide a theoretical model to describe the dynamic response of an ARALL beam. The second study summarizes possible composite materials as well as different smart

Material-Integrated Intelligent Systems: Technology and Applications, First Edition.
Edited by Stefan Bosse, Dirk Lehmhus, Walter Lang, and Matthias Busse.
© 2018 Wiley-VCH Verlag GmbH & Co. KGaA. Published 2018 by Wiley-VCH Verlag GmbH & Co. KGaA.

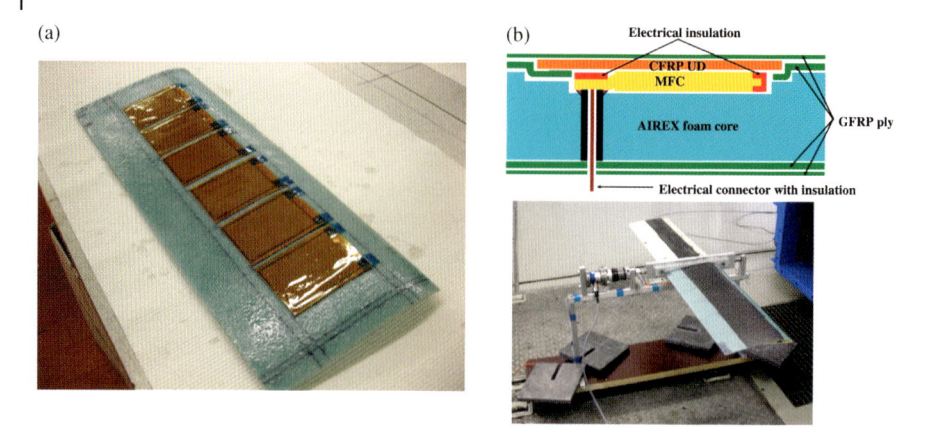

Figure 8.1 MFC applied on foam core (a) [2]; stack sequence and fabricated active wing structure (b) [2].

materials, such as piezoelectric transducer elements, that can be used in aircraft applications.

An alternative way to integrate piezoelectric transducers directly into sheet metals was discussed in Ref. [6]. A piezoceramic ring with steel end caps is positioned inside an aluminum tube with a larger diameter. Except for the region with the piezoceramic ring, the tube diameter is reduced by rotary swaging. Thus, the piezoceramic element with end caps is joined with the tube, which generates a pressure prestress in the piezoceramic ring (Figure 8.2).

Using another joining by forming process, piezoceramic fibers are integrated into aluminum sheet metal surfaces in Ref. [7]. The sheet metal is first structured by micromilling, which leads to cavities with a length of 10 mm, a depth of 312 μm, and a width of 301 μm. The contacted fibers, which have smaller dimensions, are positioned in the cavities. In a next process step, the height of the protruded webs between the cavities is reduced by microforming with a planar die. The cavity walls contact the piezoceramic fibers and a pressure prestress is

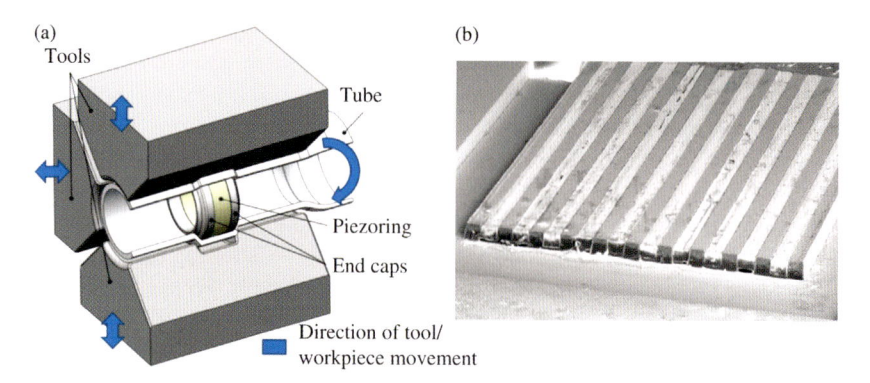

Figure 8.2 (a) Piezoceramic ring integrated in an aluminum tube [6]. (b) Piezoceramic fibers integrated in a microstructured aluminum sheet metal [7].

generated in the fibers. After unloading, the fibers are clamped in the cavities and joined form-locked with the sheet metal. If the web height is further reduced, it is possible to generate a form-fit due to lateral enclosing of the fibers with aluminum material.

Sandwiches with a local reinforcement were investigated in Ref. [8]. Two outer metallic sheets encapsulate an inner-bonded polyolefin foil, which acts as connector between the sheets. Furthermore, the adhesive acts as carrier for a flat metallic reinforcement element. The sandwiches were formed into circular cups in a deep-drawing process. The circular metallic inlays with different sizes are located in the bottom and the side walls of the cups. Although it was not realized, the possibility was discussed to integrate sensor elements instead of metallic inlays into the sandwiches.

Subsequent application of laminar piezomodules is state of the art for enabling a sensor and actuator functionality, for example, for sheet metal parts. This technology is appropriate for the fabrication of a small number of parts. It was widely used within performance studies considering parts with piezoelectric transducers that are joined with the previously shaped structure by adhesive bonding. Discalea *et al.* manufactured two exemplary carbon fiber-reinforced plastics (CFRPs) wing structures bonded with a central spar for a structural health monitoring application in Ref. [9]. Ultrasonic waves were emitted with an MFC actuator bonded with the wing structure. A corresponding MFC acts as sensor and is situated on the other side of the joined spar. The traveling ultrasonic wave crosses the spar bond, which includes artificial bond defects. It was demonstrated that different bond conditions could be monitored with analyses of the resulting measured waveforms. An impact location detection of bonded MFC was studied by Salamone *et al.* in [10]. Sensor signals of two MFC rosettes, each consisting three MFCs, are used in an algorithm that locates an impact onto an aluminum plate regardless of the wave speed in the material. Park *et al.* detected a bolt loosening of a bolt-jointed specimen using impedance–frequency curves measured with a surface-bonded MFC [11]. A repeated bolt loosening and retightening influences the stiffness of the structure and is detected via observation of resulting changes in the impedance plot. The sensor application clearly identifies healthy and damaged conditions of the bolted joint. An example of the reduction of noise by antioscillation at eigenfrequency modes is given in Ref. [12]. Several MFCs are bonded onto a car-body roof structure, acting either as sensor or as actuator. The first five bending eigenmodes were damped with a regularization scheme. As a result, the vibration amplitudes in a broadband excitation test were reduced by 20–30 dB. Kunze *et al.* [13] studied the vibration control ability of an automotive transmission shaft including surface-bonded MFCs. Bending deformations were measured using sensor MFCs. In parallel, the measured signal was used to excite the structure with corresponding actuator MFCs. The authors reached a sound pressure reduction of 12 dB with an automatic identification of regularization parameters.

Aiming at higher quantities, an increased level of automation within a new integration concept is necessary. The direct integration of piezoceramic foil transducers into fiber-reinforced structures can be adapted to generate a highly efficient production process for active sheet metal parts with integrated piezoceramic sensors and actuators. The proposed process chain developed and applied by the authors includes the integration of a laminar piezomodule between two metal sheets using an adhesive. Forming of the sandwich material takes place in a viscous condition of the adhesive, which allows a relative movement between piezomodule and sheet metals. Due to the temporary soft coupling, forming-induced elongations of the piezomodule are drastically reduced in comparison to a fully cured adhesive. Another advantage of the proposed method is the suitability for parts with a complex 3D surface shape. A subsequent application of piezomodules on such structures is rarely possible due to the high module stiffness and afterward bonding techniques.

8.2 Integration Technology

The basic idea is the integration of piezomodules in a semifinished part in a way that allows a subsequent forming without loss of its functionality. Therefore, a relative movement between piezomodule and sheet metals has to be ensured during the forming operation. A sandwich compound was designed, including an MFC (Smart-Material Corp.), which is encapsulated in an adhesive layer between two aluminum sheets. Figure 8.3 shows the different layers of a bending specimen manufactured with this approach of temporary soft coupling of the piezomodule during the forming operation.

In order to achieve a good sensor and actuator functionality, the piezomodule is placed at a distance of 0.25 mm to the neutral fiber. It is important that the piezomodule is not placed in the neutral fiber, so that strains of the piezomodule can be transformed into a deflection of the bending specimen (bimetal effect). Another advantage of this integration technology is a good protection of the MFC against chemical and mechanical influences.

Fabrication of sheet metal compounds with an integrated sensor and actuator functionality includes the following steps. At first an adhesive layer is applied on the basic sheet with a coating knife. Then the MFC is applied in the adhesive and covered by an additional adhesive layer. The contacting of the piezomodules is performed with flexible printed circuits, because of their low thickness and a good electrical insulation. The use of conventional circular electric wires could lead to visible marks on the surface of the part, caused by the soldered joint. Finally, the

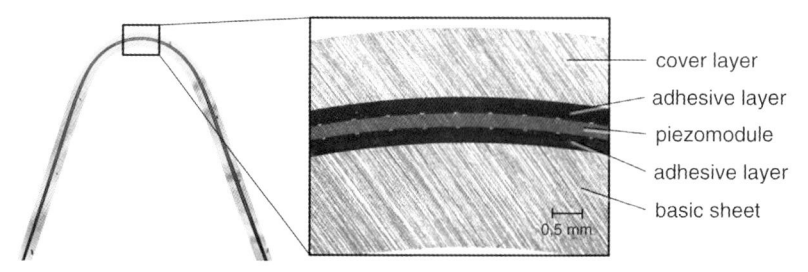

Figure 8.3 Bending specimen with integrated MFC M8528P1, bending radius 10 mm, cross section.

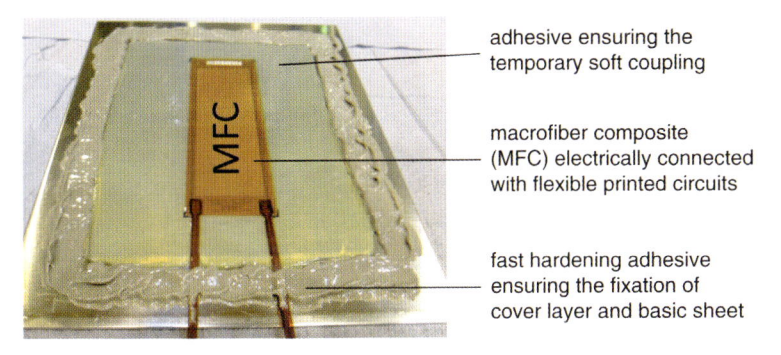

Figure 8.4 Fabrication of piezometal compounds. Basic sheet metal with adhesive and macrofiber composite.

Labels in figure:
- adhesive ensuring the temporary soft coupling
- macrofiber composite (MFC) electrically connected with flexible printed circuits
- fast hardening adhesive ensuring the fixation of cover layer and basic sheet

compound is closed with the cover layer. During the fabrication, it is important that a fixed bond between basic sheet and cover layer is ensured in the outer area. This outer fixation avoids a leakage of adhesive during the forming operation. For the fixation of the cover layer, a fast curing adhesive is appropriate. Figure 8.4 shows a basic sheet metal with the two different adhesives and an integrated MFC.

8.3 Forming of Piezometal Compounds

Piezomodules contain brittle piezoceramic fibers. Hence, small elongations can lead to an overload of the piezomodule. Forming operations induce strains in the sheet metal. A fixed joint between piezomodule and sheet metal would lead to a damage of the piezoceramic fibers in this case. Therefore, a temporary soft coupling of the piezomodule in the adhesive layer is realized during the forming process. The viscous adhesive surrounds the MFC when forming operations are performed. Thus, a relative movement between piezomodule and the sheet metals is possible and the transfer of critical elongations from the sheet metals to the MFC is drastically reduced. The adhesive acts as a lubricant during the forming operation. The fast curing adhesive at the outer area ensures the fixation of cover layer and basic sheet and is cured during the forming operation. This prohibits a leakage of adhesive and maintains clean tool surfaces. After the forming, the adhesive inside the piezometal compound completely cures.

Forming tests were performed using an aluminum alloy (AlMg4.5Mn0.7) for basic and cover sheet metal. The two-component epoxy-based structural adhesive 3 M DP 410 ensures the temporary soft coupling during the forming. The thicknesses of basic and cover sheet metals were varied between 0.8 and 1.5 mm, resulting in compound thicknesses between 2.3 and 3.0 mm with varying distances of the MFC from the neutral fiber. It was found that bending down to a radius of 10 mm did not lead to a functional degradation of the integrated MFC. Further investigations include geometries with 3D surface shapes. Deep-drawing of rectangular cups was performed with two different punch radii. Figure 8.5 shows the forming tool, the specimen type fabricated with a punch radius of

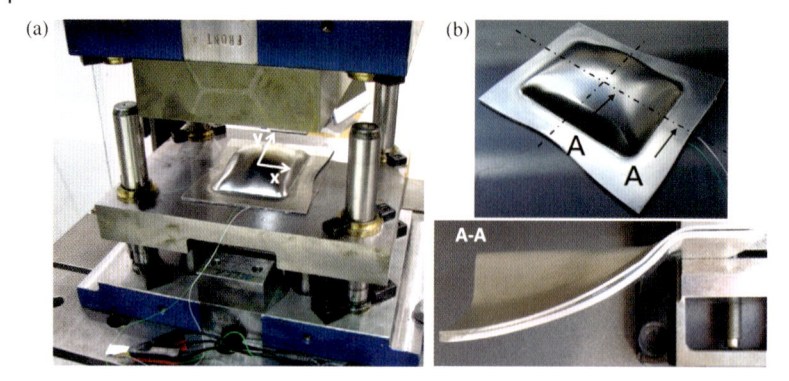

Figure 8.5 Deep-drawing of a rectangular cup with a double curvature: forming tool (a); finished part and cross section (b).

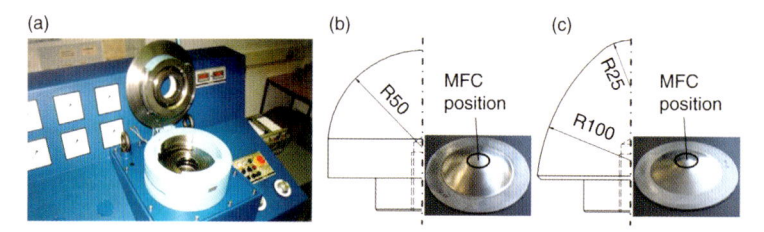

Figure 8.6 Stretch-drawing of rotationally symmetric cups with radii of 50 and 25 mm: testing machine (a), punch geometry and part R50 (b), and punch geometry and part R25 (c).

100 mm in x- and y-directions, and an exemplary cross section of the parts. The other available punch has a double-curved punch radius of 250 mm.

Increased forming loads were achieved by stretch-drawing of rotationally symmetric cups. In addition to higher elongations of the sheet metals induced by the stretch-drawing process, the radius of the punch was reduced down to 50 mm in one case and down to 25 mm in the second case. Figure 8.6 shows the Erichsen testing machine (part (a)) and the geometry of the 50 mm punch and the associated shaped part (part (b)). In Figure 8.6c the geometry of the 25 mm punch and the finished part are depicted.

Despite the temporary soft coupling, during the piezometal compound forming steps with small punch radii, overloads may occur, which can lead to a reduction of the functionality of the compound. Numerical simulations were used to investigate the MFC loading amount. The MFC contains piezofibers, electrodes, and epoxy embedding and covering polyimide layers. The consideration of the exact geometry in the numerical simulation would cause extremely high computation times. Hence, a homogenization strategy with a representative volume element (RVE) was pursued (Figure 8.7). The amount of elastic energy that is stored in the RVE for several load scenarios is used to retrieve elastic properties of the homogenized transversal isotropic material (Ref. [14] and [15]). The inelastic MFC properties were identified in uniaxial tension tests with load direction parallel and transverse to the piezofiber direction. The inelastic material behavior, as a result, is very similar for loading parallel and transverse to the fiber direction.

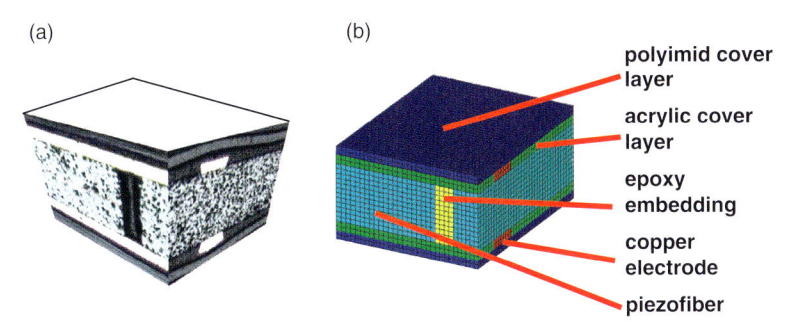

Figure 8.7 RVE for the MFC homogenization approach: geometry from microscopic sections (a) and RVE FEM model (b).

Table 8.1 Elastic and inelastic parameters of the MFC used in the numerical simulation.

Elastic parameter	Value
E_{11} (MPa)	32.14E+03
E_{22} (MPa)	19.50E+03
ν_{12} (−)	0.313
G_{12} (MPa)	6.33E+03
Inelastic parameter	Value
A_0 (MPa)	24.78
K_t (MPa)	120.89

Hence, in the numerical simulation, the inelastic material behavior is described with a simple isotropic hardening law (Eq. (8.1) with yield stress σ_y, starting yield stress A_0, tangent modulus K_t, and true strain φ). The identified elastic and inelastic parameters used in the simulation are summarized in Table 8.1.

$$\sigma_y(\varphi) = A_0 + K_t \varphi. \tag{8.1}$$

The MFC loads during the forming steps were investigated with a numerical model, using four different punch geometries. The four simulation cases include rectangular punch geometries with radii of 250 mm (R250) and 100 mm (R100) and two circular punch geometries with radii of 50 mm (R50) and 25 mm (R25). The resulting MFC strains directly rise with decreasing punch radius. The encapsulating sheet metals cause a double curvature (Gauss curvature) of the MFC, which induces relevant membrane strains. Figure 8.8 shows the von Mises stress distribution for the simulations with punch radii R100 and R50 of the upper sheet metal. Figure 8.9 gives the corresponding punch force/punch displacement curves of experiment and simulation. The exact punch force match for the four punch radii simulations (R250 and R25 have similar conformances with the

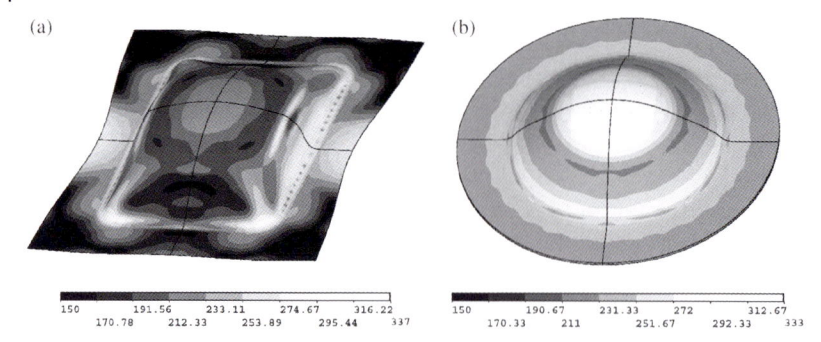

Figure 8.8 von Mises stress distribution in MPa in the numerical simulations of the forming step for deep drawing: rectangular cup R100 (a) and circular cup R50 (b).

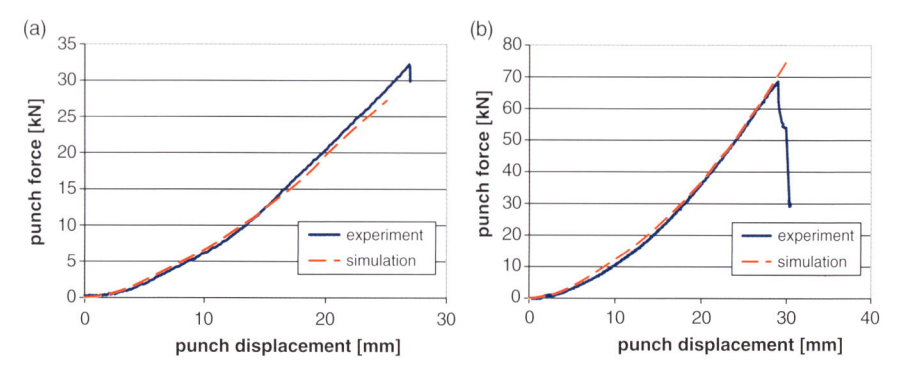

Figure 8.9 Punch force versus punch displacement, comparison between experiment and numerical simulation: rectangular cup R100 (a) and circular cup R50 (b).

experiment) leads to the conclusion that the load boundary condition for the MFC is very well reflected by the numerical simulation.

8.4 Characterization of Functionality

The functionality of piezometal compounds depends on the buildup, the material selection, the coupling between sheet metals and piezomodule, and the functionality of the piezomodule itself. Bending operations or even forming operations of a 3D surface shape with a double curvature of a higher radius do not lead to a damage of MFC, when forming is performed in a viscous condition of the adhesive. For forming operations with a double curvature of lower radii, a test of the MFC functionality is required. X-ray analysis allows the detection and the localization of fiber cracks. However, this examination method is not always easy to apply and leads to high inspection costs. Hence, a method was used, which allows getting information about the functionality of the integrated piezomodule in a simpler way.

MFC consists of piezoceramic rods sandwiched between layers of adhesive and encapsulated between polyimide films with interdigitated electrodes [16]. This

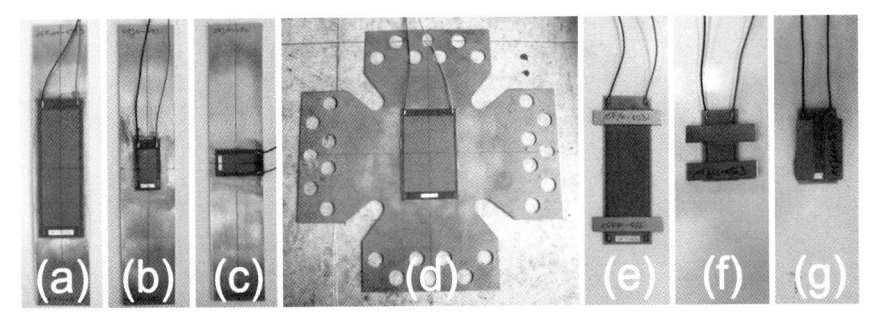

Figure 8.10 Several specimen types were defined to realize different load strain combinations parallel and perpendicular to the fiber direction.

buildup leads to a network of capacitance cells connected in parallel. The sum of the capacitances of all capacitance cells results in the total capacitance of the MFC [17]. If fiber cracks occur as a result of too high forming loads, the total capacitance of the MFC is decreased. In this way a reduced capacitance after the forming operation is an indicator for fiber cracks and a reduction of the MFC functionality. If the capacitance tends to zero, a rupture of the cords is probable.

The measurement of capacitance was performed directly before and after the forming operation. A measurement during the forming operation is possible, but is falsified by the electric charge resulting from the direct piezoelectric effect. In order to get additional information about a possible functional degradation of shaped piezometal compounds, the load levels of MFC were retrieved within forming simulations for different load scenarios with the punch radii of R250, R100, R50, and R25. To evaluate the load level and to compare it with experimental values, a function degradation model based on the maximum strain combination (strain in fiber and perpendicular direction) in the simulation was defined. With basic tension experiments, several strain combinations were realized (Figure 8.10). Specimen types (a) to (c) cause mainly strain in load direction superimposed by moderate perpendicular pressure strain due to lateral contraction of the underlying sheet metal tension strip (the MFC is bonded on the strip). The biaxial specimen type (d) was used to realize several combinations of MFC strains. Tension tests with types (e) to (g) produce pure strain in the MFC in loading direction.

With a loading profile (Figure 8.11), the MFC actuator function was determined for increasing load levels. Figure 8.11 shows the resulting test machine forces and displacements over time for an exemplary tensile test. After a loading/unloading step, the MFC is actuated with a sinusoidal voltage signal in a displacement-controlled holding stage. The resulting forces were measured and an upper and a lower hull curve defined. Subtraction of the hull curves and averaging for each holding stage gives the decreasing functionality of the MFC. All function values were then normalized with the maximum function value at test start. Finally, for several strain combinations, a maximum load level-dependent function degradation is available, which is defined between 0 and 1 (1 ... full performance, 0 ... complete performance loss). The experimental results were fitted in a next step with an analytical model with usage of the

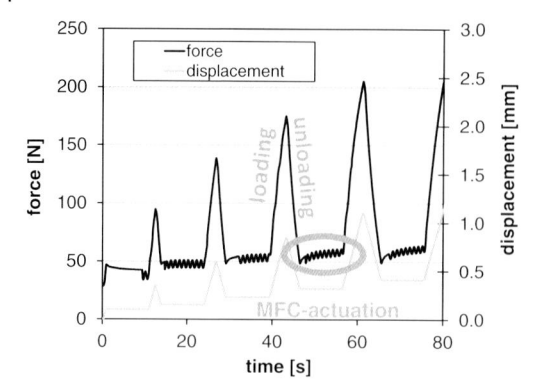

Figure 8.11 Load regime with loading/unloading phase followed by a holding stage in which the MFC is actuated.

mean squared error method (Eq. (8.2)) for the MFC performance degradation (Figure 8.12, $\varepsilon_1, \varepsilon_2$ fiber and transverse strains, Q_0–Q_5 fitting parameters, and π-number Pi).

$$p(\varepsilon_1, \varepsilon_2) = Q_0 + Q_1 \tan\left(\frac{\pi(\varepsilon_1 - Q_2)}{Q_4}\right) \tan\left(\frac{\pi(\varepsilon_2 - Q_3)}{Q_5}\right). \tag{8.2}$$

Evaluation of Eq. (8.2) with the maximum fiber and perpendicular strain combination results in a prediction of the local performance loss distribution in the simulation. To obtain the residual performance P of the MFC, a final summation step over the MFC area A with all MFC elements elem is necessary (Eq. (8.3)).

$$P(\varepsilon_1, \varepsilon_2) = \frac{1}{A} \sum_{\text{elem}} p_{\text{elem}}\left(\varepsilon_{1,\text{elem}}^{\max}, \varepsilon_{2,\text{elem}}^{\max}\right) A_{\text{elem}}. \tag{8.3}$$

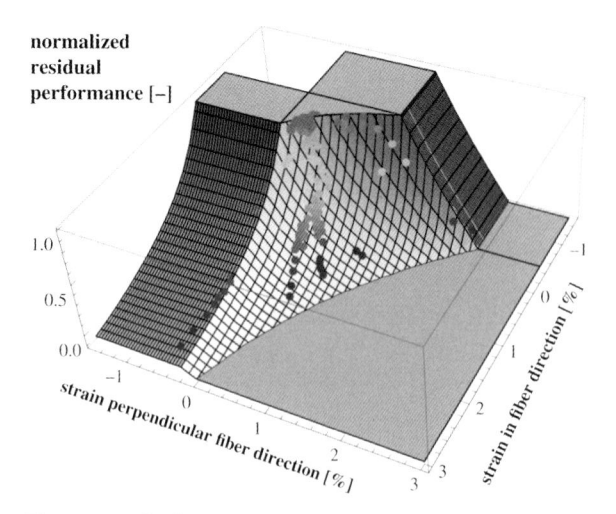

Figure 8.12 Performance degradation model in strain space. Experimental values are plotted with points that are partly covered by the analytical surface.

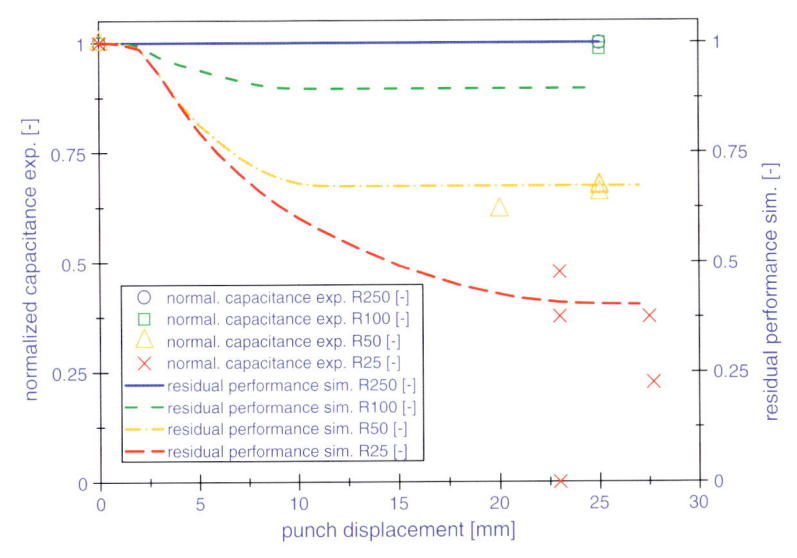

Figure 8.13 Comparison between experimentally determined normalized capacitance and computed residual performance for load scenarios R250, R100, R50, and R25.

Figure 8.13 compares the experimentally determined capacitances before and after the forming experiment with the computed residual performance curves for the different load scenarios with punch radii R250, R100, R50, and R25. The capacitances were also normalized with the maximum starting value. The simulation results are in very good agreement with the final experimental values. As already mentioned, the determination of capacitance curves leads to difficulties during the experiments because the measured capacitance that correlates with the performance is superimposed by additional parts due to electric charge generation (forming loads induce charges in the piezoceramic material).

8.5 Fields of Application

Lightweight construction and functional integration become increasingly part of modern engineering. By the integration of piezoceramic materials in sheet metal structures, an active system with sensor and actuator functionality is achieved. Several industrial branches such as aerospace industry, automotive industry, or medical industry can profit from the specific features of such parts. Sensor applications, like a crash detection or a detection of vibrations, are possible as well as the combination of sensor and actuator functionality for vibration reduction or health monitoring. Many studies deal with health monitoring, energy harvesting, or active shape control [18–20] achieved by a subsequent assembling of piezomodules. Using the presented method of the temporary soft coupling, piezometal compounds are fabricated, which offer comparable functionalities.

The health monitoring functionality is applied to an exemplary assembly of a wing structure. It consists of two shaped sheet metals bolted together on the overlapping area. Both sheets are furthermore bolted to a framework. In one sheet

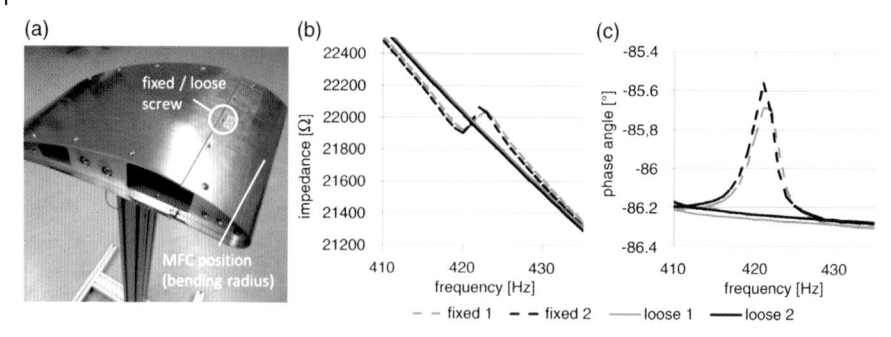

Figure 8.14 Health monitoring of a multiple bolted joint: assembly (a), comparison of impedance signal (b), and phase angle (c) for the two cases fixed and loose screws.

an MFC was integrated using the method of temporary soft coupling during the forming operation. Monitoring of the bolted joint is possible with a measurement of the MFC impedance. The fixed bond between the MFC and the sheet metal surface leads to a coupling between the mechanical impedance of the structure and the electrical impedance of the MFC. Because of the coupling, the measurement of the electrical impedance of the piezomodule gives information about the mechanical properties of the structure [21,22]. Figure 8.14 depicts the assembly of the wing structure (left), the signals for impedance (middle), and the phase angle (right) determined with an impedance meter. Two measurements were performed for the fixed and the loose condition of one screw of the bolted joint. The curves of impedance and phase angle show a significant signal change at a frequency of 422 Hz. At this frequency, the structure has a resonance frequency in case of a fixed screw. If the screw is loose, the resonance frequency has changed and is not visible in this frequency range, and hence a monitoring of the bolted joint is possible.

Another demonstrator part was designed to show the functionality of noise reduction. Therefore, a deep drawn cup with four integrated piezomodules, manufactured with the method of temporary soft coupling, was stimulated with a shaker at different resonance frequencies. A drive of the MFC at the same frequencies, but with a phase shift, resulted in a significant reduction of acoustic noise. Figure 8.15a shows the deep drawn cup, clamped and stimulated with a shaker and a microphone used for the detection of the acoustic noise. Figure 8.15b shows an FFT analysis at a shaker frequency of 233 Hz for the two cases: (a) without an additional piezomodule actuation and (b) with an additional piezomodule actuation at the same frequency with the phase shift. The drive of the four integrated MFC type M5628P1 leads to a reduction of the sound emission of 48 dB.

8.6 Conclusion and Outlook

A production technology is presented, which allows the fabrication of shaped sandwich parts with an integrated sensor and actuator functionality. The specific

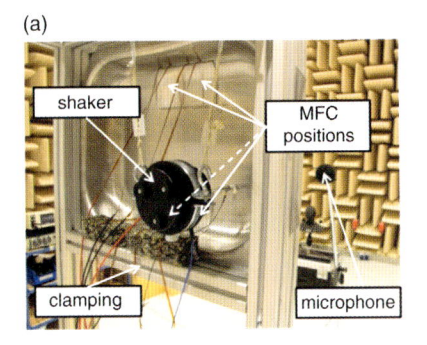

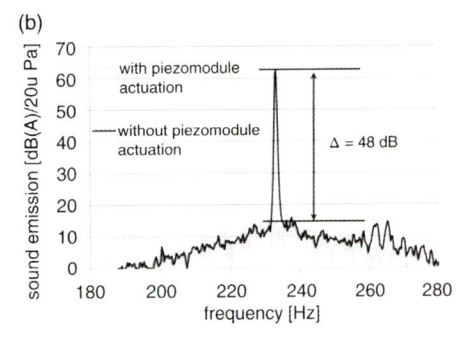

Figure 8.15 Rectangular deep drawn cup with four integrated MFC: setup for vibration reduction and the measurement of acoustic noise (a); reduction of the sound emissions of 48 dB at a frequency of 233 Hz (b).

feature is the approach of temporary soft coupling, which allows a forming of the piezomodules with brittle fibers. Forming takes place in a viscous condition of the adhesive so that a relative movement between sheet metals and piezomodule is facilitated. This method offers the possibility for process automation and the advantage of applicability on 3D surface shapes with a double curvature. The sandwich design leads to a good protection of the integrated piezomodules against mechanical and chemical influences. In experimental investigations, the forming limits have been determined. Furthermore, the fields of applications were pointed out and the potential for health monitoring was demonstrated for a multiple bolted joint. A reduction of vibrations and noise by use of the actuator functionality of the integrated piezomodules was shown for a deep drawn cup stimulated with a shaker. A way of integrating sensors into sheet metal parts is proposed, which enhances the functionality of such structures. Previous work focused on increasing mechanical properties of sheet metal compounds with a similar layup and additional inactive patches or layers [23–25]. The novel layup including an integrated active piezoelectric transducer allows for additional sensor and actuator modes of use in sheet metal parts.

The forming of piezometal compounds into a 3D surface shape with a double curvature can cause an overload in the piezoceramic module inside the compound, which can lead to a reduction of the performance of the compound. Hence, a predictability of the amount of performance loss is essential for a successful part design strategy. With basic tension tests and a specific load profile (increasing loading/unloading), the performance reduction can be measured and is associated with a certain load level. With different combinations of loads in fiber and transverse directions, the load history is also included. The experimental performance degradation results are used to define an analytical model with the mean squared error method. A detailed simulation of the forming process provides the maximum load level, which is used to calculate the residual performance with the analytical model. The elastic material properties of MFC can be retrieved in a homogenization step. The inelastic behavior can be evaluated with tension tests. The comparison between experimentally and numerically determined performances shows a good accordance. The approach facilitates a secure prediction of the residual performances dependent on the load maximum occurred in the fabrication procedure for complex shaped parts.

Nowadays, the cost for a piezoelectric foil transducer is the main obstacle that prevents the mass production of active piezometal structures in efficient process chains. The current production process consists of manufacturing steps under laboratory conditions. However, several studies showed the feasibility of series production methods for piezoelectric foil transducers with reduced costs [26]. Other difficulties lie in insufficient reparability and recycling approaches of such parts. An important point especially for development of recycling strategies is environmental compatibility. PZT ceramics as the state-of-the-art highly efficient transducer materials contain lead, which causes metal poisoning. Previous studies showed that the PZT can be substituted by several lead-free systems if an elaborate material choice according to the appropriate use case takes place (Ref. [27] gives an overview). Although the feasibility of producing sheet metal compounds with an integrated sensor and actuator functionality is given, several aspects for a sustainable and efficient process chain still have to be investigated within future work.

References

1 Rodgers, J. and Hagood, N.W. (1998) Design, manufacture and testing of an integral twist-actuated rotor blade. Proceedings of the Eighth International Conference on Adaptive Structures and Technologies, pp. 63–72.
2 Paradies, R. and Ciresa, P. (2009) Active wing design with integrated flight control using piezoelectric macro fiber composites. *Smart Materials and Structures*, **18** (3), 035010.
3 Tien, C.M.T. and Goo, N.S. (2010) Use of a piezocomposite generating element in energy harvesting. *Journal of Intelligent Material Systems and Structures*, **21** (14), 1427–1436.
4 Braga, A.M., Gama, A.L., and De Barros, L.P. (1998) Models for the high frequency response of active piezoelectric composite beams. Smart Materials and Structures: Proceedings of the 4th European and 2nd MIMR Conference, Harrogate, UK, July 6–8, 1998, CRC Press, pp. 115–122.
5 Surowska, B. (2008) Materiały funkcjonalne i złożone w transporcie lotniczym (Functional and hybrid materials in air transport). *Eksploatacja i niezawodność (Maintenance and Reliability)*, **3**, 30–40.
6 Brenneis, M. and Groche, M. (2012) Smart components through rotary swaging. *Key Engineering Materials*, **504–506**, 723–728.
7 Schubert, A., Wittstock, V., Koriath, H.-J., Jahn, S.F., Peter, S., Müller, B., and Müller, M. (2013) Smart metal sheets by direct functional integration of piezoceramic fibers in microformed structures. *Microsystem Technologies*, **20** (6), 1131–1140.
8 Sokolova, O., Carradó, A., and Palkowski, H. (2010) Production of customized high-strength hybrid sandwich structures. *Advanced Materials Research*, **137**, 81–128.

9 Discalea, F.L., Matt, H., Bartoli, I., Coccia, S., Park, G., and Farrar, C. (2007) Health monitoring of UAV wing skin-to-spar joints using guided waves and macro fiber composite transducers. *Journal of Intelligent Material Systems and Structures*, **18** (4), 373–388.

10 Salamone, S., Di Bartoli, I., Leo, P., Lanza Di Scala, F., Ajovalasit, A., D'Acquisto, L., Rhymer, J., and Kim, H. (2010) High-velocity impact location on aircraft panels using macro-fiber composite piezoelectric rosettes. *Journal of Intelligent Material Systems and Structures*, **21** (9), 887–896.

11 Park, S., Lee, J.-J., Yun, C.-B., and Inman, D.J. (2007) Electro-mechanical impedance-based wireless structural health monitoring using PCA-data compression and k-means clustering algorithms. *Journal of Intelligent Material Systems and Structures*, **19** (4), 509–520.

12 Weyer, T. and Monner, H.P. (2003) PKW-Innenlärmreduzierung durch aktive Beruhigung der durch die Motorharmonischen erregten Dachblech-Schwingungen, in Motor- und Aggregateakustik, Tagungsband zum 3. Symposium Motor- und Aggregateakustik, Magdeburg, Germany (eds H. Tschöke and W. Henze), Expert Verlag, Renningen, Germany, pp. 252–262.

13 Kunze, H., Riedel, M., Schmidt, K., and Bianchini, E. (2003) Vibration reduction on automotive shafts using piezoceramics. *Proceedings of SPIE*, **5054**, 382–386.

14 Kranz, B. and Drossel, W.-G. (2005) Homogenisierung von Materialkennwerten und Lokalisierung von Beanspruchungen bei Kompositen mit Piezohohlfasern, in *37. Tagung des DVM-Arbeitskreises Bruchvorgänge "Technische Sicherheit, Zuverlässigkeit und Lebensdauer"* (ed. M. Kuna), DVM e.V., Hamburg-Harburg, Germany.

15 Drossel, W.-G., Hensel, S., and Kranz, B. (2010) Simulation models for design and production of active structural parts with deformed piezoceramic-metal compounds, in Society of Photo-Optical Instrumentation Engineers (Proceedings of SPIE 7644), Bellingham/Wash.: Behavior and Mechanics of Multifunctional Materials and Composites, Paper 764428, San Diego.

16 Smart Material Corp. (2012) Macro Fiber Composite – MFC. Datasheet SMART DOC CPRO-V2.0en-0311.

17 Sodano, H., Lloyd, J., and Inman, D. (2006) An experimental comparison between several active composite actuators for power generation. *Smart Materials and Structures*, **15** (5), 1211–1216.

18 Di Scalea, F.L. *et al.* (2007) Health monitoring of UAV wing skin-to-spar joints using guided waves and macro fiber composite transducers. *Journal of Intelligent Material Systems and Structures*, **18** (4), 373–388.

19 Song, H.J., Choi, Y.-T., Wereley, N.M., and Purekar, A.S. (2010) Energy harvesting devices using macro-fiber composite materials. *Journal of Intelligent Material Systems and Structures*, **21** (6), 647–658.

20 Kim, D.-K. and Han, J.-H. (2006) Smart flapping wing using macro-fiber composite actuators. SPIE 13th Annual Symposium Smart Structures and Materials, vol. 6173.

21 Peairs, D.M., Park, G., and Inman, D.J. (2004) Improving accessibility of the impedance-based structural health monitoring method. *Journal of Intelligent Material Systems and Structures*, **15** (2), 129–139.

22 Ayres, J.W., Lalande, F., Chaudhry, Z., and Rogers, C.A. (1998) Qualitative impedance-based health monitoring of civil infrastructures. *Smart Materials and Structures*, **7** (5), 599–605.
23 Bothelo, E.C., Silva, R.A., Pardini, L.C., and Rezende, M.C. (2006) A review on the development and properties of continuous fiber/epoxy/aluminium hybrid composites for aircraft structures. *Materials Research*, **9** (3), 247–256.
24 Kopp, R., Abratis, C., and Nutzmann, M. (2004) Lightweight sandwich sheets for automobile applications. *Production Engineering Research and Development*, **11** (2), 55–60.
25 Sokolova, O., Kühn, M., and Palkowski, H. (2011) Lokale Verstärkungen in blechförmigen Metall/Polymer/Metall-Sandwichverbünden. *EFB Tagungsband*, **T32**, 111–126.
26 Hufenbach, W., Gude, M., Modler, N., and Kirvel, C. (2007) Development of novel piezoceramic actuator modules for the embedding in intelligent lightweight structures. *Journal of Achievements in Materials and Manufacturing Engineering*, **24** (1), 390–396.
27 Takenaka, T., Nagata, H., and Hiruma, Y. (2008) Current developments and prospective of lead-free piezoelectric ceramics. *Japanese Journal of Applied Physics*, **47** (5), 3787–3801.

9

Sensor and Electronics Integration in Additive Manufacturing

Dirk Lehmhus[1] and Matthias Busse[1,2]

[1]*University of Bremen, ISIS Sensorial Materials Scientific Centre, Wiener Str. 12, 28359 Bremen, Germany*
[2]*Fraunhofer IFAM, Shaping and Functional Materials, Wiener Straße 12, 28359 Bremen, Germany*

9.1 Introduction to Additive Manufacturing

The term additive manufacturing, short AM, describes a class of processes characterized by almost unparalleled geometrical flexibility and a direct path from the digital part model to its physical realization without the need for tooling. Economically, this straightforward approach in conjunction with the recent, rapid technology development has created significant growth over the last decades, and a firm conviction that this trend will continue: According to Wohlers *et al.*, market volume will increase from $ 4 billion in 2014 at double-digit rates to an expected $ 21 billion in 2020 [1,2].

The common approach behind these impressive figures is defined by ASTM F2792-12a by contrasting it to subtractive manufacturing processes like milling or turning: Whereas the latter remove material where it is not needed, these technologies add material where it is required – hence the term additive manufacturing. The full definition, then, describes AM as the "process of joining materials to make objects from 3D model data, usually layer upon layer, as opposed to subtractive manufacturing methodologies" [3].

Although the ASTM definition leaves room for alternative processes, building up a part layer by layer has in many cases turned out to be the easiest way of guaranteeing full access to any coordinate within the part, and thus clearly dominates the scene: Most AM processes follow this approach in one way or another, and are, therefore, based on virtually "slicing" the digital model of the object to be built into a multitude of individual, parallel layers. The slicing of the part volume typically builds on a surface-based geometry representation contained in a so-called stl file, which can be generated directly from 3D CAD data and as of today represents the most common format understood by AM systems. The alternatives are mostly proprietary and issued by equipment manufacturers, while the AMF format explicitly developed to become a future standard including additional information besides geometry, has not yet penetrated the market. The complete sequence of steps from product idea to final AM part is visualized in

Material-Integrated Intelligent Systems: Technology and Applications, First Edition.
Edited by Stefan Bosse, Dirk Lehmhus, Walter Lang, and Matthias Busse.
© 2018 Wiley-VCH Verlag GmbH & Co. KGaA. Published 2018 by Wiley-VCH Verlag GmbH & Co. KGaA.

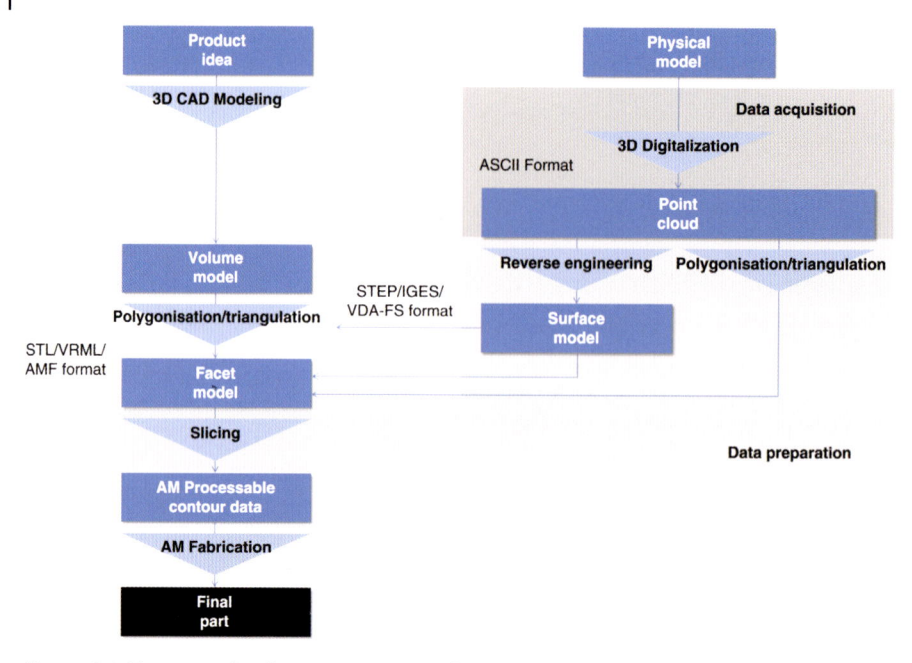

Figure 9.1 The generalized process sequence from product idea to AM manufactured part depicted here, based on Refs [4,5].

Figure 9.1. In it, both the possibilities of conceiving an original part and the alternative of replicating an existing one according to a reverse engineering approach are reflected.

Historically, AM emerged as a prototyping method. While this application has retained its relevance, additional scenarios have become accessible through technological advances. Figure 9.2 qualitatively depicts the timeline and the change in perspective on the general concept of AM since its first introduction.

Early processes like stereolithography, which uses focused laser energy to locally cure a thermoset polymer, were typically not suited for the processing of many engineering materials. Thus originally, additively manufactured prototypes were appropriate for judging appearance and basic geometry-controlled aspects of the product only (Figure 9.2, I a). Progress in processing, introduction of new methodologies, and adaptation to further types of materials with more performant characteristics facilitated the transition from purely design-oriented to functional prototypes, mimicking the characteristics expected from the final part (Figure 9.2, I b/c).

Later generation processes were used not to create the mirror image of the final component, but rather its negative – effectively the tool to produce it from the same material it was meant to be finally made of. Rapid tooling is the term attributed to this strategy, and it is used, for example, in plastic injection molding. Depending on the process and materials employed, such tools can extend from suitability for a limited number of actual prototypes to sufficient durability for smaller scale series production (Figure 9.2, II).

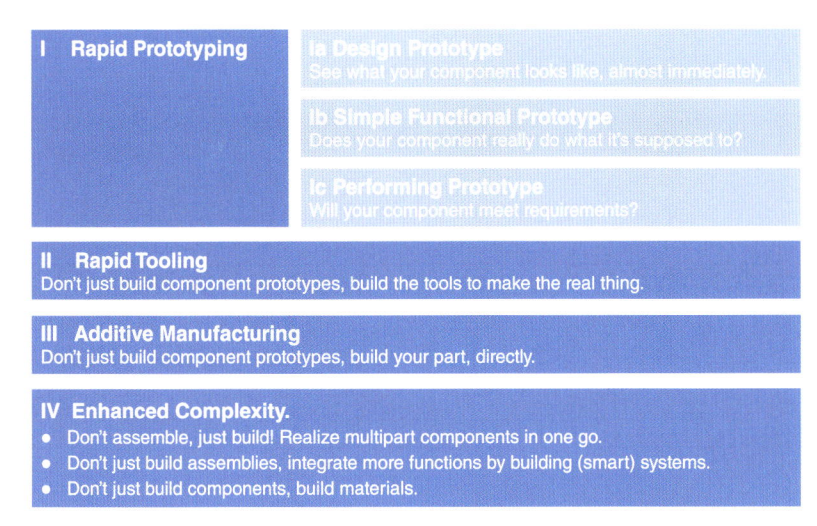

I	Rapid Prototyping	Ia Design Prototype
		See what your component looks like, almost immediately.
		Ib Simple Functional Prototype
		Does your component really do what it's supposed to?
		Ic Performing Prototype
		Will your component meet requirements?

II Rapid Tooling
Don't just build component prototypes, build the tools to make the real thing.

III Additive Manufacturing
Don't just build component prototypes, build your part, directly.

IV Enhanced Complexity.
- Don't assemble, just build! Realize multipart components in one go.
- Don't just build assemblies, integrate more functions by building (smart) systems.
- Don't just build components, build materials.

Figure 9.2 Rapid prototyping to rapid tooling to additive manufacturing (AM): a qualitative timeline extended by some technological visions currently under development.

The following third development stage employs AM as a true production process yielding final rather than prototype parts or prototyping molds (Figure 9.2, III). In 2014, such direct manufacturing accounted for a global market volume of $ 1.75 billion, or a share of roughly 45% of the total AM market in that year [2].

The piecemeal addition of materials behind AM has several advantages: At the foremost, it does not require tooling to shape a part, but just the information where the material is to be added. Thus in principle, the digital model of the part is all that is needed to start production. Lead times can thus be significantly reduced, and design changes accommodated at minimum cost – two factors that act as door-opener toward customization and individualization of products: Think in contrast of the effort of designing a forming tool, and the cost that goes with changing it, should this become necessary.

However, there is a price to pay, and that is, typically, the apparent productivity: Building up a part from scratch and layer-by-layer affords time. Typically, different AM processes are characterized in this respect by specifying the build rate, that is, the part volume or weight created per unit time. This makes sense because in a first approximation, for many processes neither the geometry nor the part size has a dominant influence on this rate, which thus allows a transfer of the value from one component to another. Table 9.1 lists an arbitrary selection of approximate build rates retrieved from literature for different processes and materials.

Naturally, the list can only provide hints at real production scenarios. Actual build rates depend on many side aspects such as material and powder characteristics, allowable layer thicknesses, and so on. This is the case if the focus is on the actual materials consolidation or fusion process. Besides, from a part rather than a material perspective, placement of the specified geometry within the building space has major effect on productivity, as it determines the number of material feed cycles relative to the fused material volume: Imagine a slender cuboid produced either in a horizontal position, or standing upright in this respect.

Table 9.1 Overview of AM processes (for more details on individual processes, see Section 9.2) for polymers and metals comparing typical build rates based on supplier specifications and/or scientific publications.

Materials	Process	System	Build rate (cm³/h)	Reference
Polymers				
Unspecified	Stereolithography[a]	3D Systems SLA 3500	15–36	[6]
ABS	FDM	Denford Up!	10–100	[g]
Unspecified	FDM	Stratasys Prodigy Plus	2.5–9.5	[6]
Unspecified	Polyjet	Objet Geometries EDEN330	10–120	[6]
Unspecified	SLS	EOS EOSINT P385	15–415	[6]
For example, PA	SLS	ProMaker P4000 X	3000–3600	[7]
Metals				
Inconel®	LENS	Optomec LENS® MR-7	up to 12	[h]
TiAl6V4	EBM	Arcam A2X	55–80	[f]
Unspecified	EBM	Arcam Q 10, Q 20	up to 80	[8]
Unspecified	SLM	unspecific	5–20	[c]
Unspecified	SLM	SLM Solution SLM 500 HL	up to 70	[8]
Unspecified	SLM/LaserCusing	ConceptLaser xline 2000R	up to 120	[d]
Unspecified	LMD-p	unspecified	5–25	[9]
Unspecified	LMD-w	unspecified	60–130	[9]
Unspecified	UAM	unspecified	about 500	[e]

Aluminum	WAAM[b]	unspecified/Fronius CMT	about 370–3700	[10]
Titanium	WAAM[b]	unspecified/Fronius CMT	about 220	[10]
Steel	WAAM[b]	unspecified/Fronius CMT	about 380–1270	[10]

Note that due to the rapid development in the field, all figures are exemplary only. Ceramics are not included due to limited relevance in the present context and a comparative lack of data.

a) DLP as further variant of the vat photopolymerization process class achieves higher built rates by using an area projection rather than point-to-point scanning approach. The CLIP variant of the DLP process offers additional advantages in speed [11].
b) Typically, parts produced by WAAM require further machining to achieve final geometry and shape.
c) http://www.rolandberger.com/media/pdf/Roland_Berger_Additive_Manufacturing_20131129.pdf (accessed February 2, 2016).
d) http://www.concept-laser.de/fileadmin/branchenbilder/PDFs/1509_X%20line%202000R_DE.pdf (accessed February 2, 2016).
e) Hehr, A. (June 2016) Fabrisonic LLC, private communication.
f) http://www.ifam.fraunhofer.de/content/dam/ifam/en/documents/dd/Infobl%C3%A4tter/additive_manufacturing-electron_beam_melting_fraunhofer_ifam_dresden.pdf (accessed February 2, 2016).
g) http://www.lboro.ac.uk/research/amrg/amrg-group/facilities/equipment/denfordlimitedup/ (accessed February 2, 2016).
h) http://www.optomec.com/wp-content/uploads/2014/04/LENS_MR-7_Datasheet_WEB.pdf (accessed February 3, 2016).

Not surprisingly, the comparison shows that build rates are best expressed in terms of cm^3/h, with the occasional combination of material and process reaching the dm^3/h level – though specifically for metal parts, the respective processes, such as WAAM, typically only generate the approximate part geometry with machining and thus subtractive processes a mandatory second step. The WAAM example as such is thus also a good example of the message implicitly contained in the listing that accepting lower resolution and/or surface quality may effectively buy a productivity advantage. However, productivity is naturally an area of continuous progress as machine suppliers constantly improve it to increase competitiveness of their products and thus their market share. A further aspect the table underlines is the fact that transferring a basic process concept like SLS/SLM from polymer to metal usually means loosing performance in (volume based) build rates.

From a broader economic perspective, the relatively low speed of production is somewhat compensated by extended process chains in conventional manufacturing with fast cycle times per individual step, but at the same time the need for several such steps to realize complex parts. Thus, the higher the part complexity, the more likely it is that the scales will tip in favor of AM processing.

Besides, if the perspective is not only on manufacturing itself, but on the full product generation process and the time it takes from first idea to series production, the necessity to develop tools, and sometimes even dedicated manufacturing systems, in parallel with products for many conventional processes, which is eliminated by the AM alternative, favors the latter by allowing significant reductions in lead time, investment, and even personnel.

Such advantages open up a number of business cases in which additive manufacturing is competitive. In a recent study, Conner *et al.* have discussed this issue in much detail, categorizing various production scenarios in terms of high or low complexity, customization, and production volume. The ensuing eight categories are then scrutinized with respect to their economic viability. The final finding of this comparison is that almost all fields offer niches for successful implementation of additive manufacturing, with the one exception of traditional, low customization/low complexity mass production – and even to the latter, some secondary access may exist via the rapid tooling approach. Table 9.2 summarizes these conjectures.

With growing maturity of and interest in additive manufacturing, a constant, rapid evolution of system capabilities can be observed (Figure 9.2, IV). A common aim in this respect is managing increasing levels of complexity. The general trend is from homogeneous materials via uniform composites and functionally graded materials (FGM) to full spatial control of material and/or material properties in multimaterial deposition combined with controlled local variation of processing characteristics.

In this context, sensor and sensor system integration as a separate type of complexity is a further research focus, which in the long run may build entirely on multimaterial approaches, realizing the full level of complexity within a single process. Currently, however, embedding of functional structures takes its first

9.1 Introduction to Additive Manufacturing

Table 9.2 AM business cases discussed in terms of production scenarios as characterized by production volume, customization, and complexity according to Conner *et al.* [12].

	Low production volume	
	Customization	
Complexity	Low	High
High	*Complexity advantage:* Manufacturing of the few extended to high complexity parts, for example, with extended functionality. *Products:* GE LEAP engine fuel nozzles (part number reduction) *AM benefit:* Low cost of complexity *Development needs:* Unspecific, general and further complexity increase, for example, multimaterial capability	*Artisan products:* Complex parts for individual use cases, produced in small numbers or unique. *Products:* Track-adapted F1 race car components *AM benefit:* Low cost of complexity/design change. *Development needs:* Further complexity increase, for example, multimaterial capabilities, for improved performance (incl. multifunctionality), improved material performance in general
Low	*Manufacture of the few:* Individual parts or small series, limited customization and complexity. *Products:* Prototypes, tooling, fixtures *AM benefit:* Low lead times and tooling costs, prototyping as original AM realm *Development needs:* Unspecific, general	*Customized for the individual:* Application-specific low complexity parts at low production volume. *Products:* Part repair tasks *AM benefit:* Fast (low lead times), cost-efficient (no tooling), and flexible adaptation to individual needs *Development needs:* Unspecific, general
High	*Mass complexity:* Large numbers of highly complex parts without need for customization. *Products:* Hip implant acetabular cup *AM benefit:* Low cost of complexity *Development needs:* New complexity options like multimaterial capabilities, productivity improvements	*Complete manufacturing freedom:* Full freedom in all three dimensions (complexity, customization, prod. volume) under economic boundary conditions. *Products:* Personalized fashion/jewelry *AM benefit:* Sole economic process *Development needs:* Productivity improvements
Low	*Mass manufacturing:* Complex, high investment cost production system, and tools for optimum productivity in large series production of identical parts. Low to no customization, complexity versus cost balance in cost-driven business models. *Products:* Si-based microelectronics *AM benefit:* None *Development needs:* Dominance of dedicated mass production processes unlikely to be broken	*Mass customization:* Large numbers of parts, each of which adapted to an individual user's preferences. *Products:* Dental braces *AM benefit:* Low cost of design adaptation. *Development needs:* Unspecific, general, productivity increase

steps by merging complementary manufacturing techniques either as chain of individual processes or in a dedicated hybrid manufacturing system.

The market segments targeted by early approaches of this kind are not necessarily specific application domains, but in many cases the replacement – with greatly increased geometrical flexibility – of printed circuit boards: What had previously been restricted to two dimensions can now be transferred to three, thus opening up new possibilities in terms of packaging and making best use of the available design space. Direct production of functional units, such as batteries, without material integration may be seen as related area of research in this respect on which future material-integrated systems will build.

Seen from a higher ranking perspective, sensor integration in additive manufacturing thus brings together the following superordinate trends:

- The increased *importance of smart products*, also on consumer level, within embracing concepts like the IoT, new HMI solutions, or the gathering and exploitation of product life cycle data.
- The *advent of additive manufacturing* and its potential to take over market shares from conventional production technologies in almost all areas except for mass manufacturing.
- The growing *demand for customization* beyond color and the impact this has on suitability of production techniques, which once more suggests a shift away from mass manufacturing.
- The parallel *advent of direct write (DW) techniques,* such as functional printing, for realization of sensors, interconnects, and other sensor system components like passives.

In an attempt to cover these diverse aspects, individual AM processes will be discussed in Section 9.2, including observations on the various process classes' potential for multimaterial processing and thus, finally, single-process sensor or sensor system production. The following Section 9.3 addresses motivations for sensor integration in AM, while Section 9.4 provides selected case studies.

9.2 Overview of AM Processes

AM systems can be categorized based on

- the method used in material deposition, where deposition means any way of adding a finite material volume element (a physical voxel) in a spatially controlled manner and fusing it to the previously deposited material in its vicinity
- the physical state of the material when deposited, or
- the material classes that can actually be processed.

The initially cited terminology standard ASTM F2792-12a is primarily rooted in the "how-to" of adding voxels/layers in the build-up of the part, and here specifically in the means by which material is supplied to the point of build-up and how the provision of the energy necessary to achieve fusion between the present and the preceding voxels/layers is realized [3]. Table 9.3 reflects this perspective and provides links to reference information on usage of the various individual

Table 9.3 Overview of additive manufacturing processes according to ASTM F2792 with a special focus on studies detailing use of processes for individual material classes as well as composites, including a qualitative evaluation of multimaterial capabilities.

ASTM F2792 process class	Individual process	Materials				Potential for 3D controlled multimaterial deposition
		Polymer	Metal	Ceramic	Composites	
Binder jetting	3D Printing	—	[13]	[14,15]	[13]	+
Directed energy deposition	Laser Engineered Net Shaping LENS®	—	[16]	[17]	[18,19]	++
	Directed light fabrication, DLF	—	[20]	—	—	
	Direct metal deposition, DMD	—	[21]	—	—	
Material extrusion	Fused deposition modeling, FDM	[22]	[23,24]	[25]		+
	Multiphase jet solidification, MJS	—	[26]	—		0
	Robocasting	—	—	[27]		0
	Freeze-form extrusion fabrication, FEF	—	—	[28,29]		0
Material jetting	Multijet/polyjet modeling, MJM/PJM	[30]	—	—		++
	Direct printing, DP	[31]	[32]	[15]		
Powder bed fusion	Selective laser melting, SLM or laser beam melting, LBM	—	[33]	—	[34]	—[a]
	Selective laser sintering, SLS[b]	[35]	[36]	[15,37]	[38]	
	Direct metal laser sintering, DMLS	—	[39]	—	[40]	
	Electron beam additive manufacturing, EBAM	—	[41]	—	—	

(*continued*)

Table 9.3 (Continued)

ASTM F2792 process class	Individual process	Materials				
		Polymer	Metal	Ceramic	Composites	Potential for 3D controlled multimaterial deposition
Sheet lamination	Laminated object manufacture, LOM	[42,43]	[44,45]	[15]	—	0
	Plate diffusion brazing, PDB	—	[46]	—	—	
Vat photopolymerization	Stereolithography	[47]	[48,49]	[15,50–52]	[53]	—

a) Ref. [54] has recently shown that with some effort, at least layer-to-layer material variation can be realized in powder bed processes, such as SLM, too.
b) Use of the term SLS for processing of metals is controversial as laser-based powder bed fusion processes typically lead to complete melting of the material. However, use of the term sintering in this context is sometimes justified when only partial melting of metals is foreseen in an approach that can be likened to liquid-phase sintering.

processes for different kinds of materials, covering the basic classes of polymers, metals, and ceramics as well as (uniform) composites, plus a qualitative evaluation of the suitability of the processes for a controlled multimaterial capability that goes beyond the fabrication of homogeneous multiphase materials from previously prepared material mixtures. The latter point is of interest because it indirectly stresses technologies that could in future support a direct integration of functional and/or electronic structures and devices.

In *binder jetting* processes, a bonding agent is deposited onto a powder bed to join the particles in one layer to each other and to the preceding one. The binder may be removed in secondary processes, and the remaining porous structures brought to full density either by sintering or via melt infiltration. *Directed energy deposition* provides a continuous material feed as well as the energy needed to join the material to previously produced layers. *Material extrusion* transfers materials to the building platform by means of dispensing them through a nozzle or orifice. The well-known fused deposition modeling (FDM) process, which uses wire filaments deposited on the building platform for part generation, falls in this category. *Material jetting*, in contrast, is a discontinuous process, using droplets rather than a continuous flow of material. *Powder bed fusion* has the powder bed, which also occurs in binder jetting, but fuses, as the name has it, the particles and layers directly, without bringing in a secondary material simply through provision of energy. In *sheet lamination*, the cross-sectional geometry of each layer is not built up in a point-to-point or voxel-to-voxel manner, but it is cut from 2D materials such as paper, foil, or sheet metal. Each of these cutouts is than bonded to the previous one. *Vat photopolymerization* uses photopolymers that are cured layer by layer by means of a suitable light source. Though last in this list, with stereolithography as the most prominent example, this process really stands at the beginning of the industrialization of additive technologies. Later examples that integrate Digital Light Processing (DLP) techniques employ projection and thus area-based rather than the pointwise curing typical of the original stereolithography process: Productivity is significantly enhanced this way. Here, further improvements have been achieved by allowing oxygen permeation at the bottom of the photopolymer vat: In DLP processes, this is the place where curing actually happens, and adhesion of the built part to the vat surface as well as the time needed for resin to fill the gap between this surface and the previously built layer are rate-limiting factors. Oxygen supply locally inhibits curing of the polymer and thus facilitates maintenance of a liquid interface between part surface and vat bottom in a process variant called continuous liquid interface production (CLIP) [11].

Table 9.4 provides an alternative classification of AM processes based on the state of the material when fed to the build location and the feed geometry based on a similar representation by Isaza and Aumund-Kopp [55]. This type of disambiguation is of specific relevance for sensor integration because it distinguishes processes with a voxel-based access (0D) to the internal material volume from those offering area-based (2D) access only. Naturally, the former offer far greater design freedom when it comes to sensor system integration. This aspect will be briefly taken up again in Section 9.4.

Table 9.4 Additive manufacturing processes and studies with a special focus on controlled multimaterial capabilities.

Feedstock state	Method of construction/basic geometry of feed or consolidation		
	1D channel	1D channel array	2D area
Liquid	Stereolithography (P)	Polyjet modeling (P)	Digital Light Processing (P, M), continuous liquid interface production (P)
Particulate	Laser sintering (P) Laser melting (M, C) Laser-engineered net shaping (M, C)	3D printing (P, M, C)[a]	
Filament/wire, molten during buildup	Fused deposition modeling (P) Wire arc AM (M)		
Solid sheets/foils			Laminated object manufacturing (P, M, C) Plate diffusion brazing (M)

Letters in brackets denote the materials typically processed (P = Polymers, M = Metals, C = Ceramics).
a) "1D channel array" refers to the binder here.

9.3 Links between Sensor Integration and Additive Manufacturing

As discussed in Section 9.1, AM of smart components featuring intelligent integrated sensor or electronic systems is a trend that owes its current popularity to the parallel advent of different emerging technologies and application scenarios such as AM itself, direct write technologies, customization, and the IoT.

Another catchphrase in this context is structural electronics. The definition of this concept, which is basically application independent, stresses a twofold role of electronic components – besides providing their functional/electronic capabilities, they are meant to take over a structural role. An often cited example in this respect is the battery of an electric car, which according to the structural electronics paradigm would not only provide energy to the car, but also contribute significantly to its structural stability. Further examples of structural electronics, as cited, for example, in the introduction to a recent IDTexEx report predicting a multibillion dollar market for the concept in the timeframe 2017–2027, include smart skins for aircraft, printed OLED lighting devices on the surface of a car's body panels, and also civil engineering structures, such as bridges with concrete-embedded sensors, and electronics for health monitoring: Evidently, structural electronics as perceived in this reference is very much the equivalent, or a synonym for material-integrated intelligent systems.[1] In practice, it is not so

---

1 http://www.idtechex.com/research/reports/structural-electronics-2017-2027-applications-technologies-forecasts-000489.asp (accessed November 24, 2016).

much these high-tech applications that are nearest to commercialization, but consumer products in which, for example, structural electronics based on AM and/or DW techniques allow better utilization of a limited design space. The simplest example in this respect are applications that transform the typically flat PCB that provides the interconnects between electronic components into a three-dimensional object adapted to product shape. What additive manufacturing offers in this respect is – besides new geometrical options – the promise of a simplified approach that realizes functionality based on a single process or at least a single manufacturing system, thus dispensing both with several processing steps and complicated tooling needed in other processes suitable for production of 3D PCB substitutes such as the MID technique. To what degree this still remains a challenge will be addressed in the following sections.

Further benefits of material level sensor integration in additive manufacturing become apparent when stepping back from the detail view of a single manufacturing process to the full panorama of manufacturing systems and organization on a larger, even global scale.

Cloud-based manufacturing has introduced shared data that is easily accessible from any place in the world, usually describing the object to be made and facilitating decentralized, distributed manufacturing solutions, though not necessarily outside the original developer's enterprise. Cloud-based design and manufacturing as, for example, described by Wu *et al.* widens and extends this original vision by assuming that most steps of the production processes can be outsourced, or acquired as a service. The range is from Software-as-a-Service (SaaS) all the way to Infrastructure-as-a-Service (IaaS) [56–58].

The result is an almost limitless global market of manufacturing capabilities, resources, and opportunities: With the digital model of the product at hand, the information system in the background (in the cloud?) will tell you where to most economically manufacture it at this moment in time and in relation to current customer requests and orders. The processes in the background can, but do not have to be of an AM type, though AM greatly facilitates this approach based on the assumption that availability of a suitable AM system and the digital product model will suffice to allow production of quality parts. Alternatively, the anticipated low level of preparatory work prior to production can be utilized to produce the part not where this will be cheapest, but where and when it is actually needed – spare part and repair solutions typically follow this approach.

Now what does sensor integration add here, and in which way is it linked to AM in this respect?

Sensor integration in general endows products with the ability to gather life cycle data. This feature is discussed in the relevant literature under the headline product life cycle monitoring, from which product life cycle management or PLM is derived [59–61]. PLM has been discussed by many authors in the context of products stemming from conventional, nonadditive manufacturing processes [62,63] and in this context, it is taken up in this book, too [64].

Abramovici as well as Lachmayer *et al.* see collection of life cycle data as enabler for generation-to-generation product optimization [63,65]. Already, this longer term perspective would profit from implementation via AM processes, based on their inherent design freedom and the limited cost of design modifications.

However, to maximize profit from life cycle information, shorter optimization cycles could be useful, as has been discussed by Lehmhus et al. [66]. Effectively, a process like AM allowing design changes to be brought in virtually without need for physical adaptations along the much shortened manufacturing chain, could go as far as eliminating all product generations, making each part produced an individual that reflects the full set of usage information accessible to the maker at the very moment of its production. AM could thus allow an economically viable realization of continuous product evolution (Figure 9.3) [66].

This generic example is in itself a strong support for developing sensor integration in AM, where the benefits are reaped from the combination of both approaches. There are, however, many more products which profit from both AM and sensor or electronic system integration independently. One such example is prostheses: Individualization through AM allows customization for the individual bearer and thus constitutes an added value in its own right [67]. Sensor integration, on the other hand, can be employed to gather knowledge about real-world loads, to monitor the state of the component in general, or even to act as prerequisite of a sensory feedback to the bearer. The first aspect would again facilitate product evolution, on a personalized level, while the last has no direct link to AM – except for the fact that AM is a much discussed or even preferred process in this application scenario and is deliberately used to realize complex, structurally optimized, and individualized designs that do not lend themselves easily to subsequent sensor integration. Thus, already from this simple manufacturability perspective, concepts for in-process sensor integration would be beneficial.

Switching the point of view toward an application perspective looking at use cases that benefit from sensor integration in general and could be realized via AM techniques, a broad overview has recently been published by Lehmhus et al. [4] and is reproduced in a slightly extended form in Table 9.5.

In most of the examples already listed, the use of AM is not a prerequisite for sensor integration, or for the general viability of the scenario. Instead, the link to AM is based on the assumption that the manufacturing technology as such will continue to penetrate many markets, and thus in the long run, a hitherto employed conventional manufacturing approach may be replaced by AM for reasons not related to sensor integration. However, whenever sensor integration may add to performance or attractiveness of the product, its AM-based realization potential may speed up the switch of manufacturing approaches.

Practical implementation of some such examples by means of AM manufacturing approaches will be discussed in the following section, where the main focus, however, is not on the application, but on the manufacturing concept utilized for sensor integration.

9.4 AM Sensor Integration Case Studies

Practical incorporation of sensors, sensor systems, and their elements in components produced via AM can be distinguished (a) based on the location of the sensor within the component and (b) with respect to the realization of both the

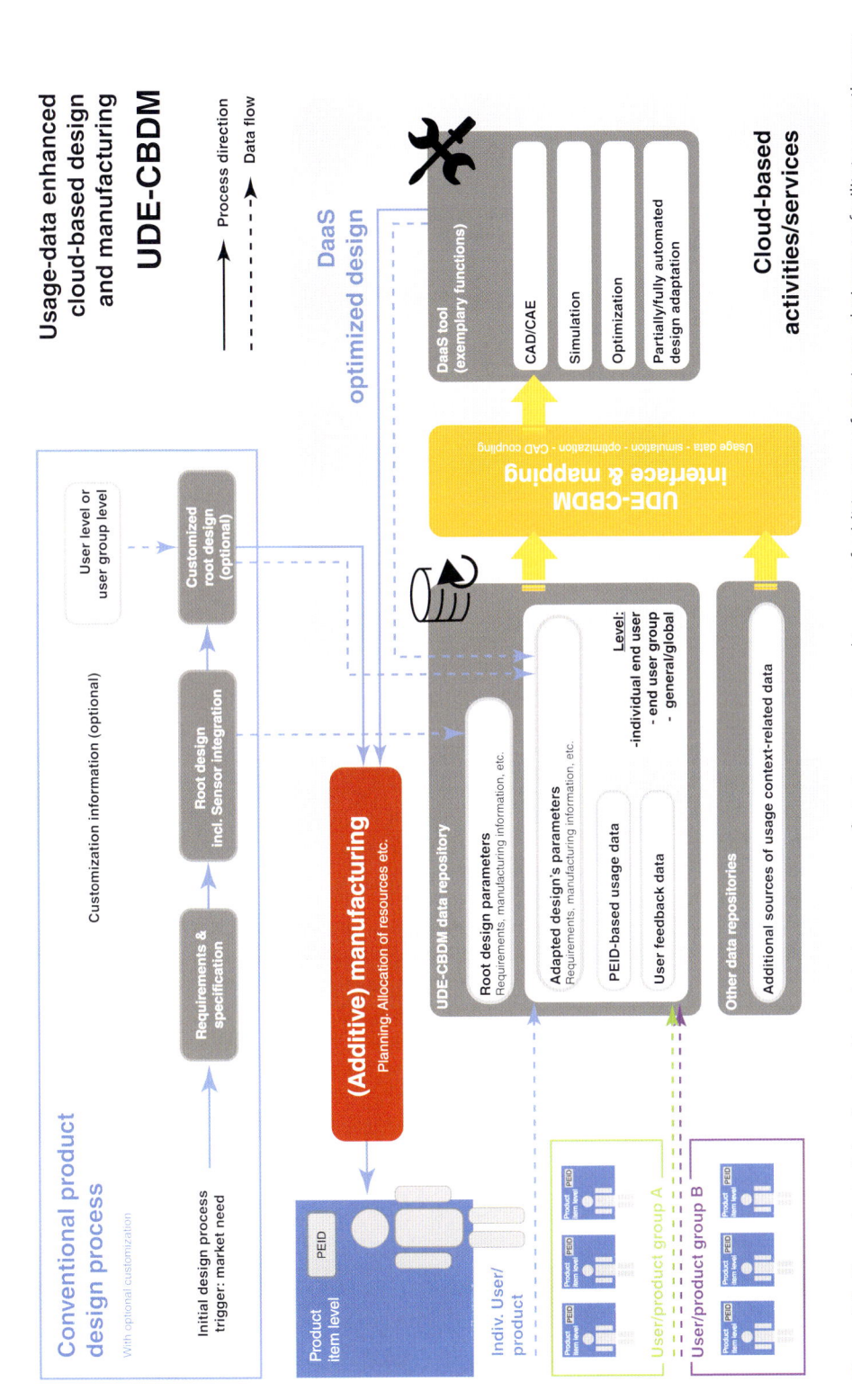

Figure 9.3 General principle of a cloud-based design and manufacturing scenario making use of additive manufacturing techniques to facilitate a continuous product optimization process beyond generation-to-generation updates [66].

Table 9.5 Examples of sensor integration use cases as recently summarized and published by Lehmhus et al. [4].

Application scenario/ motivation	Scenario description *primary part production process (examples)*	Type of sensor *Host material*	References
Anticounterfeiting	Secure identification of objects, for example, as countermeasure against plagiarism. *light metal casting (high pressure die casting/HPDC)*	RFID *Al alloy*	[68,69]
Object tracking	Tracking and localization of objects in production logistics, operating theater, and so on *light metal casting (HPDC), selective laser melting*	RFID *Al alloy, EOS IN718 Ni alloy*	[70,71]
Vibration damping	Monitoring and control of structural vibrations through material-embedded piezoelectric modules. *light metal casting (HPDC)*	piezoel. sensor/ actuator *Al alloy*	[72,73]
Structural health monitoring	Damage detection/localization/ classification up to remaining lifetime prediction in aerospace applications. *Resin transfer molding (RTM)*	fiber optical sensors *CFRP*	[74]
Fly-by-feel	Sensitized airfoil supporting autonomous flight. *composite lamination techniques (envisaged)*	for example, strain gauges, piezoelectric sensors *CFRP (envisaged)*	[75,76]
Production process monitoring	Material-integrated sensors to monitor curing processes. *Resin transfer molding (RTM)*	dielectric IDE *CFRP*	[77]
Robotic tactile sensing	Increased dexterity in robot manipulation, enhanced and safer interaction, for example, with humans. *part production processes match sensor system production processes as the sensor system and substrate is the e-skin*	PVDF piezoelectric sensors *PVDF carrier film, PDMS top layer*	[78–80]
Usage data collection	Sensors integrated in soles of shoes provide usage data, allowing production of user-adapted new versions	pressure sensors *rubber*	[66,81]
Advanced user interfaces	Sensitive materials and surfaces as means for object-user or human-computer interaction. *diverse (abstract concepts discussed)*	for example, mechanical, thermal sensors *diverse*	[82,83]

Note that these scenarios do not rely on AM techniques, but are listed here to illustrate the wide variety of use cases.

Part surface: 1 – surface-integrated conductive path, planar and 3D surfaces
2 – surface-mounted pre-feb. sensor/electr. device, planar and 2½D surf.
3 – surface-integrated sensor/organic electr. device, planar and 3D surf.

Part volume: 4 – volume-integrated conductive path, on build planes
5 – volume-integrated conductive path across build planes
6 – volume-integrated pre-fabricated sensor/electronic device
7 – volume-integrated in situ-fabricated sensor/organic electr. device

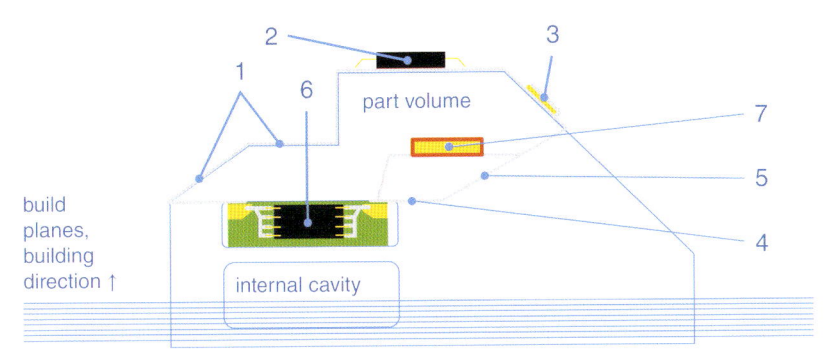

Figure 9.4 Classification of approaches for sensor and electronic system integration in additive manufacturing. (Image originally based on Hoerber et al. [84].)

sensor system and its physical embedding in terms of the manufacturing processes involved and their integration. Figure 9.4 provides an overview based on Lehmhus et al. [4,66] as well as Hoerber et al. [84].

As already stated, a typical characteristic of many AM processes is access to any material volume element (voxel) of a certain, process-dependent size and accordingly the theoretical possibility to manipulate, influence, and modify any such voxel. The prerequisite for full access of this kind is a 0D consolidation strategy, which is not realized in all AM systems, as has already been demonstrated in Table 9.4: DLP, for example, uses area-based consolidation and thus limits modification capabilities to part cross sections. Other processes, such as LOM, do not consolidate but rather join materials during shape generation. As a consequence, cross-plane volume integration is hindered, but foils used as semifinished materials lend themselves easily to external functionalization and equipment with sensor systems in a physically separated process.

The aforementioned examples as well as the distinction provided in Figure 9.4 together indicate that not all types of integration match all AM processes and manufacturing concepts. Table 9.6 classifies sensor integration scenarios according to the positioning of the sensor system on or within the part, and its production. The latter can be done

- in a separate process and manufacturing environment, limiting the actual integration to a pick-and-place operation plus attachment and provision of interconnects,
- through a separate production process realized within the same, hybrid manufacturing system that also hosts the AM process, or
- using the same process that is used for part production to create any required functional substructure.

Table 9.6 Exemplary realizations of sensor integration using additive manufacturing techniques.

Type of integration by geometry		Sensor system versus part production scenario[a]		
		Separate process, separate manufacturing system	Separate process, same (hybrid) manufacturing system	Same process, same manufacturing system
Surface integration	Planar surfaces	Inkjet printing, electroless Cu plating/3DP (binder jetting) [84,85] (Figure 9.4, No. 1, 2)	(Figure 9.4, No. 3)	
	Nonplanar surfaces	Aerosol Jet™/FDM [86] Aerosol Jet, Dispensing/FDM [4,87] Aerosol Jet/SLS [84] (Figure 9.4, No. 1, 2 & Section 9.4.2)	Aerosol Jet, µDispensing/FDM [4,87] (Figure 9.4, No. 3 & Section 9.4.2)	

Volume integration	On build planes	Fiber-optic sensor/UAM [88] Screen printing/UAM, LOM [89,90] μDispensing/DLP [91] Aerosol Jet/PolyJet [92]	Dispensing/Polyjet [93,94] 3 Wire embedding, Dispensing/FDM [95] Wire embedding/FDM [96,97] Dispensing/SLS [84] (Figure 9.4, No. 4, 7)	Dispensing/Disp. [98] Inkjet printing/3DP (binder jetting) [84] c), d) (Figure 9.4, No. 4, 7 & Section 9.4.3)a)
	Transgressing build planes	Dispensing/3DP (binder jetting) SMT devices/SLS [84] RFID system/LBM [99] (Figure 9.4, No. 6 & Section 9.4.1)	Dispensing/SLA [100] Dispensing/FDM [95] Dispensing/photo curing [101]b) [84] (Figure 9.4, No. 5 & Section 9.4.2)	To the authors' knowledge, no such solution exists yet. For polymers, DP and dispensing, for metals, DED processes have the potential for realization. (Figure 9.4, No. 5 & Section 9.4.3)a)

For each individual case, processes used are named (sensor production/AM process).

a) Readers should note that processes are usually added in the field that represents the maximum achievable complexity – that is, a process or process combination that can deposit a sensor within the part volume on or across build planes will also allow surface integration on planar as well as nonplanar surfaces.
b) The study by Muth et al. presents an approach that allows production of 3D sensor structures within a component made from elastomeric materials and consolidated by photocuring, however, this component's geometry is not defined through additive build-up, but requires a mold and is thus in a strict sense not an AM process.
c) The study by Hoerber et al. uses a binder jetting setup in which provision of binder and conductive/functional material are realized via separate inkjet printing systems. Integration using a single, multimaterial printhead would be possible, though. Similarly, the combined DPC/Polyjet process suggested by Perez et al. and Vatani et al. uses separate deposition systems for structural and functional material.
d) The approach discussed by Hoerber et al. has some capability of producing interconnects transgressing build planes, but so far it seems limited to creating interlayer vias rather than true 3D paths within the material volume.

For the various combinations, concrete examples are provided by specifying the respective process or process combination as well as a suitable reference for it. Despite its scope, the table should be seen as providing examples rather than a complete picture. It is supplemented by Table 9.7 providing information on sensors realized in conjunction with or directly through AM processes and giving details on both materials and processes.

In the following section, three distinct scenarios are discussed that represent different approaches to sensor integration in AM, as classified in Figure 9.4 and Table 9.6. These are

- cavity-based full system integration (RFID integration as example) – Section 9.4.1
- multiprocess hybrid manufacturing systems – Section 9.4.2
- dual process material jetting platforms and their potential development toward single process solutions – Section 9.4.3

The focus is not on application scenarios, of which many have already been named in the preceding section, but on materials and processes.

9.4.1 Cavity-Based Sensor and Electronic System Integration

Potentially the easiest way of integrating an intelligent sensor system in an AM component is to foresee a cavity in the component into which this system could then be inserted. This would entail interrupting the AM process prior to producing the cavity's cover layers, and potentially and temporarily removing excess material from it. The electronic system itself could then be manufactured separately, for example, using microelectronics and microsystems technology approaches and thus allowing for several different types of sensors as well as high-performance computation.

In general, all AM processes are suited for this approach, as it assumes an interruption of the building sequence in any case. However, assuming that the integrated system may be temperature sensitive, processes that require a thermal treatment of the prefabricated component following shape generation must be considered critical. This is the case, for example, for 3D printing of metals or ceramics, which as binder jetting process relies on thermal removal of the binder phase and subsequent sintering or melt infiltration. In contrast, processes with local or at least layerwise rather than global, volume-based introduction of energy, such as powder bed fusion, material extrusion, material jetting, or laminated object manufacture, are favored in this respect. Among the vat photopolymerization processes, based on the location of the build-up zone, conventional stereolithography has advantages over DLP and CLIP processes. Nevertheless, even for DLP, solutions of this type have been suggested in which the partially built structure is removed from the building environment, equipped externally with the respective systems, and immersed again in the vat for continuation of the DLP process at the exact point at which it had been stopped.

Basically, this integration strategy involves some pick-and-place operation. Many case studies rely on manual placement of systems to be integrated instead. While this is possible, automation implies that all necessary operations can be

Table 9.7 Examples of sensors and functional structures either integrated in AM parts or produced using AM processes.

Functional component	Material(s)	AM/DW process	References
Interconnect	CuNi alloys, various compositions	Aerosol Jet printing furnace sintering	[102] [103]
Interconnect	Constantan (CuNi44 Mn1 alloy)	Aerosol Jet printing furnace sintering	[104]
Interconnect	Ag	Aerosol Jet printing furnace/el. sintering	[105]
Interconnect	Carbon black particles/silicon oil suspension	Dispensing	[101]
Interconnect	MWCNT/polymer composite	DP/DPC	[106]
Interconnect	Ag (DuPont 5021)	Paste extrusion	[107]
Interconnect	Isotropic conductive adhesive, Ag	Inkjet printing	[84]
Interconnect	Cu wire	Wire embedding in FDM	[97]
Antenna	Ag	Aerosol Jet printing	[108]
Strain gauge	Ag (Adv. Nano Products Co., Ltd)	Aerosol Jet printing	[109]
Strain sensor	Carbon black particles/silicon oil suspension, additional suggestions incl. eutectic GaIn alloys, ionics liquids	Dispensing	[101]
Tactile sensor	MWCNT/photopolymer composite	DP/DPC	[94]
Pressure sensor	MWCNT/photopolymer composite	Combined screen printing and molding, curing	[110]
Thermoelectric generator	ZnO/PEDOT:PSS nanocomposite	Inkjet printing	[111]

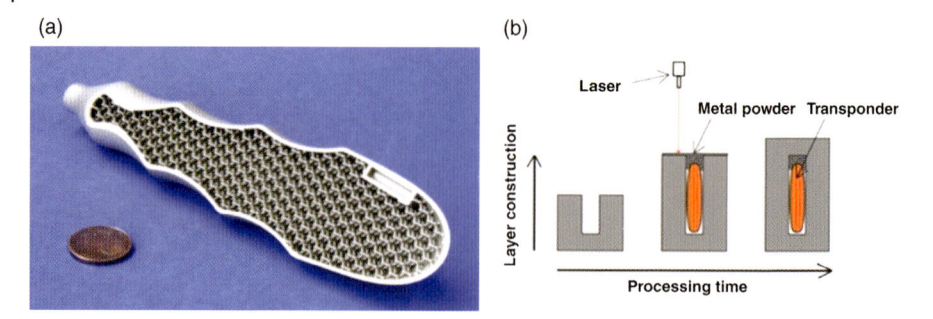

Figure 9.5 (a) Metallic AM part (lightweight surgical handle) with internal structuring and provisions for near-surface volume integration of an RFID transponder. (b) Schematic processing sequence. (Images courtesy of Fraunhofer IFAM, see also Ref. [88].)

performed in a closed environment, which is an important aspect where processes are concerned that, for example, require protective atmospheres, like the SLM process when applied to reactive materials such as aluminum or titanium. In terms of productivity, the integration process can have detrimental effects on cycle times if the electronic system is temperature sensitive and thus affords additional interruptions of the building process to allow for cooling of the already consolidated material volume. As secondary effect, a processing sequence of this kind could also affect part properties, as it might influence cohesion between the last layer manufactured prior to system insertion, and the first one produced afterward.

Case studies following this basic approach for metallic parts have been conducted at the Fraunhofer Institute for Manufacturing Technology and Advanced Materials (IFAM): Among the main objectives in these were part identification and tracking relying on material-integrated RFID systems [4]. Application scenarios include production logistics, anti-counterfeiting measures, PLM, or tracking of medical/surgical instruments [70,99]. A specific product example of this kind is depicted in Figure 9.5.

A primary characteristic of cavity-based approaches is the fact that the integrated system usually has no mechanical connection to the part itself. This can be an advantage if that system needs to be decoupled from mechanical and potentially also thermal loads acting on the component. It is, however, a drawback for the very same reason if the system in question is meant to measure mechanical or thermal loads within the respective component, from which the loose embedding would detach it. A partial remedy may be devices or systems produced on substrates that allow at least one-side formation of bonds to the host material via the substrate material employed. While such approaches have not yet been reported for metallic structures, polymer-based solutions of remotely similar kind have been evaluated: One example in this respect is provided by Chang *et al.*, who have printed functional structures on temperature-resistant substrates showing low strength of bonding to the printed lines, cured the latter and transferred them to the current build plane of a PolyJet-processed part using a print-stick-peel-approach, that is, by making use of the higher adhesive forces between printed structure and the AM part as the new compared to the original substrate [92]: For metallic parts and production processes like LBM, it would seem possible to create the sensor or interconnect on a substrate that could effectively be laser-welded to the AM part. In both cases, however, the prerequisite is that the integrated system be thinner than the typical build layer thickness of the respective AM process. For LBM processes, this requirement would translate into a thickness of system and substrate combined well below 100 µm.

A secondary drawback of the cavity-based approach is the added difficulty of producing interconnects for data and/or energy transmission to and from the integrated system. RFID systems pose fewer problems in this respect as they allow wireless communication. Metallic parts, however, generally complicate the issue, as they either electromagnetically shield the system from communication in wireless approaches unless the system is embedded adjacent to the part surface [4], or require additional insulation layers in case of solutions based on physical, wired interconnects. In polymers, provision of channels that are filled with a suitable conductive medium after the AM process represents an elegant way of circumnavigating this challenge [112].

9.4.2 Multiprocess Hybrid Manufacturing Systems

This chapter provides examples of systems that tackle the challenge of sensor integration in AM parts on the level of surface and volume integration and based on approaches that combine several different manufacturing techniques in a single setup, that is, a hybrid manufacturing system. Primary examples in this respect stem from Fraunhofer IFAM (surface integration) and the University of Texas at El Paso (UTEP, surface and volume integration) [95,96].

Surface integration on nonplanar, 3D surfaces is a general challenge in sensorization of engineering components, consumer goods, and so on, since most processes for sensor and sensor system manufacture rely on a planar buildup. In principle, direct write methods can deal better with this challenge than, for example, film deposition or mask-based techniques. In conjunction with AM processes for part manufacture, the potentially higher geometric complexity of parts as well as their usually limited surface quality raise further difficulties.

To meet the specific requirements of sensor surface integration on complex surfaces, a so-called modular manufacturing platform (MMP) has been developed and set up at Fraunhofer IFAM that combines several surface coating and functionalization as well as supporting pre- and postprocessing modules [4,113,114]. The general concept is depicted in Figure 9.6. The primary advantage of this concept is its high versatility, as well as the possibility to adapt the overall concept as realized on lab scale to requirements of application-specific production environments by adding or removing individual modules. This flexibility is achieved via the transfer system, which links the individual processes using trays pin-coded for identification by the robotic handling system as well as the various processing units. The tray-based identification also facilitates the realization of a simple overall control concept that currently covers workpiece flow from station to station via the processing units communicating their status and availability to the transfer system. Programming of the individual processing

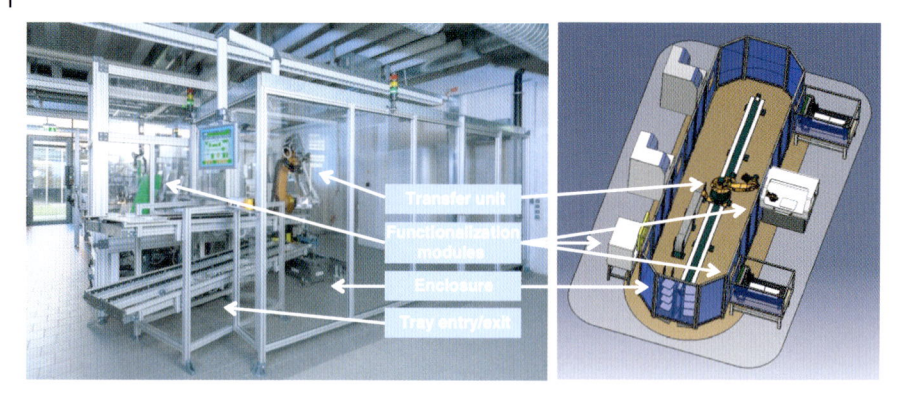

Figure 9.6 Modular manufacturing platform for 3D/nonplanar surface functionalization and sensorization using a broad portfolio of direct write processes and focusing specifically on processing of complex shape AM parts: Photograph and schematic representation of setup at Fraunhofer IFAM (Images courtesy of Fraunhofer IFAM) [4].

stations has to be done separately on unit level, with the individual devices executing the pre-programmed sequence whenever a new tray comes in. For a production environment, association of processing parameters to the identification of either parts or workpiece carriers can easily be envisaged.

In its current setup, the MMP comprises three main functionalization units, namely, the jetting testbed module (JTM), the inkjet printing module (IPM), and the screen printing module (SPM). Of these, the JTM integrates several functionalization as well as characterization steps, including Aerosol Jet™ printing, microdispensing, rotary microvalve dispensing, precision spraying for defined area coverage, and high-accuracy stylus-based geometry and topography capture. Details of these, plus the added capabilities of IPM and SPM, are given in Table 9.8, covering among others (where applicable) the achievable resolution for each of them. Complementing this choice are solutions for postprocessing such as furnaces, for example, for sintering or curing of conductive tracks deposited via any of the aforementioned processes.

Thanks to the flexibility of the MMP, selection of functionalization processes, and above all process combinations, can be adapted to the dominating requirements of each individual element of a specific sensor system and application. These practical capabilities in conjunction with AM processes are illustrated in Figure 9.7, which shows part of a leg prosthesis produced by means of FDM from PLA, a thermoplastic polymer. This part has a complex, nonplanar surface on which a strain gauge has been printed using the Aerosol Jet™ printing process (lower region of the part). This strain gauge is connected to two contact pads in the part's upper region, which like the conductive paths were realized via microdispensing. Choice of the different processes was influenced by the need for better control of geometrical features as well as material and thus sensor properties (resistance) in the case of the strain gauge.

In contrast to surface integration, a *volume integration* either with or without transgression of build planes tends to afford a tighter coupling of processes for economically viable realization of part and sensor, leading from an integrated

Table 9.8 Functionalization processes integrated in the modular manufacturing platform and their main characteristics [4].

Functionalization process	Module	Description	3D capability (qualitative)	Resolution (typical range) [μm]
Aerosol Jet printing	JTM	Optomec Inc., Albuquerque, USA, aerosol jet printhead.	+[a]	10–100
Microdispensing	JTM	Vermes Microdispensing GmbH MDS 3000	+[a]	100–1000
Rotary microvalve dispensing	JTM	Techcon Systems, used for deposition of higher viscosity materials in contact with the substrate.	+[a]	100–1000
Precision spraying valve	JTM	Techcon Systems, used for homogeneous surface coatings incl. coverage of larger areas.	+[a]	n. a. area-based
3D scanning	JTM	Stylus tip-based geometry/topography capture, used in track planning for direct write surface functionalization. Accuracy (repeat) 15 μm, max. positioning fault<8 μm.	+[a]	3 μm measuring
Inkjet printing	IPM	Inkjet printing module with variable printheads	○	40–200
Screen printing	SPM	Screen printing module suitable for structured and area coating.	–/○[a]	40 → area coating

a) Determined by current three-axis JTM setup, industrial processes allow even better (++) 3D capability.
b) Special device integrated for printing on rotationally symmetric surfaces (via part rotation) available.

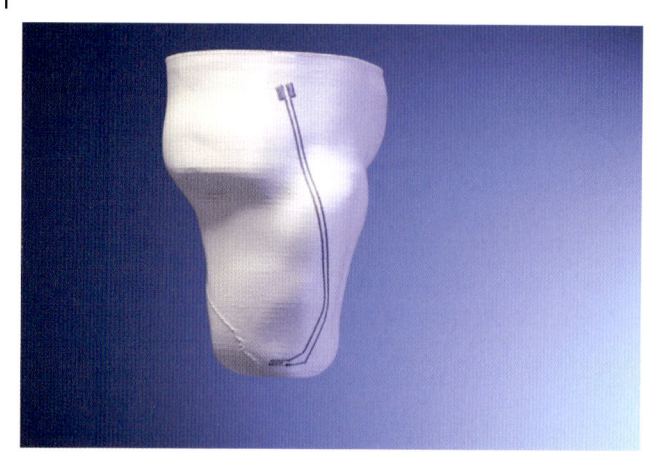

Figure 9.7 Example of a component produced via the modular manufacturing platform. Part of a leg prosthesis made from PLA using the FDM process. Strain gauges shown on the nonplanar 3D surface were Aerosol Jet printed, while the interconnects and contact pads were deposited via microdispensing techniques (Images courtesy of Fraunhofer IFAM) [4].

manufacturing cell as described above to an integrated hybrid manufacturing system: A major point in this respect is that volume integration implies switching of processes during part build-up, while surface integration can be realized in a process sequence, during which volume generation is finished first. A prominent example that follows the hybrid manufacturing system approach has been developed by the University of Texas at El Paso [95].

The system as such is described in its original configuration by Espalin *et al.* [95]. This solution includes two FDM systems that allow production of bimaterial components without having to exchange feedstock, as would have been necessary in a solitary FDM device. The build platform is transferred between these using a linear axis, which also facilitates exact positioning for application of other technologies. Among these are processes for creation of conductive paths such as microdispensing or embedding of metal wires within the polymer material laid down by FDM. To realize very fine geometrical features and to create trenches in which functional materials can be deposited at greater dimensional accuracy, a micromachining system has also been added. The setup is complemented by a pick-and-place device, which facilitates automated addition of externally produced microelectronic devices or microsystems. Contacting these embedded devices can then again rely on the aforementioned processes, following which operation the build-up of the part – by means of FDM – can be continued.

A synopsis of the various systems is given in abbreviated form in Table 9.9, which thus corresponds to Table 9.8 detailing the modular manufacturing platform.

The reader may note that the solution proposed by UTEP includes, as one of the many integrated processes, cavity-based integration as already described in the preceding section of this chapter using the pick-and-place system.

Even though the system has been conceived from the start as modular and allowing adaption via choice of modules in relation to specific product

Table 9.9 Building, functionalization, and secondary processes integrated in the hybrid manufacturing system available at UTEP [95].

Subsystem functionality	Description Maker/type
Component build-up	2FDM systems
System-to-system transfer	Sliding workpiece/building platform with pneumatic actuation
Feature generation (subtractive)	Micromachining system for realizing high resolution and surface quality features/CNC router LC 3024, Techno, Inc., New Hyde Park, NY, USA
Component integration	Pick-and-place system for integration of externally fabricated microelectronic components or microsystems
Conductive path generation	Microdispensing system, for lower conductivity requirement interconnects/Ultra 2400 Series, Nordson EFD, Westlake, OH
Conductive path generation	Wire embedding system, also capable of building plane transgression, for high conductivity requirements, embedding based on ultrasonic or Joule heating energy supply and thermoplastic substrates, wire-Ø 80–320 µm

requirements, more recent developments at UTEP seem to point toward lower levels of integration and thus a manufacturing cell concept similar to the aforementioned MMP. The respective system has recently been described by MacDonald and Wicker: It links once again two AM systems with a wire embedding station. Like in the MMP approach, transfer of workpieces is realized via a robotic system, which can also be tasked with component assembly [96].

9.4.3 Toward a Single AM Platform for Structural Electronics Fabrication

The fact that some additive manufacturing and direct write technologies are closely related has already been mentioned. Inkjet printing as used in the PolyJet and related processes, but also in direct write approaches, is just as obvious an example as the similar link that exists between the LENS™ AM and the Aerosol Jet™ printing process. In practice, this means that there is a medium-term chance at least for realizing the structural, sensors, interconnects, and (simple) passives part of material-integrated intelligent systems or structural electronics components using a single building process.

Today, systems are commercially available that combine no more than two processes to implement this capability. Prominent among these are the devices offered by Voxel8, which are based on a single platform combining two extrusion systems for FDM and microdispensing and originate from the activities of Jennifer A. Lewis group at Harvard University [115].[2] In its standard configuration marketed today as Developer's Kit, the system is delivered with PLA filament and conductive silver ink, but extending the range of materials is among Voxel8's primary

---

2 http://www.voxel8.com/technology/ (accessed November 29, 2016).

development aims. Additions to the portfolio include, on the structural side, rigid epoxies, polycarbonates or PET as well as soft silicones or polyurethanes, while on the electronic side, different types of silver, carbon, and dielectric particle-based inks are offered. Besides integrating two building systems, Voxel8's device is configured for simple removal and exact repositioning of the building platform to allow external (automated or manual) pick-and-place operations.²

In contrast to Voxel8, Stratasys' PolyJet systems build on a process that combines material jetting techniques like inkjet printing with photo polymerization. So far, to the knowledge of the authors, no structural electronics component has been produced in the latter way, though the process itself clearly would offer great potential for this: Stratasys has introduced a digital material concept that facilitates setting of the local final material composition and properties by mixing the final ink's components immediately before deposition, and in varying ratios. This allows coloring of materials as well as locally varying properties like stiffness, but besides, given that the filler added with an ink's second component would provide conductivity, it should also enable deposition of conductive paths and simple sensors at any place within the part, showing any orientation that is asked for, even transgressing build planes. Some examples of current products making use of the color and/or transparency option may serve to highlight these possibilities. However, not even the white paper on PolyJet materials published by Stratasys contains any hints at conductive or sensor materials [116].

Further methods technologically similar to the PolyJet process have been suggested by other researchers, too, using alternative designations like multijet modeling (MJM), direct printing (DP), or direct print/cure (DPC). Common principle in all cases is that an inkjet process is used to deposit fluidic materials that are then solidified by photo or UV curing.

In conjunction with sensor integration, one such example is presented by Perez and Williams, who discuss a further two-process system in which PolyJet technology provides the structure and thus the substrate for conductive traces that are created via paste extrusion. The respective study is not so much on specific applications, but focuses instead on the development of process and material combinations comparing different PolyJet standard materials in their role as substrates [107].

Lu *et al.* have followed the opposite path in their study, again looking at separate processes for sensor/interconnect and substrate production: While the former originate from a dispenser-based DPC process, the latter is generated via a polymer-based AM technique not discussed in any detail in the respective study. The result may serve as replacement for conventionally produced MID, with the advantage of added geometrical complexity both in terms of substrate shape and routing of interconnects and thus the potential to make better use of the available design space [106]. How such extended geometrical freedom can be exploited for system miniaturization has been demonstrated by Espalin *et al.* based on several consecutive generations of a three-axis magnetic flux sensor system [95].

Vatani *et al.* have moved further toward a single process by using DLP-like projection-based curing both for the consolidation of the part itself and the conductive traces. While the former emerges from an SLA or rather DLP process, the latter are once again deposited using a dispensing system [94].

An interesting alternative is the application of 3D printing approaches, that is, a binder jetting and thus powder bed process, to create conductive paths via the binder supply system. Usually, the role of the inkjet-printed binder is simply to selectively join the particles in the powder bed. Depending on the process and application, removal of the binder and subsequent sintering can yield pure ceramic or metal parts [117]. Hoerber *et al.* in contrast suggest using inkjet printing not only for deposition of the binder but also for creating conductive paths using, for example, silver nanoparticle-based inks. In their study, material for conductive structures uses a secondary inkjet printing system, however, in principle using a multimaterial printhead both binder and functional material could be provided through the same system. Since the metal content realizable through direct printing in the powder bed is limited, variants of the method suggest removal of unbound powder prior to printing the conductive ink, which would then fill the channel created by powder removal. The need for such an intermediate process step currently also forbids creating true 3D conductive paths within the part volume. However, 3D connections can nevertheless be made on the basis of interlayer vias [84].

In conclusion, it must be conceded that apparently and up to now, there is no single processing platform that would create the structural and the electronics part of a material-integrated intelligent system using the same manufacturing process, even though DP or DPC processes seem to combine all the necessary characteristics, not the least because of their close relation to inkjet printing, one of the most prominent DW processes used in product functionalization.

When it comes to optimum performance, the fact that materials for the DP class of processes naturally come as inks often incorporating the functional components as (nano-) particulate filler may introduce certain thresholds:

The production of such inks is necessarily a multiobjective optimization problem in which requirements like processing characteristics (e.g., viscosity ranges, which are limited especially for inkjet printing) or stability (e.g., need for stabilizing agents) have to be balanced with the achievable product performance.

Despite the fact that this conflict may limit the applicability of single-process sensor integration solutions (as is illustrated by the development of the copper wire embedding technique described earlier [97]), the apparent ease of the DP route is likely to secure it an appreciable market share.

Even if realized, it is equally clear that any single structural electronics manufacturing process could not dispense with foreseeing either integrated or added pick-and-place operations: For true material-integrated intelligence, like all other approaches, a solution of this kind will have to build on conventional microelectronics to realize fast and complex data processing. Interesting in this respect will be solutions that integrate the separately provided data evaluation systems more tightly than is possible today.

9.5 Conclusion and Outlook

In a sense, when applied to sensor integration, AM techniques may seem to blur the boundaries between material and component integration. This, however, is mainly due to the fact that material and component are effectively built in parallel where AM is concerned. Aside from this specific aspect of an emerging manufacturing technology, the basic capability of several AM processes to deliver sensors and even sensor systems tightly embedded in a material is out of question.

The specific appeal of sensorization in an AM context is rooted in the fact that the increasing relevance of customization experienced today in the field of consumer products, which is just discovering AM as an extraordinarily powerful tool to meet new customer demands, can be taken to completely new levels if the step from passive to intelligent, interactive products is made.

To achieve this, suitable materials and processes or process combinations are mandatory. Until today, many steps in this direction have already been taken. However, like other methodologies that closely integrate components and sensor/electronics equipment, AM will also require codesign methodologies that cover the primary part functionality plus that of the electronic system and reflect the link between both: Only then can the specific benefits of sensor system integration be fully exploited. Currently, awareness of this issue is on the rise with the first related publications available [118,119] – this at a time when the notion that AM in general may require new design philosophies is just about to penetrate the thoughts of design engineers [120].

Whether advanced concepts like usage-data-enhanced cloud-based manufacturing and design will see broad implementation will largely depend on the overall development of the manufacturing sector. Currently, sensor integration is very much linked to high-value goods, and the scope of such products which benefit sufficiently from regular updates with associated optimization is narrower still. Besides, it is specifically these high-value products that are typically based on the highest levels of engineering competence and thus represent the core competencies and the competitive edge of their maker – thus it is not at all clear whether the distributed, cloud-based manufacturing concept can apply to them, unless it is used for intracompany organization and performance optimization.

With this in mind, it seems that high-level consumer products like sporting goods and potentially also medical devices like prostheses may develop into the first products that apply such methodologies – unless manufacturing costs can be reduced by such a degree that low-cost consumer goods sporting lower requirements in terms of lifetime and reliability become economically attractive.

Abbreviations

Abbreviation	Meaning
3DP	3D printing, a powder bed/binder jetting AM process
ABS	acrylonitrile butadiene styrene, a thermoplastic polymer
AM	additive manufacturing
CBDM	cloud-based design and manufacturing
CLIP	continuous liquid interface production

(*Continued*)

Abbreviation	Meaning
DP	direct printing, similar to the PolyJet process
DPC	direct print/cure process, synonymous to DP
DLP	Digital Light Processing/Direct Light Printing/Digital Light Projector
DMLS	direct metal laser sintering
FDM	fused deposition modeling
FFF	fused filament fabrication, synonym to FDM (Second use: free-form fabrication, synonym for AM)
FOS	fiber optic sensor
HMI	human–machine interface
IDE	interdigitated electrode
IGES	initial graphics exchange specification
IoT	Internet of Things
LBM	laser beam melting, generic name for processes like SLM
LENS	laser-engineered net-shaping, designation of a directed energy deposition process commercialized by Optomec
LMD	laser material deposition
MID	molded interconnect device
MJM	multijet modeling, similar to PolyJet Modeling
MMP	modular manufacturing platform
MWCNT	multiwall carbon nanotubes
PA	polyamide thermoplastic polymer
PCB	printed circuit board
PJM	PolyJet Modeling, the PolyJet process as offered by Stratasys, based on inkjet printing and subsequent curing of photopolymers
PLA	polylactic acid, a thermoplastic polymer
PLM	product life cycle management
RFID	radio-frequency identification
SHM	structural health monitoring
SLA	stereolithography (the AM process)
SLM	selective laser melting
SLS	selective laser sintering
SMD	surface-mount(ed) device, microelectronic system suitable for surface mounting and contacting and so on, for example, on a PCB
SMT	surface-mount technology, the approach behind SMD
STEP	STandard for the Exchange of Product model data
STL	stereolithography (describes the process, but also the common AM file format that was originally introduced for stereolithography processes and thus bears their name)
UAM	ultrasonic additive manufacturing
UAV	unmanned aerial vehicle
VRML	virtual reality modeling language
WAAM	wire arc additive manufacturing

References

1 Wohlers, T. (2014) Wohlers Report 2014: 3D Printing and Additive Manufacturing State of the industry: Annual Worldwide Progress Report. Wohlers Associates, Inc., Fort Collins, CO, USA.
2 Wohlers T, Caffery T. (2015) Wohlers Report 2015: Additive Manufacturing and 3D Printing State of the Industry: Annual Worldwide Progress Report. Wholers Associates, Inc., Fort Collins, CO, USA.
3 Active Standard ASTM F2792 (2012) Standard Terminology for Additive Manufacturing Technologies, ASTM Int., West Conshohocken.
4 Lehmhus, D., Aumund-Kopp, C., Petzoldt, F., Godlinski, D., Haberkorn, A., Zöllmer, V., and Busse, M. (2016) Customized Smartness: A Survey on links between Additive Manufacturing and Sensor Integration. *Procedia Technology*, **26**, 284–301.
5 VDI-Guideline 3404 (2009) Additive Fabrication – Rapid Technologies (rapid prototyping) – Fundamentals, terms and definitions, quality parameters, supply agreements. Beuth Verlag, Berlin, Germany.
6 Brajlih, T., Valentan, B., Balic, J., and Drstvensek, I. (2011) Speed and accuracy evaluation of additive manufacturing machines. *Rapid Prototyping Journal*, **17**, 64–75.
7 ProMaker 4500 Series Product Data Sheet. Available at http://www.prodways.com/en/wp-content/uploads/sites/2/2016/11/ProMaker-P4500-EN-V21.10.2016.pdf (accessed September 6, 2017).
8 Bhavar, V., Kattire, P., Patil, V., Khot, S., Gujar, K., and Singh, R. (2014) Proceedings of the 4th International Conference and Exhibition on Additive Manufacturing Technologies – AM, September 1–2, 2014, Bangalore, India.
9 Hauser, C. (2012) Laser additive manufacturing as a key enabler for the manufacture of next generation jet engine components – technology push. EU Innovation Forums, May 9th, Aachen, Germany.
10 Williams, S. W., Martina, F., Addison, A. C., Ding, J., Pardal, G., Colegrove, P. (2016) Wire + arc additive manufacturing. *Materials Science and Technology* **32** 641–647.
11 Tumbleston, J.R., Shirvanyants, D., Ermoshkin, N., Janusziewicz, R., Johnson, A.R., Kelly, D., Chen, K., Pinschmidt, R., Rolland, J.P., Ermoshkin, A., Samulski, E.T., and DeSimone, J.M. (2015) Continuous liquid interface production of 3D objects. *Science*, **347**, 1349–1352.
12 Conner, B.P., Manogharan, G.P., Martof, A.N., Rodomsky, L.M., Rodomsky, C.M., Jordan, D.C., and Limperos, J.W. (2014) Making sense of 3-D printing: creating a map of additive manufacturing products and services. *Additive Manufacturing*, **1–4**, 64–76.
13 Michaels, S., Sachs, E., and Cima, M.J. (1994) 3-Dimensional printing of metal and cermet parts, in (eds C. Lall , A.J. Neupaver), *Advances in Powder Metallurgy and Particulate Materials*, vol. 6: Advanced Processing Techniques. Metal Powder Industries Federation, Princeton, 103–112.
14 Moon, J., Grau, J.E., Knezevic, V., Cima, M.J., and Sachs, E.M. (2004) Ink-jet printing of binders for ceramic components. *Journal of the American Ceramic Society*, **85**, 755–762.

15 Travitzky, N., Bonet, A., Dermeik, B., Fey, T., Filbert-Demut, I., Schlier, L., Schlordt, T., and Greil, P. (2014) Additive manufacturing of ceramic-based materials: additive manufacturing of ceramic-based materials. *Advanced Energy Materials*, **16**, 729–754.

16 Atwood, C., Griffith, M, Harwell, L, Schlinger, E, Ensz, M, Smugeresky, J, Romero, J.A, Greene, D, and Reckaway, D. (1998) Laser engineered net shaping (LENS^{TM}): a tool for direct fabrication of metal parts. ICALEO '98. International Congress on Applications of Lasers and Electro-Optics, Orlando FL, p. 801.

17 Balla, V.K., Bose, S., and Bandyopadhyay, A. (2008) Processing of bulk alumina ceramics using laser engineered net shaping. *International Journal of Applied Ceramic Technology*, **5**, 234–242.

18 Liu, W. and DuPont, J.N. (2003) Fabrication of functionally graded TiC/Ti composites by laser engineered net shaping. *Scripta Materialia*, **48**, 1337–1342.

19 Liu, W. and DuPont, J.N. (2004) Fabrication of carbide-particle-reinforced titanium aluminide–matrix composites by laser-engineered net shaping. *Metallurgical and Materials Transactions A*, **35**, 1133–1140.

20 Lewis, G.K. and Schlienger, E. (2000) Practical considerations and capabilities for laser-assisted direct metal deposition. *Materials & Design*, **21**, 417–423.

21 Mazumder, J., Dutta, D., Kikuchi, N., and Ghosh, A. (2000) Closed loop direct metal deposition: art to part. *OPTICS and Lasers in Engineering*, **34**, 397–414.

22 Masood, S.H. (2014) Advances in fused deposition modelling, in *Comprehensive Materials Processing*, Elsevier, pp. 69–91.

23 Masood, S.H. and Song, W.Q. (2004) Development of new metal/polymer materials for rapid tooling using fused deposition modelling. *Materials & Design*, **25**, 587–594.

24 Wu, G., Langrana, N.A., Sadanji, R., and Danforth, S. (2002) Solid freeform fabrication of metal components using fused deposition of metals. *Materials & Design*, **23**, 97–105.

25 Bellini, A., Shor, L., and Guceri, S.I. (2005) New developments in fused deposition modeling of ceramics. *Rapid Prototyping Journal*, **11**, 214–220.

26 Greulich, M., Greul, M., and Pintat, T. (1995) Fast, functional prototypes via multiphase jet solidification. *Rapid Prototyping Journal*, **1**, 20–25.

27 Cesarano, J. (1998) A review of robocasting technology. *Proceedings of MRS*, **542**, 133–139.

28 Huang, T., Mason, M.S., Zhao, X., Hilmas, G.E., and Leu, M.C. (2009) Aqueous-based freeze-form extrusion fabrication of alumina components. *Rapid Prototyping Journal*, **15**, 88–95.

29 Mason, M.S., Huang, T., Landers, R.G., Leu, M.C., and Hilmas, G.E. (2009) Aqueous-based extrusion of high solids loading ceramic pastes: process modeling and control. *Journal of Materials Processing Technology*, **209**, 2946–2957.

30 Singh, R. (2011) Process capability study of polyjet printing for plastic components. *Journal of Mechanical Science and Technology*, **25**, 1011–1015.

31 de Gans, B.-J., Duineveld, P.C., and Schubert, U.S. (2004) Inkjet printing of polymers: state of the art and future developments. *Advanced Materials*, **16**, 203–213.

32 Ladd, C., So, J.-H., Muth, J., and Dickey, M.D. (2013) 3D printing of free standing liquid metal microstructures. *Advanced Materials*, **25**, 5081–5085.

33 Kruth, J.P., Van Froyen, L., Vaerenbergh, J., Mercelis, P., Rombouts, M., and Lauwers, B. (2004) Selective laser melting of iron-based powder. *Journal of Materials Processing Technology*, **149**, 616–622.

34 Attar, H., Bönisch, M., Calin, M., Zhang, L.-C., Scudino, S., and Eckert, J. (2014) Selective laser melting of *in situ* titanium–titanium boride composites: processing, microstructure, and mechanical properties. *Acta Materialia*, **76**, 13–22.

35 Goodridge, R.D., Tuck, C.J., and Hague, R.J.M. (2012) Laser sintering of polyamides and other polymers. *Progress in Materials Science*, **57**, 229–267.

36 Kong, C.Y. and Soar, R. (2005) Method for embedding optical fibers in an aluminum matrix by ultrasonic consolidation. *Applied Optics*, **44**, 6325–6333.

37 Qian, B. and Shen, Z. (2013) Laser sintering of ceramics. *Journal of Asian Ceramic Societies*, **1**, 315–321.

38 Vaucher, S., Carreno-Morelli, E., Andre, C., and Beffort, O. (2003) Selective laser sintering of aluminium and titanium-based composites: processing and characterization. *Physica Status Solidi(a)*, **199**, R11–R13.

39 Murr, L.E., Martinez, E., Amato, K.N., Gaytan, S.M., Hernandez, J., Ramirez, D.A., Shindo, P.W., Medina, F., and Wicker, R.B. (2012) Fabrication of metal and alloy components by additive manufacturing: examples of 3D materials science. *Journal of Materials Research and Technology*, **1**, 42–54.

40 Manfredi, D., Calignano, F., Krishnan, M., Canali, R., Ambrosio, E.P., Biamino, S., Ugues, D., Pavese, M., and Fino, P. (2014) *Additive Manufacturing of Al Alloys and Aluminium Matrix Composites (AMCs), Light Metal Alloys Applications* (ed. Dr. W. A., Monteiro), InTech.

41 Gong, X., Anderson, T., and Chou, K. (2012) Review on powder-based electron beam additive manufacturing technology. In ASME, p. 507.

42 Mueller, B. and Kochan, D. (1999) Laminated object manufacturing for rapid tooling and patternmaking in foundry industry. *Computers in Industry*, **39**, 47–53.

43 Park, J., Tari, M.J., and Hahn, H.T. (2000) Characterization of the laminated object manufacturing (LOM) process. *Rapid Prototyping Journal*, **6**, 36–50.

44 Chiu, Y.Y., Liao, Y.S., and Hou, C.C. (2003) Automatic fabrication for bridged laminated object manufacturing (LOM) process. *Journal of Materials Processing Technology*, **140**, 179–184.

45 Yi, S., Liu, F., Zhang, J., and Xiong, S. (2004) Study of the key technologies of LOM for functional metal parts. *Journal of Materials Processing Technology*, **150**, 175–181.

46 Strasser, C., Prihodovsky, A., and Ploshikhin, V. (2014) Plate press brazing for the production of large metallic tools. In Proceedings of the 1st International Symposium Materials Science and Technology of Additive Manufacturing; DVS Berichte; DVS Media GmbH: Bremen, Germany, Vol. 308, pp. 12–16.

47 Hull, C.W. (1986) Apparatus for production of three-dimensional objects by stereolithography.

48 Bartolo, P.J. and Gaspar, J. (2008) Metal filled resin for stereolithography metal part. *CIRP Annals –- Manufacturing Technology*, **57**, 235–238.

49 Lee, J.W., Lee, I.H., and Cho, D.-W. (2006) Development of micro-stereolithography technology using metal powder. *Microelectronic Engineering*, **83**, 1253–1256.

50 Brady, G.A. and Chu, T.M. (1996) Curing Behavior of Ceramic Resin for Stereolithography.

51 Chartier, T., Chaput, C., Doreau, F., and Loiseau, M. (2002) Stereolithography of structural complex ceramic parts. *Journal of Materials Science*, **37**, 3141–3147.

52 Doreau, F., Chaput, C., and Chartier, T. (2000) Stereolithography for manufacturing ceramic parts. *Advanced Engineering Materials*, **2**, 493–496.

53 Karalekas, D.E. (2003) Study of the mechanical properties of nonwoven fibre mat reinforced photopolymers used in rapid prototyping. *Materials & Design*, **24**, 665–670.

54 Hengsbach, F., Kopp, P., Holzweissig, M.J., Aydinöza, M.E., Taube, A., Hoyer, K.-P., Starykov, O., Tonn, B., Niendorf, T., Tröster, T., and Schaper, M. (2017) Inline additively manufactured functionally graded multi-materials – microstructural and mechanical characterization of 316L parts with H13 layers (submitted to *Progress in Additive Manufacturing*).

55 Isaza, P.J.F. and Aumund-Kopp, C. (2013) Additive manufacturing approaches, in *Structural Materials and Processes in Transportation* (eds D. Lehmhus, M. Busse, A.S. Herrmann, and K. Kayvantash), Wiley-VCH Verlag GmbH, pp. 549–566.

56 Wu, D, Thames, J.L., Rosen, D.W., and Schaefer, D. (2012) Towards a cloud-based design and manufacturing paradigm: looking backward, looking forward. In ASME, International Design Engineering Technical Conferences & Computers and Information in Engineering Conference (IDETC 2012), pp. 315–328.

57 Wu, D., Rosen, D.W., Wang, L., and Schaefer, D. (2015) Cloud-based design and manufacturing: a new paradigm in digital manufacturing and design innovation. *Computer-Aided Design*, **59**, 1–14.

58 Xu, X. (2012) From cloud computing to cloud manufacturing. *Robotics and Computer-Integrated Manufacturing*, **28**, 75–86.

59 Wellsandt, S., Hribernik, K., and Thoben, K.-D. (2015) Content analysis of product usage information from embedded sensors and web 2.0 sources. Proceedings of the 20th International Conference on Concurrent Enterprising (ICE) & IEEE TMC Europe Conference, Belfast, United Kingdom.

60 Luetzenberger, J., Klein, P., Hribernik, K., Thoben, K.-D. (2016) Improving product-service systems by exploiting information from the usage phase. *Procedia CIRP* **47**, 376–381.

61 Wellsandt, S., Thoben, K.-D. (2016) Approach to describe knowledge sharing between producer and user. *Procedia CIRP* **50**, 20–25.

62 Kiritsis, D. (2011) Closed-loop PLM for intelligent products in the era of the Internet of things. *Computer-Aided Design*, **43** (5), 479–501.

63 Lachmayer, R., Mozgova, I., Reimche, W., Colditz, F., Mroz, G., and Gottwald, P. (2014) Technical inheritance: a concept to adapt the evolution of nature to production engineering. *Procedia Technology*, **15**, 178–187.

64 Wuest, T., Hribernik, K., and Thoben, K.-D. (2017) New markets/opportunities through availability of product life cycle data, in *Material-Inegrated Intelligent Systems* (eds S. Bosse, D. Lehmhus, W. Lang, and M. Busse), Wiley-VCH Verlag GmbH, Heidelberg.

65 Abramovici, M., Krebs, A., and Lindner, A. (2013) Exploiting service data of similar product items for the development of improved product generations by using smart input devices, in *Smart Product Engineering* (eds M. Abramovici and R. Stark), Springer, Berlin/Heidelberg, pp. 357–366.

66 Lehmhus, D., Wuest, T., Wellsandt, S., Bosse, S., Kaihara, T., Thoben, K.-D., and Busse, M. (2015) Cloud-based additive manufacturing and automated design: a PLM-enabled paradigm shift. *Sensors*, **15**, 32079–32122.

67 Cronskär, M., Bäckström, M., and Rännar, L. (2013) Production of customized hip stem prostheses – a comparison between conventional machining and electron beam melting (EBM). *Rapid Prototyping Journal*, **19**, 365–372.

68 Altimus, J.C., Johnson, V.D., Luegge, M.A., and Ramrattan, S.N. (1998) Remote identification of metal castings. *AFS Transactions*, **44**, 605–608.

69 Pille, C. (2009) Produktidentifikation, Intralogistik und Plagiatschutz - RFID-Integration in Gussbauteile. Proceedings of the BDG-Fachtagung Gussteilkennzeichnung - Methoden und Datenmanagement - Praxisberichte, Essen (Germany).

70 Isaza-Paz, J., Wilbig, J., Aumund-Kopp, C., and Petzoldt, F. (2014) RFID transponder integration in metal surgical instruments produced by additive manufacturing. *Powder Metallurgy*, **57**, 365–372.

71 Pille, C. (2010) In-process embedding of piezo sensors and RFID transponders into cast parts for autonomous manufacturing logistics. Proceedings of the Smart Systems Integration (SSI), Como (Italy), March 23–24, 2010.

72 Mayer, D., Melz, T., Pille, C., and Woestmann, F.-J. (2008) CASTRONICS – Direct integration of piezo ceramic materials in high pressure die casting parts for vibration control. Proceedings of Actuator, 11th International Conference on New Actuators & 5th International Exhibition on Smart Actuators and Drive Systems, Bremen (Germany), June 9–11th, 2008.

73 Schwankl, M., Kimme, S., Pohle, C., Drossel, W.-G., and Körner, C. (2015) Active vibration damping in structural aluminum die castings via piezoelectricity – technology and characterization. *Advanced Engineering Materials*, **17**, 969–975.

74 Di Sante, R. (2015) Fibre optic sensors for structural health monitoring of aircraft composite structures: recent advances and applications. *Sensors*, **15**, 18666–18713.

75 Salowitz, N., Guo, Z.Q., Kim, S.J., Li, Y.H., Lanzara, G., and Chang, F.-K. (2012) Bio-inspired intelligent sensing materials for fly-by-feel autonomous vehicles. Proceedings of the 11th IEEE Sensors Conference, Taipei (Taiwan), October 28–31, pp. 363–365.

76 Salowitz, N., Guo, ZQ., Li, Y.-H., Kim, K., Lanzara, G., and Chang, F.-K. (2013) Bio-inspired stretchable network-based intelligent composites. *Journal of Composite Materials*, **47**, 97–105.

77 Boll, D., Schubert, K., Brauner, C., and Lang, W. (2014) Miniaturized flexible interdigital sensor for *in situ* dielectric cure monitoring of composite materials. *IEEE Sensors Journal*, **14**, 2193–2197.

78 Correll, N., Önal, CD., Liang, H., Schoenfeld, E., and Rus, D. (2010) Soft autonomous materials – using active elasticity and embedded distributed computation, in *Experimental Robotics*. Springer Tracts in Advanced Robotics, Springer.

79 Hughes, D. and Correll, N. (2015) Texture recognition and localization in amorphous robotic skin. *Bioinspiration & Biomimetics*, **10**, 055002.

80 Seminara, L., Pinna, L., Ibrahim, A., Noli, L., Caviglia, S., Gastaldo, P., and Valle, M. (2015) Towards integrating intelligence in electronic skin. *Mechatronics*. doi: 10.1016/j.mechatronics.2015.04.001

81 Kaihara, T., Kokuryo, D., and Kuik, S. (2015) A proposal of value co-creative production with IoT-based thinking factory concept for tailor-made rubber products, in *Advances in Production Management Systems: Innovative Production Management Towards Sustainable Growth, IFIP Advances in Information and Communication Technology*, vol. 460 (eds S. Umeda, M. Nakano, H. Mizuyama, H. Hibino, D. Kiritsis, and G. von Cieminski), Springer, pp. 67–73.

82 Gross, S., Bardzell, J., and Dardzell, S. (2014) Structures, forms and stuff: the materiality and medium of interaction. *Personal and Ubiquitous Computing*, **18**, 637–649.

83 Ishii, H., Lakatos, D., Bonanni, L., and Labrune, J.-B. (2012) Radical atoms: beyond tangible bits, toward transformable materials. *Interactions*, **XI**, 38–51.

84 Hoerber, J., Glasschroeder, J., Pfeffer, M., Schilp, J., Zaeh, M., and Franke, J. (2014) Approaches for additive manufacturing of 3D electronic applications. *Procedia CIRP*, **17**, 806–811.

85 Johander, P., Haasl, S., Persson, K., and Harrysson, U. (2007) Layer, manufacturing as a generic tool for microsystem integration. Third International Conference on Multi-Material Micro Manufacture; Borovets, Bulgaria.

86 Paulsen, J.A., Renn, M., Christenson, K., and Plourde, R. (2012) Printing conformal electronics on 3D structures with Aerosol Jet technology, in Future of Instrumentation International Workshop (FIIW), pp. 1–4.

87 Runge, D. (2016) 3D Printing und gedruckte Elektronik für die Medizintechnik. M.Sc. thesis, University of Applied Science Bremerhaven.

88 Monaghan, T., Capel, A.J., Christie, S.D., Harris, R.A., and Friel, R.J. (2015) Solid-state additive manufacturing for metallized optical fiber integration. *Composites: Part A*, **76**, 181–193.

89 Li, J., Monaghan, T., Bournias-Varotsis, A., Masurtschak, S., Friel, R.J., and Harris, R.A. Exploring the mechanical performance and material structures of integrated electrical circuits within solid state metal additive manufacturing matrices. Proceedings of the 25th Annual International Solid Freeform

Fabrication (SFF) Symposium – An Additive Manufacturing Conference, The University of Texas in Austin, Austin, Texas, USA, pp. 857–864.

90 Li, J., Monaghan, T., Masurtschak, S., Bournias-Varotsis, A., Friel, R.J., and Harris, R.A. (2015) Exploring the mechanical strength of additively manufactured metal structures with embedded electrical materials. *Materials Science and Engineering A*, **639**, 474–481.

91 Li, J., Wasley, T., Nguyen, T.T., Ta, V.D., Shephard, J.D., Stringer, J., Smith, P., Esenturk, E., Connaughton, C., and Kay, R. (2016) Hybrid additive manufacturing of 3D electronic systems. *Journal of Micromechanics and Microengineering*, **26**, 105005.

92 Chang, Y.-H., Wang, K., Wu, C., Chen, Y., Zhang, C., and Wang, B. (2015) A facile method for integrating direct-write devices into three-dimensional printed parts. *Smart Materials and Structures*, **24**, 10.1088/0964-1726/24/6/065008.

93 Perez, K.B. and Williams, C.B. (2014) Design considerations for hybridizing Additive Manufacturing and Direct Write Technologies. Proceedings of the ASME 2014 International Design Engineering Technical Conferences & Computers and Information in Engineering Conference IDETC/CIE, August 17–20, 2014, Buffalo/NY, USA.

94 Vatani, M., Lu, Y., Engeberg, ED., and Choi, J.-W. (2015) Combined 3D printing technologies and material for fabrication of tactile sensors. *International Journal of Precision Engineering and Manufacturing*, **16**, 1375–1383.

95 Espalin, D., Muse, D.W., MacDonald, E., and Wicker, R.B. (2014) 3D printing multifunctionality: structures with electronics. *International Journal of Advanced Manufacturing Technology*, **72**, 963–978.

96 MacDonald, E. and Wicker, R. (2016) Multiprocess 3D printing for increasing component functionality. *Science*, **353**, aaf2093-1–aaf2093-8.

97 Shemelya, C., Cedillos, F., Aguilera, E., Espalin, D., Muse, D., Wicker, R., and MacDonald, E. (2015) Encapsulated copper wire and copper mesh capacitive sensing for 3-D printing applications. *IEEE Sensors Journal*, **15**, 1280–1286.

98 Lind, J.U., Busbee, T.A., Valentine, A.D., Pasqualini, F.S., Yuan, H., Yadid, M., Park, S.-J., Kotikian, A., Nesmith, A.P., Campbell, P.H., Vlassak, J.J., Lewis, J.A., and Parker, K.K. (2016) Instrumented cardiac microphysiological devices via multimaterial three-dimensional printing. *Nature Materials Letters*. doi: 10.1038/NMAT4782

99 Weise, J. and Aumund-Kopp, C. (2014) Selective laser melting – opportunities and challenges for complex-shaped parts for transport applications. KMM-VIN 3rd Industrial Workshop: Current Research on Materials and Technologies for Transport Applications, Dresden, Germany, November 3–4, 2014.

100 Lopes, A.J., Lee, I.H., MacDonald, E., Quintana, R., and Wicker, R. (2014) Laser curing of silver-based conductive inks for *in situ* 3D structural electronics fabrication in stereolithography. *Journal of Materials Processing Technology*, **214** (9), 1935–1945.

101 Joseph, T., Muth, J.T., Vogt, D.M., Truby, R.L., Mengüç, Y., Kolesky, D.B., Wood, R.J., and Lewis, J.A. (2014) Embedded 3D printing of strain sensors within highly stretchable elastomers. *Advanced Materials*, **26**, 6307–6312.

102 Pal, E., Kun, R., Schulze, C., Zöllmer, V., Lehmhus, D., Bäumer, M., and Busse, M. (2012) Composition-dependent sintering behaviour of chemically synthesised CuNi nanoparticles and their application in aerosol printing for preparation of conductive microstructures. *Colloid and Polymer Science*, **290**, 941–952.

103 Zöllmer, V., Pál, E., Maiwald, M., Werner, C., Godlinski, D., and Lehmhus, D. (2012) Functional materials for printed sensor structures. 1st Joint International Symposium on System-Integrated Intelligence 2012; Hannover, Germany.

104 Pal, E., Zöllmer, V., Lehmhus, D., and Busse, M. (2011) Synthesis of $Cu0.55Ni0.44Mn0.01$ alloy nanoparticles by solution combustion method and their application in aerosol printing. *Colloids and Surfaces A: Physicochemical and Engineering Aspects*, **384**, 661–667.

105 Werner, C., Godlinski, D., Zöllmer, V., Busse, M. (2013) Morphological influences on the electrical sintering process of aerosol jet and ink jet printed silver microstructures. *Journal of Materials Science: Mater Electron* **24** 4367–4377.

106 Lu, Y., Yun, H.-Y., Vatani, M., Kim, H.-C., and Choi, J.-W. (2015) Direct-print/cure as a molded interconnect device (MID) process for fabrication of automobile cruise controllers. *Journal of Mechanical Science and Technology*, **29**, 5377–5385.

107 Perez, K.B. and Williams, C.B. (2014) Characterization of *in situ* conductive paste extrusion on PolyJet Substrates. 25th Annual Solid Freeform Fabrication (SFF) Symposium, University of Texas at Austin, Austin, Texas, USA, August 4–8, 2014.

108 Maiwald, M., Werner, C., Zoellmer, V., and Busse, M. (2010) INKtelligent printing® for sensorial applications. *Sensor Review*, **30**, 19–23.

109 Maiwald, M., Werner, C., Zoellmer, V., and Busse, M. (2010) INKtelligent printed strain gauges. *Sensors and Actuators A Physical*, **162**, 198–201.

110 Vatani, M., Vatani, M., and Choi, JW. (2016) Multi-layer stretchable pressure sensors using ionic liquids and carbon nanotubes. *Applied Physics Letters*, **108**, 061908.

111 Besganz, A., Zöllmer, V., Kun, R., Pal, E., Walder, L., Busse, M. (2014) Inkjet printing as flexible technology for the deposition of thermoelectric composite structures. *Procedia Technology* **15**, 99–106.

112 Swensen, J.P., Odhner, L.U., Araki, B., and Dollar, A.M. (2015) Printing three-dimensional electrical traces in additive manufactured parts for injection of low melting temperature metals. *Journal of Mechanisms and Robotics*, 7, 021004.

113 Godlinski, D., Taubenrauch, E., Werner, C., Wirth, I., Zöllmer, V., and Busse, M. (2012) Functional integration on three-dimensional parts by means of direct write technologies. Fraunhofer Direct Digital Manufacturing Conference (DDMC2012), Berlin (Germany), March 14–15th, 2012.

114 Godlinski, D., Haberkorn, A., Kluge, O., Kohl, M., Zöllmer, V., and Busse, M. (2014) Modular manufacturing platform for 3D functionalized parts. Fraunhofer Direct Digital Manufacturing Conference (DDMC2014), Berlin, Germany, March 12–13, 2014.

115 Lewis, J.A. and Ahn, B.Y. (2015) Device fabrication: three-dimensional printed electronics. *Nature*, **518**, 42–43.
116 Stratasys (2016) PolyJet Materials White Paper. Available at http://www.stratasys.com/~/media/Main/Secure/White-Papers/WP_PJ_PolyJetMaterials.ashx (accessed November 29, 2016).
117 Wieland, S. and Petzoldt, F. (2016) Binder jet 3D-printing for metal additive manufacturing: applications and innovative approaches. *Ceramic Forum International DKG*, **93**, E26–E30.
118 Panesar, A., Brackett, D., Ashcroft, I., Wildman, R., and Hague, R. (2015) Design framework for multifunctional additive manufacturing: placement and routing of three-dimensional printed circuit volumes. *Journal of Mechanical Design*, **137**. doi: 10.1115/1.4030996.
119 Yang, S. and Zhao, Y.F. (2015) Additive manufacturing-enabled design theory and methodology: a critical review. *International Journal of Advanced Manufacturing Technology*, **80**, 327–342.
120 Thompson, M.K., Moroni, G., Vaneker, T., Fadel, G., Campbell, R.I., Gibson, I., Bernard, A., Schulz, J., Graf, P., Ahuja, B., and Martina, F. (2016) Design for additive manufacturing: trends, opportunities, considerations, and constraints. *CIRP Annals – Manufacturing Technology*, **65**, 737–760.

Part Five

Signal and Data Processing: The Sensor Node Level

10

Analog Sensor Signal Processing and Analog-to-Digital Conversion

John Horstmann,[1] Marco Ramsbeck,[1] and Stefan Bosse[2,3]

[1]TU Chemnitz, Professorship Electronic Components of Micro- and Nanotechnology, Reichenhainer Str. 70 (Weinholdbau), 09107 Chemnitz, Germany
[2]University of Bremen, Dept. of Mathematics and Computer Science, 28359 Bremen, Germany
[3]University of Bremen, ISIS Sensorial Materials Scientific Centre, Wiener Str. 12, 28359 Bremen, Germany

A smart sensor node should be able to operate autonomously and independently of the environment and other sensor nodes. It consists of different modules and units:

1) Digital signal processing and data processing units (DSP/DPU: microprocessor, virtual machine, and digital logic and register–transfer level architectures)
2) Communication units (CMU: physical, link, and data level)
3) Digital storage (DS)
4) Energy management units (EM: hardware and software control)
5) Analog signal processing (ASP: analog electronics)
6) Analog–digital conversion (ADC)
7) Energy supply and energy storage (ES)

The relation of the single units and the data flow is illustrated schematically in Figure 10.1. Typically, a sensor node processes low-level sensor data, improving the quality and accuracy using noise and fusion filter techniques. The sensor data or preprocessed information is passed from this information source to an information sink node by using message passing route in some kind of communication network.

This chapter outlines the principles of different components of a sensor node and the challenges to integrate sensor nodes in materials and mechanical structures.

Data acquisition systems build the interface between the physical parameter of the real world, which are analog and which are sensed by sensors, and the world of digital computation [1]. The devices that perform the interface function between analog and digital worlds are analog-to-digital converters (ADC). ADCs are key components in today's modern applications and are essential parts of sensing systems. In data acquisition systems, the analog signal from the sensor has to be adjusted to suit the sampling frequency and the amplitude range needed by the analog-to-digital converter. The two main functions of an ADC are sampling and

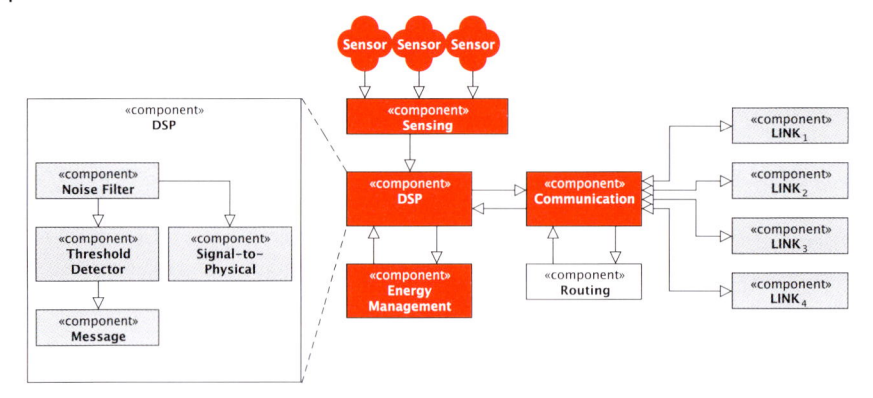

Figure 10.1 Architecture of a smart sensor node.

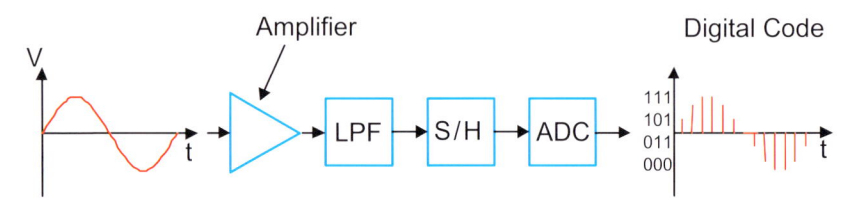

Figure 10.2 Signal characteristics and functional overview.

quantization. These two are converting the analog sensor signal from the continuous time and voltage into digital numbers having discrete values at discrete times. To convert analog signals continuously in time and at every voltage would take an infinite amount of storage. Therefore, every data acquisition system has to be designed for an appropriate sampling rate and quantization, better described as resolution. In Figure 10.2, a functional overview of a data acquisition system is given. The sensor output is first amplified to suit the input range of the following ADC. Furthermore, this functional block is used to convert the sensor output into voltages. The signal is then filtered by an antialiasing filter to remove high-frequency components that may cause an effect known as aliasing. The signal is sampled and held by a circuit and converted into a digital representation.

A deeper and rigorous introduction in measuring systems, their properties, and the design of measuring systems can be found in Refs [1,2].

10.1 Operational Amplifiers

The front end of the data acquisition system amplifies and filters the signal of the sensor node. An amplifier and filter are crucial components in the initial signal processing. The amplifier must perform at least one of the following functions: amplifying the signal, buffer the signal, convert a signal current into a voltage, or extract a differential voltage from a common mode noise [3]. To accomplish these functions the most popular amplifier will be used. It is called operational amplifier (OPAMP) which is a general purpose gain block having differential inputs.

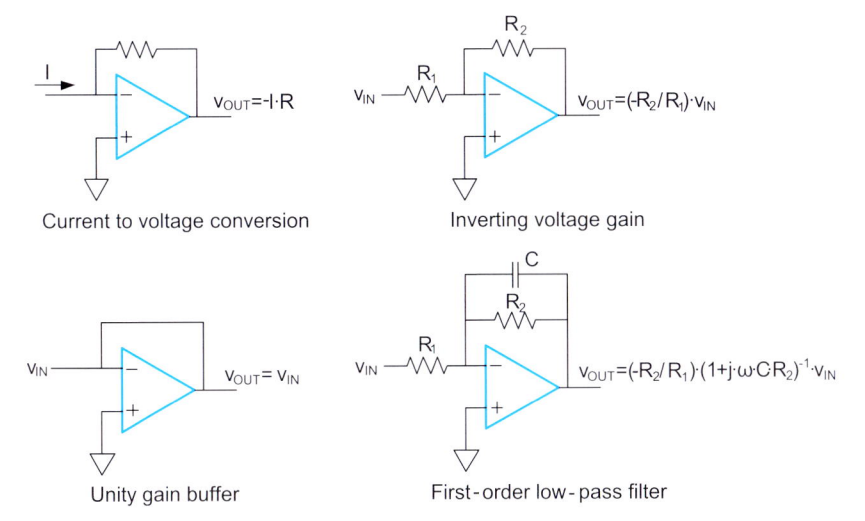

Figure 10.3 Operational amplifier circuits.

By connecting the OPAMP with different closed-loop configurations a wide variety of functionalities can be realized. A selection of configurations is shown in Figure 10.3.

The gain of the circuits depends on the external devices around the amplifier. In case of differential signal processing, the instrumentation amplifier is a better choice since both inputs having high impedances and the gain is set with one precision resistor, as shown in Figure 10.4.

Each input is directly connected to an OPAMP working as a unity gain buffer if R_{gain} is left out. Since the inputs are internally connected to a differential stage, meaning a gate of a MOSFET, the impedance is very high. The rightmost amplifier is working as a standard differential amplifier circuit with a gain equal to R_3/R_2. By including R_{gain} the transfer function is to be described as

$$\frac{v_{out}}{v_{in2} - v_{in1}} = \left(1 + \frac{2R_1}{R_{gain}}\right) \frac{R_3}{R_2}. \tag{10.1}$$

A further advantage of using an instrumentation amplifier is its high common-mode rejection ratio (CMRR). An ideal differential input amplifier responds only

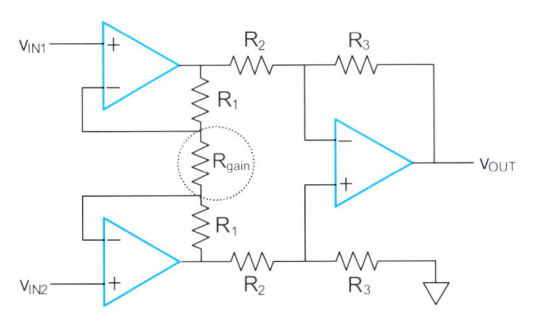

Figure 10.4 Instrumentation amplifier for high input impedance.

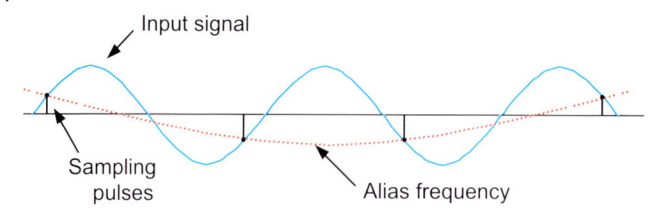

Figure 10.5 Alias frequency caused by undersampling.

to the voltage difference between its input terminals. An increase or decrease of the common-mode voltage at the input terminal should not change the output voltage. In nonideal amplifiers, the common-mode input signal causes an output response so small compared to the response of the differential input voltage. The common-mode rejection ratio is the ratio of differential voltage gain to common-mode voltage gain and can be described as

$$\text{CMRR} = 20 \log \left(\frac{A_\text{D}}{A_\text{CM}} \right), \quad (10.2)$$

where A_D is the differential voltage gain and A_CM the common-mode gain. Besides amplifying voltages, the frequency behavior in the front-end data acquisition system is important. Figure 10.2 shows a low-pass filter to prevent an effect called aliasing and removing noise components at higher frequencies. The effect of inadequate sampling rate on a sinusoid is illustrated in Figure 10.5.

In this case, sampling at a rate slightly less than twice per cycle gives a low-frequency sinusoid shown as the dotted waveform. This alias waveform can be significantly different from the original frequency. In this figure, it is easy to see that if the sinus waveform is sampled at least twice per cycle, as required by the sampling theorem, the original frequency is preserved.

10.2 Analog-to-Digital Converter Specifications

Sampling The task of the sample-and-hold (S/H) circuit is to lock the value of an input signal at a certain time and to hold it while the ADC is processing the data. Figure 10.6 shows the behavior of an ideal S/H circuit and a simplified circuit. The input signal is periodically captured and held at a constant voltage level until the

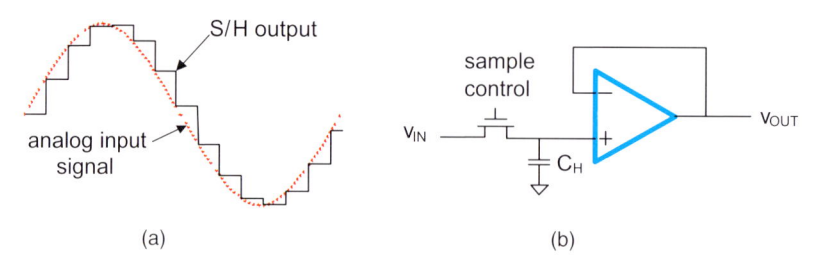

Figure 10.6 (a) Output of an ideal S/H Circuit. (b) Track-and-hold circuit using a buffer.

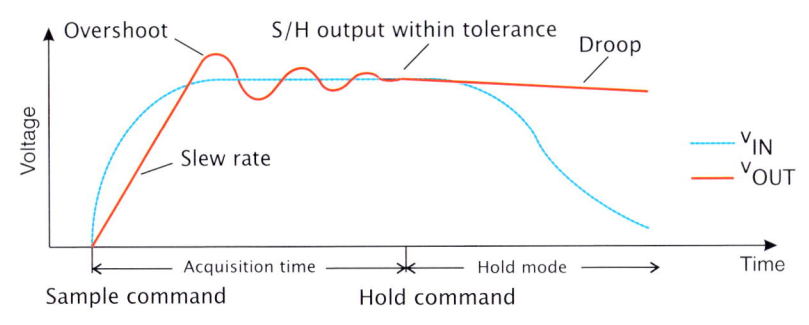

Figure 10.7 Typical errors caused by a sample-and-hold circuit.

next time step. As the performance of the S/H circuit is essential for the quality of the data conversion, it is important to determine its characteristics in detail (see Ref. [4] for technical aspects).

Most S/H circuits employ operational amplifiers and thus nonideal op-amp behavior has to be taken into account (see Figure 10.7). If the input voltage is changing rapidly, then the output of the amplifier could be limited by its slew rate. Further issues may occur if the amplifier is not correctly compensated, so that the phase margin at the transit frequency is too small. This would result in a large overshoot and a longer settling time would be required until the output of the S/H circuit approaches the signal value within the necessary tolerances. Consequently, the circuit would require a larger acquisition time, that is, the time interval between sample command and subsequent hold command. Additionally, the op-amps DC error should be considered. The overall output error depends on the amplifiers offset, linearity, and gain error. Ideally, the gain of a S/H circuit is 1 and constant over the input voltage range.

Once the S/H circuit is in hold mode further deviations from the ideal behavior of the S/H circuit may arise. The parasitic impedances associated with the following ADC as well as with the amplifier of the S/H circuit may cause leakage currents, which in turn may lead to a drop of the voltage level. A larger hold capacitor can mitigate this effect. However, this will result in a longer acquisition time, as the time needed to charge the capacitor increases.

A transient effect, which is inducing an error, is called aperture error. Since the transistor varies its switching time steered by the control signal in relation to the analog input value, a so-called aperture uncertainty is applied.

Quantization While sampling is working in the time domain, quantization affects the amplitude domain. During the hold time of the S/H the analog signal is converted into a digital representation. The ADC quantizes the analog signal by using one value out of a finite list of integer values to represent a given analog voltage level.

Figure 10.8 depicts the transfer function of an ideal 3-bit quantizer that maps the input signal into eight (2^3) digital output values. The transfer function corresponds to a staircase with equidistant steps. The width of the steps is referred to as code width. The code width signifies the smallest possible difference between two digitized values: the so-called least significant bit (LSB). In this

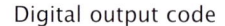

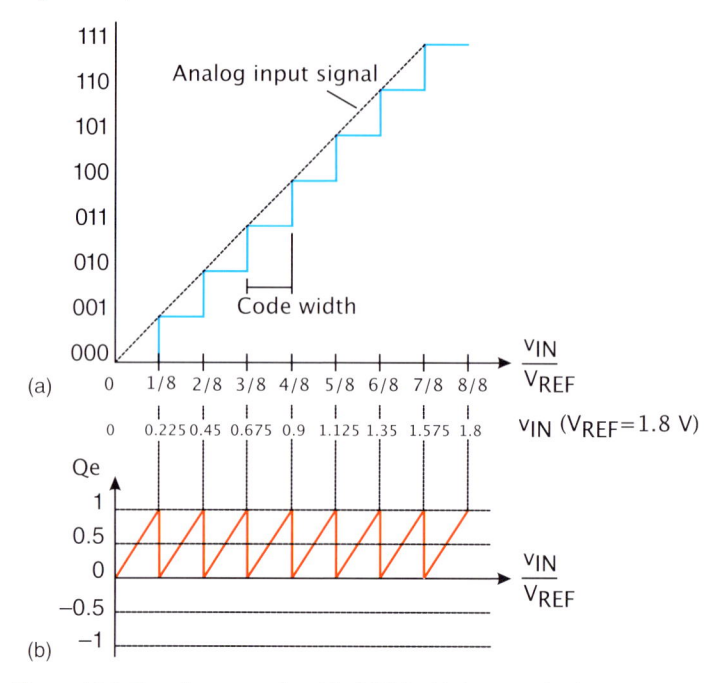

Figure 10.8 Transfer curve of an ideal ADC with its quantization error.

example 1 LSB equals 225 mV, as the full analog input range of 1.8 V is divided into eight LSB intervals each represented by a unique digital code.

Figure 10.8b shows the quantization error. It is obtained by subtracting staircase from the input graph. The average error is 1/2 LSB. The error maximal values are found at integer multiples of the code width. The standard deviation error is $1/\sqrt{12}$ LSB.

Range and Resolution Naturally, only input values within certain voltage range, that is, the input range, can be represented by a given ADC in a meaningful way. The maximum value that can still be converted is referred to as +full scale, the respective minimum value as −full scale. ADCs with a −full scale of 0 V are called unipolar. Conversely, bipolar ADCs have a symmetrical input meaning that the absolute values of +full scale and −full scale are identical. Nevertheless, most ADCs offer reference inputs (V_{REF}) that allow the full-scale range to be adjusted (within the range of the power supply voltages).

Adjusting the input range of an ADC entails that its resolution, that is, the voltage represented by the LSB, is subject to change, as well. Consequently, the resolution of an ADC is typically provided in terms of the number of bits offered by the ADC. A 16-bit ADC, for example, can resolve one part in $2^{16} = 65\,536$ of the set input range. However, other measures are frequently used to represent the ADC resolution (see Table 10.1). For example, the resolution of digital voltmeters is often given in terms of a number of digits: a 5-1/2 -digit voltmeter on a 2 V range resolves a range of +1.99 999 V to −1.99 999 V into a resolution of 0.00 001 V.

Table 10.1 Overview of different resolution and their corresponding values like "digits" and dynamic range.

Bits	Digits at a "voltmeter"	Steps in full-scale range	Step size (ppm)	Dynamic range (dB)
24		16 777 216	0.060	146
21.9	6 1/2	4 000 000	0.25	133
20.9	6	2 000 000	0.5	127
20		1 048 576	0.9	122
18.6	5 1/2	400 000	2.5	113
18		262 144	3.8	110
17.6	5	200 000	5	107
16		65 536	15	98
15.2	4 1/2	40 000	25	93
14.2	4	20 000	50	87
14		16 384	61	86
12		4 096	244	74
11.9	3 1/2	4 000	250	73
10.9	3	2 000	500	67
10		1 024	976	61
8.6	2 1/2	400	2500	53
8		256	3906	49
7.6	2	200	5000	47

Linear Errors A systematic deviation of the ADC transfer function from the behavior of the ideal ADC that can be described by a linear function across the entire input range is referred to as linear error. Linear errors are a common occurrence and can be of substantial magnitude.

Generally two types of linear errors are distinguished: offset errors and gain errors. An offset error occurs as a result of a mismatch between the LSB and the first code transition, as shown in Figure 10.9a. Because offset errors are constant across the input range, they can easily be corrected for by adequate calibration procedures. A gain error denotes the difference between the slope of the ADC transfer function and the behavior of the ideal ADC, which has an average slope of 1 (see Figure 10.9b). Since this error is systematic by definition, a correction could be implemented by a microcontroller.

Nonlinear Errors The term differential nonlinearity (DNL) of an ADC denotes the difference between the actual code width of each step of the transfer function compared to the LSB of the ideal ADC. This is illustrated in Figure 10.10. If the DNL exceeds 1 LSB, it causes a nonmonotonic transfer function that is known as a missing code.

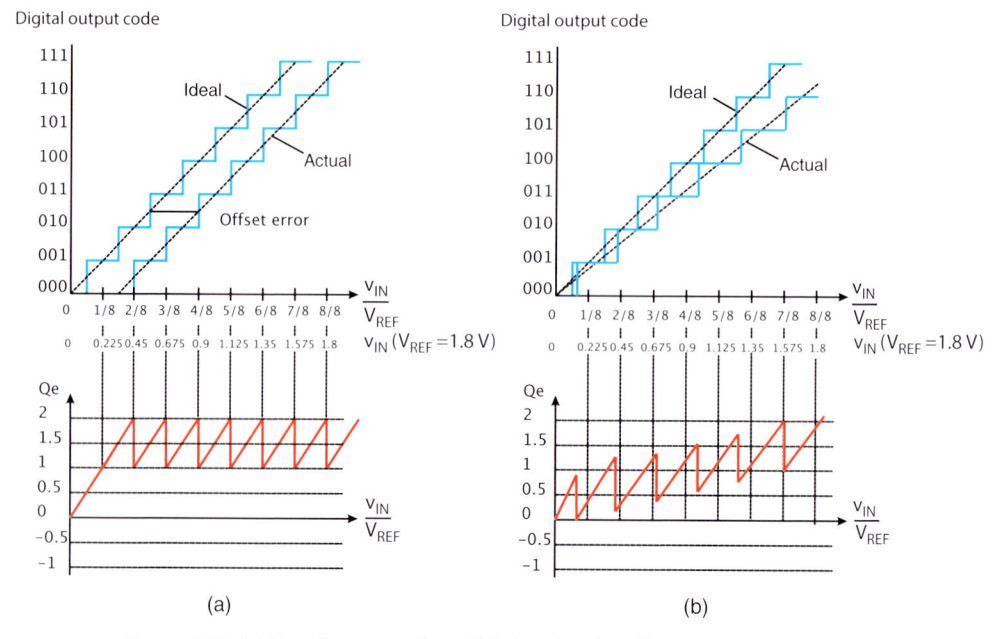

Figure 10.9 (a) Transfer curve of an ADC showing the offset error and the corresponding quantization error. (b) Transfer curve of an ADC showing the gain error and the corresponding quantization error.

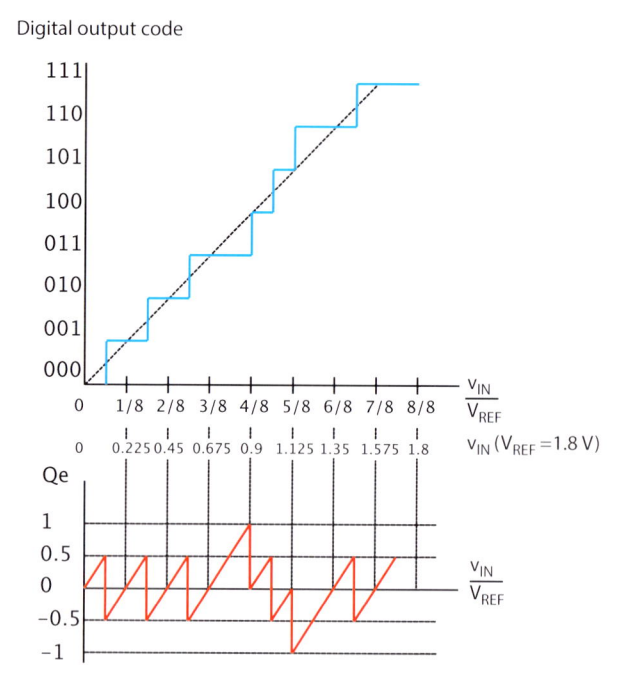

Figure 10.10 Transfer curve of an ADC indicating a nonlinear error and the corresponding quantization error.

The integral nonlinearity is defined as the difference between the date converter code transition points and the straight line with all other errors set to zero. Therefore, a straight line is drawn through the end points of the first and the last code transition. Another possibility to determine the INL is to take a look at the quantization error. There the INL will have the value of the highest peak minus the 1/2 LSB of the quantization error of an ideal ADC.

Dynamic Range The signal-to-noise ratio (SNR) of an ADC represents the ratio between the largest RMS input signal, v_{in}, and the RMS value of the noise added by the ADC, v_{noise}:

$$\text{SNR} = 20 \log\left(\frac{v_{in}}{v_{noise}}\right). \tag{10.3}$$

If the input signal is a sine wave with a peak-to-peak value equal to +full scale and −full scale, the v_{in} is given by

$$v_{in} = \frac{v_{fullscale}}{2\sqrt{2}} = \frac{V_{ref}}{2\sqrt{2}} = 2^N \frac{V_{LSB}}{2\sqrt{2}}, \tag{10.4}$$

where V_{LSB} indicates the voltage value of the least significant bit. The RMS of the noise voltage of an ideal ADC is given by its quantization error. In respect to Figure 10.8, the value can be calculated as

$$v_{noise} = \sqrt{\frac{1}{V_{LSB}} \int_{-0.5 V_{LSB}}^{0.5 V_{LSB}} V_{LSB}^2 \, dV_{LSB}} = \frac{V_{LSB}}{\sqrt{12}}. \tag{10.5}$$

The SNR of the ideal ADC is thus given by:

$$\text{SNR} = 20 \log\left(\frac{2^N V_{LSB}}{2\sqrt{2}}\right) \frac{\sqrt{12}}{V_{LSB}} = 6.02N + 1.76. \tag{10.6}$$

This equation denotes an important relation between the resolution and the SNR of an ideal ADC. An ideal 16-bit ADC thus has a signal-to-noise ratio of 98.08 dB. For nonideal ADCs this concept is often reversed to obtain a meaningful measure for the resolution of the device. In this approach, the SNR of the ADC is measured and the so-called effective number of bits (ENOB) is calculated as

$$\text{ENOB} = \frac{\text{SNR} - 1.76}{6.02}. \tag{10.7}$$

Power Dissipation The power dissipation of ADC depends on multiple parameters. Commonly, the mean of the power dissipation has a proportional dependency of the sampling rate, mainly basing on transistor switching effects (considering CMOS circuit designs), but not limited to. The power dissipation depends significantly on the data converter architecture and the particular circuit implementation, discussed in the next section. Apart from the usable resolution and the source-to-noise ratio, the power consumption of ADC components is a central design criteria in the deployment in low-power and very low-power sensing systems.

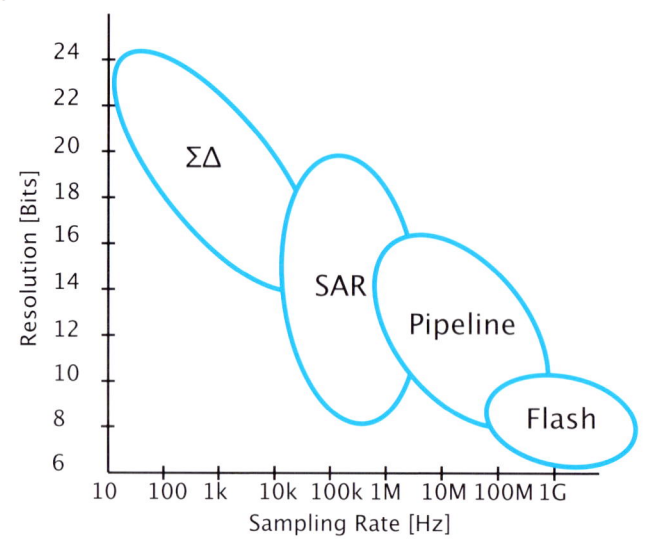

Figure 10.11 Different ADC architectures and their common usage in terms of resolution and sample rate.

10.3 Data Converter Architectures

Considering that currently thousands of converters are available on the market, the selection of the architecture most suited for the given task is an essential part of circuit and sensor node design. In this selection process, several aspects have to be considered. Sigma-delta ADCs, for example, achieve a high resolution while the corresponding sample rate is comparatively low. Therefore, sigma-delta ADCs are ideal for application in sensor signal processing, where conversion speed is not essential. Conversely, flash converters and pipeline architectures are the fastest ADC available, yet due to aspects such as chip area consumption they are limited to lower resolution. In Figure 10.11, the most common ADC architectures are compared in terms of their resolution and sample rate. Some more details can be found in Ref. [4], and design aspects are discussed in Ref. [5]. As shown in the next sections, ADC designs consist of analog and digital parts. A deeper discussion on mixed-signal circuit design can be found in Ref. [6].

The basic component of every ADC is the comparator. It can be seen as a 1-bit ADC as it has two analog inputs and one digital output. The output is a digital 1, if the voltage at the noninverting input of the comparator exceeds the voltage level at the inverting input. Otherwise, the output is a digital 0. Another functional part is the reference circuit that determines the relationship between an analog input value and the corresponding digital output. The performance of the reference circuit directly impacts on the magnitude of nonlinear errors, such as DNL and INL. Furthermore, every ADC offers the possibility to set the input range by means of a reference voltage.

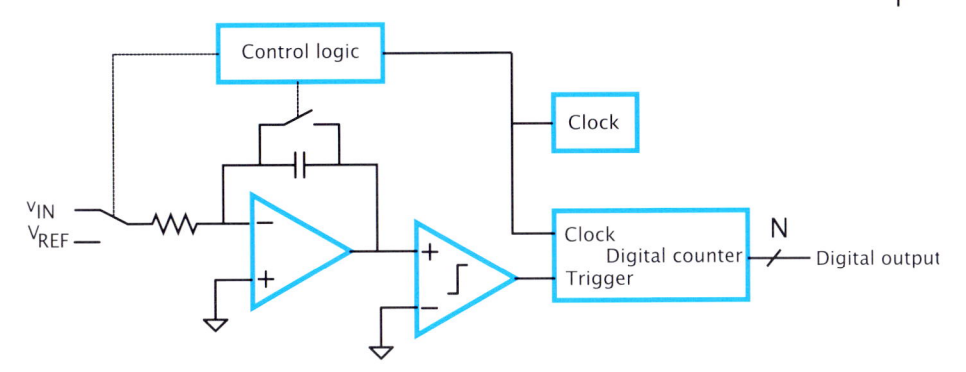

Figure 10.12 A dual-slope integrating architecture.

Integrating ADC Integrating converters are used for high-resolution and low-speed applications. A schematic overview of an integrating ADC implemented in the so-called dual-slope architecture is illustrated in Figure 10.12.

The basic components of this architecture are an integrating amplifier followed by a comparator and a digital counter. At the beginning of the conversion process, the capacitor of the integrator is discharged by the control logic. At $t=0$ the analog input signal V_{in} is connected to the input of the integrator and the capacitor is charged. After a constant time the input is switched to the reference voltage, V_{REF}. The voltage at t_1 is directly proportional to the analog input voltage. Because the reference voltage is negative, the capacitor will discharge at a rate proportional to the reference voltage. The digital counter measures the time from t_1, until the capacitor is fully discharged and the comparator will switch its output at t_2. An overview of the charging and discharging process is given by Figure 10.13.

Assuming an ideal capacitor, the ratio between the discharge time and the charge time is proportional to the ratio of the input voltage to the reference voltage. Consequently, the digital counter value is proportional to the analog input voltage, as well.

It is important to note that the input voltage is averaged during the charging process of integrating ADCs. Therein lays an important difference with respect to ADCs that employ sample-and-hold techniques. The averaging can be advantageous as it tends to reject periodic noise. This is best utilized by using an

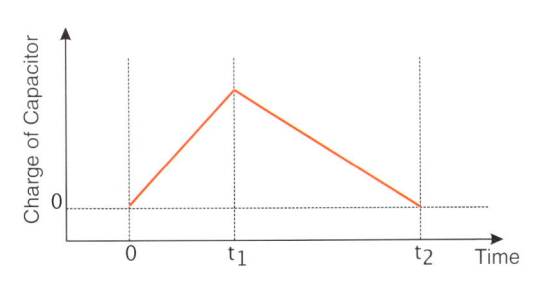

Figure 10.13 Charge of the integrating capacitor versus time.

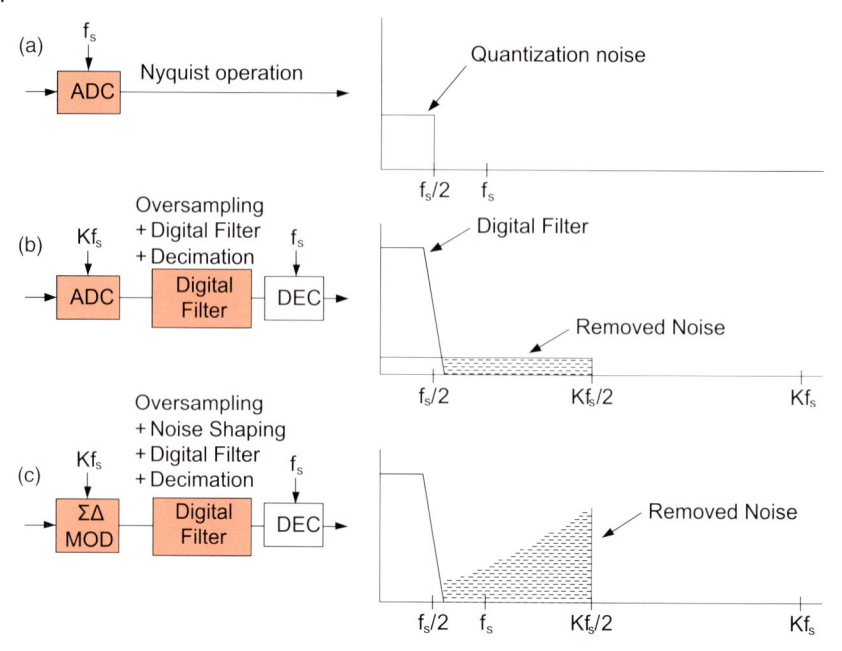

Figure 10.14 Overview of the principles of oversampling, digital filtering, noise shaping, and decimation.

integration time that is an integer multiple of the AC line period of 50 Hz or 60 Hz, so that interferences from the power line are canceled.

Sigma-Delta Converters The sigma-delta ($\Sigma\Delta$) ADCs have become a popular architecture choice. $\Sigma\Delta$-ADCs often have resolutions above 16-bit at sample rates below 100 kS/s. The high resolution is achieved by oversampling, noise shaping, and decimation. These three working principles are illustrated in Figure 10.14.

Figure 10.14a shows the distribution of quantization noise in the frequency domain for a sampling ADC operating at sampling rate of f_s to fulfill the Nyquist criteria for signals with frequencies up to $f_s/2$. If a (much) larger sampling rate is used than strictly required to fulfill the Nyquist criteria, the rms quantization noise is spread over a wider bandwidth; that is, from DC to $Kf_s/2$, where K is the oversampling factor and f_s the required sampling rate. This entails that the application of a digital low-pass filter removes a fraction of the quantization noise thus increasing the SNR as well as the ENOB (see Figure 10.14b).

However, if one were to only use oversampling to increase the resolution, the oversampling factor K would need to be 2^{2N} for a gain of N-bits of resolution. $\Sigma\Delta$-ADCs mitigate the requirement for unreasonably large oversampling factors by shaping the quantization noise to higher frequencies (see as Figure 10.14c). This is achieved by a so-called $\Sigma\Delta$ modulator. Based on the state of a 1-bit ADC, that is, a comparator, a 1-bit DAC signal is generated. The digital analog converter (DAC) signal is summed with the input and fed to an integrator, which in turn is directly connected to the comparator thus creating a negative feedback loop (see Figure 10.15).

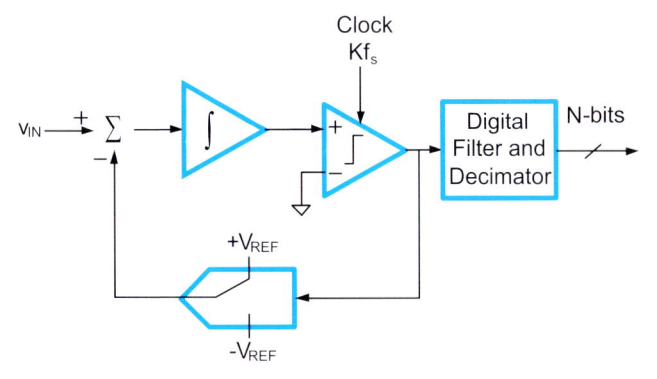

Figure 10.15 First-order sigma-delta ADC.

The input voltage is represented by the serial output stream of the comparator as the feedback loop forces the average DC voltage on the output of the 1-bit DAC to be equal to V_{IN}. If one, for example, assumes a DC of $V_{IN} = 0$ at the input of a bipolar ΣΔ-ADC, the integrator is constantly ramping up and down and the ratio between high and low states at the output of the comparator will be 1. If the input signal increases toward $+V_{REF}$, the number of high states in the serial output stream of the comparator will increase, as illustrated in Figure 10.16. Conversely, if the input signal shifts toward $-V_{REF}$, the number of low states will increase, thus reducing the average voltage.

The subsequent digital filter and the decimator process the bit stream to the final output at the sample frequency of f_s. The fact that the digital output filter reduces the bandwidth entails that the output data rate can be reduced by the decimator, without violating the Nyquist criteria. The decimator only passes every Xth result to the output while the rest is discarded. X can be any integer value as long as the output data rate is still larger than twice the input signal bandwidth.

To obtain a meaningful result at the output of the ΣΔ-ADC, a large number of samples should be averaged as the dynamic range increases as more samples are averaged. For example a digital low-pass filter that averages four samples can only assume four states indicating a resolution of 2-bit. If the filter is averaging every 16 samples, the resolution yields 4-bit and so forth. Figure 10.16 shows the output of the integrator and comparator for an input DC voltage of $V_{IN} = V_{REF}/2$. The respective output values for the 2-bit and 4-bit (not shown) ΣΔ-ADC are 3/4 and 12/16, respectively.

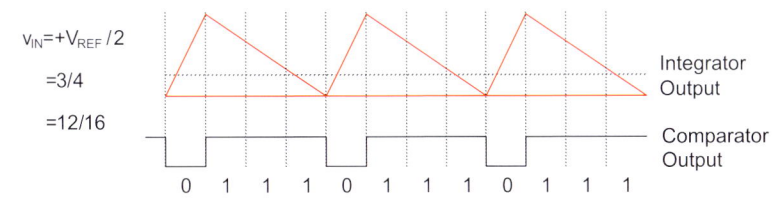

Figure 10.16 Sigma-delta waveforms at a certain analog input voltage.

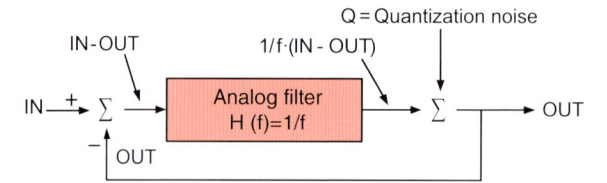

10 Analog Sensor Signal Processing and Analog-to-Digital Conversion

Figure 10.17 Simplified frequency domain model of a sigma-delta modulator.

The concept of the ΣΔ-ADCs noise shaping is best explained by considering the frequency domain. The integrator of the ΣΔ modulator is working as an analog low-pass filter with the transfer function $H(f) = 1/f$. The 1-bit DAC generates the quantization noise Q which is fed back to the input of the integrator as indicated in Figure 10.17.

For a given input signal IN, the signal after the summing block (i.e., at the input of the integrator) is IN−OUT. This signal is multiplied by the transfer function of the integrator. Finally, the quantization noise is added resulting in the following equation:

$$\text{OUT} = \frac{1}{f}(\text{IN} - \text{OUT}) + Q. \tag{10.8}$$

Solving this equation for OUT results in

$$\text{OUT} = \frac{\text{IN} + Qf}{1 + f}. \tag{10.9}$$

If the frequency of the input signal approaches zero, the output voltage consists entirely of the input voltage without any quantization noise. At higher frequencies the amplitude of the input signal will decrease while the quantization noise approaches the value of Q. In essence, the integrator has a low-pass effect on the input signal while having a high-pass effect to the quantization noise. Therefore, the integrator working as an analog filter is performing the noise shaping function of a ΣΔ-ADC. To achieve even higher resolutions at the same oversampling frequency, higher order ΣΔ modulators can be implemented. By using higher order analog filters the effect of the noise shaping is increased due to the higher attenuation of the filter. An example for a second-order ΣΔ-ADC is given in Figure 10.18.

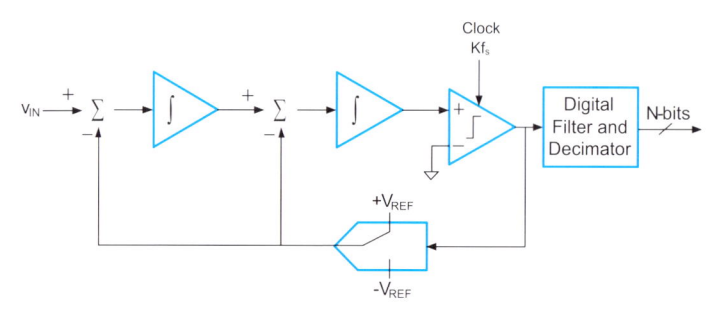

Figure 10.18 Second-order sigma-delta ADC.

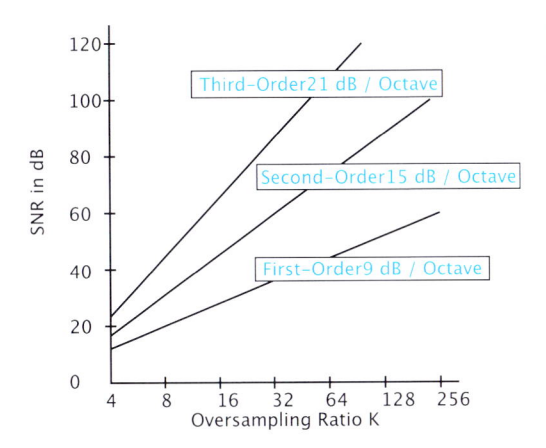

Figure 10.19 SNR versus oversampling ratio for different sigma-delta modulator orders.

In Figure 10.19, the relationship of a ΣΔ-ADC to the achievable SNR is shown. If a resolution of 12-bit is required, the necessary signal to noise ratio (SNR) of the ADC has to be 74 dB. This is, for example, achieved, using third-order ΣΔ modulator with an oversampling factor of at least 16. In order to achieve the same resolution with a first-order ΣΔ modulator, the factor would need to be approximately 512. Since high clock frequencies in a mixed-signal integrated circuit can cause severe difficulties regarding the performance of the analog circuit blocks, a trade-off between more complicated higher order ΣΔ architectures and the oversampling ratio has to be found.

Successive Approximation ADC The successive approximation ADC is by far the most popular architecture concerning data acquisition systems. Having resolutions between 8- and 18-bit while the sample rate can be greater than 1 mega sample per second (MSPS). In addition, they are low in cost due to the low chip-area consumption.

Figure 10.20a shows the basic components of a SAR-ADC. At the start of the conversion process, the S/H amplifier (SHA) is in hold mode. All bits in the successive approximation register (SAR) are set to 0 except the MSB, which is set to 1. The SAR output drives the DAC to apply the first reference voltage to the comparator. Therefore, analog voltage at the output of the DAC is always $V_{DA} = V_{REF}/2$ at the beginning of a conversion cycle. If the output of the DAC is greater than the analog input, the MSB in the register will be cleared, otherwise it is left set. The next most significant bit is set and the process continuous until the LSB is reached. Consequently, N comparison operations are required, in order to convert the analog input signal into an N-bit digital word. An example for this conversion process is shown in Figure 10.20b.

The conversion time and the sample rate of a SAR-ADC, are given by

$$T_{conversion} = NT_{setting} = \frac{1}{f_s}. \tag{10.10}$$

It is evident that the speed of the ADC is limited by the settling time, $T_{settling}$, of the DAC. It is important to note that the DAC may need a longer time to settle its

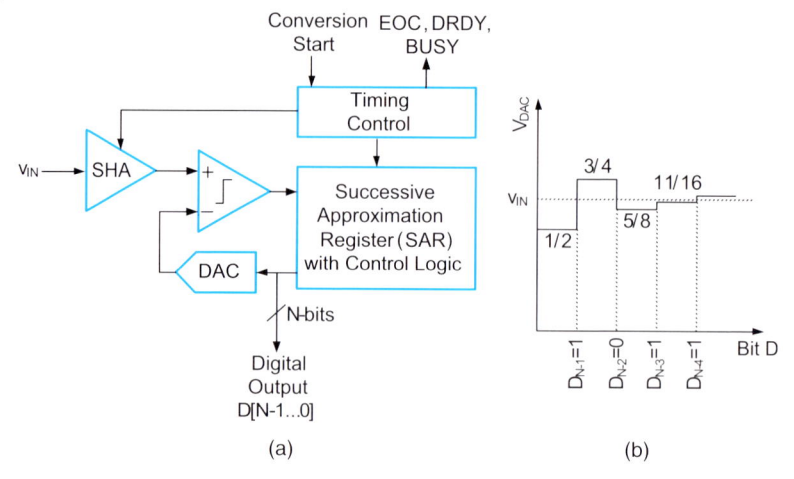

Figure 10.20 (a) Basic architecture of a successive approximation ADC. (b) Weighting process for a given analog input signal.

output voltage within the tolerances of 1/2 LSB, if the resolution is high. This entails, that the conversion time does not necessarily scale linearly with N.

One of the most popular successive approximation architectures uses the binary-weighted capacitor array as its DAC. The reference voltage of the DAC is based on the amount of charge on each of the DAC capacitors. Figure 10.21 shows a N-bit architecture.

The binary-weighted capacitor array is sampling the analog input voltage such that no sample-and-hold circuit is required. Before the conversion starts the reset switch is closed and the capacitor array is discharged. As long as the reset switch is closed the comparator acts as a unity gain buffer. In this state, the capacitor array charges to the offset voltage of the comparator performing an offset cancellation. In order to start the conversion process, V_in is connected to the input nodes of the entire array. The array is charged while the reset switch is still closed, holding its side of the capacitor array at virtual ground. The reset switch is opened while the input of the capacitors is switched from back to ground, simultaneously. As a result, the voltage at the comparator input is now $V_\text{offset} - V_\text{in}$.

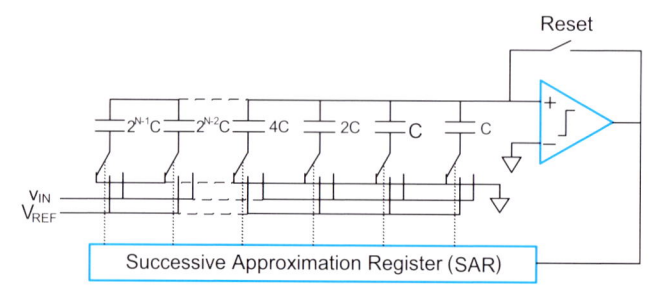

Figure 10.21 A charge redistribution architecture using a binary-weighted capacitor array.

To initiate the first conversion step, the input of the MSB capacitor (i.e., the capacitor with the largest capacitance) is switched to V_{REF}. The current voltage at the comparator input is now

$$V_{incomparator} = V_{offset} - V_{in} + D_{N-1}\frac{V_{REF}}{2}, \qquad (10.11)$$

where D_{N-1} is the output of the comparator, that is, the MSB. If the comparator output is high, the input of the MSB capacitor remains connected to V_{REF}, otherwise it is switched back to ground. Subsequently, the second largest capacitor is switched to V_{REF} and the voltage at the comparator input array becomes

$$V_{incomparator} = V_{offset} - V_{in} + D_{N-1}\frac{V_{REF}}{2} + D_{N-2}\frac{V_{REF}}{4}. \qquad (10.12)$$

The conversion process continuous until every bit in the successive approximation register is tested. The input voltage of the comparator at the end of the conversion process can be described as

$$V_{incomparator} = V_{offset} - V_{in} + D_{N-1}\frac{V_{REF}}{2} + D_{N-2}\frac{V_{REF}}{4} + \cdots + D_1\frac{V_{REF}}{2^{N-1}}$$
$$+ D_0\frac{V_{REF}}{2^N}, \qquad (10.13)$$

where N represents the resolution of the ADC.

Flash ADC Flash ADCs, or parallel ADCs, use a large number of comparators simultaneously for the signal conversion, which enables extremely fast conversion times. The schematic of a 3-bit Flash ADCs is shown in Figure 10.22. It can be

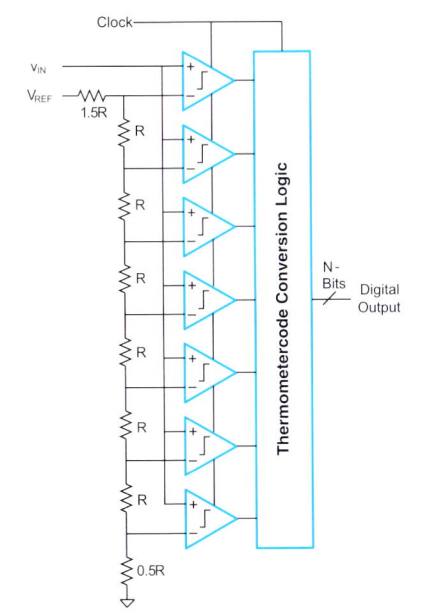

Figure 10.22 A 3-bit parallel flash converter.

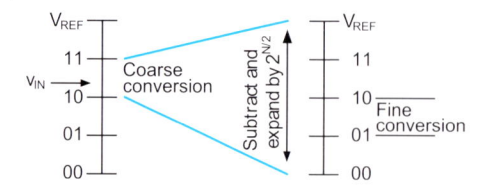

Figure 10.23 Principle of coarse and fine conversion using a two-step flash ADC.

deduced from this figure that an N-bit flash ADC requires $2^N - 1$ comparators. Furthermore, 2^N resistors are required, as each comparator needs its own reference voltage, which is supplied to be the string of resistors. The resistors are equal, with the exceptions of the first and the last resistor in the string. Thus, all neighboring nodes differ by 1 LSB.

The output of all comparators is often referred to as a thermometer code, because a certain input voltage causes the output of all comparators with a lower reference voltage to transition to a higher state. A digital encoder converts this data into the corresponding binary code. Because only the comparators and the digital encoder are limiting the sample-time fast converters can be realized. This advantage is counterbalanced by the need to double the chip-area to gain 1 bit of resolution. Therefore, flash converters rarely exceed 8-bit resolution.

One approach to limit the number of necessary comparators is the implementation of a two-step flash ADC. The working principle of a two-step flash ADC is illustrated in Figure 10.23. The converter consists of two flash ADCs where one converter estimates the most significant bits (MSB). The result of this *coarse conversion* is converted back into an analog signal and subtracted from the input signal. The remainder is converted by the second flash ADC. This ADC delivers the least significant bits, as its full range corresponds to the LSB of the coarse converter. The amount of comparators used by a two-step flash converter is given by $2(2^{N/2} - 1)$. For example a 10-bit flash converter requires 1023 comparators while a two-step flash ADC requires only 62.

10.4 Low-Power ADC Designs and Power Classification

In Figure 10.11 different converter architectures were classified by resolution and sampling rate domains. But power dissipation is the third central parameter in the design of material-integrated sensing systems demanding for low-power devices. Though there is a relation between the resolution, sampling rate, converter architecture, circuit technology, and the power dissipation of an ADC, it is difficult to give estimations of expected power dissipation by these parameters. There are some rough estimations to classify and quantify the power dissipation of different ADC architectures and in relation to the bit-resolution and sample rate. In Ref. [7] it was stated that the minimum power dissipation (lower bound) for data sampling of Nyquist ADCs (no oversampling) that can be achieved is proportional to the sampling frequency f_S and the power of the bit-resolution N:

$$P_S \sim f_S 2^{2N}. \tag{10.14}$$

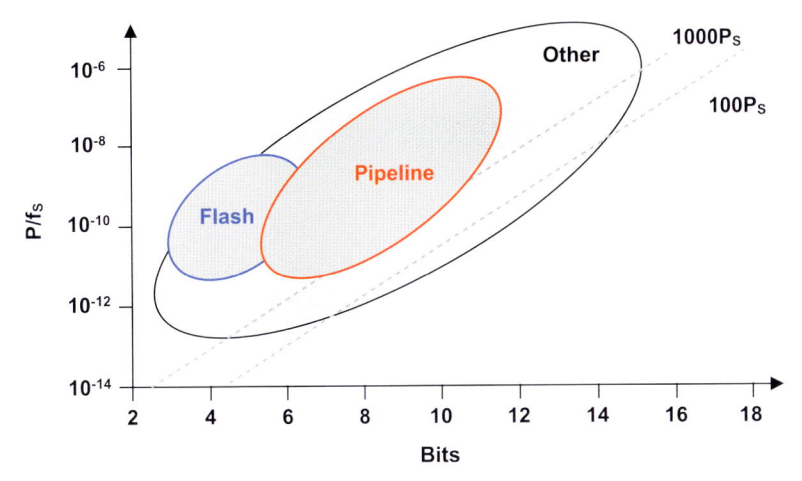

Figure 10.24 Normalized power dissipation of different ADC architectures depending on the effective resolution (SNR bits) and comparison with the estimated minimal power dissipation bound from Eq. (10.14). (Based on Ref. [7] and data from Ref. [9].)

Similar discussion about the lower limits of ADC are discussed in Ref. [8], analyzing the data sampling circuit of ADCs.

In Figure 10.24 the normalized power dissipation of different ADC architectures is shown, depending on the effective resolution (SNR bits) and giving a comparison with the estimated minimal data sampling power dissipation bound from Eq. (10.14) (based on data from Ref. [9]). As expected, there is a monotonic increase of the power dissipation in relation to the bit resolution. It is interesting that the current ADC designs have much higher power requirements than the predicted minimum power bounds. Surprisingly, the normalized power dissipation of flash ADC architecture scales better than for pipelined or SAR (other) architectures. Due to the power dependency from bit-resolution attempts were made to reconfigure the bit-resolution at runtime, for example, Ref. [10], proposing a 5–10-bit SAR ADC. Furthermore, the supply voltage can be varied between 0.4 and 1 V, further lowering the power dissipation with low sample rates. Leakage effects of the electronic circuit become dominant at sample rates below 1–2 kS/s.

In Ref. [11], an ultralow power ADC architecture and implementation was proposed using the SAR conversion method. They achieve a power consumption in the µW range by providing 8-bit resolution and sample rates up to 100 kS/s (1 V supply voltage). The standby power consumption could be achieved in the pW range. A conversion requires energy of about 30 pJ. High-speed ADC can achieve power consumptions down to the mW range, based on SAR and pipelined SAR principles (e.g., a 9-bit 150MS/s SAR ADC [12]).

10.5 Moving Window ADC Approach

Resistive sensors, such as strain gage sensors, provide only a small relative change in resistance on the order of 1% resulting from a change of applied load in the considered operating range of the sensor. Using bridge configuration, providing a

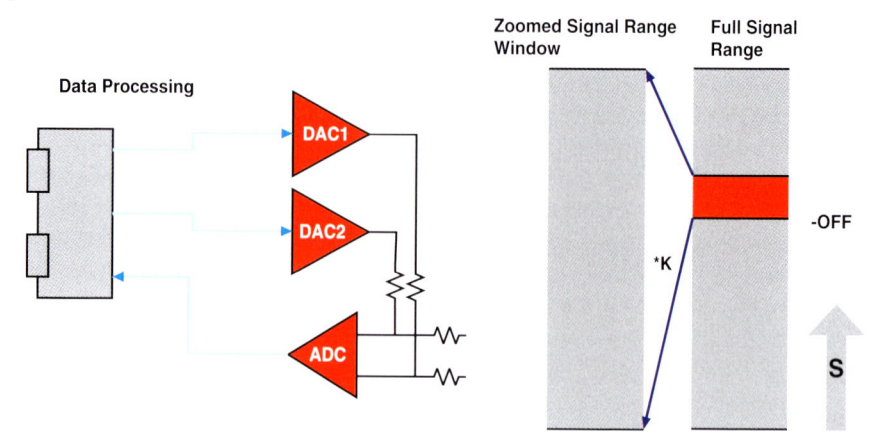

Figure 10.25 Zooming and moving window ADC approach, used, for example, with resistive strain gage sensors (ADC: analog-to-digital converter with differential input, DAC: digital-to-analog converter).

differential signal, require compensated sensors with small tolerances in strain and zero-load resistance parameters, actually not applicable to sensorial materials using, for example, printed sensors.

Assuming only one noncalibrated and uncompensated resistive sensor, a zooming window approach (see Eq. (10.15)) can be used to match an initially unknown sensor to the measurement system preserving a high and full-range resolution, shown in Figure 10.25.

$$W(s) = k(s - \text{off}), \tag{10.15}$$

with W being the window, k the zoom factor, and *off* the shift of the window.

The data processing performs an initial (or periodically repeating) autocalibration finding the center of the operational window by using fast settling successive approximation, shown in Algorithm 10.1. The zoom factor k is kept fixed (determined by the amplifier gain settings of the ADC/DAC sections), and only the offset of the window is calibrated.

Algorithm 10.1 Autocalibration Using Successive Approximation

```
1  sar ← DIGITALRANGE/2;
2  DAC1 ← GAIN0, DAC2 ← 0;
3  WHILE sar <> 0 do begin
4    IF ADC > DIGITALRANGE/2 THEN
5       DAC2 ← DAC2 + sar
6    ELSE
7       DAC2 ← DAC2 - sar;
8    sar ← shift_right(sar,1);
9  end;
10 off ← DAC2-DAC1;
```

Reconfigurable ADC supporting different resolutions can support the window approach significantly in presence of varying bandwidth and dynamic ranges. For example, in Ref. [10] a reconfigurable SAR-ADC architecture was proposed that provides a bit-resolution range from 5 to 10 bits, supporting an optimal dynamic adaption if used in conjunction with the window approach. Hence, the extended optimized dynamic range adaption using resolution configurable ADC with moving and zooming windows enables the design of very low-power sensor networks, a prerequisite for, for example, material-integrated sensing systems and self-supplied sensing systems with very limited available energy.

References

1 Webster, J.G. and Eren, H. (2014) *Measurement, Instrumentation, and Sensors Handbook*, 2nd edn, CRC Press.
2 Sydenham, P.H. and Thorn, R. (2005) *Handbook of Measuring System Design*, John Wiley & Sons, Inc, New York.
3 Johns, D. and Martin, K.W. (1997) *Analog Integrated Circuit Design*, John Wiley & Sons, Inc, New York.
4 Kester, W. (2005) *The Data Conversion Handbook*, Analog Devices.
5 Plassche, R.J.d. (2005) *CMOS Integrated Analog-to-Digital and Digital-to-AnalogConverters*, 2nd edn, Springer, New York.
6 Baker, R.J. (2010) *CMOS Mixed-Signal Circuit Design*, 3rd edn, John Wiley &Sons, Inc., New York.
7 Sundström, T., Murmann, B., and Svensson, C. (2009) Power dissipation bounds for high-speed Nyquist analog-to-digital converters. *IEEE Transactions on Circuits and Systems—I: Regular Papers*, **56** (3), 509–518.
8 Murmann, B. (2006) Limits on ADC power dissipation, in Analog Circuit Design, pp. 351–367.
9 Murmann, B. (2007) ADC Performance Survey 1997–2007. Available at http://www.stanford.edu/murmann/adcsurvey.html.
10 Yip, M. and Chandrakasan, A.P. (2011) A resolution-reconfigurable 5-to-10b 0.4-to-1V power scalable SAR ADC, in Digest of Technical Papers - IEEE International Solid-State Circuits Conference.
11 Scott, M.D., Boser, B.E., and Pister, K.S.J. (2003) An ultralow-energy ADC for smart dust. *IEEE Journal of Solid-State Circuits*, **38**, 1123–1129.
12 Lin, S.J.C., Ying, Z., Liu, C., Cheng, H., Guan Y., and Shyu, Y.T. (2013) A 9-bit 150-MS/s 1.53-mW subranged SAR ADC in 90-nm CMOS, in IEEE Symposium on VLSI Circuits, Digest of Technical Papers.

11

Digital Real-Time Data Processing with Embedded Systems

Stefan Bosse[1] and Dirk Lehmhus[2]

[1]*University of Bremen, Department of Mathematics and Computer Science, Robert-Hooke-Str. 5, 28359 Bremen, Germany*
[2]*University of Bremen, ISIS Sensorial Materials Scientific Centre, Wiener Str. 12, 28359 Bremen, Germany*

11.1 Levels of Information

On a long-term scale, material-integrated sensor systems will become common. Structural materials thus equipped will be able to judge their own state irrespective of the loads they experienced, and will not lose this ability even if damage has been incurred. Structures of this kind will show autonomy at least in the evaluation of their structural state, and the adaptation of their self-perception will base on this evaluation.

The field of structural monitoring and control has been evolved in 20 years significantly, leading to a classification of systems originally proposed by Rytter [1] and recently summarized by Lehmhus *et al.* [2] (see Figure 11.1)

The deployment of embedded systems for sensor data processing can be found in the following areas:

Structural Monitoring and Control

Automated gathering of information on mechanically loaded structures has originated in civil engineering, with bridges or skyscrapers as the primary objects of supervision [3–5]. Sensor node size is typically macroscopic here, and outward attachment to structural members is sufficient. However, the "need for speed" is clearly there. Miniaturization in combination with new developments in aerospace engineering, not the least the broader introduction of composite materials in replacement of metals, has fueled interest in the technology in this area for more than a decade [6]. At the very end of these trends are morphing structures that facilitate fly-by-feel concepts in unmanned aerial vehicles (UAV). These again strongly stress the interest in immediate availability of information. Recently, maintenance issues associated with offshore wind energy plants, which also rely heavily on composite materials, for example, for rotor blades, have created similar interest here.

Material-Integrated Intelligent Systems: Technology and Applications, First Edition.
Edited by Stefan Bosse, Dirk Lehmhus, Walter Lang, and Matthias Busse.
© 2018 Wiley-VCH Verlag GmbH & Co. KGaA. Published 2018 by Wiley-VCH Verlag GmbH & Co. KGaA.

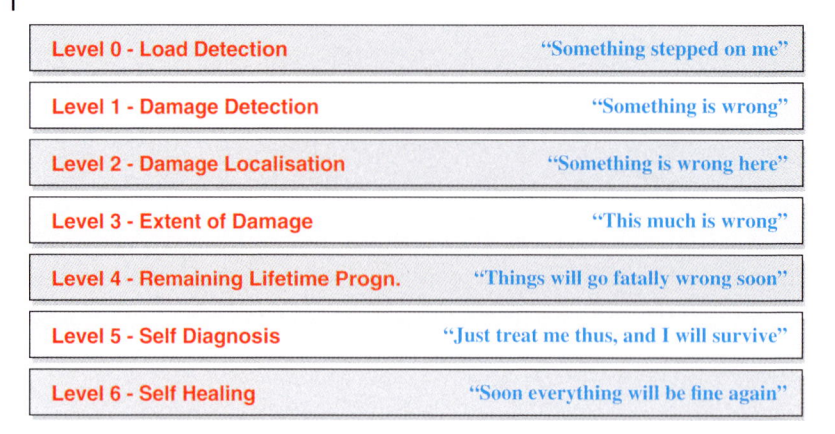

Level 0 - Load Detection	"Something stepped on me"
Level 1 - Damage Detection	"Something is wrong"
Level 2 - Damage Localisation	"Something is wrong here"
Level 3 - Extent of Damage	"This much is wrong"
Level 4 - Remaining Lifetime Progn.	"Things will go fatally wrong soon"
Level 5 - Self Diagnosis	"Just treat me thus, and I will survive"
Level 6 - Self Healing	"Soon everything will be fine again"

Figure 11.1 Algorithmic and information level hierarchy in SHM and LM systems.

Structural Health Monitoring

Structural health monitoring (SHM) adds the ability to derive not just loads, but also their effects from sensor data. Boller [7] gave a definition of a SHM system: *A SHM is the integration of sensing and possibly also actuation devices to allow the loading and damaging conditions of a structure to be recorded, analyzed, localized, and predicted in a way that nondestructive testing (NDT) becomes an integral part of the structure and a material.*

Load Monitoring

A load monitoring system (LM) can be considered as an incomplete subclass of a full SHM system, which provides spatial resolved information about loads (forces, moments, etc.) applied to a technical structure. When implemented in robot grippers [8], such systems can improve grasping performance via feedback control. Another application can be found in long distance robot manipulator structures to improve position accuracy [9]. In aerospace industry, load monitoring systems as first step toward structural health monitoring were initially implemented in military aircraft such as the Eurofigther TYPHOON [10], but have meanwhile entered the civilian market in parallel to their ongoing maturing into true SHM systems [11,12]. Real-time capabilities of the load and structural monitoring system can be essential.

Structural control, adaptive and morphing structures

The term structural control implies a material-inherent capability of changing structural characteristics in response to, and countering, the effects of external loads. It is thus beyond mere sensing and beyond the scope of systems addressed within this text. These, however, form the necessary basis of farther-reaching adaptive systems.

Besides alleviating structural loads, structural control-like systems are foreseen to facilitate a fly-by-feel approach for autonomous flight of UAVs [12]. At a second glance, this scenario can be seen as linked to robotic tactile sensing: In both cases, a major extension of perceptive capabilities at the interface between system and environment is foreseen to support interaction with the latter, and based on it, allow for increased autonomy.

Tactile Sensing

Tactile sensing systems provide extrinsic perception for robots and robotic applications [13], via systems commonly designated as smart or artificial skin. Basically they deliver spatial resolved information on forces applied to an extended but limited surface region, for example, of robot connection elements or finger tips of a robot hand [14,15]. The fingertip example is probably what comes to mind first when thinking of robotic tactile sensing. However, covering areas other than a fingertip with a smart skin capable of interpreting tactile sensor data bears additional advantages: Sensor networks of this kind can provide the robotic system with much finer information about – voluntary as well as involuntary – contacts with its surroundings than state-of-the-art joint-integrated load monitoring ever will. This can be exploited to enhance safety levels in robot–robot or human–robot cooperation, but just as well to endow the robot with the capability to monitor its own actions, profiting from their success or failure via reinforcement learning. The concept has been applied to humanoid robots in the form of tactile sensor arrays to help monitor, control, and ultimately improve their strategy to dynamically stand up from lying on their back. Such examples are merely a glimpse at the full potential of material-integrated sensor systems. From the above, many development trends can easily be extrapolated, like the simple change of perspective that turns robotic smart skins into new kinds of tactile user interface for countless purposes. The real-time capability of the tactile sensing system is fundamental.

11.2 Algorithms and Computational Models

The raw sensor output of a structural monitoring or tactile sensing system reflects the lowest level of information. Both use cases incorporating sensor networks with sensor nodes acquiring sensor data of different dimension, regarding the spatial and the temporal dimension, starting from a single-strain gage sensor delivering a scalar sensor value up to temporal distributed wave receiver data. Beside technical aspects of sensor integration of the main issue in those applications is the derivation of a mapping function $F_m(S)$, which basically maps the raw sensor data input S, in general a n-dimensional vector consisting of n sensor values (usually sampled from electromechanical transducers like piezoelectric sensors, or from electro-optical transducers like fiber-optic sensors [16]), on the desired information I, in general a m-dimensional result vector:

$$F_m(S) : \vec{S} \rightarrow \vec{I}, \quad (11.1)$$
$$\vec{S} = (s_1, s_2, \ldots), \vec{I} = (i_1, i_2, i_3, \ldots).$$

The function F_m can be considered as a feature extraction function, and the output vector I can be classified in the following:

- A condensed senor data vector retrieved by sensor data fusion, that means $S = (s_1, s_2, \ldots, s_n) \rightarrow I = S' = (s'_1, s'_2, \ldots, s'_m)$ with $m < n$, to improve the confidence of the measured sensor data.

- A spatially resolved load vector with an estimation of the load (forces) applied to a structure under test, for example, used in artificial skin [17] or robot manipulator [18] applications, providing the estimated center location (coordinates) of the region and the strength information of the load, for example, $I = (x,y,z,v)$.
- At higher abstraction level, a proposition vector of detected features and patterns f_i, for example, a damage-sensitive feature (structural defect) or a decision for maintenance, for example, $I = (f_1, f_2, f_3, \ldots)$ with $f_i = [0,1]$.

This mapping function is usually built up by functional composition, with different functions ranging from low-level to complex high-level signal processing algorithms:

$$F_m(S) = f(g(h(\ldots(m(S))))) : S \to I$$
$$\rightleftharpoons S \mapsto m \mapsto \cdots \mapsto h \mapsto g \mapsto f : \quad (11.2)$$
$$(S \to \alpha) \to (\alpha \to \beta) \to \cdots \to S \to I.$$

That means the signal processing system is composed of initially independent functions $f, g, h, \ldots$ (which can be considered as higher order functions treating arguments as functions and allowing partial evaluation) arranged in a pipeline chain, each performing the calculation of an intermediate result based on either the original sensor data S or already computed intermediate data ($\alpha \to \beta$). From a data processing point of view, these functions are operational units that can be modeled with processes with data dependencies, requiring synchronized interprocess communication, as discussed in Chapter 6.

Basically there are two different information extraction approaches: (I) Those based on an accurate mechanical and numerical model of the technical structure and the sensor, and on the other side (II) those without any or with a partial physical model. The first class comprises, as most common variant, inverse finite element methods (FEM). These provide a fast path from local system reactions such as mechanical strain derived from distributed sensors to the loads which cause them – in terms of strength level and location. This capability matches the description of a load monitoring system. Derivation and classification of damage introduced by these loads is a further issue asking for additional models describing, for example, failure mechanisms and propagation. Reintroduction of structural change effected by such local damage to the structure requires another step, and a replacement of the original (inverse) structural model: This goes to show that the method has deficiencies when it comes to providing SHM systems with the necessary flexibility for real-time data evaluation that parallels damage evolution. Nevertheless, the method has its values where immediate model adaptation is less important and is thus object of ongoing research [19–22]. The second class usually bases on classification or pattern recognition algorithms derived from supervised machine learning [23], discussed in Chapter 13. A common widely used method is the correlation analysis of measured data resulting from induced stimuli at runtime (system response) with data sets retrieved from an initial (first-hand) observation, which make it more difficult to select damage relevant features from the measurement results [7]. These

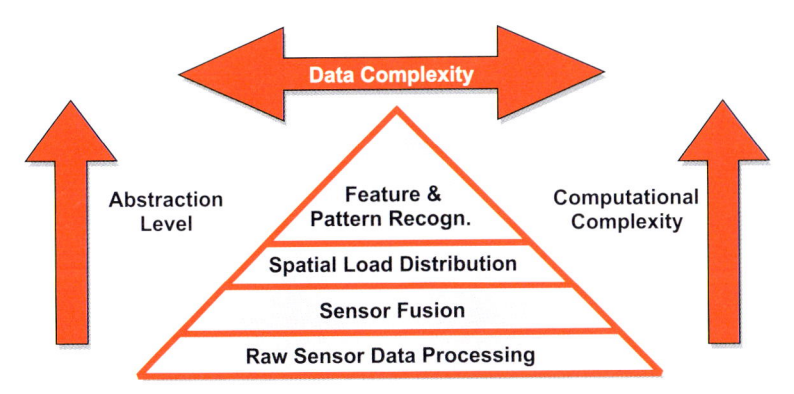

Figure 11.2 Data processing classes common to SHM, LM, and TS systems.

measurements are based on signal correlation [23] or waveform and spectral analysis, for example, from vibration analysis [16,24]. Finite element model-based simulation (FEM) can support and improve model-based sensor signal processing, for example, by identifying regions of interest that are damage-sensitive or by selecting appropriate regions to place sensors. Feature extraction using machine learning techniques with at least a model of the structure (but not necessarily of the sensor and sensor-material interaction) can be improved by FEM by substituting supervised learning under real conditions with a simulation-driven learning approach [25].

Figure 11.2 shows the different information extraction classes representing different data processing levels and their relationship to the abstraction, as well as computational and data complexity.

Several different algorithm classes and levels of data processing are used in sensor networks, varying in computational and communication complexity, memory resources, and the kind of data they process. These are summarized in Figure 11.3.

Basically computation and data processing in sensor networks can be classified by the following:

- **Number of sensors:** Algorithms are performing computations based on one sensor value (local scope) or multiple sensor values (multiple-local or global scope).
- **Temporal computation model:** The computation is performed online or offline in that sense that measuring data is either processed continuously at runtime or a limited set of data is only stored and processed after a measurement, with the first case requiring real-time capable data processing platforms.
- **Temporal processing model:** Processing of a computation is either strict sequentially (via a single process at time) or parallel processed (via multiple concurrent processes).
- **Spatial computation model:** Local (centralized) or global (distributed) computation is employed based on local or global distributed data, with either short- or long-range data dependencies.

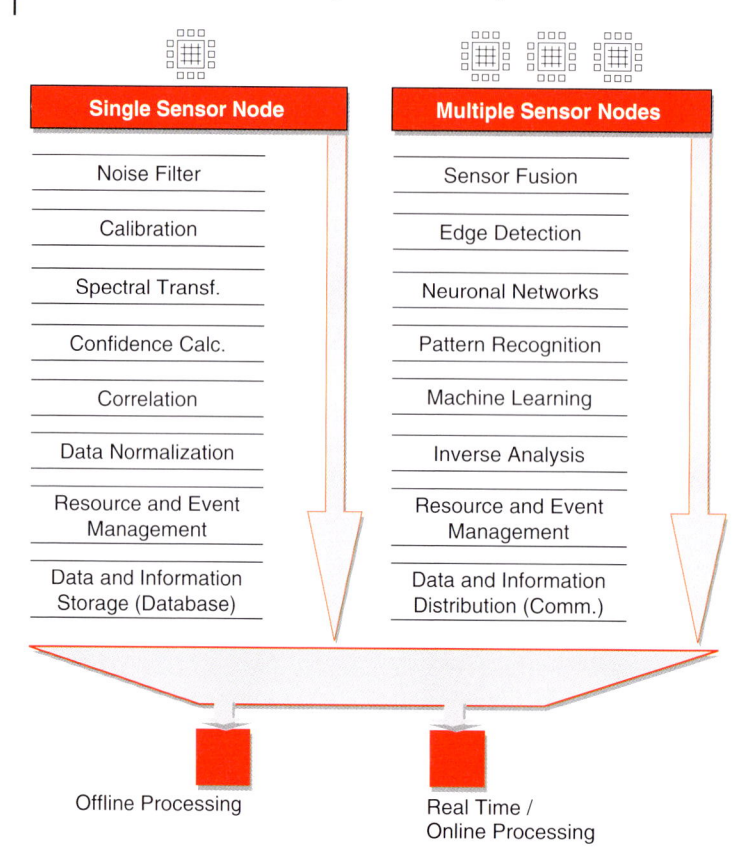

Figure 11.3 Different data processing algorithm classes used with single and multiple sensor nodes (i.e., deployed in distributed sensor networks).

- **Spatial network model:** Computations can be performed *in situ* (in-network) or outside of the sensor network (off-network).
- **Communication model:** Local computation is usually rooted in the shared memory model, whereas global computation uses message passing to exchange data.

Single-sensor-based computations are mainly dealing with filtering, scaling, and calibration, whereas multiple-sensor-based computations perform some kind of data fusion to improve the measuring quality and feature extraction producing higher level condensed information.

A deeper discussion of signal processing in sensor networks and actual and future challenges can be found in Ref. [26], pointing out the importance of distributed algorithms and data processing due to the inherent geometrical and distributed features of sensor networks, especially regarding material-embedded networks.

Algorithms can differ in their computational complexity significantly, which is usually increasing with increasing information level (i.e., abstraction). Therefore, available computational resources of processing elements (concerning instruction complexity, average throughput, and data storage) of the sensor networks (e.g., the sensor nodes itself) limit the possible online/real-time capable implementations of higher level algorithms, such as machine learning, inverse numerical approaches, and complex pattern recognition. An offline/online partitioning and codesign of algorithms is usually required to optimally satisfy resource and service constraints (hybrid data processing approach).

Scaling of algorithms (commonly used in information processing) to microchip level requires (i) simplification and (ii) partitioning in programmable software tasks using microprocessors and nonprogrammable application-specific hardware tasks using finite-state machines and parallel multi-RTL architectures (for example [27], with offline/off-network but real-time machine learning and online autonomous distributed data processing in a sensor network performing sensor acquisition, calibration, preprocessing, and communication). Parallel data processing improves computational latency by preserving energy constraints and can be a prerequisite for real-time capable data processing platforms.

Many signal processing algorithms that can be partitioned in parallel functional networks can be parallelized, for example, communication (routing), data fusion [28], transformations (FFT), and many more.

Most current work of signal processing in sensor networks still implements sequential data processing on programmable machines (microcontrollers) for the sake of simplicity and flexibility of the design flow (rapid prototyping), thus failing to exploit concurrency and the advantages of application-specific hardware design (examples can be found in Refs [29–31]). The gap between the ongoing technological progress enabling increased miniaturization and node densities, and the implementations themselves become larger with those traditional serialized software-based approaches. Traditional generic parallel microprocessor-based architectures consume too much hardware (size and costs) and power resources that inhibit material integration.

For example, in many SHM applications, correlation algorithms are used to identify signal deviations with external excitation as probing signal source (vibration analysis), thus providing an estimate of the change of state of the structure. Correlation functions generally have low computational complexity and offer the possibility for fine-grained parallelization, well suited for RTL data processing architectures.

Agent- and machine learning-based approaches for SHM gain more importance in the future (see, for example, Ref. [32]).

11.3 Scientific Data Mining

Data mining is the process of computing meaningful information from raw data sets. More precisely, it reduces the size of the input compared to the output data vector. For example, a set of strain gage sensors embedded in a technical structure is used to derive the health statements of the structure, which can be a simple Boolean value of the set {true, false} that is the output vector of the data processing. Now, let us assume we have 100 sensors that are processed by autonomous sensor nodes capable of converting the sensor values with a 12-bit digital value. If time series evaluation is neglected, the input vector consists of 100 integer values ($s_1, s_2, \ldots, s_{100}$), resulting in a 1200-bit vector, mapped finally on 1 bit! If history is relevant, the reduction in dimensionality is much higher. The information retrieval is quite easy if there is a model function defining a relationship between the sensor input data and the output information, and the process can be considered as analytical. But commonly this analytical function is missing, also if there is a mechanical model of the structure. In this case, the data mining process is nontrivial, and uses, for example, supervised machine learning methods or neural network approaches to approximate the mapping function. The sensor data can be multivariate, with data from sensors monitoring the same process or physical variable, and the input data can be at different spatial and temporal scales. Real-world sensed data can contain missing values and the sensor data can be noisy and of low quality. Information retrieval must deal with this disturbed input data producing still correct and trustful data.

Data analysis is commonly an iterative and partial interactive process partitioned in multiple steps, summarized in Figure 11.4 (details can be found in Ref. [33]). Each step depends on the kind of data to be processed and the relationship model between data and computed information. First of all, the objects of interest must be localized or identified. In robotics, for example, image analysis is used to identify geometric objects that are the specific objects to be found in an image with a semantic meaning. The features to be extracted can be related to a matching index (a probability to recognize a specific geometric shape), or the geometrical (center) position of an object to be tracked. Back propagation of data computed in one step to a previous step can be used for refinement of the analysis to improve analysis results and quality.

The analysis process is commonly nonlinear and discrete, mostly it will succeed to classify different analysis outcomes and objects. This is a classical separation problem, which is well known in machine learning techniques. For example, an image analysis should give an answer with a significant evidence of the shape of an object detected in an image belonging to a small set of geometrical classes, that is,

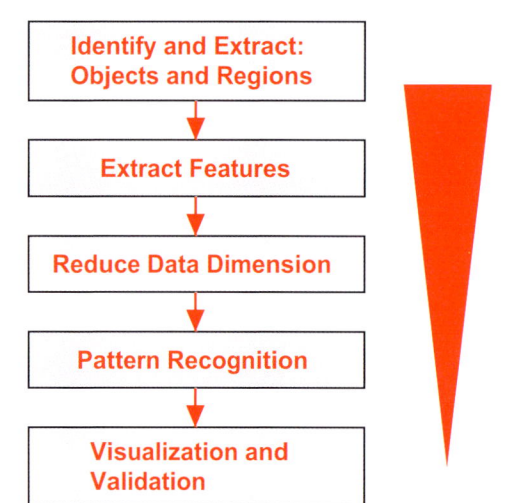

Figure 11.4 Levels of scientific data analysis used in data mining processes.

{rectangular, circular, triangle}. Again we have a significant data reduction problem. Traditionally, image analysis is related with high-resolution camera data. But these approaches are not limited to camera data. Assume again the integration of a spatially distributed sensor network integrated in materials surface of a technical structure. The sensor data sampled by the sensor network can be compared and represented with a two-dimensional matrix such as data sampled from a single image camera.

Feature Selection Data reduction can be performed by using feature selection and transformation techniques [33]. For example, damage sensitive features should be extracted in a SHM application. Feature selection is used to discriminate a subset of features that are more relevant than others in determining the values of the variables composing the mapped information. These selected variables can be highly correlated with the output. Common functional feature selection techniques, beside machine learning approaches, are distances filters, chi-squared filter, stump filter, and quality estimation by the ReliefF method [34].

Correlation Techniques Another commonly used analysis technique in SHM applications is based on temporal or spatial correlation analysis that identifies similarities between a pattern $pat(x)$ and a template $temp(x)$, which can reduce a data vector to a scalar feature index value, for example, the normalized peak value of the correlation function $\Gamma(\delta) = \langle pat(x)temp(x+\delta) \rangle$.

The autocorrelation function $\Gamma(\delta) = \langle XX \rangle$ describes the self-similarity of a continuous function X in relation to the lag δ of the function. The normalized correlation function has an output value interval $[-1,1]$ (real output value). The cross-correlation function $\Gamma(\delta) = \langle XY \rangle$ computes the similarity of two continuous functions X and Y with a displacement lag δ, shown in Eq. (11.3). The normalized cross-correlation function is scaled with the square root of the product of the autocorrelation functions of X and Y at lag 0.

X^* denotes the complex conjugate of the function X.

$$\begin{aligned}
\Gamma_Y(\delta) &= \langle Y^*(x)Y(x+\delta) \rangle = \lim_{X \to \infty} 1/X \int_{-X/2}^{X/2} Y^*(x)Y(x+\delta)dx, \\
\Gamma_{YZ}(\delta) &= \langle Y(x)Z(x+\delta) \rangle = \lim_{X \to \infty} 1/X \int_{-X/2}^{X/2} Y^*(x)Z(x+\delta)dx, \\
\Gamma_{YZ}^{\text{norm}}(\delta) &= \frac{\langle Y(x)Z(x+\delta) \rangle}{\sqrt{\langle Y(0) \rangle \langle T(0) \rangle}}, \\
\overline{\Gamma}_{AB}(d) &= \langle A^*(i)B(i+d) \rangle = 1/X \sum_{i=-X/2}^{X/2} A^*(i)B(i+d).
\end{aligned} \quad (11.3)$$

The discrete correlation function $\overline{\Gamma}(d) = \langle AB \rangle$ can be used for discrete finite data sets of functions A and B, for example, $A = \{a_1, a_2, \ldots\}$, $B = \{b_1, b_2, \ldots\}$, respectively, with a discrete lag value d.

The following Ex. 11.1 shows the application of the discrete cross-correlation analysis to sensor data sampled from accelerometer (vibration) sensors attached to bearing housings.

In the first time of operation there is a significant correlation $\Gamma(0)$ of about 0.6 between the temporal resolved sensor data set (consisting of 2000 values sampled in one second), decreasing only slightly over the time. But at the end of the bearing lifetime, the correlation decreases significantly and finally vanishes, indicating a mechanical failure. The correlation analysis reduces, therefore, a vector of dimension 2000 to a scalar value (only Γ at a lag of zero is relevant for this SHM system).

Ex. 11.1 Cross-correlation analysis of accelerometer sensor data from bearing operation comparing the time resolved sensor signal sample from a new bearing with sensor data from the aged bearing until a bearing failure occurs. [x-axis: displacement lag d, arb. units, at center position $d=0!$, y-axis: normalized cross-corr. function, arb. units, sensor data: NSF I/UCR Center for Intelligent Maintenance Systems] (a): 3 min, (b): 1 month, (c): 1 month and 3 days, and (d): 10 h later with bearing failure.

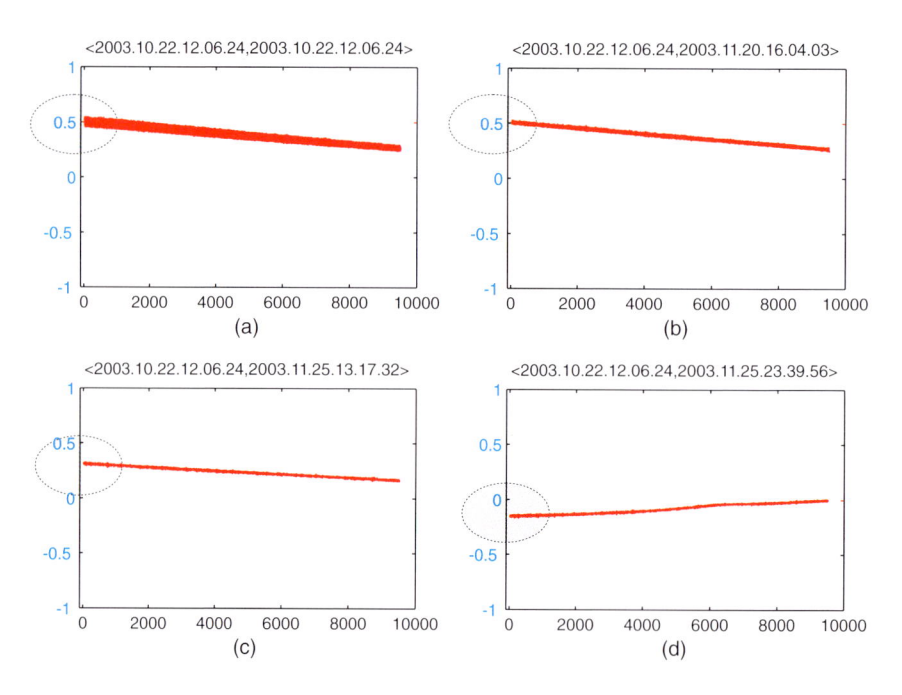

Correlation techniques can also be used for the spatial localization of sensors (proximity) toward intelligent and self-configuring distributed sensor networks by comparing the sensor data with sensor data from some sensors with a known sensor position using the cross-correlation approach [35].

Feature Transformation In contrast to feature selection algorithms, the feature transformation maps a high-dimensional input data vector on a lower dimensional output vector space. Prominent algorithms are principal component analysis (PCA) that can be combined with DNA-based algorithms, Isomap [36–38]. The Laplacion Eigenmaps algorithm is based on the k-nearest neighborhood approach, which is also applied in machine learning, resulting in a lower dimensional representation of the input data by giving the distances between data points.

11.4 Real-Time and Parallel Processing

Real-time signal processing systems must compute correct response of sensor data within a definite time limit. That means the correctness of a computation in a real-time system depends on the correctness of the results and the time in which the results are produced. A violation of the specified time bounds either decreases system performance (soft real-time system) or results in malfunction (hard real-time system). The strictness of a deadline for a given computational task depends on the application to be performed.

There are different approaches to satisfy time bounds. First of all, algorithms with constant runtime order are preferred, allowing runtime prediction at design-time, for example, using hash tables instead of linear linked lists for resource management. Usually a computational task can be composed of several subtasks, which can be processed independently and in parallel, leading to a lower overall computation latency compared with the sequential processing case, supporting the satisfaction of runtime time bounds with safety windows. Parallel data processing can be classified in spatial parallelism and decomposition applied to one data set to be computed, and temporal parallelism and decomposition applied to a stream of data sets, aka pipeline processing architectures. The latter one can improve processing element utilization and the data throughput of stream-based applications, but cannot improve the computation latency of one data set.

Parallel data processing requires the following steps [39]:

- Decomposing computation into tasks on different levels of parallelism: subprogram, procedure, loop, expression, and bit level with decreasing granularity and increasing degree of parallelism from left to right
- Assigning tasks to processes (independent computational units)
- Specification of interprocess communication (IPC) and synchronization (data access, exchange, and communication)
- Mapping processes to processing units (processors) for execution

It is usually not possible to map all processes of a computation to processing units with a 1:1 mapping due to resource constraints, resulting in simulated parallel processing by dynamically assigning processes to a processing unit, which executes several processes or part of them sequentially.

The performance of software executed on processors depends significantly on the instruction fetch–execute cycle and the load–execute–store model with the extensive use of register or memory variables by many high-level programming languages [40]. Direct hardware synthesis of programs and algorithms can overcome these performance limitations.

The possible speedup that can be achieved by subtask decomposition and parallel execution is limited by the fraction of serial execution r_s, which is considered as the nonparallelizable part of the program. This is addressed by Amdahl's law that expresses the limit of speedup by using N parallel processes in relation to the serial and parallel execution time fractions of a program, shown in Eq. (11.4) (with $r_s + r_p = 1$).

$$\text{Speedup}(N) = \frac{1}{r_s + \frac{r_p}{N}} < N. \tag{11.4}$$

Amdahl's law uses the serial program as the reference. In contrast, Gustafson's law uses the parallel program as the reference, and measures the serial fraction r'_s of the parallel program execution, and this is mainly a result from synchronization between parallel executing processes, shown in Eq. (11.5).

$$\text{Speedup}(N) = r'_s + r'_p N < N. \tag{11.5}$$

Amdahl's assumes a fixed total problem size, independent of the number of the processing units, whereby Gustafson's law assumes an increasing problem size with an increasing number of processing units, which can be more realistic in the

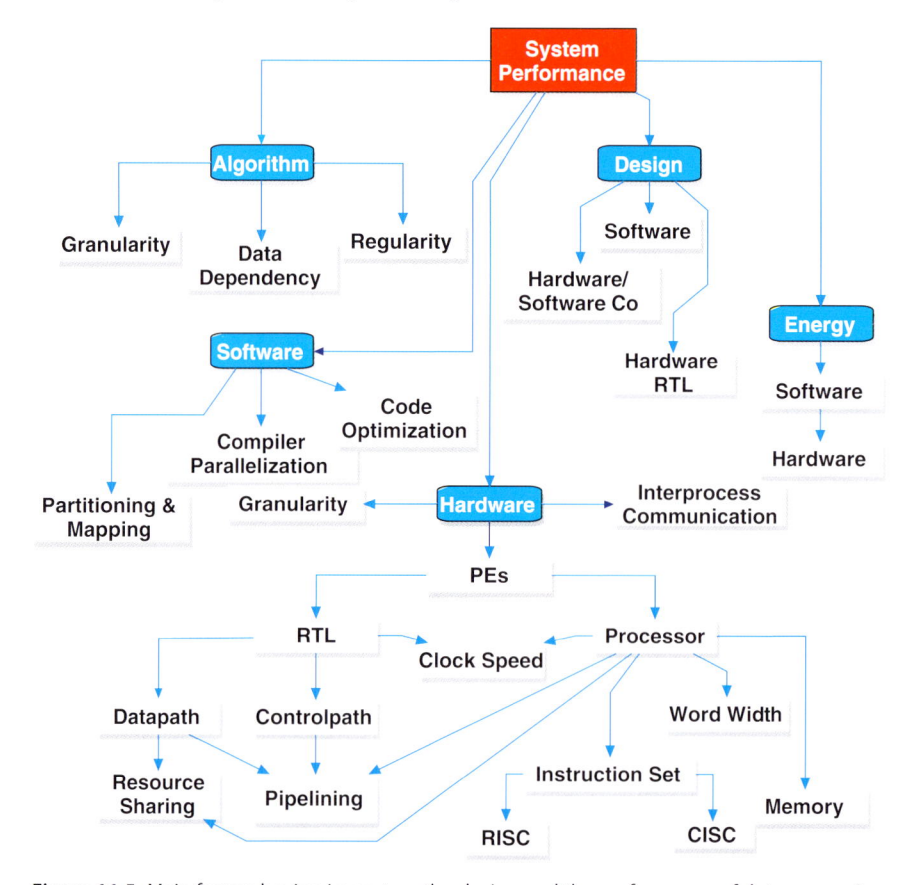

Figure 11.5 Main factors having impact on the design and the performance of data processing systems in sensor networks. (Adapted from Ref. [39].)

context of cloud computing [41]. That means that the parallel fraction increases with the number of processing units (and is not fixed).

The design of the data processing architecture used in material-integrated sensor networks requires the consideration and weighting of different impact factors, summarized in Figure 11.5, finding a merit for the satisfaction of resource requirements and computational time bounds.

Several sensor data fusion algorithms are based on integral and sum terms (see Ref. [28], for example), which can be easily parallelized in the spatial dimension by partial or complete loop unrolling and datapath parallelization, a powerful high-level optimization technique, which can be fully exploited by using RTL architectures, shown in Algorithm 11.1. Additionally, the application (calling) of functions can be parallelized by evaluating function argument expressions concurrently (due to inherent data independence). These techniques can help to reduce the computation time and to meet time bounds significantly (or at least increasing the safety window).

Data pipeline processing can be applied to continuous data streams and provides a way to reduce the temporal spacing between single computation results of different data sets and improves the utilization of chained computational blocks (temporal parallelization), which can be fully exploited by using RTL architectures, shown in principle in Algorithm 11.2.

Algorithm 11.1 Parallelization Rules for Sum and Product Terms and Function Evaluation (∥: Subprocesses executed concurrently, I: partial expression computation or instruction sequence, →: operational sequence, $[e_1, \ldots, e_n] \otimes / \oplus \rightarrow$:operational cluster block with n-ary operation)

$\sum_{i=a \text{ to } b} I(i) \Rightarrow I(a) \rightarrow+ I(a+1) \rightarrow+ \ldots \rightarrow+ I(b)$

$[I(a) \parallel I(a+1) \parallel \ldots \parallel I(b)] \rightarrow+$

$\prod_{i=a \text{ to } b} I(i) \Rightarrow I(a) \rightarrow* I(a+1) \rightarrow* \ldots$

$[I(a) \parallel I(a+1) \parallel \ldots \parallel I(b)] \otimes \rightarrow$

$\sum_i \sum_{j=a \text{ to } b} I(i,j) \Rightarrow I(a,a) \rightarrow+ I(a,a+1) \rightarrow+ \ldots$

$[\sum_j I(i,a) \parallel \sum_j I(i,a+1) \parallel \ldots \parallel \sum_j I(i,b)] \otimes \rightarrow$

$F(\varepsilon_1, \varepsilon_2, \varepsilon_3, \ldots \varepsilon_n) \Rightarrow v_1 = \varepsilon_1 \rightarrow v_2 = \varepsilon_2 \rightarrow \ldots \rightarrow F(v_1, v_2, \ldots, v_n)$

$[v_1 = \varepsilon_1 \parallel v_2 = \varepsilon_2 \parallel \ldots \parallel v_n = \varepsilon_n] \rightarrow F(v_1, v_2, \ldots v_n)$

Algorithm 11.2 Temporal parallelization rule with pipelining (simplified, s: input data, o: output data)

$\{\ldots, s_i, \ldots\} \Rightarrow F(G(H(\ldots(X(s_i)))) \Rightarrow \{\ldots, o_i, \ldots\}$

$\{\ldots, s_{i-m}, s_{i-m+1}, \ldots, s_i, \ldots\} \Rightarrow$
$\parallel [X(s_i) \rightarrow \ldots \rightarrow H(s_{i-m+2}) \rightarrow G(s_{i-m+1}) \rightarrow F(s_{i-m})] \Rightarrow$
$\{\ldots, o_{i-m}, o_{i-m+1}, \ldots, o_i, \ldots\}$

Data dependency classes

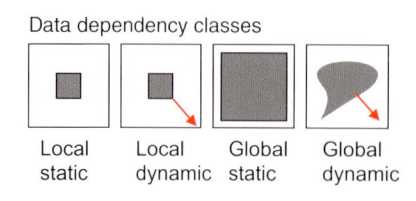

Local static | Local dynamic | Global static | Global dynamic

Figure 11.6 Different data dependency classes of algorithms.

One major feature of parallel and distributed algorithms is their data dependency, which can be classified in local short-distance and global long-distance dependency regarding tasks or partitions of a program, as shown in Figure 11.6. Intertask data dependency causes communication during the parallel or distributed data processing, which can significantly lower the available speed-up. Commonly, global data dependency causes the highest communication overhead. Communication is usually synchronized, too, and therefore, increases the sequential part of a parallel program.

Register-Transfer Level Architectures Spatial parallelism leads to an increase of resources by replication, whereby temporal parallelism leads to an increase of the utilization of a sequentially composed system. This pipelining can be very versatile in data flow computations, allowing the design and the proper control of pipelines at design time (e.g., RTL architectures with datapath pipelining). In contrast, control flow-based computations like machine programs executed on microprocessors make the design of pipelines difficult with a lower ratio of real speed-up to resource costs due to the nonpredictability of branches in the control flow and the expensive reset logic.

Application-specific designed digital logic circuits with RTL architectures allow the implementation of data latency, throughput, and resource-optimized data processing systems on microchip level. The principal structure of RT architectures is shown in Figure 11.7. Basically, combinatorial logics perform functional computations $f(o_1, o_2, \ldots)$ of data stored in registers. The datapath that performs sequential data processing consists of multiple levels with alternating registers and functional blocks between these registers levels. The input operands o_i of the nth level function f_n are the input data from registers of level n, and the results of the computation is transferred to registers of level $n + 1$. This datapath architecture offers parallel data processing on expression level and temporal pipelining of different datasets processed in different levels (stages) of the pipeline. Combinatorial logic introduces disadvantages concerned with metastability and path delays leading to increased switching activity and power consumption.

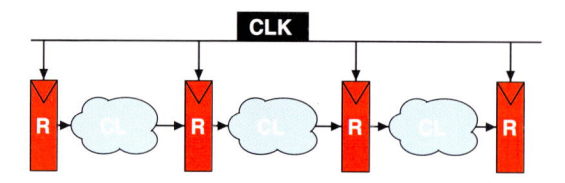

Figure 11.7 Principle data processing with register-transfer architectures (synchronous system) (R: Register, CL: Combinatorial logic, CLK: Common clock signal).

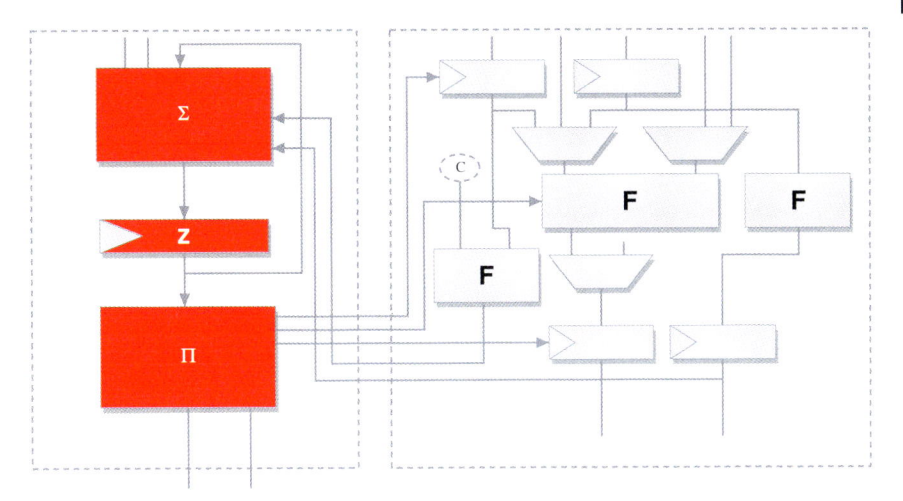

Figure 11.8 RTL data processing architecture. (Left) Finite state machine controller. (Right) Datapath and computation.

In a synchronous system with a central master clock, the highest possible clock frequency is determined by the longest path delay in a combinatorial block. Refactoring of the datapath in small functional blocks weaved between register levels increases the overall system performance. Registers are barriers for (invalid) metastable digital logic levels between combinatorial blocks.

The data processing in RTL architectures is performed sequentially and controlled by a finite state machine (FSM) meeting today's still common formulation of algorithms and communication in a sequential order, shown in Figure 11.8. The FSM partitioned the computation in multiple steps, defining the control path, with states that activates specific register stages of the datapath part of the RTL system. The datapath is usually composed of the multiple subpaths and consists of functional blocks, datapath selectors (multi- and demultiplexer), registers, and addressable cell memories (RAM). Path selectors are used for switching input and output ports of shared resources like functional blocks (adders, multipliers, ...) The datapath is bidirectionally connected with the control machine and the outside world, interfacing, for example, sensor acquisition or communication devices.

The finite state machine controller consists of a state register storing the encoded state representation (i.e., binary coding), a state transition network Σ computing the next state, and an output data generator Π providing the control signals for the datapath.

Virtual Machine Architectures Virtual machine concepts are increasingly employed in sensor networks supporting abstraction and virtualization (discussed in Ref. [42]) of the technological platform and offering advanced flexibility and programming features, including on the fly compilation of program code at runtime [43]. A virtual machine not only offers an abstraction layer of the processing system, it also offers the virtualization of resources, such as storage, sensor, or network resources, too. They enable the distribution of programs and

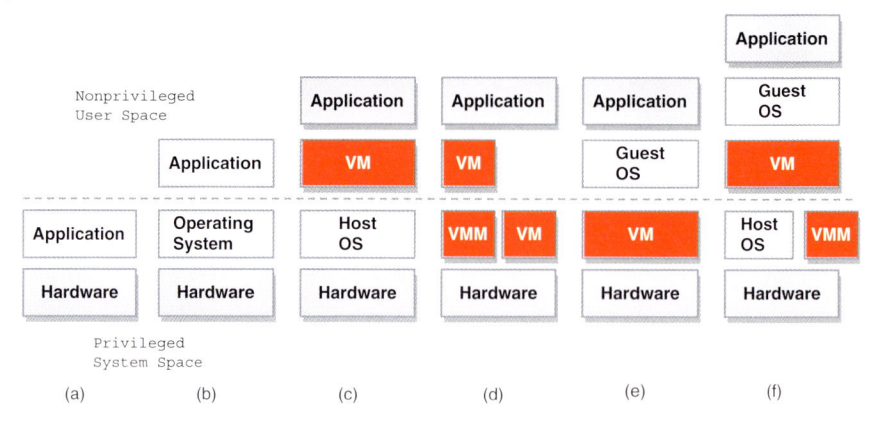

Figure 11.9 Different virtual machine (VM) architectures with and without a virtual machine monitor translating between system and nonprivileged user processing space. (a) Direct applications executed by hardware, (b) with operating system (OS), (c) with virtual machine executed in user space, (d) with VM in system and user space bridged with a VMM, (e) with a guest OS, and (f) a guest OS executed on a user space VM.

migration of mobile processes in an architecture-neutral format, which can easily be interpreted or compiled [44], usually stored in and directly executed from byte code formats. There are two different classes of virtual machines, which are capable to execute byte code: register-based and stack-based processing machines. Register-based machines commonly perform computation by modifying local register or shared memory using multiple-operand instructions, whereby stack-based machines operate on one or multiple stacks using zero-operand instructions, which access or modify few top elements of the stack. Zero-operand instruction formats offer an in-system programmable code morphing capability, the ability of program code to modify itself (in-place or out-of-place) [45], which can be used, for example, for agent behavior modification at runtime. Stack-based machine designs, originally closely related to the **FORTH** programming language invented by Chuck Moore, enable advanced resource and power optimizations [46]. Due to the functional paradigm, stack-based machines are well suited for the implementation of massive parallel systems [47]. But common programming languages are based on a random access memory model, which does not match the computation model of a stack-based machine, requiring the transformation of register-based to stack-based computation (examples of efficient transformation rules are explained in Ref. [48]).

Virtual machines (VM) can be executed in restricted user space (nonprivileged) or in unrestricted system space (privileged mode), shown in Figure 11.9. Nonprivileged applications have restricted access to system and hardware resources. Commonly software is executed directly on hardware (RTL architectures and simple embedded microcontroller systems) in the unrestricted execution space, or on the top of an operating system (OS) in restricted mode, for example, with additional memory and code protection. A virtual machine, executing an application, can be directly executed on hardware or on top of an operating system, too. In server applications, a VM can be used to execute a guest OS that executes applications. Virtual machine monitors can be used to translate between the privileged and the nonprivileged execution space. In distributed sensor network applications, there are commonly embedded systems with low resources that prohibit the use of operating systems, delegating the abstraction of hardware and the system services to the application software. Virtual machines can provide a meaningful middle layer incorporating operating system services required by application software and providing the necessary hardware abstraction, for example, network communication, sensor representation, and memory management, to make the application independent from the hardware layer. Furthermore, virtual machines can provide security and robustness by monitoring and automatic memory management based on garbage collection. There are several VM architectures, mostly based on byte-code execution, which are suitable for low-resource embedded systems. FORTH stack-based machines are one example, another is the OCaML byte-code interpreter, which was successfully implemented on a low-resource 8-bit microcontroller [49], or the Agent Forth VM [50], suitable for pure RTL microchip hardware implementations. In contrast to stack-based architectures, the tinyRef architecture [51] promotes a register-based VM for wireless sensor nodes, arguing with lower code size and increased execution speed (the program execution costs), but which cannot be held generally, as shown in Ref. [50].

References

1 Rytter, T. (1993) Vibration based inspection of civil engineering structure, Ph.D. dissertation, Department of building technology and structure engineering, Aalborg University, Denmark.
2 Lehmhus, D., Brugger, J., Muralt, P., Pané, S., Ergeneman, O., Dubois, M.-A., Gupta, N., and Busse, M. (2013) When nothing is constant but change: adaptive and sensorial materials and their impact on product design. *Journal of Intelligent Material Systems and Structures*, **24** (18), 2172–2182.
3 Chong, K.P. (1999) Health monitoring of civil structures. *Journal of Intelligent Materials Systems and Structures*, **9** (11), 892–898.
4 Peeters, B., Couvreur, G., Razinkov, O., Kündig, C., Van der Auweraer, H., and De Roeck, G. (2009) Continuous monitoring of the Øresund Bridge: system and data analysis. *Structure and Infrastructure Engineering: Maintenance, Management, Life-Cycle Design and Performance*, **5** (5), 395–405.
5 Sun, M., Staszewski, W.J., and Swamy, R.N. (2010) Smart sensing technologies for structural health monitoring of civil engineering structures. *Advances in Civil Engineering*, **2010**. doi: 10.1155/2010/724962
6 Renton, W.J. (2001) Aerospace and structures: where are we headed? *International Journal of Solids and Structures*, **39** (17), 3309–3319.
7 Boller, C. (2009) Structural health monitoring – an introduction and definition, in *Encyclopedia of Structural Health Monitoring*, John Wiley & Sons Ltd., Chichester.
8 Tracht, K., Hogreve, S., and Bosse, S. (2012) Intelligent interpretation of multiaxial gripper force sensors. Proceedings of CIRP Conference on Assembly Technologies, CATS 2012.

9 Mavroidis, C., Dubowsky, S., and Thomas, K. (1997) Optimal sensor location in motion control of flexibly supported long reach manipulators. *Transactions of the ASME, Journal of Dynamic Systems, Measurement and Control*, **119**, 718–726.

10 Hunt, S.R. and Hebden, I.G. (2001) Validation of the Eurofighter Typhoon structural health and usage monitoring system. *Smart Materials and Structures*, **10**, 497.

11 Rulli, R.P., Dotta, F., and da Silva, P.A. (2013) Flight tests performed by EMBRAER with SHM systems. *Key Engineering Materials*, **558**, 305–313.

12 dos Santos, L.G. (2013) EMBRAER perspective on the challenges for the introduction of scheduled SHM (S-SHM) applications into commercial aviation maintenance programs. *Key Engineering Materials*, **558**, 323–330.

13 Cannata, G., Dahiya, R., Maggiali, M., Mastrogiovanni, F., Metta, G., and Valle, M. (2010) Modular skin for humanoid robot systems. CogSys 2010 Conference Proceedings (Vol. 231500).

14 Dahiya, R. and Valle, M. (2007) Tactile sensing arrays for humanoid robots. Research in Microelectronics and Electronics Conference 2007. PRIME 2007 (pp. 201–204).

15 Vidal-Verdú, F., Barquero, M.J., Castellanos-Ramos, J., Navas-González, R., Sánchez, J.A., Serón, J., and García-Cerezo, A. (2011) A large area tactile sensor patch based on commercial force sensors. *Sensors (Basel, Switzerland)*, **11** (5), 5489–507.

16 Balageas, D., Fritzen, C., and Güemes, A. (eds) (2006) *Structural Health Monitoring*, ISTE.

17 Dahiya, R.S., Lorenzelli, L., Metta, G., and Valle, M. (2010) POSFET devices based tactile sensing arrays. Proceedings of 2010 IEEE International Symposium on Circuits and Systems, pp. 893–896. doi: 10.1109/ISCAS.2010.5537414

18 Franke, R., Hoffmann, F., and Bertram, T. (2008) Observation of link deformations of a robotic manipulator with fiber Bragg grating sensors. 2008 IEEE/ASME International Conference on Advanced Intelligent Mechatronics, pp. 90–95. doi: 10.1109/AIM.2008.4601640

19 Friswell, M.I. and Mottershead, J.E. (2001) Inverse methods in structural health monitoring. *Key Engineering Materials*, **204–205**, 201–210.

20 Vazquez, S.L., Tessler, A., Quach, C.C., Cooper, E.G., Parks, J., and Spangler, J.L. (2006) Structural Health Monitoring using high-density Fiber Optic Strain Sensor and Inverse Finite Element Methods.

21 Dennis, B.H., Jin, W., Dulikravich, G.S., and Jaric, J. (2011) Application of the finite element method to inverse problems in solid mechanics. *International Journal of Structural Changes in Solids*, **3** (2), 11–22.

22 Gherlone, M., Cerracchio, P., Mattone, M., Di Sciuva, M., and Tessler, A. (2011) Dynamic shape reconstruction of three-dimensional frame structures using the inverse finite element method. Proceedings of the COMPDYN 2011 Conference, Corfu, Greece, May 25–28th, 2011.

23 Farrar, C.R. and Worden, K. (2013) *Structural Health Monitoring – A Machine Laerning Perspective*, John Wiley & Sons Ltd, Chichester.

24 Carden, E.P. (2004) Vibration based condition monitoring: a review. *Structural Health Monitoring*, **3** (4), 355–377.

25 Pantke, F., Bosse, S., Lehmus, D., and Lawo, M. (2011) An artificial intelligence approach towards sensorial materials, Future Computing Conference.
26 Trigoni, N. and Krishnamachari, B. (2012) Sensor network algorithms and applications. *Philosophical Transactions. Series A, Mathematical, Physical, and Engineering Sciences*, **370** (1958), 5–10.
27 Bosse, S., Pantke, F., and Edelkamp, S. (2013) *Robot manipulator with emergent behaviour supported by a smart sensorial material and agent systems.* Proceedings of the Smart Systems Integration 2013, Amsterdam, March 13–14, 2013, NL.
28 Mitchell, H.B. (2012) *Data Fusion, Concepts and Ideas*, Springer.
29 Roscher, K.-U., Fischer, W.-J., Landgraf, J., Pfeifer, G., and Starke, E. (2007) Sensor networks for integration into textile-reinforced composites. TRANSDUCERS 2007-2007 International Solid-State Sensors, Actuators and Microsystems Conference, pp. 1589–1592. doi: 10.1109/SENSOR.2007.4300451
30 Hedley, M., Hoschke, N., Johnson, M., Lewis, C., Murdoch, A., Price, D., and Prokopenko, M. (2004) Sensor network for structural health monitoring. Proceedings of the 2004 Intelligent Sensors, Sensor Networks and Information Processing Conference, 2004. (pp. 361–366). doi: 10.1109/ISSNIP.2004.1417489
31 Ghezzo, F., Starr, A.F., and Smith, D.R. (2010) Integration of networks of sensors and electronics for structural health monitoring of composite materials. *Advances in Civil Engineering*, **2010**, 1–13.
32 Zhao, X., Yuan, S., Yu, Z., Ye, W., and Cao, J. (2008) Designing strategy for multi-agent system based large structural health monitoring. *Expert Systems with Applications*, **34** (2), 1154–1168.
33 Rajan, K. (2013) *Informatics for Materials Science and Engineering*, Elsevier.
34 Robnik-Šikonja, M. and Kononenko, I. (2003) Theoretical and empirical analysis of ReliefF and RReliefF. *Machine Learning*, **53** (1–2), 23–69.
35 Hu, F. and Hao, Q. (eds) (2013) *Intelligent Sensor Networks*, CRC Press. ISBN: 9781420062212.
36 Tenenbaum, J.B., de Silva, V., Langford, J.C. (2000) A global geometric framework for nonlinear dimensionality reduction. *Science*, **290** (5500), 2319–23.
37 Roweis, S. T. and Saul, L. K. (2000) Nonlinear dimensionality reduction by locally linear embedding. *Science*, **290**, 2323–2326.
38 Belkin, M. and Niyogi, P. (2002) Laplacian eigenmaps and spectral techniques for embedding and clustering, in T. K. Leen, T. G. Dietterich, and V. Tresp (eds), *Advances in Neural Information Processing System 14* MIT Press, Cambridge, MA.
39 Tokhi, M.O., Hossain, M.A., and Shaheed, M.H. (2003) *Parallel Computing for Real-time Signal Processing and Control*, Springer.
40 Hanna, D.M. and Haskell, R.E. (2006) Flowpaths: compiling stack-based IR to hardware. *Microprocessors and Microsystems*, **30** (3), 125–136.
41 Morales, G.D.F. (2012) *Big Data and the Web: Algorithms for Data Intensive Scalable Computing*, IMT Institute for Advanced Studies, Lucca.
42 Smith, J.E. and Nair, R.N.R. (2005) The architecture of virtual machines. *Computer*, **38** (5). doi: 10.1109/MC.2005.173

43 Müller, R., Alonso, G., and Kossmann, D. (2007) A virtual machine for sensor networks. Proceedings of the 2nd ACM SIGOPS/EuroSys European Conference on Computer Systems 2007, pp. 145–158. ACM. Available at http://doi.acm.org/10.1145/1272996.1273013.

44 Gregg, D., Casey, K., Ertl, M.A., and Shi, Y. (2008) Virtual machine showdown. *ACM Transactions on Architecture and Code Optimization*, doi: 10.1145/1328195.1328197

45 Bosse, S., Pantke, F., and Kirchner, F. (2012) Distributed computing in sensor networks using multi-agent systems and code morphing, ICAISC Conference, Poland, Zakapone.

46 Chuck Moore *et al.* (2010) WP002 GreenArrays Energy Conservation. Available at www.greenarraychips.com.

47 Chuck Moore *et al.* (2011) PB002 GreenArrays Architecture. Available at www.greenarraychips.com.

48 Park, J., Park, J., Song, W., Yoon, S., Burgstaller, B., and Scholz, B. (2011) Treegraph-based instruction scheduling for stack-based virtual machines. *Electronic Notes in Theoretical Computer Science*, **279** (1), 33–45.

49 Vaugon, B., Wang, P., and Chailloux, E. (2011) Les microcontrôleurs PIC programmés en Objective Caml, in JFLA, Vingt-deuxièmes Journées Francophones des Langages Applicatifs, pp. 177–208.

50 Bosse, S. (2015) Design and simulation of material-integrated distributed sensor processing with a code-based agent platform and mobile multi-agent systems (article). *Sensors*, **15** (2), 4513–4549.

51 Marques, I.L., Ronan, J., and Rosa, N.S. (2009) TinyReef: a register-based virtual machine for wireless sensor networks, in IEEE SENSORS 2009 Conference, pp. 1423–1426.

12

The Known World: Model-Based Computing and Inverse Numeric

Armin Lechleiter and Stefan Bosse

University of Bremen, Department of Mathematics and Computer Science,
Robert-Hooke-Str. 5, 28359 Bremen, Germany

In this chapter we consider parameter identification problems and discuss several deterministic techniques for the extraction of information on searched-for parameters from sensor data. These techniques will rely, on the one hand, on some physical model for the data acquisition process and, on the other hand, on deterministic mathematical algorithms to stably compute searched-for quantities from measured sensor data in the context of ill-conditioned or underdetermined problems. In engineering, physics, or other sciences, the process of inverting measured data in order to approximate time- and/or space-dependent model parameters typically lacks stability whenever quantities of interest can only be accessed indirectly via remote measurements or via measurements of other physical quantities. A prominent example in this class of problems is the approximation or characterization of material properties from boundary measurements of the material's behavior in nondestructive testing or structural health monitoring.

If only remote or indirect data is available, there typically exist quite different parameters that lead to rather similar sensor measurements such that their stable and accurate inversion becomes a challenging problem. In particular, unavoidable noise in the sensor measurements implies that any inversion scheme has to be stabilized in some suitable way, a concept which is also known as regularization in the terminology of applied mathematics. Once a computable physical model linking model parameters with sensor measurements is at hand, regularization theory offers inversion algorithms that stably compute model parameters from noisy sensor data by, roughly speaking, balancing the amount of information extracted from the noisy data with the level of stability of the inversion process.

In this chapter, we aim to highlight several parts of linear and nonlinear regularization theory adapted to applied scientists familiar with concepts from basic calculus; consequently, we simplify settings and results to ease access to the material. Textbooks and monographs on regularization theory that go far beyond what can be sketched in the overview on the following pages include Refs [1,2,3] and Refs [4,5], respectively.

Material-Integrated Intelligent Systems: Technology and Applications, First Edition.
Edited by Stefan Bosse, Dirk Lehmhus, Walter Lang, and Matthias Busse.
© 2018 Wiley-VCH Verlag GmbH & Co. KGaA. Published 2018 by Wiley-VCH Verlag GmbH & Co. KGaA.

12.1 Physical Models in Parameter Identification

Taking the viewpoint of an applied scientist, any parameter identification problem searches for an element in some parameter set that is able to explain the measured data. The fundamental advantage of a computable physical model for some data acquisition process is that the implicit link between model parameters and sensor measurements becomes predictable, typically at the expense of a numerical simulation. If measured sensor data are available, such a model offers, at least in principle, the possibility to test for a variety of parameters whether numerically simulated sensor measurements for test parameters match these measurements. In other words, a physical model offers the opportunity to check whether a given parameter explains the sensor measurements under the chosen physical model.

Taking a more abstract viewpoint, a physical model for some data measurement process aiming to characterize unknown parameters provides a mapping Φ from the set of admissible parameters $\mathcal{P}$ into the space of sensor measurements $\mathcal{S}$,

$$\Phi : \mathcal{P} \to \mathcal{S},$$

such that we will in the following identify Φ with the corresponding model. Disregarding possibly time-dependent measurements, $\mathcal{S}$ can most often simply be identified with some Euclidean space $\mathbb{R}^N$ of dimension N, corresponding to N sensor measurements. To make this text easily accessible for applied scientists, we will always take this viewpoint in the sequel and set $\mathcal{S} = \mathbb{R}^N$ for some $N \in \mathbb{N}$.

The above-described parameter identification task can then be – naively – rephrased as follows: Given sensor measurements $s \in \mathcal{S}$, determine suitable parameters $p \in \mathcal{P}$ such that $\Phi(p) = s$. The model Φ is usually called the forward operator within the mathematical community; the parameter identification problem hence consists in stably inverting this forward operator, in order to find parameters explaining sensor data.

Example 12.1 (Inverse Problems in Linear Elasticity)

We illustrate the above concept with an example from parameter identification in linear elasticity. In structural health monitoring, one is interested in computing material parameters of a body $\Omega \subset \mathbb{R}^3$ from measurements of its material behavior. Let us for simplicity assume that the material behavior of Ω is linearly elastic, such that there is a stiffness tensor $C : \Omega \to \mathbb{R}^{3 \times 3 \times 3 \times 3}$ characterizing the material behavior of Ω. More precisely, for each $x \in \Omega$, $C(x)$ is fourth-order tensor that linearly links the Cauchy stress tensor $\sigma : \Omega \to \mathbb{R}^{3 \times 3}_{\text{sym}}$ with the infinitesimal strain tensor $\varepsilon : \Omega \to \mathbb{R}^{3 \times 3}_{\text{sym}}$ via Hooke's law:

$$\sigma(x) = C(x) : \varepsilon(x), \quad x \in \Omega,$$

that is, $\sigma_{ij}(x) = \sum_{k,l=1}^{3} C_{ijkl} \varepsilon_{k,l}$. Without going into details, the stiffness tensor C is a positive definite linear mapping on symmetric 3×3 matrices, which possesses in total 21 degrees of freedom at each point x in Ω, due to symmetries [6,7]. The infinitesimal strain tensor is explicitly determined by the deformation field $u : \Omega \to \mathbb{R}^3$,

$$\varepsilon(x) = \varepsilon(u(x)) = \frac{1}{2} \left[\nabla u(x) + (\nabla u(x))^\mathrm{T} \right], \quad x \in \Omega, \tag{12.1}$$

such that the continuum equation yields the following governing equation for the deformation field due to some applied body force $f : \Omega \to \mathbb{R}^3$:

$$\operatorname{div}(C : \varepsilon(u(x))) = f(x), \quad x \in \Omega. \tag{12.2}$$

Suitable boundary conditions include displacement boundary conditions for the restriction of u to some subset Γ_1 of the boundary $\partial\Omega$:

$$u(x) = g_1(x), \quad x \in \Gamma_1 \subset \partial\Omega, \tag{12.3}$$

stress boundary conditions for the traction $\sigma\nu = (C : \varepsilon(u))\nu$ (where ν denotes the exterior unit normal to $\partial\Omega$):

$$\sigma(x)\nu(x) = g_2(x), \quad x \in \Gamma_2 = \partial\Omega\setminus\Gamma_1, \tag{12.4}$$

or linear combinations between these two types. If Γ_1 is a nonempty subset of $\partial\Omega$, and if $g_{1,2}$ are sufficiently regular, then there exists a unique deformation field solving the corresponding boundary value problem. For simplicity, we assume in the following that $g_{1,2} = 0$.

In principle, inversion of C from sensor measurements can be based on surface strain measurements on Γ_1 or on surface deformation measurements on Γ_2. Components of the surface strain $\varepsilon(u(x))$ at boundary points $x \in \partial\Omega$ can be measured by, for example, applied strain gauges or fiber Bragg gratings, while surface deformation can be measured optically using interferometry. Neglecting that these measurements should be adequately modeled as well, the forward operator Φ then maps parameters in the set $\mathcal{P}$ consisting of positive-definite and symmetric fourth-order tensors to a finite number N of sensor measurements of surface strain and/or surface deformation of the field u from (12.2). The inversion problem for the material parameter C can then be formulated as follows: Determine the fourth-order symmetric and positive-definite tensor C from spatially distributed sensor measurements $s \in \mathcal{S} = \mathbb{R}^N$ of surface strains and/or surface deformation.

It is important to note that linearity of the differential equation in (12.2) means that the deformation field u as well as its surface strains and surface deformation depend linearly on the applied body force f. However, as u clearly also depends on the stiffness tensor C, the application of C to $\varepsilon(u)$ in (12.2) implies that u does *not* depend linearly on C, such that the forward operator Φ is nonlinear. This fact makes the introduced inversion problem challenging, even for stiffness tensors that are spatially constant or piecewise constant.

As more sensor measurements yield a larger amount of (implicit) information on the stiffness tensor, the stability of any inversion algorithm is enhanced by collecting more data either by relying on more sensors or by applying several forces f to the body. (Of course, the analogous inversion problem can be posed and the same properties hold if the deformation field is excited by a deformation $g_1 \neq 0$ or surface traction $g_2 \neq 0$.)

A somewhat different identification problem arises if the material coefficients of Ω are known but forces acting on Ω are the searched-for quantity; thus, while the stiffness tensor C in (12.2) is known, the volume force f is the searched-for parameter. This setting is often met in load monitoring where one is interested in identifying loads on structures from sensor information on the material behavior under load. The forward mapping Φ hence maps forces f in some function space $\mathcal{P}$ on Ω (e.g., in the space of piecewise linear functions on some triangular or tetrahedral mesh decomposing Ω) to sensor measurements s_m of the surface deformation and/or surface strain caused by the deformation field due to f. As discussed above, the measurements s_m depend linearly on the deformation field u, which itself depends linearly on f, such that $\Phi : f \mapsto s_m$ is a linear mapping and the inversion problem to find a force f such that $\Phi(f) = s_m$ for given s_m is linear, too. We will see in the sequel that linear identification problems are somewhat easier to tackle compared to nonlinear ones; in particular, their analysis is significantly simpler.

12.2 Noisy Data Due to Sensor and Modeling Errors

Unfortunately, real-world sensor measurements are never ideal since the usual sensor noise together with various sensor errors perturb the sensor signal ahead of any data processing (sensor errors include, for example, digitalization, sensitivity, or drift errors as well as possible environmental influences as temperature or moisture). Similarly, no physical model can exactly reproduce real-world phenomena, such that modeling errors are further error sources. Thus, any measured signal comprises a certain level of noise $\delta > 0$ coming from the two mentioned sources. Notationally, we take this noise level into account by writing s_m^δ for noisy data in $\mathcal{S}$, while $s_m \in \mathcal{S}$ from now on denotes noise-free data according to the chosen model Φ; for $s_m \in \mathcal{S}$, there hence exists $p \in \mathcal{P}$ such that $\Phi(p) = s_m$. The absolute noise level δ is, by definition, the magnitude of the (unknown) difference of s_m and s_m^δ in $\mathcal{S} = \mathbb{R}^N$,

$$\delta = \left\| s_m^\delta - s_m \right\|_{\mathbb{R}^N} = \left(\sum_{j=1}^{N} |s_{m,i}^\delta - s_{m,i}|^2 \right)^{1/2} , \text{ for } s_m^\delta = \left(s_{m,i}^\delta \right)_{i=1}^{N}, \ s_m = \left(s_{m,i}^\delta \right)_{i=1}^{N}.$$

We will later on consider noise levels $\delta > 0$ that tend to zero and denote this as $\delta \to 0$; in this case, the definition of δ in the last equation implies that $s_m^\delta \to s_m$ in $\mathbb{R}^N$.

Note that noisy data s_m^δ does, in general, not belong to the range of Φ, roughly speaking since noise usually does not feature smoothness; even more, it is rarely the case that one finds some model parameter p that satisfies the equation $\Phi(p) = s_m^\delta$ exactly. For this reason, it is *not* advisable to attempt to solve the latter equation for given sensor data s_m^δ exactly, as any resulting solution would be required to explain the noisy data, despite we are *not* interested in explaining the noise $s_m^\delta - s_m$.

Taking a more theoretic viewpoint, whenever data are gained by remote or indirect measurements, the mapping Φ usually fails to possess a continuous inverse Φ^{-1} (if ever this inverse mapping exists at all). Even if Φ^{-1} exists, its discontinuity implies that merely computing $\Phi^{-1}(s_m^\delta)$ fails to provide sensible parameter reconstructions in practice, as $\Phi^{-1}(s_m^\delta)$ is far from $\Phi^{-1}(s_m)$ even for tiny noise levels δ. Instead, one should attempt to stabilize the inversion process by solving the operator equation approximately up to the noise level δ of the sensor data. We will illustrate several numerical techniques to do so in the subsequent sections.

Before, let us demonstrate by a simple linear problem in a finite-dimensional setting, motivated by an inverse problem in heat conduction, what can go wrong if one tries to exactly solve operator equations $\Phi(p) = s_m^\delta$ for parameter identification problems.

Example 12.2 (Inverse Heat Conduction)

Assume we are given sensor measurements $s_m^\delta \in \mathbb{R}^N$ of the temperature of a thin, homogeneous bar of length one at some time $t_0 \gg 0$; the inversion problem under consideration is to compute the initial temperature of the bar at time $t = 0$.

As a simplistic physical model for heat conduction in the bar, we choose the one-dimensional heat equation with constant thermal diffusivity $\alpha > 0$ for the temperature u:

$$\frac{\partial u}{\partial t} - \alpha \frac{\partial^2 u}{\partial x^2} = 0, \quad \text{for} \quad x \in [0,1], \quad t > 0,$$

subject to initial and (isolating) boundary conditions:

$$u(0,x) = g(x), \quad \text{for} \quad x \in [0,1], \quad \text{and} \quad \frac{\partial u}{\partial x}(t,0) = \frac{\partial u}{\partial x}(t,1) = 0, \quad \text{for} \quad t > 0.$$

Solutions to this initial boundary-value problem are of the form

$$u(t,x) = \sum_{n \in \mathbb{N}_0} \hat{u}_n \cos(\pi n x) e^{-\alpha \pi^2 n^2 t}, \quad \text{for} \quad x \in [0,1], \quad t > 0,$$

for coefficients $\hat{u}_n \in \mathbb{R}$.

(12.5)

For N sensor positions $0 \leq x_1 < x_2 < \cdots < x_N \leq 1$, the forward operator Φ maps the initial value g, a function of x, to the vector $s_m = \left(u(t_0, x_j)\right)_{j=1}^N \in \mathbb{R}^N$ containing point values of the temperature at time $t_0 > 0$. Thus, s_m^δ contains noisy sensor measurements of the temperature at time $t_0 > 0$ at the sensor positions x_j, such that $\delta^2 = \sum_{j=1}^N |s_{m,j}^\delta - u(t_0, x_j)|^2$. The operator equation to be tackled hence reads $\Phi(g) = s_m^\delta$ for given measurements s_m^δ and a searched-for initial temperature g. A straightforward solution to the operator equation $\Phi(g) = s_m^\delta$ exists once we compute a trigonometric interpolation:

$$u_{t_0,N}^\delta(x) = \sum_{n=0}^{N-1} \hat{u}_n^\delta \cos(\pi n x) \quad \text{such that} \quad u_{t_0,N}^\delta(x_j) = s_{m,j}^\delta, \quad \text{for} \quad 1 \leq j \leq N.$$

(12.6)

This interpolation is particularly easy in case that $x_j = (2j-1)/(2N)$ for $j = 1, \ldots, N$, since the discrete cosine transform of the sensor values then yields the coefficients $\hat{u}_n^\delta$:

$$\hat{u}_0^\delta = \frac{1}{\sqrt{N}} \sum_{k=0}^{N-1} s_{m,k}^\delta, \quad \hat{u}_n^\delta = \frac{\sqrt{2}}{\sqrt{N}} \sum_{k=0}^{N-1} s_{m,k}^\delta \cos(\pi x_k n), \quad \text{for} \quad 1 \leq n \leq N-1.$$

For simplicity, we hence assume from now on that $x_j = (2j-1)/(2N)$. Then (12.6) takes the explicit form

$$u_{t_0,N}^\delta(x) = \frac{1}{\sqrt{N}} \sum_{k=0}^{N-1} s_{m,k}^\delta + \frac{\sqrt{2}}{\sqrt{N}} \sum_{n=1}^{N-1} \sum_{k=0}^{N-1} s_{m,k}^\delta \cos(\pi x_k n) \cos(\pi n x), \quad x \in [0,1].$$

(12.7)

Matching the expression of $u_{t_0,N}^\delta$ at time $t = t_0$ in (12.7) with the solution u in (12.5) shows that the coefficients $\hat{u}_n$ of u from (12.7) are given by

$$\hat{u}_0 = \frac{1}{\sqrt{N}} \sum_{k=0}^{N-1} s_{m,k}^\delta, \quad \hat{u}_n = \frac{\sqrt{2}}{\sqrt{N}} \left(\sum_{k=0}^{N-1} s_{m,k}^\delta \cos(\pi x_k n) \right) \exp(\alpha \pi^2 n^2 t_0), \quad 1 \leq n \leq N-1,$$

and $\hat{u}_n = 0$ for $n \geq N$. The solution to the heat equation that equals $u_{t_0,N}^\delta$ at time $t = t_0$ thus equals

$$u_N^\delta(t,x) = \frac{1}{\sqrt{N}} \sum_{k=0}^{N-1} s_{m,k}^\delta + \frac{\sqrt{2}}{\sqrt{N}} \sum_{n=0}^{N-1} \left(\sum_{k=0}^{N-1} s_{m,k}^\delta \cos(\pi x_k n) \right) \cos(\pi n x) e^{\alpha \pi^2 n^2 (t_0 - t)},$$

and takes initial values

$$u_N^\delta(0,x) = \frac{1}{\sqrt{N}} \sum_{k=0}^{N-1} s_{m,k}^\delta + \frac{\sqrt{2}}{\sqrt{N}} \sum_{n=0}^{N-1} \left(\sum_{k=0}^{N-1} s_{m,k}^\delta \cos(\pi x_k n) \right) \cos(\pi n x) e^{\alpha \pi^2 n^2 t_0}$$

(12.8)

for $x \in [0,1]$. Consequently, $\Phi(u_N^\delta(0,x)) = (u_N^\delta(t_0,x))_{j=1}^N = s_m^\delta$, that is, $u_{0,N}^\delta$ is an exact solution to the operator equation we are considering.

However, for all but the constant term in (12.8), the noise in s_m^δ into u_n^δ is multiplied with the exponentially growing factors $\exp(\alpha \pi^2 n^2 t_0)$, such that any attempt to find reasonable initial temperatures will completely fail due to a blow-off of the exponential terms in (12.8), which is also clearly visible in Figure 12.1.

12.3 Coping with Noisy Data: Tikhonov Regularization and Parameter Choice Rules

Since parameter identification problems typically lack stability, one should refrain from attempting to solve the operator equation $\Phi(p) = s_m^\delta$ exactly. As Example 12.2 shows, unavoidable measurement errors lead to completely wrong solutions

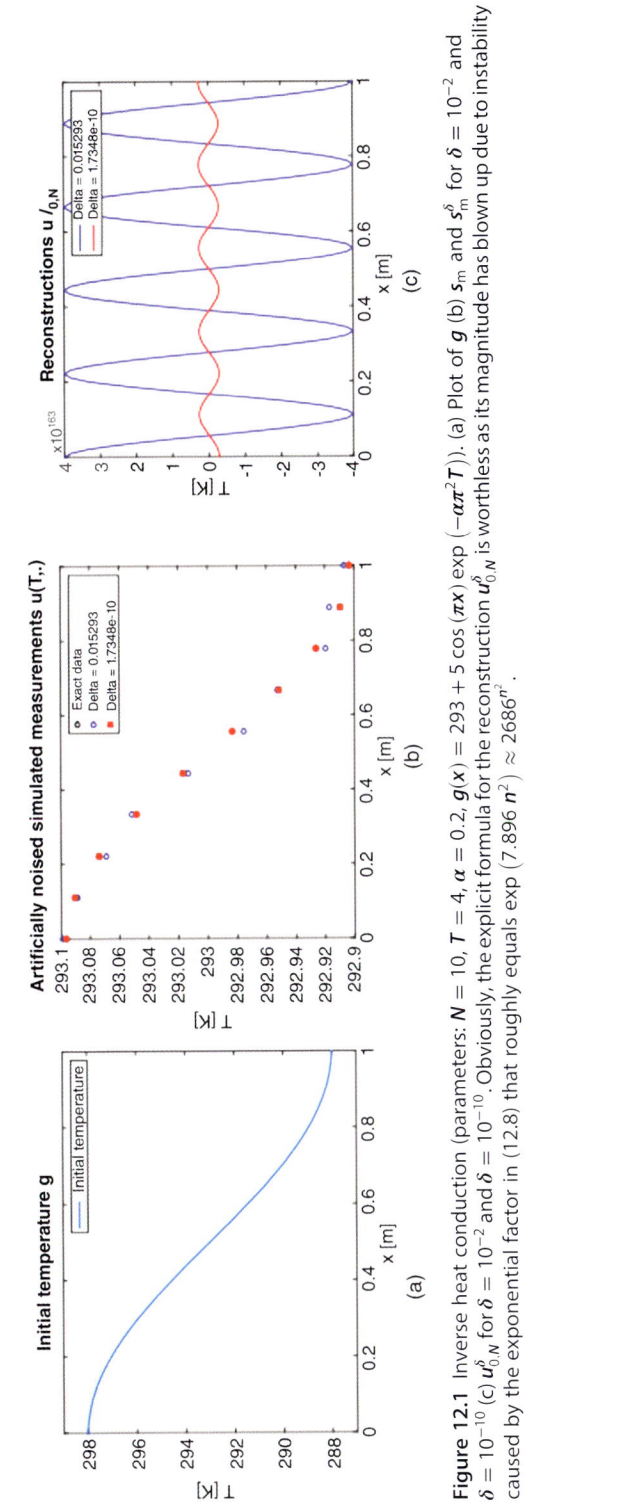

Figure 12.1 Inverse heat conduction (parameters: $N = 10$, $T = 4$, $\alpha = 0.2$, $g(x) = 293 + 5\cos(\pi x)\exp(-\alpha\pi^2 T)$). (a) Plot of g (b) s_m and s_m^δ for $\delta = 10^{-2}$ and $\delta = 10^{-10}$ (c) $u_{0,N}^\delta$ for $\delta = 10^{-2}$ and $\delta = 10^{-10}$. Obviously, the explicit formula for the reconstruction $u_{0,N}^\delta$ is worthless as its magnitude has blown up due to instability caused by the exponential factor in (12.8) that roughly equals $\exp(7.896\, n^2) \approx 2686^{n^2}$.

that suffer in particular from a blow-up in magnitude. A first idea is hence not to try to find a parameter $p \in \mathcal{P}$ such that $\Phi(p)$ equals s_m^δ, but to relax this criterion and merely attempt to find $p \in \mathcal{P}$ such that, on the one hand, the discrepancy $\left\|\Phi(p) - s_m^\delta\right\|_{\mathbb{R}^N}$ is small but, on the other hand, the magnitude of the reconstructed parameter p is not too large.

This idea requires to formalize the concept of the "magnitude" of a parameter mathematically (having, for example, the stiffness tensor $C: \Omega \to \mathbb{R}^{3\times3\times3\times3}$, the force $f: \Omega \to \mathbb{R}^3$, or the initial temperature $g: [0, 1] \to \mathbb{R}$ as parameters in Examples 12.1 and 12.2 in mind). Such a formalization is, for instance, achieved by basic functional analytic concepts such as infinite-dimensional Banach or Hilbert spaces, and norms and operators on such spaces; this mathematical theory is introduced in detail in standard textbooks as, for example, Refs [8,9]. To avoid the need for a detailed exposition of relevant infinite-dimensional theory, we opt to restrict ourselves to finite-dimensional parameter spaces X with scalar product $(\cdot,\cdot)_X$ and norm $\|\cdot\|_X = (\cdot,\cdot)_X^{1/2}$, such that $\mathcal{P} \subset X$ is subset of a finite-dimensional vector space. This situation occurs, for instance, automatically after discretization of an infinite-dimensional problem and eases the presentation considerably. In the simplest case, $X = \mathbb{R}^L$ for $L \in \mathbb{N}$, such that $(p_1, p_2)_X = p_1 \cdot p_2$ is the usual scalar product of two vectors. If X is a function space spanned by finitely many basis functions $\{b_1, \ldots, b_L\}$, then each $p = \sum_{j=1}^L v_j b_j \in X$ is represented by a unique vector $v = v_p \in \mathbb{R}^L$ containing the coefficients v_j of p for the basis functions b_j; a scalar product on X can be defined, for example, via the standard scalar product on $\mathbb{R}^L$ of the coefficient vectors, $(p, p')_X = v_p \cdot v_{p'}$.

12.4 Tikhonov Regularization

Coming back to our above idea to find parameters $p \in \mathcal{P}$ that balance the discrepancy $\left\|\Phi(p) - s_m^\delta\right\|_{\mathbb{R}^N}$ with the magnitude of p, we exploit the X-norm $\|\cdot\|_X$ to measure that magnitude and rely on a relaxation parameter $\alpha > 0$ to define suitable parameters as solutions to the minimization problem:

$$\min_{p \in \mathcal{P}} \left[\left\|\Phi(p) - s_m^\delta\right\|_{\mathbb{R}^N}^2 + \alpha \|p\|_X^2 \right].$$

As it is rather frequent that not all elements in X belong to the set of admissible parameters $\mathcal{P} \subset X$ due to, for example, sign conditions coming from the non-negativity of several physical quantities whenever those are searched-for parameters, the minimum in the last equation is indeed searched for over $\mathcal{P}$ instead of X. The penalty term $\|\cdot\|_X^2$ can further be exploited to incorporate *a priori* knowledge on the searched-for parameter p by considering the functional

$$\mathcal{J}_{\alpha, s_m^\delta}(p) = \left\| W\left(\Phi(p) - s_m^\delta\right) \right\|_{\mathbb{R}^N}^2 + \alpha \|D(p - p^*)\|_Z^2, \quad \text{for} \quad p \in \mathcal{P}, \qquad (12.9)$$

some invertible $N \times N$ matrix W, an element $p^* \in \mathcal{P}$ (or in X), and some injective linear mapping D from the finite-dimensional space X into another finite-dimensional vector space Y. The matrix W modifies the norm on $\mathbb{R}^N$ and is typically chosen as inverse of the Cholesky factor of the covariance matrix of s_m^δ (if a statistical distribution of s_m^δ is known) [10]; the mapping D is often chosen as a (discretized) differential operator that forces (piecewise) smoothness of the minimizer. We denote any minimizer of the latter functional on $\mathcal{P}$ by p_α^δ,

$$p_\alpha^\delta = \underset{p \in \mathcal{P}}{\operatorname{argmin}}\ \mathcal{J}_{\alpha, s_m^\delta}(p). \tag{12.10}$$

Note that a minimizer p_α^δ of the functional $\mathcal{J}_{\alpha, s_m^\delta}$ is not necessarily unique for a nonlinear model Φ, such that we cannot speak of the minimizer to (12.10).

The functional $\mathcal{J}_{\alpha, s_m^\delta}$ from (12.10) is the so-called Tikhonov functional with regularization parameter α, and the technique to determine stabilized solutions to $\Phi(p) = s_m^\delta$ by (12.10) is called Tikhonov regularization. (For linear mapping Φ, we later on consider a simple way to compute the minimizer by solving a linear equation.) Obviously, the choice of α crucially determines properties of resulting minimizers p_α^δ. Thus, to obtain a convergent regularization method, we further need to choose this parameter – this is typically and most easily done by a parameter choice rule in dependence on the noise level $\delta > 0$. Indeed, if the regularization parameter $\alpha = \alpha(\delta)$ is chosen suitably in dependence on the noise level δ, then several important results on the convergence of minimizers to $\mathcal{J}_{\alpha, s_m^\delta}$ have been shown [11,12]. In detail, if α is chosen as a scalar function of δ that satisfies

$$\alpha(\delta) \to 0 \quad \text{and} \quad \frac{\delta^2}{\alpha(\delta)} \to 0, \quad \text{as} \quad \delta \to 0, \tag{12.11}$$

then it is guaranteed that the family of minimizers $p_{\alpha(\delta)}^\delta$ of $\mathcal{J}_{\alpha(\delta), s_m^\delta}$ possesses at least one convergent subsequence as $\delta \to 0$. Moreover, the limit $p^\dagger \in \mathcal{P}$ of any convergent subsequence of these minimizers solves $\Phi(p^\dagger) = s_m$ and furthermore minimizes the functional $\|\cdot - p^*\|_X$ among all solutions to that equation, that is, $\|p^\dagger - p^*\|_X \leq \|p - p^*\|_X$ for all solutions $p \in \mathcal{P}$ to $\Phi(p) = s_m$; $p^\dagger$ is henceforth called a minimal norm solution. If there exists only one minimal norm solution $p^\dagger$, then $p_{\alpha(\delta)}^\delta$ converges to $p^\dagger$ as $\delta > 0$ tends to 0 (no matter which of the possibly several minimizers for fixed $\delta > 0$ is chosen); if $p^\dagger$ further satisfies source conditions [1,12], then this convergence further satisfies an algebraic convergence rate less than 1.

We emphasize that it is unrealistic to expect more than convergence of the regularizers $p_{\alpha(\delta)}^\delta$ to a solution of the nonlinear equation as the noise level $\delta > 0$ tends to zero – for fixed $\delta > 0$, it is, in general, impossible to deduce exact information from noisy data. Even if the parameter identification problem was stable, the latter convergence result is the best one can hope for.

12.5 Rules for the Choice of the Regularization Parameter

Two problems of the above formulation of Tikhonov regularization and the related convergence results naturally arise: As the convergence result holds asymptotically as $\delta \to 0$, it is unclear how to choose $\alpha = \alpha(\delta)$ in practice for some known noise level (or, even worse, if the noise level δ is unknown). Second, it is not clear how to tackle the minimization problem (12.10) numerically. While the second point is treated in the subsequent Section 12.4, we discuss in the following a number of techniques that cope with the problem of suitably choosing the regularization parameter from the first point.

A typical parameter choice method to determine $\alpha = \alpha(\delta)$ if the noise level $\delta > 0$ is known (or can at least be estimated or guessed) is the discrepancy principle of Morozov [13]: Fix some monotonically decreasing sequence $\alpha_n \to 0$, compute for each $n = 1, 2, \ldots$, a minimizer $p^\delta_{\alpha_n}$ of (12.10), and pick the first minimizer $p^\delta_{\alpha_{n^*}}$ with discrepancy $\left\| W\left(\Phi\left(p^\delta_{\alpha_{n^*}}\right) - s^\delta_m\right) \right\|_{\mathbb{R}^N}$ less than $\tau\varepsilon$ for some fixed parameter $\tau > 1$, that is,

$$\left\| W\left(\Phi\left(p^\delta_{\alpha_{n^*}}\right) - s^\delta_m\right) \right\|_{\mathbb{R}^N} \leq \tau\delta,$$

whereas $\left\| W\left(\Phi\left(p^\delta_{\alpha_n}\right) - s^\delta_m\right) \right\|_{\mathbb{R}^N} > \tau\delta,$ for $n = 1, 2, \ldots, n^* - 1.$

(12.12)

Typically, the sequence α_n is chosen as power of $1/n$ or as t^n for some t in the interval $(0, 1)$, according to the degree of instability of the identification problem, and τ is chosen in between 1.1 and 4. However, let us mention that in a specific application, optimal values of τ in terms of the distance of $p^\delta_{\alpha_{n^*}}$ to the true parameter might sometimes be smaller than 1.

The parameter choice rule (12.11) is the so-called discrepancy principle, which basically attempts to solve the equation up to the noise level [13]. The above-sketched convergence results for the minimizers $p^\delta_{\alpha(\delta)}$ hold as well for $p^\delta_{\alpha_{n^*}}$ as $\delta \to 0$. Note that a modified version of the above-described principle attempts to find $\alpha(\delta) > 0$ such that the discrepancy satisfies $\left\| W\left(\Phi\left(p^\delta_{\alpha(\delta)}\right) - s^\delta_m\right) \right\|_{\mathbb{R}^N} = \delta$ [2]; this variant is particularly attractive for linear problems, as we sketch later on.

If the noise level δ is unknown, several heuristic parameter choices have been proven to work well. We refer first to the L-curve criterion [14] that picks α by maximizing the curvature of the so-called L-curve, a curve in two dimensions defined by

$$\alpha \mapsto \left(\log \left\| W(\Phi(p^\delta_\alpha) - s^\delta_m) \right\|_{\mathbb{R}^N}, \left\| D(p - p^*) \right\|_Z^2\right) \in \mathbb{R}^2. \tag{12.13}$$

Second, the quasi-optimality criterion [15] picks α as the minimizer of

$$\alpha \mapsto \alpha \left\| D(p^\delta_\alpha - p^*) \right\|_Z. \tag{12.14}$$

Third, generalized cross-validation [16] is a heuristic parameter choice rule that applies for linear Φ and $q = 2$. For simplicity, we present this rule in the case $X = \mathbb{R}^L$ and assume first that F is a matrix representation of Φ with transpose matrix F^T, that is, $\Phi(p) = Fp$ and $Fp \cdot s = p \cdot F^T s$ for all $p \in X = \mathbb{R}^L$ and all $s \in \mathcal{S} = \mathbb{R}^N$, and that both W and D are the identity and can hence be omitted in (12.10). Under these assumptions, generalized cross-validation picks α as minimizer of

$$\alpha \mapsto \frac{\left\| Fp^\delta_\alpha - s^\delta_m \right\|^2_{\mathbb{R}^N}}{\left[\operatorname{trace}(I - F(F^T F + \alpha^2 I)^{-1} F^T)\right]^2}. \tag{12.15}$$

(Here, I denotes the identity matrix and trace is the sum of the diagonal elements of a matrix.) Both the L-curve and the quasi-optimality criterion can be interpreted as a tool to balance the discrepancy against the norm of the reconstructed parameter; the generalized cross-validation technique stems from statistics and can be interpreted as yielding a risk-minimizing ridge estimator for a solution to $\Phi(p) = s^\delta_m$ [17].

12.6 Explicit Minimizers for Linear Models

If Φ is a linear mapping, if $q = 2$, and if the set of admissible parameters $\mathcal{P}$ equals the entire linear space X, then the minimizer to (12.10) is unique and can be computed explicitly by solving a linear matrix–vector equation. To this end, let us assume that the minimization problem under consideration is already discretized, such that $X = \mathbb{R}^L$ and that $F : \mathbb{R}^L \to \mathbb{R}^N$ is a matrix representation of Φ, that is, $\Phi(p) = Fp$ for all $p \in X = \mathbb{R}^L$. The Tikhonov minimization problem (12.10) hence is to find a solution $p_\alpha^\delta \in \mathbb{R}^L$ to the minimization problem:

$$p_\alpha^\delta = \min_{p \in \mathbb{R}^L} \mathcal{J}_{\alpha, s_m^\delta}(p), \quad \mathcal{J}_{\alpha, s_m^\delta}(p) = \left\| W\left(Fp - s_m^\delta\right) \right\|_{\mathbb{R}^N}^2 + \alpha \| D(p - p^*) \|_{\mathbb{R}^M}^2, \tag{12.16}$$

where, by abuse of notation, we assume that D is an injective matrix in $\mathbb{R}^{M \times L}$ with $M \geq L$. Computing the Lagrange equation of the functional on the right of (12.10) shows that the minimizer p_α^δ solves the linear system of equations [1,2]:

$$\left[F^T W^T W F + \alpha D^T D\right] p_\alpha^\delta = F^T W^T s_m^\delta + \alpha D^T D p^*, \tag{12.17}$$

The latter matrix–vector equation is always uniquely solvable because the matrix $F^T W^T W F + \alpha D^T D$ is positive definite and hence invertible on $\mathbb{R}^L$ for all $\alpha > 0$.

Remark 12.1 If X is any finite-dimensional linear vector space, not necessarily equal to $\mathbb{R}^L$, and if Φ is a linear mapping, then a similar linear equation as (12.17) can also be set up for the minimizer of p_α^δ of $\mathcal{J}_{\alpha, s_m^\delta}$. As X possesses a finite basis $\{b_1, \ldots, b_L\}$, every $p \in X$ can be represented as

$$p = \sum_{j=1}^{L} v_j b_j, \quad \text{for a unique vector } v = v_p = (v_j)_{j=1}^{L} \in \mathbb{R}^L. \tag{12.18}$$

(When working with, for example, a finite-element discretization, the b_j are the finite element basis functions defined on a mesh.) The matrix F is then defined by $Fv = \Phi\left(\sum_{j=1}^{L} v_j b_j\right)$ for all $v \in \mathbb{R}^L$. However, expressing the norm of $D(p_\alpha^\delta - p^*)$ in Z via the norm of the coefficient vector of $p - p^*$ requires introducing a suitable Gram matrix, which we omit here, referring to Chapter 9 in Ref. [1] or Chapter 6 in Ref. [18].

Remark 12.2

1) Many results on linear models Φ in this section can be reformulated using the singular value decomposition of the matrix $WF \in \mathbb{R}^{N \times L}$, see, for example, Refs [1,2]; we skip details to avoid the need to introduce this matrix decomposition.
2) Several iterative solution methods for linear systems as (12.17) provide fast alternatives to solving these systems by a direct solver as, for example, Gaussian elimination and its variants. Such algorithms include the conjugate gradients method, the GMRES (generalized minimal residual) method and the steepest descent method [19].

In the same latter linear setting, there is an efficient way to compute $\alpha = \alpha(\delta)$ such that the modified discrepancy principle $\left\|W\left(F\left(p^\delta_{\alpha(\delta)}\right) - s^\delta_m\right)\right\|_{\mathbb{R}^N} = \delta$ introduced above is satisfied: Since the scalar function

$$\gamma(\alpha) = \left\|W\left(F\left(p^\delta_\alpha\right) - s^\delta_m\right)\right\|^2_{\mathbb{R}^N} - \delta^2 \qquad (12.19)$$

is continuously differentiable and monotonously decreasing, Newton's method allows to find a positive zero of γ. The derivative of γ equals

$$\gamma'(\alpha) = 2\left(F^T W^T \left(WF\left(p^\delta_\alpha\right) - s^\delta_m\right)\right) \cdot \frac{dp^\delta_\alpha}{d\alpha} = -2\alpha\left(D^T D\left(p^\delta_\alpha - p_0\right)\right) \cdot \frac{dp^\delta_\alpha}{d\alpha}, \qquad (12.20)$$

where we denoted the derivative of the vector p^δ_α with respect to α by $dp^\delta_\alpha/d\alpha$. This derivative is again a vector in $\mathbb{R}^L$ and satisfies a linear equation with the same matrix on the left-hand side as in (12.17):

$$\left[F^T W^T WF + \alpha D^T D\right] \frac{dp^\delta_\alpha}{d\alpha} = -D^T D\left(p^\delta_\alpha - p_0\right),$$

such that computing this derivative allows evaluating $\gamma'(\alpha)$ via (12.20). The Newton iteration to determine $\alpha = \alpha(\delta) > 0$ such that $\left\|W\left(F\left(p^\delta_{\alpha(\delta)}\right) - s^\delta_m\right)\right\|_{\mathbb{R}^N} = \delta$ hence reads

$$\alpha_{n+1} = \alpha_n - \frac{\gamma(\alpha_n)}{\gamma'(\alpha_n)}, \qquad n = 1, 2, 3, \ldots.$$

If $p_0 = 0$ and $\|s^\delta_m\|_{\mathbb{R}^N} > \delta$, then a suitable initiation is $\alpha_0 = \delta\|WF\|^2_{\mathbb{R}^L \to \mathbb{R}^N}/\left(\|s^\delta_m\|_{\mathbb{R}^N} - \delta\right)$; if $\|s^\delta_m\|_{\mathbb{R}^N} \leq \delta$, then obviously $p = 0$ already satisfies the discrepancy principle.

12.7 The Soft-Shrinkage Iteration

If one considers, in the same linear setting, the following Tikhonov functional

$$\mathcal{J}_{\alpha, s^\delta_m}(p) = \left\|W\left(Fp - s^\delta_m\right)\right\|^2_{\mathbb{R}^N} + \alpha \sum_{j=1}^L \left|p_j - p_{0,j}\right|^q, \qquad \text{for } p \in \mathbb{R}^L \text{ and fixed } q \geq 1, \qquad (12.21)$$

with different penalty term, then explicit formulas for minimizers $p^\delta_\alpha \in \mathbb{R}^L$ to this functional do not exist. Still, one can write down a fixed-point equation for p^δ_α that is particularly popular in the case $q = 1$, where it is known as the soft-shrinkage iteration [20], particularly attractive if the searched-for parameter differs from p_0 merely in few entries. The announced fixed-point iteration relies on the scalar function $f_{\alpha, q} : \mathbb{R} \to \mathbb{R}$, defined by

$$f_{\alpha, q}(t) = t + \alpha q \operatorname{sign}(t) |t|^{q-1}, \qquad \text{for } t \in \mathbb{R}, \quad q \geq 1, \quad \alpha > 0.$$

If $q > 1$, the function $t \mapsto \operatorname{sign}(t)|t|^{q-1}$ is monotonically increasing and bijective from $\mathbb{R}$ into $\mathbb{R}$, such that the function $f_{\alpha, q}$ is invertible. For simplicity, we denote its inverse function by $s_{\alpha,q}$. This inverse function determines the q-shrinkage operator $S_{\alpha,q} : \mathbb{R}^L \to \mathbb{R}^L$, defined for $p \in \mathbb{R}^L$ element-wise by

$$S_{\alpha,q}(p) = w \in \mathbb{R}^L, \quad w_j = \sum_{j=1}^{L} s_{\alpha,q}(p_j), \quad \text{for} \quad j = 1, \ldots, N.$$

The shrinkage operator is also well defined for $q = 1$, where one can even express its action on p explicitly, using the function

$$s_{\alpha,1}(t) = \begin{cases} t + \alpha/2, & \text{if } t \leq -\alpha/2, \\ 0, & \text{if } |t| < \alpha/2, \\ t - \alpha/2, & \text{if } t \geq \alpha/2. \end{cases}$$

For $q \geq 1$, one can then show that any minimizer p_α^δ to the functional in (12.21) satisfies

$$p_\alpha^\delta - p_0 = S_{\alpha,q}\left(p_\alpha^\delta - p_0 + F^T W^T \left(s_m^\delta - W F p_\alpha^\delta\right)\right).$$

Consequently, the corresponding fixed-point iteration reads

$$p_{n+1} = p_0 + S_{\alpha,q}\left(p_n - p_0 + F^T W^T \left(s_m^\delta - W F p_n\right)\right), \quad n \in \mathbb{N}, \quad (12.22)$$

for an arbitrary starting vector $p_0 \in \mathbb{R}^L$, and converges to a minimizer p_α^δ of $\mathcal{J}_{\alpha,s_m^\delta}$ defined in (12.21). If one couples the regularization parameter $\alpha = \alpha(\delta)$ with the noise level δ as in (12.11), the minimizers again converge to the minimum-norm solution to $Fp = s_m$ (respectively, to $\Phi(p) = s_m$) as $\delta \to 0$.

Of course, one can also stop this iteration by the discrepancy principle whenever the nth iterate p_n from (12.22) satisfies

$$\|Fp_n - s_m^\delta\| \leq \tau\varepsilon, \quad \text{for some fixed } \tau > 1. \quad (12.23)$$

The discrepancy principle is particularly suited for iterative regularization schemes such as (12.22), as little extra work apart from the iteration itself is necessary to set up the corresponding regularization scheme. Alternatively, heuristic parameter choice rules can also be employed; we refer to Refs [21,22] for the quasi-optimality criterion applied to iterative schemes and a comparison of several parameter choice rules for both iterative and noniterative regularization methods.

12.8 Iterative Regularization Schemes

If the model Φ is nonlinear, then there are in general no explicit representations for the minimizer of the Tikhonov functional given in (12.10). If one aims to rely on Tikhonov regularization in a nonlinear setting, one needs to compute this minimizer by some minimization scheme suitable for functionals as given in (12.9). Typically, such schemes are of iterative nature and frequently they rely on the differentiability of Φ with respect to the parameter p: By definition, $\Phi : \mathcal{P} \subset X \to \mathcal{S} = \mathbb{R}^N$ is differentiable at $p \in \mathcal{P}$ if there is a linear mapping $\Phi'(p) : X \to \mathbb{R}^N$ such that

$$\frac{\|\Phi(p+h) - \Phi(p) - \Phi'(p)h\|_{\mathbb{R}^N}}{\|h\|_X} \to 0, \quad \text{as } \|h\|_X \to 0.$$

In a setting where X (and possibly also $\mathcal{S}$) are infinite-dimensional spaces, this concept is called Fréchet differentiability; in finite-dimensional spaces, this is nothing but the well-known differentiability of functions that map $\mathbb{R}^L$ into $\mathbb{R}^N$.

12.9 Gradient Descent Schemes

Let us assume from now on that Φ is differentiable at all parameters $p \in \mathcal{P}$. Then, the derivative $\mathcal{J}'_{\alpha,s^\delta_m}(p)$ of the Tikhonov functional

$$\mathcal{J}_{\alpha,s^\delta_m}(p) = \left\| W\left(\Phi(p) - s^\delta_m\right) \right\|^2_{\mathbb{R}^N} + \alpha \| D(p - p^*) \|^2_Z$$

from (12.10) with $q = 2$ at p can be explicitly computed by the chain rule:

$$\mathcal{J}'_{\alpha,s^\delta_m}(p)h = 2\left(\Phi'(p)h, W^T W\left(\Phi(p) - s^\delta_m\right)\right)_X + 2\left(h, \alpha D^T D(p - p^*)\right)_X.$$

(12.24)

Here, $D^T : Z \to X$ is the transpose mapping to D (i.e., the transpose matrix, if $D \in \mathbb{R}^{M \times L}$ is a matrix), defined by $(p, D^T z)_X = (Dp, z)_Z$ for all $p \in X$ and $z \in Z$. If we analogously denote the transpose mapping of $\Phi'(p)$ by $\Phi'(p)^T : \mathbb{R}^N \to X$, then (12.24) implies that $-\Phi'(p)^T W^T W\left(\Phi(p) - s^\delta_m\right) - \alpha D^T D(p - p^*)$ is a descent direction of the functional $\mathcal{J}_{\alpha,s^\delta_m}$ at p. This motivates the following iteration for the minimization of $\mathcal{J}_{\alpha,s^\delta_m}$:

$$\begin{aligned} p^\delta_{n+1} &= p^\delta_n - \omega_n \left[\Phi'(p^\delta_n)^T W^T W\left(\Phi(p^\delta_n) - s^\delta_m\right) + \alpha D^T D(p^\delta_n - p^*) \right] \\ &= (I - \omega_n \alpha D^T D) p^\delta_n - \omega_n \left[\Phi'(p^\delta_n)^T W^T W\left(\Phi(p^\delta_n) - s^\delta_m\right) - \alpha D^T D p^* \right], \quad n \in \mathbb{N}, \end{aligned}$$

(12.25)

for some starting value (or initial guess) $p_0 \in \mathcal{P}$. (Here, $I : p \mapsto p$ is the identity mapping on X, that is, the identity matrix in a discretized setting.) The numbers $\omega_n > 0$ can be interpreted as a step size control or a damping parameter that should be chosen such that both linear mappings $\omega_n \Phi'(p^\delta_n)^T W^T W$ and $\omega_n \Phi'(p^\delta_n)^T W^T W$ are contractive, that is,

$$\omega_n \left\| \Phi'(p^\delta_n)^T W^T W s \right\|_X < \|s\|_{\mathbb{R}^N} \quad \text{and} \quad \omega_n \alpha \left\| D^T D p \right\|_X < \|p\|_X, \quad \text{for all } s \in \mathbb{R}^N, \ p \in \mathcal{P}.$$

In other words, the spectral matrix norm (i.e., the 2-norm) of the matrix representations of both mappings needs to be strictly less than 1 for each $n \in \mathbb{N}$ [5,23]. Typically, the latter condition is far from being easy to guarantee in applications, in particular if it is costly to set up the entire matrix representation of $\Phi'(p^\delta_n)^T$.

A related method is the so-called nonlinear Landweber iteration, where one chooses the update direction as a descent direction of the squared discrepancy $\left\| W\left(\Phi(p) - s^\delta_m\right) \right\|^2_{\mathbb{R}^N}$ (neglecting the penalty term of $\mathcal{J}_{\alpha,s^\delta_m}$), such that the resulting iteration reads

$$p^\delta_{n+1} = p^\delta_n - \omega_n \Phi'(p^\delta_n)^T W^T W\left(\Phi(p^\delta_n) - s^\delta_m\right), \quad n \in \mathbb{N}.$$

(12.26)

Again, this iteration scheme starts with some initial guess $p_0 \in \mathcal{P}$. The damping parameter $\omega_n > 0$ has to be chosen so small that $\omega_n \Phi'(p^\delta_n)^T W^T W$

is contractive, that is,

$$\omega_n \|\Phi'(p_n^\delta)^T W^T W s\|_X < \|s\|_{\mathbb{R}^N}, \quad \text{for all } s \in \mathbb{R}^N, \quad n \in \mathbb{N}.$$

Both iterations (12.25) and (12.26) are connected, which becomes evident if one replaces the term $D^T D(p_n^\delta - p^*)$ in (12.25) by $D^T D(p_{n+1}^\delta - p^*)$ and chooses $p^* = 0$. The resulting iteration then reads

$$(I - \omega_n \alpha D^T D) p_{n+1}^\delta = p_n^\delta - \omega_n \Phi'(p_n^\delta)^T W^T W (\Phi(p_n^\delta) - s_m^\delta), \quad n \in \mathbb{N},$$

which is a Landweber iteration subject to an additional smoothing by multiplying the Landweber iterate with the inverse of $I - \alpha D^T D$:

$$p_{n+1}^\delta = (I - \omega_n \alpha D^T D)^{-1} \left[p_n^\delta - \omega_n \Phi'(p_n^\delta)^T W^T W (\Phi(p_n^\delta) - s_m^\delta) \right], \quad n \in \mathbb{N}.$$

(The inverse $(I - \omega_n \alpha D^T D)^{-1}$ exists if ω_n is for all $n \in \mathbb{N}$ so small that the matrix representation of $\omega_n \alpha D^T D$ has a (spectral) matrix norm less than 1.) Typically, the numbers ω_n are either all fixed to $\omega > 0$ small enough (this requires to know the magnitude of Φ' in advance), or chosen adaptively by some step size rule that aims to minimize the value of the Tikhonov functional (for (12.25)) or of the discrepancy (for (12.26)) along the descent direction. Examples for step size rules include the bisection and the Armijo rule [24,25].

As (12.25) attempts to minimize the Tikhonov functional, whereas (12.26) attempts to minimize the discrepancy, a crucial difference between the iterations (12.25) and (12.26) is that the fixed-point equations satisfied by any limit $p^\dagger$ of the sequence of iterates $\{p_n^\delta\}$ differ:

$$0 = \Phi'(p^\dagger)^T W^T W (\Phi(p^\dagger) - s_m^\delta) + \alpha D^T D(p^\dagger - p^*), \quad \text{for } (12.25),$$

$$\text{but } 0 = \Phi'(p^\dagger)^T W^T W (\Phi(p^\dagger) - s_m^\delta), \quad \text{for } (12.25) \text{ and } (12.26).$$

If we assume that $\Phi'(p^\dagger)^T$ is injective, then any limit of the Landweber iteration (12.25) satisfies the equation $\Phi(p^\dagger) = s_m^\delta$; obviously, this does not hold for the limit to (12.25). Thus, computing a minimizer p_α^δ of $\mathcal{J}_{\alpha, s_m^\delta}$ via the iteration (12.25) must be accompanied by a parameter choice rule to determine a value of α that ensures both stability and accuracy of the resulting parameter reconstruction (see (12.12–12.15)). (Several of these rules require computing several minimizers, such that computation time might become a crucial point here; whenever runtimes become an issue, experience on suitable choices of α is typically required to reduce computation times.)

By construction, the Landweber iteration (12.26) attempts to compute a solution to the equation $\Phi(p) = s_m^\delta$ instead of a minimizer of $\mathcal{J}_{\alpha, s_m^\delta}$. We discussed in Section 12.1 that solving ill-conditioned operator equations exactly leads to incorrect results due to instability; thus, the Landweber iteration must in any case be stopped by some stopping rule that ensures stability and accuracy of the resulting approximation to the parameter p. (Such a stopping rule hence plays the same role as the regularization parameter choice rule for Tikhonov regularization.) If the noise level δ is known, a particularly attractive stopping rule is the discrepancy principle that we already discussed for Tikhonov regularization in (12.12). For an iterative regularization method, Morozov's discrepancy principle stops the iteration whenever the discrepancy is below the tolerance $\tau\delta$ for the first time: The resulting approximation to the true parameter is hence $p^\delta_{n^*}$, determined by

$$\left\| W\left(\Phi(p^\delta_{n^*}) - s^\delta_m\right) \right\|_{\mathbb{R}^N} \leq \tau\delta,$$
$$\text{whereas} \quad \left\| W\left(\Phi(p^\delta_n) - s^\delta_m\right) \right\|_{\mathbb{R}^N} > \tau\delta, \quad \text{for} \quad n = 1, 2, \ldots, n^* - 1. \tag{12.27}$$

Again, $\tau > 1$ is a fixed parameter, typically chosen between 1.1 and 4. (Suitable values for a specific application should be tested and optimized.) Other stopping rules for known noise level suitable for the Landweber iteration include the quasi-optimality criterion and Lepskii's balancing principle; if we assume that $X = \mathbb{R}^L$ such that $\Phi'(p^\delta_n)^T$ is a matrix in $\mathbb{R}^{L \times N}$, then these two rules stop the iteration at the nth iterate p^δ_n if

$$\left[\Phi(p^\delta_n) - \Phi(p^\delta_{n+1})\right] \cdot \Phi'(p^\delta_n)^T W^T W\left(\Phi(p^\delta_n) - s^\delta_m\right) \leq 2\delta \left\| \Phi'(p^\delta_n)^T W^T W\left(\Phi(p^\delta_n) - s^\delta_m\right) \right\|_X \tag{12.28}$$

and if

$$\left\| p^\delta_n - p^\delta_{n+1} \right\| \geq \tau \max\left\{ \omega_1^{1/2}, \ldots, \omega_n^{1/2} \right\} n^{1/2} \delta, \tag{12.29}$$

respectively. These noise level-free rules hence stop the Landweber iteration whenever the discrepancy is small enough or if the difference between two iterates becomes too large compared to the iteration number. Notably, both rules require to compute $n^* + 1$ iterates for the determination of the stopping index n^*, that is, one iterate more than required by the discrepancy principle. A further heuristic parameter choice rule applicable to the Landweber iteration is the Hanke–Raus rule. This stopping rule determines the stopping index rule n^* as a minimizer of the sequence of numbers $\sqrt{n}\left\| W\left(\Phi(p^\delta_n) - s^\delta_m\right) \right\|_{\mathbb{R}^N}$ in n; the resulting parameter approximation is hence $p^\delta_{n^*}$ [21,22,26].

Both iterative schemes (12.25) and (12.26) can also be used to regularize *linear* problems, where Φ can be represented by some matrix F. In this case, one usually fixes the damping parameter α_n to $\alpha > 0$; the upper bound $\alpha < 2/\|WF\|^2$ involves the spectral matrix norm $\|WF\|^2$ of WF that is crucial for convergence of the iteration.

There are numerous works concerned with the convergence and regularization properties of the Landweber iteration for both linear and nonlinear models Φ [23,27]. In particular, the following convergence result holds for nonlinear Φ if the iteration is stopped by the discrepancy principle, returning $p^\delta_{n^*}$ as reconstruction: If $\Phi'(p)$ is injective for all $p \in \mathcal{P}$ and if the initial value of the iteration is close enough to any solution to $\Phi(p) = s^\delta_m$, then $p^\delta_{n^*}$ converges to a solution $p^\dagger$ to that equation as $\delta \to 0$. For linear mappings Φ, this convergence result holds independent of the choice of the initial value of the Landweber iteration, at least if Φ is injective (see Ref. [1] for details). Similar as for Tikhonov regularization, convergence rates in terms of δ can be shown if the starting value p_0 solution satisfies a source condition [28]. (For nonlinear Φ, convergence of Landweber-type regularization methods typically rely on nonlinearity conditions for Φ; if $\Phi'(p)$ is injective for all $p \in \mathcal{P}$, these conditions are automatically satisfied in our setting as X is finite dimensional.)

12.10 Newton-Type Regularization Schemes

The Landweber iteration has the advantage to be rather robust against data noise; on the downside, this method is rather slow and cannot be advertised for small noise levels (for linear models, the number of iteration steps for constant damping parameter behaves like δ^{-2}). For this reason, a different class of iterative Newton-like methods is usually preferred, in particular for rather accurate data. Instead of trying to find stable solutions to an equation for a searched-for parameter by minimizing a functional, these methods try to approximately solve the equation $\Phi(p) = s_m^\delta$ by successive linearization: At each iterate p_n^δ, one considers the linearization of Φ at p_n^δ:

$$\Phi(p) \approx \Phi(p_n^\delta) + \Phi'(p_n^\delta)h, \quad \text{if} \quad \|h\|_X \text{ is sufficiently small.}$$

Aiming to determine an update $h_n^\delta \in X$ that defines the next iterate $p_{n+1}^\delta = p_n^\delta + h_n^\delta$ such that $\Phi(p_{n+1}^\delta) = s_m^\delta$ leads to the following linear equation for $h_n^\delta \in X$:

$$\Phi'(p_n^\delta)h_n^\delta = s_m^\delta - \Phi(p_n^\delta). \tag{12.30}$$

For parameter identification problems, this linearized equation typically cannot be stably inverted either, such that a linear regularization scheme must be applied to (12.10). Let us opt here for Tikhonov regularization and denote, as above in (12.25), by $\Phi'(p_n^\delta)^T : \mathbb{R}^N \to X$ the transpose mapping of the linear map $\Phi'(p_n^\delta)$. Furthermore, let us first choose the *a priori* information p^* from the Tikhonov functional in (12.10) or (12.16) to be zero. Due to the results from the last section, see (12.16), the minimizer $h_n^\delta \in X$ of the Tikhonov functional

$$\mathcal{J}_{\alpha_n, s_m^\delta - \Phi(p_n^\delta)}(h) = \left\| W\left(\Phi'(p_n^\delta)h - s_m^\delta + \Phi(p_n^\delta)\right) \right\|_X^2 + \alpha_n \|Dh\|_Z^2 \quad \text{on } X$$

for some regularization parameter $\alpha_n > 0$, a matrix $W \in \mathbb{R}^{N \times N}$ and an injective map D from X into the finite-dimensional space Z satisfies the linear equation:

$$\left[(\Phi'(p_n^\delta))^T W^T W \Phi'(p_n^\delta) + \alpha_n D^T D\right] h_n^\delta = (\Phi'(p_n^\delta))^T W^T (s_m^\delta - \Phi(p_n^\delta)), \quad n \in \mathbb{N}, \tag{12.31}$$

which allows for a straightforward computation of h_n^δ if $X = \mathbb{R}^L$, since all involved quantities then become matrices or vectors. (We omitted the dependence of h_n^δ on α_n.) In particular, h_n^δ is the *unique* minimizer of that functional. The resulting iterative regularization technique with iterates $\{p_j^\delta\}$ is the Levenberg–Marquardt method [29,30].

Again, we need to discuss the choice of the regularization parameters α_n, as well as the choice of a stopping rule for the Levenberg–Marquardt iteration. Concerning the first issue, it is simplest to choose α_n as a sequence of positive numbers that tends strictly monotonically to zero, such as $\alpha_n = n^{-r}$ for some $r > 0$ or $\alpha_n = r^n$ for some r in the interval $(0, 1)$. Computationally more involved are inexact Newton methods that determine $\alpha_n > 0$ such that the solution $w_n^\delta = w_{n,\alpha_n}^\delta$ to (12.31) satisfies the linearized equation up to a tolerance $\mu \in (0,1)$ [31],

$$\left\| W\left(\Phi'(p_n^\delta) h_{n,\alpha_n}^\delta - s_m^\delta + \Phi(p_n^\delta)\right) \right\|_{\mathbb{R}^N} \leq \mu \left\| \Phi(p_n^\delta) - s_m^\delta \right\|_{\mathbb{R}^N}, m \quad (12.32)$$

for some fixed $\mu \in (0,1)$.

One typically chooses μ in between 0.01 and 0.3; the techniques we used in (12.22) or (12.19) to find a suitable regularization parameter by the discrepancy principle can be used here to find a suitable $\alpha_n > 0$ such that w_{n,α_n}^δ satisfies (12.32). Of course, this implies extra computational work as w_{n,α_n}^δ possibly needs to be computed for several choices of α_n until (12.32) is satisfied.

To determine the stopping index n^* of the sequence of iterates p_n^δ of the Levenberg–Marquardt method, one can either use the discrepancy principle (12.27), the quasi-optimality criterion (12.28), or Lepskii's balancing principle (12.29) if the noise level $\delta > 0$ is known; otherwise, the heuristic stopping rule indicated above for the Landweber iteration can be exploited: Determine n^* as the index where the sequence of numbers $\sqrt{n} \left\| W(\Phi(p_n^\delta) - s_m^\delta) \right\|_{\mathbb{R}^N}$ assumes its minimal value.

Termination and convergence of the Levenberg–Marquardt method combined with the discrepancy principle can, for instance, be shown if $\Phi'(p)$ is injective for all $p \in \mathcal{P}$ (see Ref. [31] or the monographs [5]). Of course, other techniques instead of Tikhonov regularization can also be used for computing a regularized step h_n^δ; we refer to Ref. [32] for several other applicable linear schemes as the steepest descent or the conjugate gradients method.

While the Levenberg–Marquardt method chooses the *a priori* information p^* from the Tikhonov functional in (12.16) equal to zero, the iteratively regularized Gauss–Newton method [33,34] introduces *a priori* guess $p^* \in \mathcal{P}$ and forces the iterates p_n^δ to remain close to p^*. This can be interpreted as incorporation of *a priori* information on the searched-for parameter, or as a tool to increase stability of the iteration. The update $h_n^\delta \in X$ in the nth step $p_{n+1}^\delta = p_n^\delta + h_n^\delta$ of this algorithm is, to this end, defined as the unique minimizer of the functional:

$$\mathcal{J}_{\alpha_n, s_m^\delta - \Phi(p_n^\delta)}(h) = \left\| W(\Phi'(p_n^\delta) h - s_m^\delta + \Phi(p_n^\delta)) \right\|_{\mathbb{R}^N}^2 + \alpha_n \left\| D(h + p_n^\delta - p^*) \right\|_Z^2 \text{ on } X.$$

Equivalently, $h_n^\delta \in X$ satisfies the linear equation:

$$\left[(\Phi'(p_n^\delta))^T W^T W \Phi'(p_n^\delta) + \alpha_n D^T D \right] h_n^\delta = (\Phi'(p_n^\delta))^T W^T (s_m^\delta - \Phi(p_n^\delta)) + \alpha_n (p^* - p_n^\delta),$$

that becomes a matrix–vector equation upon discretization, that is, once $X = \mathbb{R}^L$ is a Euclidean space. If α_n is a sequence of positive numbers tending to zero, such that the quotient α_n / α_{n+1} is bounded in $n \in \mathbb{N}$ (the above choice $\alpha_n = t^n$ for $0 < t < 1$ satisfies this requirement), then convergence results for the iteratively regularized Gauss–Newton method can be shown, for instance, if $\Phi'(p)$ is injective for all $p \in \mathcal{P}$.

12.11 Numerical Examples in Load Reconstruction

In this last section, we present load reconstructions using two of the above-introduced regularization techniques in the context of load reconstruction, as introduced in Example 12.1. More precisely, we consider Tikhonov regularization and the Landweber iteration for load monitoring using sensors attached to the lower surface of a thin horizontal plate Ω. The plate Ω of size $0.5\,\text{m} \times 0.5\,\text{m} \times 0.02\,\text{m}$ is assumed to consist of homogeneous construction steel with elastic modulus E of $210\ \text{kN}/\text{mm}^2$ and Poisson's ration ν of 0.3. The Lamé parameters hence equal $\lambda = \nu E/[(1+\nu)(1-2\nu)] = 63/52\ \text{kN}/\text{mm}^2$ and $\mu = E/(2+2\nu) = 1050/13\ \text{kN}/\text{mm}^2$. Due to homogeneity of Ω, the rather involved stiffness tensor in (12.2) can be characterized by these two constants.

We additionally assume that one of the four horizontal sides of Ω, denoted by Γ_0, is fixed, and that the top surface Γ_+ is loaded by some space-dependent loading described by a force of the form $-(0,0,f)^T$ for a nonnegative scalar function $f : \Gamma_+ \to \mathbb{R}$. This loading function is further assumed to be small enough for that the equations of linear elasticity provide a valid model. Under these assumptions, the deformation field $u : \Omega \to \mathbb{R}^3$ is governed by the following boundary value problem for the Cauchy stress tensor $\sigma(u) = 2\mu\varepsilon(u) + \lambda\,\text{div}\,u\,I_3$:

$$\text{div}\,\sigma(u) = 0 \quad \text{in } \Omega, \quad u = 0 \quad \text{on } \Gamma_0, \quad \sigma(u)\nu = (0,0,-f)^T \quad \text{on } \Gamma_+,$$

$$\sigma(u)\nu = 0 \quad \text{on } \partial\Omega \setminus (\Gamma_+ \cup \Gamma_0). \tag{12.33}$$

(Recall that $2\varepsilon(u) = \nabla u + (\nabla u)^T$ and that ν is the exterior unit normal field to the boundary $\partial\Omega$.) Due to the Dirichlet boundary condition on Γ_0, there is a unique deformation field that solves the latter boundary value problem. Abbreviating $A : B = \sum_{i,j=1}^{3} A_{i,j}B_{i,j}$, an integration by parts implies the following associated variational formulation for the solution u:

$$\int_\Omega \left[2\mu\varepsilon(u) : \varepsilon(u) + \lambda \int \text{div}\,u\ \text{div}\,v \right] x = \int_{\Gamma_+} f \cdot v_3 S \tag{12.34}$$

for all sufficiently smooth functions $v : \Omega \to \mathbb{R}^3$. This variational formulation allows to approximate u by a finite element method, replacing u and v in (12.34) by, for example, piecewise polynomial and globally continuous vectorial finite element functions on a triangulation of Ω. For our examples presented below, we simulated loadings by piecewise quadratic elements on a regular tetrahedral mesh with mesh width $1/10$ mm.

Before explaining details on the simulated measurements, let us remark that the proceedings paper [35] and the preprint [36] contain several numerical example of loads reconstructed from sensor network data in a similar setting. Extending the examples below, these references further include particular inversion algorithms for linear problems as the conjugate gradients method and reconstructions from highly imprecise data due to incomplete data propagation through a self-organizing sensor network.

We next assume that the sensors attached to the lower surface of the plate measure surface strain in x- and y-directions at 64 sensor positions x_j on a grid of sensors of size 8×8. Such measurements can, for instance, be obtained from strain gauges; we crudely model them by the first two diagonal entries $\partial u/\partial x$ and $\partial u/\partial y$ of the strain tensor $\varepsilon(u)$ at the sensor positions.

Thus, we obtain a physical model Φ linking the applied loading with surface strain measurements to (12.33):

$$\Phi f = \left(\frac{\partial u}{\partial x}(x_j), \frac{\partial u}{\partial y}(x_j) \right)_{j=1}^{64} \subset \mathbb{R}^{128}.$$

For simplicity, we choose f in the space of piecewise constant functions on a regular quadrilateral mesh of 20×20 squares covering Γ_+ completely; see Figure 12.2a for a plot of the midpoints of these squares. The forward mapping Φ thus maps the parameter space $S = \mathbb{R}^{400}$ into $\mathbb{R}^{128}$. u and $\varepsilon(u)$ depend linearly on f through (12.33) such that Φ is a linear mapping that can be represented by a matrix F in $\mathbb{R}^{128 \times 400}$. Using the above-described finite element discretization, we simulate the 400 loadings corresponding to the standard basis in $S = \mathbb{R}^{400}$ to get an explicit approximation to the matrix F representing Φ. More precisely, the forces corresponding to these loadings vanish on all but one square of the mentioned surface quadrangulation of Γ_+; on that square, they equal $(0, 0, -1)^T$ N/cm². As the plate Ω is fixed at one vertical side parallel to the x-axis, strain in x-direction is generally smaller than strain in y-direction, which can also be observed from the plot of the matrix F in Figure 12.2b.

To obtain simulated data for inversion by Tikhonov regularization and the Landweber iteration, we further simulate (ideal) surface strain sensor measurements s_m for two forces $f_{1,2}$ plotted in Figure 12.2c and d. Figure 12.2e and f shows the piecewise constant interpolations on the 20×20 quadrangulation of Γ^+. By checking the convergence of the simulated strain data for different finite element meshes, we determined that the relative error of the simulated surface strains is less than 3%. Nevertheless, we further increase this unknown noise level due to numerical approximation by adding a scaled random matrix containing uniformly distributed numbers in $[-1, 1]$ to the simulated data, and choose the scaling such that the resulting relative noise level equals a prescribed relative noise level in between 0.0033 and 0.3. For 100 independent instances of such noise data, we compute reconstructions and present the mean reconstruction and the variance of those 100 reconstructions, to give an impression on the behavior of both regularization techniques.

For all examples computed by Tikhonov regularization, we choose both mappings W and D as the identity, such that we compute the corresponding regularizers by $p_\alpha^\delta = (F^T F + \alpha I)^{-1} F^T s_m^\delta$; the regularization parameter is computed by the discrepancy principle (12.12) for $\alpha_n = 10^{2-hn}$ with $h = 0.75$ and $\tau = 1.1$. Note that we work from now on with relative data errors, which are somewhat easier to interpret as misleading scaling effects cannot as easily occur as for absolute errors. The relative noise level of artificially noised data s_m^δ hence equals $\|s_m^\delta - s_m\|_{\mathbb{R}^{128}} / \|s_m\|_{\mathbb{R}^{128}}$, and the relative error of the corresponding reconstruction $f^\delta \in \mathbb{R}^{400}$ to the true (interpolated) load $f \in \mathbb{R}^{400}$ equals $\|f^\delta - f\|_{\mathbb{R}^{400}} / \|f\|_{\mathbb{R}^{400}}$. Consistently, we also employ the relative discrepancy when checking whether the nth Landweber iterate f_n^δ satisfies the discrepancy principle, that is, whether $\|F(f_n^\delta) - s_m^\delta\|_{\mathbb{R}^{128}} / \|s_m^\delta\|_{\mathbb{R}^{128}} < \tau \delta$.

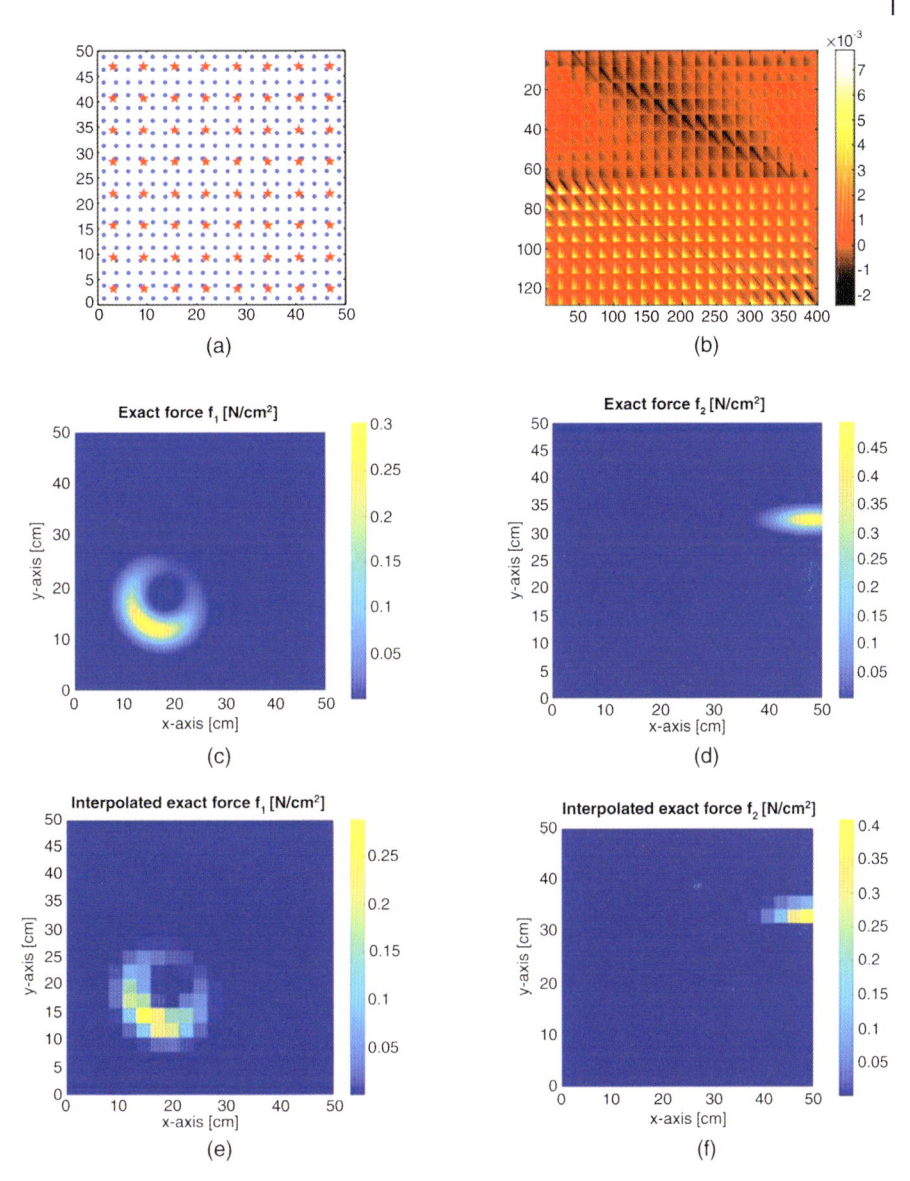

Figure 12.2 (a) Positions of the 400 midpoints of the squares defining the quadrangulation of Γ_+ for discretized loads f are indicated using blue dots; red crosses indicate the positions of the 64 sensors attached to the lower surface of the plate Ω. As for (c–f), units of x- and y-axes are cm. (b) Plot of the load–strain matrix $F \in \mathbb{R}^{128 \times 400}$. (c and d) The loadings $f_{1,2}$. (e and f) Piecewise constant interpolations of loadings $f_{1,2}$ in the space of piecewise constant functions on the regular 20×20 grid of the top surface of the plate.

As we aim to exploit the *a priori* knowledge that the exact forces correspond to a loading, and hence point in direction $(0, 0, -1)^T$, we replace any negative value of any reconstruction by Tikhonov regularization by zero, and proceed in the same way with any iterate of the Landweber iteration.

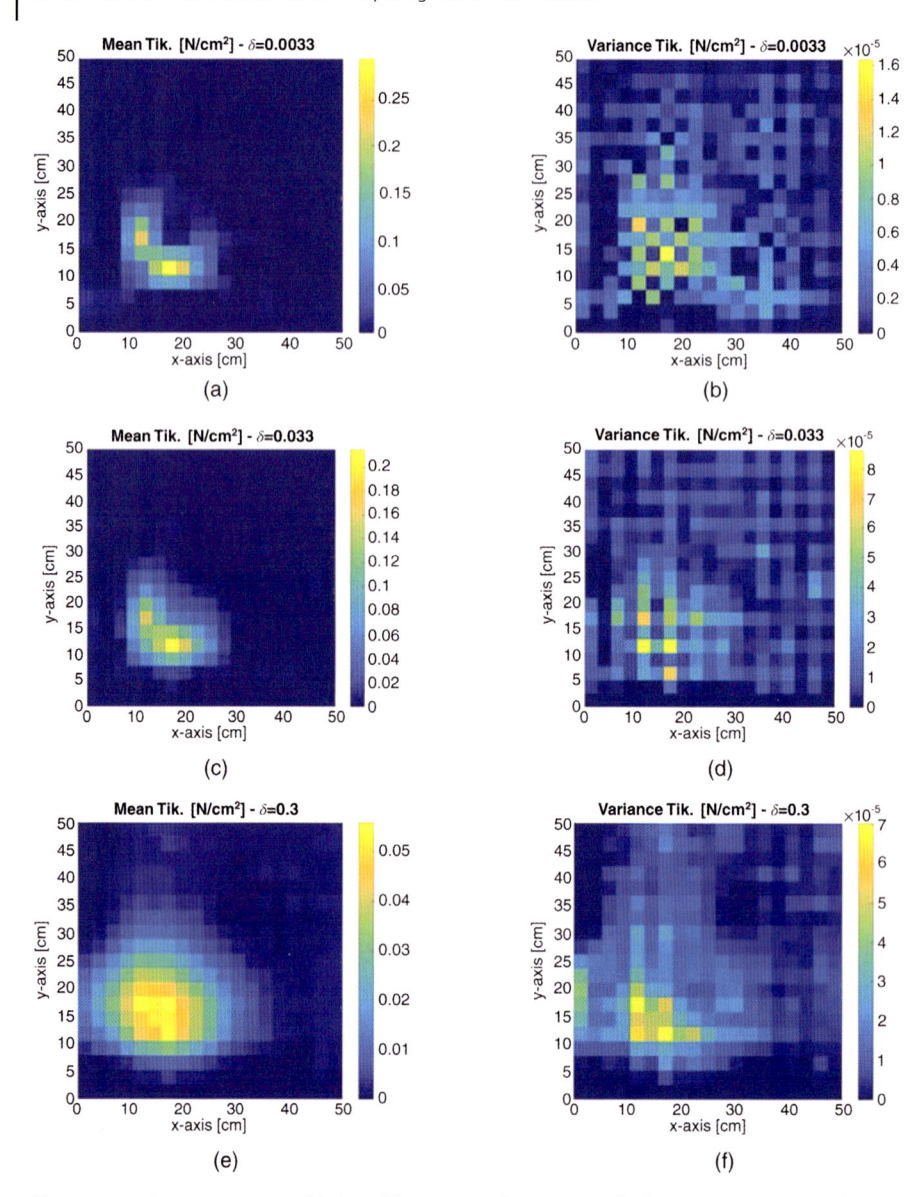

Figure 12.3 Reconstruction of f_1 by Tikhonov regularization with discrepancy principle for different relative noise levels. Reconstructions in [N/cm^2], units of **x**- and **y**-axes are cm. Noise levels $\delta = 0.0033$ (a and b), $\delta = 0.033$ (c and d), and $\delta = 0.3$ (e and f). (a, c, and e) Mean reconstruction of 100 reconstructions from independent noisy data sets. (b, d, and f) Variance of the 100 reconstructions.

Figures 12.3–12.6 show the results of Tikhonov regularization and Landweber iteration with regularization parameter and stopping index chosen by the discrepancy principle, respectively (see (12.12) and (12.28)). Each figure contains plots for the three different relative noise levels 0.0033, 0.033, and 0.3, ordered row-wise in increasing order. The left column of each figure contains plots of the mean

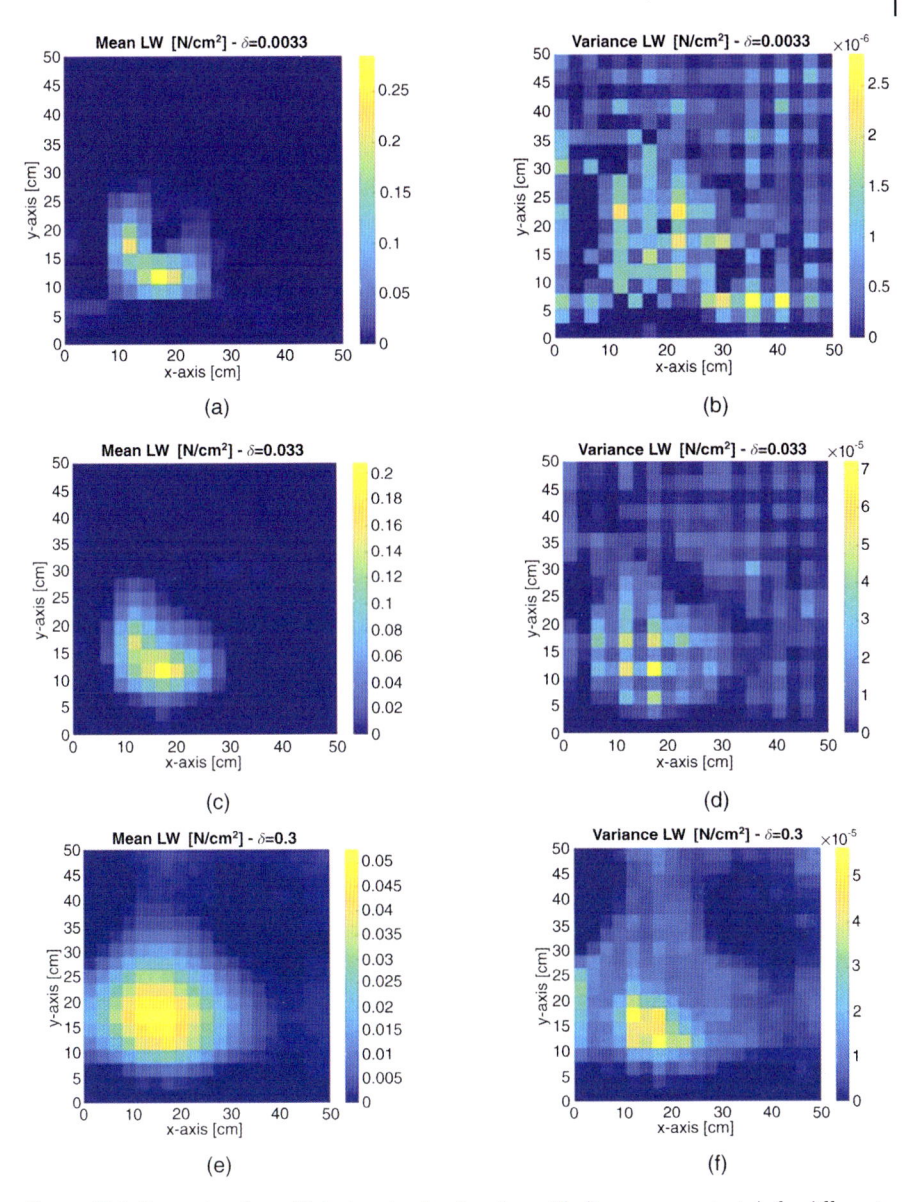

Figure 12.4 Reconstruction of f_1 by Landweber iteration with discrepancy principle for different relative noise levels. Reconstructions in [N/cm^2], units of **x**- and **y**-axes are cm. Noise levels $\delta = 0.0033$ (a and b), $\delta = 0.033$ (c and d), and $\delta = 0.3$ (e and f). (a, c, and e) Mean reconstruction of 100 reconstructions from independent noisy data sets. (b, d, and f) Variance of the 100 reconstructions.

reconstruction f_{mean} computed from all reconstructions due to 100 independent instances $f^{(k)}$ of simulated noisy data, defined by $f_{\text{mean}} = \sum_{k=1}^{100} f^{(k)}/100$. The right column contains plots of the variance of the 100 reconstructions, defined element-wise by $\text{var}(f^{(k)})_j = \sum_{k=1}^{100} \left| f_j^{(k)} - f_{\text{mean},j} \right|^2 /100$. Notably, the variances of both

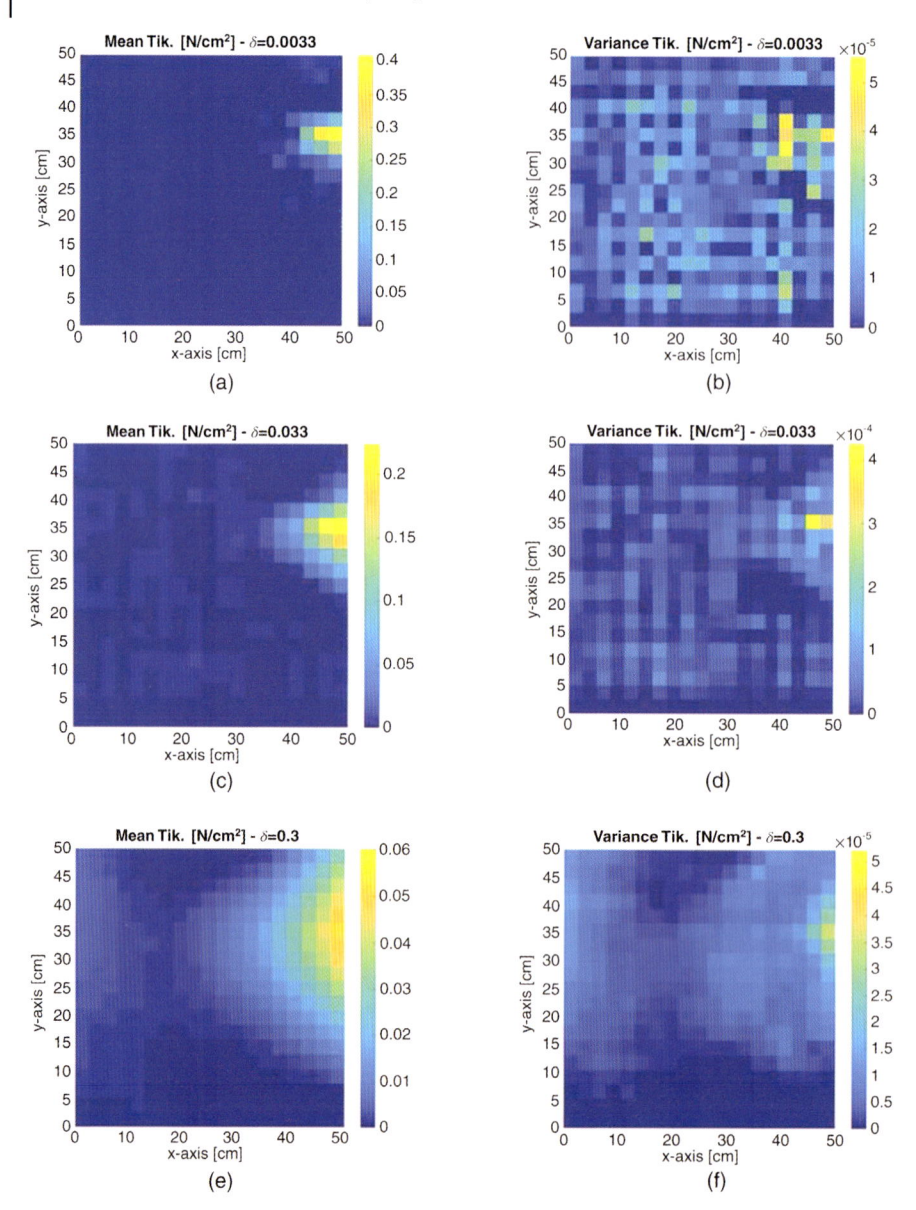

Figure 12.5 Reconstruction of f_2 by Tikhonov regularization with discrepancy principle for different relative noise levels. Reconstructions in [N/cm^2], units of **x**- and **y**-axes are cm. Noise levels $\delta = 0.0033$ (a and b), $\delta = 0.033$ (c and d), and $\delta = 0.3$ (bottom row). (a, c, and e) Mean reconstruction of 100 reconstructions from independent noisy data sets. (b, d, and f) Variance of the 100 reconstructions.

regularization schemes are so small that it is actually unnecessary to plot to best or the worst reconstruction, as those agree in the visual norm with the mean reconstruction. Furthermore, as one expects, the reconstruction quality degrades when the noise level is increased.

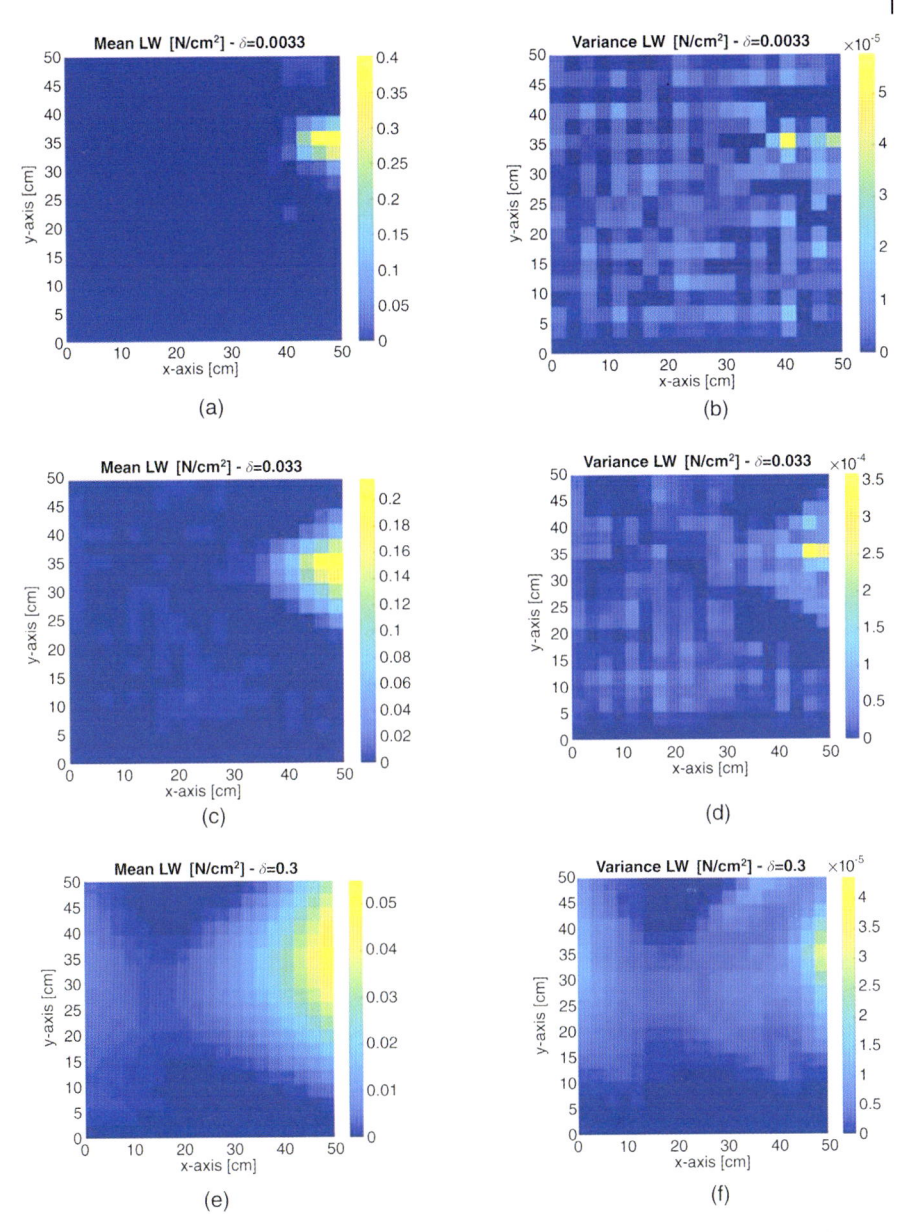

Figure 12.6 Reconstruction of f_2 by Landweber iteration with discrepancy principle for different relative noise levels. Reconstructions in [N/cm^2], units of **x**- and **y**-axes are cm. Noise levels $\delta = 0.0033$ (a and b), $\delta = 0.033$ (c and d), and $\delta = 0.3$ (e and f). (a, c, and e) Mean reconstruction of 100 reconstructions from independent noisy data sets. (b, d, and f) Variance of the 100 reconstructions.

For $\delta = 0.0033$, one notes, for instance, in Figures 12.3a and 12.4a the corner due to the nonconvex part of the original load in Figure 12.2c. When increasing the noise level to $\delta = 0.033$, this detail disappears and further increasing δ to 0.3, merely the location of the load is correctly identified. Furthermore, the peak values of the

Table 12.1 Mean computation times and variances for Tikhonov and Landweber iteration combined with the discrepancy principle (computed on an 2.4 GHz Intel Core 2 Duo processor).

Relative noise level	0.0033	0.01	0.033	0.1	0.3
Mean runtime Tikhonov regularization	1.43	1.09	1.01	0.965	0.858
Variances	0.0541	0.0001	0.0001	0.0001	0.0004
Mean runtime Landweber iteration	62.8	18.6	5.09	1.06	0.178
Variances	17.4	0.739	0.238	0.0219	0.0012

reconstructions considerably decrease for higher noise levels due to a smearing effect caused by the higher degree of regularization required to ensure stability.

Considering reconstructions of the second load f_2 from Figure 12.2d in Figures 12.5 and 12.6, the same conclusions on the reconstruction quality as for the load f_1 apply – variances of the reconstructions are generally rather small and the reconstruction quality decreases when the noise level increases.

Let us finally compare computation times for Tikhonov regularization and the Landweber iteration, both using the discrepancy principle for determining the regularization parameter or the stopping index of the scheme (12.12) and (12.28)). Table 12.1 shows that Landweber regularization should not be used for small noise levels: One iteration basically requires two matrix–vector multiplications such that the runtime essentially depends on the number of iterations required to satisfy the discrepancy principle. (There are far better iterative regularization methods for small noise level, such as the conjugate gradients method (see Ref. [35] or [1]).) As mentioned in the last section, this iteration number behaves like δ^{-2}, such that the Landweber iteration becomes infeasible for small noise level. On the other hand, for large noise levels 0.1 or 0.3, the Landweber iteration is an efficient regularization scheme that keeps up with or even outperforms Tikhonov regularization in terms of speed and reconstruction quality. The runtime of Tikhonov regularization is governed by the sequence α_n from (12.23), as that sequence determines how often one has to compute solutions $p_{\alpha_n}^\delta$ to the linear system (12.17). Note, however, that the conclusion to decrease runtime by choosing a rapidly decreasing sequence is misleading, due to increasing instability.

References

1 Engl, H.W., Hanke, M., and Neubauer, A. (1996) *Regularization of Inverse Problems*, Kluwer Academic Publishers, Dordrecht, The Netherlands.
2 Kirsch, A. (1996) *An Introduction to the Mathematical Theory of Inverse Problems*, Springer.
3 Mueller, J. and Siltanen, S. (2012) *Linear and Nonlinear Inverse Problems with Practical Applications. Computational Science and Engineering*, vol. 10, SIAM, Philadelphia, PA.

4 Schuster, T., Kaltenbacher, B., Hofmann, B., and Kazimierski, K.S. (2012) *Regularization Methods in Banach Spaces*, de Gruyter, Berlin.
5 Kaltenbacher, B., Neubauer, A., and Scherzer, O. (2008) *Iterative Regularization Methods for Nonlinear Ill-Posed Problems*, de Gruyter, Berlin.
6 Ciarlet, P.G. (1988) *Mathematical Elasticity. Vol. I: Three-Dimensional Elasticity*, North-Holland, Amsterdam.
7 Gould, P.L. (2013) *Introduction to Linear Elasticity*, Springer, New York.
8 Rudin, W. (1991) *Functional Analysis*, 2nd edn, McGraw-Hill.
9 Aubin, J.-P. (2000) *Applied Functional Analysis*, 2nd edn, John Wiley & Sons, Inc., New York.
10 Zha, H.Y. and Hansen, P.C. (1990) Regularization and the general Gauss–Markov linear model. *Mathematics of Computation*, **55**, 613–624.
11 Seidman, T.I. and Vogel, C.R. (1989) Well posedness and convergence of some regularization methods for non-linear ill posed problems. *Inverse Problems*, **5**, 227–238.
12 Engl, H.W., Kunisch, K., and Neubauer, A. (1989) Convergence rates for Tikhonov regularisation of non-linear ill-posed problems. *Inverse Problems*, **5**, 523–540.
13 Morozov, V.A. (1968) The error principle in the solution of operational equations by the regularization method. *USSR Computational Mathematics and Mathematical Physics*, **8**, 63–87.
14 Hansen, P.C. (1992) Analysis of discrete ill-posed problems by means of the l-curve. *SIAM Review*, **34** (4), 561–580.
15 Morozov, V.A. (1984) *Methods for Solving Incorrectly Posed Problems*, Springer, New York.
16 Wahba, G. (1990) *Spline Models for Observational Data*, SIAM.
17 Golub, G.H., Heath, M., and Wahba, G. (1979) Generalized cross-validation as a method for choosing a good ridge parameter. *Technometrics*, **21**, 215–223.
18 Rieder, A. (2003) *Keine Probleme mit Inversen Problemen*, 1st edn, Vieweg.
19 Kelley, C.T. (1995) *Iterative Methods for Linear and Nonlinear Equations*, SIAM, Philadelphia.
20 Daubechies, I., Defrise, M., and De Mol, C. (2004) An iterative thresholding algorithm for linear inverse problems with a sparsity constraint. *Communications on Pure and Applied Mathematics*, **57**, 1413–1457.
21 Hämarik, U., Palm, R., and Raus, T. (2009) On minimization strategies for choice of the regularization parameter in ill-posed problems. *Numerical Functional Analysis and Optimization*, **30**, 924–950.
22 Hämarik, U., Palm, R., and Raus, T. (2011) Comparison of parameter choices in regularization algorithms in case of different information about noise level. *Calcolo*, **48** (1), 47–59.
23 Hanke, M., Neubauer, A., and Scherzer, O. (1995) A convergence analysis of the Landweber iteration for nonlinear ill-posed problems. *Numerische Mathematik*, **72** (1), 21–37.
24 Armijo, L. (1966) Minimization of function having Lipschitz continuous first partial derivatives. *Pacific Journal of Mathematics*, **16**, 1–13.

25 Barzilai, J. and Borwein, J.M. (1988) Two-point step size gradient methods. *IMA Journal of Numerical Analysis*, **8**, 141–148.

26 Hanke, M. and Raus, T. (1996) A general heuristic for choosing the regularization parameter in ill-posed problems. *SIAM Journal on Scientific Computing*, **17**, 956–972.

27 Landweber, L. (1951) An iteration formula for Fredholm integral equations of the first kind. *American Journal of Mathematics*, **73**, 615–624.

28 Hanke, M. (1995) *Conjugate Gradient Type Methods for Ill-Posed Problems*, Pitman Reserch Notes in Mathematics, CRC Press.

29 Levenberg, K. (1944) A method for the solution of certain problems in least squares. *Quarterly of Applied Mathematics*, **2**, 164–168.

30 Marquardt, D. (1963) An algorithm for least-squares estimation of nonlinear parameters. *SIAM Journal on Applied Mathematics*, **11**, 431–441.

31 Hanke, M. (1997) A regularizing Levenberg–Marquardt scheme, with applications to inverse groundwater filtration problems. *Inverse Problems*, **13**, 79–95.

32 Lechleiter, A. and Rieder, A. (2009) Towards a general convergence theory for inexact Newton regularizations. *Numerische Mathematik*, **114**, 521–548.

33 Bakushinskii, A.B. (1992) The problem of the convergence of the iteratively regularized Gauss–Newton method. *Computational Mathematics and Mathematical Physics*, **32**, 1353–1359.

34 Blaschke, B., Neubauer, A., and Scherzer, O. (1997) On convergence rates for the iteratively regularized Gauss–Newton method. *IMA Journal of Numerical Analysis*, **17**, 421–436.

35 Bosse, S. and Lechleiter, A. (2014) Structural health and load monitoring with material-embedded sensor networks and self-organizing multi-agent systems. *Procedia Technology*, **15**, 669–691.

36 Bosse, S. and Lechleiter, A. (2015) A hybrid approach for structural monitoring with self-organizing multi-agent systems and inverse numerical methods in material-embedded sensor networks. *Mechatronics*, **34**, 12–37.

13

The Unknown World: Model-Free Computing and Machine Learning

Stefan Bosse[1,2]

[1]*University of Bremen, Department of Mathematics and Computer Science, Robert-Hooke-Str. 5, 28359 Bremen, Germany*
[2]*University of Bremen, ISIS Sensorial Materials Scientific Centre, Wiener Str. 12, 28359 Bremen, Germany*

13.1 Machine Learning – An Overview

An advantage of machine learning (ML) compared with numerical algorithms is the ability to handle problems with a nonpolynomial (NP) complexity class [1], that is, an exponential increase of computation time with respect to the data size. Basically, learning algorithms can be used to improve a system behavior either at runtime or at design time (or a hybrid approach of both) by optimizing function parameters to increase mainly estimation accuracy. Machine learning is based either on example data (training with labeled data) or on past experience with reward feedback retrieved at runtime of a system. The fields of application range from the optimizing and fitting of parameters of evaluation functions to full feature extraction classifiers. Machine learning is often used if there is no or only an incomplete world model specifying on behavioral or functional level of the output change of a module (the system or a part of it) in response to a change of the input stimulus.

A practically oriented classification of machine learning schemes (based on Ref. [2]), suitable for information extraction in sensor networks and supporting model-free information retrieval is given by the following list:

Supervised learning
  A process that learns a model, basically a function mapping input data (of the past) to an output data set using data comprising examples with an already given mapping, that is, using labeled data (see Figure 13.1). Two typical examples are classification and regression. Supervised learning produces a database requiring a high demand of storage resources, but the resulting learned model could be considerably small (e.g., a decision tree).

Unsupervised learning
  This is generally a process that seeks to learn structure of data (see Figure 13.1) in the absence of an identified output (like in supervised learning

Material-Integrated Intelligent Systems: Technology and Applications, First Edition.
Edited by Stefan Bosse, Dirk Lehmhus, Walter Lang, and Matthias Busse.
© 2018 Wiley-VCH Verlag GmbH & Co. KGaA. Published 2018 by Wiley-VCH Verlag GmbH & Co. KGaA.

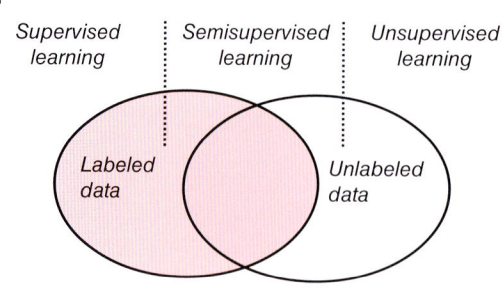

Figure 13.1 Transition from supervised to unsupervised learning.

with labeled data) or a feedback (reinforcement learning). Examples are clustering or self-organizing maps trying to group similar unlabeled data sets.

Semisupervised learning

This is the combination of supervised and unsupervised learning techniques creating a hybrid learning architecture (see Figure 13.1).

Reinforcement learning

In some situations, the output of a system (the impact of the environment in which the system operates) is a sequence of actions, and the sequence of correct actions is important, rather than one single action. Reinforcement learning seeks to learn a policy mapping to actions that optimize a received reward (the feedback), finally optimizing the behavior of a system (the reactivity). It is different from data classifiers and relates closely to the autonomous agent behavior model, discussed in Chapter 18. There are no trained example situations for correct or incorrect behavior, only rules evaluating and weighting actions and their impact on the environment and the system.

Association learning

The goal of association learning is to find conditional probabilities of association rules between different items of data sets. An association rule has the form $X \rightarrow Y$, where X and Y are item sets. The association rule belongs to a conditional probability $P(Y|X)$ giving the probability that if X occurs, then set Y is also likely to occur. Giving user-specified support and confidence thresholds, the *a priori* algorithm developed by Agrawal and Shafer [3] can find all association rules between two sets X and Y. This proposed algorithm can be parallelized.

Classification Classifier systems can be considered as modeling tools. Given a real system without a known underlying dynamics model, a classifier system can be used to generate a behavior that matches the real system. The classifier offers rules-based model of the unknown system. There is a very large number of classification algorithms, for example, commonly used, decision trees (i.e., C4.5 algorithm), instance-based learners (i. e., nearest-neighbor (NN) methods), support vector machines (basically linear classifiers), rule-based learners, neural networks, and Bayesian networks. The main purposes of classification in sensor networks are knowledge extraction and pattern recognition. For example, in Ref. [4], C4.5 and k-NN algorithms were used to classify a 18-dimensional sensor vector of a flat rubber plate to a (F,X) vector providing a strength classification of an applied load (F) with an estimation of the spatial position (X).

Decision trees as one outcome of a machine learning classifier have low computational resource requirements and can be implemented directly on microchip level, for example, used for energy management in Ref. [5].

Clustering Clustering is basically an unsupervised learning with the goal to group data sets sharing similar characteristics in clusters. The main goal is finding structure in the given set of data. A common clustering algorithm is the k-mean algorithm, which quantifies vectors and computes the distance between vectors, finally grouping nearest vectors in cluster segments (see, for example, Ref. [6],). A self-organizing map is another well-known clustering algorithm.

Regression Like classification, regression is a supervised learning approach. The goal is to learn an approximation of a real-valued function mapping an input data vector (real-valued input variables) to an output vector (the mean of response variables).

Supervised machine learning and *statistical pattern recognition* can be used to compute the response of a technical structure due to load or damage situations based only on the measured sensor data [7]. Machine learning-based *classification* can be used in the information levels 0 to 3 of Figure 11.1, but requires damage or load quantification to enable classification. Figueiredo *et al.* [8] proposed machine learning methods for the damage detection under varying operational and environmental conditions using a hybrid approach with neural network algorithms. *Neural network algorithms* can be used in conjunction with self-organizing multiagent systems and offering limited computational and data complexity promising microchip level implementations.

Unsupervised learning can be used for novelty detection. The algorithm simply indicates if the data come from a normal operating condition or not (something is wrong). *Regression* algorithms provide the output of several continuous variables, that is, the diagnosis outputs the Cartesian coordinates of the fault, or the length of a crack. The regression problem can be nonlinear and is suitable for neural networks [7].

Machine learning is well suited for the analysis process and the feature extraction (i.e., the damage estimation of a technical structure) in SHM applications [9].

13.2 Learning of Data Streams

Commonly, supervised machine learning consists of two phases: the learning and the analysis (predictive classification). The learning phase collects a number of experimental or simulated data sets assigned to a label vector given by the supervisor, that is, a training data set (X, y) is acquired off-line, mapping input data X on labels y. After the classifier was trained, real-time data can be processed, that is, at runtime, the classifier is fed by data sampled at runtime. Incremental learning adds new training data sets (X', y') at runtime, trying to improve the classifier results, though additional training sets can degrade the classification results, too. The deployment of ML in stream-based data processing requires incremental learning machines. In sensing applications, the sensor data can be

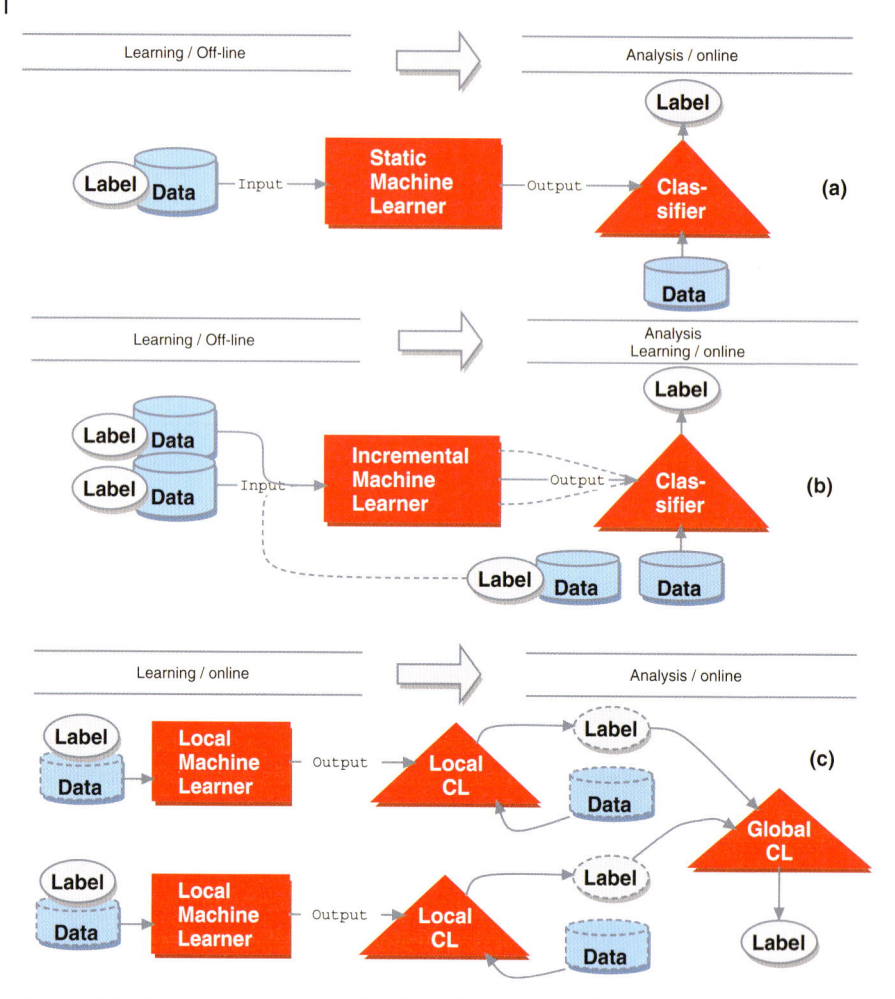

Figure 13.2 Comparison of (a) static, (b) dynamic incremental learning, and (c) distributed learning.

continuously over time and some kind of incremental and adaptive learning is required. There are different ML algorithms suitable for incremental learning, for example, decision trees (DT) or neural networks (NN), overlapping and interleaving learning and analysis phases (see Figure 13.2b).

For example, support vector machines (SVM) [10,11], NN, or decision tree classifiers are used in incremental learning environments, combined with agent-based distributed learning [12]. But SVMs are basically linear classifiers not suitable for high-dimensional data- and label vectors like they are present in sensing applications. It is important in incremental learning that the learner not relies on the storage of old data sets already learned with a full recomputation, that is, the learning algorithm must avoid relearning of the entire data set each time an iteration step is performed. Decision tree and rule (DR) learners seem to be the most promising ML methods for real-time stream-based learning. They can be learned on sequential scans. DTs are robust to errors and drift in the data, though some effort is required to perform incremental learning efficiently with low computational power using, for example, very fast decision tree (VFDT) learning applied to data streams [13].

13.3 Learning with Noise

One major issue in learning a classification model or performing clustering is the uncertainty of sensor data, that is, the random noise supposed to the original sensor data. Noise is always present, related either to a statistical variation (based on a physical effect) or being sporadic based mostly on failures of the sensor or electronics. Noise can significantly reduce the prediction quality and accuracy of a learned model. Different learning algorithms have a different sensitivity to noise. Fuzzy learning by adding uncertainty intervals to data can improve the classification stability significantly, shown in Section 13.5.

13.4 Distributed Event-Based Learning

Distributed learning divides a spatial distributed data set in clustered regions and applies learning to the limited local regions, based on a divide and conquer approach; that is, there are multiple learning instances for learning a local model (using spatially bounded data) that is finally merged in a global model, shown in Figure 13.2c. Event-based learning activates learners if there is a local stimulus detected. Decision trees are simple models derived from learning with training data, and well suited for agent-based learning. A learned model is used to map data set vectors on class values. The tree consists of nodes testing a specific attribute variable, that is, a particular sensor value, creating a path to the leaves of the tree containing the classification result, for example, the load situation class. Among the distribution of the entire learning problem, event-based activation of learning entities can improve the system efficiency significantly. Commonly, the locally sampled sensor values are used for an event prediction.

The event-based regional learning leads to a set of classification results, which can differ significantly, that is, the classification set can contain wrong predictions. To filter out and suppress these wrong predictions, a global major vote election is applied. All nodes having performed a regional classification send their result to the network collecting all votes and perform an election. This election result is finally used for the global prediction. The variance of different votes can be an indicator for the trust of the election giving the right prediction.

The following Eq. (13.1) summarizes the spatial distribution of learning. The global model learner M is divided in multiple local learner m, computing a local model k based on local data d, a tuple set consisting of input data s (e.g., sensor data) and classification labels l. The local classification function k is applied to unknown local data s predicting a specific label l, finally merged in a global prediction.

$$
\begin{aligned}
&M : D \to K(S) \qquad\qquad\qquad m_{i,j} : d_{i,j} \to k_{i,j}(s) \\
&l \in L \qquad\qquad\qquad\qquad\quad k_{i,j} : s_{i,j} \to l_{i,j} \\
&K : S \to l \qquad\qquad\qquad\qquad K : (l_{1,1}, l_{1,2}, \cdots) \to l \\
&D : \{(S^1, l^1), (S^2, l^2), \cdots\} \xrightarrow{\text{Distribution}} d_{i,j} : \{(s^1_{i,j}, l^1), (s^2_{i,j}, l^2), \cdots\} \\
&S : \begin{pmatrix} x_{1,1} & \cdots & x_{n,1} \\ \vdots & \ddots & \vdots \\ x_{1,m} & \cdots & x_{n,m} \end{pmatrix} \qquad s_{i,j} : \begin{pmatrix} x_{i-u,j-v} & \cdots & x_{i+u,j-v} \\ \vdots & \ddots & \vdots \\ x_{i+u,j-v} & \cdots & x_{i+u,j+v} \end{pmatrix}
\end{aligned}
$$

(13.1)

13.5 ε-Interval and Nearest-Neighborhood Decision Tree Learning

Traditional decision tree learner (DTL) (e.g., using the C4.5 algorithm) select data set attributes (feature variables) for decision-making only based on information–theoretic entropy calculation to determine the impurity of training set columns (i.e., the gain), which is well suited for nonmetric symbolic attribute values such as color names, shapes, and so on. The distinction probability of two different symbols is usually 1. Numerical sensor data is noisy and underlies variations due to the measuring process and the physical world. Two numeric (sensor) values a and b have only a high distinction probability if the uncertainty intervals $[a - \sigma, a + \sigma]$ and $[b - \sigma, b + \sigma]$ due not overlap. That means, not only the entropy of a data set column is relevant for numerical data, the standard deviation σ and value spreading of a specific column must be considered, too. To improve attribute selection for optimized data set splitting, a column ε-interval entropy computation was introduced that extends each value of a column vector with an uncertainty interval $[v_i - \varepsilon, v_i + \varepsilon]$. Values with overlapping intervals are considered to be nondistinguishable, lowering the entropy (with $\lfloor x$: lower bound of a value/interval, $\lceil x$: upper bound, and $|v|$ as the size of a vector), with the computation given by Eq. (13.2).

$$
\begin{aligned}
\text{entropy}_\varepsilon(\text{cols}, \varepsilon) &= \sum_{i=1\ldots|\text{cols}|} -\text{prob}_\varepsilon(\text{col}_i, \text{cols}, \varepsilon) \log_2(\text{prob}_\varepsilon(\text{col}_i, \text{cols}, \varepsilon)), \\
\text{prob}_\varepsilon(v, \text{cols}, \varepsilon) &= \frac{\sum_{i=1\ldots|\text{cols}|} \begin{cases} 0 : \text{overlap}([\text{cols}_i - \varepsilon, \text{cols}_i + \varepsilon], [v - \varepsilon, v + \varepsilon]) \\ 1 : \text{otherwise} \end{cases}}{|\text{cols}|}, \\
\text{overlap}(v_1, v_2) &= \begin{cases} \text{true} : (\lceil v_1 \rceil \geq \lfloor v_2 \rfloor \wedge \lceil v_1 \rceil \leq \lceil v_2 \rceil) \vee \\ \qquad\quad (\lceil v_2 \rceil \geq \lfloor v_1 \rfloor \wedge \lceil v_2 \rceil \leq \lceil v_1 \rceil) \\ \text{false} : \text{otherwise} \end{cases}, \\
\text{distance}(v_1, v_2) &= \left| \frac{\lfloor v_1 \rfloor + \lceil v_1 \rceil}{2} - \frac{\lfloor v_2 \rfloor + \lceil v_2 \rceil}{2} \right|.
\end{aligned}
$$

(13.2)

13.5 ε-Interval and Nearest-Neighborhood Decision Tree Learning

The ε-entropy is calculated for all data set columns, and the attribute (feature variable) for the column with highest entropy value is selected. The column can still contain nondistinguishable values with overlapping 2ε intervals. All overlapping 2ε values are grouped in partitions that cannot be classified (separated) by the currently selected attribute variable. Only partitions – ideally containing only one data set value – are used for a classification selection. All data sets in one partition create a subtree of the current decision tree node. If there is only one partition available (containing more than one class target), a data set attribute selection is based on the column with the highest standard deviation, but the 2ε separation cannot be guaranteed in this case, lowering the prediction accuracy. The basic principle of the learning algorithm, which is an adaptation of a common discrete C4.5 decision tree learner, is shown in Algorithm 13.1. It creates a model with attribute value interval selection, e.g., $x \in [500 \ldots 540]$, instead the commonly used and simplified relational value selection, e.g., $x<540$, which is an inadmissible extrapolation beyond the training set boundaries and prevent recognizing totally nonmatching data.

Algorithm 13.1   Principle Learning and Classification Algorithms

The entropy computation applying the 2ε interval to values is shown in Eq. (13.2).

```
type value = number | number range
The learned model is a decision tree with nodes and leaves
type model = Result (name: string) |
             Feature (name:string, featvals: model array) |
             Feature Value(val: value, child: model)
function createTree(datasets, target, features)
  1. Select all columns in the data set array
     with the target key
  2. If there is only one column, return a result leaf
     node with the target
  3. Determine the best features by applying entropy
     and value deviation computation
  4. Select the best feature by maximal entropy
  5. Cretae partitions from all possible column values
     for this feature
  6. If there is only one partition holding all values,
     go to step 10
  7. For each partition create a child feature value node
  8. For each child node apply the createTree function with the
     remaining reduced data set by filtering all data rows
     containing at least one value of the partition in the
     respective feature column of the data set, and by
     using a reduced remaining feature set w/o the
     current feature
  9. Return a feature node with previously created feature
     value child nodes.
     Finished.
```

```
10. Select the best feature by maximal value deviation
11. For each possible value create a feature value node
12. For each child node apply the createTree function
    with the remaining reduced data set by filtering all
    data rows containing at least one value of the
    partition in the respective feature column of the data
    set, and by using a reduced remaining feature set w/o
    the current feature
13. Return a feature node with previously created
    feature value child nodes.
    Finished.
end
function classify (model,dataset)
  I. Iterate the model tree until a result leaf is found.
  II. Evaluate a feature node by finding the best matching
      feature value node for the current feature attribute
      by finding the feature value with minimal distance to
      the current sample value from the data set.
end
```

The prediction (analysis and classification) algorithm is a hybrid approach, too. It consists of the tree iteration, but uses a simple nearest-neighborhood estimation for selecting the best matching feature value with a given sample sensor value.

The learned DT is composed of result leaves and feature/feature value selection nodes. A learned DT model can be easily transformed in a table representation, enabling the implementation of learned DT models on hardware level using simple linear lookup tables (LUT), illustrated in Figure 13.3.

13.6 Machine Learning – A Sensorial Material Demonstrator

A first attempt to investigate new smart materials was performed with experiments using a functional mock-up consisting of a flat rubber plate (equipped with nine biaxial strain-gauge sensors and the sensor network previously introduced, 70 mm sensor distance) and an experimental setup shown in Figure 13.4 with circular weights.

Figures 13.4 and 13.5 show the analysis of the difference between measured and predicted load positions (position accuracy) retrieved by machine learning with two different sensor array configurations. The plots show the spatial vector difference between the predicted and observed position with a mean value below 25 and 50 mm, respectively. The training set consisted of two different masses (103 and 306 g) and 150 positions. During learning mode, the position (of the weight) was monitored with a camera mounted above the rubber plate together with acquired sensor data delivered by the sensor network.

A system with reduced number of sensors (Figure 13.5, 100–150 mm sensor distance) results in a decrease of prediction position accuracy in the boundary

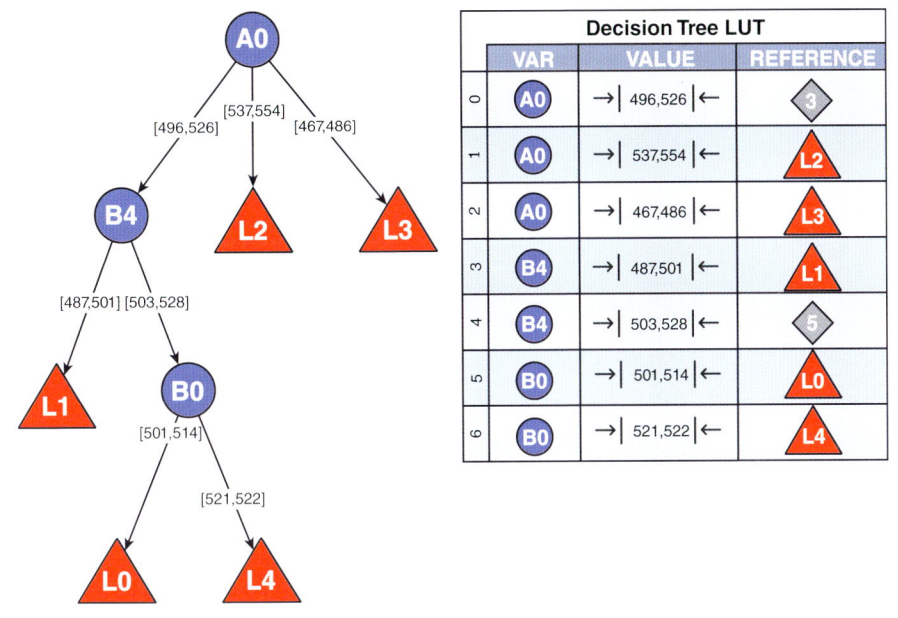

Figure 13.3 A learned decision tree and its linear lookup table representation (A_i: Sensor A on node i, B_i: Sensor B on node i, L_i: Load case i).

region, but is still usable for structure load monitoring, especially in the middle area spawned by the edges of sensors.

The sensor network attached below the rubber plate consisted of nine single sensor nodes, each equipped with analog–digital conversions and digital processing units (FPGA). Each sensor node was attached to one biaxial-aligned strain-gage sensor pair. The sensor nodes were arranged in two-dimensional network, shown in Figure 13.6.

The sensor data was collected periodically from the network of smart sensor nodes, which already performed the signal preprocessing. The sensor data was finally processed by a computer to derive load information (interpolated position and load strength classification) by using supervised machine learning. Different machine learning methods were investigated, showing different quality and stability results of the predicted loads, and differing in the required number of training experiments to reach a certain level of quality and robustness [14]:

- k-Nearest-neighbor → Numerical regression of the load position, mass, and displacement vectors
- C4.5 Decision trees → Mass classification
- Neural networks → Numerical regression of load position and mass

Finally, the k-NN classificator showed the best overall results. The original proposed approach included FEM simulation to enable simulated virtual training of the ML classifier, finally used in the load classification with data from the real sensor network sampled at runtime, can be found in Ref. [15].

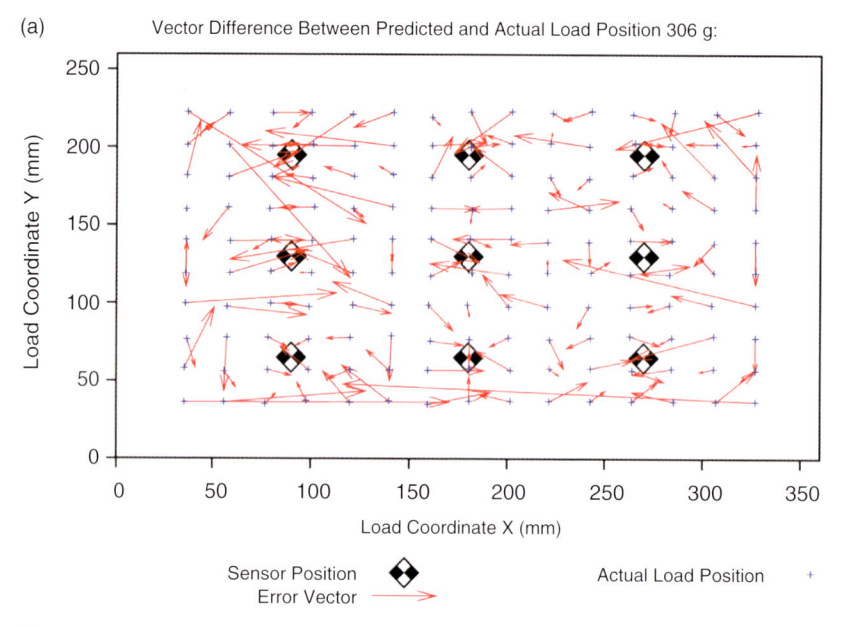

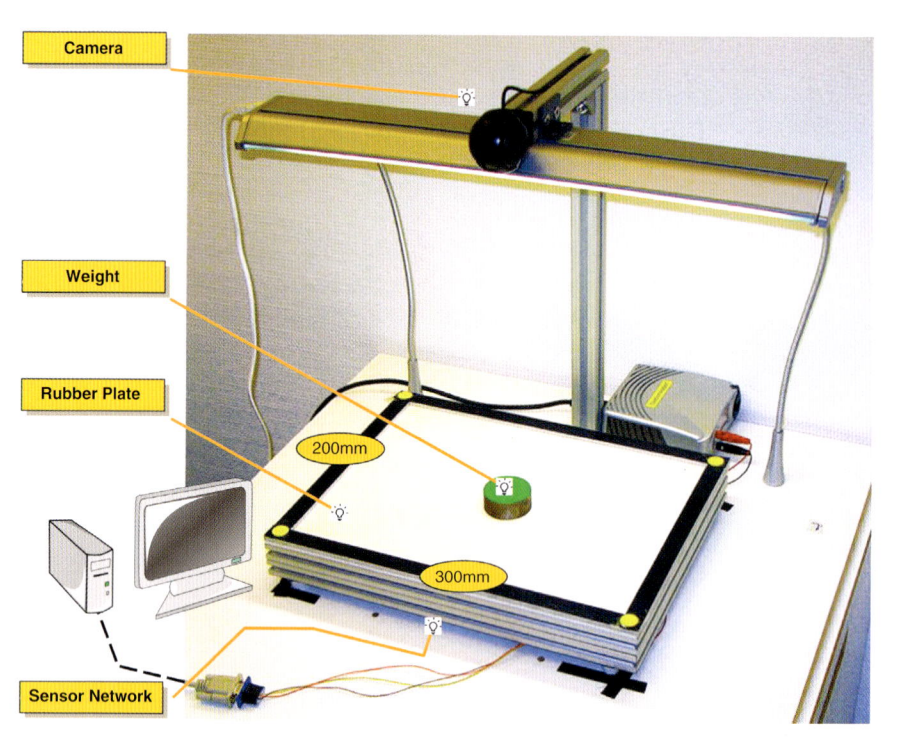

Figure 13.4 Experimental results of predicted load positions (306 g weight) with nine strain-gauge sensor pairs mounted on backside of a rubber plate (experimental test setup shown in part b) [14].

13.6 Machine Learning – A Sensorial Material Demonstrator

Vector Difference Between Predicted and Actual Load Position for 306 g:

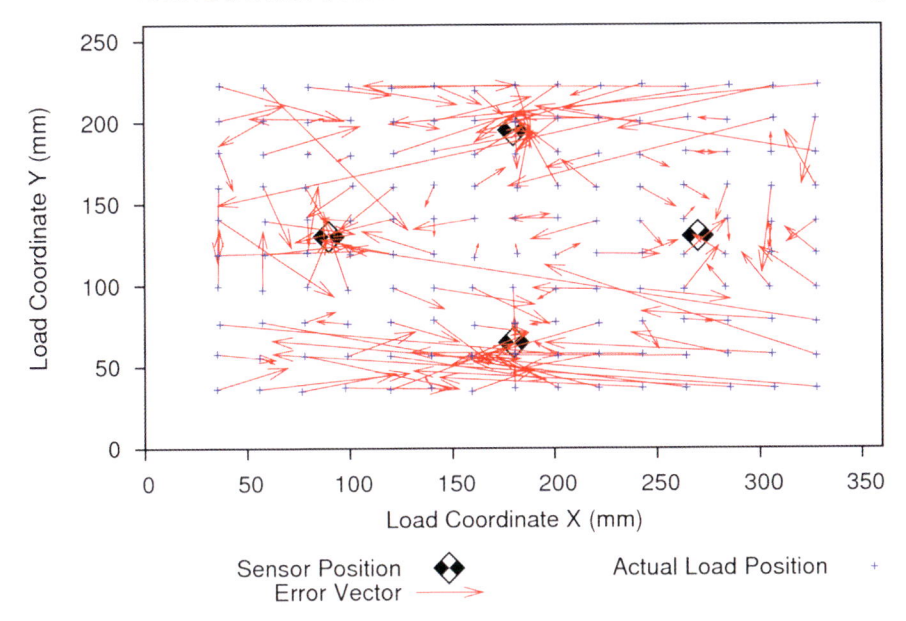

Sensor Position ◆   Actual Load Position +
Error Vector →

ISIS Demonstrator

1. Place one weight at a time on the rubber sheet.
2. Choose **"Memorize Load Case"**.
3. Repeat this procedure several times with different weights and positions.
4. Switch to **"Query Mode"** when done.

Automatic load case capturing is active: Current and memorized load cases:

You are in Training Mode.   [Memorize Load Case]   [Query Mode]

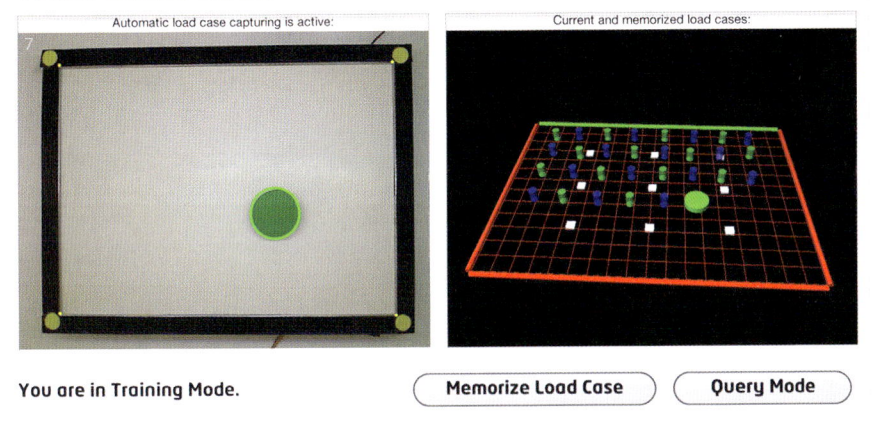

Figure 13.5 Experimental results of predicted load positions (306 g weight) with only four strain-gauge sensor pairs and ML graphical user interface (GUI, right side) [14].

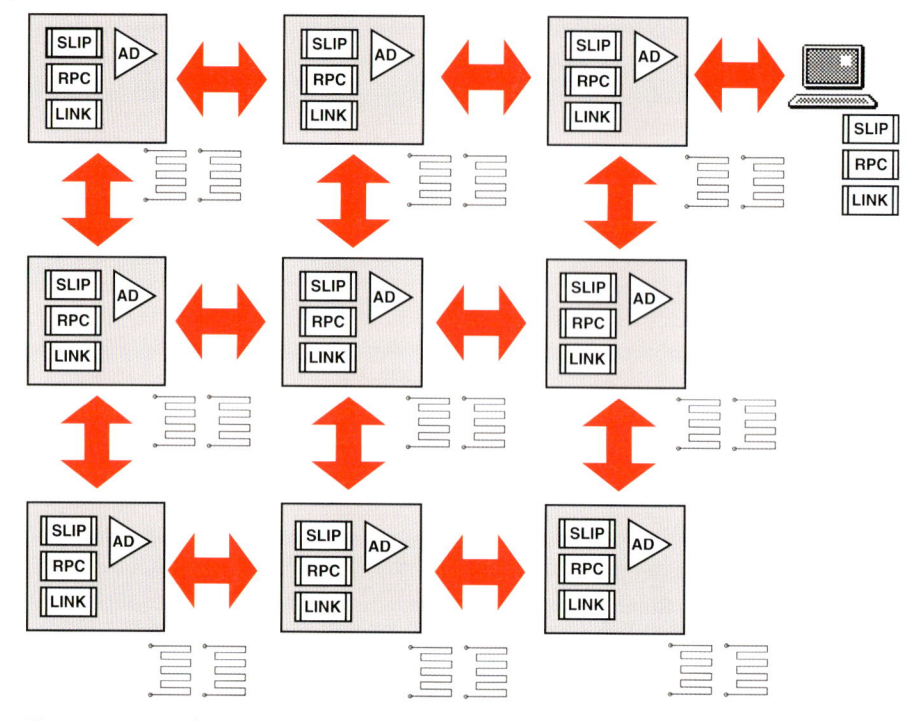

Figure 13.6 Two-dimensional sensor network with autonomous sensor nodes attached to a computer performing advanced data analysis using machine learning. Each smart sensor node consists of signal and digital data processing and communication.

Messaging was performed in the sensor network by using an adaptive Δ-distance routing protocol (SLIP) based on backtracking, which will be discussed in Chapter 16. This adaptive routing scheme enables the delivery of messages even in the case of incomplete (missing communication connections) and irregular (missing or down nodes in the two-dimensional mesh grid) networks.

References

1 Wuest, T., Weimer, D., Irgens, C., and Thoben, K.-D. (2016) Machine learning in manufacturing: advantages, challenges, and applications. *Production & Manufacturing Research*, **4** (1), 23–45.
2 Sammut, C. and Webb, G.I. (eds) (2011) *Encyclopdia of Machine Learning*, Springer.
3 Agrawal, R. and Shafer, J.C. (1996) Parallel mining of association rules. *IEEE Transactions on Knowledge and Data Engineering*, **8** (6), 962–969.
4 Bosse, S., Pantke, F., and Edelkamp, S. (2013) *Robot manipulator with emergent behaviour supported by a smart sensorial material and agent systems*. Proceedings of the Smart Systems Integration 2013, Amsterdam, NL.
5 Bosse, S. and Behrmann, T. (2011) Smart energy management and low-power design of sensor and actuator nodes on algorithmic level for self-powered

sensorial materials and robotics. Proceedings of the SPIE Microtechnologies 2011 Conference, Prague, Session EMT 101 Smart Sensors, Actuators and MEMS. doi: 10.1117/12.888124

6 Bell, J. (2015) *Machine Learning – Hands-On for Developers and Technical Professionals*, John Wiley & Sons, Ltd, Chichester.

7 Farrar, C.R. and Worden, K. (2013) *Structural Health Monitoring: A Machine Laerning Perspective*, John Wiley & Sons, Ltd, Chichester.

8 Figueiredo, E., Park, G., Farrar, C.R., Worden, K., and Figueiras, J.A. (2010) Machine learning algorithms for damage detection under operational and environmental variability. *Structural Health Monitoring*, **10** (6), 559–572.

9 Worden, K. and Manson, G. (2007) The application of machine learning to structural health monitoring. *Philosophical Transactions of the Royal Society A*, **365**, 515–537.

10 Laskov, P., Gehl, C., and Krüger, S. (2006) Incremental support vector learning: analysis, implementation and applications. *Journal of Machine Learning Research*, **7**, 1909–1936.

11 Ralaivola, L. and D'Alché-Buc, F. (2001) Incremental support vector machine learning: a local approach, in *Artificial Neural Networks — ICANN 2001, Georg Dorffner* (eds H. Bischof and K. Hornik), Springer, pp. 322–330.

12 Choi, J., Oh, S., and Horowitz, R. (2009) Distributed learning and cooperative control for multi-agent systems. *Automatica*, **45** (12), 2802–2814.

13 Gama, J. and Kosina, P. (2011) Learning decision rules from data streams, in Proceedings of the Twenty-Second International Joint Conference on Artificial Intelligence.

14 Pantke, F., Bosse, S., Lehmhus, D., Lawo, M., and Busse, M. (2011) *Combining simulation and machine-learning for real-time load identification in sensorial materials*. 2nd International Conference on Simulations in Bio-Science and Multiphysics (SimBio-M 2011), Marseille/France.

15 Pantke, F., Bosse, S., Lehmhus, D., and Lawo, M. (2011) *An artificial intelligence approach towards sensorial materials*. Proceedings of the Future Computing 2011 Conference.

14

Robustness and Data Fusion

Stefan Bosse[1,2]

[1]University of Bremen, Department of Mathematics and Computer Science, Robert-Hooke-Str. 5, 28359 Bremen
[2]University of Bremen, ISIS Sensorial Materials Scientific Centre, Wiener Str. 12, 28359 Bremen, Germany

Reliability is a key factor in the design and operation of distributed sensor networks integrated in materials and technical structures, which in current works is often not considered in a rigorous way on system design level. Failures in DSN can be categorized as follows:

1) Sensor failures
2) Communication link failures
3) Data processing failures with different sublevels
4) Algorithmic design errors
5) Race conditions, dead-locks, and live-locks in concurrent systems.

This study is not concerned with the various sources of physical failure nor with their relative importance. Instead, we will discuss potential remedies in case of failure as well as measures that help maintain functionality even after failure has occurred.

Process failures (of the operational units or the sensor node itself) can be arbitrary (i.e., due to electronic failures), part of a crash with possible recovery, a crash without recovery, or related to omissions (not all operations of a process were performed successfully or finished) [1]. Arbitrary faults are the most difficult to handle, and can be a result of a wrong design on algorithmic (programming) level. Omission faults are mainly related to incomplete communication (e.g., a process is not responding to a communication request). In contrast, process crashes can be detected and "healed" with a process recovery, either restoring the state of the process (consisting of a control and data state) to an older backup state or just by restarting the process (restoring the initial state).

Communication failures and robustness aspects are discussed in Section 16.8.

There are different approaches to achieve programming fault tolerance. Basically, there are forward and backward error recovery (assuming the correct operation of the host PE of a process). In backward recovery strategies, an older backed-up process state (consisting of the data and control state) is restored if an error occurs. Forward recovery tries to correct the wrong behavior of the process more precisely by identifying the error and modifying the system state containing the error. Exception handling provided on programming level by high-level software languages like Java, JavaScript, ML, or ADA, or by high-level hardware SoC design languages like ConPro [2] can help to handle and to detect run-time errors in a deterministic and efficient way.

If errors cannot be safely identified, which is always the case in defective PEs, voting-based approaches can be used. Redundancy can be provided by executing the same computation by different PEs (or at least processes) and performing a major voting. N-version programming is a similar but more powerful and robust approach. In this approach, the computation is performed by different algorithms executed in different processes with a final major voting [3]. But redundant computation always increases the resource demands significantly (for storage, operations, and communication).

Communication failures can be compensated with network topologies providing alternatives for path routing leading to path redundancy, discussed in Section 16.2.

Sensor fusion (also known as multisensor data fusion) is another well-known technique to increase the robustness and confidence of a sensor signal processing system by using multiple sensor outputs to improve a decision-making process. Sensor fusion can be used to

- improve the overall signal-to-noise ratio;
- increase the robustness and reliability;
- extend the coverage to supply a more complete picture of the environment and system under perception;
- improve the resolution of a measurement;
- increase the confidence of sensor information and decision-making processes; and
- reduce the measurement time.

It is a process of combining observations from different sensors. A multisensor data fusion system can be seen as a distributed system of autonomous modules, well fitting in the concept of distributed sensor networks.

The fusion process can be classified in different levels [4]:

- Raw sensor-level fusion before any preprocessing
- Feature-level fusion after sensor signal preprocessing
- Pattern-level fusion combining feature vectors
- Decision-level fusion combining multiple decisions or classifications.

There are different fusion architectures, shown in Figure 14.1. The single sensor processing architecture (a) has one fully expanded processing chain for each sensor, finally producing multiple results. Each processing chain consists of a sensor preprocessing stage, feature extraction (or information computation), postprocessing, pattern recognition, and finally the decision-making stage. A centralized sensor fusion architecture (b) has multiple sensor preprocessing chains combined in a centralized fusion stage (raw sensor-level fusion). A

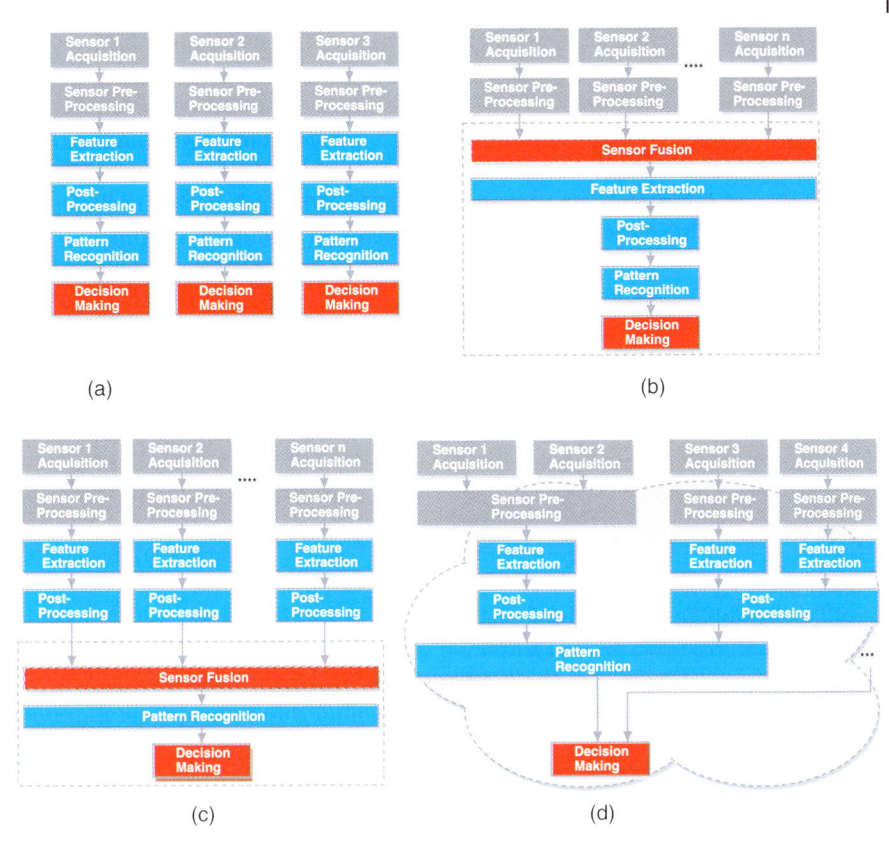

Figure 14.1 Different strategies in sensor data fusion and fusion architectures. (Adapted from Ref. [4].) (a) Single-sensor processing. (b) Centralized sensor fusion. (c) Distributed sensor fusion. (d) Heterogeneous and hierarchical sensor fusion architecture.

distributed sensor fusion architecture (c) performs the sensor preprocessing, feature extraction, and the postprocessing distributed, and finally the postprocessed data are combined in a centralized fusion stage. A heterogeneous and hierarchical architecture (d) is composed of a tree with multiple distributed processing and combination stages with one centralized decision-making stage.

14.1 Robust System Design on System Level

Serial and parallel system composition are the most common basic structures in data processing system design, which can be applied on software and hardware architecture level, illustrated in Figure 14.2. Data processing nodes in a mesh-like sensor network can be considered as a parallel composition. Furthermore, parallel data processing on microchip level can be easily exploited by using multi Register Transfer architectures [2]. On the other hand, a chain or bus network topology is a serial composition of initially independent systems. The functional composition

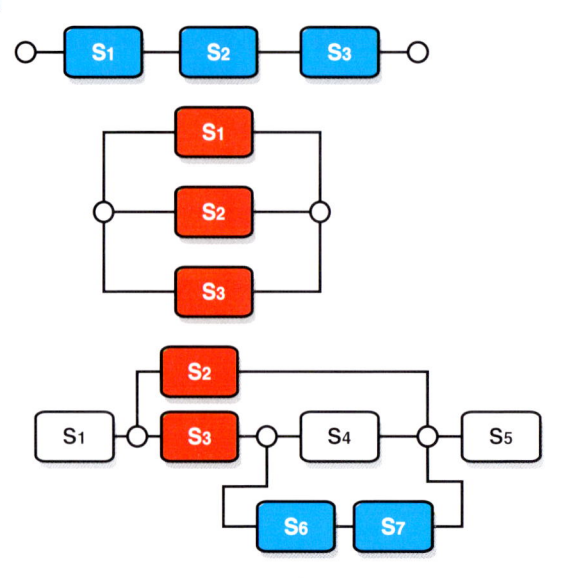

Figure 14.2 Serial, parallel, and mixed system designs.

of programs and algorithms leads to a serial system behavior, too, without providing any robustness against partial failures.

The reliability R (probability to operate without a failure at certain time t) and the mean time to failure ($MTTF$) of such a serial system compared with a parallel system is given by Eq. (14.1a) and 14.1b (λ is the failure rate of a module, one part of the system) [5]:

$$R_{\text{serial}}(t) = \prod_{i=1}^{N} R_i(t), \quad \text{MTTF}_{\text{serial}} = \frac{1}{\sum \lambda_i} = \frac{1}{\lambda_s}. \tag{14.1a}$$

$$R_{\text{parallel}}(t) = 1 - \prod_{i=1}^{N}(1 - R_i(t)), \quad \text{MTTF}_{\text{parallel}} = \sum \frac{1}{k\lambda_i}. \tag{14.1b}$$

$$R_{\text{serial+parallel}}(t) \leq 1 - \prod_{i=1}^{N}(1 - R_{\text{path},i}(t)). \tag{14.1c}$$

This means the failure of a serial system is an accumulation of all independent failures R_i of one module with a resulting constant failure λ_s rate, whereas a parallel system is an accumulation of liveliness with a nonconstant failure rate decreasing with each failure of a part. The reliability (or $MTTF$) of mixed systems consisting of weaved serial and parallel structures is difficult to calculate. An upper bound (Eq. (14.1c)) of the overall system reliability can be given, assuming a decomposition of the system in serial paths operating parallel (and independently) with a single reliability R_{path}. This implies that in a mixed system some modules can be removed without compromising the overall system operation.

M-of-N Systems Majority voting can be used to isolate modules producing faulty results, which can be implemented with agents (see Chapter 18). A major voting assumes that at least M-of-N independent modules, performing the same computation, are working properly. This technique offers fault tolerance by redundancy. If $R(t)$ is the reliability of one independent module (the probability of being operational at time t), then the reliability of the whole system is (Eq. (14.1a)) [5]

$$R_{M-of-N}(t) = \sum_{i=M}^{N} \binom{N}{i} R^i(t)(1 - R(t))^{N-i}, \qquad (14.2)$$
$$R_{M-of-N}^{corr}(t) = (1 - Q_{corr})R_{M-of-N}(t).$$

If there is even a slight correlation Q (e.g., a probability for a common fault) between the individual modules, then the overall system reliability decreases dramatically (Eq. (14.1b)), forcing a proper system design and the identification of correlation effects, for example, practically speaking, in the case of sensor networks there are shared resources like power supplies or communication structures that can fail! A wrong computation result must be considered as a partial fault of a module.

In addition to the previously described static redundancy based on voting and employing a large number of independent modules at the same time, dynamic redundancy that enable spare modules in the presence of detected faults of one actually active module can improve system efficiency (power, communication, utilization of processing elements). Dynamic redundancy is obtained by using fault detection and reconfiguration. If there is only one active module at the same time, the estimation and detection of failures or incorrect result is critical, decreasing the system reliability. The system reliability can be increased by using a hybrid approach of voting (static redundancy) and fault detection (dynamic redundancy) [5].

Autonomous agent systems can reduce computational correlation effects on system level significantly. Replication and reconfiguration features of agents can offer efficient and dynamic M-of-N system implementations.

N-Version Systems These can be considered as a more general approach of M-of-N system designs. In an N-version system, the same computation is performed by different algorithms or implementations of algorithms, increasing the independence of parallel computing modules significantly. This approach covers programming and implementation errors at design time, too, in contrast to an M-of-N system with identical modules, which all can have the same design error and produce faulty results under some (rare) circumstances, basically depending on the input data. Version independence is important; otherwise, correlated data processing errors can decrease the system reliability significantly (as already discussed with M-of-N systems).

Distributed Neighborhood Detection Failures of single sensors are hard to distinguish from random measuring errors. A well-known technique is data fusion of different independent sensor signals detected in a bounded neighborhood region around a center node, assuming nearby sensors measuring similar signals and the measurements are stochastically noncorrelated. This technique can improve fault tolerance by detecting defective modules (sensors) and can increase the quality-of-service for signal monitoring significantly, as suggested in Ref. [6].

Dynamic Self-Organizing Systems   In contrast to computational and communication robustness related often to a particular subsystem or sensor node, the liveliness of the entire systems in presence of single and multiple failures on different levels is more important. The system liveliness is mostly defined by the goals of a system, rarely on the state of components composing the entire system. For example, it is pointless if there is a node or senor failure in a spatially distributed strain gauge network that does not effect the overall load computation given by a specific numeric or machine learning mapping function. But it is difficult to predict the overall system behavior from particular failures. Due to these difficulties, self-organizing systems (SoS), commonly using multiagent systems, can overcome this lack of a complete failure model. An SoS is characterized by some degree of autonomy of individual parts, adaptability, and reconfiguration of system and subsystem configurations, for example, proposed in Ref. [7] using an agent-based SoS for feature detection in a load monitoring network.

References

1. Guerraoui, R. and Rodigues, L. (2006) *Introduction to Reliable Distributed Programming*, Springer.
2. Bosse, S. (2011) Hardware-software-co-design of parallel and distributed systems using a unique behavioural programming and multi-process model with high-level synthesis. Proceedings of the SPIE Microtechnologies 2011 Conference, April 18–20, 2011, Session EMT 102 VLSI Circuits and Systems, Prague.
3. Wu, J. (1999) *Distributed System Design*, CRC Press.
4. Farrar, C.R. and Worden, K. (2013) *Structural Health Monitoring: A Machine Learning Perspective*, John Wiley & Sons, Inc., New York.
5. Koren, I. and Krishna, C.M. (2007) *Fault Tolerant Systems*, Morgen Kaufmann.
6. Luo, X., Dong, M., and Huang, Y. (2006) On distributed fault-tolerant detection in wireless sensor networks. *IEEE Transactions on Computers*, **55** (1), 58–70.
7. Bosse, S. (2015) Design and simulation of material-integrated distributed sensor processing with a code-based agent platform and mobile multi-agent systems. *Sensors*, **15** (2), 4513–4549.

Part Six

Networking and Communication: The Sensor Network Level

15

Communication Hardware

Tim Tiedemann[1,2]

[1]*Hamburg University of Applied Sciences (HAW), Faculty of Engineering & Computer Science, Department of Computer Science, Intelligent Sensing, Berliner Tor 7, 20099 Hamburg, Germany*
[2]*Deutsches Forschungszentrum für Künstliche Intelligenz (DFKI), Robotics Innovation Center, Robert-Hooke-Straße 1, 28359 Bremen, Germany*

When designing "networking and communication" bottom-up, the hardware to be used for communication becomes the starting point. In case of material-integrated intelligent systems, special demands and limits are given by the applications. However, there is still a broad spectrum of different communication hardware types, and even standards, which could be applicable.

As it is true for other components of material-integrated intelligent systems, too, a source of potential communication hardware solutions is given in the wide field of embedded systems. Therefore, this section gives an overview of communication hardware options used in embedded applications today.

In Section 15.2 an overview of potential requirements for communication hardware in material-integrated intelligent systems is given. A list of requirements should be made for the specific target application as a first design step. These requirements determine the physical communication media and the potential communication options and standards being eligible for the further communication design process.

The Sections 15.3, 15.4, 15.5, and 15.6 describe the different types and especially the respective standards of communication hardware together with example applications.

A final summary is given in Section 15.7 and a short motivation showing communication hardware in specific applications follows in the next section. Not covered is CPU communication hardware like standard buses for CPU-peripheral or CPU-memory data transfer. Furthermore, only digital communication is dealt with here. Analog properties are described only where necessary and as physical side effects of digital communication.

15.1 Communication Hardware in Their Applications

Communication can be applied in material-integrated intelligent systems in various ways. One field is the integration in textiles (sometimes called smart fabrics), for example, with electronics for health monitoring. An early overview of research projects and applications is given by Lymberis and Paradiso [1]. They see smart fabrics and interactive textiles (SFIT) being part of an interactive communication network. Embedded into textile are potential sensors, actuators, computing, and power sources – needing communication within the textile.

For the application field of firefighters, van Torre *et al.* presented a communication solution using textile antennas [2]. Communication is used here for data exchange with external, potentially nonembedded devices. In this case radio frequency (RF) transmission is used as communication media. RF communication needs two specific problems to be solved. One is the integration of antennas the other is the power consumption that is usually very limited in material-integrated intelligent systems. Magno *et al.* reviewed different power management methodologies for wireless sensor networks [3]. This article is outlined in Section 15.5.

Finally, a completely different way of communication is presented by Jacobs with the research project "Reach" (as part of the interdisciplinary research platform "IT + Textiles") [4,5]. Here, embedded optoelectronics is used to generate (potentially dynamical) patterns on textiles for external, interhuman communication.

15.2 Requirements for Embedded Communication Hardware

The requirements for communication hardware in material-integrated intelligent systems can be derived from the requirements generally given for embedded systems. The latter are summarized by Marwedel [6].

The requirements for specific material-integrated intelligent systems have to be identified depending on the respective applications. As a basis to define requirements, metrics are needed. Examples for such metrics used by Marwedel are

- data throughput rate (including delays needed for the communication), measured, for example, in bits per second (bit/s, bps),
- latency (delay between communication start at the sender side and receiving of the first data at the receiver – notice that latency and throughput can both be high at the same time), measured, for example, in seconds (s),
- noise/interference parameters, for example, error rate in percent,
- power consumption,
- dimensions and/or weight, and
- costs.

Important requirements of communication as indicated by Marwedel [6] are as follows:

- *Efficiency*: An application can give certain (and strict) efficiency requirements in different respect. Efficiency is determined as a ratio of output (e.g., data throughput rate) by needed resource (e.g., power consumption, weight, costs). Typically for embedded systems are dimension, power, and weight requirements since they are usually limited by the structures they are embedded in. To improve the efficiency for the communication, the network topology can be selected appropriately (e.g., using bus instead of star topology, see Chapter 16). On the communication hardware level, an appropriate standard can be chosen (e.g., serial instead of parallel-wired connection, wireless instead of wired).

- *Data throughput and/or latency*: In most cases the application gives specific requirements for a minimum data throughput. These requirements can vary a lot. In some applications, the minimum data throughput and/or the maximum latency are fixed limits and need to be respected. Such cases are sometimes called real-time capability.
- *Polling versus event-triggered data transfer*: The communication protocol and/or the application can give specific requirements regarding the communication triggering. One option is to let one system (sometimes called "master") ask the others (sometimes referred to as "slaves") if communication is needed, that is, if they need data or if they have data to be sent. The slaves cannot initiate communication in this "polling" called mode. Another option is event-triggered data transfer. In this case, an event (usually application-related) initiates communication. Here, slaves can also start the communication.
- *Error protection*: In many applications, communication can be disturbed leading to single or multiple bit errors. In cases where the application cannot tolerate such errors, countermeasures need to be taken. One usual method is the transfer of additional special data leading to redundancy in the transferred data. By error detection or error correction codes (EDC/ECC) the occurrence of an error can be detected. In some applications it might be needed to introduce error protection measures on the protocol level as well (e.g., to trigger a repetition of data transfers or an interruption of current processes).
- *Robustness regarding environmental conditions*: Especially in embedded applications, the communication sometimes needs to cope with harsh environmental conditions. Examples are extreme temperature ranges, radiation, pressure, chemicals, and so on that the communication hardware needs to withstand. The most often occurring environmentally caused requirement is the robustness against electromagnetic interference (EMI). EMI can cause disturbances of the transferred data and, therefore, lead to bit errors (see Error protection).
- *Security*: Depending on the application, the data to be transferred might be classified as sensitive. In this case, special attention should be drawn on the environment and to what extent a third party could be interested in obtaining or modifying the data. To avoid this, security measures like encryption and/or authentication can be used.
- *Maintainability*: Finally, a maintainability of the communication software or hardware can be required. Needed might be the option to repair or exchange the communication hard-/software itself (e.g., to improve the data throughput). Furthermore, it might be desired to add communication means to enable *system* maintenance, for example, software updates or debugging.

In the case of material-integrated intelligent systems, the general requirements can be weighted and have to be extended depending on the specific application.

15.3 Overview of Physical Communication Classes

To get an overview of the different types of communication at first the layers involved have to be understood. As defined in Ref. [6], communication is set up in so-called channels. Channel is an abstract term that is linked to (a) an abstract relation between one sender (component) and one or multiple recipients (components) and (b) linked to communication properties. Such communication properties can be data throughput or noise/interference parameters.

The physical basis the communication in a channel is based on is called a medium. Important classes of media are wireless electromagnetic transfer, wireless optical transfer, connection via optical waveguides, and transport of charges in electrical wires.

Electrical Wires The latter class is the communication medium used in the vast majority of communication cases. It can be further subdivided into (a) a single-ended (SE) and (b)differential signal or differential pair (DS/DP) connections. SE connections use usually one reference connection between sender and recipient(s) (called ground) and an arbitrary number of additional wires transferring usually one data bit per point in time per wire (see Figure 15.1).

The names of the latter vary a lot and depend on the application. In the following descriptions they are called data lines (D_0, D_1). In most cases bit data ("0" or "1") are defined by voltage thresholds to be measured between ground and a data line. The thresholds are fixed in communication standards. The sender applies a voltage (e.g., 3.3 V, above the given threshold) to send a "1" and another voltage (e.g., 0 V, below another given threshold) to send a "0". The recipient(s) measures the voltage between their wire ends of ground and data line to identify a

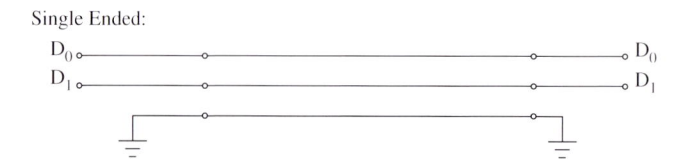

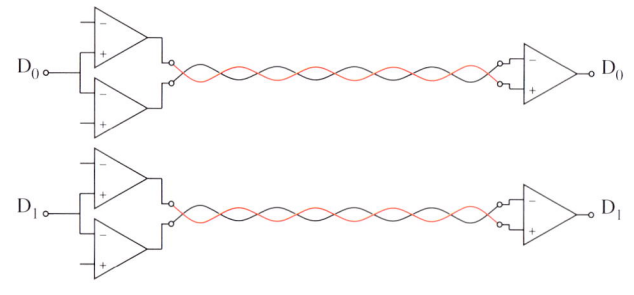

Figure 15.1 In wired electrical communication, single-ended (SE) or differential pair (DP) connections can be chosen. SE has the advantage of a simple and less expensive setup while DP can realize more robust and, thus, higher data throughput or longer distance connections. In the figure, two bits D_0 and D_1 are transferred in parallel.

bit value of "0" or "1," respectively. In some standards, the wires are connected via a resistor at (one of) the recipient(s).

In this case, the sender applies a voltage between two wires to generate a constant current through the first wire, the recipient's resistor, and back through the second wire. In this case, "0" and "1" are defined using current thresholds. One advantage of using current compared to voltage is an independence of the wire length (within limits defined by the sender driving electronics and usually defined by the standards.

In DP connections, pairs of wires are always used with one dedicated positive (+) and one dedicated negative (−) wire. The sender applies a voltage to both wires.

Often, a ground connection is given, too, and standards limit the voltage between ground and positive/negative wire. Again, bit values "0"/"1" can be defined via voltage thresholds (now measuring between positive and negative wire). The sender applies, for example, 3.3 V between ground and positive wires and 0 V between ground and negative wires to send a "1". For a "0" bit value, the sender applies 0 V between ground and negative wires and 3.3 V between ground and positive wires. Consequently, the recipient measures +3.3 V between negative and positive wires in the former case and −3.3 V in the latter situation. Hence, the difference between both measurements (bits) is 6.6 V now.

One benefit is an increased robustness against the induced disturbances. This is especially the case for disturbances that induce an erroneous offset voltage to the almost same degree on both positive and negative wire. Thus, this constant component (direct component, DC) part applied to both wires does not change the difference signal measured between both wires.

Additionally, if possible, both wires are twisted to equalize the electromagnetic influence for both. One application field of differential signaling is environments with large EMI, thus, with high noise levels.

Another application is high-speed connections. Here, the voltage levels that the sender applies are lowered compared to standard speed SE connections in the same environment. This is called low-voltage differential signaling (LVDS). The advantage of lower voltage levels is a possible reduced switching time between low- and high-voltage level on the two (positive and negative) wires. This leads to shorter times needed per bit and, therefore, a higher bit rate.

In general, the bit length (duration) can be defined in different ways. Possible are fixed lengths/bit rates, a switching between different (but fixed) lengths/bit rates (e.g., to support low-speed and high-speed communication), or the lengths dependent on the data bit to be transferred (e.g., short period between previous and next edge to transfer a "0", long period for a "1"). Furthermore, in addition to the payload data (determined by the application) often further bits are added to compensate for transfer errors by introduced redundancy or to ensure specific signal conditions (e.g., start/end marker, maximum time spans between edges). See Chapter 16 for details.

Wireless Electromagnetic Transfer In cases where a wired connection is not possible, for example, because sender and recipients have a varying and potentially larger distance, data transfer via electromagnetic waves is chosen most often. This is called radio-frequency (RF) transfer.

When an electrical current flows in a wire an electromagnetic field is generated circularly surrounding it. Furthermore, changes of such an electromagnetic field induce a voltage in electrically conducting materials.

Therefore, if a sinusoidal electrical signal is applied to a wire, a sinusoidal electromagnetic wave of the same frequency is generated. If a (electrically conducting) wire is applied to this field, a sinusoidal electrical signal of the very same frequency is generated in this receiver antenna.

In the most simple (yet still efficient) case, a straight wire with one end open (and length matching 1/4 of the wavelength) can be used at the transmission and at the receiving side. This is called a monopole antenna. On the transmission side, the sinusoidal voltage is applied between the monopole antenna and ground (e.g., the earth). On the receiving side the signal between ground and the receiving monopole antenna is measured.

The energy received is much less than the energy spent to generate the electromagnetic field. However, depending on the transmission power, the antennas, the dielectric medium between the two antennas (usually air), and the chosen frequency the receiving signal can still be measured with limited effort for distances of meters (low-power applications) or distances up to hundreds or thousands of kilometers (high power on transmission side).

Further Physical Communication Classes Besides the most often used electrically or electromagnetically based communication methods, two other physical media are used in some off the shelf communication devices. One is the data transfer using light. Here, a light transmitter (usually a light emitting diode, LED, or a laser diode) emits visible or invisible light and a light receiver transforms receiving light back in electrical signals (usually a photo diode or photo transistor). As in electrical communication, two types of media are used: Either the light is transferred via optical waveguides (flexible optical fibers or fixed solid media on a carrier) or an optical transfer in air is used between transmitter and receiver.

A third physical communication class is using acoustic data transfer. The transmitter emits mechanical oscillations (depending on the transfer medium usually in the Hertz or kilohertz, subsonic, sonic, or ultrasonic range) and as receiver special microphones or hydrophones can be used. The transfer media is usually air or water.

15.4 Examples of Wired Communication Hardware

Wired communication can be subdivided into (a) parallel communication with multiple bits being transferred in parallel on multiple wires and (b) serial communication with a transfer of one bit per time on one (SE or DP) connection (see Section "Electrical Wires"). In the case of material-integrated intelligent systems, serial communication is often preferred to reduce the effort needed for the usually complex hardware implementation. A parallel wired communication standard is IEEE-488 (renamed "IEEE 488.1, 2004", renamed "IEEE 60488.1:2004", in parallel developed by IEC as IEC-60625-1, short IEC-625, called HP-IB and GPIB, and with extensions IEEE 488.2/IEEE 60488.2:2004 and IEEE 488.1-2003, short HS-488, and as combined "Dual Logo" IEEE/IEC standard IEC-60488-1 and IEC-60488-2).

For serial wired communication, several standards of a variety of different types have been designed. Very often used in industry and embedded applications are the serial communication interface (SCI) standards RS-232, RS-422, and RS-485. The ANSI EIA/TIA-232 standard is referred to most often by "RS-232" with the current version "ANSI EIA/TIA-232-F" issued in 1997. It is a point-to-point interface connecting two devices with up to 115.2 kbit/s. The bit rates are not defined in the standard but the given rate is the usual recommended maximum in current implementations. Higher bit rates can sometimes be chosen but may lead to high error rates or even be impractical.

The interfaces defined by standard RS-422 use DS instead of SE compared to RS-232. This leads to a higher noise immunity and, therefore, can allow longer cable lengths. The second difference between RS-232 and RS-422 is that RS-422 can be used as a bus with up to 10 devices instead of a one-to-one connection only. RS-485 extends RS-422 with up to 32 devices and is a superset of RS-422. Further field bus standards are the CAN-Bus, the LIN-Bus, and PROFIBUS.

On the standard computer systems (PC and embedded systems) as serial interface, the Universal Serial Bus (USB) is one of the most often used interfaces. It uses one DP (up to USB 2.0) or three DPs (for full-duplex communication in SuperSpeed mode of USB 3.0) and allows transfer rates of up to 10 Gbit/s (USB 3.1).

A second group of communication interface standards are designed for short distance and simple chip-to-chip communication. These are, for example, synchronous serial communication interface standards I^2C (I^2S), (Inter-Integrated Circuit, up to 5.0 Mbit/s), SPI (Serial Peripheral Interface, no formal standard, max. frequency device dependent up to 100 MHz). Several microcontrollers and peripheral integrated circuits (IC) are available that support one or both of the interfaces.

For nonembedded (and especially not material-integrated) communication, Ethernet (100BASE-TX, IEEE 802.3 Clause 24, or GbE 1000BASE-T) is heavily used for communication between computer systems or between sensors and computer systems. However, especially for material-integrated applications the needed hardware is not suited and, thus, 100BASE-TX and 1000BASE-T will not be described here.

However, for use of Ethernet inside a vehicle, special solutions needed to be found that might be applicable to material-integrated communication, too. The resulting standard for automotive Ethernet Open Alliance BroadR-Reach™ (OABR)[1] focuses on the reliability of the data transmission, as well as on a cost-efficient transmission medium (two-wire line, see Figure 15.2). The standard for the physical layer was developed by the industry OPEN Alliance (One-Pair Ether-Net) special interest group (SIG)[1] and became IEEE standard (IEEE

---

1 OPEN Alliance SIG website, http://www.opensig.org (accessed June 25, 2016).

Figure 15.2 Automotive Ethernet is designed for single unshielded twisted pair cables. The photo shows two converters to standard 100BASE-TX interfaces using NXP PHY devices.

802.3bw, 100BASE-T1 with 100 Mbit/s based on BroadR-Reach, 1000BASE-T1 with 1 Gbit/s). So far, 100BASE-T1 Ethernet was designated for the transmission of image data inside the vehicle. The description of further applications (e.g., implementation of a backbone system) is new and subject of current research projects [7].

For the application in material-integrated intelligent systems, the reliability and especially the transmission medium of a single unshielded twisted pair seems to be interesting.

15.5 Examples of Wireless Communication Hardware

Standards For wireless (RF) communication, several interface standards have been developed. One of the most often used groups of standards is the WLAN (Wi-Fi) standard IEEE 802.11. In industry and private use, the standards 802.11a, b, g, and n are very common. Especially for the connection of a vehicle to other vehicles (V2V, Car-2-Car) or to the infrastructure (V2I, V2X, Car-2-X), the IEEE 802.11p standard was designed.[2] IEEE 802.11p describes the lower layers of dedicated short-range communications (DSRC) to enable wireless access in vehicular environments (WAVE). One main problem of DSRC in Car-2-X communication (addressed in IEEE 802.11p) are the short communication periods, especially when vehicles are passing each other in opposite directions.

---

2 IEEE 802.11p, http://standards.ieee.org (accessed May 28, 2016).

A higher layer based on IEEE 802.11p is IEEE 1609. One feature designated for the vehicle/traffic application (but interesting for other embedded applications as well) is the service announcement. On the so-called control channel service announcement messages (SAM) can be sent to announce services and service channels. The receiver can switch to the service channel where the contents of the services are transmitted. This is part of the higher layer standard IEEE 1609. However, here are differences between different standards, for example, between the IEEE 1609 and the European ETSI ITS-G5.

IEEE 802.11 standards offer a high data throughput but the power consumption of available devices is relatively high. Therefore, especially for embedded applications other standards like Bluetooth, Bluetooth LE, and ZigBee are often better suited.

Bluetooth is a standard for wireless short-distance communication and was invented by the telecom company Ericsson. IEEE standardized Bluetooth as IEEE 802.15.1 but does not maintain the standard. It is managed by the Bluetooth Special Interest Group (SIG).[3]

ZigBee is a specification developed by the ZigBee Alliance[4] and based on the standard IEEE 802.15.4. It is intended to be simpler and less expensive compared to Bluetooth. The ZigBee Alliance proposes ZigBee as a foundation of the Internet of Things (IoT).[4]

Bluetooth low energy (or Bluetooth LE, BLE, Version 4.0+, or Bluetooth Smart) was designed by the Bluetooth SIG.[3] It is intended to provide a reduced power consumption and reduced cost compared to classic Bluetooth.

Application Examples Several research activities related to material-integrated systems and using RF communication were conducted and published.

Kennedy *et al.* present body-worn e-textiles for the application of extravehicular activity (EVA) suits for space missions. They propose (and focus on the) e-textile antennas and give some more information on the positioning of the body-worn antennas [8]. Hertleer *et al.* focus on the antenna only and present a flexible patch antenna entirely made out of textile material [9]. Both target on the communication of the body-worn system with a base station and not on intra-garment communication.

This is also the case for van Torre *et al.*, who present a solution for off-body RF communication for firefighters and other rescue workers using textile antennas. Their proposal uses multiple input/multiple output (MIMO) communication with 2×2 or 4×4 links realized by two dual-polarized antennas, one at the front and one at the back of the firefighter jacket. As they show, the bit error rates (BER) improve with increasing diversity (while keeping the transmitted power equal) [2].

And finally, Magno *et al.* present a review of power minimization techniques. They target on wireless sensor networks but their results can clearly be relevant for many kinds of embedded and especially material-integrated applications, too.

---

3 Bluetooth Special Interest Group, http://www.bluetooth.com (accessed June 27, 2016).
4 ZigBee Alliance, http://www.zigbee.org (accessed June 27, 2016).

Besides well-established power optimization methods they discuss newer technologies with a focus on wake-up radio receivers [3].

15.6 Examples of Optical Communication Hardware

The industry "Infrared Data Association" (IrDA) provides several specifications for wireless infrared communication. They cover different layers and data rates of up to 1 Gbit/s. The physical layer IrPHY (Infrared Physical Layer Specification) defines optical link, modulation, and coding. Different data rates and modulation/coding schemes are defined in IrPHY ranging from SIR (serial infrared, up to 115.2 kbit/s) up to Giga-IR with up to 1~Gbit/s. Further layers are IrLAP (representing the data link layer of the OSI model), IrLMP (Infrared Link Management Protocol, implementation of logical channels and service registration), and further but nonmandatory protocols.

The Li-Fi Consortium (founded in 2011) is a nonprofit organization, devoted to introducing optical wireless technology (www.lificonsortium.org). After academic exchange of knowledge in the first years they focus on technology and product development. The members of the Li-Fi Consortium developed scenarios and solutions that might be relevant for different embedded communication applications. One example is GigaDock (combination of 10 Gbit/s data transfer and parallel inductive charging for plug-free data and energy transfer) another one is GigaHotspots (ceiling-mounted device communicate with 1–5 Gbit/s over distances of 3–5 m). Finally, free-space optical communication (FSO) is usually used for data transfer over longer distances of tens of meters up to a few kilometers. However, the devices used for FSO are too large and too heavy for most applications of material-integrated intelligent systems

15.7 Summary

To realize communication and networking options for material-integrated intelligent systems, one can avail oneself of many existing solutions, most of them fixed in standards. One first design decision to make is the question if a solid, tethered connection should be used (wires or optical fibers) or if a data transfer through air (or water) is preferred.

The second decision is to answer the question of using electrical (or electromagnetic) transfer or optical data transmission (or in very specific cases acoustic transmission). This point might depend on the environmental conditions, induced noise, needed data rates, and available power. Therefore, a definition of the application's communication requirements (e.g., using the parameters and metrics already listed) is an important prerequisite.

Eventually, most options of off-the-shelf products are available for electrical or electromagnetic communication. Consequently, these standards and products are often the first choice in material-integrated intelligent systems, too. However, with the specific requirements in this field there might be adaptations of the communication hardware and/or software and, thus, violations of the existing standards needed.

References

1 Lymberis, A. and Paradiso, R. (2008) Smart fabrics and interactive textile enabling wearable personal applications: R&D state of the art and future challenges. 30th Annual International Conference of the IEEE Engineering in Medicine and Biology Society, pp. 5270–5273.
2 Torre, P.V., Vallozzi, L., Hertleer, C., Rogier, H., Moeneclaey, M., and Verhaevert, J. (2011) Indoor off-body wireless MIMO communication with dual polarized textile antennas. *IEEE Transactions on Antennas and Propagation*, **59** (2), 631–542.
3 Magno, M., Marinkovic, S., Srbinovski, B., and Popovici, E.M. (2014) Wake-up radio receiver based power minimization techniques for wireless sensor networks: a review. *Microelectronics Journal*, **45** (12), 1627–1633.
4 Jacobs, Margot and Worbin, Linda (2005) Reach: dynamic textile patterns for communication and social expression. CHI'05 Extended Abstracts on Human Factors in Computing Systems, ACM, pp. 1493–1496.
5 Interactive Institute and Newmad Technologies (2016) IT+Textiles. Available at http://dru.tii.se/reform/projects/itextile/ (accessed June 27, 2016).
6 Marwedel, Peter (2011) *Embedded System Design*, Springer.
7 Tiedemann, T., Backe, C., Vögele, T., and Conradi, P. (2016) An automotive distributed mobile sensor data collection with machine learning based data fusion and analysis on a central backend system. Proceedings of the 3rd International Conference on System-integrated Intelligence SysInt, Procedia Technology, Elsevier.
8 Kennedy, T., Fink, P., Chu, A., Champagne, N., Lin, G., and Khayat, M. (2009) Body-worn e-textile antennas: the good, the low-mass, and the conformal. *IEEE Transactions on Antennas and Propagation*, **57** (4), 910–918.
9 Hertleer, C., Tronquo, A., Rogier, H., Vallozzi, L., and Van Langenhove, L. (2007) Aperture-coupled patch antenna for integration into wearable textile systems. *IEEE Antennas and Wireless Propagation Letters*, **6**, 392–395.

16

Networks and Communication Protocols

Stefan Bosse[1,2]

[1]*University of Bremen, Department of Mathematics and Computer Science, Robert-Hooke-Str. 5, 28359 Bremen, Germany*
[2]*University of Bremen, ISIS Sensorial Materials Scientific Centre, Wiener Str. 12, 28359 Bremen, Germany*

Material-integrated sensing systems like sensor networks embedded in technical structures entail new data processing and communication architectures. With increasing system complexity increased reliability and robustness of the entire commonly heterogeneous sensing system are required to maintain a specified quality of service. Failure is not an exception, it is treated as a part of the design process and the runtime behavior. Reliable networking and communication is one major part contributing to safe and useful sensing system.

Cloud computing introduces dynamics and scalability of distributed computing with virtualized resources. Though cloud computing is concerned for big data problems, it can be attractive for large-scale sensor networks, especially considering the integration of sensor networks in sensor clouds and the Internet (of Things), discussed in Chapter 17.

A communication network is basically represented with a network connectivity graph (NCG) with nodes (computers, subnetworks, mobile devices, sensors) and edges connecting the nodes, representing the technical communication links, discussed in the Chapter 15. The edges can be directed or not directed (bidirectional communication), and the graph can have cycles resulting in closed paths. A path is a part of a linear list of nodes of the graph, starting with a source node and ending on a destination node.

Communication is a central part of distributed data processing in sensor networks, influencing performance and operation significantly. A connection between two nodes is usually a uni- or bidirectional (serial) link which is capable to transfer data encapsulated in messages using protocols for coordination and coding. Message passing is the preferred communication method for large-scale networks compared with switched networks requiring crossbar or multistage banyan (butterfly) switches. Path finding and routing, the process of forwarding a message in the network with the goal to deliver messages from the sender to the receiver is required in message passing networks, and for economic reasons and the sake of simplicity there are usually no dedicated routers in sensor networks. Thus, a sensor or computational node is a service end point and a message router, too. Traditional computer networks use unique node identification not suitable for (especially loosely or *ad hoc* coupled) sensor networks. Path finding algorithms can be classified in those using routing tables managed by each network node and storing information about the network topology (usually limited to the neighborhood of the respective node), and those not relying on routing tables. Routing tables provide information for path planning in advance. Managing routing tables can be critical in resource-constrained and real-time systems, because they allocate a fairly high amount of local storage and require computational and communication activities, which should be minimized in sensor networks. Routing tables can only reflect an outdated view of the network (the network configuration within a certain time interval) and hence dealing with network changes is complicated by using routing tables.

16.1 Network Topologies and Network of Networks

The network topology is a specific structure of the network connectivity graph, introducing some regularity in the arrangement of nodes and their connectivity to neighbor nodes. Some common topologies are shown in Figure 16.1a, and the topology and technology classifications in Figure 16.1b.

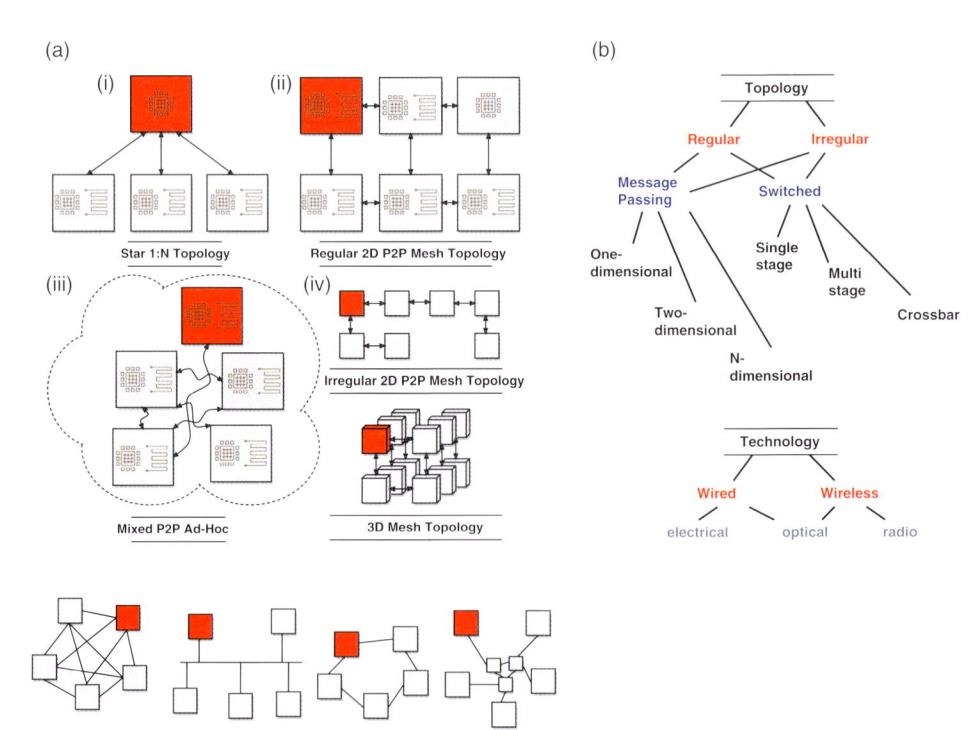

Figure 16.1 (a) Common network topologies. (b) Summary of network topology and technology classification.

One of the simplest and still widely used topologies is master–slave 1:N star networks. These are mainly used in centralized processing architectures. There is only one connection from each client to a server node (connectivity degree 1, distance 1). The failure of the server is critical. A different situation is shown in Figure 16.1a(vi)) using a broadcast medium (bus, Ethernet) that connects N individual nodes, with a connectivity degree (N-1) and a maximal distance of 1. Here, the failure of the broadcast medium (the interconnect) is critical.

Two-dimensional mesh-like networks, either fully (Figure 16.1a(ii)) or partially (Figure 16.1a(iv)) connected and using point-to-point links, are the preferred topologies for wired planar material-integrated sensor networks having the lowest interconnect complexity but still providing robustness in the presence of missing or defective links due to path redundancy (there is more than one possible path from a source to a destination node). They have a connectivity degree of 4 and a maximal distance of $\log_2 N$. Furthermore, the logical network topology corresponds to the geometrical placement order of nodes.

In fully connected networks (Figure 16.1a(v)), each node is connected with each other node, leading to a connectivity degree of N-1 and a distance of 1 with the highest degree of robustness and lowest message passing latency, but requiring the highest resource demands. Cube networks (three-dimensional topology, Figure 16.1a(iv)) provide a good compromise between the aforementioned networks, having a connectivity degree of 6 and a distance of $\log_3 N$ (resulting in a lower message passing latency compared with two-dimensional networks). They are a special case of a generic hypercube networks (n-dimensional), with very limited practical use in material-integrated sensor networks due to the complex interconnect structure.

Network topologies (Figure 16.1b) can be basically classified in regular and irregular structures. Mobile *ad hoc* networks with peer-to-peer communication are prominent examples for networks with a massive irregular structure that changes over time. The connectivity in mobile networks is dynamic in the spatial and temporal dimension and is determined by the overlapping of (mostly) circular radio transmission ranges, different from wired connectivity.

In chain or ring networks (Figure 16.1a(vii), the connectivity is 2, the maximal distance N-1, providing no (chain) or low (ring) robustness. Here, failure of a single node can already be critical: The chain is only as strong as its weakest link.

Hierarchical networks (Figure 16.1a(viii)) provide node partitioning in spatial bounded subnetworks connected with each other by using dedicated routers (for example, substar networks arranged in chain networks applied in Ref. [1]).

16.2 Redundancy in Networks

A chain network has no redundancy, but two- and three-dimensional mesh and cube networks (with four and six neighbor node connections, respectively) provide an increasing number of possible path alternatives without introducing additional high connectivity, complexity, and costs (number of connections are in the order of nodes). The possible number of paths P connecting the (upper) left

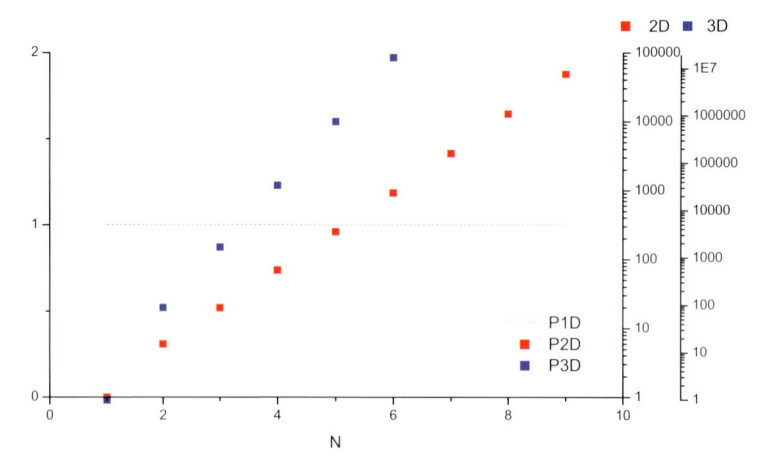

Figure 16.2 Redundancy in mesh-like networks (chain, grid, cube): possible paths from node $(0, \ldots)$ to $(n-1, \ldots)$ depending on the network size (n nodes in each dimension).

and (lower) right nodes and the connectivity costs C (number of all links in a regular configuration) of the network depend on the number of nodes (n, m, o) in each dimension, respectively, and can be calculated from (assuming a connecting path visits each node at most one time)

$$P_1(n) = 1, P_2(n,m) = \frac{(n+m)!}{n!m!}, \quad P_3(n,m,o) = \frac{(n+m+o)!}{n!m!o!},$$
$$C_1(n) = n-1, C_2(n,m) = 2nm - m - n,$$
$$C_3(n,m,o) = 3mno - m - n - o. \tag{16.1}$$

Figure 16.2 compares the redundancy of 1d, 2d, and 3d grid networks in relation to the number of nodes. Naturally, a linear chain has no redundancy (i.e., there is only one possible path between a source and destination node), but a ring has already a redundancy degree 2.

16.3 Protocols

A communication or network protocol is basically a specification. A protocol defines as follows:

- The message format and structure
- A sequence flow between communication end points or routers assigning states to communication steps, commonly defined by a finite state machine
- The operational and synchronization semantic
- Access control and security

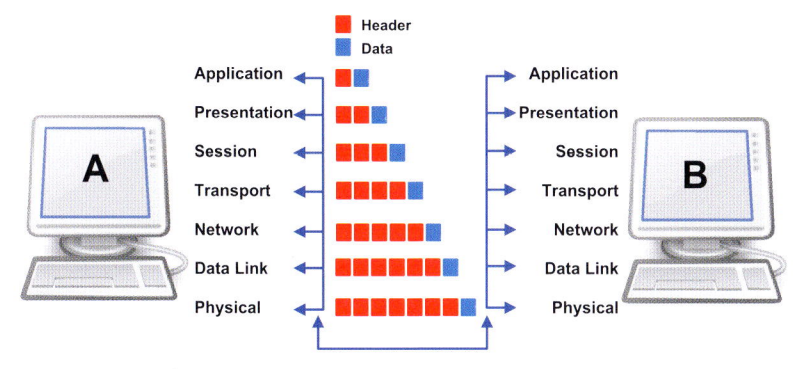

Figure 16.3 OSI layer model.

- Structural organization of networks
- Kind of node or communication end point addressing (if any)

Communication start and end point can be computers, routers, programs (or more precisely spoken processes of programs), other protocol layers, and so on, and depends on the particular communication protocol.

Therefore, there is commonly a set of protocols with different operational and organizational semantics involved in end-to-end message communication, assigned to different layers of the Open Systems Interconnection (OSI) model, shown in Figure 16.3. The layers are as follows:

1) Physical layer (technology-dependent raw data transfer)
2) Data link layer (manages specific communication devices, message fragmentation)
3) Network layer (routing of messages, flow, and congestion control)
4) Transport layer (manages end-to-end message delivery)
5) Session layer (control of logical/virtual links and sessions)
6) Presentation layer (device-independent data representation)
7) Application layer (interface to user processes)

Each protocol layer requires its own data and message encapsulation, adding each time a protocol specific header. For example, the application layer adds a process identifier, the transport layer adds computer addresses.

There are still many sensor networks based on traditional computer networking approaches like TCP/IP and remote procedure call (RPC) communication systems [2], resulting in an increased overhead and requiring a large amount of computational, power, and microchip resources, being contraindicated for energy-aware and microchip-level sensor network designs. But a case study showed that application-specific FPGA implementations using the RTL architecture of a reduced UDP/IP protocol stack requires about 200 K logic gates, which is reasonably low and suitable for microchip scale [3].

Still most communication protocols require unique addressing of nodes (or service end points), commonly human managed, able to compose an organizational structure of the network. With an increasing number of nodes beyond the trillion boundaries, this addressing scheme is questionable up to impossible. Why should two sensors communicate with each other?

Different path finding schemes propagating messages and data rely on data-based and event-based routing, that is, data sources and data sinks publish or negotiate the interest of specific data and events, for example, sensor data of a specific kind or originating from a specific bounded region of interest. The data-centric approach avoids the node addressing requirement, and, for example, replacing unique node addresses by spatial coordinates or spatial neighborhood connectivity, like in wireless networks. Bioinspired networking focuses on self-organizing and self-adaptive structures without any external organizer, for example, implementing some kind of swarm intelligence.

16.4 Switched Networks versus Message Passing

Switched networks are peer-to-peer networks connecting nodes directly. A switched network uses line multiplexer to connect a node with multiple other nodes. A full crossbar switch for a network consisting of N nodes requires N^2 multiplexer resources, and it is conflict free (each node can be connected with any other node at any time). A multistage switch network using, for example, binary switches, has lower resource requirements (usually $N\log(N)$ switches), but is not conflict free. A network switch commonly does not provide synchronization between nodes, in contrast to message passing networks (Figure 16.4).

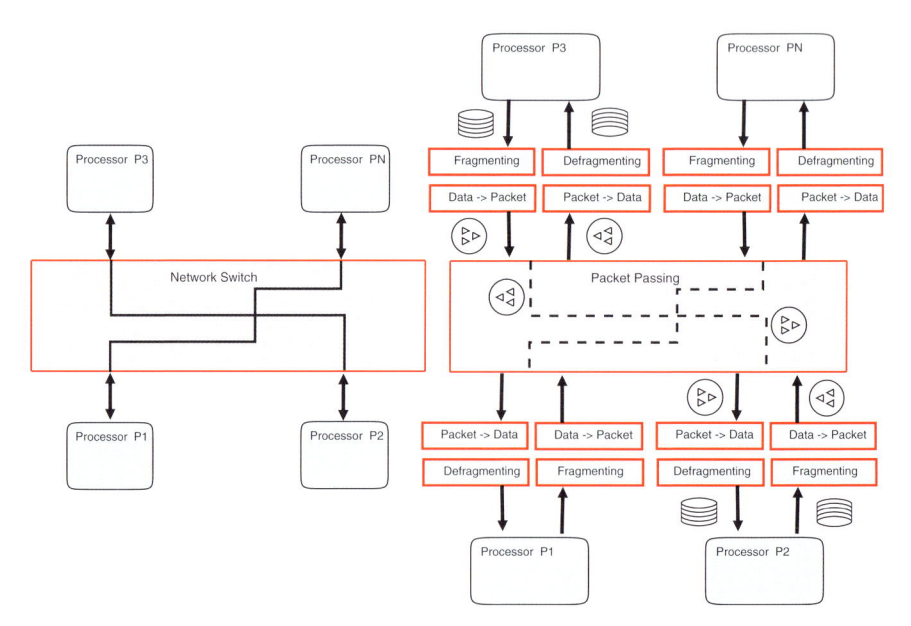

Figure 16.4 (a) Switched circuit network. (b) Message passing network.

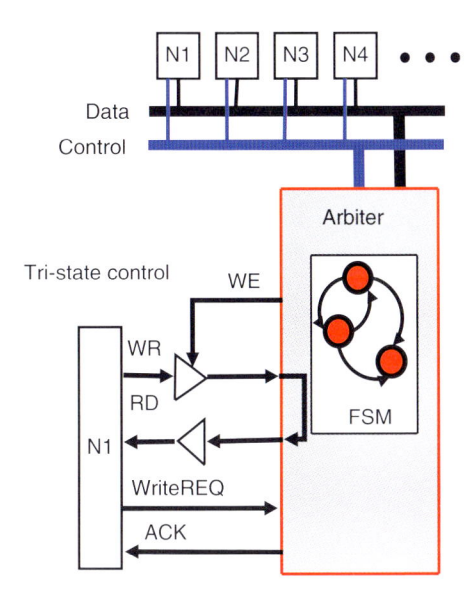

Figure 16.5 Bus request-acknowledge protocol handled by an arbiter.

16.5 Bus Systems

Bus systems are very popular in network designs, especially their deployment in material-integrated systems. A bus system has linear hardware costs of the order $\Theta(N)$, if N is the number of network nodes connected by the bus. But they have the disadvantage of a total bandwidth of 1, that is, only one connection between two nodes is possible at any time. Furthermore, one node failure can block the entire bus by a defective receiver, sender, or interconnect wire.

A bus is a switched network without node and communication process synchronization. The access of the bus is hence exclusive and requires a mutual-exclusive scheduling performed by a shared arbiter, shown in Figure 16.5. A node wanting to access the bus, that is, the data lines, will request the access by raising a request signal. If the access is granted by the arbiter, an acknowledgment is sent to the requesting node. Higher level of node synchronization can be introduced by simple messages sent by the nodes on dedicated control lines or by using the data lines with a special coding. A minimal set of messages consists of (adapted from the IEEE 488/GPIB protocol) the following:

LISTEN <dstaddr>
　　The current owner of the bus sends a control command to notify the receiver of the following data messages.
UNLISTEN
　　All nodes wait for control commands.
TALK <srcaddr>
　　The current owner of the bus publishes his address.
UNTALK
　　This control command indicates that no node should be a sender of data messages.

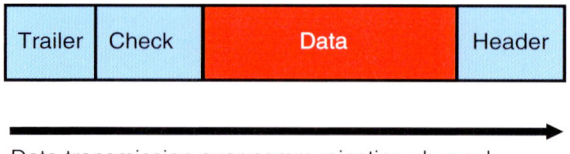

Figure 16.6 A typical message format consisting of header with information required for routing, the data payload, and optional check fields and a trailer marking the end of a message.

Each node, the interconnect wires, and the arbiter can be considered as single point of failures.

16.6 Message Passing and Message Formats

A network message is used to transport data between network nodes and processes executed on the nodes. Usually processes are communication end points, rather than computers. Message passing can involve multiple protocols belonging to different communication layers (physical, logical link, or transport layers). Each protocol adds a header to the message, shown in Figure 16.6. The header contains protocol-specific information and is used for the message routing in the network, discussed in the next section.

Message passing requires a significant amount of time, depending on the network topology (number of hops between a source and destination node), the communication hardware (bandwidth and latency), and the amount of data (N bytes):

$$T(N)_{S \to D} = T_{\text{Overhead}} + T_{\text{Routing delay}} + T_{\text{Transfer}}(N) + T_{\text{Conflict}}. \tag{16.2}$$

The message passing time adds an overhead to the data processing latency in distributed and parallel systems. Communication channels are usually shared by different "virtual" paths (connecting a set of source with a set of destination nodes). Hence, the conflict time T_{Conflict} can be a significant overhead if there is a high-channel utilization.

16.7 Routing

A network node is connected to a network via communication ports. At least one communication port is required. A network node can be a service end point or a message passing unit (or both), the router. A routing service determines the path of a message from an incoming to an outgoing port. Usually, routing bases on the source–destination parameters of a message.

In macroscale sensor network nodes are often connected by using wireless technologies with dynamic *ad hoc* network topologies. Protocol-based routing in changing networks (regarding communication connections, node position, and node failures) is one main challenge in the design of robust and energy-aware sensor networks. Most traditional routing protocols rely on an address space uniquely identifying nodes as well as routing tables storing network information usually limited to a bounded spatial scope. Path finding is required for the process of forwarding of a message, and is the main goal of routing by minimizing the number of passed routers (e.g., nodes). Commonly, there is a set of different paths connecting nodes (see Section 16.6). In the past, many routing algorithms were developed (for wireless and wired networks) finding (1) alternative paths in the case of missing or nonoperating connectivity in parts of the networks; (2) the shortest path (concerning the distance in hop counts and the delivery latency). One example is junction-based routing (discussed in detail in Ref. [4]). Routing in large-scale wireless sensor networks is discussed in detail in Ref. [5].

Geographic routing and addressing bases on geometrical relations between nodes, for example, relative addressing specifying the relative distance between nodes, can be used to avoid absolute unique node addressing, which is not applicable in large-scale miniaturized networks with an initially unknown configuration such as smart dust networks [6].

The scaling of networking down to material-integrated level requires simplified and robust approaches. Some of them capable for microchip level implementations satisfying low-resource constraints are summarized further. They all not only rely on routing tables (saving memory and computation) as well as they are able to adapt to network changes immediately.

Store and Forward Routing The store and forward (SF) algorithm divides message passing in three steps, shown in Figure 16.7a:

1) Receive and store the entire message from an input port CP_I
2) Compute the output port CP_O by commonly using look up tables giving a destination-port mapping.
3) If the output port CP_O is not busy, send the entire message to the output port, else wait for the output becoming ready and send message.

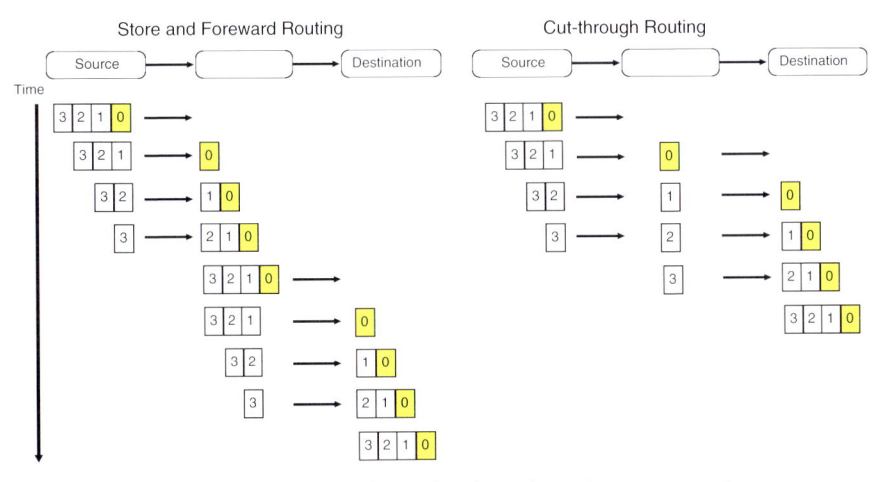

Figure 16.7 Comparison of store-and-forward and cut-through message routing.

The SF algorithm requires $N \cdot P$ full size message buffers, with N being the maximal number of buffered messages per port and P the number of communication ports per node.

The routing time for a message (or packet) of size N bytes is given by

$$T_{\text{Routing SF}}(N, H) = H\left(\frac{N_{\text{Packet}}}{B} + \Delta\right), \tag{16.3}$$

and is the product of the message transfer time (N/B) with a given communication bandwidth B and the number of hop nodes (H). An additional routing overhead (step 2) Δ applies on each hop, too.

Cut-Through and Wormhole Routing Cut-through (Ct) and wormhole (Wh) routing splits messages in parts (so-called flits). A network message consists of a (small) header and a (large) data part. If the header fits in one flit packet, only the header flit must be received and stored, which can be immediately evaluated. Hence, Ct/Wh routing is divided in different steps, shown in Figure 16.7b:

1) Receive and store the header of a message
2) Compute the output port CP_O by commonly using look up tables giving a destination-port mapping.
3) If the output port CP_O is not busy, send the header and all following flit packets to the output port.
4) Ct: If the output port is busy, store the entire message and send it if the output port is ready.

   Wh: If the output port is busy, notify the sending network node to stop sending further flit packages. The notification is propagated up to the source node.

Ct and Wh algorithms behave only differently if the output port is blocked (i.e., busy). The Ct algorithm requires P full size message buffers, whereby the Wh algorithm requires only P flit size buffers.

The routing time for a message (or packet) of size N bytes is given by

$$T_{\text{Routing CT}}(N, H) = \frac{N_{\text{Packet}}}{B} + H\Delta \ll T_{\text{Routing SF}}(N, H), \tag{16.4}$$

that is, the sum of the transfer time for the entire message (or packet) N/B with the communication bandwidth B, and H times of the routing overhead Δ occurring on each hop node.

Δ-Distance Routing In this scheme, a path from a source to a destination node in a grid-like network (of arbitrary dimension) is given by its relative distance in each Cartesian direction, the delta vector $\Delta = (\delta_1, \delta_2, \ldots, \delta_n)$. Though Δ-distance addressing and routing is one of the simplest approaches it is very well suited for microchip level implementation.

Commonly, a message is forwarded to the destination node by choosing a path with the goal to minimize one distance value after each other by passing the message along nodes in the path (hopping), each time decrementing the respective distance value Δ_i in the ith direction, until the Δ vector is finally zero (Δ^0 is the original distance vector and M is the message content). This is shown in Algorithm 16.1, considering a message as a mobile process. The route function tries to find a suitable direction to forward the message using the moveto(DIR) operation. After the message was moved to the new neighbor node, the routing function is applied again until the destination is reached.

Alg. 16.1 Standard and Simple Adpative Δ-Distance Routing (in Pseudocode Notation)

```
1   TYPE DIM = {1,2,..,m} Enumeration of network topology dimensions
2   TYPE DIR = {NORTH,SOUTH,EAST,WEST,UP,DOWN,..} Symbolic directions
3   DEF dir = fun (i) → match i with   Numerical mapping of directions
4     ... -3 → UP | -2→NORTH | -1→WEST | 1→EAST | 2→SOUTH  3 → DOWN ...
5
6   DEF route_xy = fun (Δ,Δ⁰,M) →
7     if Δ ≠ 0 then
8       for first i ∈ DIM with Δ_i ≠ 0 do:
9         if Δ_i > 0 then
10          moveto(dir(i));
11          route_xy(Δ with Δ_i:Δ_i−1,Δ⁰,M);
12          return
13        else if Δ_i < 0 then
14          moveto(dir(-i));
15          route_xy(Δ with Δ_i:Δ_i+1,Δ⁰,M);
16          return
17      discard(M)
18    else
19      deliver(M)
20
21  DEF route_xyvar = fun (Δ,Δ_0,M) →
22    if Δ ≠ 0 then
23      for each i ∈ DIM with Δ_i ≠ 0 try:
24        if Δ_i > 0 ∧ link?(dir(i)) then
25          moveto(dir(i));
26          route_xyvar(Δ with Δ_i:Δ_i−1,Δ⁰,M);
27          return
28        else if Δ_i < 0 ∧ link?(dir(-i)) then
29          moveto(dir(-i));
30          route_xyvar(Δ with Δ_i:Δ_i+1,Δ⁰,M);
31          return
32      discard(M)
33    else
34      deliver(M)
```

This trivial routing protocol requires no routing table and is suitable for low-resource hardware SoC implementations. There is only one possible path that can be traveled if Algorithm 16.1 is applied strictly ordered (route$_{xy}$), limiting this simple Δ-routing protocol to complete and regular network topologies without missing or broken links. If the routing direction can be chosen based on actual detected node connectivity (using the link? operation and route$_{xyvar}$), there is a set of disjoint paths that can be chosen alternatively.

Adaptive Δ-Distance Routing with Backtracking Robustness and support of incomplete mesh-like networks with missing and nonoperating links and nodes can be achieved by extending the simple Δ-routing protocol with a selection mechanism of the routing scheme providing the ability to overcome missing network connectivity in a specific direction [7]. This approach bases on a sink attraction and backtracking behavior. There are different routing behaviors selected by the available local link connectivity (tested by the function `link(DIR)`) and information stored in the message (Δ,Γ), shown in Algorithm 16.2, again considering a message as a mobile process. An optional preferred routing direction ω can be specified. The Γ vector marks back or opposite traveling.

Algorithm 16.2 Advanced Adaptive Δ-Routing with Backtracking (in Pseudocode Notation)

```
1    DEF route_s = fun (Δ,Δ⁰,Γ,ω,M) →
2      d^in := incoming-direction-of-message(M)
3      if Δ=0 then deliver(M,Δ⁰)
4      else if not route_normal(Δ,Δ⁰,Γ,ω,M) then
5        if not route_opposite(Δ,Δ⁰,Γ,ω,M,d^in) then
6          if not route_backward(Δ,Δ⁰,Γ,ω,M,d^in) then
7            discard(M)
8    WITH
9      route_normal = fun (Δ,Δ⁰,Γ,ω,M) →
10       for each i ∈ DIM starting with ω and with Δ_i ≠ 0 try:
11         rev := Γ ≠ 0 ∧ Δ_i ≠ Δ_i0
12         if Δ_i > 0 ∧ link(dir(i)) & not rev then
13           moveto(dir(i));
14           route_s(Δ with Δ_i:Δ_i-1,Δ⁰,Γ,ω,M);
15           return true
16         else if Δ_i < 0 ∧ link(dir(-i)) ∧ Γ_i=0 then
17           moveto(dir(-i));
18           route_s(Δ with Δ_i:Δ_i+1,Δ⁰,Γ,ω,M);
19           return true
20       Reached here: No success! All failed!
21       return false
22
23     route_opposite = fun (Δ,Δ⁰,Γ,ω,M,d^in) →
24       for each i ∈ DIM starting with ω try:
25         if Γ_i = 0 ∧ d^in ≠ dir(-i) ∧ link(dir(-i)) then
26           moveto(dir(-i));
27           route_s(Δ with Δ_i:Δ_i+1,Δ⁰,Γ with Γ_i :-1,ω,M);
28           return true
29         else if Γ_i = 0 ∧ d ≠ dir(i) ∧ link(dir(i)) then
30           moveto(dir(i));
31           route_s(Δ with Δ_i:Δ_i-1,Δ⁰,Γ with Γ_i :-1,ω,M);
32           return true
33       Reached here: No success! All failed!
34       return false
35
36     route_backward = fun (Δ,Δ⁰,Γ,ω,M,d^in) →
37       for each i ∈ DIM starting with ω try:
38         if Γ_i = 0 ∧ d^in < 0 ∧ link(dir(-i)) then
39           moveto(dir(-i));
```

```
40          route_s(Δ with Δ_i:Δ_i+1,Δ°,Γ with Γ_i :-1,ω,M);
41          return true
42        else if Γ_i = 0 ∧ d^in > 0 ∧ link(dir(i)) then
43          moveto(dir(i));
44          route_s(Δ with Δ_i:Δ_i-1,Δ°,Γ with Γ_i :-1,ω,M);
45          return true
46        else if Γ_i = -1 ∧ link(dir(-i)) then
47          moveto(dir(-i));
48          route_s(Δ with Δ_i:Δ_i+1,Δ°,Γ,ω,M);
49          return true
50        else if Γ_i = 1 ∧ link(dir(i)) then
51          moveto(dir(i));
52          route_s(Δ with Δ_i:Δ_i-1,Δ°,Γ,ω,M);
53          return true
54   Reached here: No success! All failed!
55   return false
```

For a first communication attempt, the normal routing strategy is tried with the goal to minimize the Δ-vector in each dimension in any order. If this is not possible due to a lack of required connectivity, the message is tried to send in one opposite direction increasing the Δ-vector temporarily. An opposite travel is marked in the Γ entry of the message. Finally, if this strategy also fails, the message is sent backward, finally reaching the source node again if there are no other routing alternatives found on the back path. Some care must be taken to avoid sending a message back to the direction from which it originally comes (resulting in a ping-pong forwarding with a live lock). To summarize, this simple routing algorithm uses decision-making based on information stored in the message, and neighborhood connectivity retrieved by the current node. It is suitable for peer-to-peer networks with a limited number of neighbors, as found in wired material-integrated sensor networks and benefits from low computational requirements and complexity (a minimized hardware implementation requires less than 100 K logic gates resources [8]).

The smart backtracking-based routing algorithm is well suited for path migration of mobile agents in mesh-like networks (of any dimension), requiring no host routing tables and no path planning in advance (Figure 16.8).

Data-Centric and Event-Based Routing Commonly, stream-based data communication is used in (distributed) signal processing systems, transferring data from sensors (information sources) to computational nodes (information sinks) continuously, mostly based on a service request–reply architecture with sensors as a service (SaaS). Directed diffusion [9,10] is a data-centric routing and path finding approach suitable for sensor networks, which promises energy efficient delivery of messages based on events and the interpretation of their content and the demand of the data, in contrast to peer-to-peer approaches only considering geographical ontologies of the network. There are two different event-based sensor data distribution strategies that can be applied: (1) the sensor node (the data producer) detects an event and decides to notify the neighborhood (the data consumer) or the entire network of the availability of new data by using message flooding (i.e., rumor routing in Ref. [11], a compromise between

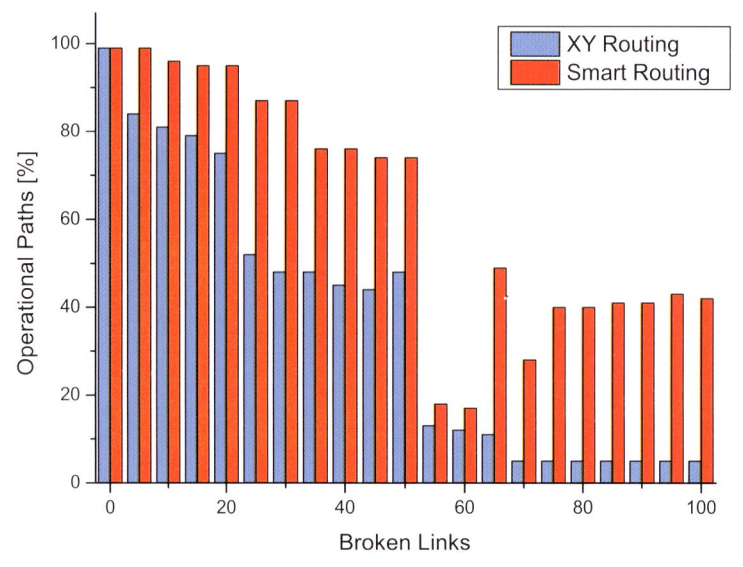

Figure 16.8 Simulation results from a reachability analysis in a two-dimensional mesh sensor network arranged in a 10 × 10 matrix with up to 180 links. Each node was connected with up to four neighbor nodes. The percentage of the operational paths in dependency of the number of broken node-to-node links was compared for the traditional nonadaptive XY and the adaptive smart routing strategy. Beyond 50 broken links the smart routing outperforms, and up to 40% of paths are still reachable [7].

flooding queries and flooding event notifications), and (2) the data consumer (recipient) notifies all neighborhood nodes in a bounded region for the interest of specific data, and the sensor node detecting an event can use this information for delivering the sensor data to a specific node on a shortest path (i.e., directed diffusion routing).

Agent-Based Routing The simplest forwarding strategy in a network is flooding the network by using multi- or broadcasting strategies performed on the link layer. An incoming message is replicated and sent to all neighbor nodes. Wireless networks have the advantage that a message can be received by many nodes around the neighborhood of the sending node efficiently implementing 1:N multi and broad-casting immediately on the link layer, but still requiring computational activity by all N nodes, decreasing the energy efficiency significantly. In wired networks, 1:N multicasting requires the composition of N unicast messages, resulting in the worst routing algorithm in terms of energy efficiency.

Usually, messages can be treated as passive units. The routing strategy depends entirely on the algorithm and knowledge provided by each forwarding node along the path. Instead, treating messages as active units – like mobile agents can do – leads to a decoupling from the capabilities and knowledge of each node and provides increased autonomy. Routing decisions are now made by the agent itself carrying the message. The goal of the agent or a collection of agents is to deliver the message carried by the agent to a destination node.

Replication can be used to increase the delivery probability and decrease the latency. Learning agents can improve path finding by considering their travel history. Data-centric directed diffusion algorithms can be easily implemented with autonomous mobile agents.

The aforementioned backtracking routing protocol was developed with an agent-based system and implemented entirely on microchip level, using neighborhood hopping and sink attraction behavior, described in detail in Ref. [14].

In Ref. [11], an event notification algorithm is proposed using notification agents installing routing information of an event, for example, a sensor value is exceeding a threshold, in all the nodes they are visiting along a path (rumor routing). This approach does not flood the network. Search agents are used to query and find an event installed by the notification agents. The root location of an event can be found by following the path the notification agents have marked, offering backtracking routing.

Directed diffusion can also be implemented with different agents. Agent-based routing is used in several works to improve communication robustness, reliability, latency, and energy efficiency, but not generally using autonomous agents for message delivery. Such examples for agent-based design methodologies can be found in Refs [15–17].

Biologically inspired routing in conjunction with agents and path marking is another approach to find optimal paths in *ad hoc* and wireless networks, for example, ant colony algorithms described in Ref. [18], which is used to find out optimal and efficient migration paths for mobile agents.

16.8 Failures, Robustness, and Reliability

Communication failures can originate from failures of the technological interconnect, the link receiver, or sender unit, and finally from process failures performing communication on protocol level. The runtime behavior of communication (with failures) and communication networks can be distinguished by different properties and behavior, all having influence of the overall system stability of the sensor network (based on Refs [19,20]):

- *Fair-loss and finite duplication*: A message from a sender process p_a directed to a destination process p_b is eventually delivered with a probability $P<1$ if none of the both processes crashes and the link is capable of retransmission (compensating lost messages). Furthermore, if a message was sent a finite number of times, it can only be delivered a finite number of times. This link behavior can be assumed to be always present in any technical system.
- *No creation/causality*: If a message was delivered at least one time to a process p_b, then there was a process p_a sending the message to p_j at least one time. If this property is violated, either by link or protocol failures, ambiguities and data corruption can occur.
- *Stubborn delivery*: A process p_a sends a message to a destination process p_b at most a finite number of times. It is possible that the message is delivered an

infinite number of times, which can be system-critical in energy autonomous networks. A retransmit-forever behavior of the sending process or node can be one cause of this event.

- *No delivery/live-lock*: A message was sent by a process p_a and requires routing through a finite number of network nodes to reach the destination process p_b. There is at least one path connecting the nodes of both processes. It is possible that a message never reaches the destination node, but circulates in the network and is routed an infinite number of times. This situation mainly has its origin in (simplified and incomplete) protocol design, rarely caused by technical failures. For example, the simple but smart routing-based communication protocol SLIP [7] used in microchip-level sensor networks suffered from live-locks occurring under rare certain network topology situations caused by (over) simplification of the protocol implementation. Live-locks are system-critical.
- *Disorder*: The order of messages is not preserved and is critical for proper communication.
- *Data corruption*: A message gets corrupted during its transmission or during routing. Nondetected data corruption can result in process or total system failures and is system-critical.

Reliability in distributed systems can be achieved by the following [20]:

1) Programming fault tolerance
2) Communication fault tolerance
3) Self-organization and replication of autonomous and mobile program (process) units, which is discussed in Chapter 18.

16.9 Distributed Sensor Networks

A distributed sensor network as the central part of a sensing system consists of multiple active sensor nodes, a set $N = \{n_1, n_2, \ldots\}$, connected and arranged in a network forming a graph $G(N,C)$ with edges C, a set $C = \{c_1, c_2, \ldots\}$, connecting nodes and providing communication (and eventually energy transfer) between nodes. Each sensor node provides signal and data processing for a set of sensors $S = \{s_1, s_2, \ldots\}$, communication, some kind of energy supply and management, and sensor interfaces. The connectivity of the sensor nodes can be used for data and power exchange, too. Each sensor node should provide a minimal degree of autonomy and independence from other nodes in the network.

From the computer science point of view, a sensor network can be composed of a set of processes $P = \{p_1, p_2, \ldots\}$ performing parallel data processing and interaction using messages $M = \{m\}$, shown in Figure 16.9. The processes communicate by exchanging data by using messages, and messages are exchanged by the nodes hosting the processes by using communication links, discussed in Chapter 15. Message-based communication, which takes place between processes, requires protocols for source and destination process synchronization and message routing (forwarding). Thus, in this sense and assuming the assumption of the above definition a distributed sensor network can be considered as a distributed computer

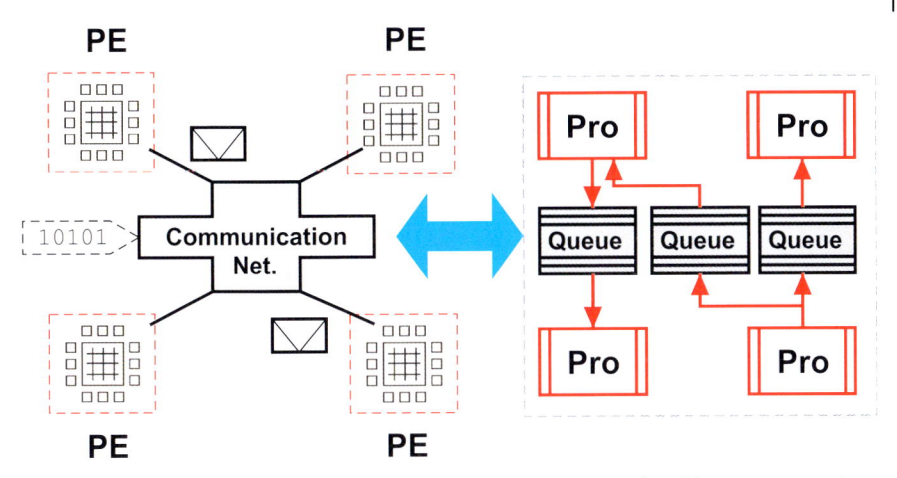

Figure 16.9 Abstraction of distributed computing in sensor networks with processes and message passing using queues.

performing distributed data processing. Interaction between nodes is required to manage and distribute data and to compute information.

The properties and operational behavior of a distributed system can be summarized as follows:

1) Processing elements (PE) represent the physical computational resources and are part of the nodes of the sensor network.
2) Processes (P) are logical resources that perform the data processing for a specific task.
3) Cooperative communication and interprocess communication handled by message passing (instead of a centralized master–slave model).
4) Communication links (the interconnection network) are physical resources and connects PEs.
5) Processes are cooperative by interaction.
6) Communication delay does not affect the overall distributed program behavior (immutability).
7) Robustness in the presence of resource failures (failures of single logical or physical resources do not affect overall system behavior).
8) Runtime reconfiguration: adaptation of single resources or the whole system in the presence of resource failures

Of these physical and logical elements, specifically 1, 2, and 4 are prone to fail.

In Ref. [21], a general discussion of the architectures, communication, and data processing issues in sensor networks can be found.

Sensor network applications introduce application-specific network domains with specific requirements and constraints. These are primarily (see Ref. [22] for details) the following:

Personal Area Networks (PAN)

Personal area networks connect primarily mobile devices or laptops with sensor networks, for example, body area networks (BAN) and wireless body area networks (WBAN) that connect smart sensors with data acquisition devices like mobile phones.

A (W)BAN can incorporate different classes of sensor types, for example, heart beat and EEG monitors, temperature, moisture, pressure, and many more.

The fields of application for BANs are medical diagnostics and smart clothing, but not limited to these main fields.

Sensor networks as a basic part of PANs and the below discussed APNs have different communication requirements. They are equipped with nodes having limited computational resources and storage capacity, limiting the communication capabilities significantly. Energy consumption can be a major challenge in the design of sensor networks.

Ambient and Pervasive Networks (APN)

Ambient and pervasive networks (APN) are strong heterogeneous networks consisting of network nodes ranging from sensors over mobile devices to home computer. They equip buildings inside and ambient environments outside with sensors and computation with the goal to adapt these environments personally and to meet personal needs. Data inference, data mining, and data modeling are the major challenges in pervasive networks. Ontologies as a tool for a common understanding between different platforms, messages, users, and data are required to provide interoperability and a high degree of adaptivity to environmental changes. Personal area networks can be considered as part of these APNs.

Wide Area Networks (WAN)

Wide area networks connect different networks on a long spatial distance scale, basically providing the connectivity for the Internet and all its public and private subnetworks. In contrast to APNs and PANs, they are equipped with network nodes and routers having enough computational power and storage capacity to support high traffic and high bandwidths. The communication over WANs is mostly client–server driven, in contrast to APNs and PANs with data and event driven communication. WANs rely on a unique node and process addressing scheme (IP-4/IP-6) and communication devices with a unique identifier (e.g., MAC address).

WANs are hierarchically organized and there exist usually a lot of paths from a source to a destination node (process) crossing different subnetworks, offering increased reliability based on redundancy and adaptive path finding.

There are basically three different layers of data processing in sensing applications, which require processing platforms with different computational power and storage capacity:

Sensing

Localized acquisition and preprocessing of sensor data, sensor data fusion.

Aggregation

Distribution and collection of sensor data, distributed sensor fusion, and globalized sensor information mapping (using Map&Reduce methods from Big data mining).

Application

Presentation of condensed sensor information, storage, visualization, interaction.

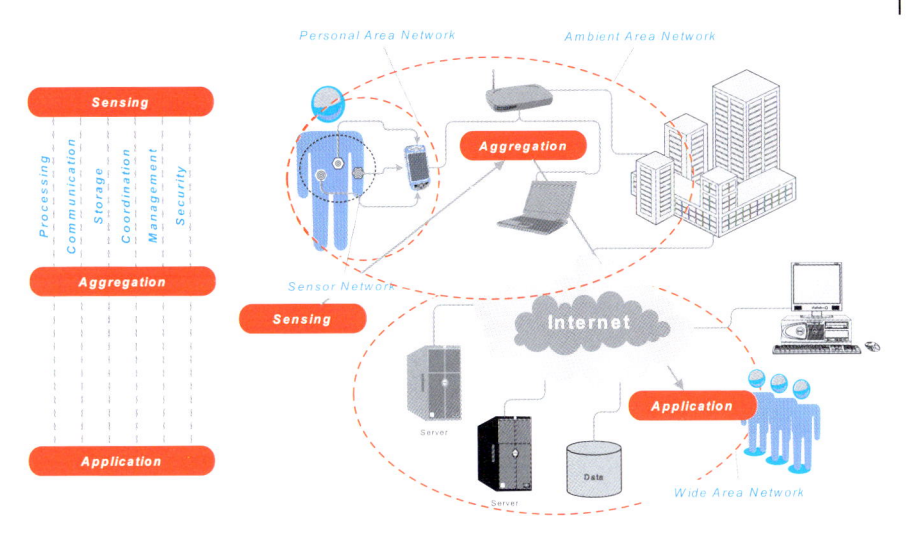

Figure 16.10 A heterogeneous network environment with PAN, APN, and WAN domains (on network level) and sensing, aggregation, and application domains on the use-case level.

These different layers can be scattered in different network domains. The sensing layer is usually located in sensor networks, for example, body area networks, the aggregation layer can be found in personal and ambient area networks, and finally the application layer can be found in ambient area and wide area networks.

The characterization of sensor network features and their operational capabilities can be further divided into the following classes and terms, handled by all three layers of the sensing application, shown in Figure 16.10:

- Processing
- Communication
- Messaging
- Storage
- Ontologies and Data Models
- Manageability
- Security

16.10 Active Messaging and Agents

There are several works using agents to implement active and robust messaging like the directed diffusion behavior, discussed in Ref. [23]. The multiagent model, a collection of individual agents acting locally but following a global goal to be satisfied, is discussed in Chapter 17. Commonly, communication tasks use passive messages that are forwarded from a source to destination node, or more precisely spoken processes as communication endpoints, by active processing units (routers, computers). In the agent model world, a message is active and the routing and transport of the message content is performed by active agents.

References

1 Ghezzo, F., Starr, A.F., and Smith, D.R. (2010) Integration of networks of sensors and electronics for structural health monitoring of composite materials. *Advances in Civil Engineering*, **2010**, 1–13.
2 Baglini, E., Cannata, G., and Mastrogiovanni, F. (2010) Design of an embedded networking infrastructure for whole-body tactile sensing in humanoid robots. 2010 10th IEEE-RAS International Conference on Humanoid Robots, pp. 671–676. doi: 10.1109/ICHR.2010.5686834
3 Löfgren, A., Lodesten, L., and Sjöholm, S. (2005) An analysis of FPGA-based UDP/IP stack parallelism for embedded Ethernet connectivity. IEEE, 23rd NORCHIP Conference.
4 Badri, S. (2011) Junction based routing: a novel technique for large network on chip platforms, Masters thesis, Jönköping Institute of Technology.
5 Li, C., Zhang, H., Hao, B., and Li, J. (2011) A survey on routing protocols for large-scale wireless sensor networks. *Sensors (Basel, Switzerland)*, **11** (4), 3498–3526.
6 Warneke, B., Last, M., and Liebowitz, B. (2001) Smart dust: communicating with a cubic-millimeter computer. *Computer*, 34, 44–51.
7 Bosse, S. and Pantke, F. (2012) Distributed computing and reliable communication in sensor networks using multi-agent systems. *Production Engineering Research and Development*. doi: 10.1007/s11740-012-0420-8
8 Bosse, S. (2011) Hardware-software-co-design of parallel and distributed systems using a unique behavioural programming and multi-process model with high-level synthesis. Proceedings of the SPIE Microtechnologies 2011 Conference, April 18–20, 2011, Prague, Session EMT 102 VLSI Circuits and Systems.
9 Intanagonwiwat, C., Govindan, R., and Estrin, D. (2000) Directed diffusion: a scalable and robust communication paradigm for sensor networks. Proceedings of the 6th annual international conference on Mobile computing and networking, 56–67.
10 Zhao, S., Yu, F., and Zhao, B. (2007) An energy efficient directed diffusion routing protocol. 2007 International Conference on Computational Intelligence and Security (CIS 2007). doi: 10.1109/CIS.2007.70
11 Braginsky, D. and Estrin, D. (2002) Rumor routing algorthim for sensor networks. Proceedings of the Workshop on Wireless sensor networks and applications, pp. 22–31.
12 Haas, Z., Halpern, J., and Li, L. (2006) Gossip-based *ad hoc* routing. *Networking, IEEE/ACM Transactions On*, **14** (3), 479–491.
13 Farooq, Muddassar (2009) *Bee-Insrpied Protocol Enginnering*, Springer.
14 Bosse, S. (2013) Intelligent microchip networks: an agent-on-chip synthesis framework for the design of smart and robust sensor networks. Proceedings of the SPIE 2013, Microtechnologie Conference, Session EMT 102 VLSI Circuits and Systems, 24-26 April 2013, Alpexpo/Grenoble, France, SPIE. doi: 10.1117/12.2017224012
15 Yang, Y., Zincir-Heywood, A.N., Heywood, M.I., and Srinivas, S. (2002) Agent-based routing algorithms on a LAN. IEEE CCECE2002. Canadian Conference on

Electrical and Computer Engineering. Conference Proceedings (Cat. No.02CH37373), 3, pp. 1442–1447.

16 Shakshuki, E., Malik, H., and Xing, X. (2007) Agent-based routing for wireless sensor network. *Advanced Intelligent Computing Theories and Applications. With Aspects of Theoretical and Methodological Issues*, Lecture Notes in Computer Science, 4681, 68–79. doi: 10.1007/978-3-540-74171-8_8

17 Ebrahimi, M., Daneshtalab, M., Liljeberg, P., Plosila, J., and Tenhunen, H. (2011) Agent-based on-chip network using efficient selection method. IEEEIFIP 19th International Conference on VLSI and SystemonChip. IEEE, pp. 284–289. doi: 10.1109/VLSISoC.2011.6081593

18 Zhang, S., He, Z., and Yang, H. (2012) Mobile agent routing algorithm in wireless sensor, in Jin, D., Lin, S. (Eds.), *Advances in Computer Science and Information Engineering*, AISC 169, vol. 2, pp. 105–113.

19 Guerraoui, R. and Rodigues, L. (2006) *Introduction to Reliabe Distributed Programming*, Springer.

20 Wu, J. (1999) *Distributed System Design*, CRC Press.

21 Akyildiz, I.F., Su, W.S.W., Sankarasubramaniam, Y., and Cayirci, E. (2002) A survey on sensor networks. *IEEE Communications Magazine*, 40 (8). doi. 10.1109/MCOM.2002.102442

22 McGrath, M.J. and Scanaill, C.N. (2014) Sensor Technologies, ApressOpen, ISBN: 9781430160134.

23 Malik, H., Shakshuki, E., Dewolf, T., and Denko, M.K. (2007) Multi-agent system for directed diffusion in wireless sensor networks. 21st International Conference on Advanced Information Networking and Applications Workshops (AINAW'07). 2. doi: 10.1109/AINAW.2007.260

17

Distributed and Cloud Computing: The Big Machine
Stefan Bosse[1,2]

[1]*University of Bremen, Department of Mathematics and Computer Science, Robert-Hooke-Str. 5,*
28359 Bremen, Germany
[2]*University of Bremen, ISIS Sensorial Materials Scientific Centre, Wiener Str. 12,*
28359 Bremen, Germany

Cloud computing is a new computation paradigm used to process a large amount of data (Big Data), commonly in a distributed environment. A cloud offers the abstraction of the underlying data storage and computing platforms commonly with a central control instance. Cloud computing pursue multiple goals:

- Virtualization of resources in heterogeneous environments, for example, storage and processors.
- Offering a service-oriented data processing approach, rather than a data-centric.
- Computing of meaningful information from raw data, known as data mining.
- Scalability of data processing,
- Load balancing of processors in a distributed computing environment.
- Storage balancing of data.

A cloud in terms of storage and computation is characterized by and composed of the following [2]:

- A parallel and distributed system architecture.
- A collection of interconnected virtualized computing entities that are dynamically provisioned.
- A unified computing environment and unified computing resources based on a service-level architecture.
- A dynamic reconfiguration capability of the virtualized resources (computing, storage, connectivity, and networks).

The taxonomy of data in large-scale sensor network applications can be split into three categories (based on the WEB data taxonomy in Ref. [1]), each having its own complexity and characteristics):

1) *Bags:* Unordered collections of data with items consisting of scalar, textual, vectors, and multiset values.

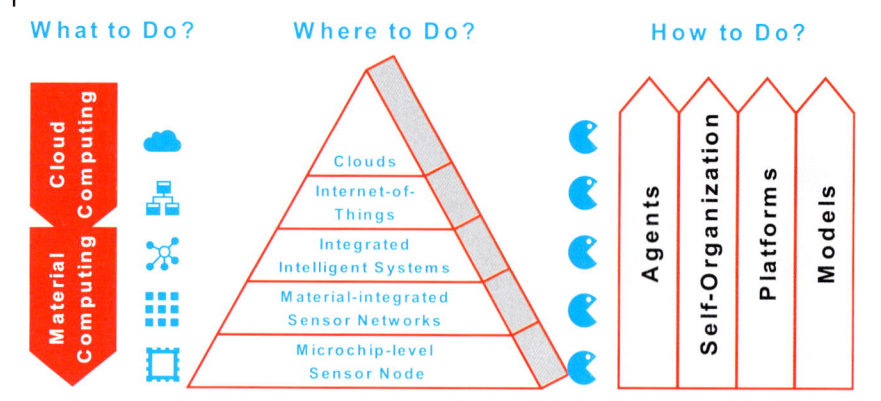

Figure 17.1 From material-integrated sensing systems (MISS) to clouds.

2) *Graphs:* Graph ordered collections of data consisting of nodes associated with data and vertexes defining the relation, that is, the structure, of data. Graphs can be used to compose similarity or query (search) graphs.
3) *Streams:* Unbounded sequences of data items ordered by time. In contrast to low-level signal data streams, the streams in sensor and WEB clouds are string heterogeneous.

Originally, cloud-based computing was primarily used in business processes and consumer–producer scenarios offering distributed databases (storage) and services (computation). In the meantime, a cloud can be any decentralized computing architecture, recently extended by sensor clouds integrating and connecting smart sensors in distributed services, data processing, and data storage. The Internet-of-Things (IoT) will heavily rely on cloud-based services and infrastructures, and hence material-integrated sensing systems (MISS). The IoT can be seen as the bridge technology between the high-density and low-resource MISS and computing and storage clouds, illustrated in Figure 17.1 [2].

Programming consists of computation and communication. This is still the case in cloud-based environments. Mobile agent-based computation and communication can be used to satisfy requirements in such a distributed and loosely coupled machine. Furthermore, the deployment of agents can enable a paradigm shift from central control instances toward self-organizing systems not requiring a central control instance. Mobile agents can be considered as an enabling technology for the seamless information processing ranging from materials to clouds, and cover both material computing (distributed computing in materials) and cloud computing (distributed computing in clouds).

Reference

1 Morales, G.D.F. (2012) *Big Data and the Web: Algorithms for Data Intensive Scalable Computing*, IMT Institute for Advanced Studies, Lucca, Italy.
2 Lehmhus, D. *et al.* (2015) Cloud-based automated design and additive manufacturing: a usage data-enabled paradigm shift. *Sensors MDPI*, **15** (12), 32079–32122.

18

The Mobile Agent and Multiagent Systems
Stefan Bosse[1,2]

[1]*University of Bremen, Department of Mathematics and Computer Science, Robert-Hooke-Str. 5, 28359 Bremen, Germany*
[2]*University of Bremen, ISIS Sensorial Materials Scientific Centre, Wiener Str. 12, 28359 Bremen, Germany*

One major question to be answered in large-scale and wide-area sensing networks is the strategy of decentralized data processing, data delivery, and information reduction. It is a similar issue arising in the Internet-of-Things (IoTs) domain that can be extended by sensorial materials integrated in the IoT domain. New unified data processing and communication methodologies are required to overcome different computer architecture and network barriers. One common distributed and coordinated data processing model is the mobile agent, a self-contained and autonomous virtual processing unit, which can be organized in groups and societies of agents. The mobile agent represents a mobile computational process that can migrate in networks, for example, the Internet domain, as well as in sensor networks. Multiagent systems (MASs) represent self-organizing societies consisting of individuals following local and global tasks and goals including the coordination of information exchange in the design and manufacturing process. A MAS is a collection of autonomous or semiautonomous problem solvers, often solving problems proactively. The agent's action and behavior depends on an abstract model of the environment founded by ontologies. A MAS relies on communication, negotiation, collaboration, and delegation with specialization including instantiation of new agents.

Multiagent systems introduce a paradigm shift from passive messaging performed by active processes to active messages with mobile processes.

Summary of the most important agent behaviors and features that can be used for data processing in distributed sensor networks is as follows:

Autonomy
  Each agent is an autonomous operation unit, independent of other agents, which uses either a virtual or a physical execution platform for data processing.
Mobility
  An agent is capable of migration of its data and processing state (eventually including its program code) from one processing platform to a spatial or logical

different platform by using containers (containing a snapshot of the agent state). After migration the agent process continues operation on the new platform at (or after) the breakpoint of the control flow at which the agent was interrupted.

Creation

An agent can be created at runtime. Agent creation is related with resource allocation. This includes usually a data container for private agent storage and a control container handled by the platform (code) for agent processing.

Inheritance and replication

An agent can be created by another (parent) agent inheriting the control and data state of the parent agent, called forking. This feature enables replication behavior in a group of agents.

Reconfiguration

Reconfiguration can be used to adapt the agent behavior based on learning and environmental perception to improve the system response to agent actions.

Reconfiguration of agents can be performed down to microchip level, for example, proposed in Ref. [1] or [2]. But commonly adaptation of agents is limited to conditional behavior, activity, or action selection at runtime [1,3]. In Ref. [4], code morphing (ability of a program to modify code at runtime) was applied by the agent itself to directly modify its behavior. Code morphing with code executed on a virtual machine is attractive and suitable for creating new behavior of agents by other agents. Code morphing techniques can be implemented efficiently on hardware level.

Agent interaction

Communication of agents is required to implement cooperating and synchronized multiagent systems. Interaction can be performed by exchanging simple messages, which requires an understanding of data types and structures of the participating agents. A tuple space database approach with synchronized access based on pattern matching is both a simple and powerful technique providing advanced fault tolerance at runtime [5], especially suited for heterogeneous agent systems.

The autonomous behavior model of mobile agents provides the capability of performing and unifying
- voting and negotiation,
- M-of-N and N-version systems,
- parallel operation,
- neighborhood computations,
- replication and self-organization, and
- adaptability based on learning,

all contributing to the design of fault-tolerant systems.

Multiagent systems with mobile autonomous agents provide an advanced and more intuitive methodology to model and implement distributed data processing systems including communication compared with traditional immobile program and messaging-related approaches, suitable for sensor networks, proposed in Ref. [6].

Distribution and parallelization

Parallelized image processing was successfully implemented with agent-based systems relying on the inherent computational independence of autonomous agents, the factorization capabilities of images, and agent cooperation with learning to improve the subtask scheduling and planning [7]. In Ref. [8], a distributed and parallel SHM system was implemented with agents consisting of different agents (agent behavior classes) performing different computations in parallel. The agents operate in a wireless sensor network, those sensors were embedded in composite materials. Castanedo *et al.* [9] used a multiagent architecture based on the belief–desire–intention (BDI) agent model for processing the data in a distributed visual sensor network by using data fusion and local information processing instead exchanging raw image data.

In the network section, three different processing layers of a distributed sensing application were introduced:
- Sensing (localized acquisition and preprocessing of sensor data including sensor data fusion)
- Aggregation (collection and distribution of sensor data and preprocessed information, globalized sensor information mapping, and sensor data fusion)
- Application (presentation of condensed sensor information, computation, storage, visualization, and interaction)

These different layers can be occupied by mobile agents performing the described tasks and offering a communication interface between the layers by providing these required features on all layers:
- Processing
- Communication
- Messaging
- Storage
- Ontologies and data models
- Manageability
- Security

18.1 The Agent Computation and Interaction Model

Traditional software is composed of objects and functions specified at design time. Although there are concepts of extending programs at runtime by dynamically loading libraries, these programs are mainly static. Usually, the execution of software is strongly coupled to a specific execution environments and operating systems. Interpreted script languages, for example, *JavaScript* (JS), are more loosely coupled and can be executed on a wider variety of host architectures, operating systems, and specific computers. Finally, a multiprogram system requires communication regulated by protocols. Two programs participating in communication must have significant knowledge of the understanding, meaning, and most of all the encoding of information encapsulated in communication protocols and messages.

Software agent technologies can overcome the limiting barriers in heterogeneous systems that can be affected by technical unreliability. An agent is an operational unit, usually a software process, that satisfies the following properties:

- *Autonomy*, that is, the execution of agents does not require continuously intervention of humans and machines, and agents have the control over their inner state and the actions they perform.
- *Reactivity*, that is, agents respond to an environmental perception that can be the information of the state of a communication network and the connectivity, sensor input, or platform-related properties (like available storage, architecture, and other agents).
- *Proactiveness*, that is, the actions agents can perform do not depend only on the perception; they are planned based on goals an agent should reach, for example, the delivery of sensor data from a source to a sink node in a network.
- *Social ability*, that is, different agents can interact with each other to reach a group goal, based on the progress and actions of other agents.

A MAS is a collection of individual agents capable to communicate and interact with each other, mainly by exchanging messages and acting following a collaborative goal. Self-organizing is a common approach for a MAS to operate in unreliable environments with missing or incomplete world models.

An agent can be considered as a computational unit situated in an environment and world, which performs computation, basically hidden for the environment, and interacts with the environment to exchange basically data. A common computer is specialized to the task of calculation, and interaction with other machines is encapsulated by calculation and performed traditionally by using messages. An agent behavior can be reactive or proactive, and it has a social ability to communicate, cooperate, and negotiate with other agents. Proactiveness is closely related to goal-directed behavior including estimation and intentional capabilities.

Agents record information about an environment state $e \in E$ and history $h: e_0 \to e_1 \to e_2 \to \ldots$. Let $I = S \times D$ be the set of all internal states of the agent consisting of the set of control states S related with activities and internal data D. An agent's decision-making process is based on this information. There is a perception function *see* mapping environment states to perceptions, a function *next* mapping internal states and perceptions $p \in$ Per to internal states (state transition), and the action–selection function *action* that maps internal states to actions $a \in$ Act, shown in Definition 18.1.

Actions of agents modify the environment, which is seen by the agent, thus the agent is part of the environment. Software agents modify data of the environment; hardware agents like robots or cyber-physical systems (CPS) modify the physical world, too. Learning agents can improve their performance to solve a given task if they analyze the effect of their action on the environment. After a performed action, the agent gets a feedback in form of a reward $r(t) = r(e_t, a_t)$. There are strategies $\pi: E \to A$ which map environment states on actions. The goal of learning is to find optimal strategies π^* that is a subset of π. The strategies can be used to modify the agent behavior. Rewarded behavior learning was addressed in Ref. [10], for example, based on Q-learning.

18.1 The Agent Computation and Interaction Model

Definition 18.1   Agent Processing and State Change by Applying Three Basic Functions in a Service Loop

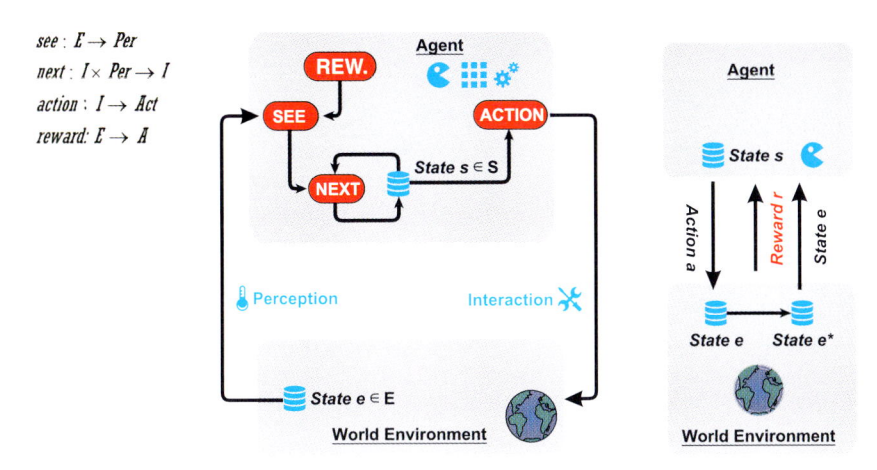

$see: E \rightarrow Per$
$next: I \times Per \rightarrow I$
$action: I \rightarrow Act$
$reward: E \rightarrow A$

Agent behavior models basically consist of perception, reasoning, computation, and acting by performing discrete actions [11], shown in Figure 18.1. Agent computation models are attractive in the context of sensorial and reactive systems found in manufacturing processes due to the inherent processing of sensing data that causes actions. Actions modify the environment, that is, commonly data stored and computations performed outside of the agent (including data and activities of other agents), but they can affect the physical world, too, regarding CPS and machine interaction.

There are basically three major behavior model classes suitable for industrial agents [11]:

Reactive agents

This agent model maps sensory input directly on a set of actions, that is, the agents react immediately to perceptive data. This (*sensor, action*) mapping function can be ambiguous and conflicts and misbehavior can result. To overcome the incomplete sensor data reactivity and to provide runtime adaptability, *subsumption* architectures were proposed [12], shown in Figure 18.1 on the upper right side. The agent behavior is composed of different behavior that can inhibit (block) lower prioritized behavior, resulting in the selection of different (*sensor, action*) mapping functions. In this work, the agent behavior can be transformed at runtime, which is a similar but more efficient and dynamic approach than inhibition used in the subsumption architecture.

Deliberative agents

They use explicitly the more natural formulation of beliefs, desires, and intentions to select plans finally performing actions by a reasoner merging all these parts of a goal-oriented behavior, shown in Figure 18.1 on the lower left side. In the most common BDI architecture [13], the beliefs rely on the sensory input accumulated over time, in contrast to pure immediate reactive behavior. The desires represent the goals of the agent, and the intentions influence future plans and actions.

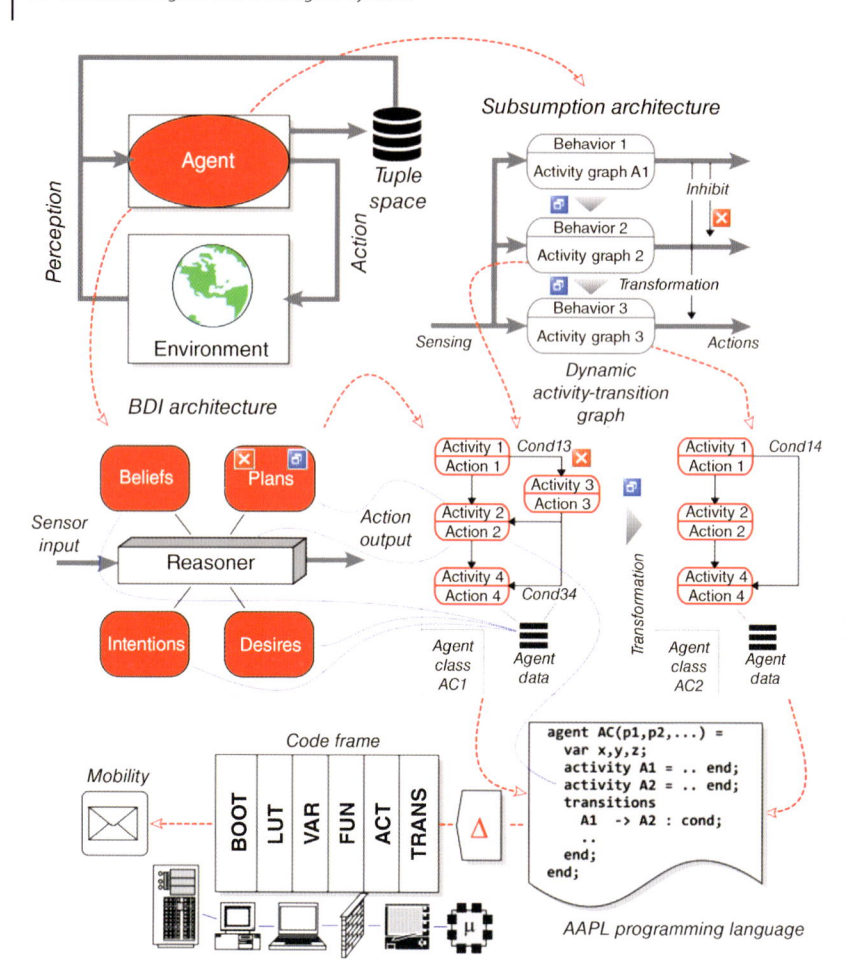

Figure 18.1 Agent models: (left, top) basic agent model; (right, top) subsumption architecture composed of different behavior; (right, middle) reconfigurable (dynamic) activity-transition graphs (DATG) used in this work; (left, middle) relation to the belief–desire–intention (BDI) architecture; and (bottom) programming model and code frame.

Hybrid agents

They combine methods and methodologies of reactive and deliberative agents with stacked layer architectures. For example, the *InteRRap* architecture [14] is composed of a world interface, behavior, plan, and cooperation *layers* propagating sensory input upward and actions downward.

Actions of agents modify the environment, which is seen by the agent, thus the agent is part of the environment. The change of the environment due to agent actions can occur immediately or be delayed, evaluating the history and past. The actions of pure reactive agents base only on the current perception, and usually an action has an effect on the environment immediately.

An agent behavior model can be partitioned into the following tasks, which must be reflected by an agent programming language model by providing suitable statements, types, and structures:

Computation

One of the main tasks and the basic action is computation of output data from input data and stored data (history). Principally, functional and procedural (or object-oriented) programming models are suitable, but history incorporating computed data and storage is handled only by the procedural programming model consequently.

Communication

Communication as the main action serves two canonical goal tuples: (i) (data exchange, synchronization), and (ii) (interaction with the environment, interaction with other agents). The latter goal tuple can be reduced to agent interaction only if the environment is handled by an agent, too. Communication between agents can link single agents to a multiagent system, by using peer-to-peer or group communication paradigms.

Mobility

Mobility of agents increases the perception and interaction environment significantly. Mobile agents can migrate from one computing environment to another finally continuing there their processing. The state of an agent, consisting of the control and data state, must be preserved on migration.

Reconfiguration

Traditional computing systems get a fixed behavior and operational set at design time. Adaption in the sense of behavioral reconfiguration of a system at runtime can significantly increase the reliability and efficiency of the tasks performed.

Replication

These are the methods to create new agents, either created from templates or by forking child agents, which inherit the behavior and state of the parent agent, finally executing in parallel. Replication is one of the major agent behaviors to compose distributed computational and reactive systems.

Agents and objects

Modern data processing is often modeled based on object-oriented programming paradigms. But there is a significant difference between agents and objects. Objects are computational units encapsulating some state and are able to perform actions (by applying methods) and communicate commonly by message passing. Objects are related with object-oriented programming and are not (or less) autonomous in contrast to agents, and they are commonly not mobile. The common object model has nothing common with proactive and social behaviors. But agents can be implemented on top of the object model with methods acting on objects. The modification of the behavior engine is basically not supported by the object programming model. Agents decide for themselves, in contrast objects require external computational units, like operating systems or users, for the decision-making process. Although object-oriented programming can be extended by parallel and concurrent processing (multithreading), multiagent systems are inherently multithreaded.

To summarize, agents are characterized by their autonomy (without or less intervention of users), reactivity, social ability, and po-activeness, combined with behavioral adaptivity based on learning or subclassing. Subclassing selects a particular behavior from a larger set of behaviors, for example, to fulfill a specific task only. This is basically reflected by the well-known divide–conquer approach that splits a large problem and its solving in smaller subtasks.

18.2 Dynamic Activity-Transition Graphs

The behavior of an activity-based agent is characterized by an agent state, which is changed by activities. Activities perform perception, plan actions, and execute actions modifying the control and data state of the agent. Activities and transitions between activities are represented by an activity-transition graph (ATG), suitable for the behavioral modeling of reactive agents. The transitions start activities commonly depending on the evaluation of agent data (body variables), representing the data state of the agent, as shown in Figure 18.2.

An activity-transition graph, related to an agent behavior class, discussed later, consists of a set of activities $\mathcal{A} = \{A_1, A_2, \ldots\}$, and a set of transitions $\mathcal{T} = \{T_1(C_1), T_2(C_2), \ldots\}$, which represent the edges of the directed graph. The execution of an activity, composed itself of a sequence of actions and computations, is related with achieving a subgoal or a satisfying a prerequisite to achieve a particular goal, for example, sensor data processing and distributions.

Usually agents are used to decompose complex tasks into simpler ones, based on the composition of MAS. Agents can change their behavior based on learning and environmental changes, or by executing a particular subtask with only a subset of the original agent behavior.

The ATG behavior model is closely related to the interaction of agents with the environment, here mainly by exchanging data by using a tuple space database and migration. Message passing between agents is available by passing signals that execute signal handler on the destination agent asynchronously. The execution of

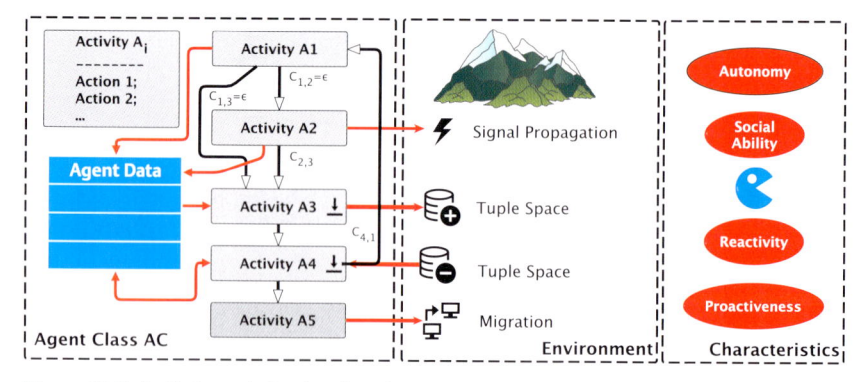

Figure 18.2 (Left) Agent behavior given by an activity-transition graph and the interaction with the environment performed by actions executed within activities. (Right) Agent characteristics.

signal handlers changes the agent data and hence has an impact on activity transitions.

The characteristics of agents can be classified in autonomy, social ability and social interaction, reactivity with respect to changes of the environment and learning based on history and rewards, and finally proactiveness by making assumptions about the estimated change of the environment resulting from actions performed by the agent.

An ATG describes the complete agent behavior. Any subgraph and part of the ATG can be assigned to a subclass behavior of an agent. Therefore, modifying the set of activities $\mathcal{A}$ and transitions $\mathcal{T}$ of the original ATG introduces several sub-behaviors implementing algorithms to satisfy a diversity of different goals.

The reconfiguration of activities $\mathcal{A} = \{\mathcal{A}_1 \subseteq \mathcal{A}, \mathcal{A}_2 \subseteq \mathcal{A}, \ldots\}$ derived from the original set $\mathcal{A}$ and the modification or reconfiguration of transitions $\mathcal{T} = \{T_1, T_2, \ldots\}$ enables dynamic ATGs and agent subclassing at runtime, shown in Figure 18.3.

18.3 The Agent Behavior Class

Behavior. A particular agent class AC_i is related with the previously introduced ATG that defines the runtime behavior and the computational action.

Perception. An agent interacts with its environment by performing data transfer using a unified tuple space with a coordinated database-like interface. Data from the environment influence the following behavior and action of an agent. Data passed to the environment (e.g., the database) influence the behavior of other agents.

Memory. State-based agents perform computation by modifying data. Since agents can be considered as autonomous data processing units, they will primarily modify private data, and a computational outcome using this data will be transferred to the environment. Therefore, each agent and agent class will include a set of body variables $V = \{v_1 : DT, v_2 : DT, \ldots\}$, which are modified by actions in activities and read in activities and transitional expressions (with DT: set of supported data types).

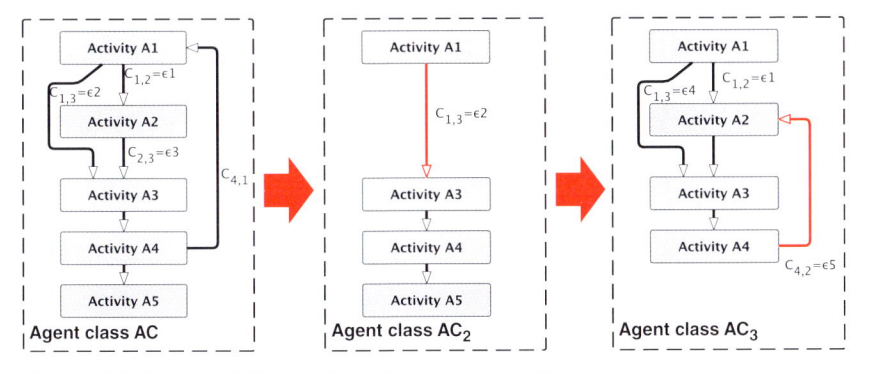

Figure 18.3 Dynamic ATGs: transformation and composition.

Parameter. Agents can be instantiated at runtime from a specific agent class creating agents with equal initial control and data states. To distinguish individual agents (creating individuals), an external visible parameter set $\mathbb{P} = \{p_1 : \text{DT}, p_2 : \text{DT}, \ldots\}$ is added, which get argument values on instantiation, enabling the creation of different agents regarding the data state. Inside an agent class, parameters are handled like variables.

To summarize, an agent class is fully defined by the *ATVPFH* tuple:

$$\begin{aligned}
\text{AC}_i &= \langle \mathbb{A}, \mathbb{T}, \mathbb{V}, \mathbb{P}, \mathbb{F}, \mathbb{H} \rangle \\
\mathbb{A} &= \{a_1, a_2, \ldots, a_n\} \\
\mathbb{T} &= \{t_{ij} = t_{ij}(a_i, a_j, \text{cond}) : a_i \xrightarrow{\text{cond}} a_j ; i,j \in \{1, 2, \ldots, n\}\} \\
a_i &= \{i_1, i_2, \ldots, i_u \in \text{ST}\} \\
\mathbb{V} &= \{v_1, v_2, \ldots, v_m\} \\
\mathbb{P} &= \{p_1, p_2, \ldots, p_l\} \\
\mathbb{F} &= \{f_1, f_2, \ldots, f_j\}, \quad \text{with } f_i : (x_1, x_2, \ldots) \to y \\
\mathbb{H} &= \{h_1, h_2, \ldots, h_k\}, \quad \text{with } h_i : (v?) \to ()
\end{aligned} \qquad (18.1)$$

with *F* as a set of computational or procedural functions, and H a set of signal handlers, discussed in the next section.

18.4 Communication and Interaction of Agents

Communication and interaction is one major paradigm in distributed computing systems and enables synchronization and negotiation between parallel and concurrent processes. One suitable approach used in the agent behavior model is the interaction of agents by exchanging data using a tuple database as a shared object supporting synchronized and atomic read, test, remove, and write operations. Agents can communicate and synchronize peer-to-peer by using signals, which can be delivered to remote execution nodes, too.

A tuple space is basically a shared memory database used for synchronized data exchange among a collection of individual agents as a suitable MAS interaction paradigm. The scope and visibility of a tuple space database can be unlimited and visible and distributed in the whole network, or limited to a local scope, for example, network node level. A tuple space provides abstraction from the underlying platform architecture, and offers a high degree of platform independence, vital in a heterogeneous network environment.

A tuple database stores a set of *n*-ary data tuples, $\text{tp}_n = (v_1, v_2, \ldots, v_n)$, a *n*-dimensional value tuple. The tuple space is organized and partitioned in sets of *n*-ary tuple sets $\nabla = \{\text{TS}_1, \text{TS}_2, \ldots, \text{TS}_n\}$. A tuple is identified by its dimension and the data type signature. Commonly, the first data element of a tuple is treated as a key. Agents can add new tuples (output operation) and read or remove tuples (input operations) based on tuple pattern templates and pattern matching, $\text{pat}_n = (v_1, x_2?, \ldots, v_j, \ldots, x_j?, \ldots, v_n)$, a *n*-dimensional tuple with actual and formal parameters. Formal parameters are wild-card place holders, which are replaced with values from a matching tuple and assigned to agent variables. The input operations can suspend the agent processing if there is actually no matching tuple is available. After a matching tuple was stored, blocked agents are resumed and can continue processing. Therefore, tuple databases provide interagent synchronization, too. This tuple space approach can be used to build distributed data structures and the atomicity of tuple operations provides data structure locking. The tuple space represents the knowledge of agents.

One of the simplest interaction models are signals. They are sent from one source agent and delivered to one or more destination agents. An agent receiving a signal, which can carry simple data or being pure symbolic, will process a signal by a signal handler, that is, a function. Signal propagation is asynchronous, that is, the sender of a signal is not aware of the successful receiving of a signal. In contrast to tuple space interaction signals are not reliable, which must be considered by the agent behavior design.

18.5 Agent Programming Models

There are already multiple agent programming languages and processing architectures, like APRIL [15] providing tuple space like agent communication, and widely used FIPA ACL and KQGML [16] focusing on high-level knowledge representations and exchange by speech acts, or model-driven engineering (e.g., INGENIAS [3]). But required fine-grained resource and processing control is missing, preventing the deployment of agents in large-scale heterogeneous applications, which is addressed properly by the reactive *AAPL* programming language, discussed below. Furthermore, most computing algorithms are given in a conventional programming language model hard to implement with knowledge- or plan-based models.

AAPL programming language
   The ATG-based agent programming language (AAPL) was designed
1) to aid and simplify the programming of multiagent systems based on a reactive agent behavior model;
2) to enable direct synthesis of optimized hardware platforms from the AAPL programming level; and
3) to enable software and simulation model synthesis without additional effort, too.
   AAPL statements are directly related to the activity-transition graph model and offer the following:
1) The definition of an agent classes containing the agent data and behavior.
2) The instantiation (creation), management, and migration of agents;
3) The inheritance of agent behavior and state by child agents using forking;
4) The modification of the ATG agent behavior at runtime, based, for example, on learning;
5) The agent interaction using signals (simple messages), basically used in coupled parent–child groups;

6) The agent interaction using a tuple space database providing synchronized data exchange with n-ary tuples $T(v_1, v_2, \ldots)$ and patterns $P(v_1, p_2?, v_2, \ldots)$, basically used in loosely coupled and heterogeneous MAS;
7) The computational and control flow statements including modern exception handling.

The activity-graph-based agent model introduced in the previous section is attractive due to the proximity to the finite-state machine model, which simplifies the hardware implementation of the agent processing platform.

An activity is activated by a transition depending on the evaluation of (private) agent data (conditional transition) related to a part of the agents belief in terms of *BDI* architectures, or using unconditional transitions (providing sequential composition). An agent belongs to a specific parameterizable agent class AC, specifying local agent data (only visible for the agent itself), types, signals, activities, signal handlers, and transitions.

Plans are related to *AAPL* activities and transitions close to conditional triggering of plans. Table 18.1 summarizes the available language statements.

Instantiation: New agents of a specific class can be created at runtime by other agents using the new AC($v_1, v_2, \ldots$) statement returning a node unique

Table 18.1 Summary of AAPL statements.

Agent class definition	Agent interaction and control
agent AC($p_1, p_2, \ldots$) =	out($v_1, v_2, \ldots$);
variables	in($v_1, x_2?, x_3?, v_4, \ldots$); try_in..
activities	mark(*TMO*, $v_1, \ldots$);
transitions	exist?($v_1, ?, \ldots$); rm($v_1, ?, \ldots$);
functions	send(*ID*, *SIG*, *arg*);
handler	broadcast(*AC*, *DX*, *DY*, *SIG*, *arg*);
end;	timer+(*TMO*, *HANDLER*); timer-..
Agent instantiation and management	Procedural statements
var *ID* := new AC($v_1, v_2, \ldots$);	x := ϵ;
var *ID* := fork($v_1, v_2, \ldots$);	if *cond* then *stm1* else *stm2*;
kill(*ID*);	f($v_1, v_2, \ldots$);
link?(*DIR*) moveto(*DIR*);	for I = 1 to 100 do ..;
Agent activities and ATG modification[a]	Transitions and ATG modification
activity A_i =	transitions =
statements	A_i -> A_j : ϵ; .. end;
end;	transition+(A_i, A_j, ϵ);
activity+($A_1, A_2, \ldots$); [a]	transition-(A_i, A_j);
activity-($A_1, A_2, \ldots$); [a]	transition*(A_i, A_j, ϵ);
Static ATG subclassing	Dynamic ATG subclassing
subclass AC'_i =	class AC' := new [*empty*] class AC;
variables	activity+ $AC'(A_1, A_2, \ldots)$; [a]
activities	activity- $AC'(A_1, A_2, \ldots)$; [a]
transitions	transition+ $AC'(A_i, A_j, \epsilon)$;
functions	transition- $AC'(A_i, A_j)$;
handler	
end;	

a) Not supported/available on the PCSP platform architecture.

agent identifier. An agent can create multiple living copies of itself with a fork mechanism, creating child agents of the same class with inherited data and control state but with different parameter initialization, done by using the fork($v_1, v_2, ..$) statement. Agents can be destroyed by using the kill(ID) statement.

Each agent has *private data* (body variables), defined by the var and var* statements. Variables in the latter statement will not be inherited or migrated! Agent body variables in conjunction with transition conditions represent the mobile data part of the agent's beliefs database.

Statements inside an activity are processed sequentially and consist of data assignments ($x := \epsilon$) operating on agent's private data, control flow statements (conditional branches and loops), and special agent control and interaction statements.

Agent interaction and synchronization is provided by a tuple space database server available on each node (Ref. [15]). An agent can store an *n*-dimensional data tuple ($v_1, v_2, ...$) in the database by using the out($v_1, v_2, ..$) statement (commonly the first value is treated as a key). A data tuple can be removed or read from the database by using the in($v_1, p_2?, v_3, ..$) or rd($v_1, p_2?, v_3, ..$) statements with a pattern template based on a set of formal (variable,?) and actual (constant) parameters. These operations block the agent processing until a matching tuple was found/stored in the database. These simple operations solve the mutual exclusion problem in concurrent systems easily. Only agents processed on the same network node can exchange data in this way. Simplified the expression of beliefs of agents is strongly based on *AAPL* tuple database model. Tuple values have their origin in environmental perception and processing bound to a specific node location.

The existence of a tuple can be checked by using the exist? function or with atomic test-and-read behavior using the try_in/rd functions. Tuple spaces interaction is generative, that is, the lifetime of a tuple created by a process (agent) can exceed the lifetime of the process. A tuple with a limited lifetime (a marking) can be stored in the database by using the mark statement. Tuples with exhausted lifetime are removed automatically (by a garbage collector). Tuples matching a specific pattern can be removed with the rm statement.

Remote interaction between agents is provided by signals carrying optional parameters (data). Signals can be used locally, too. A signal can be raised by an agent using the send(ID, S, V) statement specifying the ID of the target agent, the signal name S, and an optional argument value V propagated with the signal. The receiving agent must provide a signal handler (like an activity) to handle signals asynchronously. Alternatively, a signal can be sent to a group of agents belonging to the same class AC within a bounded region using the broadcast(AC, DX, DY, S, V) statement. Signals implement remote procedure calls. Within a signal handler a reply can be sent back to the initial sender by using the reply(S, V) statement.

Timers can be installed for temporal agent control using (private) signal handlers, too. Agent processing can be suspended with the *sleep* and resumed with the wakeup statements.

Migration of agents (preserving the local data and processing state) to a neighbor node is performed by using the `moveto(DIR)` statement, assuming the arrangement of network nodes in a graph-like networks. To test if a neighbor node is reachable (testing connection liveliness), the `link?(DIR)` statement returning a Boolean result can be used.

Reconfiguration: Agents are capable to *change their transitional network* (initially specified in the transition section) by changing, deleting, or adding (conditional) transitions using the `transitionX(`S_1, S_2`,cond)` statements (with X = '+': add, '−': remove, and '∗': change transition). *This behavior allows the modification of the activity graph, that is, based on learning or environmental changes, which can be inherited by child agents.*

Furthermore, the set of activities can be changed at runtime using `activityX(`A_1, A_2`,...)` statements (with X = '+': add, '−': remove activities), creating, for example, subclasses on the fly. New subclass templates can also be derived from existing class templates by using the `AC' := new class AC` statement and the ATG modification statements. Note that class activities and transitions reference class variables and other functions. Therefore, only subclasses can be composed (regarding the set of activities). The new subclass can be initially empty by applying an additional `empty` keyword, but still referencing the superclass. The set of class parameters are identical in the superclass and subclass.

An example for an AAPL agent class implementation is shown in Example 18.1. The explorer agent has the goal to collect sensor data in a network region of interest and to recognize a significant stimulus using a strategy basing on a divide-and-conquer approach with child explorer agents.

This agent class uses dynamic ATG reconfiguration for the creation of child explorer agents at runtime, mainly performed in the activities percept and percept_neighbor. Child and parent agents communicate via the tuple data base and signals.

Example 18.1

AAPL Sensor Explorer Agent Class Template

```
1   type direction = {NORTH,SOUTH,EAST,WEST,ORIGIN};
2
3   type keys = {ADC,FEATURE,H,MARK};
4   val TMO := 1 sec;
5   val MAXLIVE := 15;
6   val RADIUS := 4;
7   val ETAMIN := 4;
8   val ETAMAX := 8;
9   val DELTA := 10;
10
11  signal WAKEUP,TIMEOUT;
12  agent explore(dir: direction, radius: integer[1..16]) =
13    var dx,dy:integer[-100..100]; long term data
```

```
14          live:integer[0..15];
15          h:integer[0..50];
16          backdir: direction;
17          s0: integer[0..1023];
18          group: integer[0..255];
19          b:boolean;
20
21     var* s: integer[0..1023]; short term data
22          enoughinput: integer[0..15];
23          again: boolean;
24
25     activity start =
26        dx := 0; dy := 0; h:= 0;
27        if dir <> ORIGIN then
28           moveto(dir);
29           case dir of
30              | NORTH -> backdir := SOUTH;
31              | SOUTH -> backdir := NORTH;
32              | WEST  ->  backdir := EAST;
33              | EAST  ->  backdir := WEST;
34           end;
35        else
36           live := MAXLIVE;
37           backdir := ORIGIN;
38        end;
39        group := random(integer[0..1023]);
40        transition*(move,percept_neighbour);
41        out(H,id(self),0);
42        rd(ADC,s0?);
43     end;
44
45     activity move =
46        case dir of
47           | NORTH -> backdir := SOUTH; incr(dy);
48           | SOUTH -> backdir := NORTH; decr(dy);
49           | WEST  ->  backdir := EAST;  decr(dx);
50           | EAST  ->  backdir := WEST;  incr(dx);
51        end;
52        moveto(dir);
53     end;
54
55     activity diffuse =
56        decr(live);
57        rm(H,id(self),?);
58        if live > 0 then
59           case backdir of
60              | NORTH -> dir := random({SOUTH,EAST,WEST});
```

```
61              | SOUTH -> dir := random({NORTH,EAST,WEST});
62              | WEST  -> dir := random({NORTH,SOUTH,EAST});
63              | EAST  -> dir := random({NORTH,SOUTH,WEST});
64              | ORIGIN -> dir := random({NORTH,SOUTH,EAST,WEST});
65           end;
66        else kill(ME); end;
67      end;
68
69      activity reproduce =
70        var n:integer;
71        stay here, increase feature counter
72        rm(H,id(self),?);
73        if exist?(FEATURE,?) then in(FEATURE,n?);
            else n := 0; end;
74        out(FEATURE,n+1);
75        decr(live);
76        after some time perform perception again...
77        if live > 0 then
78          for nextdir in direction do
79            if nextdir <> backdir and link?(nextdir) then
80              eval(fork(nextdir,radius));
81            end;
82          end;
83        end;
84        transition*(reproduce,stay);
85      end;
86
87      Master perception
88      activity percept =
89        Send out child agents to explore
90        the neighbourhood and calculate partial H values
91        enoughinput := 0;
92        transition*(percept,move);
93        for nextdir in direction do
94          if nextdir <> backdir and link?(nextdir) then
95            incr enoughinput;
96            eval(fork(nextdir,radius));
97          end;
98        end;
99        transition*(percept,diffuse,
100                   (h<ETAMIN or h > ETAMAX)
                       and enoughinput < 1);
101       transition+(percept,reproduce,
102                   h>=ETAMIN and h <= ETAMAX and
                       enoughinput > 1);
103       Wait for child agents delivering h values or timeout
```

```
104        timer+(TMO,TIMEOUT);
105     end;
106
107     Child perception
108     activity percept_neighbour =
109        if not exist?(MARK,group) then
110           mark(TMO,MARK,group);
111           enoughinput := 0;
112           rd(ADC,s?);
113           out(H,id(self), calc());
114           transition*(percept_neighbour,move);
115           for nextdir in direction do
116              if nextdir <> backdir and inbound(nextdir)
                     and link?(nextdir) then
117                 incr enoughinput;
118                 eval(fork(nextdir,radius));
119              end;
120           end;
121           transition*(percept_neighbour,goback,
                  enoughinput < 1);
122        Wait for child agents delivering h values or timeout
123           timer+(TMO,TIMEOUT);
124        else
125           transition*(percept_neighbour,goback);
126        end;
127     end;
128
129     activity goback =
130        if exist?(H,id(self),?) then in(H,id(self),h?);
              else h := 0; end;
131        moveto(backdir);
132     end;
133
134     activity deliver =
135        var v:integer;
136        in(H,id(parent),v?);
137        out(H,id(parent),h+v);
138        send(id(parent),WAKEUP);
139        kill(ME);
140     end;
141
142     activity stay =
143        Stay here and wait
144     end;
145
146     handler WAKEUP =
```

```
147       decr enoughinput;
148       eval(try_rd(0,H,id(self),h?));
149       if enoughinput < 1 then timer-(TIMEOUT); end;
150     end;
151
152     handler TIMEOUT =
153       enoughinput := 0;
154       again := true;
155     end;
156
157     function calc():integer =
158       if abs(s-s0) <= DELTA then return 1;
            else return 0; end;
159     end;
160
161     function inbound(nextdir:direction):bool =
162      case nextdir of
163         | NORTH -> return (dy < RADIUS);
164         | SOUTH -> return (dy > -RADIUS);
165         | WEST ->  return (dx > -RADIUS);
166         | EAST ->  return (dx < RADIUS);
167       end;
168     end;
169
170     transitions = Some transitions are changed or added
            at run-time
171       start -> percept;
172       percept -> move;
173       move -> percept_neighbour;
174       percept_neighbour -> move;
175       diffuse -> start;
176       reproduce -> start;
177       goback -> deliver;
178       diffuse -> start;
179     end;
180   end;
```

18.6 Agent Processing Platforms and Technologies

An agent processing platform must provide the following capabilities and services:

1) Execution of agents
2) Instantiation of new agents at runtime
3) Support for migration of mobile agents
4) Provision of agent interaction services

One common agent platform is JADE that bases on the Java VM bytecode platform, and using AgentSpeak/Jason as a programming framework, discussed in Refs. [17,18]. Although this platform architecture is publicly available and attractive because programmers can reuse a well-known programming language, this approach is limited to full equipped computers offering a sufficient amount of resources (Main Memory >100 MB, Computational Power >100 MIPS), and is not suitable for low-resource embedded systems. Low-resource and efficient agent platforms play a key role in distributed sensor networks that are embedded in materials. To enable the design of material-integrated sensing systems, a new low-resource agent platform is introduced. This platform can be entirely implemented in hardware (HW platform technology class). Alternatively, there are also software implementations (SW platform technology class). All technology classes (HW/SW) are fully compatible on operation level supporting agent migration between hardware and software platforms.

Agents are already deployed successfully for scheduling tasks in production and manufacturing processes [19], and newer trends pose the suitability of distributed agent-based systems for the control of manufacturing processes [20], facing not only manufacturing, but maintenance, evolvable assembly systems, quality control, and energy management aspects, finally introducing the paradigm of industrial agents meeting the requirements of modern industrial applications. The MAS paradigm offers a unified data processing and communication model suitable to be employed in the design, the manufacturing, logistics, and the products themselves.

The scalability of complex industrial applications using such large-scale cloud-based and wide area distributed networks deals with systems deploying thousands up to million agents. But the majority of current laboratory prototypes of MAS deal with less than 1000 agents [20]. Currently, many traditional processing platforms cannot yet handle big numbers with the robustness and efficiency required by industry [21,22]. In the past decade the capabilities and the scalability of agent-based systems have increased substantially, especially addressing efficient processing of mobile agents.

Multiagent systems are successfully deployed in sensing applications, for example, structural load and health monitoring, with a partition in off- and online computations [23]. Distributed data mining and Map-Reduce algorithms are well suited for self-organizing MAS. Cloud-based computing, for example, as a base for cloud-based manufacturing, means the virtualization of resources, that is, storage, processing platforms, sensing data, or generic information. Hence, agent processing platform are logical virtualized processing environments, hiding technical details of the underlying physical execution platforms (microprocessors, WEB browser, digital logic, etc.).

Agent-on-chip architectures. Traditionally, agent programs are interpreted, leading to a significant decrease in performance. Agent-on-chip processing platform can be directly implemented in standalone hardware nodes without intermediate processing levels and without the necessity of an operating system, but still enabling software implementations that can be embedded in applications (Ref. [24]). The common agent behavior and programming ATG/*AAPL* model enables the processing of large-scale multiagent systems on hardware and software platforms deployed in heterogeneous networks. Agents are mobile units and can migrate between processing nodes within the network, which must be supported by the agent processing platform. The agent behavior "program" can be changed at runtime by agents, for example, based on learning or subclassification. Very low-resource and high-efficiency platforms must implement the agent behavior application-specific by using high-level synthesis to map the agent behavior of multiagent systems entirely on microchip level.

There are programmable agent processing platforms (basing on machine virtualization) that can be deployed in strong heterogeneous network environments [25], too, ranging from single microchip up to WEB JavaScript implementations, all being fully compatible on operational and interface level, and hence agents can migrate between these different platforms. A programmable platform is more flexible than the application-specific, but requires more resources (chip area, power, and storage).

Support for Heterogeneous Networks   Commonly different Hardware and Software Agent Platforms are connected in one network or network graphs. Logical networks of nodes servicing a specific task are mapped on physical platforms differing in computational capabilities, size, and power requirements, as shown in Figure 18.4.

Agents must be capable to migrate between different physical platforms! Furthermore, the agent's goals and behavior define the task to be performed requiring generic agent processing platforms suitable for a wide range of different agent classes. Material-integrated sensor networks as part of a global sensing system will only perform a reduced set of tasks, mainly related to the sensor level. Therefore, the deployment of the application-specific platform implementing the agent behavior directly can be considered, offering low-resource platforms capable for microchip scaling.

Programmable Versus Application-Specific Platforms   There are basically two different platform architectures that can be used for the processing of large-scale MAS: programmable platforms based on mobile program code that contains the complete agent behavior and agent state, and nonprogrammable platforms

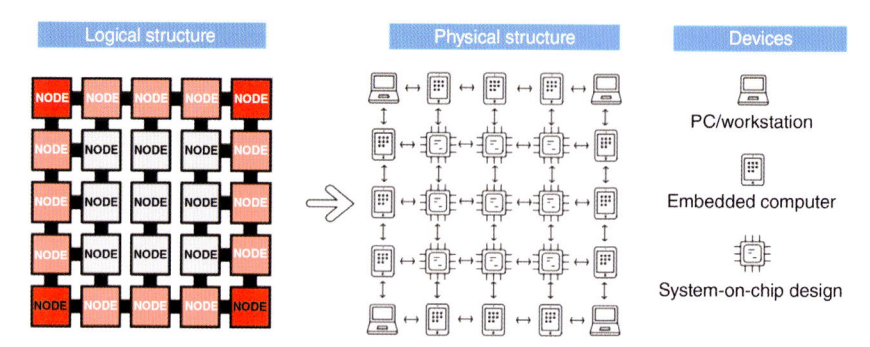

Figure 18.4 Mapping of logical on physical node structures.

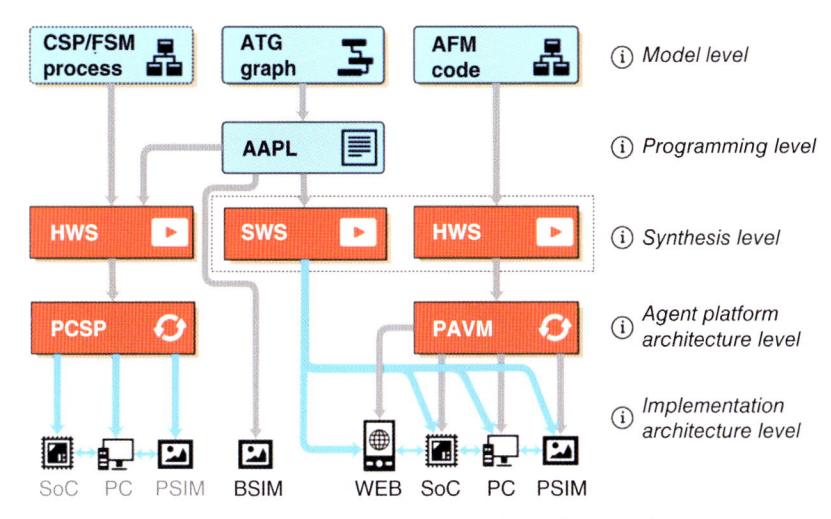

Figure 18.5 PCSP and PAVM agent processing platforms designed from a common programming level. CSP, communicating sequential processes; FSM, finite-state machine; ATG, activity-transition graph; AFM, agent forth machine; AAPL, ATG agent programming language; HWS, hardware synthesis; SWS, software synthesis; PCSP, pipelined communicating sequential processes; PAVM, pipelined agent virtual machine.

implementing the agent behavior application-specific in the platform directly, and only the agent state is mobile.

In Ref. [26] an application-specific (*PCSP*), and in Ref. [25] a programmable (*PAVM*) agent processing platform were proposed. Both are capable of supporting multiple agent behavior classes with a large number of instantiated agents and are based on a token-based agent scheduling offering advanced resource sharing and low-resource requirements (Figure 18.5).

PCSP Platform Toward the implementation of the (hardware) agent processing platform and for ATG analysis, the ATG is transformed in a state-transition (ST) petri-net (PN). Activities are mapped on states of the PN; conditional transition expressions and transition scheduling are merged with the activity states! Agents are represented by tokens passed by transitions between states of the PN. Only one token can be consumed by an activity state.

The PN is finally mapped on a communicating sequential process model architecture. The original activities are mapped on these sequential processes and the individual agents are represented with the aforementioned tokens, giving an unique agent handlers and identifiers, passed by queues between activity processes offering a pipelined CSP architecture.

In some cases process transformations must be applied to implement the correct behavior of *AAPL* statements. There are differences in pure computational and I/O-related activities. I/O statements can block (suspend) the agent processing until an event occurs.

Timed Petri-Net analysis can be used to optimize the execution of multiple agents in the same activity state and belonging to one agent class AC. Processes of computational activities with high computation time can be refactored and split into smaller units and processes with intermediate queues, and/or replicated to enable parallel agent processing to improve the overall pipeline throughput.

Hardware platform and high-level synthesis. The HW platform design is application specific with static resources implementing all supported agent classes for a maximal number of agents, which must be fixed at design-time. All parts are integrated in one system-on-chip (SoC) design on RT level. The HW platform contains all components required to create, process, and migrate agents:

- Agent manager
- Agent management and configuration tables
- Communication modules
- Agent data
- Agent processing pipelines (one for each supported agent class)

A multistage high-level synthesis approach is used to map the *AAPL* source code specification to microchip SoC level. Creation of agents at runtime is performed by creating an agent handler (token). Migration of agents is performed by transferring the data state of the agent (consisting of data of all body variables) and the control state, which is the next activity after migration.

The processing architecture consists of pipelined communicating processes. Individual agents are represented by tokens (natural numbers) that are transferred by queues from an outgoing activity to another incoming activity depending on the activity transitions specified in the *AAPL* model. An activity is implemented with a finite-state machine consuming an agent token from the input queue, processing the activity statements, and finally passing the token to an outgoing queue. There is only one in-coming transition queue for each process. There are two different kinds of activities: pure computational nonblocking activities, and event-based activities that can block the agent processing until an event happened, for example, the availability of external data or a time delay. These two activity processing kinds require different implementations.

Agent interaction is provided (locally on network node level) by a tuple space dictionary (lookup table) and globally by using signals. Pending agent signals are checked each time an agent token was removed from a queue. The tuple dictionary operations are event-based statements that can block the agent processing. Thus IO/event-based processes remove an agent token from the input queue, check the availability of requested data or the happening of an event, and in the case the request cannot be satisfied, the agent token is returned to the input queue, and the next agent token (if any) can be processed.

Local agent data are stored in a region of a memory module assigned to each agent and addressed by the agent token number. Each agent class is assigned to its own memory module. An agent is started by an agent manager that transfers the agent token into the start queue, or in the case of a migrated agent into the queue related to the last agent control state. Terminating or migrating agents are handled by a manager queue (Figure 18.6).

Software platform. For performance reasons and the lack of fine-grained parallelism, the ATG of an agent class is implemented differently:

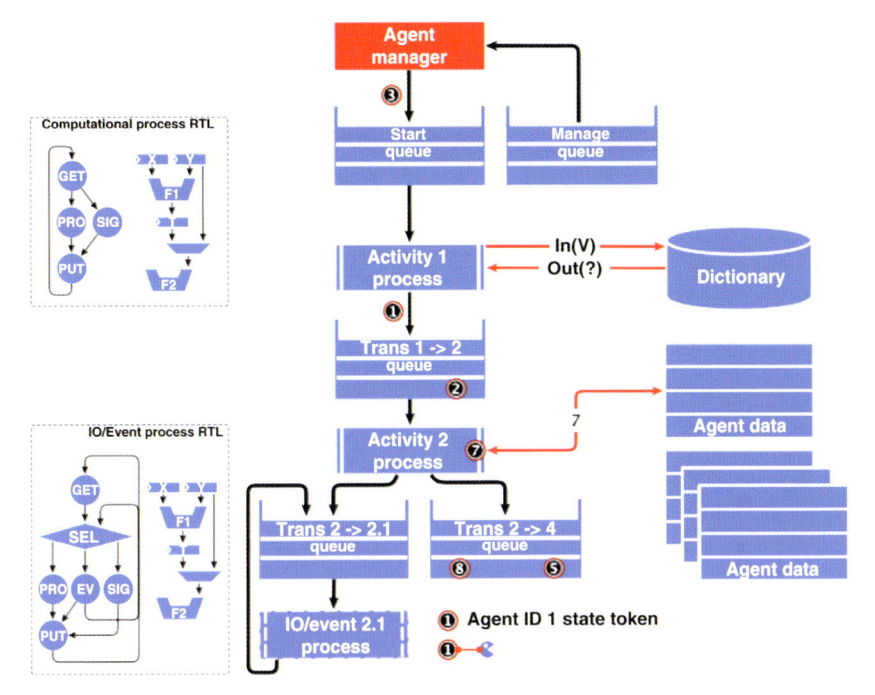

18.6 Agent Processing Platforms and Technologies

Figure 18.6 Agent implementation with a pipelined multiprocess architecture allowing operational unit sharing. Sequential processes are assigned to agent activities (representing states), queues implement state transitions. Interagent communication is performed by using a dictionary. Local agent data are stored in a region of a memory module assigned to each agent.

- Each activity of an agent class and transitions with cond. expressions are implemented with functions:

$A_i \rightarrow FA_i()\{I_1; I_2; ..; \text{return};\}$
$T_{x,y} \mid \text{Cond} \rightarrow FC_{x,y}\{\text{return cond};\}$

- Transitions are stored in dynamic lists:

```
struct tl {void (*from)(); void (*to)(); int (*cond)();
struct tl *next;};
```

- Modification of the transitional network modifies the transition lists
- Transitions between activities are handled by a transition scheduler that calls the appropriate activity functions:

```
while(!die) {next=schedule(curr,tl); curr=next; curr()}
```

- Multithreading: At runtime each agent is assigned to a separate thread executing the transition scheduler
- Operating system or multithreading primitives are used to synchronize agents (event, Mutex, Semaphore, timer).

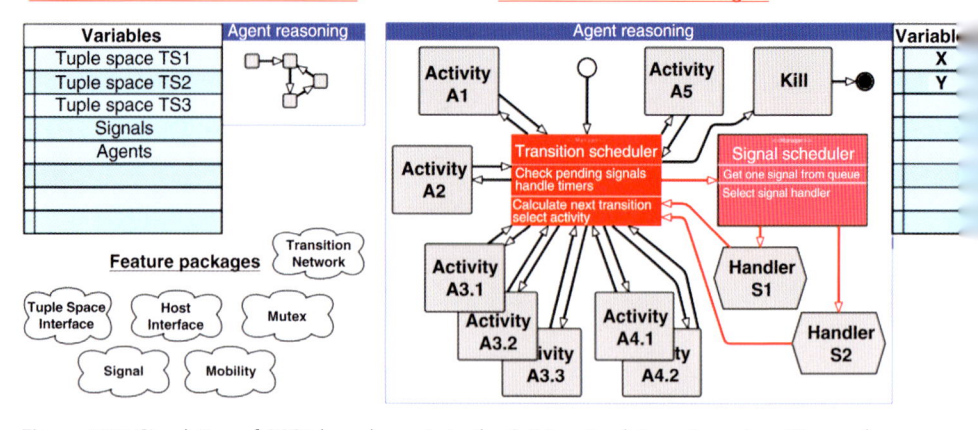

Figure 18.7 Simulation of AAPL-based agents in the SeSAm simulator using a transition and signal scheduler.

- Creation of agents at runtime is performed by creating an agent handler thread
- Migration of agents is performed by transferring the state of the agent (data state of all body variables) and control state is next activity after migration

Blocking of agent processing is handled by the thread implementation itself or by using a dedicated activity block scheduler. Further software modules implement the agent manager, tuple space databases, and networking.

Simulation platform. The SeSAm [Klügel *et al.*] simulator environment is used to perform functional testing and analysis of multiagent systems operating in distributed sensor networks, shown in Figure 18.7. The SeSAm simulator uses an agent behavior model based on ATGs, similar to the model already introduced. But (i) the SeSAm ATG model is static and fixed at simulation-time, (ii) activities may not block, and (iii) there are no signal handlers. For this reason, a transition and a signal scheduler are required that handle the agent processing. Activities with blocking statements (I/O) must be split according to the PCSP model. Migration of agents is performed by changing the spatial location!

PAVM Platform This is a pipelined agent forth virtual machine with multiple stack-based program code processors. Agent processing is performed again with tokens passed by queues to the forth processor offering fine grained agent scheduling. There are also different platform implementations: Hardware, Software, Simulation, WEB JavaScript App. They support compatibility on interface and operational level offering agent mobility crossing different platform implementations. The *AAPL* is the common and unified agent behavior model and programming source for all implementations as well architectures, shown in Figure 18.5!

PAVM Virtual Machine and Code Architecture

- The virtual machine (VM) executing tasks is based on a traditional FORTH architecture, as shown in Figure 18.8.

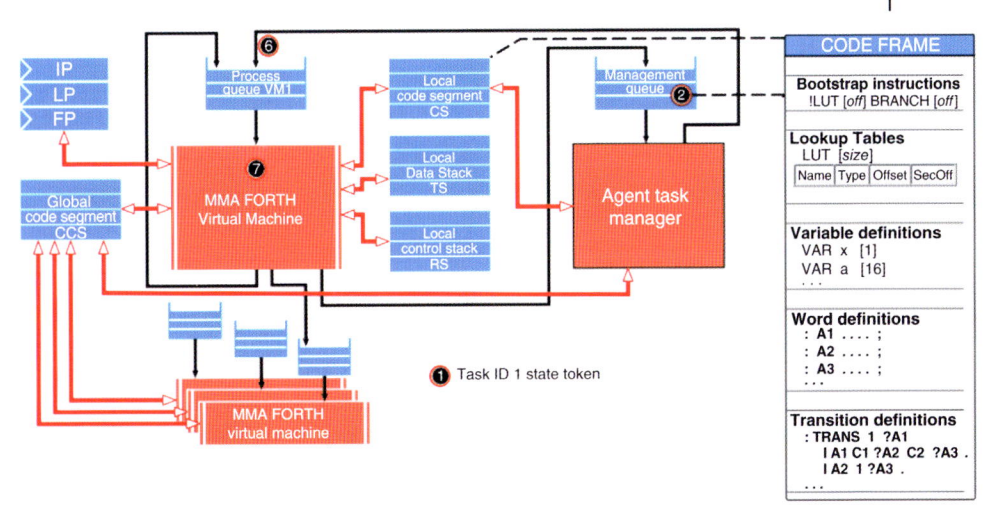

Figure 18.8 Pipelined multi-VM architecture and agent program code frame.

- It uses an extended zero-operand word instruction set (aFORTH) commonly operating on stacks rather on RAM
- Memory architecture of a VM:
  - Data (TS) stack used for data operations (computation)
  - Control (RS, return) stack used for program flow control
  - Code segment (CS) storing the program code with embedded data.
- The program is mainly organized by a composition of words (functions) stored in a code frame that embeds the agent data and control state, too.
- A word is executed by transferring the program control to the entry point in the CS; arguments and computation results are passed only by the stack(s).
- To provide fine grained granularity of task scheduling, a token-based pipelined task processing architecture was chosen.
- A task of an agent program is assigned to a token holding the task identifier of the agent program to be executed.
- The token is stored in a queue and consumed by the virtual machine from the queue.
- After a (top-level) word was executed, leaving an empty data and return stack, the token is passed back to the processing queue, and another agent task can be processed.

Comparison. Both platform architectures are compared in Table 18.2. Deploying of both architectures in one network is nontrivial due to the completely different representations of agents, basically code with embedded behavior, data, and state versus data and state only in the *PCSP* architecture. The mixing of both architectures requires agent wrapper modules transforming the data/control state of PCSP agents to an equivalent code frame that can be processed by the *PAVM* architecture and vice versa.

JAM Platform This is a *JS* implementation of the Agent Virtual Machine based on the *PAVM* architecture but supporting the direct processing of *AgentJS* agents.

Table 18.2 Comparison of the PAVM and PCSP platform architecture and their implementations.

	Programmable PAVM	Nonprogrammable PCSP
Approach	• Program code based approach • Platform is generic • Code embeds instructions, configuration (control state), and data • Migration: code transfer • Zero-operand instruction format • Stack-based data processing model	• Application-specific approach • Platform is application-specific and implements agent behavior • Activities of the ATG are mapped to processes • Token-based agent processing • Migration: data and control state transfer
Hardware	• Optimized multistack machine • Each stack processor has a local code segment and two stacks shared by all agents. There is no data segment! • Single SoC design • Multiprocessor architecture with distributed and global shared code memory • Multi-FSM RTL hardware architecture • Automatic token-based agent process scheduling and processing • Code morphing capability to modify agent behavior and program code (ATG modification) • Data- and code word sizes can be parameterized	• Pipelined communicating processes architecture composition implementing ATG and token-based agent processing • Single SoC design • Optimized resource sharing – only one PCSP for each agent class implementation required • Activity process replication for enhanced parallel agent processing • For each agent class there is one PCSP with attached data memory (agent data). • Single SoC design • LUT configuration matrix approach for ATG reconfiguration
Software	• Multithreading or multiprocess software architecture • Multi word size support • Interprocess communication: queues • Software model independent from programming language • VM sources for various programming languages: C, ML, JavaScript, and so on • Can be embedded in existing software • JavaScript platform offers WEB application integration	• Multithreading software architecture • Optimization: functional composition and implementation of ATG behavior instead PCSP • Interprocess communication: queues • Software model independent from programming language • Source code for various programming languages: C, ML, and so on • Can be embedded in existing software
Simulation	• Agent-based platform simulation • Generic simulation model – can execute machine code directly • Processor components and managers are simulated with agents	• Agent-based platform simulation • Application-specific simulation model • ATG activity processes are simulated with agents

18.6 Agent Processing Platforms and Technologies

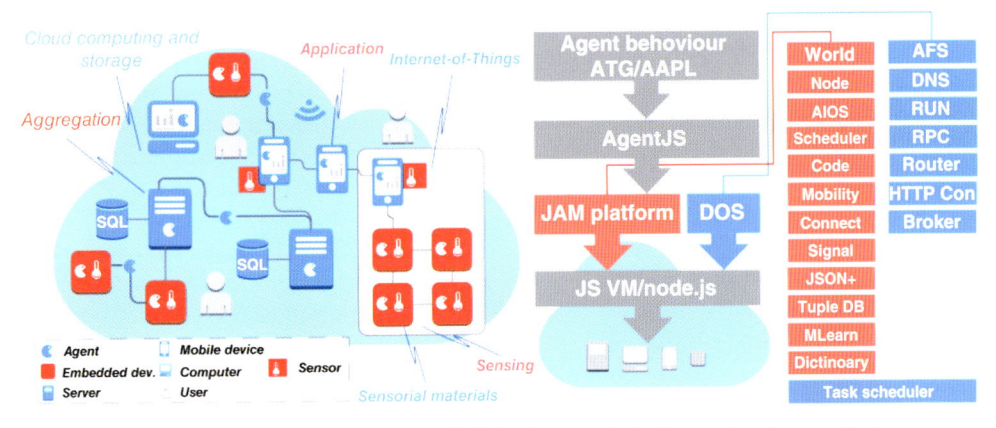

Figure 18.9 JAM AgentJS processing platform: (left) deployment in the Internet and embedded networks; (right) JAM and DOS modules.

AgentJS is the direct mapping of *AAPL* statements and the underlying programming model to *JS*. In contrast to the *PAVM* platform requires *JAM* a host platform for operation, commonly a generic computer, a mobile device (smart phone), an embedded computer, or a server. *JAM* (for details, refer to Ref. [27]) was intended for the deployment of mobile agents in large-scale and large-domain networks and environments, that is, the Internet, Clouds, and the Internet-of-Things, as shown in Figure 18.9. *JAM* completes the set of agent processing architectures, and as it bases on the *AAPL* model, it is partially interoperable with the *PAVM* and *PCSP* platform architecture by using code or state transformation modules.

JAM consists of different modules, shown in Figure 18.9, entirely implemented in *JS* that can be executed by any standalone *JS* VM or within WEB browsers. The deployment in Internet and client-side applications like browsers and hidden in virtual networks require a Distributed Co-ordination and Operation System layer (DOS) with a broker service. The main advantage of *JAM* is the unified deployment in the Internet domain.

The *AIOS* is the main execution layer of *JAM*. It consists of the sandbox execution environment encapsulating an agent process, with different privileged subsets depending on an agent role (level 0, 1, 2). Furthermore, the *AIOS* module implements the agent process scheduler and provides the API for the logical (virtual) world and node composition. The sandbox environment provides restricted access to a code dictionary based on the privilege level, enabling code exchange between agents. Level 0 agents are not privileged to replicate, create, or kill other agents and to modify their code.

Agents are either instantiated from an agent class template or forked from already existing agents. The template is genuine JS with some behavioral modifications that can be transformed in the textual *JSON+* representation, derived from the *JS* Object Notification format (*JSON*), using a modified parser and text converter. *JSON+* includes additional function code. Agents are executed always in a sandbox environment, which requires always a code-text-code transformation that is performed on agent creation or migration, discussed below.

In contrast to the *AAPL* model based on the ATG model that supports multiple blocking statements (e.g., IO/tuple space access) inside activities, *JS* is not capable of handling any kind of process blocking (there is no process and blocking concept). For this reason, scheduling blocks can be used in *AgentJS* activity functions handled by the *AIOS* scheduler. Blocking *AgentJS* functions returning a value use common callback functions to handle function results, for example, `inp(pat,function(tup){..})`.

Agent mobility, provided by the *AIOS* `moveto(dst)` statement, requires a process snapshot and the transfer of the data and control state of the agent process. The control state of an agent is stored in a reserved agent body variable `next`, pointing to the next activity to be executed. The data state of an *AgentJS* agent consists only of the body variables. Thus, the migration starts with a code-to-text transformation to the extended *JSON+* representation of the agent object, transportation of the text code to another logical or physical node, and a back text-to-code conversion with a new sandbox environment. The agent object is finally passed to the new node scheduler and can continue execution.

The JAM was extended with machine learning as a service, that is, learning agents can access basic machine learning operations provided as a platform service, offered by `model= learn(datasets, classes, features, alg?)` and `feature=classify(model, dataset)` primitives. The agent stores only the learned model, and do not carry any learning algorithms, leading to a separation of the learning algorithm (platform) from the data (agent).

SEJAM: JAM Simulator World Commonly, execution and simulation platforms are completely different environments, and simulators are significantly slower in the agent execution compared with real-world agent processing on optimized processing platforms. *SEJAM* is a *JS-APPL* simulator implemented on top of the *JAM* platform layer, executing agents with the same VM as a standalone agent platform would do.

This capability leads to a high-speed simulator, only slowed down by visualization tasks and user interaction. Furthermore, multiple simulators can be connected via a stream link (sockets, IP network connection, etc.), improving the simulation performance by supporting parallel agent processing. Furthermore, the simulator can be directly connected to any other *JAM* node.

The GUI of the simulator and the simulation world is shown for an example setup in Figure 18.10. The GUI consists of the simulation world, composed of 64 logical nodes connected with virtual circuit links. Each node shape provides information about the node name in the first row, the number of agents and tuples in the database in the second row, and some flag indicators in the last row, for example, flags signaling the existence of specific agents or sensor values.

On the right side, there is a code and data navigator. Each node can be selected including the world object. The code navigator can be used to explore node and agent information in a *JSON*-like tree presentation. The bottom part of the simulator contains a logging and message window. Agents can write messages to this window, and a compacted *JSON* can be printed from selected items in the

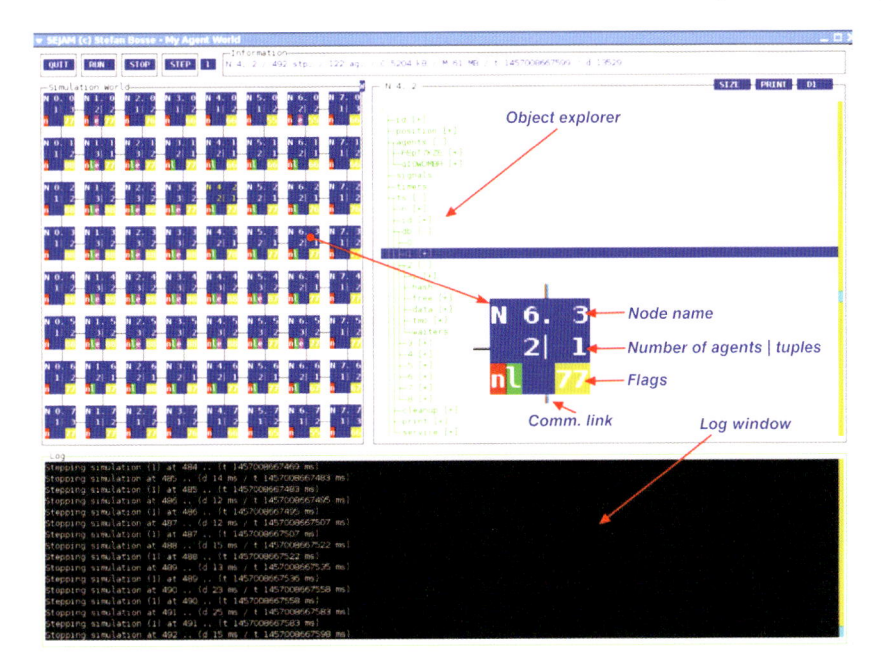

Figure 18.10 SEJAM simulation demonstration with a simulation world consisting of 8×8 logical nodes populated with mobile and nonmobile agents, indicated by markings on the bottom of the node shape (blue rectangle).

object navigator tree. Furthermore, agents executed in the simulator world inherit a special simulation object, which can be used to get specific simulation and world information, for example, the current simulation step, or support for creation of agents on a specific node, for example, used by the world agent, that is the only agent created and started at the beginning of the simulation. Multiple simulator worlds (*SEJAM* instances) can be connected enabling the composition of complex simulation worlds.

18.7 Agent-Based Learning

Traditional machine learning is partitioned in two separated actions: (i) collecting of data; (ii) learning or classification using this data. Supervised machine learning requires additional labeled (tagged) data, assigning a specific class symbol to a dataset, whereas unsupervised learning require a feedback from the environment to evaluate a given hypothesis. It is well known that unsupervised learning can be integrated in the reactive agent model (agent learning), but multiagent systems can create a bridge between traditional agent and machine learning, shown in Figure 18.11. A distributed MAS can be used to implement a multi-instance ML task, which is originally only implemented by a single ML instance. An example for a distributed multi-instance ML using MAS is outlined in the following section.

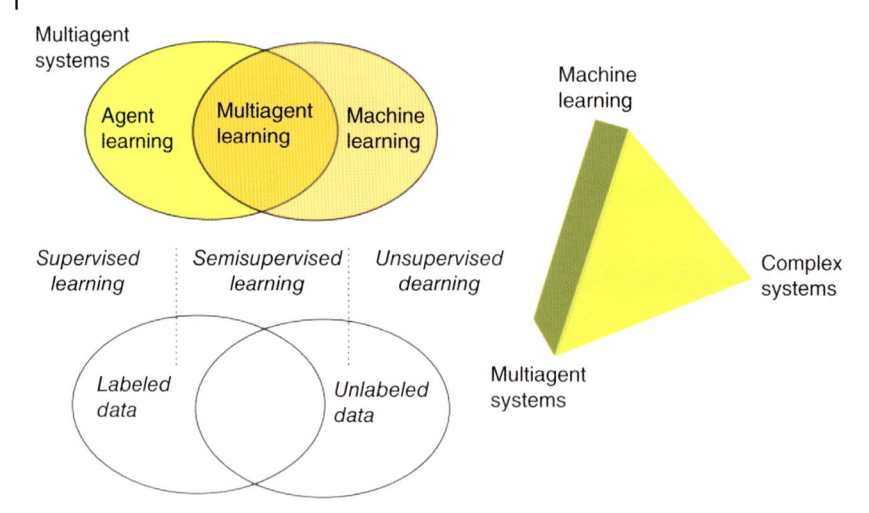

Figure 18.11 Classification of agent-based machine learning and the synergy triangle of classical machine learning, multiagent systems, and complex systems.

18.8 Event and Distributed Agent-Based Learning of Noisy Sensor Data

In Ref. [28], it was shown that three different data processing and distribution approaches can be used and implemented with agents, leading to a significant decrease in network communication activity and a significant increase of the reliability and quality-of-service:

1) An event-based sensor distribution behavior is used to deliver sensor information from source sensor to computation nodes based on local decision and sensor change predication.
2) Adaptive path finding (routing) supports agent migration in unreliable networks with missing links or nodes by using a hybrid approach of random and attractive walk behavior.
3) Self-organizing agent systems with exploration, distribution, replication, and interval voting behavior based on feature marking are used to identify a region of interest (ROI, a collection of stimulated sensors) and to distinguish sensor failures (noise) from correlated sensor activity within this ROI.

In structural monitoring applications, sensor nodes are commonly arranged in some kind of a two-dimensional grid network (Figure 18.12) and they provide spatially resolved and distributed sensing information of the surrounding technical structure, for example, a metal plate or a composite material equipped with strain sensors. Usually a single sensor cannot provide any meaningful information of the mechanical structure. One example for an information mapping discussed in Chapter 12 uses inverse numerical computation to determine load situations. An alternative approach is machine learning used to classify different learned load situations, for example, to distinguish between normal load and overload

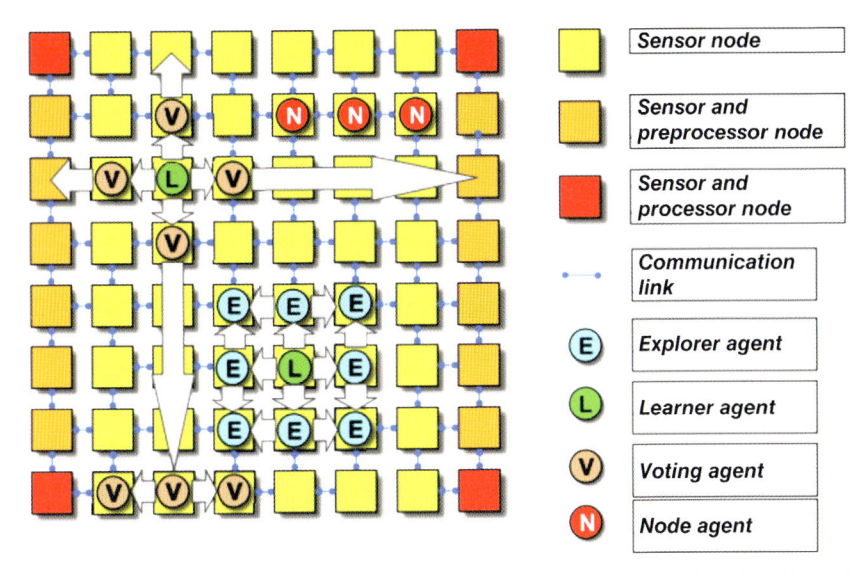

Figure 18.12 The logical view of a sensor network with a two-dimensional mesh-grid topology (left) and examples of the population with different mobile and immobile agents (right): node, learner, explorer, and voting agents. The sensor network can contain missing or broken links between neighbor nodes. Nonmobile node agents are present on each node. Pure sensor nodes (yellow nodes in the inner square) create learner agents performing regional learning and classification. Each sensor node has a set of sensors attached to the node, for example, two orthogonal placed strain gauge sensors measuring the strain of a mechanical structure.

conditions. In the following use case distributed learning is applied to spatially bound regions in the network, the regions of interest (ROI), that provide an event-based prediction and classification of the load case situation using supervised machine learning. The entire learning problem depending on the data from all sensor nodes is divided into multiple local ROI segment learners capable of predicting the local load situation based on the sensor data from the ROI, finally using a majority decision from a set of activated learners.

Again, mobile agents are used to collect (percept) and deliver sensor data to learner agents, limited to the ROI, shown in Figure 18.12, by using a divide-and-conquer approach. The advanced ϵ-Interval and NN Learning algorithm from Section 13.5 is used online with ROI data sets to compute a decision tree used for load case prediction and recognition.

Distributed and event-based machine learning is evaluated with simulated load data from Ref. [28]. Different load situations were applied to a metal plate, and the strain of the plate was computed at particular points using FEM simulation, finally mapped on artificial sensor data processed by a MAS in the *SEJAM* MAS simulator. The structure of the sensor network was already shown in Figure 18.12. A load-strain matrix T is required to compute artificial strain values.

Originally, the simulated loads used to compute the load-strain matrix T were cylindrical weights placed at $N=400$ weight positions from an equidistant rectangular grid for $N_x=N_y=20$. The force on the upper surface of the upper horizontal plate due to the loading hence vanishes outside the circle covered by the weight; inside this circle the force points in direction-z and equals $1\,\text{N/cm}^2$. Since the used deformation model is linear, the actual value assigned to this force is unimportant if it re-scale the load-strain matrix such that the magnitude of reconstructed loads matches those of the true load for noise-free data. After computing the deformation field, the surface strain in x- and y-direction at the sensor points was extracted by computing the deformation field and the extracted surface strain for a sequence of refined meshes.

Apart from the simple cylindrical loads to compute the load-strain matrix T, five spatially extended and different loads $l^{(1)}, \ldots, l^{(5)}$ with different characteristics and acting on different parts of the steel plate were simulated, shown in Figure 18.13.

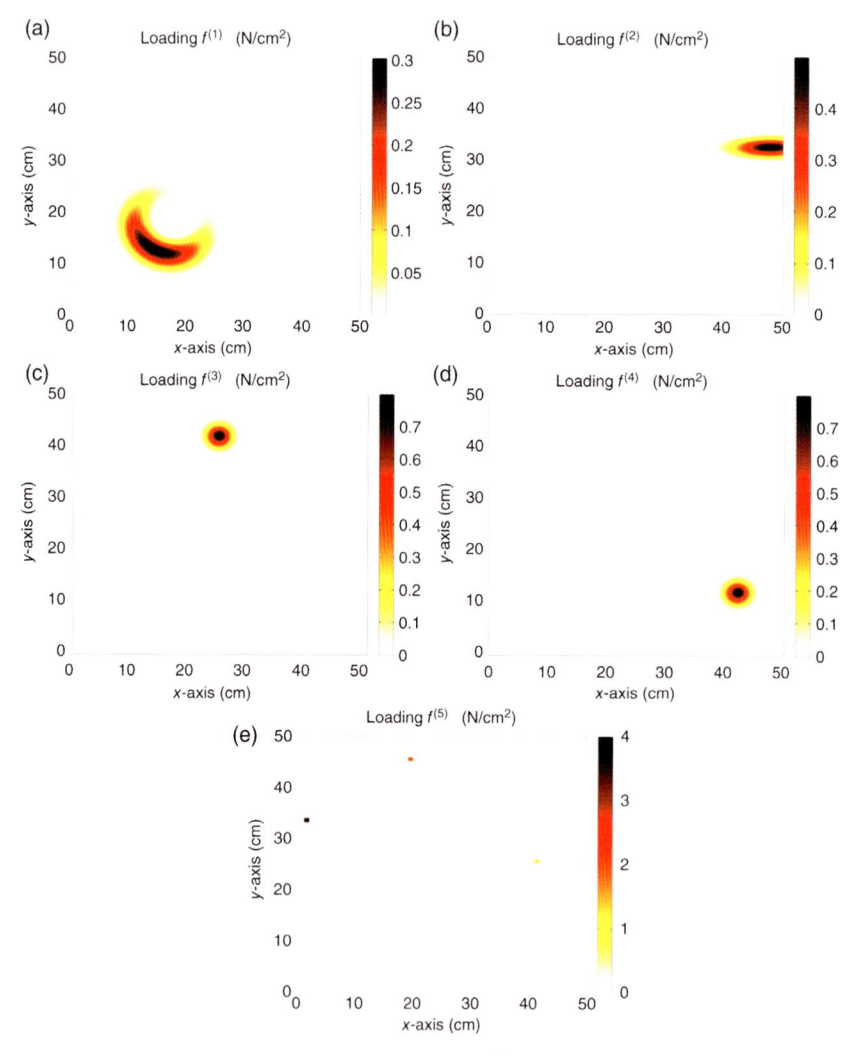

Figure 18.13 From (a) to (e): the five artificial loads $l^{(1)}, \ldots, l^{(5)}$. (From Bosse and Lechleiter [28].).

18.8 Event and Distributed Agent-Based Learning of Noisy Sensor Data

The simulated strain values are then converted to integer values in the range between 1 and 1024, with a no-load signal corresponds to the value 512. If σ_i denotes a simulated strain value, this conversion is done using the formula $s_i = b\ 512 + 10{,}000^* \sigma_i\ c$, $1 \leq i \leq 2M$. ($\lfloor a \rfloor$ denotes the largest integer smaller than or equal to $a \in \mathbb{R}$).

Thus, five strain measurement vectors $s^{(1)}, \ldots, s^{(5)}$ are computed., and these five data sets were feed consecutively as sensor values into the simulation framework for the sensor network shown in Figure 18.12. After a randomly chosen load situation was applied, the response of the LM is evaluated. Between

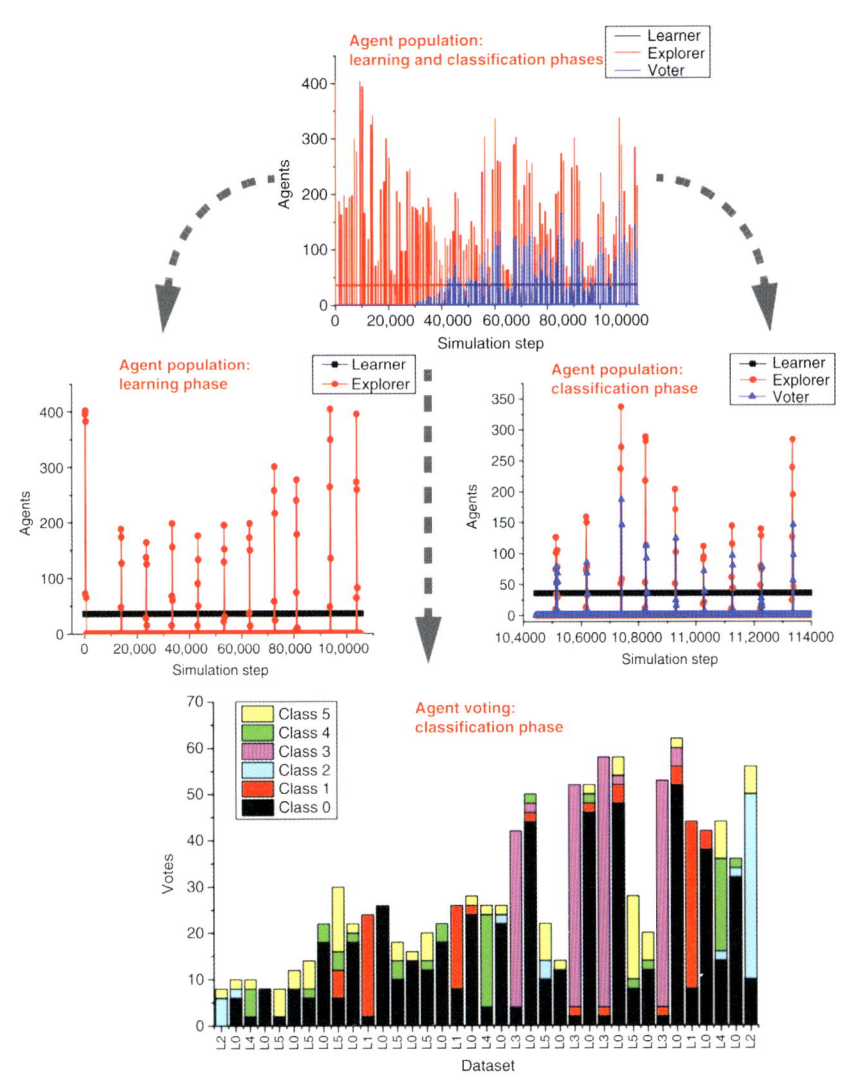

Figure 18.14 Simulation results. The top figure shows the event-based system behavior by the temporal agent population for a long-time run with a large set of single training and classification runs, with a zoom shown in the middle two figures. The bottom figure shows global classification results obtained by major voting of all event-activated regional learner agents.

two load cases there is always the null-load case that is applied to relax the system. For the learning an $\epsilon = 5$ setting was used. Monte Carlo simulation of sensor noise was applied with $\epsilon = 5$, too, adding equally distributed noise intervals $[-\epsilon, \epsilon]$ to the sensor values.

Simulation results are shown in Figure 18.14. The top figure shows the temporal agent population for a long-time run with a large set of single training and classification runs, with a zoom shown in the middle two figures. Each peak represents a particular training or classification run. In this experiment, the learner agents are nonmobile, and hence the population do not change in time. Side and edge nodes are not populated with learner agents. A learner agent covers the ROI containing its host node and eight surrounding nodes. If the host node detects a sensor event (change), it notifies the waiting learner agents by storing a `TODO` tuple in the database, consumed by the learner agent. Either a training (learning) or classification (prediction) request is send. In both operational cases, the learner agent will send out explorer agents to collect sensor data in the neighborhood, which is back delivered to the learner. In the classification modus, the learner will use the already trained and learned model to predict the load case situation. The result is sent out in the network by using voting agents, finally accumulated by the four edge nodes of the network by the major vote election.

The bottom figure shows global classification results obtained by major voting of all event-activated regional learner agents. The load sequence was randomly chosen, but always with an idle load situation (l^0) inserted between two different load cases. The learner must predict this load after a change in the load situation, too, meaning there is no load applied to the structure. The major election results show a very accurate prediction of significant distinguishable loads, that is, l^1, l^2, l^2, and l^4. The last load case l^5 is hard to distinguish from the zero load case.

References

1 Bosse, S. (2013) Intelligent microchip networks: an agent-on-chip synthesis framework for the design of smart and robust sensor networks. Proceedings of the SPIE 2013, Microtechnologie Conference, Session EMT 102 VLSI Circuits and Systems, Alpexpo/Grenoble, France, SPIE, April 24–26. doi: 10.1117/12.2017224012

2 Meng, Y. (2005) An agent-based reconfigurable system-on-chip architecture for real-time systems. Proceeding ICESS '05 Proceedings of the Second International Conference on Embedded Software and Systems, pp. 166–173.

3 Sansores, C. and Pavón, J. (2008) An adaptive agent model for self-organizing MAS, in Proceedings of the 7th International Conference on Autonomous Agents and Multiagent Systems (AAMAS 2008) (eds. L. Padgham, D.C. Parkes, J. Müller, and S Parsons), May 12–16, 2008, Estoril, Portugal, pp. 1639–1642.

4 Bosse, S., Pantke, F., and Kirchner, F. (2012) Distributed computing in sensor networks using multi-agent systems and code morphing. ICAISC Conference, Poland, Zakapone.

5 Qin, Z., Xing, J., and Zhang, J. (2010) A replication-based distribution approach for tuple space-based collaboration of heterogeneous agents. *Research Journal of Information Technology*. doi: 10.3923/rjit.2010.201.214

6 Guijarro, M., Fuentes-fernández, R., and Pajares, G. (2008) A multi-agent system architecture for sensor networks, in *Multi-Agent Systems – Modeling, Control, Programming, Simulations and Applications, IntechOpen*.

7 Lückenhaus, M. and Eckstein, W. (1997) A multi-agent based system for parallel image processing. Proceedings of the International Conference on Parallel and Distributed Methods for Image Processing at SPIE's Annual Meeting, Proc. SPIE 3166.

8 Yuan, S., Lai, X., Zhao, X., Xu, X., and Zhang, L. (2006) Distributed structural health monitoring system based on smart wireless sensor and multi-agent technology. *Smart Materials and Structures*, **15** (1), 1–8.

9 Castanedo, F., García, J., Patricio, M.a., and Molina, J.M. (2010) A multi-agent architecture based on the BDI model for data fusion in visual sensor networks. *Journal of Intelligent & Robotic Systems*, **62** (3–4), 299–328.

10 Junges, R. and Klügel, F. (2012) How to design agent-based simulation models using agent learning. Proceedings of the Simulation Conference (WSC).

11 Bussmann, S., Jennings, N.R., and Wooldridge, M. (2004) *Multiagent Systems for Manufacturing Control*, Springer.

12 Brooks, R. A robust layered control system for a mobile robot, *IEEE Journal on Robotics and Automation*, vol. 2, no. 1, pp. 14–23, 1986.

13 Wooldrige, M. (1999) Intelligent agents, in *Multiagent Systems: A Modern Approach to Distributed Artificial Intelligence* (ed. G. Weiss), MIT Press.

14 Müller, J. The Design of Intelligent Agents, vol. 1177, Lecture Notes in Artificial Intelligence, Springer, ISBN 978-3-540-62003-7.

15 McCabe, F.G. and Clark, K.L. (1995) APRIL – agent process interaction language, in *Intelligent Agents Theories Architectures and Languages LNAI*, vol. 890 (eds M. Wooldridge and N.R. Jennings), Springer.

16 Kone, M.T., Shimazu, A., and Nakajima, T. (2000) The state of the art in agent communication languages. *Knowledge and Information Systems*, **2** (3), 259–284.

17 Bordini, R.H. and Hübner, J.F. (2006) BDI agent programming in AgentSpeak using Jason, in *Computational Logic in Multi-Agent Systems*, vol. 3900, Lecture Notes in Computer Science, Springer, pp. 143–164.

18 Chunlina, L., Zhengdinga, L., Layuanb, L., and Shuzhia, Z. (2002) A mobile agent platform based on tuple space coordination. *Advances in Engineering Software*, **33** (4), 215–225.

19 Caridi, M. and Sianesi, A. (2000) Multi-agent systems in production planning and control: an application to the scheduling of mixed-model assembly lines. *International Journal of Production Economics*, **68**, 29–42.

20 Leitão, P. and Karnouskos, S. (eds) (2015) *Industrial Agents Emerging Applications of Software Agents in Industry*, Elsevier.

21 Marík, V. and McFarlane, D.C. (2005) Industrial adoption of agent-based technologies. *IEEE Intelligent Systems*, **20** (1), 27–35.

22 Pechoucek, M. and Marík, V. (2008) Industrial deployment of multi-agent technologies: review and selected case studies. *Autonomous Agents and Multi-Agent Systems*, **17** (3), 397–431.

23 Bosse, S. and Lechleiter, A. (2014) Structural health and load monitoring with material-embedded sensor networks and self-organizing multi-agent systems. Procedia Technology, Proceeding of the 2nd SysInt Conference, Bremen, Germany. doi: 10.1016/j.protcy.2014.09.039

24 Bosse, S. (2014) Distributed agent-based computing in material-embedded sensor network systems with the agent-on-chip architecture, *IEEE Sensors Journal, Special Issue MIS.* doi: 10.1109/JSEN.2014.2301938

25 Bosse, S. (2015) Design and simulation of material-integrated distributed sensor processing with a code-based agent platform and mobile multi-agent systems. *MDPI Sensors*, **2015** (2), 4513–4549.

26 Bosse, S. (2014) Design of material-integrated distributed data processing platforms with mobile multi-agent systems in heterogeneous networks (Conference). Proceedings of the 6\u2019th International Conference on Agents and Artificial Intelligence ICAART 2014. doi: 10.5220/0004817500690080

27 Bosse, S. (2015) Unified distributed computing and co-ordination in pervasive/ubiquitous networks with mobile multi-agent systems using a modular and portable agent code processing platform. The 6th International Conference on Emerging Ubiquitous Systems and Pervasive Networks (EUSPN 2015), Procedia Computer Science.

28 Bosse, S. and Lechleiter, A. (2016) A hybrid approach for Structural Monitoring with self-organizing multi-agent systems and inverse numerical methods in material-embedded sensor networks. *Mechatronics*, **34**, 12–37.

Part Seven

Energy Supply

19

Energy Management and Distribution

Stefan Bosse[1,2]

[1]*University of Bremen, Department of Mathematics and Computer Science, Robert-Hooke-Str. 5, 28359 Bremen Germany*
[2]*University of Bremen, ISIS Sensorial Materials Scientific Centre, Wiener Str. 12, 28359 Bremen, Germany*

With increasing sensor node density and the shift from centralized toward decentralized data processing architectures, we can observe that energy supply becomes a key technology and limiting factor in the design of sensor networks and their runtime stability. Centralized energy supply architectures have limited scaling ability and efficiency. Furthermore, they can introduce single-point-of-failures.

Decentralized energy supply architectures rely on local energy storage and energy harvesting, introducing hard energy constraints during runtime and the requirement for energy-aware systems. Sensor nodes in common sensing applications are typically battery-powered. There are different aims of energy-aware systems:

1) Extension of the overall operational runtime (lifetime) of (a) the system that can be considered as the network level, and (b) the sensor node considered as the node level.
2) Enabling of more powerful operations (algorithmic selection).
3) Safekeeping of the global operational system behavior with or without the partial failure of system (sensor node) components due to energy bottlenecks.

Energy-aware systems address the minimization of the energy consumption and the energy management optimizing the usage of limited energy resources to safekeeping the system vividness, which present different strategies. The first major area focuses on the design of low-power systems reducing the power consumption of electronic devices; the second area focuses on scheduling of tasks to be performed by sensor nodes and the network with the goal to minimize the consumed energy. But the problem of scheduling to minimize the energy consumption (especially on fine-grained device and component level) is a NP problem, that is, the scheduling and computational complexity increase exponentially. And each data processing requires a specific amount of energy to be performed, and hence (software-based) energy management at runtime requires a specific amount of energy, too, which must be in balance with the benefit of

Material-Integrated Intelligent Systems: Technology and Applications, First Edition.
Edited by Stefan Bosse, Dirk Lehmhus, Walter Lang, and Matthias Busse.
© 2018 Wiley-VCH Verlag GmbH & Co. KGaA. Published 2018 by Wiley-VCH Verlag GmbH & Co. KGaA.

energy reduction. Low-power design and energy management are already established in distributed sensor networks. Details can be found in Ref. [1].

The activity of a sensor node can be partitioned in different major tasks:

1) Analog signal processing
2) Digital data processing
3) Communication
4) Sleeping and being idle (i.e., waiting for events)

The second and third produce dominant contributions to the overall energy consumption of a sensor node.

19.1 Design of Low-Power Smart Sensor Systems

In electronic circuits, an electrical current flowing through a resistor causes a thermal power dissipation. Static and dynamic current flows must be distinguished. There are basically two different transistor technologies: bipolar (current-controlled current sources), and field effect-based (voltage-controlled currents sources). Modern MOS field effect transistor technologies pose a huge ratio of a static current (no switching activity) to a dynamic current (switching activity), mainly a result of charging and discharging of parasitic electrical capacities.

Assuming major CMOS technology, the power consumption of a digital circuit on chip level can be expressed by a static (leakage) and dynamic (switching) part and by the total number N_{tot} and mean number of switching transistors N_{sw} per time unit T, respectively:

$$\begin{aligned} P_{static} &\sim N_{tot} k_{tech} V_{dd} \\ P_{dynamic} &\sim f \overline{N}_{sw} C V_{dd} \end{aligned} \quad (19.1)$$

where f is the switching frequency (equal to $1/T$), C is a technology parameter (i.e., a normalized electrical charge capacity), and V_{dd} is the supply voltage of electronic circuit.

There are different optimization strategies that can be applied to a digital circuit:

1) Reduction of the switching frequency f of synchronous (clock-driven) circuits, which leads to an increase of the data processing latency (the computation time), but that can be compensated by parallelization techniques.
2) Reduction of the number of switching transistors, which requires either application-specific designs (static) or the disabling of unused digital components (dynamic at runtime); parallelization increases transistor resources and the number of switching transistors.
3) Improving the technology (e.g., minimizing the parasitic capacity C).
4) Reducing the supply voltage V_{dd} (usually requiring a lowering of the switching frequency due to an increase of the switching latency).

Typical power consumption of different microcontrollers, processors, wireless communication controllers, and analog-digital conversion circuits is shown in Table 19.1.

Table 19.1 Examples of typical power dissipation (and possible typical duration) of microcontrollers and peripheral components.

Component	Description	Power (peak)
MSP430	Low-power 16 bit microcontroller, Texas Instruments [2]	Standby: 3 µW 16 MIPS: 1 mW @ 16 MHz
Atmel ATMEGA 8L	Low-power 8 bit microcontroller, Atmel [3]	Standby: 1 mW 4 MIPS: 12 mW @ 4 MHz
Intel® Core™ i5 4310U CPU	Generic and graphics processor for mobile computers, Intel [4]	40,000 MIPS: 15 W @ 2 GHz 10,000 MIPS: 6 W @ 750 MHz
Wifi	Wireless communication, mid range [5]	20 mW/Mbps, 2 mW min.
Bluetooth®	Wireless communication, short range [5]	BR/DR: 33 mW/MBps, 300 µW min. Low En.: 100 mW/MBps, 30 µW min.
Zigbee®	Wireless communication, short range [5]	300 mW/MBps, 50 µW min.
wM-Bus, 868 MHz	Wireless communication, long range [6]	0.3 W max./40 ms
wM-Bus, 169 MHz	Wireless communication, long range [6]	1 W max./100 ms
GSM	Wireless communication, long range [6]	7 W max./0.6 ms
MAX11102	Analog-digital converter, 12 bit, SAR with S&H, Maxim [7]	5 mW @ 3 Msps

MIPS, million instructions per second; Mbps, million bits per second; Msps: million samples per second; TDP, thermal design power.
Sources: Refs [2–7].

In low-power systems, wireless communication consumes commonly the dominant part of the electrical power and can reach a fraction of 99%, caused by the analog section of a communication system and the required radio power for wireless communication. Wired communication creates a much lower contribution below 20% [1]. Low-power processors consume several order fewer energy than their desktop computer counterparts and provide more fine-grained power control at runtime.

The design of low-power systems, especially application-specific digital circuits, can profit from predesign simulations on gate level, shown in Figure 19.1, evaluating the effect of algorithm complexity on power consumption. In addition, this power analysis delivers an estimation of the power consumption, in particular computations in a sensor node, which can be used as an input for fine-grained dynamic power management.

If there is no explicit computational activity, there is still transistor switching activity, mainly resulting from clock-driven register and control state machine

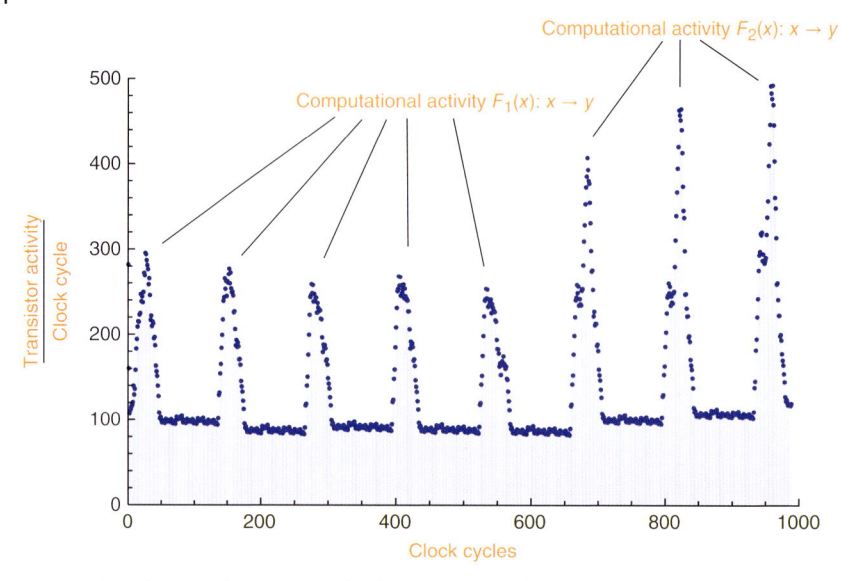

Figure 19.1 Gate-level simulation of a digital circuit analyzing the switching activity of gates in relation to algorithmic activity [8]. The peaks in the transistor switching activity accumulated of all logic gates in the device under test (DUT) result from computational activity, and the height of the peaks is related to the computational complexity of the used algorithm.

logic. The power analysis workflow is shown in Figure 19.2. On top level, an algorithm consisting of a set of functions $F = \{fn_1, fn_2, \ldots\}$ is modeled with a signal flow graph, which is transformed in a state-transition (SN) Petri net. Functional blocks fn_i are mapped to transitions, and states represent data that are exchanged between those functional blocks. The partitioning of functional blocks to transitions of the net can be performed at different composition and complexity levels [8]. Sensor data are acquired periodically and passed to the data processing system. A token of the net is equal to a data set of one computation processed by the functional blocks. Different functional blocks can be placed in concurrent paths of the net.

The Petri net is then used (1) to derive the communication architecture, and (2) to determine an initial configuration for the communication network, which is related to a multiprocess programming model, finally used to derive a hardware model for an SoC design on register-transfer level. Functional blocks with a feedback path require the injection of initial tokens in the appropriate states.

The aim of the power analysis is to retrieve an explicit assignment of the power (or the integral energy) required to execute a specific function block fn_i under specific conditions like the data processing rate, r_i = data sets/time unit, or the data set size w_i, that is, the set F is mapped on a parameterized power set $P = \{p_1(fn_1, r_1, w_1), p_2(fn_2, r_1, w_1), \ldots, p_i(fn_i, r_j, w_k), \ldots\}$. The power-function set can be used (1) at design-time for the optimization of a hardware design, and (2) at runtime for dynamic power management, for example, by selecting from a set of functions that perform basically the same computation but with different accuracy or quality and energy consumption.

19.1 Design of Low-Power Smart Sensor Systems

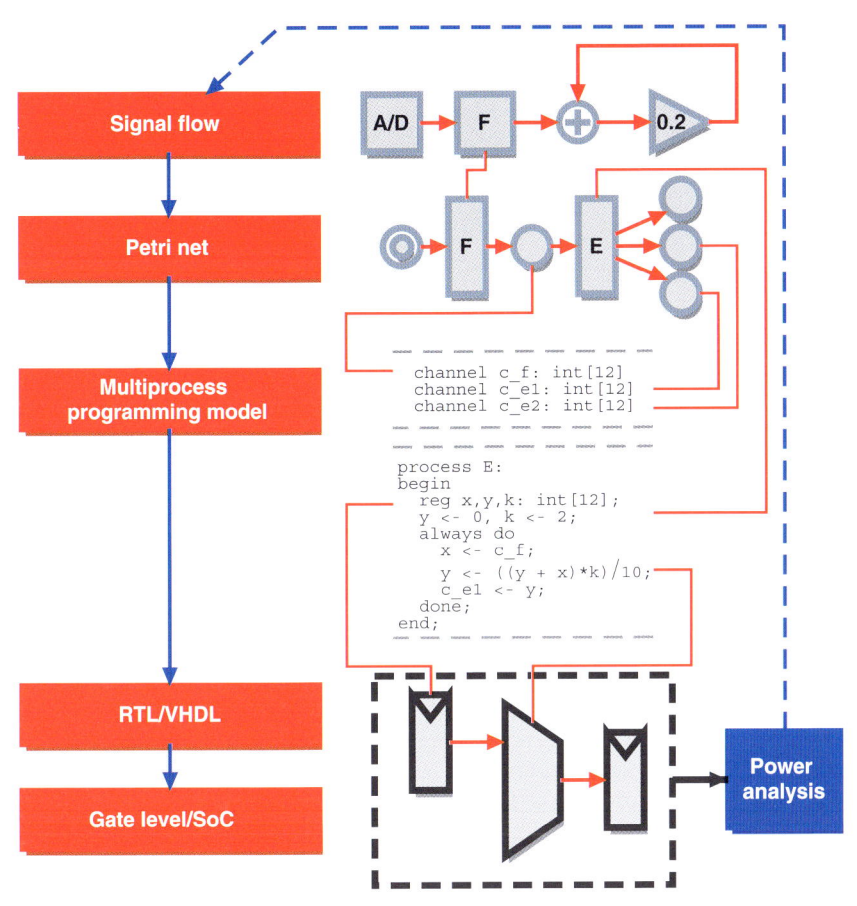

```
channel c_f: int[12];
channel c_e1: int[12];
channel c_e2: int[12];

process E:
begin
  reg x,y,k: int[12];
  y <- 0, k <- 2;
  always do
    x <- c_f;
    y <- ((y + x)*k)/10;
    c_e1 <- y;
  done;
end;
```

Figure 19.2 Power analysis workflow deriving the relation between algorithm activity and complexity with the power consumption [8].

Possible power savings on device level at runtime address [1]:

- Disabling of idle (not used) components (e.g., leakage current control techniques for memories).
- Switching between multiple low-power modes reducing the performance (e.g., data transmission bandwidth), that is, a range between full operation (active), standby, power-down, and disabled.
- Supply voltage, clock frequency, and threshold voltage scaling (but affecting the performance of the device).
- Dynamic variation of ADC sampling rates.
- Algorithmic selection from a set of computational units performing basically the same computation but differing in power consumption and accuracy or quality of service.
- Reconfiguration of device components.

19.2 A Toolbox for Energy Analysis and Simulation

In the last years, the number of available high performance but low-power embedded systems has grown exponentially, enabling application scenarios for tailored sensor systems. For many application cases, battery-powered or self-powered systems are needed, for example, in the context of wireless sensor networks. A designer has to assure that the system is provided with the appropriate amount of energy and assuring enough power to fulfill its given tasks. Often this can only be done when the analysis is carried out in the context of the real application. Consequently, a simulation has to consider the environmental condition of this context as well.

Therefore, a simulation toolbox for energy and power analysis of independent sensor nodes is required, a tool for designing modular sensor systems, proposed in Ref. [9] using the Matlab/Simulink framework. The focus of this tool will be on the economic and efficient use of power and energy performed on the embedded system level.

The base of this toolbox is a set of simulation blocks modeling the power behavior of embedded modules like energy sources, converters, storage, and load. A set of tools for observing losses, energy throughput, and power lags assists the system designer to set up an economic solution. A strong emphasis lies on the modeling of modern energy harvesting principles and the embedding physical situation.

One main goal of this toolbox is to overcome the oversizing of the power supply of electric systems for the "worst case". Instead, a situation-dependant adaptive energy management will control different operation modes of embedded systems to cope with different power supply and energy situations. Therefore, these systems can be specified more accurate and economic.

To save energy, the different operation modes will lead to a tailored sensor data processing. Besides using a software-controlled microprocessor system, dedicated and application-specific configurable hardware blocks can be added. The simulation and the energy management rely on load models that consider different implementations on work task level.

Every measurement systems designer dealing with battery- or self-powered systems have to consider the energy behavior. Most classical measuring devices were designed to work on a constant and sufficient power supply. Measurement experts often are not experienced in low-power design. A lot of optimizations can be applied using technology from mobile computing, control theory, and artificial intelligence. Using sensors not only for measuring, but also for energy harvesting, creates new perspectives in the sensor node design and deployment. Future methods for sensorial materials are consisting of low-power design, adaptation in the power consumption and energy management up to self-organization, self-localization, fault tolerance, cognition, and grid intelligence.

Sensor nodes could be implemented according the architectural model, shown in Figure 19.3. The model of a sensor node can be divided into the parts data (acquisition, signal and digital data processing, and communication) and energy (supply, storage, consumption). Although the design of low-power systems focuses on the first level on the optimization of data processing and

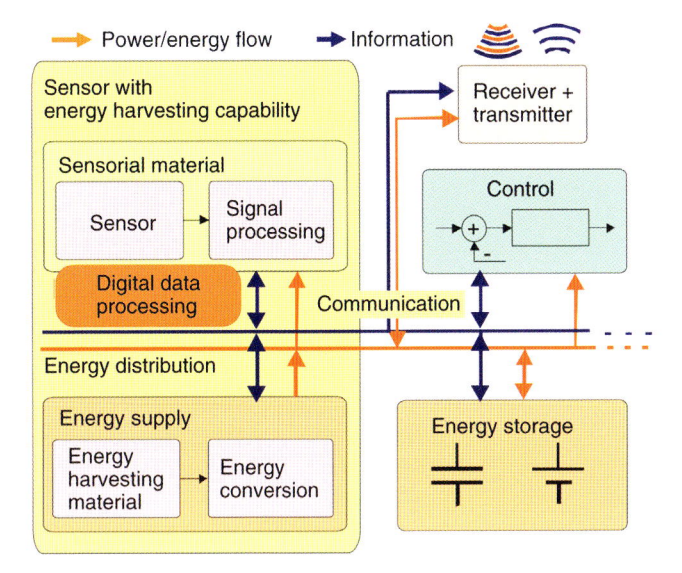

Figure 19.3 Architecture of a sensorial material embedding a sensor node with energy management and energy harvesting [9].

communication with respect to the energy consumption, this toolbox will concentrate on modeling, simulation, and analysis of the energy branch of the entire system. For most self-powered applications, the energy provided by the energy harvesting device has to get converted, shown in Figure 19.4.

The structure of the toolbox is close to the node architecture scheme according Figure 19.3 and is ordered from the energy's point of view (details can be found in Ref. [10]):

- *Energy sources*
  - Ideal sources
  - Batteries
  - Inductive/magnetic
  - Mechanical
  - Piezoelectric
  - Photoelectric

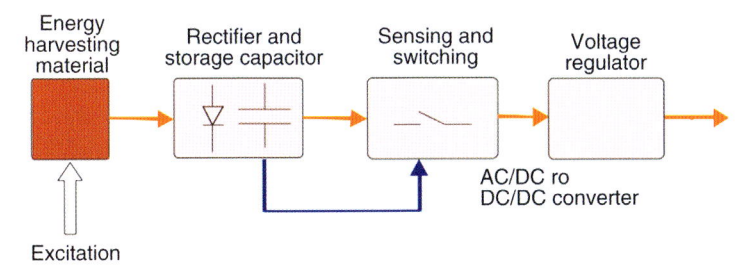

Figure 19.4 Principle energy harvesting architecture transforming some physical property into an electrical signal that can be stored.

- Thermal electric
- Transformers
- *Energy converters*
  - AC/DC
  - DC/DC
  - Voltage regulators + PWM
  - Amplifier
- *Energy storage*
  - Accumulators
  - Capacitors
- *Energy consumers*
  - Micro controllers (including memory)
  - Sensors and signal processing
  - Actuators
  - Communication modules (including radio)
  - Illumination

For each element there are generic blocks in the structure to cover the main functionality. Special parts can be derived and added to the library. According to the naming conventions in Matlab/Simulink, an input signal is connected to an "in-port" and the output of a block is fed through the "out-port.". A parameter mask in Simulink can be provided as a GUI for setting parameter values comfortably, which are then connected to constant values inside the block model implemented in Simulink [11]. All blocks of the energy toolbox are masked for convenience, for example, to parameterize the block according to the data sheet.

An example power model of a sensor node is shown in Figure 19.5, consisting of different power components (energy sources, converters, and consumers) and a power balancer performing some kind of power analysis and management (controlling the components). The power model is a parametric diagram with a constraint block (the power balancer implementing a power optimization function) and a set of constraint parameters (the sensor node components).

The application of the analysis tool delivers information about the layout of the sensor node's power system by showing the dynamic energy flows of the different components. These results and models can be incorporated into a subsequent low-power design and DPM. Most of the modeled components are dominated by a sleep schedule executing a cyclic wake-up reducing energy demand by an estimated duty cycle of 1:100 [10].

Energy Source Block The energy sources class covers the different methods for providing energy. In traditional design approaches, there is an assumption of a constant source always providing enough power. This tool enables the user to tailor the ratings for average and maximum use. For the possibility of environment-dependent energy sources, the blocks have the ability to be controlled by environmental parameters, for example, by solar radiation in the case of a photovoltaic cell. A solar cell generator converts light pulses into electrical power P_{MPP}, calculated as follows: $P_{MPP} = \eta A E = FF\ I_{sc} U_{oc} = I_m U_m$, where η is the efficiency, E is the irradiance, A is the area of the cell, FF is the fill factor, I_{sc}

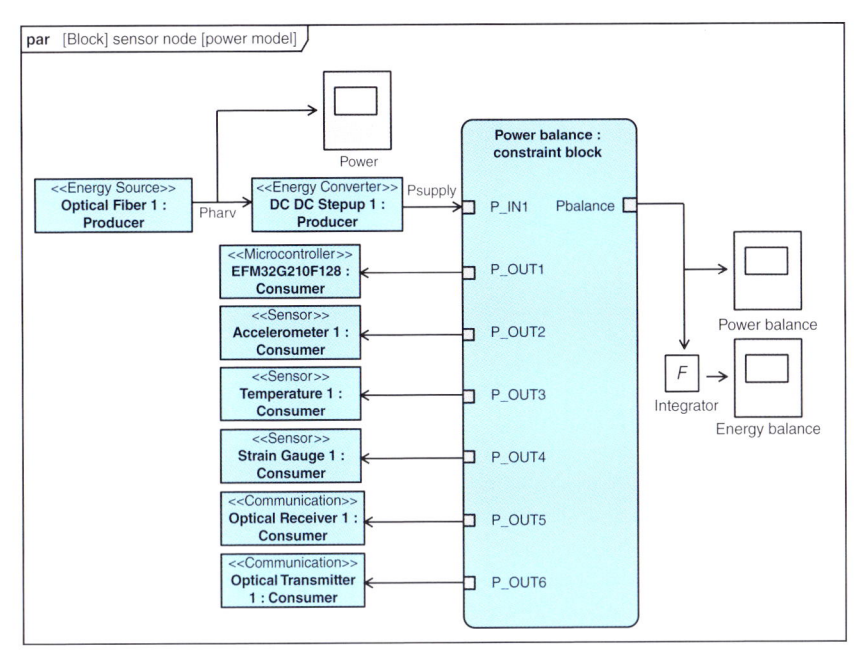

Figure 19.5 Parametric diagram of the power model of a sensor node using the toolbox components (power sources, consumers, and analyzers). (Adapted from Ref. [10].)

is the short-circuit current, U_{oc} is the open-circuit voltage, I_m is the optimum operating current, and U_m is the optimum operating voltage.

Energy Storage Block The generic capacitor block contains a simple model of a capacitor. There is a series resistance R representing effects of dielectric losses and conductor resistance. The capacitor inductance is represented by the series inductance L. A parallel resistance R_p models leakage current flow and the self-discharge. These values have to be entered into the blocks parameter mask. If the value of the series resistance is not known, the dissipation factor $\tan(d)$ and the according frequency can be entered instead. Note that the time needed for self-discharge is approximately $5 \times R \times C$. An initial voltage across the capacitor can be entered into the mask. The in-port signal represents the power to be stored on the capacitor.

Energy Converter Blocks Components for converting and regulating energy flows are technically mandatory. Focusing on the energy and efficiency, the most important fact is that these devices have electric losses and so the loss of power has to be considered. Most modern devices like direct current DC-to-DC converters use switched-power circuits and smoothing capacitors. A generic model for calculating losses without implementing the real operation is sufficient at the level of system definition and could be recalculated when the hardware layout is fixed.

Energy Consumer Block The consumer block provides generic constant power consumption. It is a good way to estimate the component's energy demand using datasheet's values, that is, useful for modeling components that have not been modeled in detail yet. A cyclic wake-up consumer block models the power consumption with an electrical switch having a resistance when in "on" position and switching losses. The power consumed is calculated as $P_{loss} = t_{switch} f_{pwm} UI + I_{avg}^2 R_{on}$, where t_{switch} is the time needed to switch between on and off, f_{pwm} is the switching frequency, U is the voltage of the power source, I is the current at maximum voltage, I_{avg} is the average current, and R_{on} is the ohmic resistance of the switch when in "on" position [12]. The output port provides information on power consumption.

Typically, energy management is performed by a central controller with limited fault tolerance and the requirement of a well-known environment world model for energy sources, sinks, and storage. Energy management in a network can additionally involve the transfer of energy between network nodes.

As already introduced, system-on-chip hardware design uses advanced high-level synthesis approaches on higher algorithmic level that can improve energy management and power efficiency based on results from toolbox analysis. Models and model parameters provided by the toolbox can be used for algorithm design, configuration, and hardware synthesis gathered by the power analysis introduced in Section 19.1.

19.3 Dynamic Power Management

The application of low-power hardware design techniques is static and cannot exploit energy reduction at runtime. Besides sensor nodes implemented with application-specific digital circuits (system-on-chip designs), sensor nodes are often operating with microcontrollers and light-weighted operating systems.

An operating system (OS) is the central instance of a computer system. They control resources (memory, devices, processors), schedule device access, implement a hardware abstraction layer, and provide unified programming interfaces for services accessing devices and data structures (e.g., file systems). Since the past decade, the OS additionally performs dynamic power management (DPM) of the entire system, originally required for mobile devices (e.g., notebooks, later smart phones), finally deployed in server computers, too.

DPM basically affects two main resources of a computer system: (1) the processor; (2) input/output devices. Runtime power control of system components is closely related to optimized scheduling of access operations.

Assuming central processing units (CPU) and peripherals supporting different power states, there are two different DPM techniques that can be combined:

- *CPU-centric DPM:* Based on Eq. (19.1), dynamic core voltage and frequency scaling (DVS/DFS) of processors based on processor workload is one major power reduction technique at runtime.
- *I/O-centric DPM:* Peripheral I/O devices supporting different power states can be shutdown or switched to standby mode if there is no activity. In contrast to

DVS/DFS techniques, it introduces power state switching of peripheral devices and a time delay affecting the access operations that compromises I/O performance.

The main action performed by DPM is the scheduling of a set of tasks, $T = \{t_1, t_2, \ldots\}$, which has to be executed. Each task is parameterized by the tuple $t_i = (a_i, d_i, l_i)$, with a_i as the arrival time, d_i the latest deadline time, and l_i the estimated duration time of the task in instruction cycles. Assuming the CPU can operate on different core voltages $V = \{v_1, v_2, \ldots\}$, associated with a set of speed levels $F = \{f_1, f_2, \ldots\}$. An algorithm is required to perform all tasks in T with minimal energy consumption, that is, assigning an appropriate speed level $f_j(v_j)$ to each task t_i to meet all (or at least as many) deadlines of the tasks:

$$E \sim \sum_{i=1}^{n} v_j^2 f_j l_i \tag{19.2}$$

19.3.1 CPU-Centric DPM

EDF Algorithm The well-known earliest deadline first (EDF) algorithm is a dynamic-scheduling algorithm primarily used in real-time operating systems. It places the task set T in a priority queue [13], that is, an ordered ready list.

LEDF Algorithm The low-energy EDF (LEDF) algorithm [14], shown in Algorithm 19.1, is an extension of the EDF algorithm. It considers different CPU speeds and adapts scheduling based on tasks missing their deadlines. The scheduling policy tries to minimize the total energy consumed by the task set and guarantees the deadline for periodic tasks. For all tasks, an absolute deadline is computed and that is recalculated at each release based on the absolute time of release and the relative deadline [1]. The task with the earliest (first) deadline is selected for execution. Energy optimization is performed by checking the task deadline and if it can be met by executing it at a lower voltage (speed). Each speed at which the processor can run is considered for deadline computation in order from the lowest to the highest. For a given speed, the approximation of the worst-case execution time of the task is calculated based on the estimated instruction count of the task. If this execution time is too high to meet the current absolute deadline for the task, then the next higher speed is considered.

The LEDF algorithm has a computational complexity of $O(N \log N)$ for a processor with two speed levels, where N is the total number of tasks to be executed. Each scheduling requires computation and hence reduces the overall energy optimization efficiency, though the LEDF algorithm introduces only a low overhead compared with other parts of an operating system. The complexity of the LEDF algorithm becomes $O(N \log N + kN)$, with k as the number of speed settings that are supported by the CPU. In Ref. [1] the computational overhead of the scheduling is reported to be below 1‰ and can be considered as negligible compared with the execution time of the tasks. The power saving of the LEDF

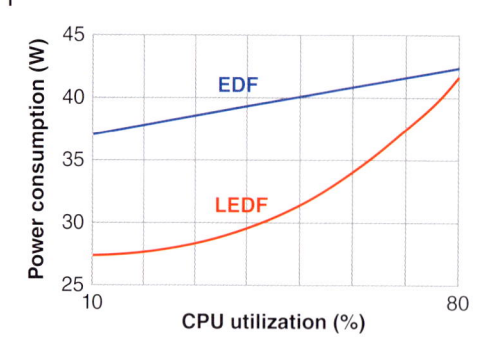

Figure 19.6 Illustration of the benefit for power saving of the LEDF compared with the EDF algorithm. (Based on data from Ref. [1].)

algorithm compared with the conventional EDF is shown in Figure 19.6, and reaches up to 40% with low CPU utilization.

One disadvantage of this approach is the requirement for task deadlines. Event-driven tasks with a short runtime can be assigned to computable deadlines, but pure computational tasks with varying instruction lengths (depending on the state of the program and input data) cannot be captured very well by this approach.

Material-integrated sensor networks will be equipped with low-resource and low-power microcontrollers supporting at least one active and one sleep state, demanding for power state transition algorithms, similar to the approaches discussed in *I/O-centric DPM*. Microcontrollers consist of a microprocessor, memory, and a large number of peripheral devices, all integrated on a system-on-chip design, hence profiting from *I/O*-centric DPM.

Algorithm 19.1 The Simplified LEDF Algorithm Adapted from Ref. [14]

```
Procedure LEDF()
Begin
  Do
    If (tasks waiting to be scheduled in ready list) Then
      Sort(deadlines) in ascending order
      Schedule(task) with earliest deadline
      Check if deadline can be met at lower speed/voltage
      If (deadline can be met) Then
        Schedule(task) to execute at lower speed/voltage
      ElseIf (deadline cannot be met) Then
        Check if deadline can be met at higher speed/voltage:
        If (deadline can be met) Then
          Schedule(task) to execute at higher speed/voltage
        ElseIf (deadline cannot be met) Then
          Task cannot be scheduled. Call exception handler!
    Else
      Do nothing.
  Done
End
```

19.3.2 I/O-Centric DPM

Scheduling of *I/O* devices is a technique that was originally introduced in non-real-time OS to minimize the access time for *I/O* requests. For example, a hard disk can only read consecutive blocks on the disk with a high throughput. Random access causes significant read/write head positioning delays. One major algorithm to schedule different *I/O* hard disk requests is a classical elevator algorithm solving multiple elevator requests on different floors and different directions. Such methods perform well in reactive systems, but a minimization of access times in real-time systems does not guarantee deadlines. With respect to energy-aware systems, the elevator algorithm minimizes the energy consumption of *I/O* devices, but cannot profit from different power states of such devices. Again an algorithm is required for managing and scheduling of multipower state *I/O* devices, with the restriction but common case of devices supporting two power states (active and sleep/standby). It is assumed that a system consists of a set of p devices $I/O = \{io_1, io_2, \ldots, io_p\}$. Different power states and power state transitions must be considered with respect to their power consumption and duration for each device io_j:

- Wake-up transition: sleep → working with $P_{wu,j} \cdot t_{wu,j}$
- Shutdown transition: working → sleep with $P_{sd,j} \cdot t_{sd,j}$
- Working state: $P_{w,j}$
- Sleep state: $P_{s,j}$

A wake-up or shutdown state transition is expensive, and hence there is a break-even time t_{be} for a sleep state between two working states consuming equal or more power than the case without the temporary shutdown of the device. This is a constraint not existing in CPU speed/voltage switching and complicates the scheduling algorithm significantly! For idle time periods shorter than the break-even time power is saved by keeping the working state.

According to Ref. [1], there is again a set of n tasks accessing the *I/O* devices $T = \{t_1, t_2, \ldots, t_n\}$, which has to be scheduled. Each task is parameterized by the tuple $t_i = (a_i, c_i, p_i, d_i, L_i)$, with a_i as the arrival time, c_i a worst-case execution time, p_i a period, d_i a deadline time, and L_i a device usage list of a task. Commonly, the deadline of the task is equal or less than the period time $d_i \leq p_i$. The energy consumed by a specific device io_i is given by:

$$E_i = P_{w,i} t_{w,i} + P_{s,i} t_{s,i} + M P_{st,i} t_{st,i} \tag{19.3}$$

with M as the number of power state transitions requiring the power $P_{st} = P_{wu} = P_{sd}$ for a time t_{st}. Associated with a task set T there is a set of l jobs $J = \{j_1, j_2, \ldots, j_l\}$ containing all tasks to be executed. Except for the job period, a job inherits all parameters of the task. A job is an instance of a task. If there is a job set J that uses a set of *I/O* devices *I/O*, a scheduling algorithm has to identify a set of start times $S = \{s_1, s_2, \ldots, s_l\}$ for the jobs such that the total energy consumed $\sum_{i=1}^{p} E_i$ is minimal and all jobs meet their deadline.

19.3.3 EDS Algorithm

From the tasks to be executed a job tree is generated. A common approach to find a scheduling solution is called pruning technique by iteratively pruning branches in the tree with respect to energy and time. A vertex v of this scheduling tree is a tuple (i, t, e), where i is the index of the job j_i, t a start time for the job, and e the energy consumed by the device until time t. A path from the root vertex to an intermediate vertex v is termed a partial schedule with some of the jobs. A path from the root vertex to a leaf vertex is a complete schedule. A complete schedule in that no job misses its deadline is suitable to solve the scheduling problem because temporal pruning eliminates all infeasible partial schedules.

The EDS algorithm, shown in Algorithm 19.2, uses the job list as an input and computes all possible nonpreemptive minimum energy schedules for the given job set. Details can be found in Ref. [1]. The algorithm shows the complexity to find an optimal energy-aware solution for I/O devices. Although this is traditionally part of an operating system only available on more powerful computer systems, low-resource embedded systems deployed in material-integrated sensing systems cannot compute the EDS algorithm due to required memory and computational resources.

Algorithm 19.2 EDS Algorithm Ref. [1]

```
Procedure EDS(J,l)
J is the Job Set with elements (i,time,energy)
l is the number of jobs
openList is a list of unexpanded vertices
currentList is a list of vertices at the current depth
t is a time counter

t:=0; d:=0; currentList:={}; openList:={(0,0,0)};
Add(vertex(0,0,0) to openList)
ForEach vertex v=(ji,time,_) ∈ openList
  t := time + completionTime(ji);
  Find all jobs J' released up to time t:
  J' := {j elem J | time(jᵢ) ≤ t}
  ForEach job j ∈ J' Do
    If (j has not been previously scheduled) Then
      V := {Find all possible scheduling instants for j};
      Computer energy for each generated vertex in V
      currentList := {currentList ∪ V};
    EndIf
  Done
  ForEach pair of vertices (v₁,v₂) ∈ currentList Do
    If v₁=v₂ and partialSchedule(v₁)=partialSchedule(v₂) Then
      If energy(v₁) > energy(v₂) Then
        Prune(v₁)
      Else
        Prune(v₂);
    EndIf
```

```
    Done
    openList := { openList ∪ {Unpruned vertices of currentList}};
    currentList := {};
    Incr(d);
    If d=1 Then Terminate;
  Done
```

The low-energy device schedule (LEDES) algorithm [15], a near-optimal, deterministic device-scheduling algorithm for two-state I/O devices that can be used for online device scheduling. It is assumed that the start time for each job is fixed and known *a priori*. Under this assumption, the energy-aware device scheduling problem P_{io} can be simplified. In contrast to EDS, the LEDES algorithm schedules devices and not jobs. The only critical check is that the idle time of a device is greater than the break-even time.

The P_{io} scheduling problem: Given the start times $S = \{s_1, s_2, \ldots, s_n\}$ and completion times $C = \{c_1, c_2, \ldots, c_n\}$ of the *n* tasks in a real-time task set *T* that uses a set $I/O = \{io_1, io_2, \ldots, io_p\}$ of I/O devices, determine a sequence of sleep/working states for each I/O device io_i such that total energy consumed $\Sigma_{i=1}^{p} E_i$ by I/O is minimal and all tasks are executed within the deadline interval. This algorithm – although not giving the best possible solution compared with EDS – is more suitable for material-integrated low-resource system due to lower memory and computational requirements. The LEDES algorithm is shown in Algorithm 19.3.

Algorithm 19.3 LEDES Algorithm Ref. [15] for Online Scheduling of jth Device io_j ($t_{0,j}$ is the device power state transition time of device *j*, *L* the device usage list, and curr the current scheduling step)

```
Procedure LEDES(io_j,τ_i,(τ_{i+1},curr)
  If (Scheduling step curr is the start of τ_i) Then
    If (state(io_j) = powered-up) Then
      If (io_j ∉ {L_i ∪ L_{i+1}}) Then
        shutdown(io_j)
      EndIf
      If (io_j ∈ L_{i+1}}) Then
        If (s_{i+1} - (s_i+c_i) >= t_{0,j}) Then
          shutdown(io_j)
        EndIf
      EndIf
    Else
      If (io_j ∈ L_{i+1} and s_{i+1} - (si-ci) < t_{0,j}) Then
        wakeup(io_j)
      EndIf
    EndIf
  ElsIf (Scheduling step curr is completion of task (τ_i) Then
    If (state(io_j) = powered-up) Then
      If (io_j nelem L_{i+1} and si+1-curr >= t0,j) Then
```

```
        shutdown(io_j)
      EndIf
    Else
      wakeup(io_j)
    EndIf
  EndIf
End
```

A significant advantage of online *I/O* device scheduling compared with offline approaches is that the DPM decision-making can exploit underlying hardware features such as buffered read and write operations. In contrast, an offline device schedule constructed as a table in system memory prevents the use of such features. Thus, the flexibility of online scheduling enhances the effectiveness of device scheduling [1].

19.4 Energy-Aware Communication in Sensor Networks

The total energy consumed in a sensor network can be classified in these contributions with a rough weighting:

- Analog data processing (ADC) [5%]
- Digital data processing (computation) [30%]
  - Microprocessor [25%]
  - Memory [5%]
- Data communication [60%]
  - On-chip [<5%]
  - Interchip (same node) [5%]
  - Internode [50%]
- Idle/standby (including auxiliary electronics and electrical losses) [5%]

In large-scale sensor network, the internode communication consumes the dominant part of the energy. Energy consumed by transmitting data using wireless communication over a 10–00-m range consumes as much energy as thousands of data processing operations [1]. Large-scale sensor networks commonly use message-based communication and routing (delivery of a message along a path) creates a significant overhead. Smart communication is required to reduce the energy consumptions. Avoiding communication is the most successful energy saving in sensor networks! Following there are some techniques shown saving energy by reducing communication and overhead.

The big volume of sensed information in large-scale networks and the requirement to aggregate data from the different sensed nodes for detecting events of interest lead to a large number of communications across the different nodes. The energy consumption of communication is proportional to the bit-size of a message, and in the case of wireless transmission proportional to the square of distance. Routing increased the energy for one message multiple times and depends on the number of intermediate nodes forwarding the message along its path.

Clustering Assuming mesh-like network structures, clustering of sensor nodes in regions of interest (ROI) and limiting communication mostly to these regions can reduce communication significantly. These clusters perform local processing of the sensed data and extract some useful information that is communicated to the nodes outside the cluster. See, for example, Ref. [16], for a deeper discussion of clustering and dedicated cluster communication protocols. The clustering concept is close to the multiagent model and self-organizing systems with entities interaction with each other in a limited region.

Caching-Based Communication In contrast to generic purpose communication networks, sensor nodes exchange mostly raw sensor and aggregated sensor data. This enables classification of communication messages. In stream-based sensor processing, the sensor data are sent multiple times without a change in the sensor value. This offers the opportunity to cache sensor data that is encapsulated and transferred in messages. Each node in a defined cluster caches the n previous data values sensed in this cluster. Assuming all the sensor nodes in the cluster have a coherent cache, then if the new sensed value that needs to be transmitted matches with one of the cached values, then only the index number of the matching cache entry is transmitted instead of the actual data value. This can reduce the size and hence the costs of communication significantly if the size of the generated sensor data is large compared with the data size of a cache index. Details can be found in Chapter 35 of Ref. [1]. If a dedicated and optimized communication protocol is used, the minimal packet size is $|p| = 1 + w + \log_2 n$, with w as the sensor data bit-size and n the number of cache entries. The first bit indicates the kind of a message (cache or data). This very simple caching approach is suitable for low-resource material-integrated sensor nodes. One disadvantage of this simple approach is the proper handling of cache inconsistency due to communication or node failures.

Data-Centric and Event-Based Communication Instead of transmitting sensor data periodically (request-based) in a data stream, the sensor nodes (as a data source) can decide if there was a significant change of sensor data that can be of interest for other nodes (as a data sink and collector). Only if an event is detected, sensor data are sent to sink nodes. This event-based communication (discussed in Section 16.7) can reduce the overall communication activity significantly and require no further synchronization, although nodes in a cluster can collectively decide if there was an event. This is discussed in Section 18.7 and applies clustering and event-based communication in distributed machine learning.

There are many popular data-centric routing algorithms for minimizing energy consumption, for example, diffusion routing, which use local spatial gradients to select paths for sending information from data source to sink nodes. Energy-aware metric and the geographical position of each node can be used to determine a route, too. Hybrid approaches can deal with constrained energy and quality-of-service considerations to improve routing. The lowest energy path is not always be the optimal path considering long-term network connectivity. Global energy minimization can result in uneven energy consumption patterns across particular sensor nodes. Consequently, some nodes could deplete their energy resources sooner than other nodes, reducing the system vividness and information quality of the entire sensor network.

19.5 Energy Distribution in Sensor Networks

With increasing miniaturization and sensor density, decentralized self-supplied energy concepts and energy distribution architectures are preferred over centralized.

Self-powered sensor nodes collect energy from local sources, for example, by using thermoelectrical energy harvester or solar cells. But additionally they can be supplied by external energy sources. Nodes in a sensor network can use communication links to transfer energy, for example, optical links are capable of transferring energy using laser or LE diodes in conjunction with photodiodes on the destination side, with a data signal modulated on an energy supply signal. An example can be found in Ref. [17]. The sensor node uses an optical link based on a fiber connection (wired communication). Two different optical wavelengths are used to transmit data and energy over the fiber link. The principle node architecture is shown in Figure 19.7, simplifying the transmission of data and communication by using one optical wavelength, which finally reduces the node complexity. In this approach, the communication signal is modulated on the (strong) energy signal on the sender side, and filtered out on the receiver side [18].

The main building blocks of a sensor node, the proposed technical implementation of the optical serial interconnect modules, and the local energy management module collecting energy from a local source, for example, a thermoelectric generator, and energy retrieved from the optical communication receiver modules. The data processing system can use the communication unit to transfer data (D) and superposed energy (E) pulses using a light-emitting or laser diode. The diode current, driven by a differential-output sum amplifier, and the pulse duration time determine the amount of energy to be transferred. The data pulses have a fixed intensity several orders lower than the adjustable energy pulses. On the receiver side, the incoming light is converted into an electrical current by using a photo diode. The data part is separated by a high-pass filter, the electrical energy is stored by the harvester module.

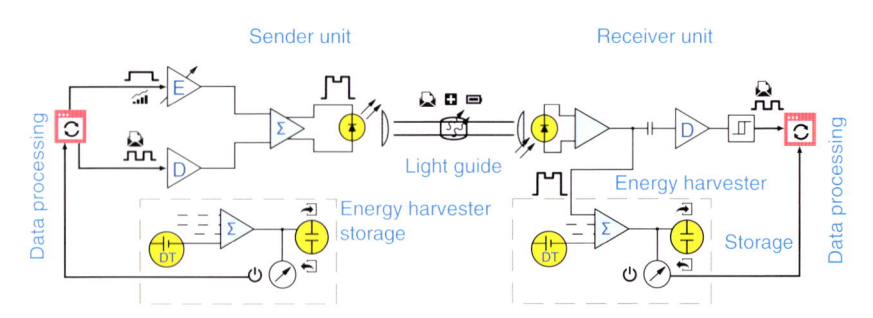

Figure 19.7 Principle sensor node architecture with energy harvesting using an optical link to transmit data and energy.

Sharing of one interconnect medium for both data communication and energy transfer significantly reduces node and network resources and complexity, a prerequisite for a high degree of miniaturization required in high-density sensor networks embedded in sensorial materials. Point-to-point connections and mesh-network topologies are preferred in high-density networks because they allow good scalability (and maximal path length) in the order of $O(\log N)$, with N as the number of nodes.

Assuming a mesh-like network topology with peer-to-peer connectivity of neighbor nodes, nodes can exchange energy based on demand and negotiation, offering energy distribution and balancing capabilities. Nodes with energy surplus can transfer energy from their local energy storage to neighbor nodes with low-energy deposit. This technical feature enables distributed and global energy management, discussed in the next section. Energy distribution capabilities of sensor nodes can eliminate a centralized energy supply infrastructure.

19.5.1 Distributed Energy Management in Sensor Networks Using Agents

Having the technical abilities explained in the previous section, it is possible to use active messaging to transfer energy from good nodes having enough energy toward bad nodes, requiring energy. An agent can be sent by a bad node to explore and exploit the near neighborhood. The agent examines sensor nodes during path travel or passing a region of interest (perception) and decides to send agents holding additional energy back to the original requesting node (action). In addition, a sensor node is represented by a node agent, too. The node and the energy management agents must negotiate the energy request.

The agents have a data state consisting of data variables and the control state, and a reasoning engine, implementing behavior and actions. Table 19.2 poses the different agent behavior required for the smart energy management.

A node having an energy level below a threshold can send a help agent with a delta-distance vector specifying the region of interest in a randomly chosen direction. The help agent hops from one sensor node to the next until the actual delta-vector is zero. If there is a good node found, with local energy above a specified threshold, the help agent persists on this node and tries to send periodically deliver agents transferring additional energy to the originator node. An additional behavior, help-on-way, changes the deliver agent into an exploration agent, too. Such a modified agent examines the energy level of nodes along the path to the destination. If a bad node was found, the energy carried with the agent is delivered to this node, instead to the original destination node. This approach is independent of the agent processing platform and the mobile agent model.

Simulation carried out in Ref. [18] with the SeSam agent simulator environment was used to show the suitability of this approach. Each sensor node is modeled with an agent, too. Energy management agents and sensor node agents negotiate energy demands and communicate with each other by using globally shared variables. If the energy deposit of a node is below a threshold $E_i < E_{\text{low}}$ (called bad node), help agents are sent out; if $E_i > E_{\text{high}} > E_{\text{good}}$, then distribute

Table 19.2 Agents with different behavior used to manage and distribute energy.

Agent type	Behavior
Request	*Point-to-point agent*: This agent requests energy from a specific destination node, returned with a `Reply` agent.
Reply	*Point-to-point agent*: Reply agent created by a `Request` agent, which has reached its destination node. This agent carries energy from one node to another.
Help	*ROI agent*: This agent explores a path starting with an initial direction and searches a good node having enough energy to satisfy the energy request from a bad node. This agent resides on the final good node for a couple of times and creates multiple deliver agents periodically in dependence of the energy state of the current node.
Deliver	*Path agent*: This agent carries energy from a good node to a bad node (response to `Help` agent). Depending on selected sub-behavior (`HELPONWAY`), this agent can supply bad nodes first, found on the back path to the original requesting node.
Distribute	*ROI agent*: This agent carries energy from and is instantiated on a good node and explores a path starting with an initial direction and searches a bad node supplying it with the energy.

ROI, region of interest.

agents are used to distribute energy to surrounding bad nodes. If $E_i > E_{good} > E_{low}$, then the node is fully functional (called good node).

Simulation carried out in Ref. [18] with the SeSam agent simulator environment was used to show the suitability of this approach. Each sensor node is modeled with an agent, too. Energy management agents and sensor node agents negotiate energy demands and communicate with each other by using globally shared variables. If the energy deposit of a node is below a threshold $E_i < E_{low}$ (called bad node), help agents are sent out; if $E_i > E_{high} > E_{good}$, then distribute agents are used to distribute energy to surrounding bad nodes. If $E_i > E_{good} > E_{low}$, then the node is fully functional (called good node).

The simulation showed that help agents with simple exploration and exploitation behavior are suitable to meet the goal of a regular energy distribution and a significant reduction of bad nodes unable to process sensor information, but additional distribute agents create no significant benefit. The multiagent implementation offers a distributed management service rather than a centralized approach commonly used. The simple agent behaviors can be easily implemented in digital logic hardware (application-specific platform approach).

The temporal development of bad and good nodes with and without smart energy management is show in Figure 19.8.

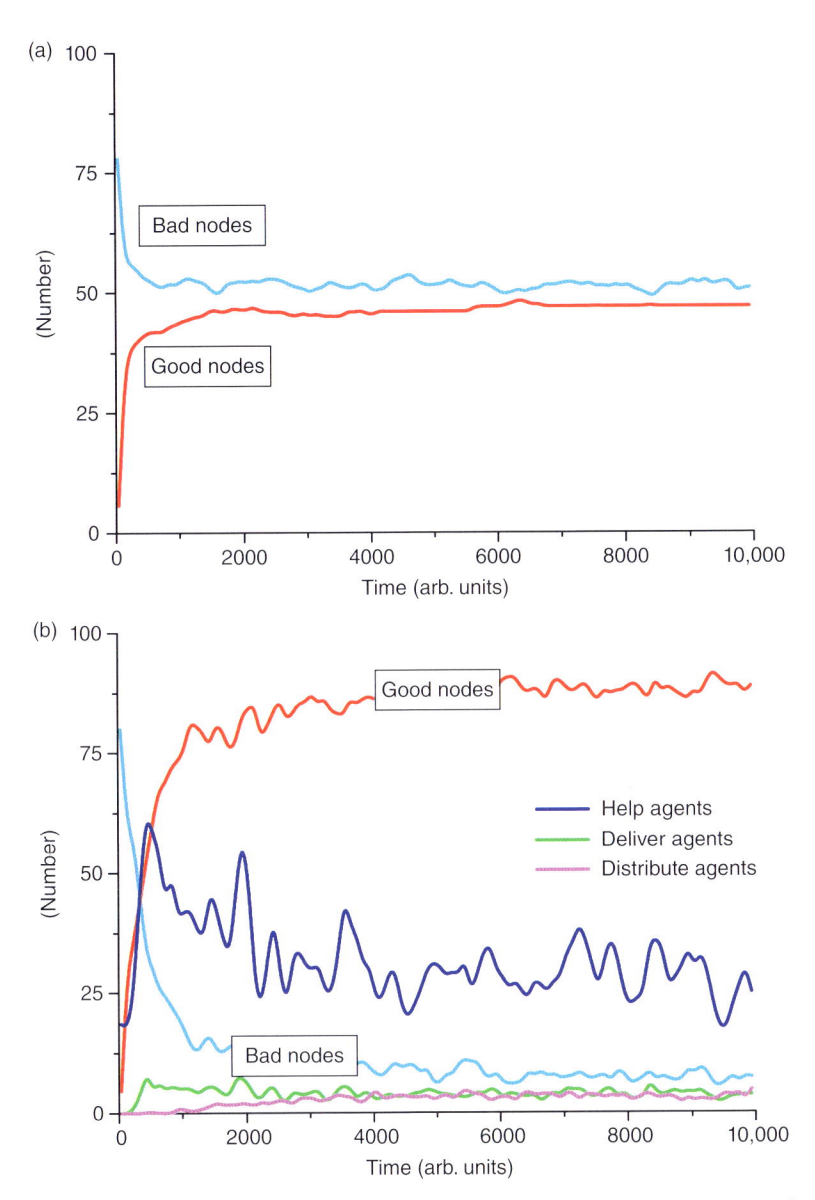

Figure 19.8 Simulation results from Ref. [18] for a distributed sensor network of self-supplying nodes without (a) and with (b) energy distribution using agents.

References

1 Iyengar, S.S. and Brooks, R.R. (eds) (2005) *Distributed Sensor Networks*, CRC Press.
2 Albus, Z., Valenzuela, A., and Buccini, M. (2009) Ultra-Low Power Comparison: MSP430 vs. Microchip XLP, Tech Brief, Texas Instruments.

3 Atmel (2013) 8-bit Atmel with 8K Bytes In-System Programmable Flash, ATmega8/ATmega8L, Rev.2486AA–AVR–02/2013.
4 Intel® (2016) Intel® Core(i5-4300U Processor (3M Cache, up to 2.90GHz), Web Specification. http://ark.intel.com/products/76308/Intel-Core-i5-4300U-Processor-3M-Cache-up-to-2_90-GHz.
5 Tochigi, M. (2013) Achieving low power wireless connectivity for battery powered healthcare sensors, Murata Europe, MURA053-D3 Technical Note.
6 Feckl, F. (2015) IoT long-life wireless sensors require ultra-low power architectures. *Electronics Component News*, May 21, 3:43 pm. http://www.ecnmag.com/article/2015/05/iot-long-life-wireless-sensors-require-ultra-low-power-architectures (retrieved May 21, 2014).
7 Maxim (2016) Low-Power ADC ICs. http://www.maximintegrated.com/en/products/analog/data-converters/analog-to-digital-converters/precision-adcs/low-power-adc-ics.html.
8 Bosse, S. and Behrmann, T. (2011) Smart Energy Management and Low-Power Design of Sensor and Actuator Nodes on Algorithmic Level for Self-Powered Sensorial Materials and Robotics. Proceedings of the SPIE Microtechnologies 2011 Conference, April 18–20, Prague, Session EMT 101 Smart Sensors, Actuators and MEMS. doi: 10.1117/12.888124
9 Behrmann, T., Zschippig, C., Lemmel, M., and Bosse, S. (2010) Toolbox for Energy Analysis and Simulation of self-powered Sensor Nodes. Proceedings of 55 IWK Internationales Wissenschaftliches Kolloquium, Workshop A3 Analysis and Synthesis to Energy Efficiency Optimisation (IWK-10), September 13–17, Ilmenau.
10 Behrmann, T., Budelmann, C., Lemmel, M., Bosse, S., and Lehmhus, D. (2013) Tool chain for harvesting, simulation and management of energy in Sensorial Materials. *Journal of Intelligent Material Systems and Structures*, **24** (18), 2245–2254.
11 Mathworks (2010) Mathworks: Simulink Reference.
12 Ma, K., Bahman, A. S., Beczkowski, S., and Blaabjerg, F. (2015) Complete loss and thermal model of power semiconductors including device rating information. *IEEE Transactions on Power Electronics*, **30** (5), 2556–2569.
13 Jeffay, K. *et al.* (1991) On non-preemptive scheduling of periodic and sporadic tasks with varying execution priority. Proceedings of the Real-Time Systems Symposium, December, p. 129.
14 Swaminathan, V. and Chakrabarty, K. (2001) Real-time task scheduling for energy-aware embedded systems. *Journal of the Franklin Institute*, **338** (6), 729–750.
15 Swaminathan, V. and Chakrabarty, K. (2003) Energy-conscious, deterministic I/O device scheduling in hard real-time systems. *IEEE Transactions on Computer-Aided Design of Integrated Circuits and Systems*, **22** (7), 847.
16 Loh, P. and Pan, Y. (2009) An energy-aware clustering approach for wireless sensor networks. *International Journal of Communications, Network and System Sciences*, **2** (2), 131–141.
17 Budelmann, C. and Krieg-Brückner, B. (2011) Sensor Network Based on Fibre Optics for Intelligent Sensorial Materials. EUROMAT 2011 – European

Conference and Exhibition on Advanced Materials and Processes, September 12–15, Montpellier, France.

18 Bosse, S. and Kirchner, F. (2012) Smart Energy Management and Energy Distribution in Decentralized Self-Powered Sensor Networks Using Artificial Intelligence Concepts. Proceedings of the Smart Systems Integration Conference 2012, Session 4, March 21–22, Zürich, Schweiz, ISBN: 978-3-8007-3423-8.

20

Microenergy Storage

Robert Kun,[1,2] Chi Chen,[3,4] and Francesco Ciucci[5,6]

[1]*University of Bremen, Faculty of Production Engineering, Innovative Sensor and Functional Materials Research Group, Badgasteiner Strasse 1, 28359 Bremen, Germany*
[2]*Fraunhofer Institute for Manufacturing Technology and Advanced Materials – IFAM, Wiener Strasse 12, 28359 Bremen, Germany*
[3]*Department of Mechanical and Aerospace Engineering, The Hong Kong University of Science and Technology, Clear Water Bay, Kowloon, Hong Kong, China*
[4]*Department of NanoEngineering, University of California San Diego, 9500 Gilman Dr, Mail Code 0448, La Jolla, California 92093-0448, USA*
[5]*Department of Mechanical and Aerospace Engineering, The Hong Kong University of Science and Technology, Clear Water Bay, Kowloon, Hong Kong, SAR China*
[6]*Department of Chemical and Biological Engineering, The Hong Kong University of Science and Technology, Clear Water Bay, Kowloon, Hong Kong, SAR China*

20.1 Introduction

The development of micro/nanosystems has increased the demand for integrating micropower modules. In portable devices and electronics, insufficient energy supply has become the main concern for users. In the 2015 CES show, Fortune and SurveyMonkey released a poll asking about the most exciting feature that a new smart phone may have. The top 1 answer is, not surprisingly, a prolonged battery life [1]. In microelectronics, this could be realized considering the energy requirements for them are usually "minimum." For example, a typical sensor requires a few microwatts in standby mode and tens of milliwatts when active. This amounts to a few watt-hours as a lifetime consumption of the sensor. As a comparison, the battery energy capacity of an iPhone 6 is about 7 Wh, which translates to a "perpetual" power source for a sensor. In essence, it would be practically attractive if the devices may be self-sustained and can operate stably for a considerable amount of time. This demand has motivated researchers to work on energy harvesting (EH) and storage, in addition to selecting energy efficient devices that minimize energy consumption [2]. However, with system size decreasing, energy solution is also increasingly difficult compared with the macroscale, with one of the difficulties being miniaturizing the energy systems while maintaining the complete functionality. In 2014, a company Wetag launched a kickstarter campaign promoting their conceptual iFind as the first "battery-free" item tag that can harvest electromagnetic energy from the environment. The most attracting feature of this tag is, without doubt, the promise of being "battery-free" and operating consistently without human monitoring.

Material-Integrated Intelligent Systems: Technology and Applications, First Edition.
Edited by Stefan Bosse, Dirk Lehmhus, Walter Lang, and Matthias Busse.
© 2018 Wiley-VCH Verlag GmbH & Co. KGaA. Published 2018 by Wiley-VCH Verlag GmbH & Co. KGaA.

Remarkable fund-raising results were achieved in the campaign: $520 000 was raised before the funding suspension, which was more than 20 times as what was requested [3]. However, the company has not delivered the product as promised. This points out not only the needs for self-powered or long-lasting devices but also the technological difficulty of achieving that. The international technology roadmap of semiconductors forecasted that the energy supplies, in terms of lifetime, volume, and mass performance, may pose technical obstacles for the semiconductor industry and limit the developments of microelectromechanical systems (MEMS) [4].

Power supplies for MEMS can be categorized into two types, regenerative and nongenerative, depending on their sources of energy [5]. The generative power supplies utilize the operating condition and generate power *in situ*, that is, EH and energy scavenging, for example, commercial desk calculators powered by solar cell. On the other hand, for nongenerative energy supply, the system needs a continuous supply of active materials or fuels.

Depending on the energy sources, the generative supplies may utilize mechanical energy [6], thermal energy, electromagnetic energy, and so on. Nongenerative energy supplies, on the other hand, usually refer to the family of energy storage and conversion systems, comprising batteries, supercapacitors, and fuel cells. Although it is possible to use either EH or energy storage alone, both are usually combined since environmental energy sources may be intermittent and the EH energy profile hardly matches the consumption profile of the device. For example, devices using solar cells alone cannot function at night and energy storage is necessary therein. A second possibility is to solely use energy storage systems to power the device. This finds applications in many battery-based electronics. Finally, in addition to EH and energy storage, smart energy management system may also be needed to coordinate the system. There have been some achievements in designing a whole system, such as the thermal system by Becker *et al.* [7], the MPPT design [8], Heliomote [9], Everlast [10], Prometheus [2], Ambimax [11], PUMA [12], and so on. Three different routes are listed in Figure 20.1.

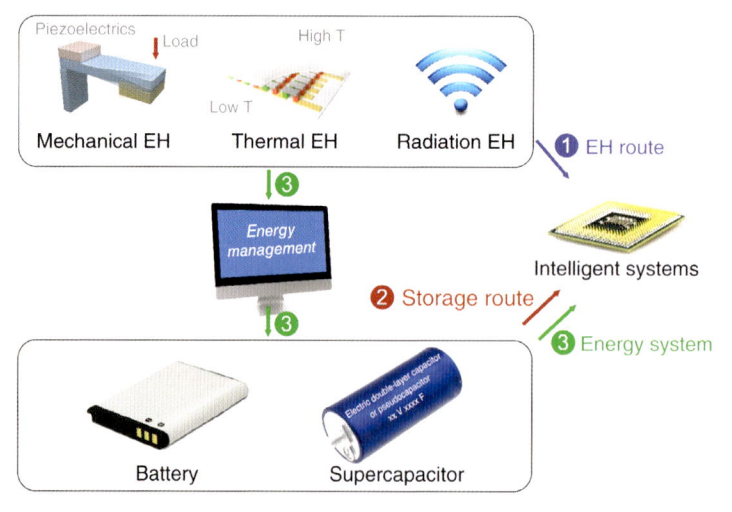

Figure 20.1 Three routes of powering intelligent systems. Route 1 harvests energy from various sources and directly outputs to the systems, 2 uses only storage system, and 3 combines EH, energy management, and storage systems.

20.2 Energy Harvesting/Scavenging

Energy harvesters transduce environmental energy sources into usable electricity that powers devices. Depending on the types of energy sources, these harvesters can be classified mainly into three categories: mechanical energy harvesters, thermal energy harvesters, and radiation energy harvesters. Often times, another term, energy scavenging, is used interchangeably, despite the subtle difference.

Mechanical EH comprises large spectrum of devices. Some of such devices are based on the materials' electromechanical or magnetomechanical coupling mechanism to convert the mechanical energy into electricity. Others use physical phenomena that do not rely on the material properties. The former category includes piezoelectric materials, such as $PbZr_xTi_{1-x}O_3$ (PZT) or $K_{1-x}Na_xNbO_3$ (KNN), which are capable of producing electricity when strained. Magnetostrictive materials change their magnetization under *Villari* effect when strain is applied and electricity can be produced by further electromagnetic induction. Moreover, this material family is not limited to ceramics, rather it also finds members in polymers. Another type of mechanical EH operates on other physical phenomena that are not based on smart materials. For example, in electrostatic generators, strain-induced thickness change of the membrane results in a change of the capacitance and thereby voltage. The strain also induces the change of chemical potential change of charges, and further concentration gradients. Based on the concentration gradient, charged particles may accumulate at electrodes, resulting in electrical potential difference. Mechanical motion can also be turned to electricity through electromagnetic induction. Recently, nanogenerators based on triboelectricity are realized by contact of two materials with different charge affinity.

Thermal EH is another attractive energy harvesting technology. Among the thermal EHs, thermoelectric generators (TEGs) are the most commonly used. TEGs generate electrical power by the Seebeck effect, whereby a spatial temperature gradient applied to a circuit at the junction of two dissimilar conductors induces electromotive force and further electrical energy.

Finally, radiation EH might be the most attractive due to the pervasiveness of radio waves. One large subclass comprises solar EHs, including silicon-based photovoltaic systems, dye-sensitized solar cells, organic solar cells, and perovskite-type solar cells. The perovskite-type solar cells [13] have sparked great interests in the field due to the fact that its efficiency has skyrocketed in past 2 years by 6% points reaching 20.1% [14], while the same achievements were made with multicrystalline solar cells in two decades [15]. In addition, it could be made transparent, and coupling with conventional silicon cell can boost the efficiency further. Another type of radiation EH harvests radio frequency (RF) waves. The RF waves can be the results of radio, TV, satellite, 3G and 4G networks, WiFi, and so on. It is attractive, since no motion is needed. However, low energy density, on the order of nW/cm^3, is intrinsic disadvantage of this EH and besides, RF fades with R^{-2} from the source.

For more detailed content about various EH technologies, the readers are referred to Chapter 7.

20.3 Energy Storage

Historical Background Mechanical separation of the electric charge in dielectric materials, called as triboelectric effect, was already known in the ancient times. However, the first practical device that could store electrical energy (i.e., due to physical charge accumulation) was invented in the late eighteenth century. This invention called as "Leyden jar" was a milestone in the history of electrical energy storage; the capacitor was born. Indeed, the "Leyden jar" can be considered as a direct predecessor of the nowadays widespread dielectric capacitors [16].

In the early nineteenth century, inspired by the earlier studies and experiments on electricity of Luigi Galvani, the Italian professor and scientist Alessnadro Volta has constructed the first device that was able to produce electrical energy. Volta has recognized that electric current would flow in an external circuit if metals with different chemical natures (e.g., zinc and copper, that is, electrodes) were immersed into an electrolyte (brine) and then connected electrically [17]. His early invention nowadays known as "Voltaic pile" was a real breakthrough in his time since this device could produce electrical current continuously for the first time and the *battery* was invented. This enabled afterward further inventions, discoveries, and scientific findings in the field of electrochemistry and material science, as well.

Accordingly, storage of the electrical energy can eventuate in direct as well as indirect fashion. In case of *direct* manner, no conversion of the electrical energy takes place during the storage process. The storage of electrical energy is based merely on electrostatic attraction–repulsion mechanisms. Considering *indirect* mode of operation, a certain conversion process between different energy forms has to be involved in the storage mechanism.

20.3.1 Capacitors

Working Principle and Classification Capacitors are electronic devices that could store electrical energy directly based on physical charge separation. Simplest capacitor is build up from two metallic plates separated by an electric insulator of any kind of dielectric material. Historically, air, vacuum, glass, (oil-soaked) paper, polymers, ceramics, and mica are the most important dielectric materials used in capacitors [18]. By contacting the two metal plates to an external electric source, charging process of the capacitor takes place; equal amounts of charge carriers with opposite charges will be gradually separated. As a result, one of the metallic plate gains negative charges and the other plate gains positive charges; consequently, a potential difference (cell voltage) between two plates will evolve. The amount of the stored energy of a capacitor can be expressed as follows (see Eq. (20.1)):

$$E = \frac{1}{2} C \times U^2. \tag{20.1}$$

where C is the capacitance of the capacitor in farad (1 F = 1 C/V) and U is the cell voltage in volts. For a simplest parallel-plate capacitor, capacitance can be expressed as

$$C = \epsilon_r \times \epsilon_0 \times \frac{A}{d}. \tag{20.2}$$

Equation (20.2) demonstrates that capacitance will be higher if electrode surface (A) and relative permittivity of the applied dielectric medium (ϵ_r) are higher and in the same time separation of the electrodes (d) is small; ϵ_0 is permittivity of the vacuum.

Based on the construction principle, modern capacitors can be classified into two main classes: (i) dielectric and (ii) electrochemical capacitors (Figure 20.2).

Due to their relatively low capacitance values (range from several pF to hundreds of µF), small-sized dielectric and electrolytic capacitors are mainly used as primary circuit board elements [19]. Nevertheless, due to the drastic reduction of the thickness of the dielectric layer (e.g., down to several tens of nanometers thickness, depending on type of electrolytic capacitor) and due to significant increase of the electrode surface area, electrolytic capacitors can provide capacitance values routinely in the 10^{-7}–10^1 F range by operating (cell) voltages up to few hundred volts DC. However, these systems cannot be considered as practical energy storage units for microelectronics. Further reduction in the length of the charge separation (down to several angstroms) and further substantial increase of the electrode surface area have been achieved in the newest generation of capacitors. These so-called *supercapacitors* (SCs) substantially differ from both previously described systems. SCs provide the highest energy density among the capacitors [20]. Due to SC technology, realization of practical gravimetric energy densities on the order of 1–10 Wh/kg and specific power density values of up to few 10 kW/kg are possible nowadays [21]. This

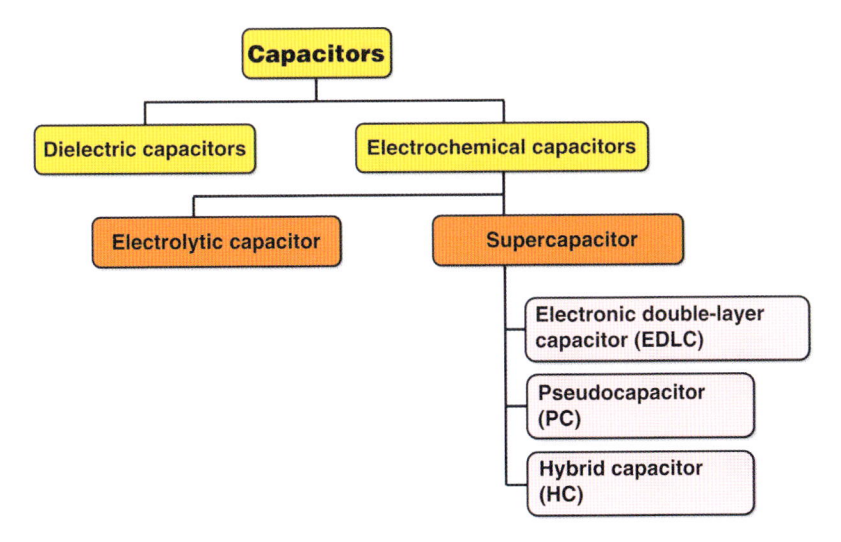

Figure 20.2 Classification of capacitors.

makes them feasible to use as stand-alone energy storage devices or they can be combined with electrochemical storage units (i.e., batteries) depending on the application scenario. Instead of (etched) metallic foils, electrode materials in SCs are carbon-based materials with high electric conductivity and remarkable high porosity; activated carbon is currently used as industrial standard [22]. However, many advanced carbon-based electrode material have been investigated and proposed [21]. Most importantly, (templated) activated carbon, single/multiwall carbon nanotubes (SWCNTs, MWCNTs) [23], graphene [24–28], and carbon (nano)fibers [29] were investigated and discussed as possible electrode materials in SCs. Such SC can operate in a broad temperature range (depending on electrolyte used) and possesses good charging/discharging efficiency (~85–98%). Storage mechanism in the pure electronic double-layer capacitor (EDLC) is based on the buildup of the Helmholtz double layer in the immediate vicinity of the outer electrode surface upon charging in the S/L (electrode/electrolyte) interface. In that manner electrode surface gains negative or positive charge, and it is compensated with the accumulation of equal number of counter ions those reside in the electrolyte. In pure EDLC, ideally no charge transfer occurs on the electrode/electrolyte interface during charging and discharging process; hence, extraordinary cycle life (~10^6 cycles) retains. In pseudocapacitors (PCs), metal oxides or conductive polymers are used as electrochemically active component composited with carbon-based electrode materials [30–32]. In PCs besides the double-layer capacitance, a faradaic current originating from a fast and highly reversible redox reaction of the pseudocapacitive material is also accountable for the further enhancement of capacitance. However, this happens at the expense of cycle life of the capacitor and reduction of power density. Since in PCs faradaic charge transfer occurs through the electrode/electrolyte interface (similar to batteries), some side effects such as electrolyte degradation, irreversible active material degradation, and mechanical stress (i.e., volume changes during cycling) of the electrode layer can result in reduced cycle life and the kinetics as well as rate of the fast reversible redox reaction limit the power density. Hybrid capacitors (HCs) are asymmetrical in construction and composed of both ELDC and PC electrodes.

Microscale Supercapacitors From application point of view, microsupercapacitors are recently attracting considerable technological and research interest. The need for reliable integrated power sources for low-power applications in microelectronics, lab-on-a-chip devices, wearables, wireless communication chips, microprocessors, and MEMS is increasing uninterruptedly [33]. If limited areal space is available for the integration of a storage device, design aspects will be more dominant. In contrast to conventional capacitors, microsupercapacitors possess planar, predominantly interdigitated structure [34–36].

Capacity and therefore energy density of a conventional sandwich-structured capacitor can be adjusted by controlling electrode layer thickness. However, due to increase of electrode layer thickness *cross-plane migration distance* of the ions will be longer and equivalent series resistance (ESR) will be higher, as well [37]. This eventuates in diminished power density, that is, slower charge/discharge capability of the device. Therefore, interdigital device structures were proposed

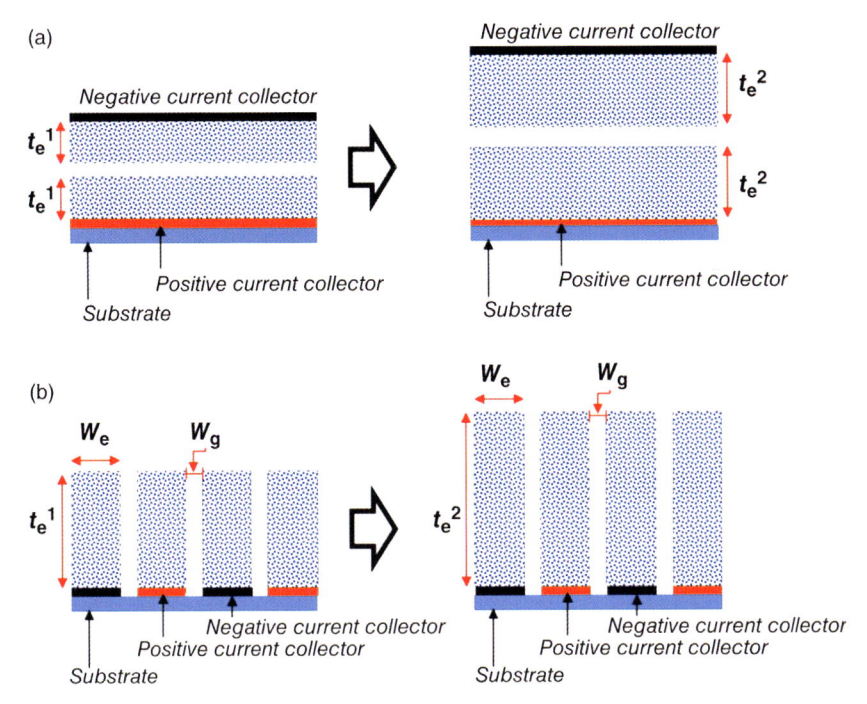

Figure 20.3 Schematic representation of increase of the electrode layer thickness in case of (a) conventional sandwich and (b) interdigitated capacitor electrode design.

for improved capacity and energy density and enhanced high-power capability of the microsupercapacitor. In Figure 20.3, sandwich-structured (*a*) and interdigitated (*b*) capacitor designs are demonstrated. In conventional sandwich-structured capacitor, the only way to improve capacity is to increase electrode layer thickness t_e ($t_e^1 < t_e^2$, Figure 20.3a) (by sustaining adjusted pore structure and using same separator thickness), which again results in higher internal resistance and longer ion diffusion pathways, which finally result in reduced high-power capability of the device. On the other hand, in interdigitated capacitor structures two adjacent electrodes (width w_e) are separated by a defined gap w_g (Figure 20.3b). In case of confined areal space on the substrate, increasing electrode layer thickness ($t_e^1 \rightarrow t_e^2$ and $t_e^1 < t_e^2$) results in higher energy density while maintaining fast ion diffusion across the electrode layer since cross-plane migration distance would not be affected in this manner. Fine control on parameters w_e, w_g, and t_e during fabrication process is possible, therefore overall device performance can be adjusted. Typical values for parameters w_e, w_g, and t_e fall between several tens to hundred micrometers. Based on current literature data, microsupercapacitors based on carbon electrode materials can reach 1–10 mF/cm^2 specific capacitance and ≤10 mWh/cm^3 specific (volumetric) energy densities [36,38–40].

Construction of Microsale Devices Fabrication of microelectrode patterns onto various substrates can be accomplished by digital printing technologies. Among various printing techniques, such as screen, transfer, aerosol jet, inkjet printing is one of the most notable techniques. Inkjet printing technology offers several advantages, such as cost-effectiveness (i.e., lower capital expenditures and production costs), easy processing, possibility to industrial scale-up, and patterning of various substrates of various forms [22,41,42]. Furthermore, being as an additive manufacturing technology the amount wasted material can be extremely minimized [43]. Inkjet printed microstructures are limited to resolutions approximately several tens of micrometers. In printed microsupercapacitors, a lateral spatial resolution of few tens of micrometers is nowadays possible [44,45]. Despite the ability to deposit ink droplets precisely without prepatterning the substrate, some problems are foreseeable and those originating from the nature of this printing technology. Drop coalescence, satellite droplets, blurred edge lines, misplaced droplets, and coffee-ring effect [46,47] are the main issues that result in low-quality printed patterns finally diminish performance on device scale.

In order to increase volumetric energy density of the microsupercapacitors, three dimensional electrode structures were proposed. The feasibility of the production of 3D interdigitated electrode structures was demonstrated using the additive manufacturing technique called selective laser melting (SLM) [48]. By the SLM process, high energy laser impacts a metal powder bed, where metal (alloy) particles will be locally melted and fused due to high energy impact and finally a three-dimensional structure will be generated in a layer-by-layer fashion. Columnar like interdigital electrodes with pillar diameter of 500 μm and pillar height of 8 mm were fabricated on 1 cm^2 device footprint. The symmetrical capacitor contained polypyrrole (PPy) electrode material and poly(vinyl alcohol)/H_3PO_4 electrolyte. Volumetric capacitance of 2.4 F/cm^3 and energy density of 0.214 mWh/cm^3 were achieved at current density of 3.74 mA/cm^3. After 1000 charge/discharge cycles, the device delivered about 80% of its initial capacity, cycled at current density of 14.98 mA/cm^3. The authors pointed out that further miniaturization of the electrode structure, therefore reduction of device footprint, is possible.

Lithography-based fabrications technique can also be used by the production of microsupercapacitor electrodes. This method has to be chosen if precise electrode separation is an important issue and electrode pattern is below the resolution limit of that of inkjet or other digital print technologies. However, depending on design and applied electrode materials in the device lithography-based processes cannot be reasonable sometimes, for example, the need for impractically extended etching times. Besides printing and lithography methods, microextrusion was proved as feasible technique in order to prepare interdigitated microsupercapacitor architectures [49]. High viscosity ink will be, during this process, extruded through a deposition nozzle. Deposition speed, nozzle size, and printing trajectory can be easily set and control, therefore influence the properties of the patterned microelectrode. This method enables patterning both two- and tree-dimensional microelectrode architectures.

Materials in Microsupercapacitors Commonly applied electrode materials in microsupercapacitors can be classified into three main groups and here is a substantial similarity to conventional supercapacitors, such as carbon materials (e.g., activated carbon, carbon nanotubes, graphene bases materials, carbon aerogels), transition metal oxide (e.g., MnO_2, RuO_2, NiO) containing composite electrode materials, and finally conducting polymers (e.g., polypyrrole, polyaniline, polythiophene). Anode and cathode materials in the microsupercapacitor can be the same (symmetrical material design); however, anode material could differ from cathode material as well (asymmetrical design). As an example, for the asymmetrical design, one electrode uses pure EDLC carbon-based electrode materials, while other contains PC material. Using asymmetric electrode arrangement extending the cell voltage above 1 V in case of aqueous electrolyte-based devices is possible, since electrodes with different composition possess high overpotentials for hydrogen and/or oxygen evolutions [50,51].

Besides electrode materials, electrolyte plays a very important role in the microsupercapacitors, since it influences device performance significantly. There are several requirements for the electrolyte material, most importantly high electrochemical stability (wide voltage window), high ionic conductivity, sufficient solvated ionic radius, low viscosity, low volatility, low toxicity, low cost, and high purity [52]. Furthermore, electrolyte should be less corrosive or noncorrosive and should provide good electrode–electrolyte interaction. Three main types of liquid electrolyte materials are applied in microsupercapacitors. These are typically aqueous electrolyte, organic solvent-based electrolyte, or ionic liquid-type electrolyte. Water-based electrolytes are very appealing being high-safety, nontoxic, and low-cost materials with good availability, high ionic conductivity, and easy handling. Despite several advantages, the poor electrochemical stability (electrochemical stability window of about 1 V), therefore low energy density on device scale in the practice, and severe corrosion issues (acid- or base-type electrolytes) limit their application potential. Using organic solvent-based (typically ethylene carbonate, propylene carbonate, acetonitrile) electrolytes, higher cell potentials up to 2.5–2.7 V can be realized. The achievable higher energy density impairs cell safety, since these types of electrolytes are toxic and flammable. Furthermore, their high sensitivity against air and humidity makes handling, packaging, cell assembly much more complicated compared with aqueous systems. The promise of almost up to 5 V cell voltage has driven considerable attention to use ionic liquids as electrolytes in (micro-)supercapacitors. The notably higher maximum cell voltage, low toxicity, nonvolatility, high-safety, high chemical stability, and high-temperature stability (for high-temperature applications) are the main advantages of such electrolyte systems. However, poor ionic conductivities especially at ambient temperatures, high production costs, complicated and complex synthesis technologies, and purity issues in the final products are the main limitations of the large-scale prevalence of ionic liquids as electrolytes.

Technological viability of liquid phase electrolytes is commonly known and accepted for conventional capacitors. In the meantime, advanced manufacturing and packaging technologies have been developed that provide safe, leak-proof, and robust encapsulation of the capacitors cells, that is, cell components. Due to the robustness and bulkiness of the established packaging technologies for conventional capacitors, they cannot be applied for microscale devices. If the storage device has to be integrated into a confined areal space or volume, state-of-the-art macroscale packaging technologies will fail. Using advanced microfabrication techniques, microscale energy storage devices with several square millimeter footprints can be routinely realized nowadays. However, packaging technologies for microscale applications are not on the same technological level. In order to be capable to realize practical microscale energy storage devices with liquid electrolytes, advanced packaging technologies on the microscale have to be first developed and established. Enabling direct integration of the microscale energy storage device onto chips, it is preferable if the storage device consists entirely solid-state materials including electrolyte. Realization of high performance solid-state microsupercapacitors for future application, high performance electrode materials and architectures and better performing solid-state electrolytes are essential. Using solid-state electrolytes in the microsupercapacitors, issues regarding packaging and sealing could be significantly reduced or even eliminated. Furthermore, application of solid-state electrolyte in the microsupercapacitor cell fabrication of all-solid-state mechanically deformable microsupercapacitor is possible. This opens novel pathways in the technological integration and endows the microsupercapacitors with added functionality such as flexibility (i.e., wearables, flexible micro/electronics) and transparency. Nevertheless, research on solid-state microsupercapacitors is in the early stages it can be considered as a dynamically emerging technology [53].

20.3.2 Batteries

The Galvanic Cell Dielectric and electrochemical capacitors are capable to store electrical energy directly, that is, in the absence of any conversion process during storage mechanism. Indeed, pseudocapacitive materials provide storage scheme that is similar to electrochemical energy storage. As it was discussed in the previous section, in pseudocapacitive materials a fast and preferably highly reversible redox reaction is responsible for the increased energy density of the storage device. If the electrode material would be composed entirely of pseudocapacitive material, this resembles that of electrodes used in rechargeable (i.e., secondary) galvanic cells.

Generally, galvanic cells are electrochemical devices that are capable to convert chemical energy into electricity. This working principle is based on the spatial separation of a redox reaction into oxidation and reduction reaction. Considering the discharge process as the spontaneous event during the galvanic cell operation, the electrode where oxidation occurs called as anode (negative electrode), while reduction takes place on the cathode side (positive electrode). The ionic contact of the electrodes is ensured by an appropriate electrolyte. During discharge process, electrodes are connected to an external load with the help of an electron conductor (i.e., electric wire) where electrons will be moved from the anode side toward cathode half-cell meanwhile they provide electrical work. For the charge neutrality ions will be diffused in the electrolyte phase. In primary cells, the electrochemical redox reaction is nonreversible; however, it must be ideally highly reversible for secondary (rechargeable) batteries.

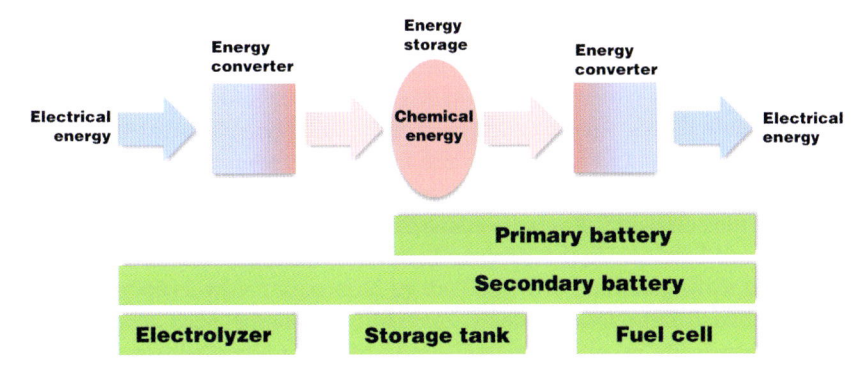

Figure 20.4 Operating schemes of different electrochemical storage devices. (Adapted from W. Weydanz, A. Jossen, Moderne Akkumulatoren richtig einsetzen.)

In order to better understand the working principle of batteries (primary cell), rechargeable batteries (secondary cell), and fuel cells (ternary cell), one should consider the storage system as technical entity composed of energy converter and energy storage (Figure 20.4). The unit energy storage is the core element of such system and it is capable to store energy in chemical form (i.e., in form of certain chemical compound). The energy stored in chemical form is converted afterward into electricity through a chemical–electrical converter. This process is known as the common discharge process of an electrochemical storage device. Primary cells or batteries cannot be recharged after the discharge process. After the full conversion of the stored chemical energy into electrical energy, they may be discarded and replaced. If the system is equipped with an electrical–chemical converter, as well, then electrical energy can be converted into chemical energy (i.e., into an electrochemically active compound). This process is well known as the charging process of a rechargeable battery. During charging of a secondary cell or battery, discharge products of the cell will be transformed ideally into the initial chemical state and the initial chemical composition prior to discharge will be recovered. In primary and secondary cells, battery electrodes represent the energy converter and energy storage units in the same time. In that sense batteries are energy converters with integrated energy storage. Redox-flow battery systems, however, represent batteries equipped with nonintegrated, external storage. In the case of fuel cells, converter (the fuel cell itself) and storage tank are not integrated, as well. Furthermore, if such system is equipped with an electrolyzer, then continuous operation can be achieved. Operation of fuel cells and common fuel cell chemistries are further discussed in Section 20.3.3.

Battery Chemistries There are numerous battery chemistries available. Historically, aqueous electrolyte-based cells (primary, secondary) were developed first. Important and technologically relevant cells include Leclanché cell, zinc–carbon, and the "alkaline" battery (Zn/MnO_2). Well-known secondary battery chemistries include lead–acid (Pb), nickel–iron (NiFe), nickel–cadmium (NiCd), and nickel–metal hydride (NiMH). An interesting technological solution, if cathode active material is not incorporated, stored in the battery cell, therefore, increases storage capacity and energy density of the battery cell using the same manufactured size. This is how metal–air (e.g., Zn/air cell) batteries are constructed. The driving force for the development of novel battery technologies is always the claim to design a cell with higher energy density with enhanced power density characteristics, increased safety, and last but not the least, low manufacturing costs. The early invention of the secondary cell was a breakthrough already; however, the extension of the cell working voltage above 3 V has opened brand new horizons. Electrochemical stability of the applied electrolyte was, and still is, a limiting factor for battery cell operating voltage. Using aqueous electrolytes (acidic or base type) cell voltage is limited to 1.5–2.0 V (electrolytic decomposition of water into hydrogen and oxygen gases begins at around 1.23 V between Pt electrodes).

Lithium and Li-Ion Technology Revolutionary technological invention of the 1960s was the application of lithium in a galvanic cell. Lithium is a lightweight metal and possesses the lowest electrochemical reduction potential among all known chemical elements. However, its high reactivity with water hindered its application in batteries. To overcome this limitation, the use of organic liquid electrolyte was proposed. As a result, in the early 1990s the first commercialized lithium battery became available. Since then vast amount of development of the lithium and lithium-ion technology has been done. The available primary "lithium" batteries contain metallic lithium as anode (e.g., Li/MnO$_2$ *CR2032 coin cell*; Li/SOCl$_2$ *D-size cylindrical cell*; Li/FeS$_2$ *AA-size mignon cell*, etc.). However, "lithium-ion" technology refers to secondary battery systems in which no elementary lithium can be found. Lithium-ion batteries working is based on the so-called rocking chair principle. Meaning that lithium ions shuttle between the two electrodes, where they take part in an intercalation (or conversion) reaction. Electrode active materials therefore act as host materials. Based on this working principle, there are a wide variety of host materials for both anode and cathode sides that can be arbitrarily combined with each other in order to obtain a fully functional lithium-ion cell. Typical anode host material in today's lithium-ion batteries is graphite. However, silicon (Si), tin (Sn), and transition metal oxides, such as lithium titanate (Li$_4$Ti$_5$O$_{12}$), have also been tested in secondary lithium-ion cells [54]. Although (nanostructured) silicon and tin-based electrodes may have very high specific capacities (almost 10 times higher than graphite – $Q_n^{graphite} = 372$ mAh/g), they suffer massively from the high volume change ($\sim$400–440%) during de/lithiation reaction [55]. This results in fast electrode degradation and finally diminishing of cell performance due to pulverization of the electrode material, loss of contacts, and increased internal electrode and cell impedance. Using lithium titanate on the anode side [56–59], on the one hand intrinsically safer battery with increased cycle life can be constructed; however, one should cope with reduced cell voltage, which, in turn, results in lower energy density of the "titanate battery." Historically, the layered lithium cobalt oxide (LiCoO$_2$) was proposed by Goodenough *et al.* as possible cathode host material in lithium-ion cell [60,61]. Since then a vast variety of transition metal oxides were suggested and verified as possible cathode active material in such systems [62]. These materials can be easily classified into three groups based on the dimensionality of the lithium-ion diffusion in the host lattice. Olivine (orthorhombic) structures, such as LiFePO$_4$ and related metal phosphate materials (1D) [63–65]; the second group consists of layered metal oxides, for instance, LiCoO$_2$, LiNiO$_2$, Li(Ni$_{1/3}$Co$_{1/3}$Mn$_{1/3}$)O$_2$ (2D) [66–69]; and the cubic spinel, such as LiMn$_2$O$_4$ (3D), constitute the third group [70–74].

Among the known and established cell chemistries lithium and lithium-ion technology possess the highest energy density values from all. Technologist and other battery experts use the specific values in order to be able to compare energy content and power capabilities of the different battery chemistries. Energy content (i.e., the product of nominal cell voltage (V) and cell capacity (Ah)) of a given battery can be specified per unit mass or even per unit volume. In the first case gravimetric or "specific energy" can be derived (Wh/kg, Wh/g), while expressions in Wh/l, Wh/cm^3 units correspond to volumetric "energy density" values. However, there is a huge difference between theoretical and practical realizable specific energy of a battery. Theoretical specific energy can be derived from the change in the Gibbs free energy of the overall electrochemical reaction involved in the storage process divided by the mass of active materials. Practical specific energy values in fact are always lower since the additional weight of other battery (electrochemically inactive) components (e.g., housing/packaging, current collectors, separator, electrolyte, safety elements) and the different sort of overvoltage which latter causes reduced cell voltage during discharge compared with the open-circuit voltage of the corresponding cell. In order to realize high specific energy values, the mass fraction of the active components must be maximized (e.g., thicker electrode layers). However, an increased mass fraction of the active material loading results in higher internal cell resistivity. This, in turn, has negative influence on power capability of the battery. Therefore, in conventional lithium-ion battery cells energy content and power characteristic are conflicting goals, and it is always a trade-off. So-called "energy cells" contain more active material (e.g., in form of thicker electrode layers), but they lack in good power capabilities. Power cells contain less amount of active material (thinner and optimized electric conductive electrode layers) and therefore have lower capacity and energy content, but they can be discharged/charged with relatively high current densities without diminishing cell performance.

Li-Ion Battery Manufacturing Construction principle of the different lithium-ion cells are common and based on the formation of thin electrode layers deposited on corresponding metallic current collectors (copper at the anode side and aluminum at the cathode side). These electrode layers will be rolled up afterward (cylindrical cells), folded or stacked (prismatic or pouch cells) using a suitable separator membrane (polyolefin-based highly porous foil) in between the two electrode layers. Then electrode pair rolls or stacks are placed in the corresponding housing (metal can, plastic container, laminated polymer pouch), soaked with liquid organic electrolyte, degassed, electrochemically formed, finally degassed again and sealed, then prepared for distribution, installation, or for example, for subsequent battery pack assembly, and so on. It should be noted that on the cell level decreasing cell size results in reduced specific energy. This is due to higher mass fraction of the electrochemically inactive components presenting in the cell [75]. For small-sized cells, housing, safety, or other passive components possess higher relative mass fraction, therefore reduce specific energy values in such smaller sized systems. This implies that construction of microscale lithium-ion batteries (and other conventional liquid electrolyte-based systems) on the footprint area of few square millimeters is hardly possible. However, the rapid and uninterruptable prevalence of autonomous sensors, wireless and self-powering microdevices, and MEMS requires energy storage solutions for such microscale application scenarios. A maintenance-free operation with excellent operational reliability and the on-board integrated energy storage devices are therefore essential. Because of the two main critical issues regarding conventional battery systems, namely, limitation by the radical downsizing and the inherent risk of electrolyte leakage/burning, completely new battery concepts are needed.

Li-Ion Microbatteries The real and effective miniaturization of the energy storage device can be achieved by construction of solid-state microbatteries. Solid-state battery technology enables more freedom and flexibility by designing and integration of the energy storage system into the microscale devices. In solid-state batteries, as their name already suggests, all battery components are in the solid phase. Since lithium-ion technology possesses the highest achievable practical specific energy (and energy density) among battery chemistries, it is obvious that microscale energy storage devices are based predominantly on this battery technology [76]. Working mechanism and the applied electrochemically active components are very similar to that of conventional, liquid electrolyte-based lithium-ion cells. In a basic configuration of a thin film (planar or 2D) solid-state battery between anode and cathode layers, a thin layer of an appropriate lithium ion conducting solid-state electrolyte is deposited. Different type of inorganic [77,78] and organic [79] solid-state electrolytes are known and well described in the literature. Important classes include ceramic [80–84], glassy [85–87], and polymer-based [88–90] lithium-ion conductive electrolytes. However, the common problem and main disadvantage of solid-state electrolyte is their relatively low room-temperature lithium-ion conductivity [91]. For instance, the very popular solid-state electrolyte used in 2D microbatteries; the glass-type lithium phosphorous oxynitride (LiPON) possesses lithium-ion conductivity on the order of 3×10^{-6} S/cm [92,93] at room temperature. Indeed, disadvantage arises from the low ionic conductivity could be circumvented by using thin electrolyte layers (on the order of about 1 μm layer thickness) between anode and cathode.

Thin Film ("2D") Microbatteries Typically, planar microbatteries are fabricated onto appropriate substrates. Cathode, electrolyte, and anode layers are subsequently deposited onto the substrate along with the corresponding current collector layers [94]. Most important requirement for anode and cathode active materials is their relatively high volumetric energy density values. Furthermore, deposited layers should possess high ionic as well as electronic conductivities for the improved power capabilities of the battery cell. Unfortunately, and based on the thin film configuration, planar microbatteries have rather low volumetric capacity and therefore energy density. Volumetric capacity of the battery cell is determined by the layer thickness of the cathode and size of the footprint area (i.e., total volume of the cathode layer). In order to provide reasonable energy density (battery capacity) of a thin film battery footprint of the cell must be drastically increased, sometimes in order of several cm^2, since cathode layer thickness is normally less than 5 µm. Therefore, in accordance with some opinions thin film planar microbatteries cannot be considered as "real microbatteries" [95]. Consequently, practical microbatteries must have high areal energy density in order to be capable to store the required energy in a typical mm^3-sized volume. If the lateral dimensions of the battery cell are strictly confined and the length scales are on the several 100 µm (up to 1 mm), high aspect ratio electrodes have to be constructed. As it was previously mentioned, cathode active materials possess generally low intrinsic ionic and electronic conductivities, therefore in order to minimize power losses this is always a limiting factor by the cell (electrode) design. The ultimate goal by the construction of a microbattery is to maximize storage capacity of the battery cell using the given confined footprint area, commonly discussed in terms of µAh/cm^2 unit.

Three-Dimensional Microbattery Concepts Instead of using 2D thin film design, in fact, a real microbattery with 3D architecture can be fabricated. In 3D microbatteries, electrode layers with much higher active area are integrated compared with 2D design. Thus, on the same footprint area significantly higher storage capacity (specific energy) can be realized. A further advantage of such systems that power capabilities of the battery cell will not diminish due to careful design of the 3D electrodes [96]. The applied electrode design allows for short lithium-ion diffusion path, and microstructures with low electrical resistivity can be made. These features together with the high specific contact area between electrode and solid electrolyte boost applicable maximum current density even higher. Conclusively, energy and power performance capabilities of the 3D microbattery cell could be fine-tuned and scaled independently.

Recently, several concepts have been suggested for 3D microbattery architectures (Figure 20.5). Different designs using periodic or aperiodic electrode arrays have been demonstrated [76,97].

Plausible 3D design is the periodic interdigitated electrode arrangement (Figure 20.5a and b). Solid-state electrolyte represents as the continuous phase, and it encloses the rod-shaped or platelet-type electrode arrays. However, solid-state electrolyte does not necessarily have to comprise the continuous phase. Rod-shaped or columnar electrode array can be conformal coated with a thin layer of solid-state electrolyte (Figure 20.5c). Remaining free volume is filled up with the material of the second electrode material (the latter acts as continuous phase) afterward. In an entirely aperiodic configuration, sometimes called as "sponge" architecture [97], solid-state electrolyte covers the hierarchical nanoparticulate network of the electrode active material. Remaining open porosity (micro- or mesoporous structure) has to be filled up with the second electroactive material (Figure 20.5d). The successful fabrication of such interpenetrated composite network is the key factor for the full functional battery cell. However, conformal coating of a random hierarchical nanoparticle network with suitable solid-state electrolyte remains a challenge – considering the high-quality requirements (e.g.,

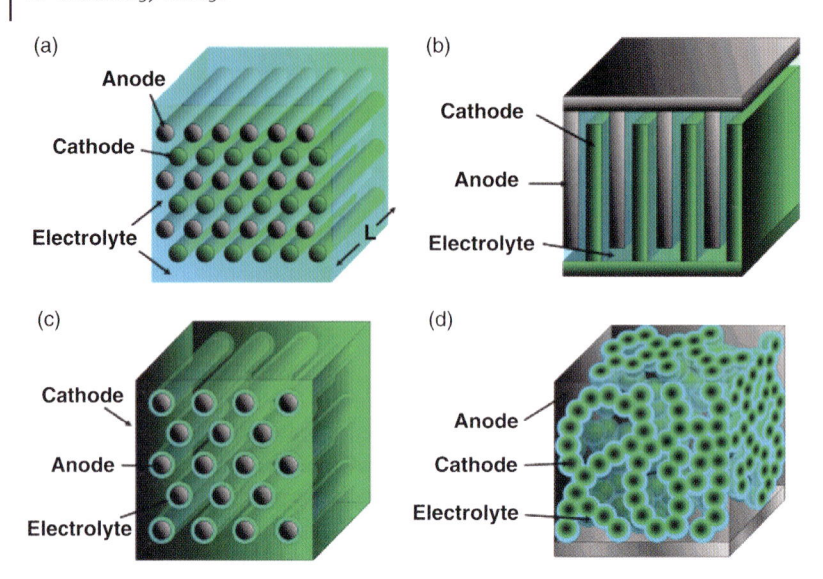

Figure 20.5 Examples of prospective 3D architectures for 3D batteries: (a) array of interdigitated cylindrical cathodes and anodes; (b) interdigitated plate array of cathodes and anodes; (c) rod array of cylindrical anodes coated with a thin layer of ion-conducting dielectric (electrolyte) with the remaining free volume filled with the cathode material; (d) aperiodic "sponge" architectures in which the solid network of the sponge serves as the charge insertion cathode, which is coated with an ultrathin layer of ion-conducting dielectric (electrolyte), and the remaining free volume is filled with an interpenetrating, continuous anode. (Reprinted with permission from Ref. [97]. Copyright 2004, American Chemical Society.)

layer thickness, layer homogeneity, absolutely pinhole free) of the electrolyte coating [98].

Although commercialization and serial industrial production of such 3D microbatteries are still not realized, successful demonstration of the above-discussed concepts under laboratory conditions have already been done. The successful utilization of active electrode area enhancement (area gain) was demonstrated in different manner. Three-dimensional substrate was fabricated by templated electrochemical deposition route [99].

On a planar substrate 3D periodic columnar structure could be generated that served as 3D current collector and substrate for the microbattery assembly (Figure 20.6). The simulated surface area enhancement can reach about 30-fold, compared with the fully flat current collector surface. Aspect ratio and packing density of the created nanorods are easily tunable; however, such 3D structures are mechanically very sensitive and fragile that could limit the practical applicability of such microbattery design. Instead of creating free-standing highly fragile rods on the substrates' surface, an "inside-out" concept is the application of microchannel plate (MCP) as 3D substrate [95].

Nathan *et al.* used disk-shaped microchannel plate as substrate, where channel dimensions were 500 µm length and 50 µm diameter. Perforated silicon substrate was used as well, which was created by inductively coupled plasma etching. Functional layers of the 3D microbattery were deposited sequentially, afterward,

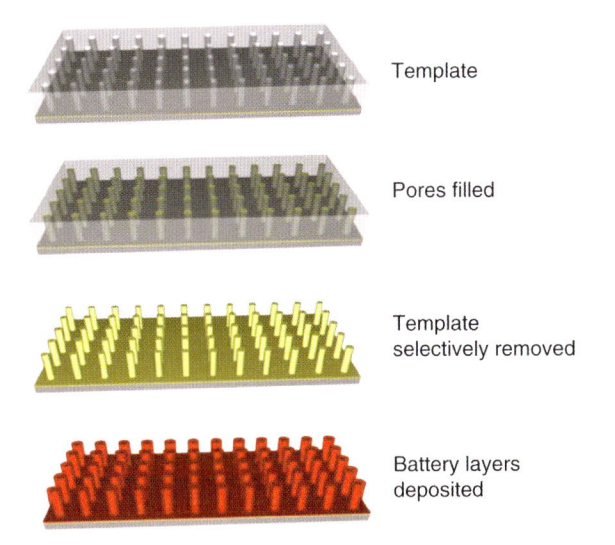

Figure 20.6 Three-dimensional battery manufacturing based on template deposition of nanorods. (Adapted with permission from Ref. [98]. Copyright 2011, Wiley-VCH Verlag GmbH & Co. KGaA.)

such as, Ni-current collector, MoO_yS_x cathode, hybrid polymer electrolyte, and lithiated graphite as bifunctional anode and current collector material. It was clearly demonstrated that area gain (about 25–30-fold) provided by the microchannels correlates well with the capacity gain of the 3D microbattery. The overall mechanical robustness of such 3D microbattery system is a further advantage over the previously discussed free-standing nanorod-type design.

Using electrochemical or reactive ion etching techniques, microscale trenches were fabricated into silicon substrate. The concept was proposed by Notten *et al.* who used step-conformal sequential coating of the silicon microtrenches for the functional battery components (Figure 20.7).

To do so chemical vapor deposition (CVD) or atomic layer deposition (ALD) can be used. As one of the advantages of the concept, the authors claim that the proposed technique is based on standardized microelectronic methods that are advantageous from the point of view of the production. On the other hand, based on the microtrench design (i.e., parallel planar thin films), the risk of the local overcharge in the microbattery is minimized, thus increasing cycle life of the storage device. On the negative end, the step-conformal coating of such dense structure remains still challenging. The above-presented microbattery concepts are common in their *periodic* electrode array architectures.

A representative example for the aperiodic, "sponge"-like architecture is based on three-dimensional aerogels. The main component of such system is an aerogel framework, which comprises a randomly aligned hierarchical nanoparticulate network with significantly large open porosity. In an ideal case, the functional material of the aerogel possesses sufficient electronic and ionic conductivity. Therefore, it can be used as bifunctional current collector and active electrode component. The available large open porosity can accommodate the further

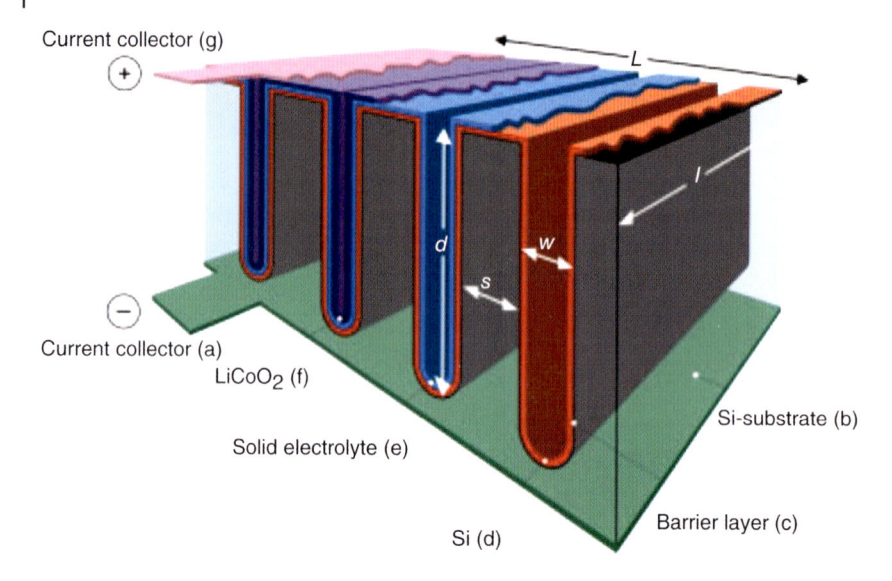

Figure 20.7 Three-dimensional integrated all-solid-state battery to be applied as energy supply unit to power autonomous devices. The various layers, denoted from (a) to (g), are deposited in high-aspect ratio trenches etched in, for example, silicon. The geometry is characterized by the footprint length (L), footprint width (l), and trench width (w), depth (d), and spacing (s). (Adapted with permission from Ref. [100]. Copyright 2008, Wiley-VCH Verlag GmbH & Co. KGaA.)

battery components (electrolyte/separator and another electrode active material) and can provide enough space for the volume changes caused by the lithium-ion extraction/intercalation into the host crystal lattice during cycling. Nevertheless, for maximal volumetric energy density an optimum porosity is needed, namely too high open porosity could result in diminished volumetric energy density. Admittedly, uniform and dense covering of such random spongy oxide framework with a thin electrolyte/separator layer is an issue. Possible formation of pinholes in the electrolyte layer causes internal short circuiting of the cell, therefore results in more intensive self-discharge of the battery cell, which eventually strongly affects and therefore limits its applicability. In the last step, fully or partial filling of the remaining open porosity of the aerogel framework with the second electrode materials has to be done. A homogeneously distributed, percolated electric conductive nanoparticle network can act as fully functional second electrode and current collector material in the same time. Such aerogel-based 3D aperiodic architecture concept is demonstrated in Figure 20.8.

Conclusion and Future Prospects Since the mass production of real 3D microbatteries is still missing, the examples above illustrate the feasibility of the fabrication of a miniaturized fully functional 3D microbattery. Batteries possess much higher energy density (volumetric as well as gravimetric) values than that of state-of-the-art supercapacitors; furthermore, solid-state battery technology allows for successful physical miniaturization of such storage devices. In that sense, high energy density batteries (e.g., Li-ion technology) offer an ideal solution

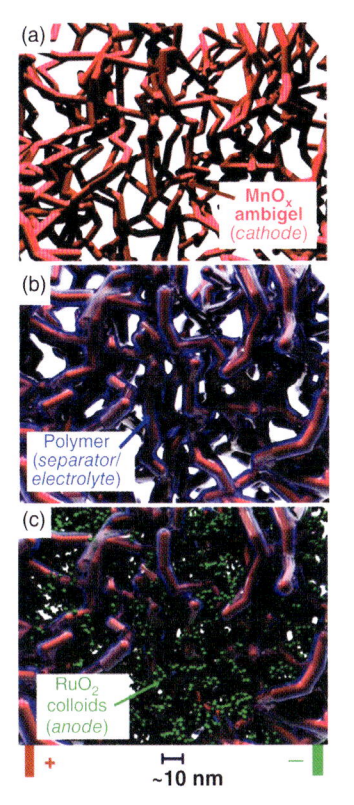

Figure 20.8 Schematic showing the three steps needed to assemble a fully interpenetrating nanoscale battery in which all three functional components (cathode–ion-conducting separator–anode) are sized on the order of 10 nm and are separated by <50 nm. (Printed with permission from Ref. [101]. Copyright 2007, American Chemical Society.)

for energy storage devices for miniaturized electronic devices, autonomous sensor nodes (equipped with a microscale energy harvester), on-board energy storage for MEMS, and so on. In order to overcome the low storage capacity per footprint area of the planar (thin film) batteries, one should significantly increase the area of the electrode active surface. Therefore, several real 3D microbattery architectures were proposed in the past and it was demonstrated by the authors that surface area gain results in the enhancement of reversible discharge capacity of the microbattery. In contrast to thin film batteries, which are already commercially available through several manufacturers, the fabrication of a 3D structured microbattery is much more complex. For instance, conformal deposition of a dense, thin, and pinhole-free solid-state electrolyte layer into the battery architecture is still a very challenging issue. However, due to recent very active worldwide research work in this scientific–technological field, the commercialization of the first real 3D solid-state microbattery will be hopefully realized soon.

20.3.3 Fuel Cells

A fuel cell is a device that converts the chemical energy in a fuel directly into electricity without combustion. Therefore, the efficiency of fuel cells can be higher than conventional combustion-based engine or even exceeds the Carnot cycle limit. Aside from the different chemistry, one difference between fuel cell and battery is that the current output from a single fuel cell (10^2–10^3 mA/cm^2) is typically orders of magnitude larger than that from a single battery (10^{-1}–10^1 mA/cm^2) [102]. Therefore, fuel cells can provide much higher power than batteries. Miniaturized fuel cell, a.k.a. microfuel cell, is alternatives for battery systems for powering MEMS, thanks to the unlimited fuel feeding. The demand for microfuel cell arises from the need of uninterrupted power supplies for electronics. Instead of charging like the batteries do, fuel cells take fuels. The refuel speed is fast, and the fuel cell can be run continuously given timely refueling.

A typical fuel cell has three layers: anode, electrolyte, and cathode. They are normally categorized by their electrolytes. For example, fuel cells are categorized as solid oxide fuel cell (SOFC), proton-exchange membrane fuel cell (PEMFC), alkaline fuel cell (AFC), molten carbonate fuel cell (MCFC), phosphoric acid fuel cell (PAFC), and so on. Due to the different operating conditions, not all fuel cells are suitable for applications in MEMS systems. Normally, the high-temperature fuel cells are not ideal candidates for MEMS, but proven suitable for large-scale power generation. For example, AFC is used in space shutters, and PAFC, MCFC, and SOFC for cogeneration. However, microscale SOFC has recently been studied and may be promising for small-scale application. The comparison of energy density and specific energy of three potential fuel cells, together with various battery technologies, can be found in Figure 20.9.

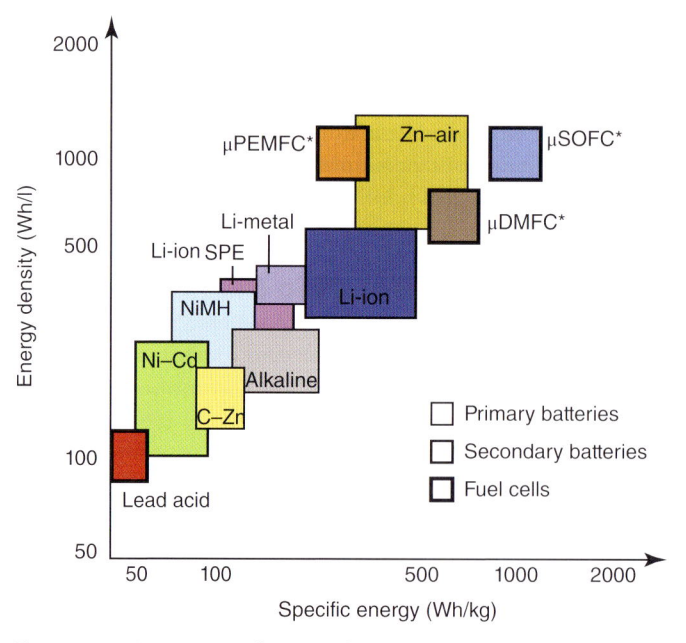

Figure 20.9 Comparison of energy density and specific energy of several micro fuel cells and battery systems. *Estimated values, as these devices are not fully developed yet. (Adapted with permission from Ref. [103]. Copyright 2009, Elsevier.)

20.3.3.1 Low-Temperature Fuel Cells

Low-temperature fuel cells have the advantage that they can work at similar temperatures as most electronic devices. The overall reaction typically involves the production of proton and proton migration across the electrolyte. The fuel feeding systems split this type of fuel cells into three categories: (i) pure hydrogen, (ii) hydrocarbon, or (ii) on-board hydrogen-derived reformed hydrocarbons [104,105]. All of these fuel cells belong to the family of PEMFCs.

The main component of a PEMFC is a positive–electrolyte–negative (PEN) structure. Fuels are oxidized at the anode, producing protons and electrons. Electrons are connected by the current collectors and directed to the out circuits generating current flows, while the protons transport directly across the electrolyte and combine with oxygen and electrons at the cathode side. PEMFC is suitable for small-scale power generation. The advantages of PEMFCs include low operating temperature, the ease of fabrication and long life. Several PEMFCs work on pure hydrogen. However, hydrogen energy is notoriously difficult for its storage transportation and additional requirements for accessory equipment. Therefore, many research works have been focusing on developing fuel cell that takes fuels other than pure hydrogen. Among these fuel cells, DMFC holds great promises. It works on methanol for small-scale [106] power generation. Methanol has an energy density of 4384 Wh/l. It is relatively cheaper and easier to store, handle, and oxidize. The drawback includes the crossover of methanol and low oxidation rate. Recently, a DMFC with $2.25\,cm^2$ active area has achieved the $32\,mW/cm^2$ with 4 M methanol at room temperature [107].

20.3.3.2 High-Temperature Fuel Cells

Owing to its high energy density and specific energy (up to four times of lithium-ion batteries), μSOFC is a promising source for MEMS devices demanding high energy [108]. Unlike PEMFCs and DMFCs, SOFCs are fuel flexible. They work on hydrogen and carbon monoxide, and also on hydrocarbons. Unfortunately, hydrocarbon reforming happens typically above $600\,°C$. This limits the operation of μSOFCs to high temperatures. However, there are other conversion pathways that are based on partial oxidation of hydrocarbons allow them to operate below $600\,°C$ [109]. In that condition, for example, power outputs from μSOFCs can be as high as $861\,mW/cm^2$ at $450\,°C$ [110]. In terms of the size of the μSOFC, the thickness of the PEN structure can be made on the order of 100 nm and the active area ranges from 0.01 to $10\,mm^2$, thin and small enough to fit to MEMS. In addition, the total power can be 0.01 mW to several watts, enabling their diversity in application selection [108]. Due to their high operating temperature that PEMFC-type of cells cannot sustain, μSOFCs may power sensors in automotives, as vehicle exhaust gas meter ($650\,°C$) or as liquid film probes ($800\,°C$).

20.3.4 Other Storage Systems

Microbial Fuel Cells MFCs take advantage of microorganisms to convert organic matter into fuel, mimicking bacterial interactions [111]. This will be suitable for medical devices. This type of fuel cell has a theoretical voltage on the order of

1.1 V, but the maximum achieved is 0.8 V and during current generation the voltage is below 0.62 V [111]. MFCs may find applications in biosensors, medical devices, agricultural applications, wastewater treatments, and waste disposal.

Microcombustors Microcombustors or gas turbine have attracted attention as alternatives for batteries due to their high energy density [112,113]. It has been partially driven by the military demand for soldier-portable devices and micropropulsion [114–116], which need high energy density and power density energy supplies [117]. Microcombustors use hydrocarbons as fuel and are therefore intrinsically superior to conventional batteries in power output. For example, the lithium-ion batteries have specific energy from 100 to 250 Wh/kg or 250–620 Wh/l of energy density, while the combustion of hydrocarbon fuels produces specific energy from 6300 to 15 400 Wh/kg, using fuels from methanol to methane [117]. In particular, the combustion of hydrogen produces around 39 500 Wh/kg. The specific power of lithium-ion batteries are around 275 W/l [118], much less compared with the microcombustors. One example is that microgas turbine built by MIT in 1990s, as the inception of this concept [119], produced already 10–50 W within 1 cm^3 volume for 7 g/h fuel consumption, equivalent to 10–50 kW/l. Recent advances have increased this number further to 100–1000 kW/l [120].

While batteries and supercapacitors are primary choices for microscale energy storages, fuel cell systems may have niche applications for their high energy density and power density. In energy and power demanding MEMS devices, these alternative energy storage/conversion devices may be preferred.

20.4 Summary and Perspectives

The selection of energy storage devices for power microscale system needs more analysis and one needs to be aware of the tradeoff between energy density, power density, and cycle life. Currently, no batteries or supercapacitors can meet all three needs. Conventional batteries have high energy density, while they are no rival to supercapacitors in terms of power density and cycle life. Fuel cells can store much more energy and the power density is also higher than batteries, but they require refueling, which may not be convenient for applications. This same is also true for microcombustors. Lastly, despite the disadvantages of almost all energy storage devices, the energy storage components are necessary. The extension of cycle life of lithium batteries, by replacing cathode to durable LiFePO$_4$ or making batteries all solid, may stimulate more interests in applying these batteries to MEMS devices.

References

1 Murray, A. (2015) What do consumers want? Better batteries, not wearables. Available at http://fortune.com/2015/01/07/what-do-consumers-want-better-batteries-not-wearables/.

2 Jiang, X., Polastre, J., and Culler, D. (2005) Perpetual environmentally powered sensor networks, IPSN 2005. Fourth International Symposium on Information Processing in Sensor Networks, pp. 463–468.

3 Kickstarter (2017) iFind – The World's First Battery-Free Item Locating Tag. Available at http://www.kickstarter.com/projects/yuansong84/ifind-the-worlds-first-battery-free-item-locating (accessed September 04, 2017).

4 Watanabe, T., Tatsumura, K., and Ohdomari, I. (2006) New linear-parabolic rate equation for thermal oxidation of silicon. *Physical Review Letters*, **96** (19), 196102.

5 Cook-Chennault, K., Thambi, N., and Sastry, A. (2008) Powering MEMS portable devices: a review of non-regenerative and regenerative power supply systems with special emphasis on piezoelectric energy harvesting systems. *Smart Materials and Structures*, **17**, 043001.

6 Bowen, C., Kim, H., Weaver, P., and Dunn, S. (2014) Piezoelectric and ferroelectric materials and structures for energy harvesting applications. *Energy & Environmental Science*, **7**, 25–44.

7 Becker, T., Kluge, M., Schalk, J., Otterpohl, T., and Hilleringmann, U. (2008) Power management for thermal energy harvesting in aircrafts. Proceedings of the IEEE Sensors Conference, pp. 681–684.

8 Alippi, C. and Galperti, C. (2008) An adaptive system for optimal solar energy harvesting in wireless sensor network nodes. *IEEE Transactions on Circuits and Systems I: Regular Papers*, **55**, 1742–1750.

9 Raghunathan, V., Kansal, A., Hsu, J., Friedman, J., and Srivastava, M. (2005) Design considerations for solar energy harvesting wireless embedded systems. Proceedings of the 4th International Symposium on Information Processing in Sensor Networks, IEEE Press, pp. 64.

10 Simjee, F. and Chou, P.H. (2006) Everlast: long-life, supercapacitor-operated wireless sensor node. ISLPED'06 Proceedings of the 2006 International Symposium on Low Power Electronics and Design, pp. 197–202.

11 Park, C. and Chou, P.H. (2006) Ambimax: autonomous energy harvesting platform for multi-supply wireless sensor nodes. 3rd Annual IEEE Communications Society on Sensor and Ad Hoc Communications and Networks, pp. 168–177.

12 Park, C. and Chou, P.H. (2004) Power utility maximization for multiple-supply systems by a load-matching switch. Proceedings of the 2004 International Symposium on Low Power Electronics and Design, pp. 168–173.

13 Liu, M., Johnston, M.B., and Snaith, H.J. (2013) Efficient planar heterojunction perovskite solar cells by vapour deposition. *Nature*, **501**, 395–398.

14 Jeon, N.J., Noh, J.H., Yang, W.S., Kim, Y.C., Ryu, S., Seo, J., and Seok, S.I. (2015) Compositional engineering of perovskite materials for high-performance solar cells. *Nature*, **517**, 476–480.

15 Green, M.A., Emery, K., Hishikawa, Y., Warta, W., and Dunlop, E.D. (2015) Solar cell efficiency tables (version 45). *Progress in Photovoltaics: Research and Applications*, **23**, 1–9.

16 Kurzweil, P. (2009) *Encyclopedia of Electrochemical Power Sources*, vol. 3 (eds J. Garche, C. Dyer, P. Moseley, Z. Ogumi, D. Rand, and B. Scrosati), Elsevier, Amsterdam, pp. 596–606.

17 Mills, A.A. (2003) Early voltaic batteries: an evaluation in modern units and application to the work of Davy and Faraday. *Annals of Science*, **60**, 373–398.

18 Jayalakshmi, M. and Balasubramanian, K. (2008) Simple capacitors to supercapacitors – an overview. *International Journal of Electrochemical Science*, **3**, 1196–1217.

19 Nishino, A. (1996) Capacitors: operating principles, current market and technical trends. *Journal of Power Sources*, **60**, 137–147.

20 Dubal, D.P., Wu, Y.P., and Holze, R. (2016) Supercapacitors: from the Leyden jar to electric busses. *ChemTexts*, **2**, 13.

21 Pandolfo, A.G. and Hollenkamp, A.F. (2006) Carbon properties and their role in supercapacitors. *Journal of Power Sources*, **157**, 11–27.

22 Pech, D., Brunet, M., Taberna, P.L., Simon, P., Fabre, N., Mesnilgrente, F., Conédéra, V., and Durou, H. (2010) Elaboration of a microstructured inkjet-printed carbon electrochemical capacitor. *Journal of Power Sources*, **195**, 1266–1269.

23 (a) Liu, C.C., Tsai, D.S., Chung, W.H., Li, K.W., Lee, K.Y., and Huang, Y.S. (2011) Electrochemical micro-capacitors of patterned electrodes loaded with manganese oxide and carbon nanotubes. *Journal of Power Sources*, **196**, 5761–5768. (b) Jiang, Y.Q., Zhou, Q., and Lin, L. (2009) Planar MEMS supercapacitor using carbon nanotube forests. IEEE 22nd International Conference on MEMS, pp. 587–590.

24 Yoo, J.J., Balakrishnan, K., Huang, J., Meunier, V., Sumpter, B.G., Srivastava, A., Conway, M., Reddy, A.L.M., Yu, J., Vajtai, R., and Ajayan, P.M. (2011) Ultrathin planar graphene supercapacitors. *Nano Letters*, **11**, 1423–1427.

25 Zhang, L.L., Zhou, R., and Zhao, X.S. (2010) Graphene-based materials as supercapacitor electrodes. *Journal of Materials Chemistry*, **20**, 5983–5992.

26 Stoller, M.D., Park, S.J., Zhu, Y.W., An, J.H., and Ruoff, R.S. (2008) Graphene-based ultracapacitors. *Nano Letters*, **8**, 3498–3502.

27 Huang, X., Qi, X.Y., Boey, F., and Zhang, H. (2012) Graphene-based composites. *Chemical Society Reviews*, **41**, 666–686.

28 Li, F., Chen, J., Wang, X., Xue, M., and Chen, G.F. (2015) Stretchable supercapacitor with adjustable volumetric capacitance based on 3D interdigital electrodes. *Advanced Functional Materials*, **25**, 4601–4606.

29 Yoon, S.H., Korai, Y., Mochida, I., Marsh, H., and Rodriguez-Reinoso, F. (eds) (2000) *Sciences of Carbon Materials*, Universidad de Alicante, p. 287.

30 Feng, L., Zhu, Y., Ding, H., and Ni, C. (2014) Recent progress in nickel based materials for high performance pseudocapacitor electrodes. *Journal of Power Sources*, **267**, 430–444.

31 Rauda, I.E., Augustyn, V., Dunn, B., and Tolbert, S.H. (2013) Enhancing pseudocapacitive charge storage in polymer templated mesoporous materials. *Accounts of Chemical Research*, **46**, 1113–1124.

32 Faraji, S. and Ani, F.N. (2014) Microwave-assisted synthesis of metal oxide/hydroxide composite electrodes for high power supercapacitors – a review. *Journal of Power Sources*, **263**, 338–360.

33 Xiong, G., Meng, C., Reifenberger, R.G., Irazoqui, P.P., and Fisher, T.S. (2014) A review of graphene-based electrochemical microsupercapacitors. *Electroanalysis*, **26**, 30–51.

34 Dinh, T.M., Mesnilgrente, F., Conédéra, V., Kyeremateng, N.A., and Pech, D. (2015) Realization of an asymmetric interdigitated electrochemical microcapacitor based on carbon nanotubes and manganese oxide. *Journal of the Electrochemical Society*, **162**, A2016–A2020.

35 Wu, Z.S., Parvez, K., Feng, X., and Müllen, K. (2013) Graphene-based in-plane micro-supercapacitors with high power and energy densities. *Nature Communications*, **4**, 2487–2494.

36 El-Kady, M.F. and Kaner, R.B. (2013) Scalable fabrication of high-power graphene micro-supercapacitors for flexible and on-chip energy storage. *Nature Communications*, **4**, 1475–1483.

37 Shen, C., Wang, X., Zhang, W., and Kang, F. (2011) A high-performance three-dimensional micro supercapacitor based on self-supporting composite materials. *Journal of Power Sources*, **196**, 10465–10471.

38 Pech, D., Brunet, M., Durou, H., Huang, P., Mochalin, V., Gogotsi, Y., Taberna, P.L., and Simon, P. (2010) Ultrahigh-power micrometre-sized supercapacitors based on onion-like carbon. *Nature Nanotechnology*, **5**, 651–654.

39 Beidaghi, M. and Wang, C. (2012) Micro-supercapacitors based on interdigital electrodes of reduced graphene oxide and carbon nanotube composites with ultrahigh power handling performance. *Advanced Functional Materials*, **22**, 4501–4510.

40 Lin, J., Zhang, C., Yan, Z., Zhu, Y., Peng, Z., Hauge, R.H., Natelson, D., and Tour, J.M. (2013) 3-Dimensional graphene carbon nanotube carpet-based microsupercapacitors with high electrochemical performance. *Nano Letters*, **13**, 72–78.

41 Wang, S., Liu, N., Tao, J., Yang, C., Liu, W., Shi, Y., Wang, Y., Su, J., Li, L., and Gao, Y. (2015) Inkjet printing of conductive patterns and supercapacitors using a multi-walled carbon nanotube/Ag nanoparticle based ink. *Journal of Materials Chemistry A*, **3**, 2707–2413.

42 Fan, L., Zhang, N., and Sun, K. (2014) Flexible patterned micro-electrochemical capacitors based on PEDOT. *Chemical Communications*, **50**, 6789–6792.

43 Lawes, S., Riese, A., Sun, Q., Cheng, N., and Sun, X. (2015) Printing nanostructured carbon for energy storage and conversion applications. *Carbon*, **92**, 150–176.

44 Ervin, M.H., Le, L.T., and Lee, W.Y. (2014) Inkjet-printed flexible graphene-based supercapacitor. *Electrochimica Acta*, **147**, 610–616.

45 Le, L.T., Ervin, M.H., Qiu, H., Fuchs, B.E., and Lee, W.Y. (2011) Graphene supercapacitor electrodes fabricated by inkjet printing and thermal reduction of graphene oxide. *Electrochemistry Communications*, **13**, 355–358.

46 Deegan, R.D., Bakajin, O., Dupont, T.F., Huber, G., Nagel, S.R., and Witten, T.A. (1997) Capillary flow as the cause of ring stains from dried liquid drops. *Nature*, **389**, 827–829.

47 Derby, B. (2010) Inkjet printing of functional and structural materials: Fluid property requirements, Feature stability, and Resolution. *Annual Review of Materials Research*, **40**, 395–414.

48 Zhao, C., Wang, C., Gorkin J III, R., Beirne, S., Shu, K., and Wallace, G.G. (2014) Three dimensional (3D) printed electrodes for interdigitated supercapacitors. *Electrochemistry Communications*, **41**, 20–23.

49 Sun, G., An, J., Chua, C.K., Pang, H., Zhang, J., and Chen, P. (2015) Layer-by-layer printing of laminated graphene-based interdigitated microelectrodes for flexible planar micro-supercapacitors. *Electrochemistry Communications*, **51**, 33–36.

50 Long, J.W., Bélanger, D., Brousse, T., Sugimoto, W., Sassin, M.B., and Crosnier, O. (2011) Asymmetric electrochemical capacitors – stretching the limits of aqueous electrolytes. *MRS Bulletin*, **36**, 513–522.

51 Wu, T.H., Chu, Y.H., Hu, C.C., and Hardwick, L.J. (2013) Criteria appointing the highest acceptable cell voltage of assymetric supercapacitors. *Electrochemistry Communications*, **27**, 81–84.

52 Wang, G., Zhang, L., and Zhang, J. (2012) A review of electrode materials for electrochemical supercapacitors. *Chemical Society Reviews*, **41**, 797–828.

53 Xiong, G., Meng, C., Reifenberger, R.G., Irazoqui, P.P., and Fisher, T.S. (2014) A review of graphene-based electrochemical microsupercapacitors. *Electroanalysis*, **26**, 30–51.

54 Goriparti, S., Miele, E., De Angelis, F., Di Fabrizio, E., Zaccaria, R.P., and Capiglia, C. (2014) Review on recent progress of nanostructured anode materials for Li-ion batteries. *Journal of Power Sources*, **257**, 421–443.

55 Kim, T.-H., Park, J.-S., Chang, S.K., Choi, S., Ryu, J.H., and Song, H.-K. (2012) The current move of lithium ion batteries towards the next phase. *Advanced Energy Materials*, **2**, 860–872.

56 Prakash, A.S., Manikandan, P., Ramesha, K., Sathiya, M., Tarascon, J.-M., and Shukla, A.K. (2010) Solution-combustion synthesized nanocrystalline $Li_4Ti_5O_{12}$ as high-rate performance Li-ion battery anode. *Chemistry of Materials*, **22**, 2857–2863.

57 Takami, N., Hoshina, K., and Inagaki, H. (2011) Lithium diffusion in $Li_{4/3}Ti_{5/3}O_4$ particles during insertion and extraction. *Journal of the Electrochemical Society*, **158**, A725–A730.

58 Bresser, D., Paillard, E., Copley, M., Bishop, P., Winter, M., and Passerini, S. (2012) The importance of "going nano" for high power battery materials. *Journal of Power Sources*, **219**, 217–222.

59 Yi, T.F., Yanga, S.Y., and Xie, Y. (2015) Recent advances of $Li_4Ti_5O_{12}$ as a promising next generation anode material for high power lithium ion batteries. *Journal of Materials Chemistry A*, **3**, 5750–5777.

60 Mizushima, K., Jones, P.C., Wiseman, P.J., and Goodenough, J.B. (1980) Li_xCoO_2 (0<x<-1): a new cathode material for batteries of high energy density. *Materials Research Bulletin*, **15**, 783–789.

61 Jo, M., Jeong, S., and Cho, J. (2010) High power $LiCoO_2$ cathode materials with ultra energy density for Li-ion cells. *Electrochemistry Communications*, **12**, 992–995.

62 Myung, S.-T., Amine, K., and Sun, Y.-K. (2015) Nanostructured cathode materials for rechargeable lithium batteries. *Journal of Power Sources*, **283**, 219–236.

63 Padhi, A.K., Nanjundaswam, K.S., and Goodenough, J.B. (1997) Phospho-olivines as positive-electrode materials for rechargeable lithium batteries. *Journal of the Electrochemical Society*, **144**, 1188–1194.

64 Deng, S., Wang, H., Liu, H., Liu, J., and Yan, H. (2014) Research progress in improving the rate performance of LiFePO$_4$ cathode materials. *Nano-Micro Letters*, **6**, 209–226.

65 Wang, J. and Sun, X. (2015) Olivine LiFePO$_4$: the remaining challenges for future energy storage. *Energy & Environmental Science*, **8**, 1110–1138.

66 Alcántara, R., Lavela, P., Tirado, J.L., Zhecheva, E., and Stoyanova, R. (1999) Recent advances in the study of layered lithium transition metal oxides and their application as intercalation electrodes. *Journal of Solid State Electrochemistry*, **3**, 121–134.

67 Dou, S. (2013) Review and prospect of layered lithium nickel manganese oxide as cathode materials for Li-ion batteries. *Journal of Solid State Electrochemistry*, **17**, 911–926.

68 Jung, S.-K., Gwon, H., Hong, J., Park, K.-Y., Seo, D.-H., Kim, H., Hyun, J., Yang, W., and Kang, K. (2014) Understanding the degradation mechanisms of LiNi$_{0.5}$Co$_{0.2}$Mn$_{0.3}$O$_2$ cathode material in lithium ion batteries. *Advanced Energy Materials*, **4**, 1300787.

69 Rozier, P. and Tarascon, J.M. (2015) Review – Li-rich layered oxide cathodes for next-generation Li-ion batteries: chances and challenges. *Journal of the Electrochemical Society*, **162**, A2490–A2499.

70 Thackeray, M.M., Johnson, P.J., de Picciotto, L.A., Bruce, P.G., and Goodenough, J.B. (1984) Electrochemical extraction of lithium from LiMn$_2$O$_4$. *Materials Research Bulletin*, **19**, 179–187.

71 Yi, T.F., Zhu, Y.R., Zhu, X.D., Shu, J., Yue, C.B., and Zhou, A.N. (2009) A review of recent developments in the surface modification of LiMn$_2$O$_4$ as cathode material of power lithium-ion battery. *Ionics*, **15**, 779–784.

72 Erickson, E.M., Ghanty, C., and Aurbach, D. (2014) New horizons for conventional lithium ion battery technology. *Journal of Physical Chemistry Letters*, **5**, 3313–3324.

73 Nitta, N., Wu, F., Lee, J.T., and Yushin, G. (2015) Li-ion battery materials: present and future. *Materials Today*, **18**, 252–264.

74 Aydinol, M.K. and Ceder, G. (1997) First-principles prediction of insertion potentials in Li-Mn oxides for secondary Li batteries. *Journal of the Electrochemical Society*, **144**, 3832–3835.

75 Andre, F., Langer, F., Schwenzel, J., and Kun, R. (2014) Energy storage options for self-powering devices. *Processing Technology*, **15**, 248–257.

76 Roberts, M., Johns, P., Owen, J., Brandell, D., Edstrom, K., Enany, G.E., Guery, C., Golodnitsky, D., Lacey, M., Lecoeur, C., Mazor, H., Peled, E., Perre, E., Shaijumon, M.M., Simon, P., and Taberna, P.L. (2011) 3D lithium ion batteries – from fundamentals to fabrication. *Journal of Materials Chemistry*, **21**, 9876–9890.

77 Knauth, P. (2009) Inorganic solid Li ion conductors: an overview. *Solid State Ionics*, **180**, 910–916.

78 Bachman, J.C., Muy, S., Grimaud, A., Chang, H.-H., Pour, N., Lux, S.F., Paschos, O., Maglia, F., Lupart, S., Lamp, P., Giordano, L., and Horn, Y.S. (2016) Inorganic solid-state electrolytes for lithium batteries: mechanisms and properties governing ion conduction. *Chemical Reviews*, **116**, 140–162.

79 Hallinan, D.T., Jr., and Balsara, N.P. (2013) Polymer electrolytes. *Annual Review of Materials Research*, **43**, 503–525.

80 Thangadurai, V., Narayanan, S., and Pinzaru, D. (2014) Garnet-type solid-state fast Li ion conductors for Li batteries: critical review. *Chemical Society Reviews*, **43**, 4714–4727.

81 Zheng, Z., Fang, H., Yang, F., Liu, Z.-K., and Wang, Y. (2014) Amorphous LiLaTiO$_3$ as solid electrolyte material. *Journal of the Electrochemical Society*, **161**, A473–A479.

82 Buschmann, H., Dölle, J., Berendts, S., Kuhn, A., Bottke, P., Wilkening, M., Heitjans, P., Senyshyn, A., Ehrenberg, H., Lotnyk, A., Duppel, V., Kienle, L., and Janek, J. (2011) Structure and dynamics of the fast lithium ion conductor "Li$_7$La$_3$Zr$_2$O$_{12}$." *Physical Chemistry Chemical Physics*, **13**, 19378–19392.

83 Logéat, A., Köhler, T., Eisele, U., Stiaszny, B., Harzer, A., Tovar, M., Senyshyn, A., Ehrenberg, H., and Kozinsky, B. (2012) From order to disorder: the structure of lithium-conducting garnets Li$_{7-x}$La$_3$Ta$_x$Zr$_{2-x}$O$_{12}$ ($x=0-2$). *Solid State Ionics*, **206**, 33–38.

84 Kokal, I., Somer, M., Notten, P.H.L., and Hintzen, H.T. (2011) Sol–gel synthesis and lithium ion conductivity of Li$_7$La$_3$Zr$_2$O$_{12}$ with garnet-related type structure. *Solid State Ionics*, **185**, 42–46.

85 Bates, J.B., Dudney, N.J., Gruzalski, G.R., Zuhr, R.A., Choudhury, A., Luck, C.F., and Robertson, J.D. (1992) Electrical properties of amorphous lithium electrolyte thin films. *Solid State Ionics*, **53–56**, 647–654.

86 Lee, Y.-I., Lee, J.-H., Hong, S.-H., and Park, Y. (2004) Li-ion conductivity in Li$_2$O-B$_2$O$_3$-V$_2$O$_5$ glass system. *Solid State Ionics*, **175**, 687–690.

87 Choi, C.H., Cho, W.I., Cho, B.W., Kim, H.S., Yoon, Y.S., and Tak, Y.S. (2002) Radio-frequency magnetron sputtering power effect on the ionic conductivities of Lipon films. *Electrochemical and Solid-State Letters*, **5**, A14–A17.

88 Maranas, J.K. (2012) Polyelectrolytes for batteries: current state of understanding, in *Polymers for Energy Storage and Delivery: Polyelectrolytes for Batteries and Fuel Cells*, ACS Symposium Series (eds K. Page *et al.*), American Chemical Society, Washington, DC (Chapter 1).

89 Agrawal, R.C. and Pandey, G.P. (2008) Solid polymer electrolytes: materials designing and all-solid-state battery applications: an overview. *Journal of Physics D: Applied Physics*, **41**, 223001.

90 Croce, F. and Scrosati, B. (2003) Nanocomposite lithium ion conducting membranes. *Annals of New York Academy of Science*, **984**, 194–207.

91 Takada, K. (2013) Progress and prospective of solid-state lithium batteries. *Acta Materialia*, **61**, 759–770.

92 Bates, J.B., Dudney, N.J., Gruzalski, G.R., Zhur, R.A., Choudhury, A., Luck, C.F., and Robertson, J.D. (1992) Electrical properties of amorphous lithium electrolyte thin films. *Solid State Ionics*, **53–56**, 647–654.

93 Yu, X.H., Bates, J.B., Jellison, G.E., and Hart, F.X. (1997) A stable thin-film lithium electrolyte: lithium phosphorus oxynitride. *Journal of the Electrochemical Society*, **144**, 524–532.

94 Patil, A., Patil, V., Shin, D.W., Choi, J.W., Paik, D.S., and Yoon, S.J. (2008) Issue and challenges facing rechargeable thin film lithium ion batteries. *Materials Research Bulletin*, **43**, 1913–1942.

95 Nathan, M., Golodnitsky, D., Yufit, V., Strauss, E., Ripenbein, T., Shechtman, I., Menkin, S., and Peled, E. (2005) Three-dimensional thin-film Li-ion microbatteries for autonomous MEMS. *Journal of Microelectromechanical Systems*, **14**, 879–885.

96 Pikul, J.H., Zhang, H.G., Cho, J., Braun, P.V., and King, W.P. (2013) High-power lithium ion microbatteries from interdigitated three-dimensional bicontinuous nanoporous electrodes. *Nature Communications*, **4**, 1732.

97 Long, J.W., Dunn, B., Rolison, D.R., and White, H.S. (2004) Three-dimensional battery architectures. *Chemical Reviews*, **104**, 4463–4492.

98 Oudenhoven, J.F.M., Baggetto, L., and Notten, P.H.L. (2011) All-solid-state lithium-ion microbatteries: a review of various three-dimensional concepts. *Advanced Energy Materials*, **1**, 10–33.

99 Perre, E., Nyholm, L., Gustafsson, T., Taberna, P.L., Simon, P., and Edström, K. (2008) Direct electrodeposition of aluminium nano-rods. *Electrochemistry Communications*, **10**, 1467–1470.

100 Baggetto, L., Niessen, R.A.H., Roozeboom, F., and Notten, P.H.L. (2008) High energy density all-solid-state batteries: a challenging concept towards 3D integration. *Advanced Functional Materials*, **18**, 1057–1066.

101 Long, J.W. and Rolison, D.R. (2007) Architectural design, interior decoration, and three-dimensional plumbing en route to multifunctional nanoarchitectures. *Accounts of Chemical Research*, **40**, 854–862.

102 Chen, D., Chen, C., Baiyee, Z.M., Shao, Z., and Ciucci, F. (2015) Nonstoichiometric oxides as low-cost and highly-efficient oxygen reduction/evolution catalysts for low-temperature electrochemical devices. *Chemical Reviews*, **115**, 9869–9921.

103 Evans, A., Bieberle-Hütter, A., Rupp, J.L., and Gauckler, L.J. (2009) Review on microfabricated micro-solid oxide fuel cell membranes. *Journal of Power Sources*, **194**, 119–129.

104 Yamaguchi, T., Zhou, H., Nakazawa, S., and Hara, N. (2007) An extremely low methanol crossover and highly durable aromatic pore-filling electrolyte membrane for direct methanol fuel cells. *Advanced Materials*, **19**, 592–596.

105 Kundu, A., Jang, J., Gil, J., Jung, C., Lee, H., Kim, S.-H., Ku, B., and Oh, Y. (2007) Micro-fuel cells: current development and applications. *Journal of Power Sources*, **170**, 67–78.

106 Sundarrajan, S., Allakhverdiev, S.I., and Ramakrishna, S. (2012) Progress and perspectives in micro direct methanol fuel cell. *International Journal of Hydrogen Energy*, **37**, 8765–8786.

107 Falcão, D., Oliveira, V., Rangel, C., and Pinto, A. (2015) Experimental and modeling studies of a micro direct methanol fuel cell. *Renewable Energy*, **74**, 464–470.

108 Garbayo, I., Pla, D., Morata, A., Fonseca, L., Sabaté, N., and Tarancón, A. (2014) Full ceramic micro solid oxide fuel cells: towards more reliable MEMS power generators operating at high temperatures. *Energy & Environmental Science*, **7**, 3617–3629.

109 Scherrer, B., Evans, A., Santis-Alvarez, A.J., Jiang, B., Martynczuk, J., Galinski, H., Nabavi, M., Prestat, M., Tölke, R., and Bieberle-Hütter, A. (2014) A thermally self-sustained micro-power plant with integrated micro-solid oxide

fuel cells, miro-reformer and functional micro-fluidic carrier. *Journal of Power Sources*, **258**, 434–440.

110 Su, P.-C., Chao, C.-C., Shim, J.H., Fasching, R., and Prinz, F.B. (2008) Solid oxide fuel cell with corrugated thin film electrolyte. *Nano Letters*, **8**, 2289–2292.

111 Logan, B.E., Hamelers, B., Rozendal, R., Schröder, U., Keller, J., Freguia, S., Aelterman, P., Verstraete, W., and Rabaey, K. (2006) Microbial fuel cells: methodology and technology. *Environmental Science & Technology*, **40**, 5181–5192.

112 Epstein, A. and Senturia, S. (1997) Macro power from micro machinery. *Science*, 276, 1211–1211.

113 Kim, N.I., Kato, S., Kataoka, T., Yokomori, T., Maruyama, S., Fujimori, T., and Maruta, K. (2005) Flame stabilization and emission of small Swiss-roll combustors as heaters. *Combustion and Flame*, **141**, 229–240.

114 Jacobs, R., Christopher, H., Hamlen, R., Rizzo, R., Paur, R., and Gilman, S. (1996) Portable power source needs of the future army: batteries and fuel cells. *IEEE Aerospace and Electronic Systems Magazine*, **11**, 19–25.

115 Hitt, D.L., Zakrzwski, C.M., and Thomas, M.A. (2001) MEMS-based satellite micropropulsion via catalyzed hydrogen peroxide decomposition. *Smart Materials and Structures*, **10**, 1163.

116 Lewis, D.H., Janson, S.W., Cohen, R.B., and Antonsson, E.K. (2000) Digital MicroPropulsion. *Sensors and Actuators A: Physical*, **80**, 143–154.

117 Kaisare, N.S. and Vlachos, D.G. (2012) A review on microcombustion: fundamentals, devices and applications. *Progress in Energy and Combustion Science*, **38**, 321–359.

118 Mohamed, M.R., Sharkh, S.M., and Walsh, F.C. (2009) Redox flow batteries for hybrid electric vehicles: progress and challenges. IEEE Vehicle Power and Propulsion Conference, pp. 551–557.

119 Epstein, A., Senturia, S., Al-Midani, O., Anathasuresh, G., Ayon, A., Breuer, K., Chen, K., Ehrich, P., Esteve, E., and Frechette, L. (1997) Micro-heat engines, gas turbines and rocket engines – the MIT microengine project. 28th Fluid Dynamics Conference, Fluid Dynamics and Co-located Conferences, American Institute of Aeronautics and Astronautics, 1997, pp. 1–12.

120 Ju, Y. and Maruta, K. (2011) Microscale combustion: technology development and fundamental research. *Progress in Energy and Combustion Science*, **37**, 669–715.

21

Energy Harvesting

Rolanas Dauksevicius[1] and Danick Briand[2]

[1]*Kaunas University of Technology, Institute of Mechatronics, Studentu 56, LT-51424 Kaunas, Lithuania*
[2]*Ecole Polytechnique Fédérale de Lausanne, Institute of Microengineering, Microsystems for Space Technologies Laboratory, Rue de la Maladière 71b, CH-2000 Neuchâtel, Switzerland*

21.1 Introduction

A process of capturing and converting ambient energy into useable electrical power is referred to as energy harvesting (EH). Though there are subtle differences [1], the term *energy scavenging* is sometimes used interchangeably. Energy harvesting is a truly broad field that encompasses technologies for harnessing ambient energy both on a large scale (e.g., wind, solar, hydro) and on a small scale. This chapter covers only the latter aspect, referred to as micro energy harvesting [2], which is associated with low-power electronics domain, where typical level of harvested power is in the range of nW/cm^2 to mW/cm^2. In general, a complete EH device encompasses three parts: an *energy harvester* (*micropower generator*), power conditioning circuit, and a storage element such as supercapacitor or rechargeable battery. This chapter focuses only on the first part.

The last two decades witnessed a substantial progress in microfabrication techniques as well as low-power microelectronics, which accelerated the development of a wide range of battery-powered portable, wearable, implantable, or embedded devices such as healthcare sensors or wireless sensor nodes. However, there are a number of adverse issues associated with the electrochemical energy sources such as low power density, short life span, generation of harmful waste, and high maintenance costs. In many applications, it is often uneconomical, impractical, or even impossible to replace or recharge depleted batteries, for example, in the case of large-scale wireless sensor networks (WSNs) or implanted biomedical devices. Furthermore, battery technology is not a feasible solution for reliable long-term powering of remote systems deployed in harsh and/or hardly accessible environments (e.g., oil/gas/mining/deep sea/space applications). As a consequence, energy harvesting technologies emerged as a result of the need to alleviate and ultimately eliminate reliance on disposable or rechargeable batteries, which impose significant constraints on device size, weight, and

Material-Integrated Intelligent Systems: Technology and Applications, First Edition.
Edited by Stefan Bosse, Dirk Lehmhus, Walter Lang, and Matthias Busse.
© 2018 Wiley-VCH Verlag GmbH & Co. KGaA. Published 2018 by Wiley-VCH Verlag GmbH & Co. KGaA.

cost. Energy harvesters as self-sufficient green energy sources are particularly needed for the following:

- Autonomous WSNs (including remote sensors) providing pervasive monitoring capabilities within industrial, civilian, health care, environmental, agri-food, security, defense, and other sectors. Power requirements of current wireless sensors implemented in low-power VLSI design are in the range of 10s to 100s of µW, which may be at least partially delivered by EH devices [3–6].
- Wearable and implantable biomedical devices [7,8], smart fabrics-interactive textiles (SFIT) [9], and wireless body area networks (WBANs) [10]. Development of flexible EH devices for wearable electronics is currently a topic of particularly intense interest since bulky and rigid battery packs constitute a serious bottleneck hindering the progress in the field.

As a whole, the field of energy harvesting is considered to be one of the most important enabling technologies (along with RFID, wireless power transfer, and green electronics) for the self-sustainable implementation of Internet of Things [11,12]. The field is rapidly advancing through continuous development of novel materials with emphasis on eco-friendly hybrid (nano)composites as well as through introduction of effective design architectures, particularly in the case of micro/nanoengineered EH devices.

There is a multitude of ambient energy sources that may be exploited for generation of usable electrical energy. This chapter covers the most common types of ambient energy: (i) mechanical energy (vibrations, pressure, static deformations), (ii) thermal energy (spatial and temporal temperature gradients), (iii) light energy (solar or indoor illumination), (iv) radio-frequency electromagnetic energy (emissions from TV/radio/mobile/WiFi transmitters). There are also other energy forms and mechanisms that have been proposed for EH purposes, but they are beyond the scope of this chapter, for example, exploiting stray magnetic fields [13], acoustic energy [14], chemical energy (e.g., enzymatic fuel cells [15]), microbial activity [16], metabolic energy [17],[1] naturally occurring pH differences [18], reverse electrowetting [19], or artificial photosynthesis [20].

21.2 Mechanical Energy Harvesters

Harvestable mechanical energy is ubiquitous in the surrounding natural and built environments, which are abundant with sources moving or oscillating with amplitudes and frequencies that typically exhibit high spatial and temporal variability. Biomechanical energy associated with humans or animals, which is characterized by ultralow frequency and large-amplitude motions, may also be harnessed in order to power wearable, embedded, or implanted biomedical devices [8]. The most common type of mechanical energy used for harvesting is mechanical vibration. Naturally occurring mechanical energy is usually associated with vibrations induced by air or water flows. Manmade sources of mechanical energy such as

---

1 http://voltreepower.com/bioHarvester.html (accessed September 22, 2017).

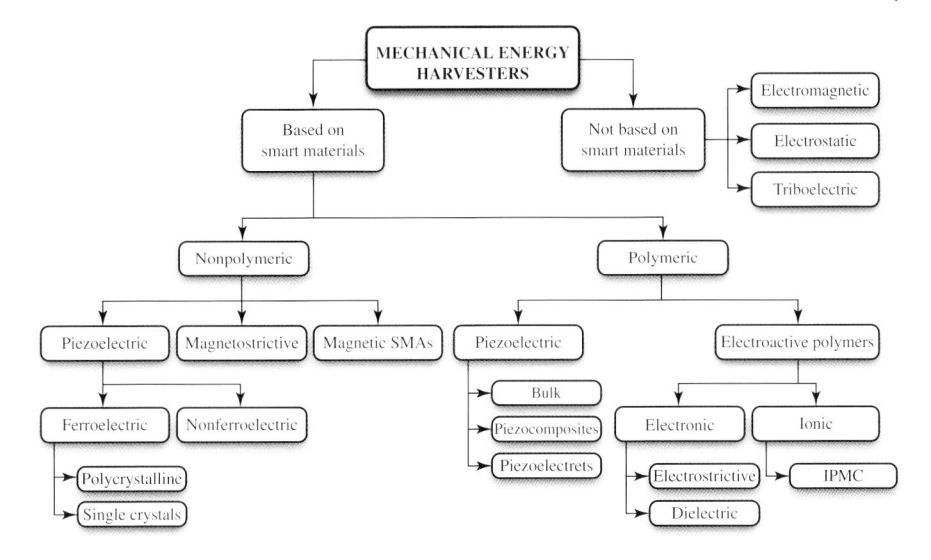

Figure 21.1 Classification of mechanical energy harvesters. SMAs: shape memory alloys, IPMC: ionic polymer metal composite.

industrial machinery, civil infrastructures, or transportation usually emit relatively low-level random vibrations that are challenging to harvest efficiently. Mechanical energy harvesters (MEHs) may be broadly subdivided into *vibration (kinetic) energy harvesters* (VEHs) relying on inertial force or direct force external excitations and *strain energy harvesters* that use surface strain fluctuations to generate electrical energy. Mechanical energy harvesting is implemented predominantly by means of VEHs. To maximize energy generation performance, the design of the harvesters has to be tailored for specific application taking into consideration nature of input excitation, which may be harmonic, random, or impulse. Physical limits of power output in VEHs are directly determined by the device volume and mass, which implies that with miniaturization the harvested power inevitably drops. Overall, the harvestable energy is limited by the energy source and, practically, only a few percent of the initially available mechanical energy can be converted into the useable electrical power [1].

Conversion of ambient mechanical energy into useable electrical power is accomplished through one or several mechanisms of electromechanical transduction. MEHs may be conditionally subdivided into two large groups depending on whether their operation relies on smart materials or not (Figure 21.1):

- Harvesters employing smart materials operate on the basis of static or dynamic strain induced in the material. There is a large variety of materials exhibiting strong electro- or magnetomechanical coupling, which can be successfully exploited for mechanical-to-electrical energy conversion. Piezoelectric energy harvesters are by far the most prominent representative in this group, followed by electroactive polymers. Application of magnetostrictive materials and magnetic shape memory alloys (SMAs) is relatively scarce at the moment. This group also encompasses hybrid generators that combine several transduction mechanisms

(at least one of them relies on smart materials) such as magnetostrictive/piezoelectric (magnetoelectric), magnetostrictive/inductive, and piezoelectric combined with electromagnetic, electrostatic, or triboelectric.
- Harvesters that do not use smart materials for electromechanical transduction operate on the basis of relative displacements occurring between elements in magnet/coil assembly (electromagnetic) or between charged capacitor plates or electrode fingers of a variable capacitor (electrostatic) as well as through contact and separation between two materials with different charge affinities (triboelectric).

Continuously growing interest in mechanical energy harvesting is reflected in a number of publications indexed in Web of Science™ database for the period of 2003–2013 [21] with obvious predominance of articles on piezoelectric generators (~2000), followed by the electromagnetic (~650) and electrostatic ones (~300). Table 21.1 provides a comparison of key advantages and disadvantages that are typically attributed to different types of MEH devices.

21.2.1 Piezoelectric Micropower Generators

Piezoelectric vibration energy harvesters (P-VEHs) commonly employ a mono/polycrystalline or polymeric piezoactive materials to generate electric charge in response to applied mechanical stress (Figure 21.2). Popularity of piezoelectric generators is attributed to inherently simple design architecture, which makes them easily amenable to miniaturization through application of relatively mature MEMS technologies. P-VEHs also benefit from higher voltage and power density levels with respect to their electromagnetic and electrostatic counterparts, respectively. Power scaling is more advantageous for piezoelectric generators ($V^{4/3}$) in comparison to their electromagnetic counterparts (V^2) [22]. Therefore, in terms of power density, piezoelectric transduction is generally preferred for generator sizes below ~0.5 cm^3 [23]. There is an extensive range of piezoelectric materials that are typically used for implementation of P-VEHs:

1) Ferroelectric piezoceramics:
   - *PZT-based:* "soft" ceramics (e.g., PZT-5 H) or "hard" ceramics (e.g., PZT-4).
   - *Containing lead:* polycrystalline (e.g., PZT, PbTiO$_3$) or single crystals (e.g., PZT, PMN-PT, PZN-PT).
   - *Lead-free [24]:* for example, BaTiO$_3$, LiNbO$_3$, K$_{1-x}$Na$_x$NbO$_3$ (KNN), or Na$_{1-x}$K$_x$NbO$_3$ (NKN).
2) *Nonferroelectric crystalline materials:* for example, AlN, ZnO, GaN, CdS, InN.
3) Piezoelectric polymers and composites [25–27].

PZT is the most widely used material due to favorable combination of piezoelectric properties (d_{31} and d_{33} reaching 320 and 650 pC/N, respectively [28])[2] and cost. Single-crystal piezoceramics potentially offer much better piezoelectric performance (for PMN-PT, d_{31} and d_{33} reaching 1800 and 3000 pC/N, respectively [28]),[3] however high cost currently precludes their wider

---
2 http://www.piezo.com/prodmaterialprop.html (accessed September 22, 2017).
3 http://www.hcmat.com/Pmn_Properties.html (accessed September 22, 2017).

Table 21.1 Comparison of main transduction mechanisms employed for mechanical energy harvesting.

Transduction mechanism	Advantages	Disadvantages
Piezoelectric	• High power density • Relatively high output voltage • Mature microfabrication technologies • External voltage source not required • Relatively simple design architectures • High electromechanical coupling in piezoceramics (particularly in single crystals)	• Low coupling in piezoelectric thin films • Microfabrication technologies are not always CMOS-compatible • High brittleness of piezoceramics • Aging effects: fatigue, depolarization • High cost of piezoceramics (single-crystal)
Electrostatic	• High output voltage • Highly mature microfabrication technologies • Tunable electromechanical coupling	• External voltage source (precharging) required • Mechanical constraints required
Electromagnetic	• High power density at macro/mesoscale • External voltage source not required • Low optimal load resistance • Efficient at low frequencies	• Low output voltage • Low power density at microscale • Inefficient at microscale (immature microfabrication methods for magnets) • Difficult to integrate with MEMS • Bulky (magnets, pick-up coil)
Magnetostrictive	• Ultrahigh coupling coefficient (<0.9) • High flexibility • Suitable for high-frequency applications • No depolarization problem	• Pick-up coil required • Magnets may be required for bias magnetic field • Difficult to integrate with MEMS • Nonlinear effects

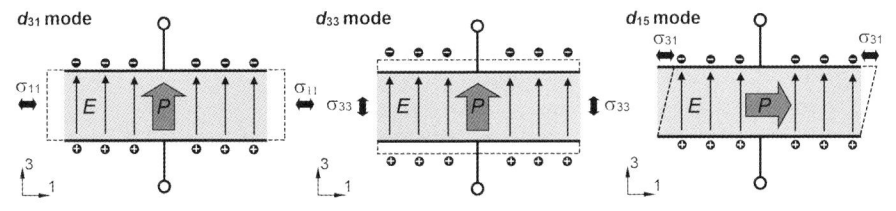

Figure 21.2 Operating modes of piezoelectric vibration energy harvesters. (a) Transverse. (b) Longitudinal. (c) Shear.

applicability [29]. Due to high Curie temperature (~400 °C) and favorable piezoelectric properties (d_{31} and d_{33} reaching 50 and 340 pC/N, respectively), KNN and its various compositions are considered as the most promising lead-free piezoceramic candidates to replace PZT [30–32]. Many ferroelectric piezoceramics including the aforementioned ones have Curie temperatures below 600 °C. Therefore, for ultrahigh temperature (>1000 °C) applications (e.g., power, transport, oil/gas/space exploration), different piezoelectric materials (with d_{31} and d_{33} typically reaching only several pC/N) are considered including wide bandgap semiconductors AlN and GaN, ferroelectric LiNbO$_3$, as well as more exotic materials such as rare-earth calcium oxyborate single crystals (e.g., GdCOB) [28,33].

Most piezoelectric VEHs are designed to be operated in one of the three coupling modes (Figure 21.2), which are distinguished in terms of direction of applied stress with respect to poling and electric field directions [28,34]:

- In the most common d_{31} (transverse) operation mode, the direction of applied tensile stress σ_{11} is perpendicular to the direction of poling and produced electric field [35]. In this case, the charges are collected by means of planar top and bottom electrodes (Figure 21.3a).
- In the less common d_{33} (longitudinal) mode, the direction of applied tensile stress σ_{33} is parallel to poling and electric field directions [37]. Interdigitated electrodes (IDEs) are typically employed for collection of charges (Figure 21.3b).

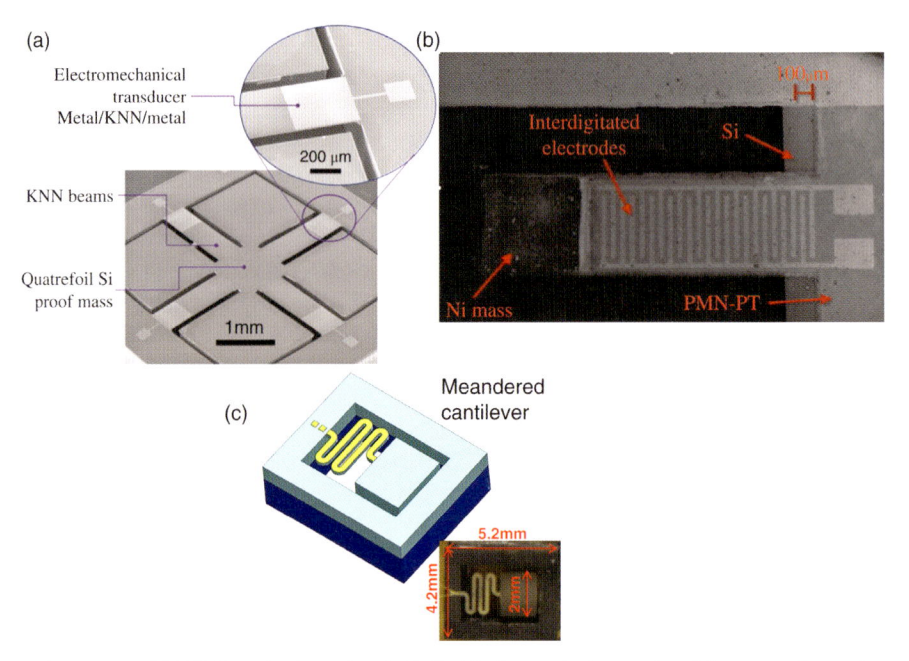

Figure 21.3 (a) MEMS-scale piezoelectric VEH with clamped–clamped multibeam configuration operating in d_{31} mode. (Reproduced with permission from Ref. [32]. Copyright 2011, IOP Publishing.) (b) Cantilever-type VEH operating in d_{33} mode. (Reproduced with permission from Ref. [37]. Copyright 2014, Elsevier.) (c) Spiral-shaped piezoelectric VEH. (Reproduced with permission from Ref. [36]. Copyright 2012, Elsevier.).

- In the least common d_{15} (shear) mode, the direction of applied shear stress σ_{31} is parallel to the poling direction, but perpendicular to the direction of electric field [38].

Values of d_{33} and d_{15} coefficients are much higher in comparison to d_{31}, but in practice it does not straightforwardly translate into appreciable improvements in energy harvesting performance. In the case of d_{33} mode, optimized design of IDEs is required in terms of electrode spacing and area in order to achieve higher electrical output with respect to d_{31} generators. Another limitation is inapplicability of IDEs configuration for nonferroelectric materials such as AlN, whose deposition via sputtering is a well-established and CMOS-compatible technique. For d_{15} harvesters, the necessity to use different electrodes for poling and charge collection entails substantial fabrication complications [39].

P-VEHs are traditionally designed as linear oscillators and are mostly implemented using cantilever-type composite transducers in unimorph or bimorph configuration with proof mass at the end [40]. Cantilever structure is preferred due to its highest responsiveness since it exhibits the highest average strain for a given excitation force [41]. Tapered cantilevers are used to homogenize strain level throughout the length of the structure, thereby enabling minimization of size and weight of a generator for a given power output [42]. P-VEHs are also implemented using stiffer structural configurations such as clamped–clamped beams (Figure 21.3a) [32] or circular diaphragms [34], which can be used in harvesting energy from slowly varying pressure fields [43] or acoustic waves [14]. Pressure energy harvesting is typically implemented using mechanically amplified stack architecture operating in d_{33} mode [44], while impact energy may be harvested using cymbal transducer [45]. More elaborate geometries such as zigzag [46], spiral (Figure 21.3c) [36], or ring-shaped [47] are attractive for designing microscale generators due to possibility to achieve lower resonant frequency, but still retain high strength.

Linear P-VEHs are highly frequency-selective since power output is maximized only when the generator is excited at (or very near) its resonant frequency. Consequently, they constitute a viable micropower source only for ambient excitations with well-defined dominant frequency that is stable in time, which is uncommon in practice. Slight deviation from the resonance leads to pronounced deterioration in performance of linear generators. Therefore, a variety of configurations exploiting nonlinearities have been proposed over the years (see Section 21.2.7).

Several mathematical models for predicting dynamic and electrical response of P-VEHs have been proposed over the last 20 years, ranging in complexity from approximate lumped parameter model of a generic VEH [48] to accurate distributed parameter model of a cantilever-type P-VEH with full electromechanical coupling [49] as well as models accounting for mechanical nonlinearities [50].

Piezoelectric VEHs are realized on different geometric scales using various fabrication methods:

- Macro/mesoscale generators are manufactured using mechanical or manual assembly.
- Micro/MEMS-scale generators are fabricated by means of various microfabrication techniques used for deposition of thin/thick piezoelectric films,

for example, sol–gel processing [51], sputtering [52], screen printing [53], epitaxial growth [54], pulsed laser deposition [55], aerosol deposition [56], or micro-pen direct-writing [57]. Electromechanical coupling of deposited piezoelectric thin films is still inferior to their bulk counterparts [34]. Therefore, as an alternative, subtractive approaches based on thinning of piezoceramic sheets are pursued as well [58,59].

- Nanoscale generators (nanogenerators) are fabricated through synthesis of nanostructures by using a multitude of chemical or physical methods such as vapor-phase deposition, chemical solution deposition, chemical vapor deposition, aqueous chemical growth, or pulsed laser deposition.

Depending on the geometric scale, power outputs from generators vary in a very wide range from nanowatts to milliwatts. In general, R&D of linear P-VEHs on a macro/meso/microscales has reached a certain level of maturity by now with several generators available on the market.[4–7] Significant efforts are still being directed toward realization of MEMS-scale P-VEHs as enablers of low-cost high-volume self-powered electronic devices. Considerable progress has been made in developing VEHs with CMOS-compatible materials such as the biocompatible AlN,[3] which is an attractive lead-free alternative to PZT in terms of piezoelectric coupling because of low dielectric constant [60].

The last decade has witnessed a remarkable upsurge in fundamental research and engineering development of nanostructured piezoelectric materials [61] in order to implement flexible, foldable, or stretchable MEH devices (nanogenerators) for efficient harvesting of (bio)mechanical energy with various potential applications, particularly in the field of implantable biosensors or personal electronics [62–64]. Piezoelectric nanogenerators are mainly fabricated using semiconducting wurtzite compounds (e.g., ZnO, GaN, InN, CdS, and CdSe), ferroelectrics (e.g., PZT, $BaTiO_3$, and $KNbO_3$) and polymers (e.g., PVDF) [61,65]. Implementations based on ZnO nanowires are the most widely studied due to favorable blend of various properties in ZnO such as relatively strong piezoelectric response (d_{31} and d_{33} reaching ∼5 and ∼27 pC/N, respectively [28]), biocompatibility, and optical transparency. Harvesting performance of nanogenerators based on GaN or InN appears promising as well, although there is noticeably less reports on nanogenerators based on the III-nitride semiconductors. The same applies to ferroelectric nanowires since it is challenging to obtain perfect crystal structure through synthesis. Piezoelectric nanostructures of different configurations and dimensions are used for mechanical-to-electrical energy conversion:

- *1D nanostructures:* for example, nanorods/nanowires of ZnO (Figure 21.4a) [66,67] or GaN [68], PZT nanofibers [69] and ribbons [70,71] (Figure 21.4b and c), PVDF nanofibers [72].

---

4 www.microgensystems.com (accessed September 22, 2017).
5 www.microstrain.com (accessed September 22, 2017).
6 http://www.mide.com/products/volture/piezoelectric-vibration-energy-harvesters.php (accessed September 22, 2017).
7 http://www.arveni.fr/en/t-rex-vibration-harvester (accessed September 22, 2017).

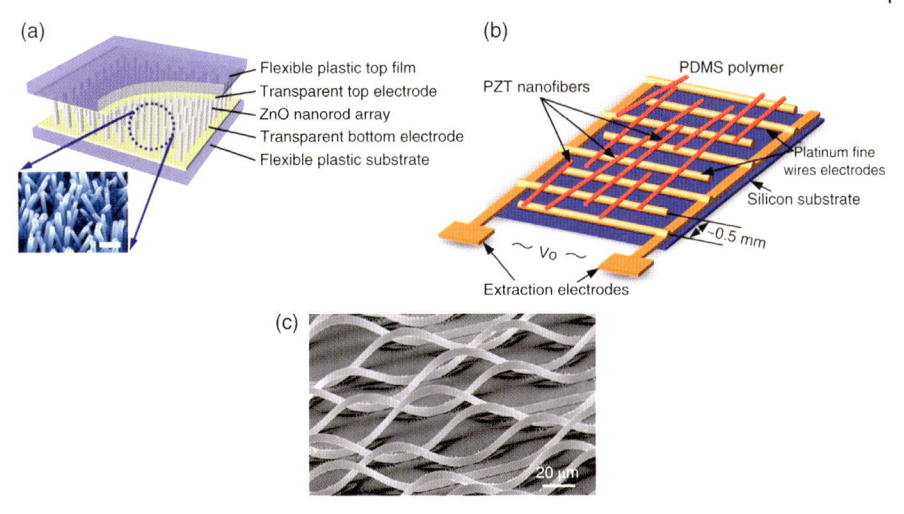

Figure 21.4 (a) Flexible nanogenerator based on ZnO nanorods. (Reproduced with permission from Ref. [67]. Copyright 2009, Wiley-VCH Verlag GmbH & Co.) (b) Nanogenerator based on PZT nanofibers. (Reproduced with permission from Ref. [69]. Copyright 2010, The American Chemical Society.) (c) Nanogenerator based on buckled PZT nanoribbons. (Reproduced with permission from Ref. [71]. Copyright 2011, The American Chemical Society.)

- *2D nanostructures:* for example, thin films of ZnO (Figure 21.5c) [73] or $BaTiO_3$ [74].

In general, nanogenerators are not commercially viable yet and issues of performance optimization, reliability, and robustness still need to be resolved [27]. To overcome some of these problems, graphene films, as a new generation material with outstanding mechanical, electrical, and optical properties, have been introduced as flexible and transparent electrodes in order to implement stretchable and transparent nanogenerators with excellent mechanical robustness and electrical functionality under high strains [77,78]. In terms of performance improvement, piezoelectric response at the nanoscale could be substantially enhanced by means of coupling between polarization and strain gradient – flexoelectric effect, which exists in all dielectrics [79]. Nanoflexoelectricity has recently gained increasing attention in mechanical energy harvesting community [61,80,81] due to tantalizing possibility to create someday a purely flexoelectric nanogenerator made of nonpiezoelectric dielectric materials.

Application of piezoelectric polymers and composites for mechanical energy harvesting continues to attract considerable scientific interest, which is fueled by the need for efficient, conformable, lightweight, biocompatible, and low-cost smart materials for self-powered wearable or implantable devices undergoing large strains during exploitation [7]. The challenge in this case is to design a piezoelectric composite with adequate piezoelectric response, compliance, and temperature tolerance. Piezoelectric polymers used in MEHs can be divided into three main categories as follows [25].

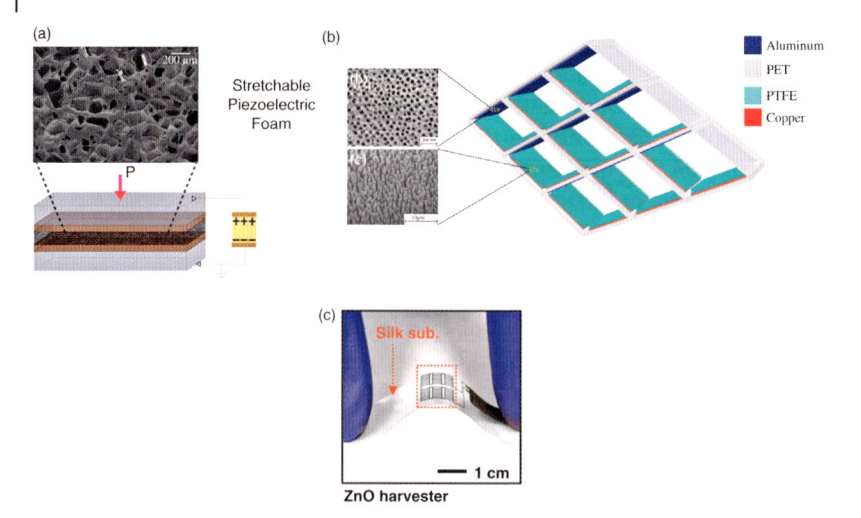

Figure 21.5 (a) Piezoelectric nanoparticle–polymer composite foam, which is based on porous PDMS with BaTiO$_3$ nanoparticles and carbon nanotubes. (Reproduced with permission from Ref. [75]. Copyright 2014, The American Chemical Society.) (b) Triboelectric nanogenerators with integrated rhombic gridding having structurally multiplied unit cells connected in parallel. (Reproduced with permission from Ref. [76]. Copyright 2013, The American Chemical Society.). (c) Biocompatible nanogenerators based on ZnO thin films fabricated on silk substrate. (Reproduced with permission from Ref. [73]. Copyright 2013, Wiley–VCH Verlag GmbH & Co.)

Bulk piezopolymers, which are solid films with piezoelectric mechanism manifesting due to their molecular structure and its arrangement. PVDF and its copolymer PVDF-TrFE, having the highest piezoelectric coefficients ($d_{33} \approx 20$–74 pC/N) out of all ferroelectric polymers, are the most common piezoelectric polymers. Polyamides and polyureas are considered to be promising materials for energy harvesting at higher temperatures [27]. This category also includes PVDF with inorganic particles of PZT [82], BaTiO$_3$ [83], or ZnO [84,85] added to enhance piezoelectric response as well as introduction of TrFE [86], carbon black [87], or nanotubes [88], which enables improvement of crystal orientation (strengthening of electric field).

Piezocomposites (piezoelectric nanocomposites) combine polymer flexibility and high coupling of piezoceramic materials, which are embedded as particles, rods, or fibers, for example, MFC®,[8] PZT/epoxy [89], SU-8/ZnO [90], PMN-PT/PDMS [91], or BaTiO$_3$/PDMS with carbon nanotubes (Figure 21.5a) [75] that enhance stress transfer from porous polymer to piezoelectric nanoparticles (achieving $d_{33} \approx 112$ pC/N, while for BaTiO$_3$ d_{33} exceeds 200 pC/N). Major part of volume of piezocomposites is occupied by the inactive polymer; therefore, the flexibility of the composite is gained at the expense of piezoelectric performance.

---

8 www.smart-material.com (accessed September 22, 2017).

Voided charged polymers (piezo/ferroelectrets) contain gas voids and are charged so as to form internal dipoles that undergo polarization under strain. Cellular polypropylene (PP) is the most popular material in this category, although materials based on fluoropolymers have been recently considered in order to improve thermal stability of piezoelectric mechanism [92]. Piezoelectrets are interesting highly compliant and lightweight polymer materials with d_{33} coefficient ranging from several hundred to several thousand pC/N depending on measurement conditions (quasi-static versus dynamic) [25,92,93]. On the other hand, they are characterized by low electromechanical coupling [28]. Piezoelectrets only recently have been considered for EH purposes [94–97].

21.2.2 Micropower Generators Based on Electroactive Polymers

A class of polymeric smart materials – electroactive polymers (EAPs) – are gaining increased interest by offering distinct advantages over the piezoceramics: flexibility, processability, and low cost [98]. Two classes of EAPs are commonly distinguished [99]: *electronic* and *ionic*. Electromechanical conversion in the former relies on the polarization-based or electrostatic mechanisms, while in the latter on solvent (water) and charge (ions) transport.

A type of *electronic* EAPs – dielectric EAPs (dielectric elastomers) – has emerged as promising smart materials for small- and large-scale harvesting of ultralow-frequency biomechanical energy [100,101] and water/air flow energy [102,103], respectively. The latter requires large stretch areas; therefore, piezoelectric and electromagnetic VEHs are hardly adaptable to these conditions. Mechanism of energy harvesting in dielectric EAPs is analogous to that in electrostatic generators presented above: when an elastomer membrane is prestretched and precharged, a reduction in the tensile force under the open-circuit condition thickens the membrane, thereby reducing the capacitance and increasing the voltage [104]. Unfortunately, high voltages of several kilovolts are required to implement the conversion. Various energy harvesting systems based on dielectric EAPs have been demonstrated so far, including mesoscale generators for harvesting sub-hertz motions for wearable applications [101]. Materials used for generators based on dielectric EAPs include acrylic- and silicone-based elastomers as well as natural rubber [105]. In general, EAP films with improved mechanical and electrical properties are still required for increasing the efficiency of these generators to make them commercially viable [106].

Another type of *electronic* EAPs – electrostrictive polymers – perform electromechanical transduction mainly through electric field-induced molecule motion or phase transition and require sub-kilovolts, which are appreciably lower than in the case of dielectric elastomers. Furthermore, these polymers are characterized by mechanical energy densities comparable to those of piezoelectric single crystals [107,108].

Feasibility of mechanical energy harvesting was only recently demonstrated for a category of *ionic* EAPs, referred to as ionic polymer metal composites (IPMC) [26,109,110], composed of a hydrated ionomeric membrane (e.g., Nafion®, Flemion®) that is sandwiched between metal electrodes. Mechanical-to-electrical energy conversion in IPMCs relies on charge concentration gradient that is induced in the composite under strain, which then leads to accumulation of charges at the electrodes and the consequent formation of the potential difference [109]. A number of models have been proposed in order to explain electromechanical transduction in IPMCs [111]. In contrast to dielectric elastomers, ionic electroactive polymers do not require high voltages and because of low stiffness, large capacitance, and ability to operate in wet conditions, they are promising for energy harvesting from complex deformations in underwater environments by extracting energy from steady flows [112] or vortices [113], although in-air harvesting is also explored [109]. It should be noted that at present power densities achieved with IPMCs are considerably lower in comparison to P-VEHs with power outputs in the nanowatt range [111,114].

21.2.3 Electrostatic Micropower Generators

Electrostatic VEHs operate on the basis of variable capacitor, where one electrode is associated with an oscillating suspended mass and the other electrode is fixed (Figure 21.6a). Excitation force acting on the proof mass induces variations in dielectric gap or in overlap area between electrodes, thereby generating additional

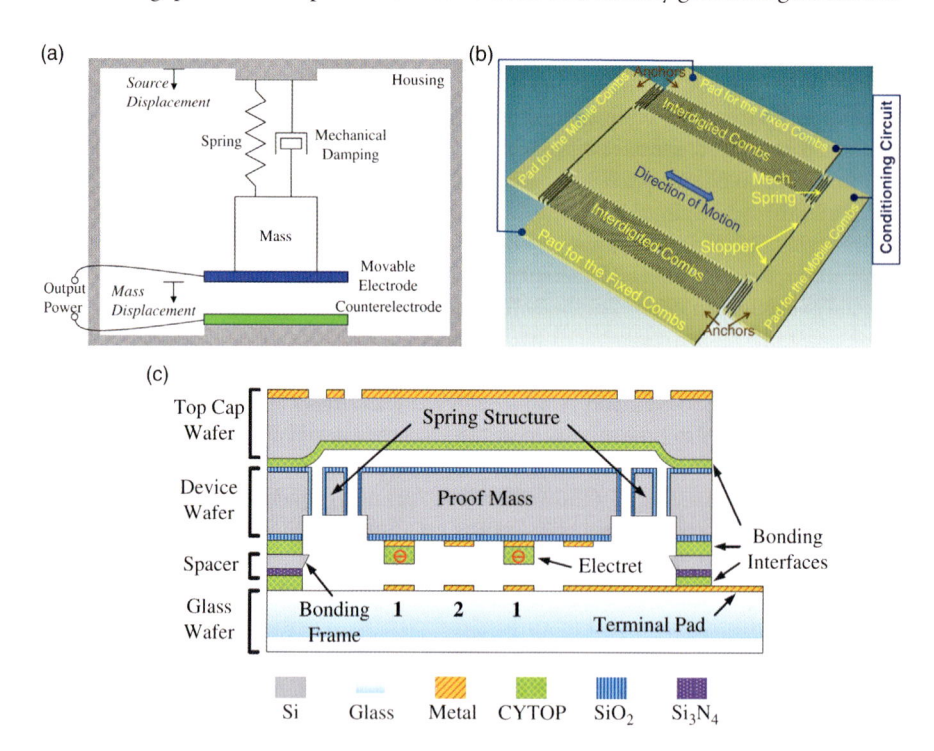

Figure 21.6 (a) Schematic representation of gap-closing electrostatic generator. (Reproduced with permission from Ref. [121]. Copyright 2008, Elsevier.) (b) In-plane gap-closing electrostatic generator exhibiting broadband response due to combined effect of electrical nonlinearities and mechanical impact. (Reproduced with permission from Ref. [115]. Copyright 2014, IOP Publishing). (c) Electret-based overlap varying electrostatic generator. (Reproduced with permission from Ref. [119]. Copyright 2013, IOP Publishing.)

Table 21.2 Comparison of basic types of electrostatic vibration energy harvesters [127].

Type	Advantages	Disadvantages
In-plane gap closing	• Larger max. capacitance	• Mechanical stops needed
Out-of-plane gap closing	• Good stability • Largest max. capacitance	• Largest mechanical damping • Surface adhesion
In-plane overlap	• Highest Q factor • No mechanical stops required	• Lowest maximum capacitance • Stability problems for large deflections

charges at the electrodes in order to balance the bias voltage. The former are termed as gap-closing converters, which may have in-plane (Figure 21.6b) [115] or out-of-plane configuration (Figure 21.6a) [116,117], while the latter are termed overlap varying converters (Figure 21.6c) [118,119]. Each type has its own characteristic advantages and disadvantages, which are summarized in Table 21.2. Hybrid gap-closing/overlap configurations have been reported as well [120]. Electrostatic VEHs are particularly suitable for wafer-level implementations since CMOS-compatible silicon-based microfabrication processes are well developed and enable realization of nonlinear springs with repeatable mechanical properties, which is beneficial for development of wideband VEHs (Figure 21.6b). The necessity of initial precharge by external voltage source for initiation of conversion process is the main drawback of electrostatic generators as this requires dedicated complex circuitry and induces high electrical losses [121,122]. The problem is circumvented (Figure 21.6c) [119] by introducing organic or inorganic dielectric materials with quasi-permanent charge polarization – electrets [123,124]. However, miniaturization of electret-based devices leads to the associated issues of low surface potential and poor long-term charge stability [125]. Neither external voltage source nor electret is needed for the recently proposed electrostatic generator, which relies on charge pumping by using two materials with different work functions [121,126].

21.2.4 Electromagnetic Micropower Generators

For generator sizes from several cm^3 upward the electromagnetic (electrodynamic) VEHs constitute the most mature and cost-effective micropower generation technology to date [28]. It was first introduced in Seiko AGS Kinetic quartz watch[9] and have recently witnessed other successful commercial deployments.[10,11] More than a hundred prototypes of electromagnetic generators

---

9 http://www.seikowatches.com/world/technology/kinetic/index.html (accessed September 22, 2017).
10 https://perpetuum.com/ (accessed September 22, 2017).
11 http://www.enocean.com/en/enocean_modules/eco-200/ (accessed September 22, 2017).

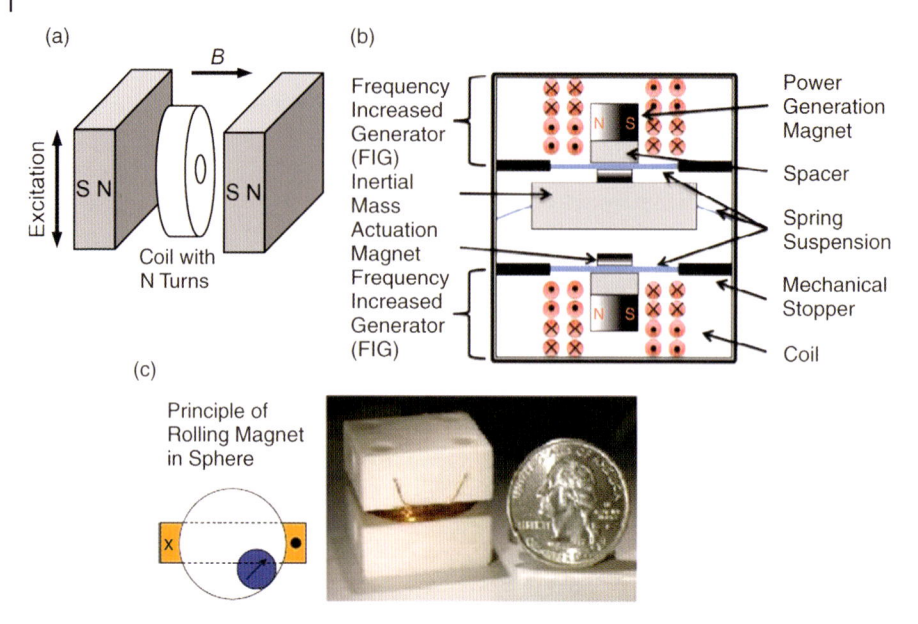

Figure 21.7 (a) Schematic representation of electromagnetic vibration energy harvester. (b) Schematic drawing of electromagnetic generator based on frequency up-conversion mechanism. (Reproduced with permission from Ref. [130]. Copyright 2009, Elsevier.) (c) Electromagnetic generator for harvesting energy from 3D motions. (Reproduced with permission from Ref. [132]. Copyright 2009, IOP Publishing.)

have been developed starting from 1995 when a seminal paper by Williams and Yates has been published [48]. They vary considerably in volume (<1 to >100 cm^3) and levels of peak power output (from submicrowatt to subwatt) [128,129]. Most of the developed electromagnetic generators are intended for 1D excitation sources (Figure 21.7b) [128,130], although 2D [131] and even 3D (Figure 21.7c) [132] configurations have also been proposed. Obviously, the use of permanent magnets makes electromagnetic generators large and bulky, which, depending on the application, is not necessarily a complication. However, performance and manufacturability of these generators deteriorate rapidly with device size shrinking below 1 cm^3. Wafer-scale implementations suffer from low power output due to limited number of planar coil turns and insufficient vibration amplitudes. Furthermore, it is challenging to integrate these harvesters with standard electronics [133].

21.2.5 Triboelectric Nanogenerators

Very recently, MEH devices, referred to as triboelectric nanogenerators (TENGs), and operating on the basis of coupled effect of contact electrification [134] and electrostatic induction, were proposed (Figure 21.5b) [76,135,136]. Mechanical energy harvesting is realized here through a periodic contact and separation of two materials with different charge affinities, thereby functioning as a charge pump that forces induced electrons to flow between the electrodes via an electrical load. Though several models have been proposed, the fundamental mechanism of contact electrification is not yet fully understood [137]. Material selection for TENG implementation is enormous due to ubiquity of triboelectricity, which provides favorable conditions to develop highly flexible and textile-embedded generators for wearable electronics that harness energy from human motion [138,139]. Several modes of operation of TENGs are distinguished [140]: vertical contact separation, lateral sliding, single-electrode, freestanding triboelectric layer, and the combinations thereof. Primary shortcoming of current TENGs is relatively low current output, while open-circuit voltages reach several hundred volts. Other challenges such as fluctuations in power output, high-output impedance, and fragility also need to be addressed before TENGs become practical micropower sources [141,142]. Deeper understanding of triboelectrification and further efforts in modification of contacting surfaces are required to improve triboelectric charge density [140]. Meanwhile, for textile-based TENGs, the issues of breathability, washability, stretchability, and softness still await to be tackled [138].

21.2.6 Hybrid Micropower Generators

Hybrid mechanical energy harvesting concepts that integrate different transduction mechanisms have been explored as a means of enhancing generated power output. For example, VEHs employing magnetostrictive materials operate on the basis of *Villari effect* and electromagnetic induction, whereby vibration-induced strain of the magnetically polarized material produces a change in its magnetic flux density, which is then converted into electrical current by means of a pick-up coil. Higher energy conversion efficiency and flexibility are key advantages of magnetostrictive/inductive VEHs over their piezoelectric counterparts (Table 21.1) [143]. Typical magnetostrictive materials used for energy harvesting include highly flexible amorphous metallic glass (Metglas) [143], crystalline rare earth–iron alloy TbDyFe (Terfenol-D) [144], and more recently discovered FeGa alloy (Galfenol) [145]. The latter has high permeability, resulting in eddy current generation even at low frequencies, which cancels some of the induced magnetic flux density change, thereby reducing generated power [145,146]. Alternatively, magnetic shape memory alloys (MSMAs) may be used to harvest mechanical energy [147], although requirement for high biasing magnetic field is a limitation [148]. MSMAs have the potential advantage of achieving reversible strain levels up to 10%, which is about two orders of magnitude higher than those of Terfenol-D, Galfenol, and piezoelectrics [146].

Vibration energy may also be harvested through magnetoelectric effect produced in magnetostrictive/piezoelectric laminates and composites [149,150]. Superior transduction efficiency can be achieved by combining piezoelectric effect with magnetostriction, which is characterized by high energy density and strong magnetomechanical coupling. More interesting representatives of this class of devices include multiaxial (Figure 21.8a) [151] and self-tunable [152] generators. Other examples of hybridization in mechanical energy harvesting include piezoelectric–electromagnetic [153,154], piezoelectric–electrostatic (Figure 21.8c) [117], or piezoelectric–triboelectric (Figure 21.8b) [155] generators.

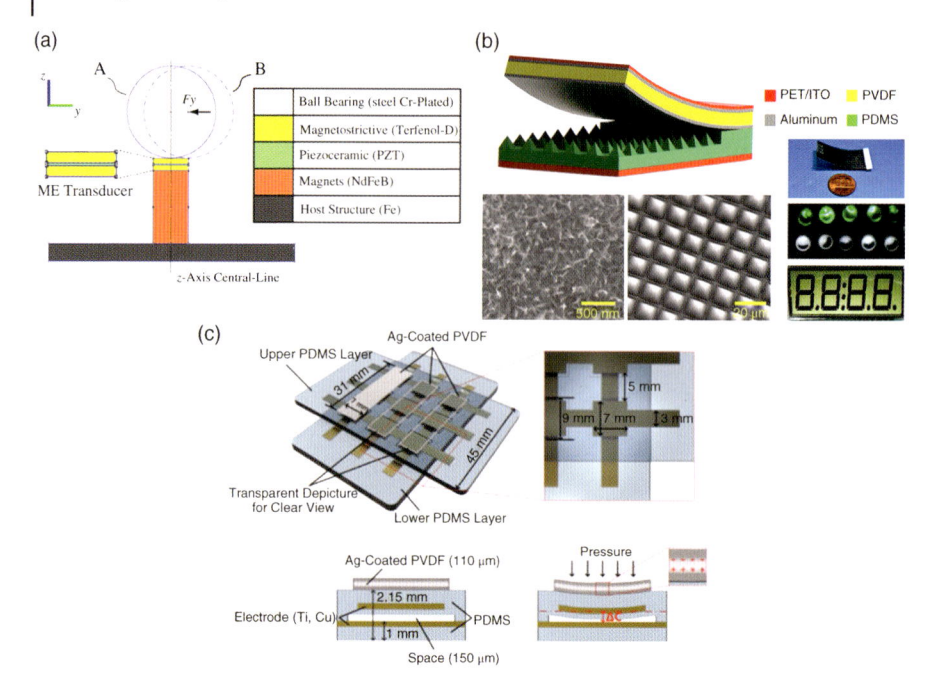

Figure 21.8 (a) Biaxial magnetoelectric generator. (Reproduced with permission from Ref. [151]. Copyright 2012, Elsevier). (b) Hybrid piezoelectric-triboelectric nanogenerator. (Reproduced with permission from Ref. [155]. Copyright 2013, American Chemical Society). (c) Hybrid piezoelectric-electrostatic generator. (Reproduced with permission from Ref. [117]. Copyright 2014, IOP Publishing.)

21.2.7 Wideband and Nonlinear Micropower Generators

Strong frequency selectivity of linear VEHs is a serious drawback leading to poor performance in many real-life application scenarios. It also entails manufacturability issues due to necessity to apply tight tolerances in order to achieve the predefined resonant frequency of the generator for a particular excitation case. Majority of realistic vibration sources are stochastic, multifrequency, or time varying. Therefore, extensive research efforts are currently directed toward investigation of methods for improving broadband response of VEH devices. Ambient vibration energy is usually distributed over wide spectrum and predominantly at (very) low frequencies (0–100 Hz), which poses major mechanical design challenges for VEHs [29,156]. Various strategies are pursued to enlarge the operational bandwidth [157,158]: resonance tuning [159], multifrequency arrays [160], frequency up-conversion [161], and use of multimodal [162] and nonlinear [163] oscillators.

Passive (intermittent) or active (continuous) tuning of operational frequency is implemented in VEHs by using mechanical (springs, screws), magnetic, or piezoelectric methods [157,158]. In general, tunable VEHs tend to underperform in the case of random or rapidly varying excitations and necessity for additional power source complicates the design and aggravates scalability [50].

Introduction of multifrequency arrays is a straightforward approach to widen the bandwidth, which is accompanied by the increase of device weight and volume resulting in reduced power density. Moreover, in this case, more complex power conditioning circuitry is required in order to avoid voltage cancelation.

Mechanical frequency up-conversion has emerged as one of the most promising solutions to increase conversion efficiency at very low input frequencies (<50 Hz). Its key benefit is decoupling of the excitation frequency and resonant frequency of the generator, which results in lower sensitivity of harvesting performance to variation of excitation frequency. This principle is typically implemented as a two-stage design, where nonresonant [164] or resonant [165] primary element responds to low-frequency excitation and impulsively triggers high-frequency oscillations of piezoelectric generator(s) through contact [165], noncontact [164], or hybrid [130] interaction (Figure 21.7b). As a general rule, resonant VEHs, which exploit resonance amplification phenomenon, are characterized by higher power output, achieved at the expense of operational bandwidth, while nonresonant devices provide larger bandwidth accompanied by lower output levels.

Beneficial application of nonlinearities in VEH devices is a very active research domain at the moment with many research challenges still ahead [29,50,156]. Nonlinearity can be intentionally introduced into VEH device through (i) piecewise-linear restoring forces implemented using stoppers [115] or (ii) nonlinear restoring forces (attained via magnetic attraction, repulsion, or using buckled structure), which is then tailored to achieve monostable [166,167], bistable [156,168,169], or even multistable [170,171] configurations. Above specific excitation amplitude, the bi/multistable VEH device switches between the two or more high-energy states in a highly nonlinear manner (aperiodic, chaotic, etc.), which under certain conditions can significantly increase both average power density and operational frequency range (particularly when bandwidths of the excitation and generator are comparable). Gains in harvesting performance in this case heavily depend on nature and intensity of input excitation as well as on initial conditions. Highly complex dynamic behavior of nonlinear VEHs compared to their linear counterparts remains the principal challenge, inhibiting performance enhancement for specific excitation scenarios. Furthermore, development of optimized power conditioning circuits for nonlinear VEHs is only in its early stages [50].

21.2.8 Concluding Remarks

After 15–20 years of extensive studies, mechanical energy harvesting still remains a topic of considerable interest, fueled by the need to implement self-charging power sources for various autonomous devices. Due to inherently simple configuration, straightforward downscaling, and attractive power density levels, piezoelectric generators based on bulk ceramics or thin/thick films have attracted the most extensive R&D efforts, which led to commercial launches of several macro/mesoscale devices. Nonetheless, electromagnetic generators currently represent the most attractive and commercially successful VEH technology at the macroscale. Meanwhile, development of electrostatic generators is also progressing since they can be conveniently integrated into self-powered microsystems using mature CMOS-compatible microfabrication techniques. Recent years have witnessed shift of research focus toward development of highly compliant energy harvesters exhibiting a favorable combination of piezoelectric, mechanical (flexibility/stretchability), optical (transparency), and other (e.g., biocompatibility) properties. Piezoelectric nanocomposites are considered to be one of the most promising candidates in this regard. Meanwhile, electroactive polymers appear to be particularly suited for water/air energy harvesting at a larger scale. In addition, the field of nanogenerators relying on synthesized piezoelectric nanostructures has been increasingly expanding over the last 10 years by promising breakthroughs in powering of wearable and implantable biomedical devices. Key issues for current nanogenerator implementations are related to insufficient power generation capability as well as biocompatibility and durability issues. Introduction of graphene films and utilization of flexoelectric effect are some of the emerging approaches for addressing these issues. Triboelectric nanogenerators, which were proposed more recently, appear as promising micropower sources in wearable applications. Major challenges for the nanogenerators include insufficient current outputs as well as various wearability aspects such as stretchability, washability, and breathability.

There remain many difficulties to be resolved in near future in order to develop useable and cost-effective self-charging micropower sources based on mechanical energy harvesters. Insufficient power density and bandwidth are the primary challenges for the current VEH technology since levels of average power harvested in real-life conditions is at least an order of magnitude lower than what is practically required for majority of applications. Undoubtedly, there has been significant progress in developing tunable and nonlinear generators with improved wideband response; however, improvements in performance are often accompanied by added design complications, implying increased manufacturing costs. Optimization of harvesting performance of nonlinear generators (including energy management circuitry) currently constitutes one of the core research challenges in the field. In addition, commercial adoption of VEH devices is also impeded by issues related to manufacturability, robustness, and reliability. Long-term durability studies are scarce, which is a point of concern that should be taken into account by research community. For example, aging effects in piezoelectric harvesters (deterioration of piezoelectric properties) or long-term charge stability in electret-based electrostatic generators have to be addressed more rigorously in order to develop alternative micropower sources that meet the expected life span requirements. Increasing number of publications in the field of mechanical energy harvesting indicates that there are many ongoing research activities focusing on the aforementioned challenges with new solutions being proposed continuously.

21.3 Thermal Energy Harvesters

21.3.1 Introduction to Thermoelectric Generators

Devices that directly convert temperature differences into electricity are referred to as thermoelectric (power) generators (TEGs). Waste thermal energy (heat) as one of the primary sources of clean, fuel-free, and low-cost energy have been used for power generation on a kilowatt-scale for many decades [172]. The end of 20th century brought significant advancements in micro/nanofabrication techniques with more efficient thermoelectric (TE) materials, which reignited research and development of small-scale TEGs that effectively operate at low temperature differences producing DC power levels on the order of milliwatts (with power densities reaching tens of $\mu W/cm^2$ for on-body applications and several mW/cm^2 for industrial heat sources [3]). Significant interest in small-scale TEGs is attributed to abundancy of low heat sources (e.g., process heat, solar heat, living body) as well as to important design-related advantages such as silent operation, absence of moving components, compactness, and ultrahigh reliability. TEGs generate electrical power by virtue of *Seebeck effect*, where a spatial temperature gradient dT/dx (T – temperature, x – coordinate) applied to a circuit at the junction of two dissimilar conductors induces electromotive force. Thermocouple (TC), as a representative TEG device (Figure 21.9a), is constructed of p-type and n-type semiconductor elements with a metallic interconnect, placing the two elements thermally in parallel and electrically in series. Open-circuit voltage V_{OC} is generated in the TEG device in proportion to the temperature difference between the hot and cold sides (ΔT), total number of TCs (n) connected electrically in series (i.e., size of thermopile) (Figure 21.9b), and Seebeck coefficient (α): $V_{OC} = n\alpha\Delta T$. Small-scale TEGs may be broadly subdivided into macroTEGs, which are based on well-established bulk material technology, and microTEGs, which are implemented using various microfabrication techniques. The former are available commercially[12–16] and are typically fabricated on ceramic substrates serving as an electrical insulation and a foundation to form a number of connected p–n junctions.

MicroTEGs attracted increased research efforts in the mid-1990s by promising a multitude of benefits over their bulk counterparts: (i) higher p–n junction density (enabling energy harvesting from lower thermal gradients), (ii) low-cost batch processing and smaller consumption of expensive TE materials (e.g., rare earth metal tellurium), and (iii) lightweight generators characterized by increased design flexibility (size/shape adaptable to specific applications). MicroTEGs are fabricated using mainly Bi_2Te_3-type compounds and applying such methods as physical vapor deposition (e.g., sputtering) [174],[12,17] chemical vapor deposition [177],[18] and electrochemical deposition [178], as well as using screen [174], spray [179], or dispenser [180] printing on foil. In contrast to macroTEGs, only few successful commercialization cases are known for microTEGs.[17–19]

---

12 www.perpetuapower.com (accessed September 22, 2017).
13 http://www.marlow.com/products/power-generators.html (accessed September 22, 2017).
14 www.customthermoelectric.com (accessed September 22, 2017).
15 https://www.tellurex.com/products/thermoelectric-power-generators/ (accessed September 22, 2017).
16 http://www.europeanthermodynamics.com/index.php/products/thermal/thermoelectrics.html (accessed September 22, 2017).
17 http://www.micropelt.com/en/energy-harvesting.html (accessed September 22, 2017).
18 http://www.lairdtech.com/products/category/646 (accessed September 22, 2017).
19 http://www.perpetuapower.com/technology.htm (accessed September 22, 2017).

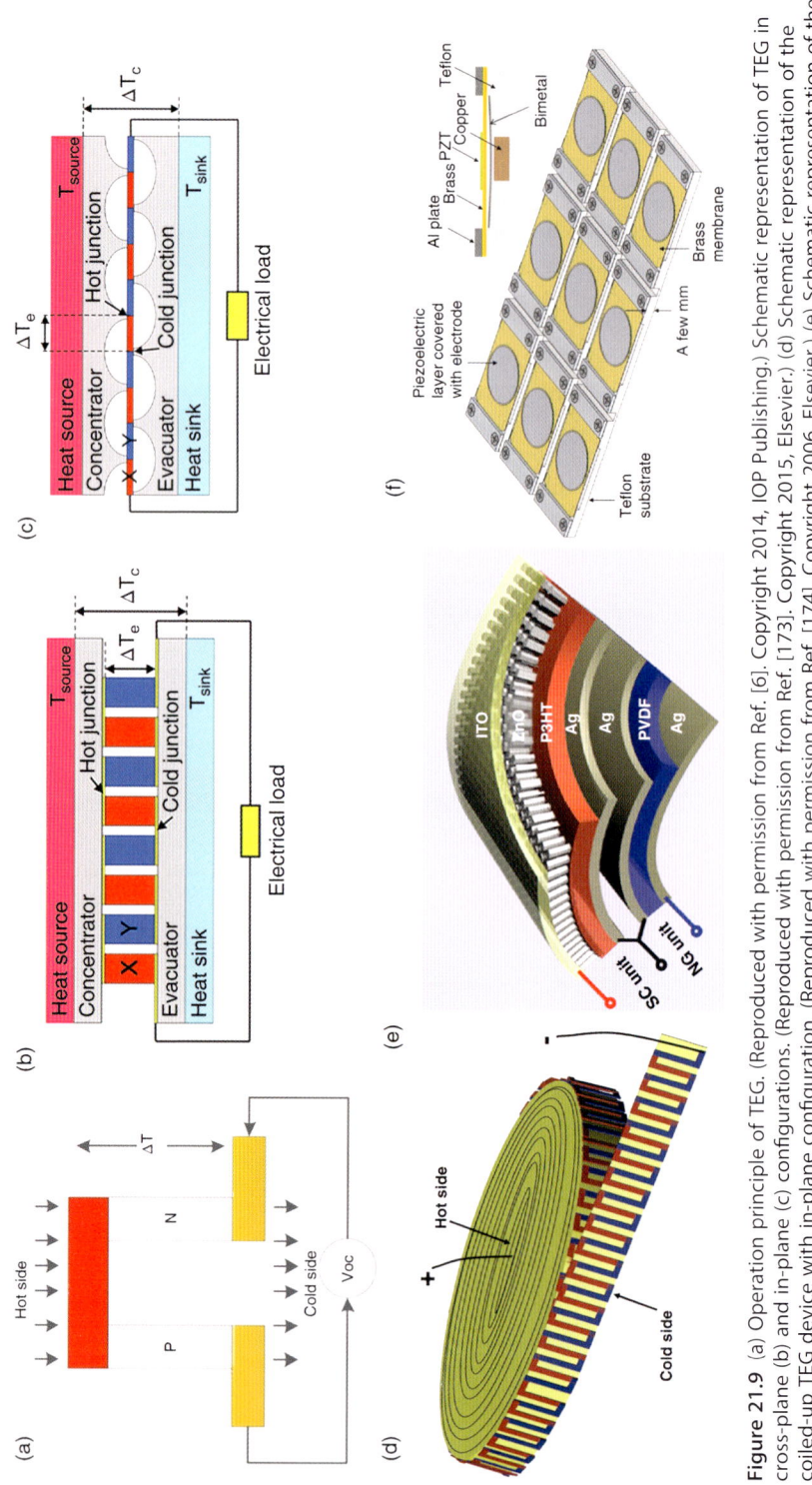

Figure 21.9 (a) Operation principle of TEG. (Reproduced with permission from Ref. [6]. Copyright 2014, IOP Publishing.) Schematic representation of TEG in cross-plane (b) and in-plane (c) configurations. (Reproduced with permission from Ref. [173]. Copyright 2015, Elsevier.) (d) Schematic representation of the coiled-up TEG device with in-plane configuration. (Reproduced with permission from Ref. [174]. Copyright 2006, Elsevier.) (e) Schematic representation of the flexible PVDF-based piezoelectric–pyroelectric–photovoltaic hybrid cell. (Reproduced with permission from Ref. [175]. Copyright 2013, American Chemical Society.) (f) 3 × 3 matrix of thermal-to-electrical energy converters based on bimetal and piezoelectric materials. (Reproduced with permission from Ref. [176]. Copyright 2014, Elsevier.)

There are two principal configurations that are used for implementation of TEGs [173]:

- Cross-plane design (Figure 21.9b), where the heat flows perpendicularly to the substrate. This vertical thermopile architecture is used in all macroTEGs and in many microTEGs, where a characteristic π-shaped pattern of thousands of TCs is formed (e.g., ~60 000 TCs within an area of 6 mm^2 [177]).
- In-plane design (Figure 21.9c), where the heat flows parallel to the substrate. This planar thermopile architecture delivers relatively higher voltage outputs due to longer TCs, allows achieving potentially higher number of TCs per unit area, and is amenable to low-cost printing techniques that are preferable for large-area TE harvesting applications [181]. An interesting representative of this category of TEGs is a circular device (Figure 21.9d) with coiled-up strip of thin-film substrate with over 5000 TCs occupying a volume of less than 0.1 cm^3 [174].

Some of the most successful applications of TEGs include radioisotope-powered energy sources for space applications [182] or cardiac pacemakers [183] as well as miniaturized bulk TEGs in wristwatches (e.g., Citizen, Seiko) [184]. The most promising novel applications include wireless sensors (e.g., in industrial setting or buildings [6]), autonomous biomedical devices, and automotive applications (e.g., using high-temperature TEGs to harvest waste heat in exhaust system of a car [185]).

21.3.2 Thermoelectric Materials and Efficiency

Performance of TEG device is essentially determined by the efficiency of thermoelectric material, which is quantified by dimensionless temperature-dependent figure of merit $zT = (\alpha^2 \sigma / \lambda) T$ (λ – thermal conductivity, σ – electrical conductivity). In general, the family of TE materials is vast and diverse, comprising semiconductors, semimetals, ceramics, and polymers, encompassing mono- and polycrystals as well as nanocomposites. The most widely used TE materials since 1950s (Figure 21.10) [186] have been alloys of chalcogenides based on heavy elements (e.g., Bi_2Te_3 or PbTe), which exhibit maximum zT values near 1 (e.g., typical conversion efficiency of Bi_2Te_3 near room temperature is in the range of ~4–6% for ΔT in the range of ~100–200 °C) [187,188]. It became a true challenge in the field of thermoelectrics to find a material with high zT (>1) since α, σ, and λ are interdependent through the electronic structure: increase in α leads to decrease in σ, while increase in σ is accompanied by increase in λ [189]. Since 1990s when concept of nanostructuring in TE materials was introduced, thermoelectrics community has been intensively pursuing a target of $zT \geq 3$, which is considered to be necessary in order to make TEGs a more competitive technology. Unfortunately, despite two decades of active research, state-of-the-art thermoelectric materials still have a very low efficiency of thermal-to-electrical energy conversion [190,191]. Historically, more appreciable improvement of TE efficiency (up to $zT \sim 1.3–1.7$ at high temperatures) was achieved predominantly through pronounced suppression of lattice thermal conductivity, which was accomplished by nanostructuring or compositing with various microstructures

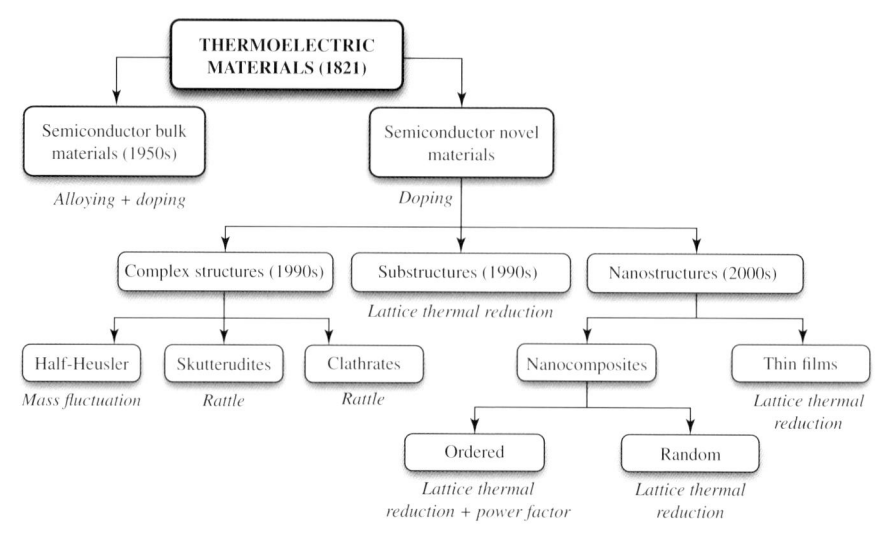

Figure 21.10 Historical progression of efforts dedicated to development of thermoelectric materials and improvement of zT. (Redrawn with permission from Ref. [186]. Copyright 2013, Elsevier.)

and additives. Introduction of low-dimensional structures in thin films (e.g., quantum wells, superlattices, quantum wires, or dots) proved to be effective in reducing lattice thermal conductivity as well as enhancing power factor ($\alpha^2\sigma$). However, growth of superlattices by vacuum-based thin film deposition techniques is neither particularly cost-effective nor suitable for large-scale production. Therefore, a significant amount of research efforts are currently directed to development of nanostructured bulk materials (nanocomposites), which can be synthesized in large amounts by means of solid-state methods [186,188,191,192]. Highly versatile colloidal synthesis techniques appear to be particularly suitable for production of novel bulk nanocomposites possessing well-defined composition, interfaces, and distribution of phases at nanometer scale, thereby promising a considerably enhanced thermoelectric performance [193]. The most recent efforts in the field of TE materials have been focused on "panoscopic" approach [188], that is, simultaneous and, ideally, synergistic integration of various advanced zT-enhancing approaches (e.g., nanostructuring, all-scale hierarchical architecturing, matrix/precipitate band alignment, intramatrix electronic structure engineering), which currently led to development of materials with zT in excess of 2.5 with predicted TEG device efficiencies of ~8–15% for ΔT in the range of ~100–200 °C. The grand challenge in raising zT till 3 or higher is associated not only with reduction of thermal conductivity even further but also with enhancement of the power factor [188,190,194].

Low abundance or toxicity of heavy elements (e.g., Te, Pb) employed in common bulk TE materials encourage thermoelectrics community to actively seek for alternative high-performing TE materials [195], for example, skutterudites and clathrates [196], half-Heusler compounds [189], silicides and pnictides [195], oxides [194], or organic semiconductors [197]. With tellurium being one of the rarest elements on earth, there is an urgent need to find a sustainable high-zT material that can be cost-effectively produced in large quantities. Unfortunately, at the moment various novel TE materials have limited commercial success due to issues related to efficiency, reliability, processability, and cost. It could be noted that in the last 15 years, polymer-based materials have undergone about three orders of magnitude increase in TE efficiency [198] with reported zT values reaching ~0.4 [199]. The interest in these materials (e.g., PEDOT:PSS, polythiophene (PTH), polyacetylene (PA), polypyrrole (PPy), polyaniline (PANI)) is understandable as they offer the benefits of low thermal conductivity, material abundance, and a prospect of relatively simple solution-process-based fabrication of environmentally benign, large-area, flexible, and lightweight TEG devices. On the other hand, they exhibit low electrical conductivity, Seebeck coefficient, and stability, which impedes their application in TE devices. Continuous efforts are made to enhance efficiency of organic TE materials by means of nanostructuring, hybridization with semiconducting stabilizers, carbon nanotubes, graphene, inorganic TE nanoparticles, and so on [197,198,200].

Apart from development of more efficient TE materials, additional improvements of TEGs performance can be achieved through optimized architectural design by integrating various novel materials such as carbon nanotubes for enhanced power dissipation properties and graphene for superior electrical conductivity at junctions [201] or exploiting unique properties of metamaterials [202]. Some of the remaining critical challenges in terms of commercialization of TEG devices are related to sublimation, oxidation, degradation of thermal and electrical interface, thermal expansion mismatch, ability to withstand high temperatures [203], as well as ensuring robustness of new high-performance TE materials through in-depth assessment of their mechanical properties [188].

21.3.3 Other Thermal-to-Electrical Energy Conversion Techniques

In addition to TEGs that rely on spatial temperature gradients (*Seebeck effect*), a less investigated alternative is pyroelectric energy generation (PEG), which uses pyroelectric materials to convert temperature fluctuations (temporal temperature gradients dT/dt) into electrical energy. Pyroelectric materials are dielectrics that undergo spontaneous electrical polarization as a consequence of temperature changes. These materials are of interest for thermal energy harvesting since (i) they do not need heat sinks to maintain temperature gradient, (ii) they could potentially operate with high thermodynamic efficiency, and (iii) many pyroelectric materials are stable up to very high temperatures [204–206]. At the moment, there are no commercialized PEG devices yet. Main challenges in this domain include (i) developing methods to generate temperature oscillations since useable thermal transients are rare in nature (e.g., thermal gradients can be transformed into time-varying temperature by means of cyclic pumping), (ii) increasing operational frequency (e.g., by generation of mechanical oscillations from a temperature gradient) since PEGs are generally restricted to extremely low-frequency operation (typically <1 Hz) due to inherently large heat capacity of materials, which severely limits the amount of power that can be harvested. There is a wide selection of pyroelectric materials that are used or could potentially be used for development of PEG devices [205,206] including ferroelectric ceramics (e.g., PZT, PMN-PT, BNT-BT, Mn:BNT-BT, LiNbO$_3$, KNN compositions, KNbO$_3$, TGS), nonferroelectric crystalline materials, which have relatively low pyroelectric coefficients compared to the ferroelectric counterparts (e.g., AlN, ZnO, GaN, CdS), PVDF-based polymers as well as various emerging nanocomposite materials (e.g., KNbO$_3$ nanowire–PDMS composite [207]). Application of micro/nanostructured pyroelectric materials opens new avenues in the development of high-performance PEGs. It was demonstrated that thin films allow increasing rate of temperature change, while tailoring of nanowire dimensions can enable "giant" pyroelectric response [205,206]. Since all pyroelectrics are piezoelectric, it is natural that attempts were made to combine these two properties into hybrid thermal/mechanical EH devices. One of key challenges in this case is to ensure that both harvesting mechanisms are operating in-phase (i.e., changes in polarization are constructive) leading to enhanced power output [206]. As an example, such hybridization has been successfully demonstrated in stretchable piezoelectric–pyroelectric nanogenerator based on micropatterned PVDF–TrFE and PDMS–CNTs composite with graphene electrodes [208] as well as in flexible PVDF-based piezoelectric–pyroelectric–photovoltaic hybrid cell (Figure 21.9e) [175].

In addition, other alternative technologies for converting waste heat into electricity are available. Thermophotovoltaics, enabling conversion of heat to electricity via photons, has been studied for decades and still remains an active research area gaining impetus from micro/nanoscale engineering, which promises more substantial efficiency improvements [209,210]. Several more exotic thermal-to-electrical energy conversion approaches have been considered recently such as magnetocaloric (thermomagnetic) generation [211], application of ferromagnetic materials [212], thermoacoustic–piezoelectric [213], or thermomechanic–piezoelectric (Figure 21.9f) [176] energy harvesting.

21.4 Radiation Harvesters

21.4.1 Light Energy Harvesters

Light energy harvesters (LEHs) use solar panels to directly convert incident natural or artificial light into electricity via *photovoltaic effect*, whereby photons generated from light are absorbed (producing proton holes), stimulating current orthogonal to the flow of proton holes (Figure 21.11a). It is the most prevalent method for harvesting energy outdoors and is now a well-established technology, which has been evolving since 1970s and still keeps improving in terms of efficiency, flexibility, and durability. Key benefits of LEHs are relatively easy integration, modularity, lack of moving parts, and noise. However, LEHs require large-area panels of proper orientation in order to collect sufficient amounts of solar energy. Variable availability of both sunlight and indoor illumination is an obvious issue to be taken into account. The amount of harvested power depends on the intensity, spectral composition, and incident angle of the light as well as on

Figure 21.11 (a) Schematic representation of operation principle of light energy harvester. (b) Wireless sensor module powered by RF energy harvester. (Reproduced with permission from Ref. [5]. Copyright 2014, Elsevier.) (c) Schematic representation of hybrid solar/vibration energy harvester based on DSSC and ZnO nanowires. (Reproduced with permission from Ref. [214]. Copyright 2011, Wiley-VCH Verlag GmbH & Co.) (d) R2R gravure printed rectenna on PET foil with schemes for printed capacitor, diode, and bottom Ag electrode. (Reproduced with permission from Ref. [215]. Copyright 2012, IOP Publishing)

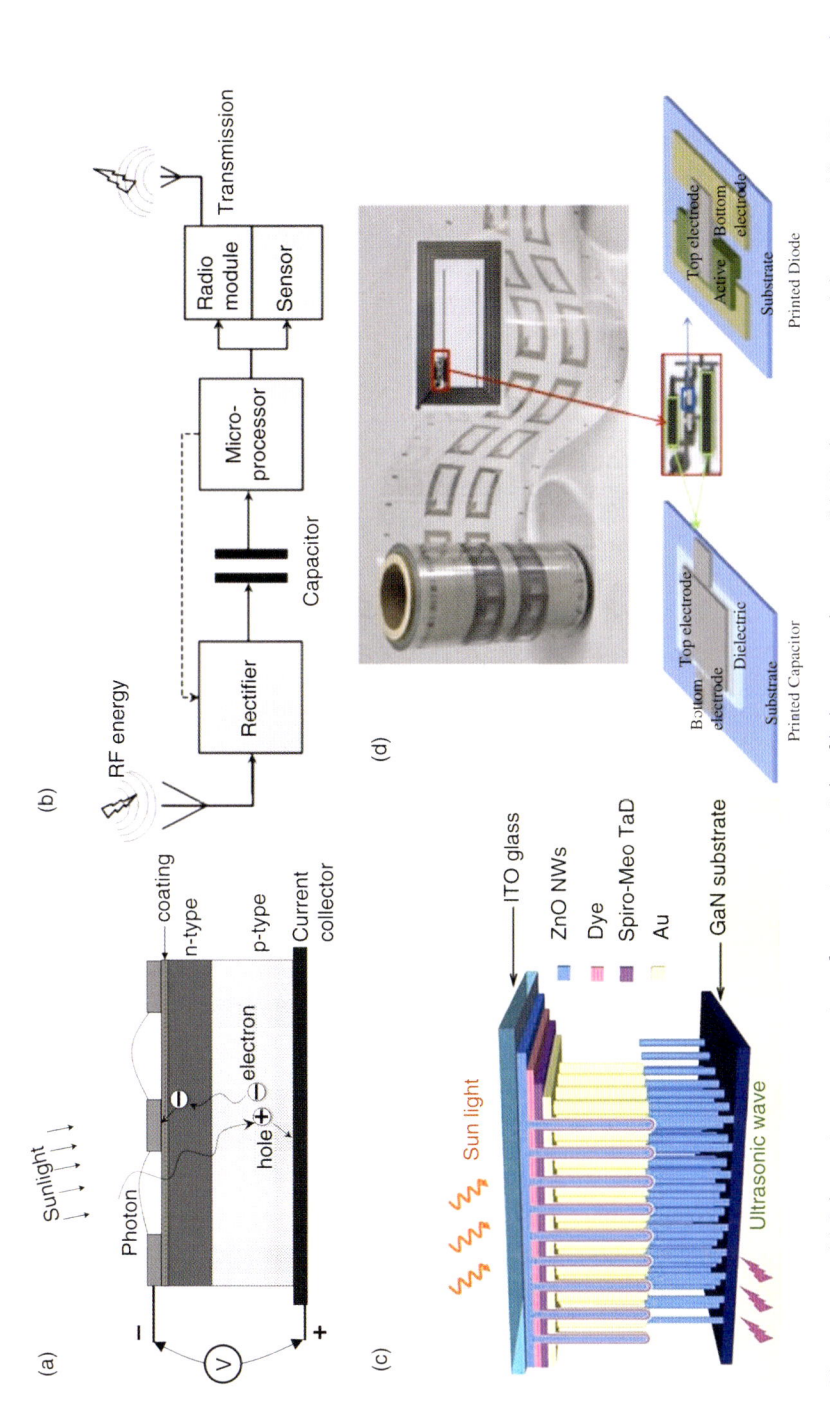

the size, sensitivity, temperature, and type of solar (photovoltaic (PV)) cells. Typical single-junction silicon PV cells generate about 0.5–0.6 V (open-circuit) and are usually connected in series to deliver higher voltages. LEHs are characterized by high harvested power density of up to $\sim 10\,mW/cm^2$ in daytime with power conversion efficiency (PCE) for single-junction PV cells ranging in $\sim$5–30% depending on the material used. PCE for the most common commercially available single-junction PV cells is $\sim$15%, while the most sophisticated multijunction cells achieve efficiencies in $\sim$32–46% range.[20] Due to significantly lower illumination levels, the harvested power densities indoors are considerably smaller ($\sim 10\,\mu W/cm^2$); therefore, thermal or vibration energy harvesters could be more suitable in buildings depending on specific application [6].

PV cells are typically classified in terms of generations depending on the underlying technology and materials as summarized below [216–219].[20]

First generation (1G) is based on thick crystalline films of V or II–VI group semiconductors (mainly Si). High-end Si-based PV cells achieve PCE of over 25%, while the theoretical limit for the single-junction PV cells is about 30%. Currently, 1G PV cells constitute the predominant part of commercial production with $\sim$85–90% market share. Major issues of 1G PV technology are high material cost, stringent purity requirements for Si crystals, high processing temperatures, and negative environmental impact.

Second generation (2G) is based on vacuum deposition of thin films on inexpensive substrates such as glass. Typical materials in 2G cells include poly/microcrystalline and amorphous Si, CIGS, and CdTe. The issue with 2G technology is that enlarged PV array areas are required to compensate for the lower PCE ($\sim$20%). In general, material/fabrication costs for 2G cells are still overly high, which limits their practical application ($\sim$10–15% market share).

Third generation (3G) is based on novel concepts, materials, and structures including hot carrier cells [220], nanomaterials (e.g., thin films, nanotubes/nanorods/nanowires/nanofibers, quantum dots, nanocrystals [193,221–223]), tandem/multijunction architectures [219], organic–inorganic heterostructures [222,224], 3D cell structures [221], optical metamaterials [202] as well as graphene-derived materials for transparent conducting electrodes [225]. Emerging 3G PV technologies that are based on low-temperature and highly scalable solution processing methods may be subdivided into three categories as follows.

Hybrid organic–inorganic perovskite cells (e.g., $CH_3NH_3PbI_3$) constitute a disruptive PV technology, which have undergone remarkable progress recently with doubling of PCE in 2010–1015 and as of 2017 reaches $\sim$22%. Essential benefit of this PV technology is its amenability to roll-to-roll (R2R) processing, which has been recently demonstrated in fabricating PV modules, thereby promising cost-effective large-scale production in the future [226].

Dye-sensitized solar cells (DSSCs) and *quantum dot-sensitized solar cells (QDSSCs)* [218,227] operate on the principle that is nearly analogous to photosynthesis. This PV technology is attractive due to added functionalities such as low-weight, flexibility, and semitransparency. The current focus in DSSCs

---

20 https://www.nrel.gov/pv/assets/images/efficiency-chart.png (accessed 21 September, 2017).

development is mainly on the modification of dyes, electrolyte, and the semiconductor oxides for a better performance.

Organic (or polymer) solar cells (OSCs) featuring two subgroups: (i) state-of-the-art architectures based on bulk heterojunctions (BHJs) consisting of blends of semiconducting polymers (e.g., P3HT, PEDOT:PSS, MEH:PPV, PCPDTBT) and fullerene derivatives (e.g., $PC_{61}BM$, $PC_{71}BM$); (ii) hybrid organic/inorganic solar cells, which combine flexibility, large-area processability, and low cost of polymeric thin films with the stability of novel inorganic nanostructures with the aim to improve charge transport and optical coupling. The principle difference between the two subgroups is that in the case of the latter the acceptors are inorganic. Employed inorganic nanostructures include semiconducting nanoparticles (e.g., CdS, CdSe, ZnO, PbS, PbSe), CNTs, graphene, metal nanoparticles (e.g., Au, Ag), metal oxides (e.g., TiO_2, ZnO, CuO_2, Nb_2O_5), and carbon–metal oxide nanohybrids [217,222,224,228]. More recently, notable improvements of the optical coupling into active layers were achieved through incorporation of plasmonic effects (e.g., using metal nanoparticles) [223]). In terms of environmental compatibility, it should be noted that Pb-containing hybrid OSCs and perovskite cells share the potential problem of Pb contamination, while OSCs based on BHJs have a potential advantage since they do not contain significant quantities of heavy metal ions [226].

Despite significant progress in raising efficiency of OSCs via incorporation of inorganic nanostructures, polymer-based LEHs are currently less efficient (PCE up to ∼6–11%) in comparison to DSSCs (PCE up to ∼11–12%). Key advantage of OSCs is easy processability, which could lead to inexpensive fabrication of LEHs by printing in a roll-to-roll fashion at high speeds. The main issue to be addressed is the poor chemical stability of OSCs when exposed to long-term sunlight under ambient conditions. In contrast, a lifetime of over 20 years has been demonstrated for DSSCs with slight degradation in dye materials. The efficiency reduction of both DSSCs and OSCs when scaling up to larger modules is still a significant problem. In the case of OSCs, this is mainly attributed to high sheet resistance of the transparent electrode at larger areas. In the case of DSSCs, light scattering layer approach constitutes an effective method to enhance efficiency, although this is technologically challenging and expensive. In general, despite noticeable progress of 3G PV technologies, significant performance improvements are still required for these technologies (particularly for OSCs) to become competitive with the previous PV generations in terms of cost per watt. Further advances in synthesis and incorporation of tailored nanostructures could lead to further improvements in the efficiency of 3G PV harvesting systems [193].

LEHs are implemented by means of two distinct approaches: (i) using hybrid assembly relying on off-the-shelf components of different sizes and types and (ii) using monolithic integration, where PV cells are incorporated above an IC chip. The latter approach offers higher efficiencies due to diminished wiring. In addition, it could be noted that LEHs combined with other energy harvesting mechanisms have also been proposed, for example, solar/RF [229], solar/thermal [230], solar/vibration (Figure 21.11c) [214], or solar/thermal/vibration [231]. Furthermore, application of resonant nanoantennas (nantennas) has recently gained attention as an alternative approach to implement solar energy harvesting [232]. However, the technology is far from fruition due to major bottleneck related to inefficient rectification of electric fields oscillating at frequencies that correspond to high-infrared and visible light.

21.4.2 RF Energy Harvesters

Present environment is abundant with radio frequency (RF) energy owing to the continuously increasing proliferation of wireless communication and broadcasting infrastructure such as radio, television, satellite, cellular networks, WiFi stations, and so on. Electronic devices that are intended to use these technologies receive only a very small part of the energy emitted by the transmitters and a significant amount of this energy is dissipated as heat or wasted. Therefore, efforts are undertaken to recycle ambient RF energy and use it for powering of stand-alone electronics (typically sensors) without compromising the signal-to-noise ratio of nearby devices employing RF signals for operation. RF (or electromagnetic wave) energy harvesting is a truly attractive concept as it requires no motion, pressure, or heat flows to generate electrical energy. RF is an electromagnetic (EM) wave that is generated when alternating voltage or current is applied to an antenna. As a consequence, the motion of free electrons in the metal of the antenna induces a magnetic field, which cause RF waves to be radiated by the antenna with a frequency the same as the applied voltage. Though the wavelength of RF is the highest among the EM waves and the corresponding energy content is the lowest ($E = hc/\lambda$), still ambient RF was demonstrated to be capable of powering wireless sensors in some cases [5,233]. The main challenge in RF energy harvesting is extremely low average power density levels, which vary considerably ($\sim 0.1\,\text{nW/cm}^2$–$1\,\mu\text{W/cm}^2$) depending on the type of RF energy source (e.g., GSM or WiFi transmitter) and its distance to the powered sensor (e.g., $\sim 0.1\,\mu\text{W/cm}^2$ at 100 m distance from GSM900 base station). Power received by RF energy harvester decreases with distance (R) by a factor of $1/R^2$ and it is much more challenging to utilize ambient RF energy for harvesting when the energy source is far away (more than 6 km) [3,233]. Instead of harvesting ambient RF energy already available in the target environment due to telecommunication transmissions (*passive* RF energy harvesting), it is also possible to use *active* RF energy harvesting concept (generally known as wireless power transfer), when a dedicated RF energy source[21,22] is intentionally placed in the vicinity of a sensor to be powered.

The principal element in RF energy harvester is a rectenna (Figure 21.11b), which comprises a receiving antenna and a rectifier circuit (typically diode-based). Impedance matching circuit is typically applied between these subcomponents in order to realize an LC circuit with fairly high quality factor. A receiving antenna intercepts a portion of the ambient RF radiation, which induces AC voltage and current in the antenna. The wavelength of the incident EM wave has to be a multiple of the antenna characteristic length in order to induce a resonant electrical current. Subsequently, a rectifier converts AC signal to DC and boosts

---

21 www.powercastco.com (accessed September 22, 2017).
22 www.witricity.com (accessed September 22, 2017).

the voltage to a level needed for the application (e.g., operation of a wireless sensor). The harvested energy is typically accumulated in a capacitor that is connected to a microprocessor used to control (dis)charging of the capacitor, output voltage, and radio transmission from the powered wireless sensor. As with many other current technologies, fabrication of antennas via R2R printing (Figure 21.11d) has become a preferred option offering the advantages of low profile and weight, easy manufacturing, and enhanced integration capabilities with different circuitries [215,234].

Two principal rectenna array architectures could be distinguished [235]:

- DC combining configuration, where each rectenna is equipped with a separate rectifier and, subsequently, DC power from each rectifier is combined in parallel, series, or hybrid manner. This architecture allows receiving radio waves from a broad angle range; however, it possesses lower RF–DC conversion efficiency due to lower RF power input to each rectifier. This configuration is suitable when dealing with large rectenna arrays.
- RF combining configuration, where multiple antenna elements are connected to an RF combiner followed by a single rectifier. In a point-to-point RF system (narrow beam), this configuration offers the most efficient power transfer scheme.

Rectifier circuits are usually designed for operation under specific conditions with respect to input power levels and loading conditions. The conversion efficiency of these circuits is about 10–30% for low RF input power levels (from −30 to −20 dBm) and reaches 60–85% for high input power levels (∼20 dBm). In general, efficiency of RF–DC conversion diminishes nonlinearly with decreasing input power levels and increasing EM wave frequencies [233,236]. Typically, a single rectenna is not sufficient for ensuring reliable powering of a device. Voltage and current delivered by the RF energy harvester can be magnified through larger effective collection area by connecting identical rectenna elements in series or in parallel, respectively. Wireless signals, such as radio/TV or GSM, are spread over multiple frequencies in urban areas and the power level of each signal could be as low as −40 dBm. Therefore, multiband or wideband RF energy harvesters have been successfully demonstrated for addressing these challenging cases [233,236].

In conclusion, it could be noted that due to very low power densities available in the surroundings harnessing ambient RF energy as a power source is currently feasible only for some niche applications operating with extremely low duty cycles (typically below ∼6–8%). Further improvements in this harvesting technology are required in terms of transmission (e.g., via beam steering) as well as receiving and conversion efficiency (through advanced rectenna designs).

21.5 Summary and Perspectives

Field of small-scale energy harvesting emerged over the last couple of decades as an alternative micropower generation technology intended to replace or augment batteries in various autonomous electronic systems with a foremost focus on ultralow power implementations. A typical example of such system is a wireless sensor network with interconnected nodes, which require some sort of self-charging power supply in order to achieve a prolonged maintenance-free service life, thereby ensuring cost-effectiveness of the network in a particular application scenario such as industrial or building automation. It is commonly acknowledged that compact, inexpensive, and highly reliable energy harvesting modules seamlessly embedded in an autonomous smart system (e.g., WSN or WBAN) constitute a prerequisite for implementation of long-term vision of intelligent world of the future, which is defined by a number of concepts such as "Internet of Things," "Ambient Intelligence," or "Smart Environments".

Energy harvesting mechanisms that have been extensively explored to date mostly rely on derivation of electrical energy from mechanical, thermal, and photonic energy sources. Field of mechanical energy harvesting has undergone probably the most appreciable progress over the last 15 years, particularly in the case of inertial-type vibration-based generators with several commercial launches of electromagnetic and piezoelectric devices in recent years. Despite considerable research efforts, the latter EH technology is still struggling to deliver performance level that could be adequate for a wide variety of practical applications. This is mainly attributed to relatively poor efficiency of common piezoelectric materials. Nevertheless, wider industrial adoption may still be achieved if compact and robust piezoelectric generators with enhanced operational bandwidth and higher power density would be implemented by applying low-cost processing techniques. In terms of materials, it appears very unlikely that near future will bring low-cost piezoelectric materials with markedly improved piezoelectric properties. However, it is anticipated that further research efforts in nanostructured materials will lead to better performing and less costly intrinsically compliant piezoelectric polymer composites with favorable blend of piezoelectric and mechanical properties. In terms of structural design, additional improvements in power density could be achieved by introducing novel physical and geometrical configurations optimized for specific applications as well as by enhancing multifunctionality of EH devices or rationally implementing hybrid architectures based on several electromechanical transduction mechanisms and/or different smart materials. These trends also apply to electrostatic energy harvesters, piezoelectric or triboelectric nanogenerators, and generators based on electroactive polymers.

Thermoelectric generators are highly reliable, compact, and noise-free micropower sources that offer the benefit of direct conversion of waste heat into useable electrical energy. However, availability of TEGs on the market is rather limited due to high cost and low efficiency of common TE materials. Significant research efforts are currently dedicated to improving performance of both inorganic and organic TE materials through multiple nanoengineering approaches. There is an obvious focus on investigation of sustainable TE materials since the traditional ones contain toxic and/or scarce constituents. In addition, issues of processability and reliability of novel TE materials (e.g., polymer-based compositions) need to be seriously addressed in order to enable cost-effective fabrication of robust and efficient TEG devices on a large scale. In addition to TEG technology, thermal-to-electrical energy converters based on pyroelectric materials gradually emerge as a promising technology to harvest thermal fluctuations. Though commercialization stage has not been reached yet, more pronounced progress for these generators is expected due to advances in nanostructured pyroelectric materials. As with mechanical energy harvesters, hybridization also opens up new avenues in the development of better performing devices for harnessing thermal energy, including pyroelectric–piezoelectric, thermoacoustic–piezoelectric, or thermomechanical–piezoelectric generators.

Photovoltaics is undoubtedly the most mature energy harvesting technology, which has witnessed several generations in its evolution but still is about four times more expensive with respect to traditional energy sources based on fossil fuels. Despite considerable efforts of fabricating PV cells using various novel materials and techniques, the market is still dominated by the traditional (1G) silicon wafer-based solar cells, followed by the 2G PV technology that uses semiconductor thin films deposited on low-cost substrates. Third-generation photovoltaics is an emerging technology that currently undergoes rapid progress due to advances in nanotechnologies, optical metamaterials, photonics, plasmonics, and semiconducting polymers. Low-temperature and highly scalable solution processing methods applied in implementation of 3G PV cells offer the prospect of truly cost-competitive technology for light energy harvesting. In particular, high-efficiency hybrid perovskite cells produced by R2R-friendly processes is a very rapidly advancing technology that has excellent commercial potential, although replacement of Pb with less toxic element may be critical to mass deployment. Meanwhile, other 3G PV technologies such as dye-sensitized and polymer-based solar cells are far from outperforming the dominant PV technologies in terms of cost and efficiency. One of the major challenges for these novel PV technologies is efficiency reduction when scaling up to larger modules as well as long-term reliability concerns for polymer-based cells when exposed to sunlight under atmospheric conditions. In terms of system integration, major complications in application of light energy harvesters are associated with intermittent availability and intensity of sunlight or indoor illumination, notably inferior generated power levels when harnessing artificial light as well as the necessity to use large-area panels of proper orientation to collect sufficient amounts of light. Nevertheless, depending on specific deployment location and target application, light energy harvesters may still be the most cost-effective solution since the achievable maximum power density levels are the highest among the available energy harvesting methods. In contrast to PV cells, RF energy harvesters are characterized by the lowest power density levels, which are sufficient only in some very specific cases such as ultralow-power sensors operating with extremely low duty cycles. Nevertheless, the concept of RF energy harvesting is very attractive since our urban environment is abundant with electromagnetic waves owing to omnipresence of public telecom services such as GSM and WiFi. Unfortunately, its implementation is challenging due to strong reduction in input power level with increasing distance to the RF transmitter and nonlinear relationship between the efficiency of RF–DC conversion and input power level or frequency of electromagnetic wave. At the moment, ambient RF energy harvesting is not feasible for powering miniature sensor modules with several cm^2 in size and power consumption of about $100\,\mu W$ (antennas of at least hundreds of cm² would be required). Moreover, there are no commercial generators intended for harvesting ambient RF energy since such devices may impede the primary function of electromagnetic waves. Meanwhile, commercial systems for wireless power transfer (*active* RF energy harvesting) are available, but in this case transmitted power levels are restricted by international regulations in order to avoid interference and health-related issues.

A micropower generator never operates in isolation; therefore, the performance of a complete EH device is heavily dependent on efficiency of energy management circuit, which performs conditioning of the converted signal in order to provide electrical energy in a form suitable for downstream system components such as sensors and wireless transceivers. Interoperability between system components is a crucial factor since there typically exists a strong two-way coupling between the micropower generator and the accompanying energy management circuit. Therefore, complete EH device has to be designed in a holistic manner as an integrated system so as to maximize energy conversion efficiency and minimize losses. Furthermore, EH device must be unobtrusive when deployed in a specific environment, which requires to take into consideration the ambient energy source itself. In some cases, energy harvesting process may have an unfavorable or detrimental impact on the energy source or the surrounding environment, for example, providing unpleasant cooling sensation when using thermoelectric generator for wearable electronics, disturbing the operation of nearby receivers when using ambient RF energy harvesters, or modifying the intended mechanical response of a host structure with mounted motion-based generator. In general, characteristics of a particular environment in terms of energy availability have a major impact when selecting the most suitable EH method for a specific application. It is very common that waste energy from ambient source is not continuously available or exhibits pronounced daily and/or seasonal variations in intensity or frequency content. Thus, prior knowledge about duty cycle of availability of ambient energy in a targeted location is generally required when designing an autonomous smart system relying on energy harvesting.

A number of various issues still need to be resolved for energy harvesters to become a viable alternative to the competing battery technology in terms of size, cost, and reliability. As an intermediate step toward large-scale adoption of EH devices, hybrid solutions integrating generators with batteries/supercapacitors prove to be effective in some cases. Current designs of energy harvesters tend to be highly tailored to specific applications, which is economically feasible only in the case of high-volume implementations (e.g., self-powered tire-pressure monitoring system). It should be noted that the overviewed EH techniques have already reached a certain level of maturity by now, which means that the resulting technology push is intensifying and should ultimately lead to an accelerated uptake of EH solutions within diverse application areas. As an emerging technological trend, energy harvesting currently faces a noticeable conservatism in terms of industrial adoption, which is associated with insufficient technological readiness of some EH solutions and the concomitant doubts regarding long-term capability of micropower generators to operate without failures. Furthermore, it is evident that for EH devices to become truly ubiquitous, they must become as versatile as batteries. This is one of the key challenges in the field, which calls for highly adaptive generator architectures suitable for a wide range of applications. Design hybridization emerges as a promising approach, which allows implementing multisource micropower generators where several EH mechanisms are constructively combined for enhanced power output under varying ambient conditions. On the other hand, introduction of advanced ultralow-power circuits and further miniaturization of sensors to micro/nanoscales leads to continuous decrease of power requirements in wireless devices, which is a favorable trend for the field of energy harvesting since the generators gradually become a viable energy source in an increasing number of applications.

It is obvious that small-scale energy harvesting as enabling technology stimulated a surge of scientific activities in the 21st century. As a consequence, this produced a significant impetus for exploration of different classes of materials, including development of novel nanotechnology-assisted synthesis routes with a recent focus on highly scalable fabrication methods such as R2R processing of large-area, conformable, lightweight, transparent, and eco-friendly composite material structures. Energy harvesting remains a topic of intense interest and will continue to evolve further in terms of generator designs, materials, and processing techniques. Concerted efforts of academia and industry will certainly lead to introduction of innovative device configurations with enhanced adaptability, synthesis of novel nanostructured materials with optimally tailored physical properties, advancement of solution-process-based fabrication methods in order to upscale the processes to mass-market production, as well as design of more efficient energy management circuitries able to fully exploit the potential of micropower generators, thereby promising many years of autonomous operation for future smart systems including material-integrated intelligent systems. When considering the latter, the aspect of reliability of material-integrated EH systems is pivotal due to increased risk of premature failure in a long run as a consequence of degrading performance of EH components (e.g., embedded smart materials) under harsh ambient conditions or excessive input loads, which are very likely to occur during the expected long operation of such systems (e.g., sensorial materials intended for structural health monitoring). Lifetime considerations become even more significant in the face of intensifying R&D activities in the domain of novel polymer-based (nano)composites targeted for mechanical, thermal, or light energy harvesting. It is acknowledged that long-term reliability of emerging polymeric PV and TE materials is a serious issue, which requires dedicated research efforts in order to ensure their robustness and functionality throughout the projected service life of the EH system. Material-integrated mechanical energy harvesters face similar reliability challenges, which are further aggravated by processability issues. For example, fragility of piezoceramics highly complicates its incorporation within material-integrated intelligent systems, which should encompass a multitude of autonomous interconnected sensors and, consequently, energy harvesting modules. Furthermore, most piezoceramics-based harvesters operate in transverse mode and can sustain only very low strain levels, which makes them susceptible to excessive mechanical loads that may be sporadically induced in many actual operation conditions. Therefore, one of the considerable challenges is to develop highly efficient compliant energy-generating smart materials that would constitute a more viable alternative in terms of manufacturability and durability of material-integrated intelligent systems. It should also be noted that lifetime issues usually receive much less attention from the scientific community. Therefore, there is a pressing need for systematic evaluation of long-term performance of flexible and/or stretchable mechanical energy harvesters based on novel piezoelectric nanocomposites, electroactive and voided charged polymers, nanogenerators exploiting triboelectric effect, and so on. For example, aging effects and charge stability issues in piezoelectric and electret-based generators, respectively, should be thoroughly studied in order to ensure that material-integrated intelligent systems meet the expected service life requirements.

References

1 Steingart, D., Roundy, S., Wright, P.K., and Evans, J.W. (2008) Micropower materials development for wireless sensor networks. *MRS Bulletin*, **33** (4), 408–409.
2 Briand, D., Yeatman, E., and Roundy, S. (eds) (2015) *Micro Energy Harvesting*, Wiley-VCH Verlag GmbH, Weinheim, Germany.
3 Vullers, R.J.M., van Schaijk, R., Visser, H.J., and Penders, J. (2010) Energy harvesting for autonomous wireless sensor networks. *IEEE Solid-State Circuits Magazine*, 29–38.
4 Kausar, A.S.M.Z., Reza, A.W., Saleh, M.U., and Ramiah, H. (2014) Energizing wireless sensor networks by energy harvesting systems: scopes, challenges and approaches. *Renewable & Sustainable Energy Reviews*, **38**, 973–989.
5 Dewan, A., Ay, S.U., Karim, M.N., and Beyenal, H. (2014) Alternative power sources for remote sensors: a review. *Journal of Power Sources*, **245**, 129–143.
6 Matiko, J.W., Grabham, N.J., and Beeby, S.P. (2014) Review of the application of energy harvesting in buildings. *Measurement Science and Technology*, **25**, 012002.
7 Leonov, V. (2011) Energy harvesting for self-powered wearable devices, in *Wearable Monitoring Systems* (eds A. Bonfiglio and D. De Rossi), Springer Science+Business Media, LLC, pp. 27–49.
8 Romero, E. (2013) *Powering Biomedical Devices*, Elsevier Inc.
9 Lemey, S., Declercq, F., and Rogier, H. (2014) Textile antennas as hybrid energy-harvesting platforms. *Proceedings of the IEEE*, **102** (11), 1833–1857.
10 Vallejos de Schatz, C.H., Medeiros, H.P., Schneider, F.K., and Abatti, P.J. (2012) Wireless medical sensor networks: design requirements and enabling technologies. *Telemedicine and E-Health*, **18** (5), 394–399.
11 Roselli, L., Carvalho, N.B., Alimenti, F., Mezzanotte, P., Orecchini, G., Virili, M., Mariotti, C., Goncalves, R., and Pinho, P. (2014) Smart surfaces: large area electronics systems for Internet of Things enabled by energy harvesting. *Proceedings of the IEEE*, **102** (11), 1723–1746.
12 Cook, B.S., Vyas, R., Kim, S., Thai, T., Le, T., Traille, A., Aubert, H., and Tentzeris, M.M. (2014) RFID-based sensors for zero-power autonomous wireless sensor networks. *IEEE Sensors Journal*, **14** (8), 2419–2431.

13 Bhuiyan, R.H., Dougal, R.A., and Ali, M. (2010) A miniature energy harvesting device for wireless sensors in electric power system. *IEEE Sensors Journal*, **10** (7), 1249–1258.

14 Horowitz, S.B., Sheplak, M., Cattafesta, L.N., and Nishida, T. (2006) A MEMS acoustic energy harvester. *Journal of Micromechanics and Microengineering*, **16** (9), S174–S181.

15 Luz, R.A.S., Pereira, A.R., de Souza, J.C.P., Sales, F.C.P.F., and Crespilho, F.N. (2014) Enzyme biofuel cells: thermodynamics, kinetics and challenges in applicability. *ChemElectroChem*, **1** (11), 1751–1777.

16 Ewing, T., Phuc, T.H., Babauta, J.T., Tang, N.T., Heo, D., and Beyenal, H. (2014) Scale-up of sediment microbial fuel cells. *Journal of Power Sources*, **272**, 311–319.

17 Himes, C., Carlson, E., Ricchiuti, R.J., Otis, B., and Parviz, B.A. (2010) Ultralow voltage nanoelectronics powered directly, and solely, from a tree. *IEEE Transactions on Nanotechnology*, **9** (1), 2–5.

18 Bhuyan, S. and Hu, J. (2013) A natural battery based on lake water and its soil bank. *Energy*, **51**, 395–399.

19 Krupenkin, T. and Taylor, J.A. (2011) Reverse electrowetting as a new approach to high-power energy harvesting. *Nature Communications*, **2**, 448.

20 Kim, D., Sakimoto, K.K., Hong, D., and Yang, P. (2015) Artificial photosynthesis for sustainable fuel and chemical production. *Angewandte Chemie, International Edition*, **54** (11), 3259–3266.

21 Toprak, A. and Tigli, O. (2014) Piezoelectric energy harvesting: state-of-the-art and challenges. *Applied Physics Reviews*, **1** (3), 031104.

22 Marin, A., Bressers, S., and Priya, S. (2011) Multiple cell configuration electromagnetic vibration energy harvester. *Journal of Physics D: Applied Physics*, **44** (29), 295501.

23 Kim, S.-G., Priya, S., and Kanno, I. (2012) Piezoelectric MEMS for energy harvesting. *MRS Bulletin*, **37** (11), 1039–1050.

24 Roedel, J., Jo, W., Seifert, K.T.P., Anton, E.-M., Granzow, T., and Damjanovic, D. (2009) Perspective on the development of lead-free piezoceramics. *Journal of the American Ceramic Society*, **92** (6), 1153–1177.

25 Ramadan, K.S., Sameoto, D., and Evoy, S. (2014) A review of piezoelectric polymers as functional materials for electromechanical transducers. *Smart Materials and Structures*, **23** (3), 033001.

26 Baur, C., Apo, D.J., Maurya, D., Priya, S., and Voit, W. (2014) Advances in piezoelectric polymer composites for vibrational energy harvesting, in *Polymer Composites for Energy Harvesting, Conversion, and Storage*, ACS Symposium Series (eds L. Li, W. WongNg, and J. Sharp), vol. 1161, American Chemical Society, pp. 1–27.

27 Crossley, S., Whiter, R.A., and Kar-Narayan, S. (2014) Polymer-based nanopiezoelectric generators for energy harvesting applications. *Materials Science and Technology*, **30** (13A), 1613–1624.

28 Bowen, C.R., Kim, H.A., Weaver, P.M., and Dunn, S. (2014) Piezoelectric and ferroelectric materials and structures for energy harvesting applications. *Energy & Environmental Science*, **7** (1), 25–44.

29 Li, H., Tian, C., and Deng, Z.D. (2014) Energy harvesting from low frequency applications using piezoelectric materials. *Applied Physics Reviews*, **1**, 041301.

30 Ringgaard, E., Wurlitzer, T., and Wolny, W.W. (2005) Properties of lead-free piezoceramics based on alkali niobates. *Ferroelectrics*, **319**, 97–107.

31 Safari, A. and Abazari, M. (2010) Lead-free piezoelectric ceramics and thin films. *IEEE Transactions on Ultrasonics Ferroelectrics and Frequency Control*, **57** (10), 2165–2176.

32 Minh, L.V., Hara, M., and Kuwano, H. (2013) Micro-energy harvesters integrated with a quatrefoil-shaped proof mass suspended by multiple (K,Na)NbO$_3$ beams. *Japanese Journal of Applied Physics*, **52** (7), UNSP 07HD08.

33 Zhang, S. and Yu, F. (2011) Piezoelectric materials for high temperature sensors. *Journal of the American Ceramic Society*, **94** (10), 3153–3170.

34 Aktakka, E.E., Peterson, R.L., and Najafi, K. (2014) High stroke and high deflection bulk-PZT diaphragm and cantilever micro actuators and effect of pre-stress on device performance. *Journal of Microelectromechanical Systems*, **23** (2), 438–451.

35 Marzencki, M., Ammar, Y., and Basrour, S. (2008) Integrated power harvesting system including a MEMS generator and a power management circuit. *Sensors and Actuators A: Physical*, **145**, 363–370.

36 Liu, H., Lee, C., Kobayashi, T., Tay, C.J., and Quan, C. (2012) Piezoelectric MEMS-based wideband energy harvesting systems using a frequency-up-conversion cantilever stopper. *Sensors and Actuators A: Physical*, **186**, 242–248.

37 Tang, G., Yang, B., Liu, J.Q., Xu, B., Zhu, H.Y., and Yang, C.S. (2014) Development of high performance piezoelectric d$_{33}$ mode MEMS vibration energy harvester based on PMN-PT single crystal thick film. *Sensors and Actuators A: Physical*, **205**, 150–155.

38 Wang, D.-A. and Liu, N.-Z. (2011) A shear mode piezoelectric energy harvester based on a pressurized water flow. *Sensors and Actuators A: Physical*, **167** (2), 449–458.

39 Kim, S.-B., Park, H., Kim, S.-H., Wikle, H.C., Park, J.-H., and Kim, D.-J. (2013) Comparison of MEMS PZT cantilevers based on d$_{31}$ and d$_{33}$ modes for vibration energy harvesting. *Journal of Microelectromechanical Systems*, **22** (1), 26–33.

40 Elfrink, R., Kamel, T.M., Goedbloed, M., Matova, S., Hohlfeld, D., van Andel, Y., and van Schaijk, R. (2009) Vibration energy harvesting with aluminum nitride-based piezoelectric devices. *Journal of Micromechanics and Microengineering*, **19**, 095005.

41 Roundy, S. and Wright, P.K. (2004) A piezoelectric vibration based generator for wireless electronics. *Smart Materials and Structures*, **13** (5), 1131–1142.

42 Mateu, L. and Moll, F. (2005) Optimum piezoelectric bending beam structures for energy harvesting using shoe inserts. *Journal of Intelligent Material Systems and Structures*, **16** (10), 835–845.

43 Deterre, M., Lefeuvre, E., and Dufour-Gergam, E. (2012) An active piezoelectric energy extraction method for pressure energy harvesting. *Smart Materials and Structures*, **21** (8), 085004.

44 Feenstra, J., Granstrom, J., and Sodano, H. (2008) Energy harvesting through a backpack employing a mechanically amplified piezoelectric stack. *Mechanical Systems and Signal Processing*, **22** (3), 721–734.

45 Kim, H.W., Priya, S., Uchino, K., and Newnham, R.E. (2005) Piezoelectric energy harvesting under high pre-stressed cyclic vibrations. *Journal of Electroceramics*, **15** (1), 27–34.

46 Karami, M.A. and Inman, D.J. (2011) Analytical modeling and experimental verification of the vibrations of the zigzag microstructure for energy harvesting. *Journal of Vibration and Acoustics: Transactions of the ASME*, **133** (1), 011002.

47 Massaro, A., De Guido, S., Ingrosso, I., Cingolani, R., De Vittorio, M., Cori, M., Bertacchini, A., Larcher, L., and Passaseo, A. (2011) Freestanding piezoelectric rings for high efficiency energy harvesting at low frequency. *Applied Physics Letters*, **98** (5), 053502.

48 Williams, C.B. and Yates, R.B. (1995) Analysis of a micro-electric generator for microsystems. Proceedings of the 8th International Conference on Solid-State Sensors and Actuators, 1995 and Eurosensors IX. Transducers, pp. 369–372.

49 Erturk, A. and Inman, D.J. (2009) An experimentally validated bimorph cantilever model for piezoelectric energy harvesting from base excitations. *Smart Materials and Structures*, **18** (2), 025009.

50 Daqaq, M.F., Masana, R., Erturk, A., and Quinn, D. (2014) Closure to "Discussion of 'On the role of nonlinearities in energy harvesting: a critical review and discussion'". *Applied Mechanics Reviews*, **66** (4), 040801 (Daqaq, M., Masana, R., Erturk, A., and Quinn, D.D. (2014) *ASME Appl. Mech. Rev.*, 66 (4), 040801)

51 Lin, Y., Andrews, C., and Sodano, H.A. (2010) Enhanced piezoelectric properties of lead zirconate titanate sol–gel derived ceramics using single crystal PbZr$_{0.52}$Ti$_{0.48}$O$_3$ cubes. *Journal of Applied Physics*, **108** (6), 064108.

52 Elfrink, R., Pop, V., Hohlfeld, D., Kamel, T.M., Matova, S., de Nooijer, C., Jambunathan, M., Goedbloed, M., Caballero, L., and Renaud, M. (2009) First autonomous wireless sensor node powered by a vacuum-packaged piezoelectric MEMS energy harvester. Proceedings of the IEEE International Electron Devices Meeting (IEDM'09) Baltimore, pp. 543–546.

53 Zhu, D., Glenne-Jones, P., White, N., Harris, N., Tudor, J., Torah, R., Almusallam, A., and Beeby, S. (2013) Screen printed piezoelectric films for energy harvesting. *Advances in Applied Ceramics*, **112** (2), 79–84.

54 Isarakorn, D., Briand, D., Janphuang, P., Sambri, A., Gariglio, S., Triscone, J.-M., Guy, F., Reiner, J.W., Ahn, C.H., and deRooij, N.F. (2011) The realization and performance of vibration energy harvesting MEMS devices based on an epitaxial piezoelectric thin film. *Smart Materials and Structures*, **20** (2), 025015.

55 Prabu, M., Banu, I.B.S., Vijayaraghavan, G.V., Gobalakrishnan, S., and Chavali, M. (2013) Pulsed laser deposition and ferroelectric characterization of nanostructured perovskite lead zirconate titanate (52/48) thin films. *Journal of Nanoscience and Nanotechnology*, **13** (3), 1938–1942.

56 Lee, B.S., Lin, S.C., Wu, W.J., Wang, X.Y., Chang, P.Z., and Lee, C.K. (2009) Piezoelectric MEMS generators fabricated with an aerosol deposition PZT thin film. *Journal of Micromechanics and Microengineering*, **19**, 065014.

57 Zhang, Y., Jiang, H., Zhang, J., Liu, W., and Jiang, S. (2013) Characterization of PZT thick films fabricated by micro-pen direct-writing. *Journal of Materials Science: Materials in Electronics*, **24** (10), 3680–3685.

58 Aktakka, E.E., Rebecca, L.P., and Najafi, K. (2011) Thinned-PZT in SOI process and design optimization for piezoelectric inertial energy harvesting. 16th International Solid-State Sensors, Actuators and Microsystems Conference (TRANSDUCERS'11), pp. 1649–1652.

59 Janphuang, P., Lockhart, R., Uffer, N., Briand, D., and de Rooij, N.F. (2014) Vibrational piezoelectric energy harvesters based on thinned bulk PZT sheets fabricated at the wafer level. *Sensors and Actuators A: Physical*, **210**, 1–9.

60 Andosca, R., McDonald, T.G., Genova, V., Rosenberg, S., Keating, J., Benedixen, C., and Wu, J. (2012) Experimental and theoretical studies on MEMS piezoelectric vibrational energy harvesters with mass loading. *Sensors and Actuators A: Physical*, **178**, 76–87.

61 Zhang, J., Wang, C., and Bowen, C. (2014) Piezoelectric effects and electromechanical theories at the nanoscale. *Nanoscale*, **6** (22), 13314–13327.

62 Wang, Z.L. and Wu, W. (2012) Nanotechnology-enabled energy harvesting for self-powered micro-/nanosystems. *Angewandte Chemie, International Edition*, **51** (47), 11700–11721.

63 Radousky, H.B. and Liang, H. (2012) Energy harvesting: an integrated view of materials, devices and applications. *Nanotechnology*, **23** (50), 502001.

64 Fang, X.-Q., Liu, J.-X., and Gupta, V. (2013) Fundamental formulations and recent achievements in piezoelectric nano-structures: a review. *Nanoscale*, **5** (5), 1716–1726.

65 Espinosa, H.D., Bernal, R.A., and Minary-Jolandan, M. (2012) A review of mechanical and electromechanical properties of piezoelectric nanowires. *Advanced Materials*, **24** (34), 4656–4675.

66 Wang, Z.L. and Song, J.H. (2006) Piezoelectric nanogenerators based on zinc oxide nanowire arrays. *Science*, **312** (5771), 242–246.

67 Choi, M.-Y., Choi, D., Jin, M.-J., Kim, I., Kim, S.-H., Choi, J.-Y., Lee, S.Y., Kim, J.M., and Kim, S.-W. (2009) Mechanically powered transparent flexible charge-generating nanodevices with piezoelectric ZnO nanorods. *Advanced Materials*, **21**, 2185–2189.

68 Gogneau, N., Chretien, P., Galopin, E., Guilet, S., Travers, L., Harmand, J.-C., and Houze, F. (2014) GaN nanowires for piezoelectric generators. *Physica Status Solidi*, **8** (5), 414–419.

69 Chen, X., Xu, S., Yao, N., and Shi, Y. (2010) 1.6V nanogenerator for mechanical energy harvesting using PZT nanofibers. *Nano Letters*, **10** (6), 2133–2137.

70 Kwon, J., Seung, W., Sharma, B.K., Kim, S.-W., and Ahn, J.-H. (2012) A high performance PZT ribbon-based nanogenerator using graphene transparent electrodes. *Energy & Environmental Science*, **5** (10), 8970–8975.

71 Qi, Y., Kim, J., Nguyen, T.D., Lisko, B., Purohit, P., and McAlpine, M. (2011) Enhanced piezoelectricity and stretchability in energy harvesting devices fabricated from buckled PZT ribbons. *Nano Letters*, **11**, 1331–1336.

72 Fang, J., Niu, H., Wang, H., Wang, X., and Lin, T. (2013) Enhanced mechanical energy harvesting using needleless electrospun poly(vinylidene fluoride) nanofibre webs. *Energy & Environmental Science*, **6** (7), 2196–2202.

73 Dagdeviren, C., Hwang, S.-W., Su, Y., Kim, S., Cheng, H.Y., Gur, O., Haney, R., Omenetto, F.G., Huang, Y., and Rogers, J.A. (2013) Transient, biocompatible electronics and energy harvesters based on ZnO. *Small*, **9** (20), 3398–3404.

74 Park, K.-I., Xu, S., Liu, Y., Hwang, G.-T., Kang, S.-J.L., Wang, Z.L., and Lee, K.J. (2010) Piezoelectric $BaTiO_3$ thin film nanogenerator on plastic substrates. *Nano Letters*, **10** (12), 4939–4943.

75 McCall, W.R., Kim, K., Heath, C., La Pierre, G., and Sirbuly, D.J. (2014) Piezoelectric nanoparticle–polymer composite foams. *ACS Applied Materials & Interfaces*, **6** (22), 19504–19509.

76 Yang, W.Q., Chen, J., Zhu, G., Yang, J., Bai, P., Su, Y., Jing, Q., Cao, X., and Wang, Z.L. (2013) Harvesting energy from the natural vibration of human walking. *ACS Nano*, **7** (12), 11317–11324.

77 Kwon, J., Sharma, B.K., and Ahn, J.H. (2013) Graphene based nanogenerator for energy harvesting. *Japanese Journal of Applied Physics*, **52**, 06GA02–06GA10.

78 Hu, C., Song, L., Zhang, Z., Chen, N., Feng, Z., and Qu, L. (2015) Tailored graphene systems for unconventional applications in energy conversion and storage devices. *Energy & Environmental Science*, **8** (1), 31–54.

79 Nguyen, T.D., Mao, S., Yeh, Y.-W., Purohit, P., and McAlpine, M. (2013) Nanoscale flexoelectricity. *Advanced Materials*, **25** (7), 946–974.

80 Jiang, X., Huang, W., and Zhang, S. (2013) Flexoelectric nano-generator: materials, structures and devices. *Nano Energy*, **2** (6), 1079–1092.

81 Deng, Q., Kammoun, M., Erturk, A., and Sharma, P. (2014) Nanoscale flexoelectric energy harvesting. *International Journal of Solids and Structures*, **51** (18), 3218–3225.

82 Van den Ende, D.A., van de Wiel, H.J., Groen, W.A., and van der Zwaag, S. (2012) Direct strain energy harvesting in automobile tires using piezoelectric PZT–polymer composites. *Smart Materials and Structures*, **21** (1), 015011.

83 Dang, Z.M., Fan, L.Z., Shen, Y., and Nan, C. (2003) Dielectric behavior of novel three-phase $MWNTs/BaTiO_3/PVDF$ composites. *Materials Science and Engineering B: Solid State Materials for Advanced Technology*, **103** (2), 140–144.

84 Dodds, J.S., Meyers, F.N., and Loh, K.J. (2012) Piezoelectric characterization of PVDF-TrFE thin films enhanced with ZnO nanoparticles. *IEEE Sensors Journal*, **12** (6), 1889–1890.

85 Mao, Y., Zhao, P., McConohy, G., Yang, H., Tong, Y., and Wang, X. (2014) Sponge-like piezoelectric polymer films for scalable and integratable nanogenerators and self-powered electronic systems. *Advanced Energy Materials*, **4** (7), 1301624.

86 Chu, B., Zhou, X., Ren, K., Neese, B., Lin, M., Wang, Q., Bauer, F., and Zhang, Q.M. (2006) A dielectric polymer with high electric energy density and fast discharge speed. *Science*, **313** (5785), 334–336.

87 Lallart, M., Cottinet, P.-J., Lebrun, L., Guiffard, B., and Guyomar, D. (2010) Evaluation of energy harvesting performance of electrostrictive polymer and carbon-filled terpolymer composites. *Journal of Applied Physics*, **108** (3), 034901.

88 Ramaratnam, A. and Jalili, N. (2006) Reinforcement of piezoelectric polymers with carbon nanotubes: pathway to next generation sensors. *Journal of Intelligent Material Systems and Structures*, **17** (3), 199–208.

89 Patel, I., Siores, E., and Shah, T. (2010) Utilisation of smart polymers and ceramic based piezoelectric materials for scavenging wasted energy. *Sensors and Actuators A: Physical*, **159** (2), 213–218.

90 Prashanthi, K., Miriyala, N., Gaikwad, R.D., Moussa, W., Rao, V.R., and Thundat, T. (2013) Vibrational energy harvesting using photo-patternable piezoelectric nanocomposite cantilevers. *Nano Energy*, **2** (5), 923–932.

91 Xu, S., Yeh, Y., Poirier, G., McAlpine, M.C., Register, R.A., and Yao, N. (2013) Flexible piezoelectric PMN-PT nanowire-based nanocomposite and device. *Nano Letters*, **13** (6), 2393–2398.

92 Von Seggern, H., Zhukov, S., and Fedosov, S. (2011) Importance of geometry and breakdown field on the piezoelectric d_{33} coefficient of corona charged ferroelectret sandwiches. *IEEE Transactions on Dielectrics and Electrical Insulation*, **18** (1), 49–56.

93 Hillenbrand, J. and Sessler, G.M. (2004) Quasistatic and dynamic piezoelectric coefficients of polymer foams and polymer film systems. *IEEE Transactions on Dielectrics and Electrical Insulation*, **11** (1), 72–79.

94 Feng, Y., Hagiwara, K., Iguchi, Y., and Suzuki, Y. (2012) Trench-filled cellular parylene electret for piezoelectric transducer. *Applied Physics Letters*, **100** (26), 262901.

95 Zhang, X., Sessler, G.M., and Wang, Y. (2014) Fluoroethylenepropylene ferroelectret films with cross-tunnel structure for piezoelectric transducers and micro energy harvesters. *Journal of Applied Physics*, **116** (7), 074109.

96 Anton, S.R., Farinholt, K.M., and Erturk, A. (2014) Piezoelectret foam based vibration energy harvesting. *Journal of Intelligent Material Systems and Structures*, **25** (14), 1681–1692.

97 Pondrom, P., Hillenbrand, J., Sessler, G.M., Boes, J., and Melz, T. (2014) Vibration-based energy harvesting with stacked piezoelectrets. *Applied Physics Letters*, **104** (17), 172901.

98 Cottinet, P.-J., Guyomar, D., and Lallart, M. (2011) Electrostrictive polymer harvesting using a nonlinear approach. *Sensors and Actuators A: Physical*, **172** (2), 497–503.

99 Biggs, J., Danielmeier, K., Hitzbleck, J., Krause, J., Kridl., T., Nowak, S., Orselli, E., Quan, X., Schapeler, D., Sutherland, W., and Wagner, J. (2013) Electroactive polymers: developments of and perspectives for dielectric elastomers. *Angewandte Chemie, International Edition*, **52** (36), 9409–9421.

100 Jean-Mistral, C. and Basrour, S. (2010) Scavenging energy from human motion with tubular dielectric polymer. SPIE Proceedings of Conference on Electroactive Polymer Actuators and Devices (EAPAD), San Diego, vol. 7642, 764209.

101 McKay, T.G., Rosset, S., Anderson, I.A., and Shea, H. (2015) Dielectric elastomer generators that stack up. *Smart Materials and Structures*, **24** (1), 015014.

102 Chiba, S., Waki, M., Kornbluh, R., and Pelnine, R. (2008) Innovative power generators for energy harvesting using electroactive polymer artificial muscles.

SPIE Proceedings of Conference on Electroactive Polymer Actuators and Devices (EAPAD), San Diego, vol. 6927, 692715.

103 Maas, J. and Graf, C. (2012) Dielectric elastomers for hydro power harvesting. *Smart Materials and Structures*, **21**, 064006.

104 Koh, S.J.A., Zhao, X., and Suo, Z. (2009) Maximal energy that can be converted by a dielectric elastomer generator. *Applied Physics Letters*, **94** (26), 262902.

105 Kaltseis, R., Keplinger, C., Koh, S.J.A., Baumgartner, R., Goh, Y.F., Ng, W.H., Kogler, A., Troels, A., Foo, C.C., Suo, Z., and Bauer, S. (2014) Natural rubber for sustainable high-power electrical energy generation. *RSC Advances*, **4** (53), 27905–27913.

106 Graf, C., Hitzbleck, J., Feller, T., Clauberg, K., Wagner, J., Krause, J., and Maas, J. (2014) Dielectric elastomer-based energy harvesting: material, generator design, and optimization. *Journal of Intelligent Material Systems and Structures*, **25** (8), 951–966.

107 Cottinet, P.-J., Guyomar, D., Lallart, M., Guiffard, B., and Lebrun, L. (2011) Investigation of electrostrictive polymer efficiency for mechanical energy harvesting. *IEEE Transactions on Ultrasonics Ferroelectrics and Frequency Control*, **58** (9), 1842–1851.

108 Ren, K., Liu, Y., Hofmann, H., Zhang, Q.M., and Blottman, J. (2007) An active energy harvesting scheme with an electroactive polymer. *Applied Physics Letters*, **91** (13), 132910.

109 Brufau-Penella, J., Puig-Vidal, M., Giannone, P., Graziani, S., and Strazzeri, S. (2008) Characterization of the harvesting capabilities of an ionic polymer metal composite device. *Smart Materials and Structures*, **17** (1), 015009.

110 Jo, C., Pugal, D., Oh, I.-K., Kim, K.J., and Asaka, K. (2013) Recent advances in ionic polymer–metal composite actuators and their modeling and applications. *Progress in Polymer Science*, **38** (7), 1037–1066.

111 Tiwari, R. and Kim, K.J. (2013) IPMC as a mechanoelectric energy harvester: tailored properties. *Smart Materials and Structures*, **22**, 015017.

112 Cellini, F., Cha, Y., and Porfiri, M. (2014) Energy harvesting from fluid-induced buckling of ionic polymer metal composites. *Journal of Intelligent Material Systems and Structures*, **25** (12), 1496–1510.

113 Peterson, S.D. and Porfiri, M. (2012) Energy exchange between a vortex ring and an ionic polymer metal composite. *Applied Physics Letters*, **100** (11), 114102.

114 Farinholt, K.M., Pedrazas, N.A., Schluneker, D.M., Burt, D., and Farrar, C. (2009) An energy harvesting comparison of piezoelectric and ionically conductive polymers. *Journal of Intelligent Material Systems and Structures*, **20** (5), 633–642.

115 Basset, P., Galayko, D., Cottone, F., Guillemet, R., Blokhina, E., Marty, F., and Bourouina, T. (2014) Electrostatic vibration energy harvester with combined effect of electrical nonlinearities and mechanical impact. *Journal of Micromechanics and Microengineering*, **24** (3), 035001.

116 Wang, F. and Hansen, O. (2014) Electrostatic energy harvesting device with out-of-the-plane gap closing scheme. *Sensors and Actuators A: Physical*, **211**, 131–137.

117 Eun, Y., Kwon, D.-S., Kim, M.-O., Yoo, I., Sim, J., Ko, H.-J., Cho, K.-H., and Kim, J. (2014) A flexible hybrid strain energy harvester using piezoelectric and electrostatic conversion. *Smart Materials and Structures*, **23** (4), 045040.

118 Basset, P., Galayko, D., Paracha, A.M., Marty, F., Dudka, A., and Bourouina, T. (2009) A batch-fabricated and electret-free silicon electrostatic vibration energy harvester. *Journal of Micromechanics and Microengineering*, **19** (11), 115025.

119 Crovetto, A., Wang, F., and Hansen, O. (2013) An electret-based energy harvesting device with a wafer-level fabrication process. *Journal of Micromechanics and Microengineering*, **23** (11), 114010.

120 Li, Y., Celik-Butler, Z., and Butler, D.P. (2015) A hybrid electrostatic micro-harvester incorporating in-plane overlap and gap closing mechanisms. *Journal of Micromechanics and Microengineering*, **25** (3), 035027.

121 Kuehne, I., Frey, A., Marinkovic, D., Eckstein, G., and Seidel, H. (2008) Power MEMS: a capacitive vibration-to-electrical energy converter with built-in voltage. *Sensors and Actuators A: Physical*, **142** (1), 263–269.

122 Cook-Chennault, K.A., Thambi, N., and Sastry, A.M. (2008) Powering MEMS portable devices: a review of non-regenerative and regenerative power supply systems with special emphasis on piezoelectric energy harvesting systems. *Smart Materials and Structures*, **17** (4), 043001.

123 Sakane, Y., Suzuki, Y., and Kasagi, N. (2007) The development of a high-performance perfluorinated polymer electret and its application to micro power generation. *Journal of Micromechanics and Microengineering*, **18** (10), 104011.

124 Wada, N., Mukougawa, K., Horiuchi, N., Hiyama, T., Nakamura, M., Nagai, A., Okura, T., and Yamashita, K. (2013) Fundamental electrical properties of ceramic electrets. *Materials Research Bulletin*, **48** (10), 3854–3859.

125 Leonov, V., van Schaijk, R., and Van Hoof, C. (2013) Charge retention in a patterned SiO_2/Si_3N_4 electret. *IEEE Sensors Journal*, **13** (9), 3369–3376.

126 Varpula, A., Laakso, S.J., Havia, T., Kyynaräinen, J., and Prunnila, M. (2014) Harvesting vibrational energy using material work functions. *Scientific Reports*, **4**, 6799.

127 Roundy, S., Wright, P.K., and Rabaey, J. (2003) A study of low level vibrations as a power source for wireless sensor nodes. *Computer Communications*, **26** (11), 1131–1144.

128 Cepnik, C., Lausecker, R., and Wallrabe, U. (2013) Review on electrodynamic energy harvesters: a classification approach. *Micromachines*, **4** (2), 168–196.

129 Moss, S.D., Payne, O.R., Hart, G.A., and Ung, C. (2015) Scaling and power density metrics of electromagnetic vibration energy harvesting devices. *Smart Materials and Structures*, **24** (2), 023001.

130 Galchev, T., Kim, H., and Najafi, K. (2009) A parametric frequency increased power generator for scavenging low frequency ambient vibrations. *Procedia Chemistry*, **1**, 1439–1442.

131 Spreemann, D., Manoli, Y., Folkmer, B., and Mintenbeck, D. (2006) Non-resonant vibration conversion. *Journal of Micromechanics and Microengineering*, **16**, 169–173.

132 Bowers, B. and Arnold, D. (2009) Spherical, rolling magnet generators for passive energy harvesting from human motion. *Journal of Micromechanics and Microengineering*, **19**, 094008.

133 Wang, P., Tanaka, K., Sugiyama, S., Dai, X., Zhao, X., and Liu, J. (2009) A micro electromagnetic low level vibration energy harvester based on MEMS technology. *Microsystem Technologies*, **15** (6), 941–951.

134 Horn, R.G. and Smith, D.T. (1992) Contact electrification and adhesion between dissimilar materials. *Science*, **256**, 362–364.

135 Hou, T.-C., Yang, Y., Zhang, H., Chen, J., Chen, L.-J., and Wang, Z.L. (2013) Triboelectric nanogenerator built inside shoe insole for harvesting walking energy. *Nano Energy*, **2** (5), 856–862.

136 Dhakar, L., Tay, F.E.H., and Lee, C. (2015) Development of a broadband triboelectric energy harvester with SU-8 micropillars. *Journal of Microelectromechanical Systems*, **24** (1), 91–99.

137 Matsusaka, S., Maruyama, H., Matsuyama, T., and Ghadiri, M. (2010) Triboelectric charging of powders: a review. *Chemical Engineering Science*, **65** (22), 5781–5807.

138 Pu, X., Li, L., Song, H., Du, C., Zhao, Z., Jiang, C., Cao, G., Hu, W., and Wang, Z.L. (2015) A self-charging power unit by integration of a textile triboelectric nanogenerator and a flexible lithium-ion battery for wearable electronics. *Advanced Materials*, **27** (15), 2472–2478.

139 Lee, S., Ko, W., Oh, Y., Lee, Y., Baek, G., Lee, Y., Sohn, J., Cha, S., Kim, J., Park, J., and Hong, J. (2015) Triboelectric energy harvester based on wearable textile platforms employing various surface morphologies. *Nano Energy*, **12**, 410–418.

140 Wang, S., Lin, L., and Wang, Z.L. (2015) Triboelectric nanogenerators as self-powered active sensors. *Nano Energy*, **11**, 436–462.

141 Du, W., Han, X., Lin, L., Chen, M., Li, X., Pan, C., and Wang, Z.L. (2014) A three-dimensional multi-layered sliding triboelectric nanogenerator. *Advanced Energy Materials*, **4** (11), 1301592.

142 Niu, S., Liu, Y., Zhou, Y.S., Wang, S., Lin, L., and Wang, Z.L. (2015) Optimization of triboelectric nanogenerator charging systems for efficient energy harvesting and storage. *IEEE Transactions on Electron Devices*, **62** (2), 641–647.

143 Wang, L. and Yuan, F.G. (2008) Vibration energy harvesting by magnetostrictive material. *Smart Materials and Structures*, **17** (4), 045009.

144 Zhao, X. and Lord, D.G. (2006) Application of the Villari effect to electric power harvesting. *Journal of Applied Physics*, **99** (8), 08M703.

145 Marin, A., Tadesse, Y., and Priya, S. (2013) Multi-mechanism non-linear vibration harvester combining inductive and magnetostrictive mechanisms. *Integrated Ferroelectrics*, **148** (1), 27–52.

146 Karaman, I., Basaran, B., Karaca, H.E., Karsilayan, A.I., and Chumlyakov, Y.I. (2007) Energy harvesting using martensite variant reorientation mechanism in a NiMnGa magnetic shape memory alloy. *Applied Physics Letters*, **90** (17), 172505.

147 Jani, J.M., Leary, M., Subic, A., and Gibson, M.A. (2014) A review of shape memory alloy research, applications and opportunities. *Materials & Design*, **56**, 1078–1113.

148 Ma, J. and Karaman, I. (2010) Expanding the repertoire of shape memory alloys. *Science*, **327** (5972), 1468–1469.

149 Ma, J., Hu, J., Li, Z., and Nan, C.-W. (2011) Recent progress in multiferroic magnetoelectric composites: from bulk to thin films. *Advanced Materials*, **23** (9), 1062–1087.

150 Yang, J., Wen, Y., and Li, P. (2011) Magnetoelectric energy harvesting from vibrations of multiple frequencies. *Journal of Intelligent Material Systems and Structures*, **22** (14), 1631–1639.

151 Moss, S.D., McLeod, J.E., Powlesland, I.G., and Galea, S.C. (2012) A bi-axial magnetoelectric vibration energy harvester. *Sensors and Actuators A: Physical*, **175**, 165–168.

152 Li, M., Wen, Y., Li, P., and Yang, J. (2013) A resonant frequency self-tunable rotation energy harvester based on magnetoelectric transducer. *Sensors and Actuators A: Physical*, **194**, 16–24.

153 Wacharasindhu, T. and Kwon, J. (2008) A micromachined energy harvester from a keyboard using combined electromagnetic and piezoelectric conversion. *Journal of Micromechanics and Microengineering*, **18** (10), 104016.

154 Tadesse, Y., Zhang, S., and Priya, S. (2009) Multimodal energy harvesting system: piezoelectric and electrodynamic. *Journal of Intelligent Material Systems and Structures*, **20** (5), 625–632.

155 Han, M., Zhang, X.-S., Meng, B., Liu, W., Tang, W., Sun, X., Wang, W., and Zhang, H. (2013) r-shaped hybrid nanogenerator with enhanced piezoelectricity. *ACS Nano*, **7** (10), 8554–8560.

156 Harne, R.L. and Wang, K.W. (2013) Review of the recent research on vibration energy harvesting via bistable systems. *Smart Materials and Structures*, **22** (2), 023001.

157 Zhu, D., Tudor, M.J., and Beeby, S.P. (2010) Strategies for increasing the operating frequency range of vibration energy harvesters: a review. *Measurement Science and Technology*, **21**, 022001.

158 Twiefel, J. and Westermann, H. (2013) Survey on broadband techniques for vibration energy harvesting. *Journal of Intelligent Material Systems and Structures*, **24**, 1291–1302.

159 Challa, V.R., Prasad, M.G., Shi, Y., and Fisher, F.T. (2008) A vibration energy harvesting device with bidirectional resonance frequency tenability. *Smart Materials and Structures*, **17** (1), 015035.

160 Liu, J.-Q., Fang, H.-B., Xu, Z.-Y., Mao, X.-H., Shen, X.-C., Chen, D., Liao, H., and Cai, B.-C. (2008) A MEMS-based piezoelectric power generator array for vibration energy harvesting. *Microelectronics Journal*, **39** (5), 802–806.

161 Gu, L. and Livermore, C. (2011) Impact-driven, frequency up-converting coupled vibration energy harvesting device for low frequency operation. *Smart Materials and Structures*, **20**, 045004.

162 Abdelkefi, A., Najar, F., Nayfeh, A.H., and Ben, A.S. (2011) An energy harvester using piezoelectric cantilever beams undergoing coupled bending-torsion vibrations. *Smart Materials and Structures*, **20** (11), 115007.

163 Gammaitoni, L., Neri, I., and Vocca, H. (2009) Nonlinear oscillators for vibration energy harvesting. *Applied Physics Letters*, **94** (16), 164102.

164 Tang, Q.C., Yang, Y.L., and Li, X. (2011) Bi-stable frequency up-conversion piezoelectric energy harvester driven by non-contact magnetic repulsion. *Smart Materials and Structures*, **20**, 125011.

165 Liu, H., Tay, C.J., Quan, C., Kobayashi, T., and Lee, C. (2011) A scrape-through piezoelectric MEMS energy harvester with frequency broadband and up-conversion behaviors. *Microsystems Technologies*, **17**, 1747–1754.

166 Mann, B.P. and Sims, N.D. (2008) Energy harvesting from the nonlinear oscillations of magnetic levitation. *Journal of Sound and Vibration*, **319**, 515–530.

167 Nguyen, D.S., Halvorsen, E., Jensen, G.U., and Vogl, A. (2010) Fabrication and characterization of a wideband MEMS energy harvester utilizing nonlinear springs. *Journal of Micromechanics and Microengineering*, **20** (12), 125009.

168 Betts, D.N., Kim, H.A., Bowen, C.R., and Inman, D.J. (2012) Optimal configurations of bistable piezo-composites for energy harvesting. *Applied Physics Letters*, **100** (11), 114104.

169 Pellegrini, S.P., Tolou, N., Schenk, M., and Herder, J.L. (2013) Bistable vibration energy harvesters: a review. *Journal of Intelligent Material Systems and Structures*, **24** (11), 1–10.

170 Kim, P. and Seok, J. (2014) A multi-stable energy harvester: dynamic modeling and bifurcation analysis. *Journal of Sound and Vibration*, **333** (21), 5525–5547.

171 Zhou, S., Cao, J., Inman, D.J., Lin, J., Liu, S., and Wang, Z. (2014) Broadband tristable energy harvester: modeling and experiment verification. *Applied Energy*, **133**, 33–39.

172 Rowe, D.M. (2006) *Thermoelectrics Handbook: Macro to Nano*, CRC Press/Taylor & Francis, Boca Raton, FL.

173 Yuan, Z., Ziouche, K., Bougrioua, Z., Lejeune, P., Lasri, T., and Leclercq, D. (2015) A planar micro thermoelectric generator with high thermal resistance. *Sensors and Actuators A: Physical*, **221**, 67–76.

174 Weber, J., Potje-Kamloth, K., Haase, F., Detemple, P., Voeklein, F., and Doll, T. (2006) Coin-size coiled-up polymer foil thermoelectric power generator for wearable electronics. *Sensors and Actuators A: Physical*, **132**, 325–330.

175 Yang, Y., Zhang, H., Zhu, G., Lee, S., Lin, Z.-H., and Wang, Z.L. (2013) Flexible hybrid energy cell for simultaneously harvesting thermal, mechanical, and solar energies. *ACS Nano*, 7, 785–790.

176 Puscasu, O., Monfray, S., Boughaleb, J., Cottinet, P.J., Rapisarda, D., Rouviere, E., Delepierre, G., Pitone, G., Maitre, C., Boeuf, F., Guyomar, D., and Skotnicki, T. (2014) Flexible bimetal and piezoelectric based thermal to electrical energy converters. *Sensors and Actuators A: Physical*, **214**, 7–14.

177 Strasser, M., Aigner, R., Lauterbach, C., Sturm, T.F., Franosch, M., and Wachutka, G. (2004) Micromachined CMOS thermoelectric generators as on-chip power supply. *Sensors and Actuators A: Physical*, **114**, 362–370.

178 Glatz, W., Schwyter, E., Durrer, L., and Hierold, C. (2009) Bi_2Te_3-based flexible micro thermoelectric generator with optimized design. *Journal of Microelectromechanical Systems*, **18** (3), 763–772.

179 Navone, C., Soulier, M., Testard, J., Simon, J., and Caroff, T. (2011) Optimization and fabrication of a thick printed thermoelectric device. *Journal of Electronic Materials*, **40** (5), 789–793.

180 Chen, A., Madan, D., Wright, P.K., and Evans, J.W. (2011) Dispenser-printed planar thick-film thermoelectric energy generators. *Journal of Micromechanics and Microengineering*, **21**, 104006.

181 Takano, H., Takubo, C., Asai, K., Sawai, Y., Fujieda, T., Naito, T., Kurata, H., and Goto, Y. (2015) Printed thermoelectric devices using vanadate glass composite materials. *Electronics and Communications in Japan*, **98** (3), 41–46.

182 Lange, R.G. and Carroll, W.P. (2008) Review of recent advances of radioisotope power systems. *Energy Conversion and Management*, **49** (3), 393–401.

183 Parsonnet, V., Driller, J., Cook, D., and Rizvi, S.A. (2006) Thirty-one years of clinical experience with "nuclear-powered" pacemakers. *Pace-Pacing and Clinical Electrophysiology*, **29** (2), 195–200.

184 Elsheikh, M.H., Shnawah, D.A., Sabri, M.F.M., Said, S.B.M., Hassan, M.H., Bashir, M.B.A., and Mohamad, M. (2014) A review on thermoelectric renewable energy: principle parameters that affect their performance. *Renewable & Sustainable Energy Reviews*, **30**, 337–355.

185 Yang, J. and Stabler, F.R. (2009) Automotive applications of thermoelectric materials. *Journal of Electronic Materials*, **38** (7), 1245–1251.

186 Alam, H. and Ramakrishna, S. (2013) A review on the enhancement of figure of merit from bulk to nano-thermoelectric materials. *Nano Energy*, **2** (2), 190–212.

187 Shakouri, A. (2011) Recent developments in semiconductor thermoelectric physics and materials. *Annual Review of Materials Research*, **41**, 399–431.

188 Zhao, L.-D., Dravid, V.P., and Kanatzidis, M.G. (2014) The panoscopic approach to high performance thermoelectrics. *Energy & Environmental Science*, **7** (1), 251–268.

189 Bos, J.-W.G. and Downie, R.A. (2014) Half-Heusler thermoelectrics: a complex class of materials. *Journal of Physics: Condensed Matter*, **26** (43), 433201.

190 Vining, C.B. (2009) An inconvenient truth about thermoelectrics. *Nature Materials*, **8** (2), 83–85.

191 Sothmann, B., Sanchez, R., and Jordan, A.N. (2015) Thermoelectric energy harvesting with quantum dots. *Nanotechnology*, **26** (3), 032001.

192 Vineis, C.J., Shakouri, A., Majumdar, A., and Kanatzidis, M.G. (2010) Nanostructured thermoelectrics: big efficiency gains from small features. *Advanced Materials*, **22** (36), 3970–3980.

193 Kovalenko, M.V., Manna, L., Cabot, A., Hens, Z., Talapin, D.V., Kagan, C.R., Klimov, V.I., Rogach, A.L., Reiss, P., Milliron, D.J., Guyot-Sionnnest, P., Konstantatos, G., Parak, W.J., Hyeon, T., Korgel, B.A., Murray, C.B., and Heiss, W. (2015) Prospects of nanoscience with nanocrystals. *ACS Nano*, **9** (2), 1012–1057.

194 Ren, G., Lan, J., Zeng, C., Liu, Y., Zhan, B., Butt, S., Lin, Y.H., and Nan, C.-W. (2015) High performance oxides-based thermoelectric materials. *JOM*, **67** (1), 211–221.

195 Goncalves, A.P. and Godart, C. (2014) New promising bulk thermoelectrics: intermetallics, pnictides and chalcogenides. *European Physical Journal B*, **87** (2), 42.

196 Kleinke, H. (2010) New bulk materials for thermoelectric power generation: clathrates and complex antimonides. *Chemistry of Materials*, **22** (3), 604–611.

197 Zhang, Q., Sun, Y., Xu, W., and Zhu, D. (2014) Organic thermoelectric materials: emerging green energy materials converting heat to electricity directly and efficiently. *Advanced Materials*, **26** (40), 6829–6851.

198 Toshima, N. and Ichikawa, S. (2015) Conducting polymers and their hybrids as organic thermoelectric materials. *Journal of Electronic Materials*, **44** (1), 384–390.

199 Kim, G.-H., Shao, L., Zhang, K., and Pipe, K.P. (2013) Engineered doping of organic semiconductors for enhanced thermoelectric efficiency. *Nature Materials*, **12** (8), 719–723.

200 McGrail, B.T., Sehirlioglu, A., and Pentzer, E. (2015) Polymer composites for thermoelectric applications. *Angewandte Chemie, International Edition*, **54** (6), 1710–1723.

201 Bonaccorso, F., Colombo, L., Yu, G., Stoller, M., Tozzini, V., Ferrari, A.C., Ruoff, R.S., and Pellegrini, V. (2015) Graphene, related two-dimensional crystals, and hybrid systems for energy conversion and storage. *Science*, **347** (6217), 1246501.

202 Chen, Z., Guo, B., Yang, Y., and Cheng, C. (2014) Metamaterials-based enhanced energy harvesting: a review. *Physica B: Condensed Matter*, **438**, 1–8.

203 LeBlanc, S., Yee, S.K., Scullin, M.L., Dames, C., and Goodson, K. (2014) Material and manufacturing cost considerations for thermoelectrics. *Renewable & Sustainable Energy Reviews*, **32**, 313–327.

204 Sebald, G., Guyomar, D., and Agbossou, A. (2009) On thermoelectric and pyroelectric energy harvesting. *Smart Materials and Structures*, **18** (12), 125006.

205 Lingam, D., Parikh, A.R., Huang, J., Jain, A., and Minary-Jolandan, M. (2013) Nano/microscale pyroelectric energy harvesting: challenges and opportunities. *International Journal of Smart and Nano Materials*, **4**, 229–245.

206 Bowen, C.R., Taylor, J., LeBoulbar, E., Zabek, D., Chauhan, A., and Vaish, R. (2014) Pyroelectric materials and devices for energy harvesting applications. *Energy & Environmental Science*, **7** (12), 3836–3856.

207 Yang, Y., Jung, J.H., Yun, B.K., Zhang, F., Pradel, K.C., Guo, W., and Wang, Z.L. (2012) Flexible pyroelectric nanogenerators using a composite structure of lead-free $KNbO_3$ nanowires. *Advanced Materials*, **24** (39), 5357–5362.

208 Lee, J.-H., Lee, K.Y., Gupta, M.K., Kim, T.Y., Lee, D.-Y., Oh, J., Ryu, C., Yoo, W.J., Kang, C.-Y., Yoon, S.-J., Yoo, J.B., and Kim, S.-W. (2014) Highly stretchable piezoelectric–pyroelectric hybrid nanogenerators. *Advanced Materials*, **26** (5), 765–769.

209 Bitnar, B., Durisch, W., and Holzner, R. (2013) Thermophotovoltaics on the move to applications. *Applied Energy*, **105**, 430–438.

210 Sundarraj, P., Maity, D., Roy, S.S., and Taylor, R.A. (2014) Recent advances in thermoelectric materials and solar thermoelectric generators – a critical review. *RSC Advances*, **4** (87), 46860–46874.

211 Vuarnoz, D., Kitanovski, A., Gonin, C., Borgeaud, Y., Delessert, M., Meinen, M., and Egolf, P.W. (2012) Quantitative feasibility study of magnetocaloric energy conversion utilizing industrial waste heat. *Applied Energy*, **100**, 229–237.

212 Lallart, M., Wang, L., Sebald, G., Petit, L., and Guyomar, D. (2014) Analysis of thermal energy harvesting using ferromagnetic materials. *Physics Letters A*, **378** (43), 3151–3154.

213 Pillai, M.A. and Deenadayalan, E. (2014) A review of acoustic energy harvesting. *International Journal of Precision Engineering and Manufacturing*, **15** (5), 949–965.

214 Xu, C. and Wang, Z.L. (2011) Compact hybrid cell based on a convoluted nanowire structure for harvesting solar and mechanical energy. *Advanced Materials*, **23** (7), 873–877.

215 Park, H., Kang, H., Lee, Y., Park, Y., Noh, J., and Cho, G. (2012) Fully roll-to-roll gravure printed rectenna on plastic foils for wireless power transmission at 13.56MHz. *Nanotechnology*, **23** (34), 344006.

216 Zhang, Q., Myers, D., Lan, J., Jenekhe, S.A., and Cao, G. (2012) Applications of light scattering in dye-sensitized solar cells. *Physical Chemistry Chemical Physics*, **14** (43), 14982–14998.

217 Jayawardena, K.D.G.I., Rozanski, L.J., Mills, C.A., Beliatis, M.J., Nismy, N.A., and Silva, S.R.P. (2013) 'Inorganics-in-organics': recent developments and outlook for 4G polymer solar cells. *Nanoscale*, **5** (18), 8411–8427.

218 Kouhnavard, M., Ikeda, S., Ludin, N.A., Khairudin, N.B.A., Ghaffari, B.V., Mat-Teridi, M.A., Ibrahim, M.A., Sepeai, S., and Sopian, K. (2014) A review of semiconductor materials as sensitizers for quantum dot-sensitized solar cells. *Renewable & Sustainable Energy Reviews*, **37**, 397–407.

219 Bahrami, A., Mohammadnejad, S., and Soleimaninezhad, S. (2013) Photovoltaic cells technology: principles and recent developments. *Optical and Quantum Electronics*, **45** (2), 161–197.

220 Koenig, D., Casalenuovo, K., Takeda, Y., Conibeer, G., Guillemoles, J.F., Patterson, R., Huang, L.M., and Green, M.A. (2010) Hot carrier solar cells: principles, materials and design. *Physica E*, **42** (10), 2862–2866.

221 Yu, M., Long, Y.-Z., Sun, B., and Fan, Z. (2012) Recent advances in solar cells based on one-dimensional nanostructure arrays. *Nanoscale*, **4** (9), 2783–2796.

222 Babu, V.J., Vempati, S., Sundarrajan, S., Sireesha, M., and Ramakrishna, S. (2014) Effective nanostructured morphologies for efficient hybrid solar cells. *Solar Energy*, **106**, 1–22.

223 Chou, C.-H. and Chen, F.-C. (2014) Plasmonic nanostructures for light trapping in organic photovoltaic devices. *Nanoscale*, **6** (15), 8444–8458.

224 Fan, X., Zhang, M., Wang, X., Yang, F., and Meng, X. (2013) Recent progress in organic–inorganic hybrid solar cells. *Journal of Materials Chemistry A*, **1** (31), 8694–8709.

225 Das, S., Sudhagar, P., Kang, Y.S., and Choi, W. (2014) Graphene synthesis and application for solar cells. *Journal of Materials Research*, **29** (3), 299–319.

226 Yan, J. and Saunders, B.R. (2014) Third-generation solar cells: a review and comparison of polymer: fullerene, hybrid polymer and perovskite solar cells. *RSC Advances*, **4** (82), 43286–43314.

227 Deepak, T.G., Anjusree, G.S., Thomas, S., Arun, T.A., Nair, S.V., and Nair, A.S. (2014) A review on materials for light scattering in dye-sensitized solar cells. *RSC Advances*, **4** (34), 17615–17638.

228 Wong, Y.Q., Wong, H.Y., Tan, C.S., and Meng, H.F. (2014) Performance optimization of organic solar cells. *IEEE Photonics Journal*, **6** (4), 8400426.

229 Collado, A. and Georgiadis, A. (2013) Conformal hybrid solar and electromagnetic (EM) energy harvesting rectenna. *IEEE Transactions on Circuits and Systems I: Regular Papers*, **60** (8), 2225–2234.

230 Tan, Y.K. and Panda, S.K. (2011) Energy harvesting from hybrid indoor ambient light and thermal energy sources for enhanced performance of wireless sensor nodes. *IEEE Transactions on Industrial Electronics*, **58** (9), 4424–4435.

231 Bandyopadhyay, S. and Chandrakasan, A.P. (2012) Platform architecture for solar, thermal, vibration energy combining with MPPT and single inductor. *IEEE Journal of Solid-State Circuits*, **47** (9), 2199–2215.

232 Bareiss, M., Krenz, P.M., Szakmany, G.P., Tiwari, B.N., Kaelblein, D., Orlov, A.O., Bernstein, G.H., Scarpa, G., Fabel, B., Zschieschang, U., Klauk, H., Porod, W., and Lugli, P. (2013) Rectennas revisited. *IEEE Transactions on Nanotechnology*, **12** (6), 1144–1150.

233 Kim, S., Vyas, R., Bito, J. et al. (2014) Ambient RF energy-harvesting technologies for self-sustainable standalone wireless sensor platforms. *Proceedings of the IEEE*, **102** (11), 1649–1666.

234 Reig, C. and Avila-Navarro, E. (2014) Printed antennas for sensor applications: a review. *IEEE Sensors Journal*, **14** (8), 2406–2418.

235 Olgun, U., Chen, C.-C., and Volakis, J.L. (2011) Investigation of rectenna array configurations for enhanced RF power harvesting. *IEEE Antennas and Wireless Propagation Letters*, **10**, 262–265.

236 Valenta, C.R. and Durgin, G.D. (2014) Harvesting wireless power. *IEEE Microwave Magazine*, **15** (4), 108–120.

Part Eight

Application Scenarios

22

Structural Health Monitoring (SHM)

Dirk Lehmhus[1] and Matthias Busse[1,2]

[1] University of Bremen, ISIS Sensorial Materials Scientific Centre, Wiener Str. 12, 28359 Bremen, Germany
[2] Fraunhofer IFAM, Shaping and Functional Materials, Wiener Straße 12, 28359 Bremen, Germany

22.1 Introduction

Structural health monitoring (SHM) is, as Farrar *et al.* have put it, "the process of detecting damage in structures" [1]. Farrar's perspective is centered on the aims of SHM. Encompassing anything from external investigation to structure- and material-integrated systems, it deliberately leaves open the question of implementation. Roughly 15 years ago by now, Renton has formulated the promise of material-integrated SHM in a more lyrical, but nevertheless very memorable way [2]: "In 20 years we will have structure that talks to us: tells us how it's feeling, where it hurts, how it changed its shape; and what loads it's experiencing."

This quotation very clearly exemplifies that SHM specifically in the aerospace industry has for a significant time been considered the most obvious area of application for material-integrated intelligent systems. However, at least for this industry, the timescale originally implied by the author of this forward-looking study must seem unrealistic today: SHM in conjunction with material-integrated intelligence is still an area of intense research and has not yet entered the stage of widespread commercialization. This said, it must be added that other industry sectors have emerged in which implementation is more easily achieved, thanks to lower technological and/or regulatory obstacles.

What the above quotation also implicitly reveals is the inspiration behind SHM – our own nervous system. Load-bearing structures in the human body like bones are known to adapt their internal structure to prevalent loads. Cyclic and nontypical, one-of-a-kind loads, however, are exempted from this natural optimization process. To a certain degree, their possible effects are covered or alleviated by the nervous system, which allows preventive reflexive action, or influences behavior, for example, during healing of injuries, to limit loads on damaged structures. Thus, the idea behind SHM is the realization of a technical nervous system – of technical pain [3–7].

Material-Integrated Intelligent Systems: Technology and Applications, First Edition.
Edited by Stefan Bosse, Dirk Lehmhus, Walter Lang, and Matthias Busse.
© 2018 Wiley-VCH Verlag GmbH & Co. KGaA. Published 2018 by Wiley-VCH Verlag GmbH & Co. KGaA.

Seen from this point of view, SHM is an application concept rather than a technology. It can be adapted to several scenarios in which engineering structures are required to safely and efficiently bear loads, and in which failure should not occur unexpectedly. Typically, automotive [8], railway [9,10], maritime [11], and aerospace structures [12] as well as civil engineering ones [13,14], not the least among them offshore structures [15] and wind energy plants [16,17], fall into this category. In recent years, the wider spread of composite materials in commercial aircraft, currently culminating in the Boeing 787 or the Airbus A350, have led to an increased interest in SHM: Composite materials show complex, brittle failure behavior and are subject to internal damage like delamination that may escape visual inspection. This characteristic is reflected in safety factors that effectively introduce a significant weight penalty [18]. Today, SHM is widely implemented in monitoring of large infrastructure like bridges. In civil engineering applications like this, both material integration and miniaturization are less critical, often allowing the realization of full SHM solutions by combination of standard modules. Nevertheless, these efforts are relevant for material-integrated systems, too, in terms of the results they deliver on aspects like data evaluation or energy management.

The basic idea behind SHM is the application of nondestructive evaluation and testing (NDE/NDT) methods to survey, over an extended period of time, the condition of load-bearing structural components, primarily for increased safety. In contrast to a conventional NDT-based evaluation performed at either regular time or usage intervals, SHM introduces continuity to the evaluation process. In practice, the availability of structural state information at any time allows scheduling of maintenance operations on a need rather than, for example, on a fixed time basis (Maintenance on Demand (MoD), predictive maintenance etc.).

The distinctive characteristics of such a system have been formulated by Worden *et al.* as the "fundamental axioms" of SHM [19]:

1) All materials have inherent flaws or defects (Axiom I).
2) The assessment of damage requires a comparison between two system states (II).
3) Identifying the existence and location of damage can be done in an unsupervised learning mode, but identifying the type of damage present and the damage severity can generally only be done in a supervised learning mode (III).
4) Sensors cannot measure damage. Feature extraction through signal processing and statistical classification is necessary to convert sensor data into damage information (IVa).
5) Without intelligent feature extraction, the more sensitive a measurement is to damage, the more sensitive it is to changing operational and environmental conditions (IVb).
6) The length- and timescales associated with damage initiation and evolution dictate the required properties of the SHM sensing system (V).
7) There is a trade-off between the sensitivity to damage of an algorithm and its noise rejection capability (VI).
8) The size of damage that can be detected from changes in system dynamics is inversely proportional to the frequency range of excitation (VII).

22.1 Introduction

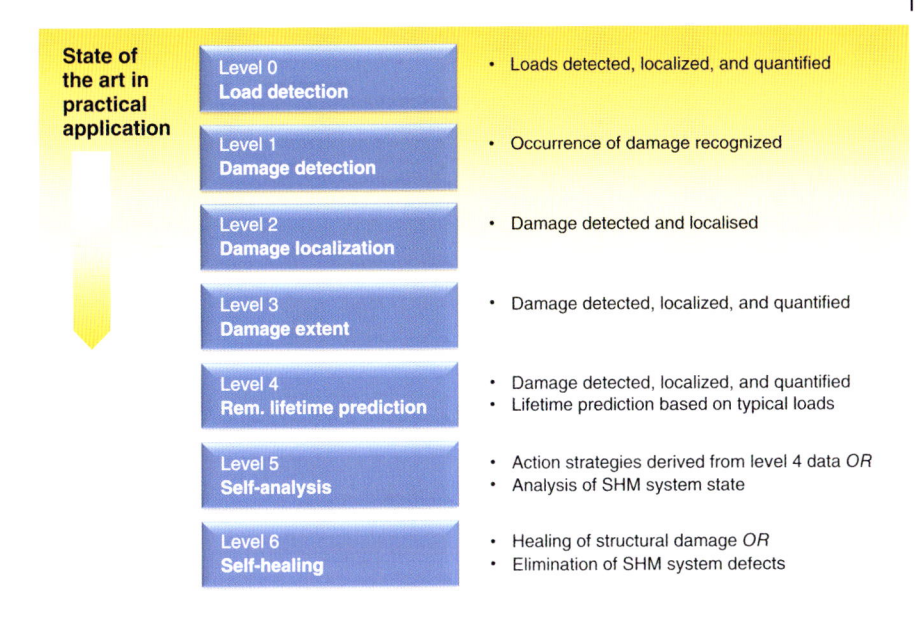

State of the art in practical application		
	Level 0 Load detection	• Loads detected, localized, and quantified
	Level 1 Damage detection	• Occurrence of damage recognized
	Level 2 Damage localization	• Damage detected and localised
	Level 3 Damage extent	• Damage detected, localized, and quantified
	Level 4 Rem. lifetime prediction	• Damage detected, localized, and quantified • Lifetime prediction based on typical loads
	Level 5 Self-analysis	• Action strategies derived from level 4 data *OR* • Analysis of SHM system state
	Level 6 Self-healing	• Healing of structural damage *OR* • Elimination of SHM system defects

Figure 22.1 SHM system levels of capability as described by Lehmhus *et al.* [22]. (Based on Rytter and Inman *et al.* [20,21].)

What these basic paradigms underline is the topical width of SHM as an area of research: Technologically, SHM incorporates all issues related to sensor networks. What is added, and this is explicitly highlighted in the above axioms, is the need for in-depth knowledge of the mechanical/structural behavior of the materials the monitored components are made of. The more sophisticated the envisaged system gets, the more detailed this knowledge has to be.

In consequence, systems addressing the SHM issue can do so on several different levels, starting from simple load monitoring and extending it up to the aforementioned capability of not only detecting but also locating and analyzing any incurred damage both as it is at the instant of measurement, and in terms of its future development. In terms of this scope of systems, Rytter [20] has offered an initial, stage-wise distinction that has been taken up by Inman *et al.* [21] and Lehmhus *et al.* [22], respectively. Figure 22.1 lists the relevant categories and provides a brief explanation of their meaning. Of these, specifically self-healing is somewhat set apart: As of today, healing is typically based on a hard-wired mechanism that allows structural materials to regain integrity and strength, but not on the evaluation of sensory data. Approaches of this kind thus do not require material-integrated intelligent systems, but depend on the buildup of the material itself. A common example in this respect are microcapsules embedded in a host material and filled with a reactant that, once their shells have been fractured by an advancing crack, release the liquid reactant, which then seals the cracks. The reactants curing process is typically supported by a catalyst likewise embedded in the matrix.

As system development progresses through the categories described in Figure 22.1, the level of detail increases from representation of linear-elastic

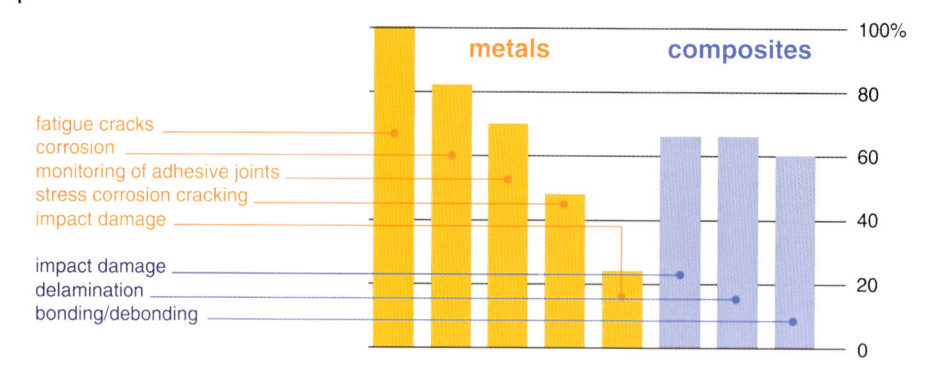

Figure 22.2 Information expected from SHM systems applied to metallic as opposed to composite structures, according to a survey performed by Bartelds [23].

behavior, which may suffice for load identification in some cases, to global and local failure mechanisms and the progressive development of local failure, as well as the latter under conditions of varying service histories and secondary external influences.

Altogether, such knowledge can roughly be grouped into three distinct categories:

- Understanding failure and failure propagation mechanisms
- Understanding the link between sensor signals and damage as well as damage accumulation
- Understanding the influence of external parameters (environmental, operational, etc.) on the aforementioned aspects

In terms of the actual materials considered, in aerospace and automotive applications as well as in many wind energy plants, fiber-reinforced composites get the main focus: In fact, it is the transition, be it a lasting one or not, from metal-dominated to CFRP-based large passenger aircraft as well as the growth and move toward offshore locations (which turns maintenance into an even more costly affair) of wind energy plants that has fueled much of the current interest in SHM. However, as such, SHM is not linked to a single material or material class, as Figure 22.2 illustrates based on an inquiry performed prior to the advent of the aforementioned technological developments.

What is naturally reflected in the expectations toward the system are the differences in the relevant failure mechanisms associated with metals on the one hand and polymer–matrix composite materials on the other.

Because structures that warrant or even necessitates monitoring, and also load bearing ones, are typically safety-relevant, SHM systems need to conform to stricter specifications than sensor networks in general in terms of

- robustness,
- reliability,
- fault tolerance, and so on.

These aspects are relevant on hardware, but also on software level. Because of such tighter requirements, developing standards that support safe implementation of SHM is a major prerequisite of a further spreading of such systems. The lack of such standards thus constitutes a major obstacle hindering their implementation in particular in the aerospace industry.

A first official document addressing this need has recently been published by the Society of Automotive Engineers (SAE): Among others, the Aerospace Recommended Practice ARP6461 "Guidelines for the Implementation of Structural Health Monitoring on Fixed Wing Aircraft" contains a definition of SHM systems, discusses motivations behind their use, and outlines abstract usage scenarios in conjunction with basic principles. Furthermore, it elaborates on validation, verification, and certification procedures for SHM systems, a topic also highlighted by Bockenheimer and Speckmann from an Aerospace perspective [24].

Following the wider understanding formulated at the onset of this chapter, according to SAE ARP6461, structural health monitoring forms a generic part of the much broader aircraft health management. The latter encompasses engine, systems, and finally structural health management, with the latter further broken down to include crew observations, maintenance records, fleetwide data analysis, and finally SHM itself as sources of information for assessing and maintaining the structural health of an aircraft. SHM is thus "the process of acquiring and analyzing data from onboard sensors to determine the health of a structure" [25].

Obviously, this definition does not incorporate the aforementioned continuity brought in by the colloquial use of the term SHM. Consequently, SAE ARP6461 distinguishes between scheduled and automated SHM. While scheduled SHM means reading out sensor data at fixed intervals, it is automated SHM that adopts the continuity aspect, as in this case the system itself is expected to call for maintenance operations when required and must thus establish an ideally constant awareness of structural state, and detect any state change. In principle and in terms of the procedure applied, scheduled SHM can thus be likened to NDT investigations performed during scheduled maintenance operations, although using onboard rather than external sensors.

Figure 22.3, which is based on Ref. [26], depicts a general scheme of an SHM system and highlights two different approaches employed in SHM: Besides the distinction between scheduled and automated SHM, SAE ARP6461 assumes a technology-based delimitation between damage monitoring and operation monitoring. Damage monitoring is understood as direct observation of the structure with the single aim of detecting damage. Operation monitoring refers to the use of sensors not necessarily deployed for SHM purposes as a source of data reflecting operational conditions over time and the usage of these data for inference of information on structural state and/or probability of failure.

As an aside, because operation monitoring does allow for use of dedicated sensors, load monitoring falls into this category.

Since ARP6461 assumes that mode of operation and technology type can be freely combined, four different SHM categories emerge. Of these, automated damage and operation monitoring are nearest to the understanding of SHM as continuously active, onboard solution – and thus most likely to see implementation as material-integrated intelligent systems.

The following section will discuss motivations for SHM in more detail and provide an exemplary overview of SHM case studies that are meant to illustrate

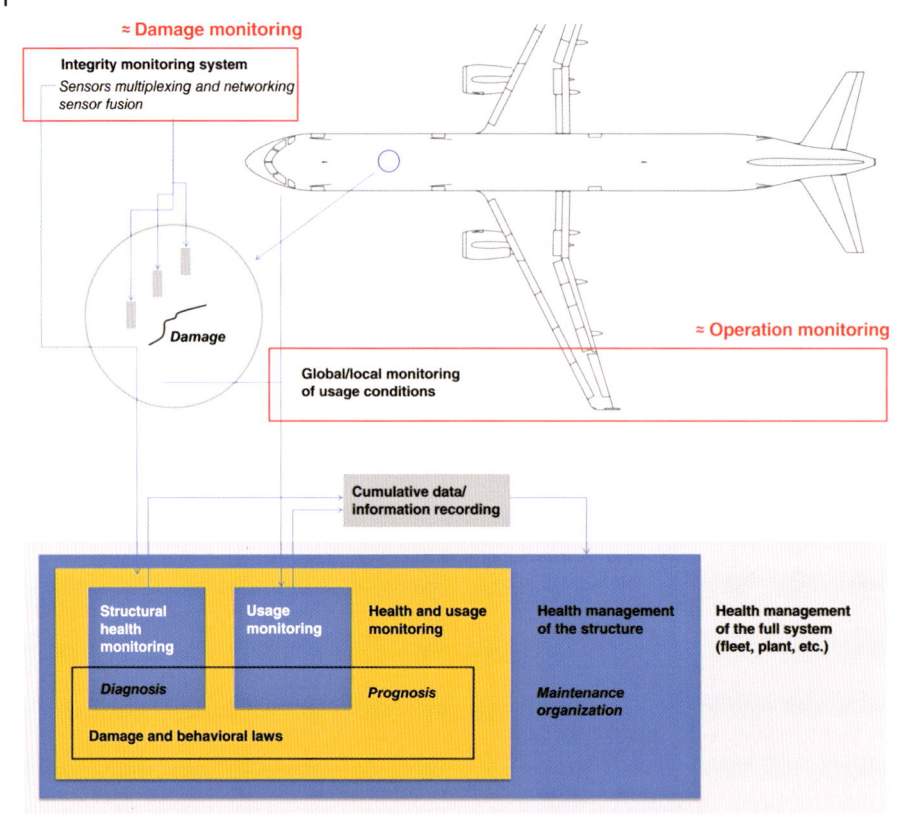

Figure 22.3 Organization of an SHM system according to Balageas *et al.* highlighting the two general approaches, damage and operational monitoring [26]. (Drawing by Julien Scavini, CC BY-SA 3.0.) https://commons.wikimedia.org/w/index.php?curid=16851855

the width of the topic in terms of application areas, technologies, and research foci. Subsequently, aspects of the individual components that make up an SHM system will be discussed in more detail. In this, the system breakdown suggested by SAE ARP6461 will be used as guideline for discussion, and thus as basis of the section's structuring [25].

22.2 Motivations for SHM System Implementation

While the general purpose of SHM is to keep track of the structural state of a technical system, when implemented, this information can be used in several different ways. The main motivation for SHM can thus be

- to increase safety levels,
- to realize weight reductions while maintaining overall safety levels, and
- to reduce maintenance costs.

The latter two aspects are highlighted in Figure 22.4, which illustrates schematically the interdependency and possible trade-off between inspection interval

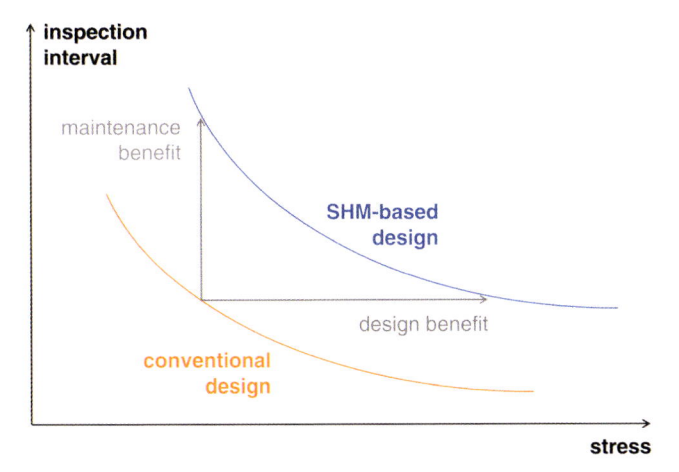

Figure 22.4 Motivations for the implementation of SHM systems: reduction of maintenance costs versus next-generation lightweight design.

and allowable stress level for a load-bearing structure. The two curves represent a conventional design approach and another one that makes use of an SHM system. The basic assumption is that the structure will ultimately reach its safety limits through accumulation of damage over time, that the amount of damage can be accurately detected, and that its accumulation is understood well enough to predict its further development. Under such circumstances, when maintaining the design stress level, the inspection interval for a given level of safety can be raised: This perspective constitutes a maintenance benefit. By shifting the point of view, a design benefit – in practice, a weight reduction – can be realized, which is illustrated by the horizontal arrow. In this case, an increased stress level is accepted, while the inspection intervals valid for a conventional design approach are retained.

Figure 22.5 further describes the principle. Here, fixed inspection intervals have to be tuned to ascertain that accumulation of damage does not exceed critical limits in between scheduled maintenance events. The figure also illustrates schematically how either a change in damage evolution rate or a singular event causing an offset of the damage size curve could counteract this approach if not detected in time by continuous monitoring.

Prerequisite of both Figures 22.4 and 22.5 is a matching design principle: Damage tolerance, which allows damage to occur on the basis that either its growth is understood well enough to maintain safety over a predictable period of time or structural fall-back solutions exist that guarantee safety even after local failure has occurred. Figure 22.6 provides an overview of the respective design philosophies as categorized by Boller and Buderath in view of design for fatigue loads [27].

Figure 22.5 thus stresses the benefit of continuous monitoring in the damage-tolerant design context: If damage accumulation follows the previously assumed path, fixed maintenance intervals may be spaced accordingly. However, if via some unexpected event, like an impact causing superimposed damage, either the

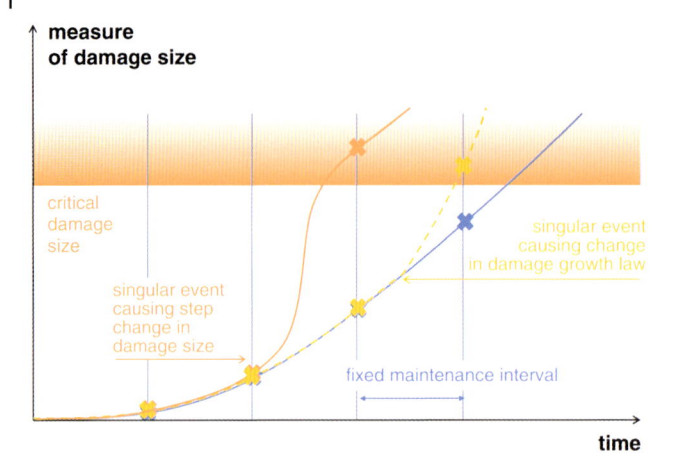

Figure 22.5 Maintenance scheduling versus the development and propagation of damage, schematic view.

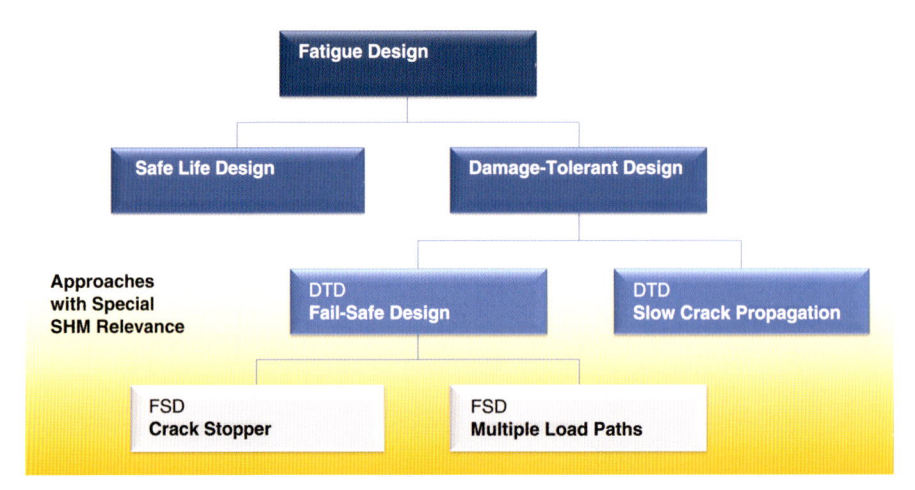

Figure 22.6 Aerospace fatigue design principles based on a similar disambiguation by Boller and Buderath [27].

level or the governing law changes without this being noticed, a critical damage size threshold might be passed prior to the next scheduled inspection.

Naturally, the importance of lightweight design and/or reduction of maintenance differs depending on the business sector. For the aerospace industry, Figure 22.7 reflects both aspects: Obviously, the relations between the various cost items may change, but the message that both reduction of weight (which directly affects fuel consumption) and maintenance effort (included as MRO, that is, maintenance, repair, and overhaul) have a major impact on the final cost figure is clear. Apart from the issue of direct costs addressed here, it is obvious that an aircraft should be airborne and thus earning revenue for as much of its service life as safely possible rather than being grounded for MRO operations. As an aside, it

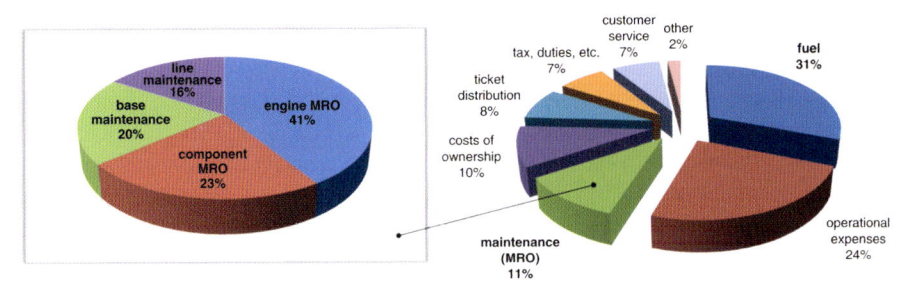

Figure 22.7 Overview of direct operating cost factors in commercial aircraft operation, according to footnote[1].

may be noted that both fuel consumption and maintenance requirements tend to rise as the aircraft ages.[2]

Keller and Poblete have discussed the business case for SHM from a generic perspective, using an aircraft application as example and discussing factors that should form part of a baseline enterprise model as well as the deviation from this starting point effected by SHM implementation [28]. More concretely, Pattabhiraman *et al.* have estimated the economic advantage of condition-based versus scheduled maintenance using an Airbus A320 as object of their study: According to their findings, supporting maintenance operations with an SHM system can save enough costs over a lifetime estimated at 60 000 flights for this type of short-haul aircraft even if balanced against additional fuel consumption caused by the SHM system increasing fuselage weight by as much as 5–10%. SHM usage concepts and associated maintenance procedures considered include the following [29]:

- *SHM-supported scheduled maintenance:* The SHM system is used during scheduled maintenance to support or replace NDE techniques. In other words, the structural state is determined offline at regular intervals using the onboard SHM system rather than external test equipment. Which components are to face a detailed investigation is not a decision taken by the SHM system.
- *Condition-based maintenance (CBM):* The SHM system ascertains the structural state of the aircraft over an extended period of time, but during scheduled maintenance only, and will call for in-depth structural maintenance operations if the evaluation results suggest that accumulated damage exceeds the relevant thresholds. The SHM system determines which components require detailed investigation.
- *CBM-skip:* The SHM system relies on onboard sensors and ground-based evaluation, reducing complexity and weight of airborne systems. It will ascertain structural states in shorter intervals, typically during A-checks (approximately 100 flight interval, 350–650 h of service), and support decisions on skipping unnecessary structural maintenance for certain components during

---

1 http://www.jetms.aero/en/press-release/gediminas-ziemelis-selecting-the-optimal-engine-for-the-a320-neo-the-pw1100g-gtf-or-the-cfmi-leap-x.html (accessed January 3, 2017).
2 http://www.ascendforairlines.com/2013-issue-no-2/size-and-age-matters (accessed December 12, 2016).

more comprehensive maintenance operations like C-checks performed every 12–18 months.

Other authors have identified further maintenance categories. Tchakoua *et al.* distinguishes between reactive and corrective maintenance, which implies running the system up to the point of failure, preventive maintenance, which matches scheduled maintenance in relying on a fixed time interval separating recurring maintenance work, and predictive maintenance, which is the only approach using knowledge about the structural condition to motivate maintenance [30]. The same authors consider corrective maintenance as being diagnostic by nature (faults are detected), whereas predictive and the narrower understanding of preventive maintenance are summarized as prognostic (faults are prevented), leading to the following, detailed classification [30]:

- *Diagnostic (fault detection):* Corrective maintenance, executed after the fault
  - Palliative maintenance (provisional rehabilitation)
  - Curative maintenance (permanent rehabilitation)
- *Prognostic (fault prevision):* Preventive maintenance, executed before the fault
  - Systematic maintenance (time-based, scheduled intervention)
  - Conditional maintenance (parameter/threshold-based intervention)
  - Predictive maintenance
    - Forecast maintenance (parameter prevision based intervention)
    - Proactive maintenance (condition-based intervention)

In the comparison by Pattabhiraman *et al.*, all three approaches led to very similar maintenance cost figures. All of these were shown to undercut conventional scheduled maintenance by roughly the same margin of almost 15% (pessimistic assumption) up to more than 20% (optimistic assumption) [29].

22.3 SHM System Classification and Main Components

SHM systems can be distinguished based on a number of different features. The listing below provides examples, but is not necessarily fully exhaustive in this respect:

- Type of material under scrutiny (i.e., for material-integrated systems, the host material)
  - fiber-reinforced polymers
  - metals
  - polymers
  - concrete
- Type of sensor(s)
  - active sensors
  - passive sensors
- Type of sensing principle employed
  - optical sensors
  - piezoelectric sensors

- piezoresistive sensors
- electromagnetic impedance
- Type of effect/influence monitored
  - corrosion
  - fatigue
  - impact
  - load, and so on.
- Concept of communication within the system
  - wired communication
  - wireless communication, and so on.
- Type of information derived, according to system classification in Figure 22.1
  - load monitoring
  - damage detection
  - damage localization, and so on up to self-analysis and healing
- Concept of data evaluation
  - model-based versus model-free approaches
  - hybrid approaches, and so on.
- Localization of data evaluation
  - distributed versus centralized data evaluation
  - fully or partially material-integrated versus external
  - for aerospace application, onboard versus ground-based
- Time of data evaluation relative to timestamp of incidents detected
  - online versus offline
  - real-time versus, for example, scheduled data evaluation
- Application scenario/environment, industrial sector
  - aerospace
  - railway
  - automotive
  - civil engineering
  - maritime/offshore industries

What is exemplified in this list linking SHM to the broader field of intelligent and specifically material-integrated intelligent systems is the fact that design and development of practically applicable concepts of this kind must consider several disciplines in parallel. Achieving an optimum combination of partial solutions thus requires an adapted multidisciplinary design methodology. Suggestions in this respect have, for example, been made by Peters *et al.* promoting a coordinated multidisciplinary design approach designated S^3 (sensor–structure–system) logic [31]. The problem as such is also known from other communities linked to smart or cyber-physical system design [32]. It also forms part of this book, where it is treated on generic as well as on detailed level in Part Two.

The component-wise breakdown of a material-integrated SHM system naturally mirrors that of a generic sensor system and includes the following hardware items and functionalities (listing partially based on SAE ARP6461 [25]):

- Transducers
  - sensor elements
  - actuator elements

- transducer interconnects
- Signal conditioning components
  - filters
  - amplifiers
  - AD converters, and so on.
- Signal/data processing components
  - microprocessors/microcontrollers
  - memory/data storage and so on
- Communication systems
  - interconnects
  - data buses
  - transmitters/receivers, and so on
- Control and interface devices
  - display, and so on
- Data storage
- Energy supply systems
  - energy storage
  - energy harvesting
  - energy transmission interconnects

Since the reference cited here refers to aerospace applications, a distinction is made between airborne and ground-based components as well as those that are not necessarily bound to either location.

The specific challenge in realizing a material-integrated intelligent SHM system is that in such a system, finally all components would not only be airborne, but also, ultimately, be integrated in the primary aircraft structure. This, basically, describes the level of ambition implied by the introductory quotation reflecting about "structure that talks to us" [2].

The actual choice of components (sensors, data acquisition, signal/information processing, etc.) for an SHM system depends on the monitoring task and field of application, since these aspects set the boundary conditions of what to track, at which level of accuracy and time resolution, and so on. To give an example that is also reflected in the selection of case studies in Table 22.1, monitoring of civil engineering structures is often based on optical sensors, namely, fiber Bragg gratings (FBGs), which allow spatially extended networks and simple multiplexing, but are typically associated with large-size evaluation units that might introduce an intolerable weight penalty in applications dominated by the need for lightweight design, like the aerospace industry.

22.3.1 Sensor and Actuator Elements for SHM Systems

The choice of sensors available for SHM is wide and naturally linked to the material under scrutiny, the nature of the measurands, and the adopted sensing/monitoring principle. The listing of SHM system disambiguation criteria has already provided an indication of the scope, which is extended in Table 22.1.

In the former, as general way of delimiting different SHM systems, the underlying principles of sensing have been named. Among them is the distinction

Table 22.1 Classes and types of sensors used in SHM with associated application areas and references.

Fundamental technology/ principle	Physical effect exploited	Materials monitored (relevant examples)	Transducer type (example)	Feature detected
Acoustic emission	Detection of acoustic waves associated with external and internal structural events like impacts, crack initiation, and growth delamination	Metals composites	Piezoelectric patch	(Impacts) Cracks delamination
Acousto-Ultrasonics	Propagation of acoustic waves, reflection/ change of propagation characteristics, attenuation, and so on at local inhomogeneities, interfaces, damage sites, and so on	Metals composites including MMC and CMC	PWAS, piezoelectric patch, phased array PWAS	Cracks delaminations
Comparative vacuum monitoring	Connection of parallel small-scale evacuated and pressurized through cracks, detection of pressure change/loss of vacuum	Metals composites (adhesive joints)	CVM patch, external pressure monitoring device	Cracks (crack propagation) corrosion debonding
Environmental degradation monitoring	Monitoring of environmental conditions inducing/supporting corrosion/(pH, temperature, humidity, and so on), associated corrosion model for data evaluation and prediction	Metals	Individual or multifunctional sensors and sensor patches covering the relevant characteristics	Corrosion
Electrical crack gauges (ECG, rip wire array)	Rip wire approach – loss of electrical connectivity for cracks passing the sensor, crack propagation accessible when using multiwire ECG patch	Metals composites	Single of multiple crack wire patches	Cracks (crack propagation)
Fiber Bragg gratings	Wavelength-dependent reflection/extinction in optical fiber with local periodic variation of refractive index (Bragg Grating), reflected wavelength controlled by grating periodicity which changed by longitudinal strain	Composites metals concrete	Optical fibers with locally inscribed gratings, multiple gratings with different constant possible per fiber, FOS-FBG patches, for example, via direct write techniques for foil-integrated sensors through integrated optics approaches	Strain impacts delaminations (temperature via thermal strain)

(*continued*)

Table 22.1 (Continued)

Fundamental technology/ principle	Physical effect exploited	Materials monitored (relevant examples)	Transducer type (example)	Feature detected
Foil-based eddy current sensors	Generation of eddy currents within the material monitored, inductive detection of their characteristics and of deviations from the ideal state caused by damage	Metals	Sensor patches incorporating coils for eddy current generation and detection	Cracks corrosion
Ultrasonic imaging	Transmission and reflection of ultrasound in structures and reconstruction of 2D images	Metals composites	Ultrasound transducer	Cracks voids delaminations
Microwave sensors	Transmission of microwaves through the material according to a pitch-catch principle, returning signal correlated with water content	Composites		
Sandwich materials, for example, with honeycomb core	Microwave antennae	Water uptake		
Strain gages	Combination of resistivity (material property) and resistance (caused by geometrical effects) change in a conductor caused by mechanical strain	Metals		
Composites ceramics concrete	Metal or semiconductor strain gauges	Strain (temperature via temperature sensitivity or thermally induced strain)		

Extension and synopsis of similar compilations provided by Speckmann and Henrich [33], Balageas et al. [26], and Sbarufatti et al. [34].

between damage and operation monitoring: Either damage is detected directly, or the operational conditions are captured in order to deduce the probability of damage having been incurred.

Operational monitoring, but even more so damage detection can in turn either be based on locally probing the structure or on gathering data about its global behavior from which, for example, the location or extent of damage can be derived.

Both systems have their own advantages and drawbacks: A damage detection system may identify, for example, location, size, and evolution of damage directly and thus more easily and potentially with higher accuracy. However, it delivers less information than an operational monitoring system in cases where either no damage is present yet or the damage remains static, that is, the system does not undergo any detectable change. Operational monitoring systems, on the other hand, are characterized by their indirect access to the damage information, as they reconstruct damage characteristics based on previously established links between their foreseen causes and their magnitude, distribution over time, and so on as captured during the operation of the structure under survey. The link itself could, for example, be the inverse solution of a deterministic system model or a classification of sensor signal combinations established via machine learning.

Provided that the underlying models permit this, operational monitoring allows tracking the state change of a structure before it is explicitly expressed in (macroscopic) damage. In principle, this capability could be used to redefine baselines for the decision of critical system states, rooting their identification in the service history of the structure. Assuming that data evaluation is flexible in its use of models, when compared to direct damage detection systems like rip wire or CVM, operational monitoring systems could even experience an improvement of predictive and damage identification capabilities once improved models have become available.

In Figure 22.8, we will distinguish between SHM sensors based on the aspects global versus local monitoring and active versus passive sensing or rather monitoring principle. The distinction between active and passive approaches has been discussed in detail for aerostructures by Staszewski et al. [35].

Passive sensing then refers to a "listening" sensor system that evaluates the effects of external stimuli. Active sensing, in contrast, would involve elements that create the stimulus themselves. As a consequence, active sensing systems are typically conceived as combinations of active and passive sensing elements.

The reader should note that the proposed disambiguation between an active or a passive nature of the act of sensing itself differs somewhat from the common notion that passive sensors are those that require a power supply (e.g., piezoresistive sensors), while active sensors may not (piezoelectric sensors): Our perspective, for example, allows piezoelectric "active" sensors to be used in an active manner as emitters of elastic waves and in a passive way for detecting mechanical strain. However, some authors, in a reversal of our own perspective, see passive sensors as those generating energy, whereas active sensors depend on some external energy supply. This alternative view is, for example, proposed by De Rossi et al., who rate, for example, piezo- or thermoelectric sensors as passive and piezo- or thermoresistive sensors as active [36].

22 Structural Health Monitoring (SHM)

Local sensing
- Rip wire-based methods
- Comparative vacuum monitoring (CVM)
- Acoustic emission techniques
- Strain monitoring using
  - Piezoresistive sensors like strain gages
  - Piezoelectric sensors
  - Fiber-optical sensors loke Fiber Bragg Gratings (FBGs)

- Piezoelectric sensor/actuator combinations for damage detection
  - Pitch-catch mode
  - Pulse-echo mode
  - Phased array approaches
- Lamb wave excitation and propagation monitoring

Passive sensors | **Active sensors**

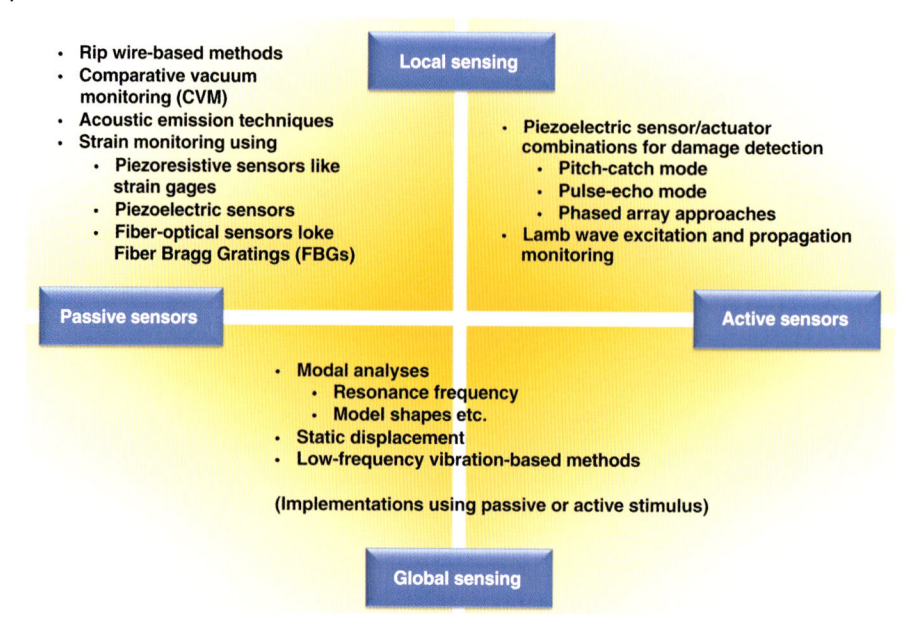

- Modal analyses
  - Resonance frequency
  - Model shapes etc.
- Static displacement
- Low-frequency vibration-based methods

(Implementations using passive or active stimulus)

Global sensing

Figure 22.8 Examples of global and local SHM system concepts using either active or passive sensing principles.

In general, direct detection of damage in this sense is more often based on active systems, while operational monitoring can more easily rely on passive systems.

In terms of the global versus local distinction, global response can be tested either actively by generating a stimulus or passively using the stimuli introduced in service. Similarly, local monitoring can use focused active as well as passive methods.

Possibly the simplest way of detecting that something is wrong with the monitored structure is to cause a conductive path to be interrupted by damage. This is the conventional rip wire approach. Comparative vacuum monitoring (CVM) can be seen as an extension of this basic concept. In this case, vacuum and controlled pressure channels are located side by side, with cracks connecting these, thus compromising the maintenance of vacuum in the respective channels or galleries. Whereas rip wires could also be envisaged as material-integrated solution, CVM is most suitable for surface application, including surfaces internal at least to the structure in the case of monitoring adhesive joints. Among the advantages of CVM is the fact that these systems can detect surface cracks situated directly below the sensor patch even if these do not damage the patch itself. Besides, self-analysis approaches have been suggested for this type of sensor, which thus fulfills an important requirement with respect to the avoidance of "false positive signals" [37].

Acoustic emission (AE) techniques typically use piezoelectric sensors to detect structure-borne sound signals originating, for example, from fracture or damaging events like impacts within the material or structure. Further to the detection of the event as such, triangulation and similar methods may be used to identify the source of the signal captured [38,39].

Common to the former approaches is the fact that damage is detected directly, meaning that basically no information can be gained as long as the structure is still healthy. Naturally, this affects predictive capabilities of the methods. AE is different in this respect since characteristics of the recorded signal may be used to derive additional information about nature and extent of damage [40,41].

In contrast to rip wire, CVM and AE techniques, strain monitoring can continuously capture information (local strain or strain states) related to the loading of the structure. Damage is not detected directly this way – instead, its occurrence, and potentially also its size and place, can be predicted via the accumulated strain history in combination with models that describe damage initiation and evolution in relation to both individual overload situations and lower level loads acting over extended periods of time. Alternatively, in case a known load is acting on the structure (e.g., a test load), deviations from the overall strain pattern expected from a healthy structure may be exploited to draw conclusions on the nature of their sources.

Practical realization of strain measurement can follow several different approaches, most common among these being the use of conventional, piezoresistive strain gauges, piezoelectric sensors, or optical strain sensors, the latter usually fiber-based and configured as interferometric sensors (Fabry–Pérot-, Mach–Zehnder-, Michelson or Sagnac interferometer), or as grating (fiber Bragg grating, long period grating, chirped or tilted FBG) or distributed (making use of Raman, Rayleigh, or Brillouin scattering) sensors [42]. Beyond conventional glass fiber integration, several such optical sensors have also been realized in polymer fibers or even foils, sometimes using direct-write techniques to realize the necessary structuring according to an integrated optics approach, as demonstrated for a Mach–Zehnder interferometer by Koerdt and Vollertsen [43]. Details about the various types of optical sensors are given in the respective chapter within this book (chapter 5).

Active techniques are very often based on excitation of a stimulus through piezoelectric elements (often called piezoelectric wafer active sensors (PWAS)) used either exclusively as actuator or both in an actuator and a sensor role [44]. The clear separation between actuator and sensor is an example of the pitch-catch method [45], in which a probing signal is emitted by one PWAS and detected by another, while the opposite method is designated the pulse-echo approach: Here, the emitter turns detector for the reflected signal. A special option facilitated by the coordinated use of a set of neighboring PWAS is the phased array approach, in which signal emission by the PWAS is controlled in order to create a directed wave front. Via this approach, the actual scanning of the monitored structure is enabled [46,47].

Besides the ultrasonic probing approach already described, PWAS elements can be used to excite Lamb waves [48]. Lamb waves are elastic guided waves observed in plate-like structures with associated particle motions perpendicular to the plate and in the plane defined by it. Depending on the excitation frequency, two fundamentally different classes of modes of Lamb wave propagation are observed, namely, the symmetric and the antisymmetric modes, where symmetry is related to particle motion perpendicular to an assumed plate center plane, which at the same time represents the symmetry plane. Among the noteworthy effects that render Lamb waves interesting for SHM applications is the fact that in fiber-reinforced composites, damages will effect wave propagation and may even cause a mode change [47–50].

A major problem associated with usage of Lamb waves for SHM is the correct understanding and accurate prediction of their interaction with damage on the one hand and complex structural features like stiffeners, changes in material thickness, or, in hybrid structures, even the material itself on the other. In addition, several authors have shown that, for example, in fiber-reinforced composites, Lamb wave propagation is greatly influenced by environmental conditions like temperature and humidity – to such a degree that at present, an actual SHM system for in-service monitoring following this approach must still be considered a vision [51–53].

While the aforementioned options either detect damage or loads directly at the sensor's location or probe the monitored structure locally, global monitoring techniques consider the response of the structure as a whole and derive information about damage from deviations from the expected behavior. One example in this respect is modal analysis, which considers resonance frequencies or modal shapes and damping: While data may be gathered locally via accelerometers and/or strain sensors in order to determine, for example, the modal shape itself, this local access is primarily used to reconstruct the global response that is then evaluated, for example, in terms of signs of damage. Depending on whether the required stimulus is provided by the monitoring system itself or is of external nature, the method can either be classified as active or as passive. A detailed overview of modal analysis principles used in damage detection has been provided by Farrar and Doebling [54]. Vibration-based solutions in general, including modal analysis, have been reviewed by Das *et al.* from a civil engineering point of view [55]. Alternatives to vibration-based analyses include static displacement analysis.

Table 22.1 lists concrete examples of sensors used in SHM scenarios, naming the physical effect that is being exploited, the fundamental sensing or transducing principle, if applicable, the transducer material or system employed, and the features/types of damage detectable. Those cases where types of sensors discussed specifically match the needs of material-integrated intelligent systems are treated in more detail in the dedicated chapters of this book. The list is not conclusive, but should be seen as an overview of technologies.

A typical example of a lab-scale evaluation of sensor and sensing principle suitability for SHM is depicted in Figures 22.9 and 22.10. In both cases, the object of study as shown in Figure 22.9 is a representative section of a helicopter fuselage, made from aluminum and featuring skin and stiffeners. The various sensors are attached to the skin as depicted in both figures, and a static test load (Figure 22.9a) is applied to a healthy and various deliberately damaged variants of this structure.

Figure 22.9 exemplifies a passive sensing approach, in which strain sensors of electrical (piezoresistive strain gauges) and optical (fiber Bragg gratings in optical fibers) types are used to sample the strain field in the structure, and detect changes caused by artificially introduced damage. The sensor information is matched with an FEM simulation of healthy and damaged structures, as shown in Figure 22.9.

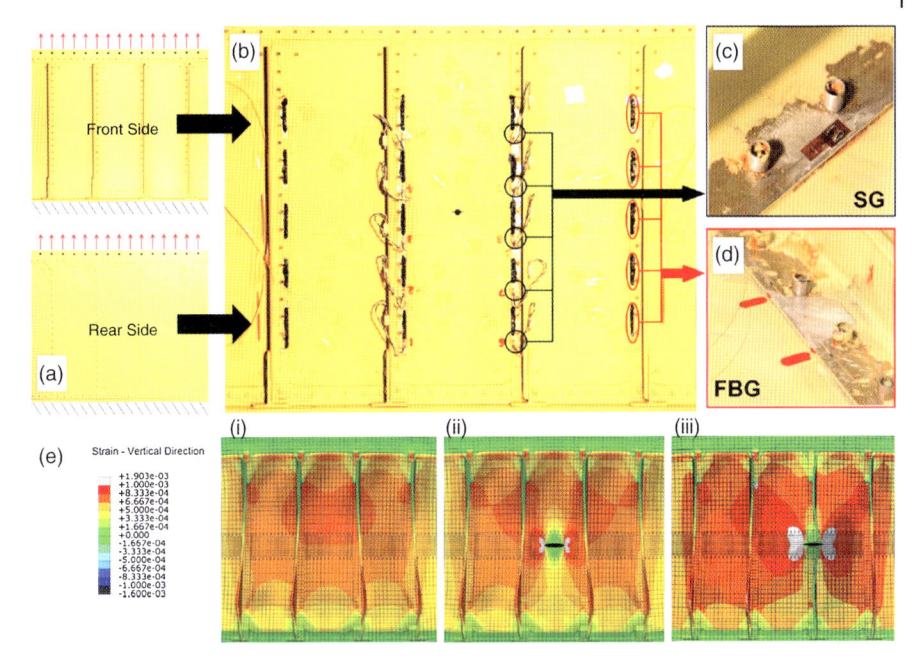

Figure 22.9 Helicopter fuselage panel with stiffeners as test case for SHM sensor evaluation. (a) Static loads applied during testing. (b–d) Application of strain gauges (SG) and fiber Bragg grating (FBG) sensors to the structure. (e) Simulated strain field under test load as benchmark for sensor output: (i) healthy structure, (ii) simulated crack at center of structure, and (iii) simulated failure of stringer. (Images by permission of John Wiley & Sons, Inc. [34]).

The healthy structure's strain distribution serves as baseline for the evaluation of the strain field, and once a deviation is observed, further simulations can support identification of its sources.

Figure 22.10 depicts the corresponding active approach, which follows a pitch-catch strategy: Individual PWAS applied to the structure are used to create a probing signal, which is captured by further PWAS devices. Again, baseline data are collected for the various ultrasonic wave transmission paths, with deviations from the baseline signifying the presence of damage. Figure 22.10c and d shows the setup as well as the results obtained for the same types of damage used in the passive sensing example.

The relative importance of the various types of sensors and sensing principles for SHM has been evaluated by Balageas *et al.* [26]. The results are semi-quantitative at best, as they are based on data sources that are not necessarily fully representative of the area (e.g., contributions to the first two International Workshops on Structural Health Monitoring held at Stanford in 1997 and 1999), but still show some interesting trends. Among these is the fact that preference for specific types of sensors depends on the application area, with fiber-optic sensors dominating the civil engineering sector, which also shows more variation, and piezoelectric ones representing the major share in the aerospace industry. When seen from a material and nature of damage perspective, a similar evaluation of composite structure delamination monitoring looking at 1150 studies published

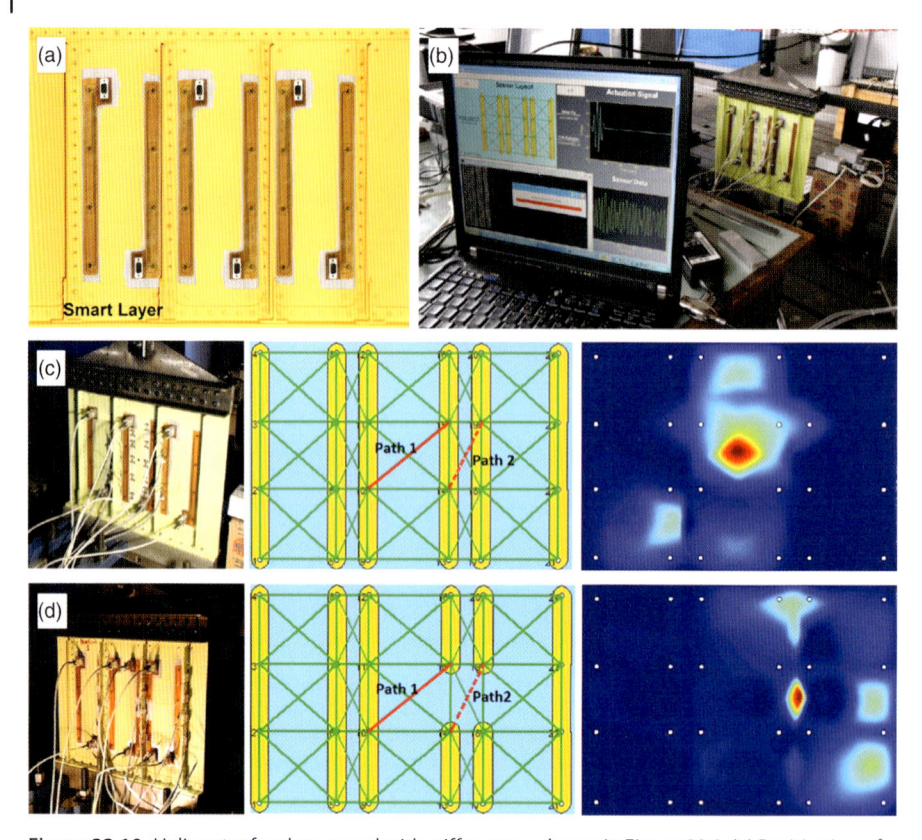

Figure 22.10 Helicopter fuselage panel with stiffeners as shown in Figure 22.9. (a) Positioning of piezoelectric sensor patches. (b) Signal processing and data evaluation unit with sensorized structure under test load. (c) Evaluation of Lamb wave propagation for damage detection for under test load with simulated crack at center of structure. (d) With simulated failure of stringer. (Images by permission of John Wiley & Sons, Inc. [34]).

between 1997 and 2003, the findings revealed a similar dominance of piezoelectric sensors (68%; of these, 65% PZT single-patch sensors, 8% SMART layer sensors, 7% other types of piezoelectric sensors, and 4% classical acoustic emission sensors), followed by fiber-optic (20%; of these, 38% fiber Bragg gratings) and electromagnetic sensors (12%). When distinguishing based on the underlying method of detection, for composite delamination, high-frequency wave propagation assumes the leading position at 28%, followed by AE (21%), static strain (17%), and AE & Lamb waves (10%). Vibration analysis, static/dynamic displacement, and electrical conductivity measurement all reach 4%, while electromechanical impedance comes in at 3%. Local methods thus apparently gain considerably higher levels of attention [26].

22.3.2 Communication in SHM Systems

Communication shall primarily refer to communication between sensor nodes here. In this sense, SHM systems as such share many aspects with sensor networks in general. However, either the typical SHM application scenarios or the boundary conditions of material integration, or the combination of both factors may require specific solutions. General remarks on sensor network communication taking into account the basic requirements of material integration are predominantly treated in chapters 15 and 16 and will not be reiterated here, where aspects with specific relevance for SHM application environments shall be highlighted instead.

Wireless as opposed to wired transmission of information is one of the most generic questions in this respect, and an important side aspect is that besides information, there might also be a need to distribute energy within the sensor network: If application requirements demand a specific solution for either of these tasks, to follow this approach in the other case may be the most economic decision. Economic in this case may not only refer to the cost of implementation but also to energy needs. Looking at communication from an energy efficiency perspective, a wired solution using electricity as medium would be preferable to the light-based option, while both could outperform a wireless approach. However, introducing conductive paths adds complexity and above all weight, which is negligible in some but extremely important in other scenarios such as the aerospace one. Besides, going wireless might improve network reliability and robustness, since structural damage would not endanger communication paths. In a wired network, the mere possibility of losing the physical links between neighboring sensor nodes affords provisions for rerouting communication in a damaged network, like the delta routing approach suggested by Bosse and Pantke [56] as well as Bosse and Lehmhus [57].

In contrast, however, a lack of conductive paths for energy distribution in conjunction with the low efficiency of wireless node-to-node energy transmission creates a need to redesign energy supply on node level toward the aim of energetic autonomy of sensor nodes. In practice, such autonomy can only be achieved through combined energy harvesting and storage. The former can utilize a large variety of effects such as structural vibration via piezoelectric elements (in this case, a combination of harvesting and active or passive sensing may sometimes be realized, as, for example, has been demonstrated by Zhao *et al.* [58,59]), temperature gradients via thermogenerators based on the thermoelectric effect, or ambient electromagnetic radiation via suitable antennae. Details of both energy harvesting and storage will be discussed in chapters 20 and 21 within this book.

The price to be paid for dispensing with interconnects for energy transmission thus is a significant increase in the required capabilities and thus the complexity of the sensor nodes. Secondary benefits in this case, however, are the elimination of interconnects as possible failure initiation sites, and the fact that no interconnects will need no dedicated repair concept. The latter typically require considerable efforts specifically for material-integrated rather than surface-mounted systems. In contrast, the ensuing difficulty or even impossibility of optimally balancing nodal energy levels in accordance with the requirements of the monitoring task throughout the network is among the drawbacks of wireless systems.

Not the least for this reason, wireless communication also requires actions aimed at reducing the amount of information passed between sensor nodes. This can be achieved through preprocessing of data at the sensor node level to facilitate transmission of higher level information only. However, such distributed data processing will once again raise the local energy demand. Possible compensating measures include the introduction of sleep states, keeping the number of active sensors low during normal operation and uncritical system health states, but waking up additional sensors whenever more detailed information is required. In a similar approach, the algorithms called upon for data preprocessing may be selected in order to either reduce overall energy consumption or deliver the highest level of information in terms of accuracy, reliability, and so on. Balancing the momentary energy and information needs thus becomes a multiobjective optimization problem, which also has to take predictions of future developments of system and monitored component state in terms of operational conditions, structural health, and energy resources into account.

As a consequence of such competing demands, hybrid communication approaches may be of interest. In these, a local subset of sensor nodes may be connected through physical paths, whereas communication between these individual subnets is handled wirelessly.

Besides the aforementioned inherent difficulties linked to fully or partially wireless systems, special application environments may create additional predicaments – These include the issue of electromagnetic compatibility (EMC) and thus potentially the need for electromagnetic shielding, as well as the general availability of suitable frequency bands for data transmission. Specifically in aerospace environments, the EMC aspect is of considerable importance and may suffice to enforce wired rather than wireless solutions.

22.3.3 SHM Data Evaluation Approaches and Principles

The primary aim of data evaluation is to provide the right information at the right time, and at a sufficient and defined level of accuracy, reliability, and so on. In addition to achieving a specific probability of detection (PoD) for all relevant types of structural damage (damage monitoring), loads, and/or environmental conditions (operation monitoring), since SHM systems may have to partake in decisions about operational fitness of the structures they survey, the task of data evaluation is not limited to correctly report critical situations, but at the same time to avoid "false positive" feedback. A false positive signal unrecognized as such would force the operator of the system under surveillance to take it out of service as a precautionary measure for detailed investigation without any actual need. In the case of an aircraft application, this means grounding the aircraft and thus effectively cutting revenue inflow. Practically this possibility implies that the system has to be able to clearly distinguish between actual deficiencies of the structure and sets of signals erroneously indicating such deficiencies. Causes of false positive messages may be several, including the malfunctioning of sensors or other system components causing a partial lack or corruption of sensory information. For this reason, an important feature of SHM systems is a self-diagnosis capability.

Depending on the monitoring principle, different levels of data evaluation can be distinguished. Assuming that an SHM system's aim is detection and characterization of damage, Sbarufatti *et al.* distinguish between direct detection (associated, for example, with electrical crack gauges), direct indication (e.g.,

22.3 SHM System Classification and Main Components

CVM), and feature extraction (e.g., strain measurement using strain gauges, FBGs, etc.) [34]. The former two approaches are typically associated with damage monitoring, while the latter can be used in both damage and operational monitoring.

Damage monitoring typically needs a model that relates the sensor signal to damage characteristics like location, size, and so on. Operational monitoring first needs to link the sensor signal to structural response, which could be local strain. A second model is needed to connect the measured set of structural response data to the structural state and thus identify, locate, and characterize potential damage sites.

Data evaluation approaches currently used in SHM can be sorted in two large groups – These are designated the model-based, deterministic methods and the model-free approaches. While the former typically rely on a system model for which a stable inverse solution has been identified prior to the deployment of the monitored structure [60,61], the latter are often based on artificial intelligence (AI) approaches like machine learning [62–64]. In this second case, depending on the exact approach chosen, some kind of classification of sensor signal patterns is undertaken and a relation determined that links these patterns, for example, to a specific structural state or a combination of external loads.

Figure 22.11 illustrates an example of the general structure of an SHM system based on machine learning using a combination of physical and virtual training to

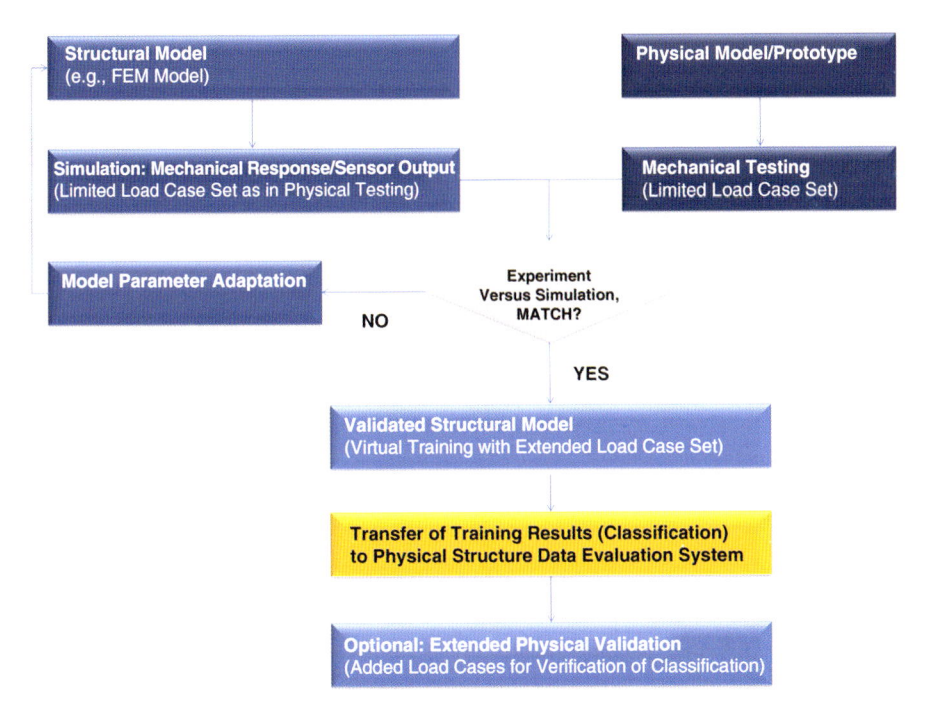

Figure 22.11 Making an SHM data evaluation system fit for service. A machine learning approach based on virtual evaluation/training using a verified FEM model of the monitored structure. (Diagram adapted from Lehmhus [65].)

achieve this classification aim. The prerequisite is a simulation environment that can accurately reflect the system behavior under all relevant conditions of service.

Figure 22.11 incorporates this aspect through the initial comparison of simulation and physical test results. The fundamental idea behind this approach is to fine-tune model characteristics via the setting of parameter values like material properties to create a sufficient match between experimental and simulation results for a limited set of loading conditions. The validated simulation model then considerably allows extending this set in the virtual world through simulation alone. The result is a large set of external loading conditions, associated internal loads, and/or damage characteristics in combination with sensor signal sets: The latter are gained from both the experimental and the simulation efforts.

For the machine learning system, this wealth of information constitutes the basis for classification of sensor signal sets in a supervised learning approach, and, thus, once the classification outcome (i.e., the classification rules derived) has been transferred to the material- or structure-integrated data evaluation system, the latter will be able to derive meaning from incoming sensor signals. This capability is not limited to the exact load cases studied in the training phase, but can also identify most likely scenarios based on a similarity consideration.

The physical location of data evaluation is another important issue – material-integrated intelligence clearly affords that data evaluation happens within the material. Renton had the same in mind when he imagined "structure that talks to us" [2]. In contrast, specifically in aerospace applications, a large proportion of suggested SHM systems even features ground-based or postponed data analysis and limits the role of the onboard part of the full SHM system to gathering the necessary data. An approach like this will necessarily forbid immediate recognition of critical damage and delay any possible response.

However, in shifting data evaluation to the individual sensor node, a balance has to be maintained, which may be tipped by the energy issue: Extensive local processing of data will bear an energy penalty linked to the computational effort – at the same time, however, it may reduce the need for data transmission, as the retrieval of information from data is effectively a reduction of its amount. Where communication is energy-intensive, like in wireless networks, this may be another motivation for local preprocessing.

Furthermore, immediacy in terms of availability of information requires not only local but also real-time data processing. Bosse *et al.* have addressed these issues. In their approach, which is based on multiagent system techniques [66,67], a further focus is on organizing data evaluation tasks in view of the condition of limited computational resources at node level, which are a key characteristic of material-integrated systems. Their solution is hybrid, as it combines model-free and model-based strategies in a distributed system that uses parallel computation to achieve soft real-time solutions through reduced computation and communication latencies – In practice, multiagent systems interact with inverse FEM. Of these, the former is primarily responsible for sensor data acquisition, aggregation, and precomputation, but also offers (decision tree type) machine learning as a service. The latter capability is used, in a simulation-based validation of the system, to predict loads acting on a model structure by drawing on local strain information gathered from a large number of virtual sensors. The structure assumed in this test case is a metal plate that is constrained on one side and is subjected to different local loads during virtual training. The results confirm the predictive capabilities of the machine learning approach even if the actual load case has not been part of the training data set. Comparison of the results with an inverse numerics approach suggests that machine learning can be used to preclassify a reconstructed load matrix. For both techniques, robustness could be demonstrated by simulating noise and drift. Adaptivity has been investigated on system level, which is covered by the SoS/MAS methods, while the necessary flexibility to cope with structural state change can at least partly be achieved through supplementing the machine learning with an incremental learning capability active during operation of the system. Future studies will look at the latter aspect in more detail [68,69]. Suggestions in this respect have previously been made by Lehmhus *et al.*, focusing on the combination of localized FEM approaches and once again multiagent systems techniques [70].

An overview of further studies focusing on machine learning in an SHM context has been offered by several authors, including Worden and Manson [62], Nick *et al.* [71], and Smarsly *et al.* [64]. Figueiredo *et al.* study machine learning approaches specifically with respect to their potential for overcoming the difficulty of separating influences of damage from those of environmental ones and other service-related boundary conditions [72]. Such a focused approach is also suggested by Melia *et al.*, who consider machine learning a possible method for detecting sensor fault, thus limiting the risk of "false positives" [73]. Besides machine learning, other techniques commonly associated with the field of artificial intelligence have been examined, among them being neural networks [74].

22.4 SHM Areas and Application and Case Studies

SHM system implementation naturally differs depending on the application area. Larger scale systems have been employed for tens of years in civil engineering to monitor the state of bridges and other large infrastructure [75]. Applications of this kind have helped to develop generic features of SHM systems in fields like data evaluation. However, there is only limited need in these applications, for example, to miniaturize sensors, sensor systems, and networks and to integrate them in structures on the level of the material.

For this reason, applications in the transport industry (railway, automotive, aerospace, and less so maritime for similar reasons as in the case of civil engineering structures) tend to be of greater interest from the present work's perspective. An exception to this rule are civil engineering and similar large-scale structures that build on composite materials, as these facilitate or even require the integration of sensors within the material due to their layered buildup and the possibility of delamination as one relevant mode of failure: Here, it is especially the rotor blades of wind energy plants that have gained attention in the past, but bridges have also been built from fiber-reinforced composites as well as ships.

Table 22.2 collects published information about individual SHM solutions for various application scenarios and describes the main features of these. Broader reviews of SHM system implementation have been provided by several authors, although not for all application areas. Examples that focus on application rather than on technological aspects are listed below:

- Aerospace industry
  - *Bartelds, 1998* [23]: Description of conventional, nonsmart system structural assessment techniques, expected progress toward smart systems, and technological as well as regulatory barriers against and needs for their implementation from an early, predominantly European perspective.
  - *Diamanti and Soutis, 2010* [12]: Short overview of NDE methods in general followed by a detailed discussion of Lamb wave techniques for SHM, including application examples for composite and metallic materials.
- Railway industry
  - *Barke and Chiu* [9]: Discussion of approaches for monitoring of vehicles to protect themselves as well as the track, identification of main parameters defining vehicle condition together with methods of determining these for actual vehicles, predominantly using wayside rather than onboard investigation methods.
  - *Hodges et al.* [10]: Introduction to wayside as well as onboard condition monitoring as used in the railway industry focusing on systems based on wireless sensor networks. Several individual application examples are given and links to the related publications provided, from wheels and pantographs to the level of bridges, tunnels, and so on, thus slightly overlapping with the SHM in civil engineering topic. Aspects like sensors, network topology, and data evaluation are discussed. Among future trends, increased sensor fusion and real-time capabilities are named.
- Civil engineering
  - *Chang et al., 2003* [82]: Critical review of SHM approaches in civil engineering, pointing out that established damage detection methods at the time still failed to correctly identify lower level damage and introducing research trends in general monitoring concepts, sensing technology, and data evaluation.
  - *Brownjohn, 2007* [13]: Description of history and motivation for SHM applied to civil infrastructures, case studies including suspension and box girder bridges as well as tall buildings and tunnels, technologies/methods including fiber-optic sensors, GPS-based solutions, and piezoelectric (PZT) patches. Development needs are identified, including the problem of guidelines and regulations also known from the aerospace industry, in this case discussed from a civil engineering perspective.
- Wind energy plants
  - *Schubel et al., 2013* [83]: Focus on composite rotor blades, other important aspects like monitoring of gearboxes not included. Discussion includes both beginning-of-life (composite matrix cure and associated residual strain monitoring) and middle-of-life phase (SHM). Sensing principles compared include dielectric, acoustic, ultrasonic, thermal, and fiber-optic methods.

Table 22.2 SHM application areas and exemplary case studies described by their main characteristics.

Application area	Description, motivation	SHM system level according to Figure 22.1	Material monitored	Sensor type	Number of sensors	Data evaluation Centralized, distributed, real time yes/no and so on	Network type Wired/wireless, topology, ...	Reference
Aerospace								
	Active damage monitoring using Lamb waves of a representative aerospace structure reflecting a door surrounding made up of skin with door cut-out and thickness variation between 2 and 8 mm, stringers and stiffeners – total size of structure approximately 5.1 m length, 3.5 m width, and 3 m outer radius. Objective of the study: Demonstration of damage detection through Lamb waves in a complex CFRP structure. No mechanical testing – damage was brought in through controlled local impacts	Detection, localization and extent of damage (1, 2, 3)	Carbon fiber-reinforced polymer (CFRP)	Disk-shaped piezoelectric sensors embedded in polymer, prestressed in production (DuraAct™ modules, thermally induced residual stresses)	584 in 126 arrays of 3, 4, or 5 sensors each	Centralized, external evaluation and control (demonstrator, on-board monitoring possible)	Wired connections within sensor arrays comprising 3–5 sensors, direct wired connection of arrays to evaluation unit (star topology)	[76]
	Active damage monitoring on a multistiffener composite plate representing a aircraft horizontal tail plane. Focus on determination of minimum number of sensor required for high PoD. Damage detection algorithm and other aspects of sensor network besides number and positioning of sensors were not covered	Detection, localization, and estimation of damage extent (1, 2, 3)	Carbon fiber-reinforced polymer (CFRP)	Piezoelectric patch, disk shaped, for Lamb wave excitation and detection	8 sensor arrays containing a linear arrangement of 4–7 sensors each	Centralized, external evaluation and control	Wired connection of sensor elements, no explicit topology information provided, apparently 4–7 sensors	[77]

(continued)

Table 22.2 (Continued)

Application area	Description, motivation	SHM system level according to Figure 22.1	Material monitored	Sensor type	Number of sensors	Data evaluation Centralized, distributed, real time yes/no and so on	Network type Wired/wireless, topology, ...	Reference
	Active damage monitoring on an aluminum aircraft wing using ultrasonic-guided waves produced by disk-shaped PZT piezoelectric transducers. Comparison of different approaches (RF and microwave) for wireless power supply and interrogation of the system, using either sensor	Detection, localization, and estimation of damage extent (1, 2, 3)	Aluminum	PZT piezoelectric disk sensor/actuator elements	8 sensors in a circular array of approximately 25 cm diameter	Centralized, external evaluation and control [58] and local data preprocessing [59]	Wired [58] and wireless using either RF or microwave systems/antennae for power supply [59]	[58,59]
Railway								
	Passive monitoring of rolling stock via accelerometers and structure-borne sound, high-frequency ultrasonic sensors, wheel–rail contact as stimulus to detect damage like wheel tread pitting. Sensor positions within wheel set hollow shafts	Detection, localization, and estimation of damage extent (1, 2, partly 3)	Steel	Ultrasonic sensors, accelerometers	8 ultrasonic sensors plus accelerometers per hollow shaft	Onboard, local signal processing, data storage and wireless communication (Bluetooth) to external data evaluation, potentially also onboard	No information provided	[78]
Automotive								
	Application by stitching of a carbon fiber sewing thread acting as resistive sensor within an epoxy-matrix glass-fiber-reinforced polymer component produced by vacuum-assisted resin infusion	Damage detection and (limited) localization, intensity estimation	Glass-fiber-reinforced polymer (GFRP)	Carbon fiber sewing thread (roving), resistance change due to consecutive individual filament breakage at increasing loads	Test of single sensor to study general feasibility	External data evaluation	Not applicable	[79]

Wind energy plants

Description	Focus	Material	Sensor type	Sensor details	Data evaluation	Network	Ref
Combined passive and active monitoring of a 40 m rotor blade under cyclic loads in a test rig, large number of sensors, optical energy, and information transmission to counter lightning strike risk, damage monitoring approach	Damage detection, localization, and estimation of damage extent (1, 2, partly 3)	Glass and carbon fiber-reinforced polymer (GFRP, CFRP)	Acoustic emission (AE) and acousto-ultrasonics (AU)	64 sensors, in monitoring arrays of 3-5 m edge length	At least signal processing at sensor node level, central node for final data evaluation and information retrieval	Wired using optical information and energy transfer, master node linked to several sensor nodes, topology not explicitly defined	[78]
Focus on SHM algorithm for monitoring of wind turbine rotor blades based on accumulated strain energy data accessible through piezoelectric elements combining sensing and energy harvesting tasks. Energetically autonomous sensor nodes transmitting information at accumulation of a set amount of vibration energy, which is linked to accumulated strain energy. FEM simulation-based evaluation using real life strain gage data and virtual piezoelectric sensor responses	Primarily damage detection, estimates of location	Fiber-reinforced composite, principle transferable to other materials	Piezoelectric combining energy harvesting and sensing functionality	Not exactly specified – for algorithm evaluation, strain data from FOS on a physical (reference) wind turbine is available, plus simulated strain date at FEM model node level	No data evaluation at individual sensor node level, signal transmission to central, onboard data evaluation node at a given threshold of accumulated energy, extended evaluation period, that is, no real-time monitoring	Wireless network, star topology with central data evaluation node	[80]

General mechanical engineering

Description	Focus	Material	Sensor type	Sensor details	Data evaluation	Network	Ref
Operation monitoring combining information from dedicated strain sensors and general onboard sensor suite to derive load information, the former in a wireless network using 2.4 GHz radio communication. Special measures like data compression and sleep cycles to limit energy consumption	Load monitoring	Steel	Strain gages, hydraulic pressure sensors, joint position sensors	17 strain gage rosettes,	Local data preprocessing (compression), centralized main evaluation (on board)	Wireless for strain gauges, others wired	[81]

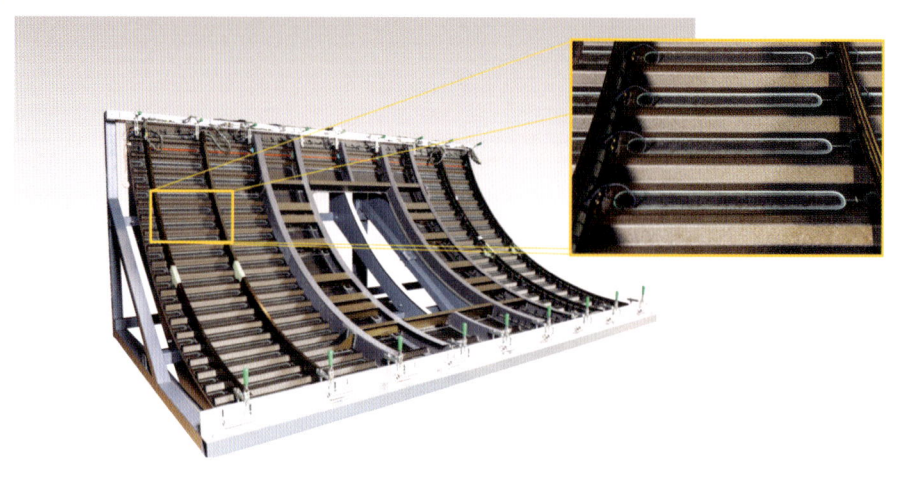

Figure 22.12 Full size representative commercial aircraft door surround structure made from CFRP and comprising altogether 584 piezoelectric sensors in 126 arrays (detail view top right) as developed and produced in the course of the SARISTU project. (See first entry in Table 22.2, as well as Schmidt *et al.* [76].)

- *Tchakoua et al., 2014* [30]: Discussion of wind energy plant condition monitoring from the viewpoint of reducing maintenance efforts and increasing availability, showing, for example, that rotor blades only represent a minor contribution to failure (although associated downtime is high) and covering solutions for all relevant wind turbine subsystems.

Figure 22.12 graphically illustrates one of the studies presented in Table 22.2: In this study, a full-size representative door surrounding structure has been build up in the course of the European project SARISTU. Other than in the case investigated by Sbarufatti summarized in Figures 22.9 and 22.10, the structure is not metallic, but made from fiber reinforced composites. The door surrounding is chosen as example because it is prone to impact during ground handling, reprovisioning and loading of the aircraft. The monitoring approach is active, analyzing Lamb wave propagation for signs of damage, using the respective data of the healthy structure as a reference or baseline. The number of sensors exceeds that of most other studies: Altogether 584 piezoelectric elements capable of operating as sensor or actuator were deployed, grouped in arrays of 3, 4, or 5 sensors, leading to a total of 126 such patches. The sensors themselves are of special type: The so-called DuraAct™ modules, which were complemented by conventional SmartLayer™ patches, are encapsulated in EPDM during fabrication at a temperature of 180 °C. During cooling, the higher coefficient of thermal expansion of EPDM compared to the piezoceramic leads to the development of tensile residual stresses in the EPDM, and accordingly compressive ones in the piezoceramic material, thus somewhat alleviating the brittle nature of the latter. The main objective of the study was to investigate experimentally whether or not the complex buildup of the door surround structure with stiffeners, frames, and changes in material thickness would hamper or even completely prevent the detection of damage via Lamb waves. Damage was artificially introduced at several locations and the structure investigated afterward. Conventional NDT techniques were used to gather reference information on extent and severity of damage.

22.5 Implications of Material Integration for SHM Systems

Material integration of SHM systems means moving one step further toward the aim of gathering data directly at the critical location. The expectation is that quality of data obtained thus should be increased in comparison to surface-mounted systems. The downside is that if integration of the sensor system means structurally weakening the host material, this deliberate move toward highly loaded areas also implies increasing the likelihood of failure of both the sensor and the component it is meant to monitor. This dilemma is pronounced in SHM applications, as these imply a critical importance of mechanical characteristics. A negative effect of the sensor in this respect would thus be felt more heavily. The aspect of reducing the sensors footprint for exactly this reason has been thoroughly discussed by Lang *et al.* and Dumstorff *et al.* [84,85].

Furthermore, in comparison with surface application, material integration may result in improved protection of the integrated system against damage. However, once damage has been incurred, the repair of material-integrated systems will at least be more difficult than that of surface-attached solutions, if not impossible. In practice, case studies have demonstrated repair solutions, for example, for material-integrated fiber-optic sensors, but in these and similar cases, it is usually only sensors and/or interconnects that are considered, and not embedded electronic systems that could only be replaced. Most likely, the layout of material-integrated systems will have to account for the possibility of damage to the monitoring equipment, for example, by providing a certain level of redundancy in terms of the number of sensors deployed, but potentially also with respect to data evaluation hardware and algorithms employed. A very important aspect in this respect will be the capability to quantify the remaining level of trust for a partially damaged system, as such information might allow a damage-tolerant design strategy not only for structural features but also for the SHM system itself.

For the same reasons, that is, the difficulty of access, material-integrated SHM systems either have to be completely energy-autonomous, or they must have access points that allow simple topping-up of system-wide energy levels. In this respect, approaches that allow wireless energy transfer into the system from the outside as suggested by Salas *et al.* may prove interesting, because in their case the leading of interconnects through all the layers of the component as well as its surface can be dispensed with [86]. Alternative approaches also suggested by Salas *et al.* even consider the powering of active sensor systems from the outside [87]. In practice, this could mean that in a structure loaded primarily in bending, the surface layer experiencing highest stress levels would not have to be pierced, and vulnerable system components could remain entirely within the neutral plane or at least in limited load areas.

Besides the downsides mentioned above, difficulty of access can also create the explicit need to revert to material-integrated systems: From an NDE/NDT point of view, critical areas that can only be reached with great difficulty (e.g., through partial disassembly) or not at all during maintenance or manufacturing on assembly level would profit from an SHM system brought in during production of the respective component on material level. This way, material-integrated SHM systems could even facilitate new design principles that are currently hampered by the nonavailability of suitable NDE/NDT approaches. Sbarufatti *et al.* report about a study on helicopter maintenance performed by the US Navy that employed surface-attached CVM sensors to detect further growth of an already known crack in an area that would have required 4 h of disassembly work had the same task been executed using conventional methods, while the CVM approach, once the sensor had been installed, reduced this effort to a mere 5 min [34].

Finally, since for material integration miniaturization is mandatory, as secondary effect, the overall weight penalty associated with an add-on SHM system could be reduced by introduction of material-integrated systems. In effect, the potential design benefit expected from an SHM system could thus be raised to a larger fraction of its full potential.

From a data evaluation point of view, SHM systems realized as material-integrated intelligent systems face two main, additional challenges:

First, the complex task of identifying either damage or loading conditions has to be rethought to make it manageable using the limited computational power a material-integrated intelligent system under energy and volume constraints can provide.

Second, the system should ideally be able to react to state change: If characteristics of the structure are altered either due to a variation of environmental or service-related boundary conditions or due to introduction or accumulation of damage, the predictive capabilities of the system should not be compromised by loss of individual sensors, communication paths, and so on.

The latter aspect, that is, the adaptation of either models or classification rules to changing states of the monitored system, is a considerable challenge specifically if such an adaptation has to be accomplished during usage of the system in real time – a capability that could be used to great benefit by adapting usage patterns of the monitored structure to a recognized state of ill health immediately (e.g., flight profiles in case of aircraft), curtailing or at least slowing down the further degradation of structural state.

22.6 Conclusion and Outlook

Structural health monitoring is a topic with an intuitively clear vision, which is summarized very graphically in the statement by Renton quoted at the onset of this chapter: "Structure that talks to us . . ." [2].

Thus, while the goal is very clear, its realization remains difficult, affording interrelated progress across several different disciplines in a field that implies optimization of safety-relevant structures and thus a critical role of the envisaged system. In this, while SHM may be the natural application scenario for material-integrated intelligent systems, and while it is representative of

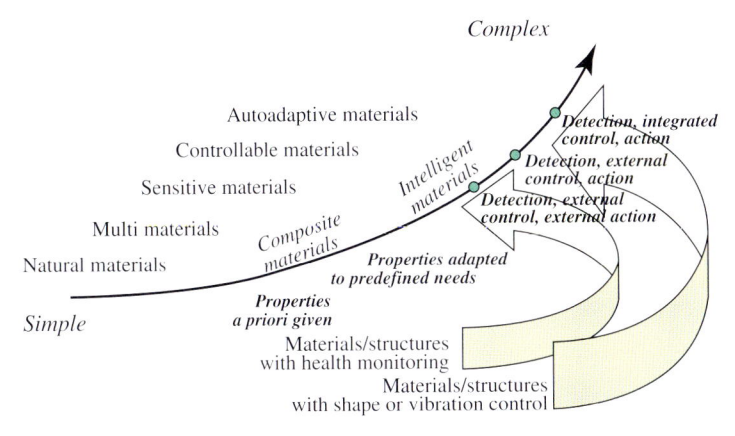

Figure 22.13 Development trends from simple to complex structures incorporating monitoring and control according to Balageas et al. [26].

many aspects of this topic in general, it is at the same time among the most challenging scenarios.

For the latter reason, it is not unlikely that SHM will turn out not to be the trailblazer for material-integrated intelligent systems as which it has been seen for quite some time, but rather leave this role to consumer-oriented applications with lower requirements in terms of accuracy, reliability, or robustness, and not the least shorter service lives. The business case for SHM, however, is actually there in many cases, as has been demonstrated by authors like Keller and Poblete [28] or Pattabhiraman et al. [29].

In the visualization of development trends offered by Balageas et al. and represented here in Figure 22.13 [26], from a perspective of commercialization, we are today still well in between the transition from composite to intelligent materials, even though on lab scale approaches for adaptive materials exist by the score, reaching up to the level of fly-by-feel scenarios [88].

For SHM of safety-critical components, be it in aerospace or other applications, once the necessary clarity in regulatory requirements has been established, two specific aspects may push commercial implementation further: One of them is the use of SHM systems not so much for in-service monitoring, but as integrated support for NDE/NDT. In the aerospace industry specifically, there is a notion that this use of SHM systems could become instrumental for realizing new weight-saving structural designs, the introduction of which is currently hampered by a lack of suitable NDE/NDT approaches. This way, material-integrated sensing could become a facilitator for a step change in lightweight design. Similarly important is the possibility of using SHM systems to already monitor the production of components: Besides enabling process optimization, the information obtained thus could be used to derive performance data specific to the individual component, and thus potentially an individual baseline for subsequent in-service SHM.

This said, further technological development is still an issue in many fields: As has been pointed out, for most SHM applications, not only the probability of detection (PoD) of damage is an issue (i.e., the return of "true positive" results in relation to the actual cases of damage), but also the minimization of "false positives": Taking the aerospace example, a false positive message from the SHM system that cannot be easily identified as being just that will lead to the grounding of the aircraft and an addition or rescheduling of maintenance operations. While the first implies a loss of revenue, the other will create direct costs.

Thus, the selectivity of the system with respect to the distinction between true and false positives is of similar importance to its PoD characteristics. Practically solving this issue may involve integrating the capability of monitoring the state of the SHM system itself. Possible approaches toward this aim may include plausibility checks based on relations between the individual sensor's data, potentially under known conditions of loading related to specific operational states. On the downside, especially in aerospace environments, components to be monitored undergo a major change in environmental conditions during service cycles and are, at the same time, subject to high levels of noise. In an aircraft, from ground to cruising altitude, parameters like temperature and humidity vary greatly, and vibrations originating, for example, from engine operation are superimposed to damage-related signals received by the SHM sensors.

Real-time data evaluation in a distributed, material-integrated monitoring system is a further challenge, especially if the ability to respond to system state change comes into play. Activities in this field include introduction of model-updating capabilities as well as the development of hybrid and completely model-free approaches.

Finally, even if all the aforementioned issues had been solved for a given type of material, advanced materials and structures might evolve, requiring a renewal of work already done in view of their altered characteristics. Hybrid materials and structures are an illustrative example in this respect: Even though both fiber-reinforced polymers (FRP) and metals are reasonably well understood in terms of SHM implementation, this is not necessarily the case for materials combining them, like fiber–metal laminates (FML) or FRP-metal transition structures [89,90].

References

1 Farrar, C.R., Park, G., Allen, D.W., and Todd, M.D. (2006) Sensor network paradigms for structural health monitoring. *Structural Control and Health Management*, **13**, 210–225.
2 Renton, W.J. (2001) Aerospace and structures: where are we headed? *International Journal of Solids and Structures*, **38**, 3309–3319.
3 Lang, W. et al. (2011) Sensorial materials – a vision about where progress in sensor integration may lead to. *Sensors and Actuators A: Physical*, **171**, 1–2.
4 Lehmhus, D., Bosse, S., and Busse, M. (2013) Sensorial materials, in *Structural Materials and Processes in Transportation* (eds D. Lehmhus, M. Busse, A.S. Herrmann, and K. Kayvantash), Wiley-VCH Verlag GmbH, Weinheim, Germany.
5 McEvoy, M.A. and Correll, N. (2015) Materials that combine sensing, actuation, computation and communication. *Science*, **347**, 1261689-1–1261689-8.

6 Mekid, S. and Kwon, O.J. (2009) Nervous materials: a new approach for better control, reliability and safety of structures. *Science of Advanced Materials*, **1**, 276–285.

7 Mekid, S., Saheb, N., Khan, S.M.A., and Qureshi, K.K. (2015) Towards sensor array materials: can failure be delayed? *Science and Technology of Advanced Materials*, **16**, 034607.

8 Herrmann, S., Wellnitz, J., Jahn, S., and Leonhardt, S. (2013) Structural health monitoring for carbon fiber resin composite car body structures, in *Sustainable Automotive Technologies 2013, Lecture Notes in Mobility*, Springer, Heidelberg, pp. 75–96.

9 Barke, D. and Chiu, W.K. (2005) Structural health monitoring in the railway industry: a review. *Structural Health Monitoring: An International Journal*, **4**, 81–94.

10 Hodges, V.J., O'Keefe, S., Weeks, M., and Moulds, A. (2015) Wireless sensor networks for condition monitoring in the railway industry: a survey. *IEEE Transactions on Intelligent Transportation Systems*, **16**, 1088–1106.

11 Okasha, N.M., Frangopol, D.M., and Deco, A. (2010) Integration of structural health monitoring in life-cycle performance assessment of ship structures under uncertainty. *Marine Structures*, **23**, 303–321.

12 Diamanti, K. and Soutis, C. (2010) Structural health monitoring techniques for aircraft composite structures. *Progress in Aerospace Science*, **46**, 342–352.

13 Brownjohn, J.M.W. (2007) Structural health monitoring of civil infrastructure. *Philosophical Transactions of the Royal Society A*, **365**, 589–622.

14 Inaudi, D. and Glisisc, B. (2008) Overview of 40 bridge structural health monitoring projects. *Bridge Maintenance, Safety Management, Health Monitoring and Informatics – IABMAS '08: Proceedings of the 4th International IABMAS Conference, Seoul, Korea, July 13–17, 2008.* (eds H.-M. Koh and D.M. Frangopol), Taylor & Francis.

15 Nichols, J.M. (2003) Structural health monitoring of offshore structures using ambient excitation. *Applied Ocean Research*, **25**, 101–114.

16 Ciang, C.C., Lee, J.-R., and Bang, H.J. (2008) Structural health monitoring for a wind turbine system: a review of damage detection methods. *Measurement Science and Technology*, **19**, 122001 (20pp).

17 Li, D., Ho, S.-C.M., Song, G., Ren, L., and Li, H. (2015) A review of damage detection methods for wind turbine blades. *Smart Materials and Structures*, **24**, 033001 (24pp).

18 Hermann, A.S. (2013) Polymer matrix composites, in *Structural Materials and Processes in Transportation* (eds D. Lehmhus, M. Busse, A.S. Herrmann, and K. Kayvantash), Wiley-VCH Verlag GmbH, Weinheim.

19 Worden, K., Farrar, C.R., Manson, G., and Park, G. (2007) The fundamental axioms of structural health monitoring. *Proceedings of the Royal Society A*, **463**, 1639–1664.

20 Rytter, A. (1993) Vibrational based inspection of civil engineering structures. Ph. D. thesis, University of Aalborg, Denmark.

21 Inman, D.J. *et al.* (ed.) (2005) *Damage Prognosis for Aerospace, Civil and Mechanical Systems*, John Wiley & Sons Ltd., Chichester.

22 Lehmhus, D. et al. (2013) When nothing is constant but change: adaptive and sensorial materials and their impact on product design. *Journal of Intelligent Material Systems and Structures*, **24**, 2172–2182.

23 Bartelds, G. (1998) Aircraft structural health monitoring, prospects for smart solutions from a European viewpoint. *Journal of Intelligent Material Systems and Structures*, **9**, 906–910.

24 Bockenheimer, C. and Speckmann, H. (2013) Validation, verification and implementation of SHM at Airbus. International Workshop on Structural Health Monitoring (IWSHM 2013), Stanford, USA, September 10th–12th.

25 Aerospace recommended practice (ARP) (2013) 6414: Guidelines for Implementation of Structural Health Monitoring on Fixed Wing Aircraft. Society of Automotive Engineers (SAE).

26 Balageas, D., Fritzen, C.-P., and Güemes, A. (2006) *Structural Health Monitoring*, ISTE Ltd., London.

27 Boller, C. and Buderath, M. (2007) Fatigue in aerostructures – where structural health monitoring can contribute to a complex subject. *Philosophical Transactions of the Royal Society A*, **365**, 561–587.

28 Keller, K. and Poblete, J. (2011) The business case for SHM, in *System Health Management: With Aerospace Applications* (eds S.B. Johnson, T.J. Gormley, S.S. Kessler, C.D. Mott, A. Patterson-Hine, K.M. Reichard, and P.A. Scandura), John Wiley & Sons, Ltd, Chichester.

29 Pattabhiraman, S., Gogu, C., Kim, N.H., Haftka, R.T., and Bes, C. (2012) Skipping unnecessary structural airframe maintenance using an onboard structural health monitoring system. *Proceedings of the Institution of Mechanical Engineers Part O: Journal of Risk and Reliability*, **226**, 549–560.

30 Tchakoua, P., Wamkeue, R., Quhrouche, M., Slaoui-Hasnaoui, F., Tameghe, T.A., and Ekemb, G. (2014) Wind turbine condition monitoring: state-of-the-art review, new trends, and future challenges. *Energies*, **7**, 2595–2630.

31 Peters, C., Zahlen, P., Bockenheimer, C., and Herrmann, A.S. (2011) Structural health monitoring (SHM) needs S^3 (sensor-structure-system) logic for efficient product development. Proceedings of the SPIE 7981, Sensors and Smart Structures Technologies for Civil, Mechanical, and Aerospace Systems. 79815N (April 18, 2011). doi: 10.1117/12.880479

32 Bombieri, N., Drogoudis, D., Gengemi, G., Gillon, R., Grosso, M., Macii, E., Poncino, M., and Rinaudo, S. (2015) Addressing the smart systems design challenge: the SMAC platform. *Microprocessors and Microsystems*, **39**, 1158–1173.

33 Speckmann, H. and Henrich, R. (2004) Structural health monitoring (SHM) – overview on technologies under development. Proceedings of the 16th World Congress on Non-Destructive Testing (WCNDT'04), Montreal, Canada, August 30–September 3.

34 Sbarufatti, C., Manes, A., and Giglio, M. (2014) Application of sensor technologies for local and distributed structural health monitoring. *Structural Control and Health Monitoring*, **21**, 1057–1083.

35 Staszewski, W.J., Mahzan, S., and Traynor, R. (2009) Health monitoring of aerospace composite structures – active and passive approach. *Composites Science and Technology*, **69**, 1678–1685.

36 De Rossi, D., Carpi, F., Lorussi, F., Mazzoldi, A., Scilingo, E.P., and Tognetti, A. (2002) Electroactive polymer fibers and fabrics for distributed, conformable and interactive systems. Proceedings of the First European Congress on Structural Health Monitoring, Cachan, France, July 10–12.

37 Roach, D. (2009) Real time crack detection using mountable comparative vacuum monitoring sensors. *Smart Structures and Systems*, **5** (4), 317–328.

38 Eaton, M.J., Pullin, R., and Holford, K.M. (2012) Towards improved damage location using acoustic emission. *Journal of Mechanical Engineering Sciences*, **226**, 2141–2153.

39 Meyendorf, N., Frankenstein, B., Hentschel, D., and Schubert, L. (2007) Acoustic techniques for structural health monitoring. IV Conferencia Panamericana de END, Buenos Aires, Brazil, October.

40 Baumgaertel, H., Kneifel, A., Gontscharov, Sergei, and Krieger, K.-L. (2014) Investigations and comparison of noise signals to useful signals for the detection of dents in vehicle bodies by sound emission analysis. *Procedia Technology*, **15**, 716–725.

41 Gontscharov, S., Baumgaertel, H., Kneifel, A., and Krieger, K.-L. (2014) Algorithm development for minor damage identification in vehicle bodies using adaptive sensor data processing. *Procedia Technology*, **15**, 586–594.

42 Di Sante, R. (2015) Fibre optic sensors for structural health monitoring of aircraft composite structures: recent advances and applications. *Sensors*, **15**, 18666–18713.

43 Koerdt, M. and Vollertsen, F. (2011) Fabrication of an integrated optical Mach–Zehnder interferometer based on refractive index modification of polymethylmethacrylate by krypton fluoride excimer laser radiation. *Applied Surface Science*, **257**, 5237–5240.

44 Giurgiutiu, V. (2008) *Structural Health Monitoring with Piezoelectric Wafer Active Sensors*, Associated Press, Burlington, ON.

45 Chang, F.K. and Ihn, J.B. (2008) Pitch catch active sensing methods in structural health monitoring for aircraft structures. *Structural Health Monitoring*, 7, 5–19.

46 Qiu, L. and Yuan, S. (2009) On development of a multi-channel PZT array scanning system and its evaluating application on UAV wing box. *Sensors and Actuators A: Physical*, **151**, 220–230.

47 Yu, L. and Giurgiutiu, V. (2007) *In situ* optimized PWAS phased arrays for Lamb wave structural health monitoring. *Journal of Mechanics of Materials and Structures*, **2**, 459–488.

48 Giurgiutiu, V. (2005) Tuned Lamb wave excitation and detection with piezoelectric wafer active sensors for structural health monitoring. *Journal of Intelligent Material Systems and Structures*, **16**, 291–305.

49 Padiyar, M.J. and Balasubramaniam, K. (2014) Lamb-wave-based structural health monitoring technique for inaccessible regions in complex composite structures. *Structural Control and Health Monitoring*, **21**, 817–832.

50 Schubert, K.J. and Herrmann, A.S. (2011) On attenuation and measurement of Lamb waves in viscoelastic composites. *Composite Structures*, **94**, 177–185.

51 Schubert, K.J. and Herrmann, A.S. (2012) On the influence of moisture absorption on Lamb wave propagation and measurements in viscoelastic CFRP using surface applied piezoelectric sensors. *Composite Structures*, **94**, 3635–3643.

52 Schubert, K.J. and Herrmann, A.S. (2013) A compensation method for environmental influences on passive Lamb wave based impact evaluation for CFRP. *Key Engineering Materials*, **569–570**, 1265–1272.

53 Schubert, K.J., Brauner, C., and Herrmann, A.S. (2014) Non-damage-related influences on Lamb wave-based structural health monitoring of carbon fiber-reinforced plastic structures. *Structural Health Monitoring: An International Journal*, **13**, 158–176.

54 Farrar, C.R. and Doebling, S.W. (1997) An overview of modal-based damage identification methods. Proceedings of the DAMAS 97 International Workshop, Sheffield, UK, June 30–July 2.

55 Das, S., Saha, P., and Patro, S.K. (2016) Vibration-based damage detection techniques used for health monitoring of structures: a review. *Journal of Civil Structural Health Monitoring*, **6**, 477–507.

56 Bosse, S. and Pantke, F. (2013) Distributed computing and reliable communication in sensor networks using multi-agent systems. *Production Engineering, Research and Development*, **7**, 43–51.

57 Bosse, S. and Lehmhus, D. (2010) Smart communication in a wired sensor- and actuator-network of a modular robot actuator system using a hop-protocol with delta-routing. Proceedings of the Smart Systems Integration Conference, Como, Italy, March 23–24. ISBN 978-3-8007-3208-1.

58 Zhao, X., Gao, H., Zhang, G., Ayhan, B., Yan, F., Chiman, K., and Rose, J.L. (2007) Active health monitoring of an aircraft wing with embedded piezoelectric sensor/actuator network: I. Defect detection, localization and growth monitoring. *Smart Materials and Structures*, **16**, 1208–1217.

59 Zhao, X., Qian, T., Mei, G., Kwan, C., Zane, R., Walsh, C., Paing, T., and Popovic, Z. (2007) Active health monitoring of an aircraft wing with embedded piezoelectric sensor/actuator network: II. Wireless approaches. *Smart Materials and Structures*, **16**, 1218–1225.

60 Kefal, A. and Oterkus, E. (2015) Structural health monitoring of marine structures by using inverse finite element method, in *Analysis an Design of Marine Structures* (eds C. Guedes Soares and R.A. Shenoi), Taylor & Francis Group, London.

61 Uhl, T. (2011) Inverse problem in structural damage identification. Proceedings of the 19th International Conference on Computer Methods in Mechanics (CMM-2011), Warsaw, Poland, May 9–12.

62 Worden, K. and Manson, G. (2007) The application of machine learning to structural health monitoring. *Philosophical Transactions of the Royal Society A*, **365**, 515–537.

63 Farrar, C.R. and Worden, K. (2013) *Structural Health Monitoring – A Machine Learning Perspective*, John Wiley & Sons, Ltd., Chichester.

64 Smarsly, K., Dragos, K., and Wiggenbrock, J. (2016) Machine learning techniques for structural health monitoring. Proceedings of the 8th European Workshop on Structural Health Monitoring (EWSHM'16), Bilbao, Spain, July 5–8.

65 Lehmhus, D. (2015) Material-integrated intelligent systems – notes on state of the art and current trends. Proceedings of the 5th Scientific Symposium of the CRC/TR 39 PT-PIESA, Dresden, Germany, September 14–16.

66 Bosse, S. (2014) Distributed agent-based computing in material-embedded sensor network systems with the agent-on-chip architecture. *IEEE Sensors Journal*, **14**, 2159–2170.

67 Bosse, S. (2015) Unified distributed computing and co-ordination in pervasive/ubiquitous networks with mobile multi-agent systems using a modular and portable agent code processing platform. *Procedia Computer Science*, **40**. doi: 10.1016/j.procs.2015.08.312

68 Bosse, S. and Lechleiter, A. (2014) Structural health and load monitoring with material-embedded sensor networks and self-organizing multi-agent systems. *Procedia Technology*, **15**, 668–690.

69 Bosse, S., Lechleiter, A., and Lehmhus, D. (2017) Data evaluation in smart sensor networks using inverse methods and artificial intelligence (AI): towards real-time capability and enhanced flexibility. *Advances in Science and Technology*, **101**, 55–61.

70 Lehmhus, D. *et al.* (2009) Simulation techniques for the description of smart structures and sensorial materials. *Journal of Biological Physics and Chemistry*, **9**, 143–148.

71 Nick, W., Shelton, J., Asamene, K., and Esterline, A. (2015) A study of supervised machine learning techniques for structural health monitoring. Proceedings of the 26th Modern Artificial Intelligence and Cognitive Science Conference (MAICS), Greensboro, NC, April 25–26.

72 Figueiredo, E., Park, G., Farrar, C.R., Worden, K., and Figueiras, J. (2010) Machine learning algorithms for damage detection under operational and environmental variability. *Structural Health Monitoring: An International Journal*, **10**, 559–572.

73 Melia, C., Cooke, A., and Grayson, S. (2016) Machine learning techniques for automatic sensor fault detection in airborne SHM networks. Proceedings of the 8th European Workshop on Structural Health Monitoring (EWSHM 2016), Bilbao, Spain, July 5–8.

74 De Fenza, A., Sorrentino, A., and Vitiello, P. (2015) Application of artificial neural networks and probability ellipse methods for damage detection using Lamb waves. *Composite Structures*, **133**, 390–403.

75 Ko, J.M. and Ni, Y.Q. (2005) Technology developments in structural health monitoring of large-scale bridges. *Engineering Structures*, **27**, 1715–1725.

76 Schmidt, D., Kolbe, A., Kaps, R., Wierach, P., Linke, S., Steeger, S., von Dungern, F., Tauchner, J., Breu, C., and Newman, B. (2015) Development of a door surround structure with integrated structural health monitoring system, in *Smart Intelligent Aircraft Structures (SARISTU): Proceedings of the Final Project Conference* (eds C.P. Woelcken and M. Papadopoulos), Springer, Heidelberg, pp. 935–945.

77 Gao, D., Wang, Y., Wu, Z., Rahim, G., and Bai, S. (2014) Design of a sensor network for structural health monitoring of a full-scale composite horizontal tail. *Smart Materials and Structures*, **23**, 055011.

78 Meyendorf, N., Frankenstein, B., and Schubert, L. (2012) Structural health monitoring for aircraft, ground transportation vehicles, wind turbines and pipes – prognosis. Proceedings of the 18th World Conference on Nondestructive Testing, Durban, South Africa, April 16–20.

79 Mitschang, P., Molnar, P., Ogale, A., and Ishii, M. (2007) Cost-effective structural health monitoring of FRPC parts for automotive applications. *Advanced Composite Materials*, **16**, 135–149.

80 Lim, D.-W., Mantell, S.C., and Seiler, P.J. (2016) Wireless monitoring algorithm for wind turbine blades using Piezo-electric energy harvesters. *Wind Energy*, **20**, 551–565.

81 Allen, W.E. and Sundermeyer, J.N. (2005) A structural health monitoring system for earthmoving machines. IEEE International Conference on Electro Information Technology, May 22–25. doi: 10.1109/EIT.2005.1627011

82 Chang, P., Flatau, A., and Liu, S.C. (2003) Review paper: health monitoring of civil infrastructure. *Structural Health Monitoring*, **2**, 257–267.

83 Schubel, P.J., Crossley, R.J., Boatend, E.K.G., and Hutchinson, J.R. (2013) Review of structural health and cure monitoring techniques for large wind turbine blades. *Renewable Energy*, **51**, 113–123.

84 Dumstorff, G., Paul, S., and Lang, W. (2014) Integration without disruption: the basic challenge of sensor integration. *IEEE Sensors Journal*, **14**, 2102–2111.

85 Lang, W., Jakobs, F., Tolstosheeva, E., Sturm, H., Ibragimov, A., Kesel, A., Lehmhus, D., and Dicke, U. (2011) From embedded sensors to sensorial materials – the road to function scale integration. *Sensors and Actuators A: Physical*, **171**, 3–11.

86 Salas, M., Focke, O., Herrmann, A.S., and Lang, W. (2014) Wireless power transmission for structural health monitoring of fiber-reinforced-composite materials. *IEEE Sensors Journal*, **14**, 2171–2176.

87 Salas, M., Focke, O., Herrmann, A.S., and Lang, W. (2016) Wireless actuation of piezo-elements for the structural health monitoring of carbon-fiber-reinforced-polymers. *Mechatronics*, **34**, 128–136.

88 Salowitz, N. and Chang, F.-K. (2012) Bio-inspired intelligent sensing materials for fly-by-feel autonomous vehicles. *IEEE Explore*. doi: 10.1109/ICSENS.2012.6411534

89 Alderliesten, R.C. (2015) Designing for damage tolerance in aerospace: A hybrid material technology. *Materials & Design*, **66**, 421–428.

90 Schimanski, K., von Hehl, A., Zoch, H.-W. (2013) Failure behavior of diffusion bonded transition structures for integral FRP-Aluminum compounds. *Procedia Materials Science*, **2**, 189–196.

23

Achievements and Open Issues Toward Embedding Tactile Sensing and Interpretation into Electronic Skin Systems

Ali Ibrahim, Luigi Pinna, Lucia Seminara, and Maurizio Valle

Electrical, Electronic and Telecommunications Engineering, Department of Naval, via Opera Pia 11A, 16145 Genoa, Italy

23.1 Introduction

The development of the electronic skin (e-skin) is a very challenging goal that should be tackled from a system perspective.

The e-skin is usually intended as a hybrid stack-wise arrangement that incorporates tactile sensing (structural and functional materials, signal conditioning and acquisition, signal processing) and interpretation. Sensory inputs similar (but not limited) to those possessed by humans are essential to provide the necessary feedback to explore the environment and interact with objects.

A specific example of this general structure is shown in Figure 23.1 (adapted from Ref. [1]).

The *bottom layer* (*substrate*) is made of a structural material that can be rigid (e.g., the robot mechanical structure) or soft. Next layer (electronic layer) *hosts the electronic circuits*. Conventional electronics is typically integrated on very hard and flat (brittle) surfaces. Here the need is to conform to curved surfaces, requiring flexibility but also stretchability, to a certain extent, to follow all movements and deformations of the parts into which the electronic layer is integrated. The adoption of a flexible substrate does not necessarily guarantee the flexibility of the entire electronic circuit, as a too dense or not well-organized layout may drastically limit the flexibility of the overall structure. Also, system flexibility does not imply stretchability. In fact, even if the substrate is stretchable, the routing lines are intrinsically not, unless a dedicated design is adopted. Some interesting concepts are related to the creation of compliant and stretchable interconnections [2,3] and a very widespread approach to materials and mechanics for stretchable electronics is contained in a complete overview [4]. Requirements on conformability and stretchability put severe constraints on the reliability of the electronics, as mechanical stress on the electronic circuits can cause faults on interconnections and circuits. The counteraction can be at a material level; for what concerns this review, we will focus on increasing

---

Material-Integrated Intelligent Systems: Technology and Applications, First Edition.
Edited by Stefan Bosse, Dirk Lehmus, Walter Lang, and Matthias Busse.
© 2018 Wiley-VCH Verlag GmbH & Co. KGaA. Published 2018 by Wiley-VCH Verlag GmbH & Co. KGaA.

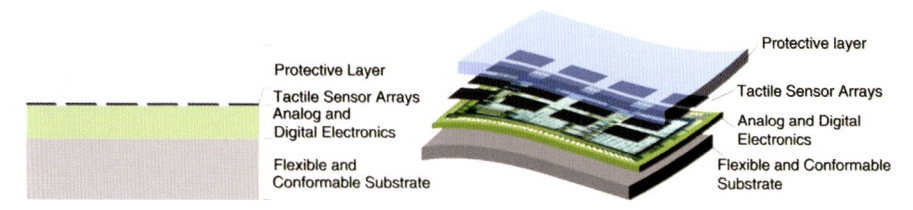

Figure 23.1 General physical structure of the e-skin system.

robustness through redundancy at functional and circuit levels: The drawback is the unavoidable increase of complexity and power consumption.

Next, *sensors* are embedded in a *protective layer*. Sensors can be multimodal in that they can measure different features of the input stimuli, for example, normal and tangential forces. The geometry of the sensor array (i.e., overall area, sensor size, sensor pitch, sensor distribution) depends on the transducer technology and on the given application requirements. The *protective layer*, which is usually polymer based (e.g., PDMS), protects the whole from damages induced by contact with objects, environmental chemical agents and water, and so on. Moreover, it implements a mechanical filtering of the input stimulus and concentrates/distributes the mechanical stimulus onto the sensor array below depending on the thickness and the compliance of the layer. As a consequence, e-skin spatial resolution depends on the sensor geometrical arrangement as well as on the features of this protective layer.

In this chapter, we first study materials and transducers, focusing on most promising sensor technologies. Thereafter, the issue of how to distill useful information from the stream of data produced by tactile sensors is addressed: Raw sensor measurements are processed and organized, trying to infer relations or learning patterns from data collected over time or from different settings. Finally, the tactile information has been interpreted to build a coherent picture of the environment and its evolution over time.

The need of extracting information from multiple sensor data streams impacts on the placement and on the type of sensors, and on the processing throughput and latency. Organizing and fusing sensor data streams so that salient information is extracted is a critical step in the process of sensor data interpretation. The time taken for sensing and interpreting limits the response time in closed-loop control systems. In general, benchmarks and metrics focus on the speed of processing, the quantity of data to be processed, the efficiency of data abstraction, and the associated error rate. As a consequence, efficient real-time embedded implementation is required.

The chapter is organized as follows: Sections 23.1 and 23.2 report a condensed assessment of the state of the art of the available transduction methods and of tactile data processing, respectively. In order to provide an estimation of the computational complexity of the hardware implementation of the processing algorithms, Section 23.3 provides a study of the computational requirements concerning two existing and sound approaches, that is, the electrical impedance tomography (EIT) algorithm [5] and a machine learning (ML) algorithm based on tensorial kernel [6]. Conclusions and future perspectives are reported in Section 23.4.

23.2 The Skin Mechanical Structure

23.2.1 Transducers and Materials

A sensing element can be seen as a structural unit (e.g., capacitive) that produces a signal as a response to a mechanical stimulus or as a material (or aggregation of materials) – for example, piezoelectric [7], piezoresistive [8], and optical [9] – which intrinsically convert the mechanical stress/strain into an optical or electrical signal [10]. When constituted by a mix of different materials having different properties, e-skin shows multimodal sensing capabilities [11], bendability, flexibility, and stretchability, hopefully shrink and wrinkle ability as human skin has [12]. On the basis on what happens in human skin, transduction technologies and corresponding sensors should enable such capabilities as normal and shear force sensing, tensile strain monitoring and vibration detection (at least up to 800 Hz) [13], and e-skin featuring a large frequency bandwidth that spans from 0 to 1 kHz is desirable. According to the application, the spatial resolution (defined as the smallest distance between two distinguishable contact points [14]) should range from a minimum of 1 mm to a maximum of 20–30 mm. Detectable force should span in a range of three orders of magnitude (e.g., 1–1000 g [13]). Even if human skin features high hysteresis, it is preferable that e-skin presents a low hysteresis, to avoid significant processing and complex electronics. The requirements outlined above together with some others (extracted and adapted from literature, in particular from Refs [14–16]) are summarized in Table 23.1. The requirements in Table 23.1 are general and can

Table 23.1 Design requirements for tactile sensing system.

Design criteria		Character guideline
Detectable force range (dynamic range)		0.01–10 N (1000 : 1)
Tactile sensing element (Taxel) pitch (for array only)		≤1 mm for small sensing areas ≥5 mm for large less sensing areas
Spatial resolution		≤1 mm for fingertips 5 mm ÷ 20–30 mm (e.g., limbs, torso, etc.)
Sensor frequency bandwidth (sensor response time)		About 1 kHz (1 ms)
Temporal variation		Both dynamic and static
Mechanical sensing detection capability		Normal and shear forces; vibrations
Sensor system characteristics	Mechanical	Flexible, stretchable, conformable and soft, robust, and durable
	Electrical	Low-power, minimal wiring, and cross talk, electrically and magnetically minimal sensitivity
Sensor response		Monotonic, not necessarily linear, low hysteresis, stable, and repeatable

Source: Adapted from Refs [14–16].

be satisfied totally or partially, according to the target application. Many of the previous requirements are satisfied by many examples reported in the literature, even if, to our knowledge, no e-skin implementation satisfies all of them.

Transduction technologies and the functional materials are the focus of this section. The commonly used are capacitive, piezoelectric, optical, and resistive/piezoresistive (based on conductive polymer films and elastomer composites), with their advantages and disadvantages [13,14,17,18]. Capacitive technology has a well-established design and fabrication technique and it is, along with resistive, the most diffused technology. Capacitive e-skin requires dielectric materials having high dielectric constant (k) to increase sensitivity, and ferroelectric polymers are usually preferred [10]. High-k thin elastomer dielectrics can be developed by an appropriate chemical design, or by addition of high-k fillers or conductive fillers [13] in the elastomer, to design stretchable and flexible capacitive tactile sensors [19,20]. Nanostructured materials like carbon nanotubes (CNT) can enhance sensitivity, dynamic range, and resolution [10]. Capacitive tactile sensors have been developed for small- [21] and large-area tactile sensing [22] and 3D pressure/force sensing [23]. Unfortunately, stray capacitances and cross talk between sensor elements, electromagnetic interference (EMI) sensitivity, and the need of relatively complex electronic circuitry are the most significant drawbacks.

Conductive polymer composite films are used for the development of flexible and compliant large-area resistive-based e-skin, which can be wrapped around curved surfaces. Elastomer composites also are used to realize stretchable resistive sensors [24], although their use is limited to pressure sensing/imaging applications [14] (e.g., electrical impedance tomography by Tawil *et al.* [25]). Stassi *et al.* [17] provides a comprehensive review on the different types of composite materials used in the development of piezoresistive sensing devices. According to Stassi *et al.* [17], e-skin based on resistive solutions and flexible composite materials could give the possibility to satisfy almost all the general requirements presented in Table 23.1. Tactile sensors based on piezoelectric materials – such as PZT, PVDF, PVDF-TrFE, to name but a few – are ideal for dynamic tactile sensing [26] due to their large frequency bandwidth and reduced response time (e.g., useful for monitoring dexterous manipulation, sensing fine surface features and textures). Moreover, piezoelectric materials are mechanically flexible, robust, and present high sensitivity. Examples of small- [27] and large-area [28] piezoelectric-based tactile sensing systems are present in literature. Piezoelectric sensors can exhibit drift in sensor response over time [13] and are mechanically not stretchable. Examples of piezoelectric tactile sensors for 3D force sensing [29,30] exist in literature.

When the number of sensing elements increases, optical-based tactile sensors can be a solution, because of a simplified and cross-talk-free wiring [14] and the possibility to have any electronic component on the sensing areas [31]. Optical tactile sensors can be used for dynamic tactile sensing as well [31]. The use of plastic optical fibers (POFs) allows overcoming sensor limitations related to fragility and rigidity [14]. Polymer-waveguide-based sensors satisfy requirements such as thin film architecture, localized force sensing and multipoint recognition, robustness to bending and fast response times [31], stretchability, stability, low hysteresis and high sensitivity (e.g., weight as low as 10 mg is detectable), and easy to fabricate [32]. For more comprehensive and exhaustive overview of materials and transduction techniques, which is not compatible with the page limits of this chapter, the readers are referred to other recent reviews, for example, Refs [10,13–18,33–35].

In Table 23.2, we report a comparison of the tactile sensing technologies described above and presented in the recent scientific literature (covered years: 2011 to present). In particular, such parameters as hysteresis, bandwidth (or response time), and spatial resolution have not been included as no complete information has been found.

23.2.2 An Example of Skin Integration into an Existing Robotic Platform

While in the next section we will deal with sensor signal processing and interpretation, here we report on how it is possible to practically couple the skin structure to a robotic platform. We illustrate a literature example, consisting in integrating capacitive sensor arrays into the iCub robot.

Effective integration of tactile sensors into real robots requires conformable structures that can be deployed on curved surfaces. Various system-level issues have to be managed, like wiring, networking, power consumption, maintenance, and lowering production costs. Figure 23.2 shows a graphical sketch of the integration of the ROBOSKIN[1] capacitive e-skin system into the iCub robot, as described in Ref. [22]. Similar approaches can be employed for the integration of different skin concepts into other robots.

A skin system for humanoids that integrates distributed pressure sensors based on the capacitive technology has been presented in Ref. [49]. It consists of triangular modules interconnected to form a system of sensors that can be deployed on nonflat surfaces. The basic functional element is a capacitor in which the dielectric deforms when pressure is applied. Patterned conductive areas on a flexible flexible printed circuit board (FPCB) form the first plate of the capacitor. On top of the FPCB, there is a deformable dielectric (3D air mesh fabric) covered by a conductive (Lycra) layer that provides the second plate of the capacitor and works as a common ground plane protecting the sensors from electromagnetic interferences. The third external layer has special hemlines with holes that host screws to keep the cover in place (see Figure 23.3a). Therefore, it can be easily substituted, if damaged, and removed to check the status of the FPCB and electronics below. The FPCB is shaped as a triangle hosting 12 sensors (i.e., taxels) and a capacitance to digital converter (CDC, AD7147 from Analog Devices) performs the AD conversion and transmits the capacitance values to a serial line. Several triangles can be interconnected to form a flexible mesh of sensors to cover the desired area. The triangles and the connections among them are flexible: The resulting mesh can, therefore, be adapted to curved surfaces. The system has been successfully integrated into different humanoid robots, for example, iCub [50],

---

1 "Skin-Based Technologies and Capabilities for Safe, Autonomous and Interactive Robots" is a FP7 STREP European project.

Table 23.2 Comparison of some tactile sensing technologies reported in international articles and conferences (covered years: 2011–2015).

Reference		Sensitivity		Force/pressure range		Bandwidth[1]/Resonant frequency[2]/Cutoff frequency[3]/Sensor response time[4]	Mechanical properties	Linearity	No. of sensors	Sensor size (mm)	Repeatability[1]/Stability[2]/Standard deviation[3]
		Normal	Shear	Normal (kPa)	Shear (kPa)						
Capacitive											
2011	[36]	—	1.967fF/N	—	±4 N	<100 Hz	Flexible	Nearly linear	1	3.5×1.5×5	High[1] 0.428 fF ±2 N range[3] 1.35 fF ±4 N range[3]
2012	[37]	0.14 kPa^{-1} 0.05 kPa^{-1}	—	0–5 5–20	—	—	Flexible	Piece-wise linear	4×4	3×3	—
2013	[23]	0.024 kPa^{-1} 6.6e-4 kPa^{-1}	2.8e-4 kPa^{-1}	0–10 10–140	0–200	10 MHz[3]	Flexible	Nearly linear	1 module of 2×2 capacitors	0.06×4 (each capacitor)	No
2014	[38]	2.24%/MPa	—	240–1000	—	—	Flexible	Almost linear	1	1.5×6×6	—
2015	[19]	0.001–0.01 kPa^{-1}	—	5–405	—	—	Stretchable	Almost linear	6	9×5	Yes[2]
Piezoelectric											
2011	[39]	42 mV/N	—	3–5 N	—	—	Flexible	Nearly linear	1	—	Good[1]
2013	[40]	0.021 V/mN	—	10–100 mN	—	—	Flexible	Yes	4×4	1 mm (radius)	Good[2]
2014	[41]	~8 nA/N	—	0.5–8 N	—	—	Flexible	Nearly linear	1	—	Low[1]
2015	[42]	430 mV/N (@200 Hz)	—	0.5–2 N	—	Up to 1200 Hz	Flexible	Yes	1	—	—
Resistive/piezoresistive											

Year	Ref									
2011	[43]	~0.004 kPa^{-1}	—	0–240	—	—	Flexible	Yes	8×8	1.5×1.5	10% (<150 kPa)[1] 20% (>150 kPa)[1]
2012	[44]	—	1.7e-6 Pa^{-1}	—	±1.8	—	Flexible	Yes	1	40×40	—
2013	[45]	1.3e-5 kPa^{-1}	1.3e-5 kPa^{-1}	0–400	0–100	—	Flexible	Yes	1	2×2×0.3	—
2014	[46]	—	—	<0.1 Pa 0.1–1.4	—	0.2 ms[4]	Flexible	—	1 (graphene array)	2 cm^2	Good[1] Good[2]
2015	[8]	−0.5 kPa^{-1}	—	~0–30	—	—	Flexible	—	8×8	1×1	High[1]

Optical

Year	Ref									
2011	[9]	−26.7 (ΔV/mV)/g	—	0–40 g	—	600 ms[4]	Flexible	Almost	1	30×30	High[2]
2012	[32]	0.02 kPa^{-1} (when stretched) 0.05 kPa^{-1} (when bent)	—	0–5 10 (const load) 10 (const load)	—	300 ms[4]	Stretchable	No	1	50×50	Good[2]
2013	[47]	0.8 pm/Pa	1.3 pm/Pa	<2.4 kPa	≤0.6	—	Soft	Almost	1	27×27×22	No
2014	[48]	~0.1 V/N	~0.1 V/N	0–8 N	0–2 N	1 ms[4]	Stretchability if stretchable wiring	Yes	6×6 (each sensor has 2×2 taxels)	6.4×6.4	6.77% (error)[1]

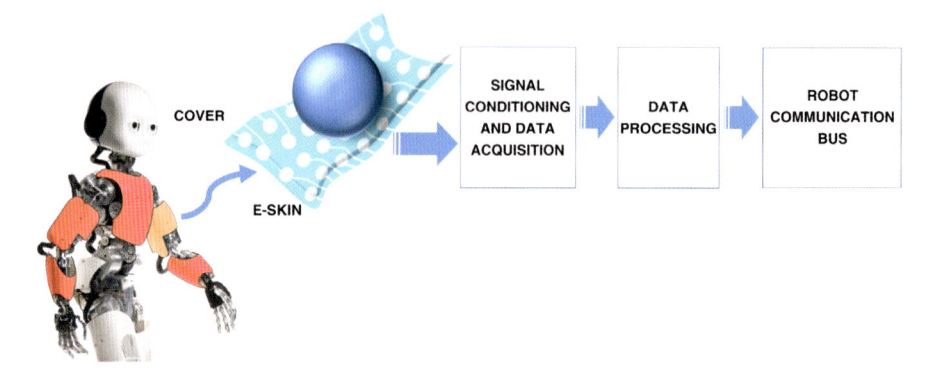

Figure 23.2 Graphical sketch of the integration of the ROBOSKIN e-skin system into iCub. (Image courtesy of the Italian Institute of Technology – IIT, Genova, Italy).

KASPAR [51], and NAO [52], and also the Schunk robotic arm as shown in Figure 23.4.

The following are the steps that were pursued for the skin integration into the iCub forearm:

- The mesh of triangles (see Figure 23.4) was glued on the cover of the iCub forearm using a bicomponent glue and with the help of a vacuum system in order to improve the adhesion on the 3D surface. In Figure 23.3c, it is possible to see the final result of this procedure.
- For the dielectric layer, different layers of fabric have been glued together, cut, and shaped to adapt to the robot part. The cover was then mounted and fixed with screws to the iCub forearm (see Figure 23.3a and b): This allows easy mounting and substitution. This procedure has been pursued for the integration of the sensor arrays on other iCub parts (the two arms, palms and torso, and also on the WAM arm from Barret Technology [22]).

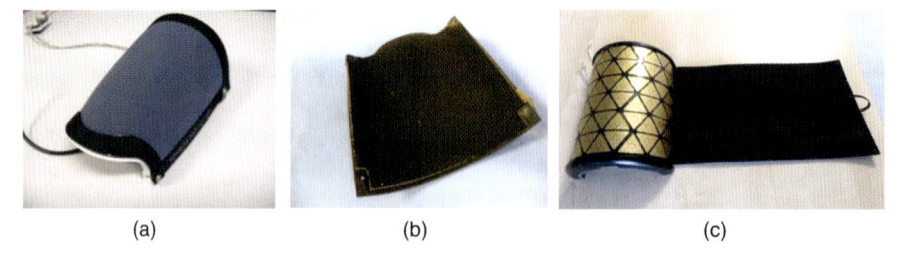

Figure 23.3 The sole dielectric layer that has been integrated into the sensor array in this version of the capacitive skin. (a) Front side. (b) Back side. (c) The complete skin patch (dielectric layer unrolled on the side) as mounted on the Schunk robotic arm. (Image courtesy of the Italian Institute of Technology – IIT, Genova, Italy).

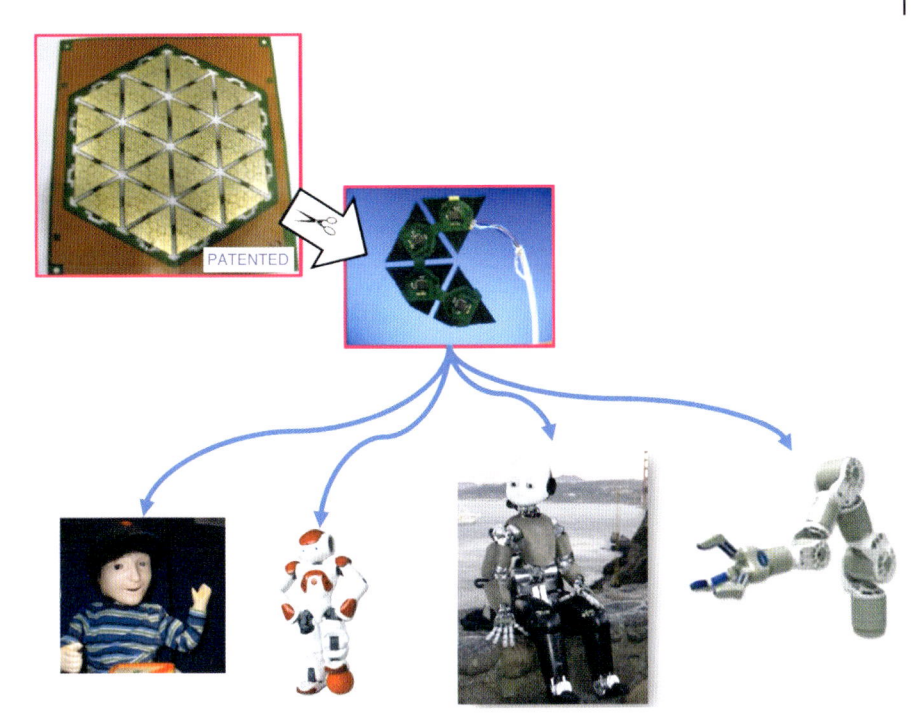

Figure 23.4 Capacitive e-skin applied to different robots. (Image courtesy of the Italian Institute of Technology – IIT, Genova, Italy).

23.3 Tactile Information Processing

Approaching tactile sensing at a *system* level means managing all issues associated with the integration of data processing and information management into a sensitive system built on structural and functional host materials. In this chapter, we limit the overview to what information is relevant and which are available methods to extract it from sensor data, discarding the wider problem of interpreting and reacting on the basis of this available structured and unstructured tactile information.

What kind of information is being propagated from the sensitized object to the robot/human? Interesting information about the minimum functional requirements for a robotic tactile sensing system mimicking human in-hand *manipulation* is contained in Ref. [14]. Those requirements shed light on what is needed to be detected during tactile contact for the complex manipulation process. An interesting review is contained in Ref. [53], in that HRI (human–robot interaction) context with the scope of summarizing (i) what is detected during tactile contact of a robot with a human and (ii) how the detected information is used by the robot. As said above, we only focus on aspect (i). Without pretending to be exhaustive, Table 23.3 extends what is reported in Ref. [53], including relevant references from journal publications only, published during 2005–2015.

Table 23.3 What information is extracted from sensor data (covered years: 2005–2015, journal publications only).

Contact	Location	[5,22,54–60]
	Area	[5,28,56,60–66]
	Static	[22,23,40,61–63,65,67–70]
	Dynamic	[28,30,43,45,54,64,69,71–73]
Force	Normal	[5,22,23,45,56–58,66,67,72,74–78]
	Shear	[23,40,45,67,68,74–77]
	Magnitude (intensity)	[23,45,56–60,66,67,70,72]
	Orientation (direction)	[23,40,45,59,60,67,70,72,74–76]
	Moment	[72,79–81]
	Pressure distribution (only normal force)	[22,24,32,54,63–65,82–84]
	Whole force distribution	[61,62,85,86]
High-Level Info	Texture/roughness	[71,87–94]
	Stiffness (contact object)	[24,54]
	Slip detection	[18,30,71,95–98]
	Vibration	[71,72]
	Touch modality	[5,6,66,99–103]

Tactile data processing concerns *lower level* information as that required for object grasping and manipulation and related to an accurate estimation of finger–object interaction such as contact location, area and duration, contact force intensity, direction and distribution, together with temperature. Touch also enables more *high-level* information for the classification of attributes of the contacting objects, for example, roughness, textures, patterns, shapes, as well as data related to object movement on the cutaneous surface, for example, slip detection, vibration, up to the discrimination of the touch modality.

Except for temperature, which is still a hot topic, which deserves a specific discussion, and which is beyond the scope of the present study, what is named *low-level* information can be faced either using *distributed* or *concentrated* approach.

In a distributed approach, complete information is attributable to the reconstruction of the force distribution at a given instant [48,104]. Either regularization methods (e.g., Tichonov) have been used in specific systems to deal with the ill-posed problem of retrieving the (continuum) complete force distribution starting from (finite) sensor data [105] or case-by-case approaches are used and specific sensor arrangements have been proposed leading to three-axis contact force distribution [61,62,85]. On the other hand, a general method for the reconstruction of the spatial distribution of contact forces as well as their intensities and directions starting from embedded sensor data has been recently proposed [86]. In this case, continuum mechanics is used as a framework for the direct problem of mechanical stimuli transmission to the sensor array and the problem is inverted using an optimization procedure and accounting for the physics of the problem. The importance of this method resides in its generality, as it is independent of the specific employed transducer and tactile application, provided that stress information comes from embedded sensor in an elastic layer.

More frequently, the distribution of the sole normal component of the force (pressure) is reconstructed, which allows for punctual information that is however far from being complete [22,24,32,54,63–65,82–84].

In a concentrated approach, either force resultants and moments are available and the contact location (centroid position at each instant) can be retrieved [79] or contact location(s), the force resultant(s), and point(s) of application are known and moments can be calculated [80].

While treating object movement on the cutaneous surface, dynamic sensors are involved (see Table 23.3). In this context, sliding systems have to be cited as they are relevant from the point of view of applications. For interesting literature about sliding and slip prevention, whose in-depth analysis is not at the core of this discussion, the reader is referred to Refs [18,30,95,96] and references therein.

Slip detection can be either inferred directly from vibration signals [30,71,106–108] or indirectly by characterizing and identifying the dynamic force variations associated with slip [97,98]. A widespread method (also used in prosthetic commercial systems[2]) to avoid sliding is to monitor the normal and shear force from a grasped object [23,45,67,74–77], while keeping the normal-to-tangential force ratio above a certain value [109]. Measurement of 3D contact forces is also critical for determining the full grasp force/torque and for handling fragile and irregular objects.

Other topics that are related to moving touches are the discrimination of textures and the interpretation of touch modalities [99], although in the context of social robotics information other than tactile (e.g., vision) probably needs to be involved – like for humans – to properly discriminate the modality of touch. To deepen these concepts, the reader is referred to Ref. [100] and references therein. If the tactile-sensing framework has to face such challenging assignments such as texture and touch modality recognition, machine learning techniques [110] may prove to be useful as explicit formalization of the input–output relationship is difficult to attain. In this case, the actual problem is to discriminate between a set of stimuli that the system is expected to recognize and empirical induction is used by learning-from-examples approach. Machine learning paradigms represent a powerful technology for tackling clustering, classification, and regression problems in complex domains and they have been widely used in robotics to retrieve partial contact information on specific systems [66,87–89,100,111]. Nevertheless, ML has its own limitations and the quality of results depends on the quality of training data and also generality is not warranted.

All these processing aspects can be integrated into overall systems that closely resembles to the human skin. Very interesting examples are reported in a recent review [13], which includes all existing research on multimodal systems that retrieve both structured and unstructured tactile information. To cite an example,

---

2 www.ottobock.com.

multimodal HEX-O-SKIN [1,71] measures normal and shear forces together with the hardness of the contact object and vibrations, enabling slip detection, impact sensation, and contact roughness sensing.

23.4 Computational Requirements

Besides the huge amount of tactile data to be processed in real time, the computation complexity poses a tough challenge in the development of the embedded electronic system. Computational requirements depend on the overall number of operations (mainly arithmetic) that the algorithm must perform and on the real-time operation.

This section presents an assessment of the computational requirements taking into account two sound and completed approaches: (i) electrical impedance tomography to classify social touch [5]; (ii) tensorial kernel approach to classify touch modality [101]. Figure 23.5 shows the algorithmic steps needed to classify the input touch: In the EIT approach, the complexity of the computation mainly lies in the first step (EIT inverse solution); while in the tensorial kernel approach, it lies in the singular value decomposition.

23.4.1 Electrical Impedance Tomography

Electrical impedance tomography is an imaging technique used to estimate the internal conductivity distribution of an electrically conductive body based on

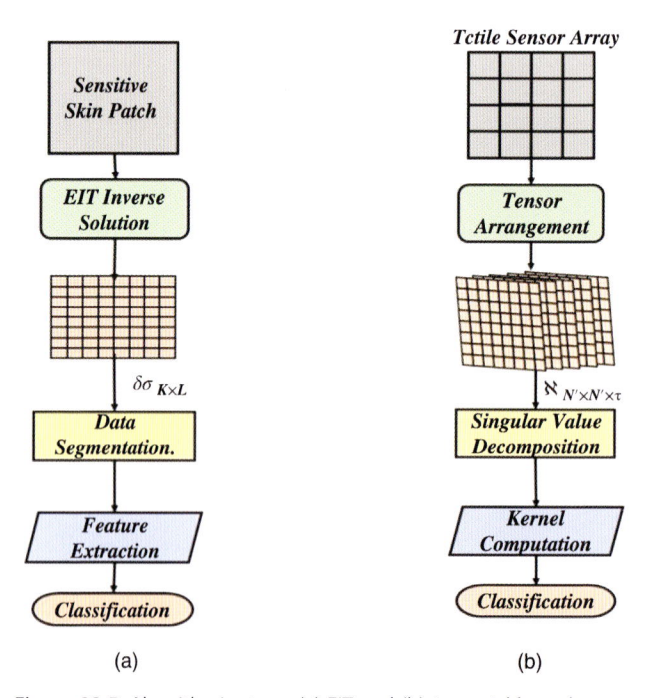

Figure 23.5 Algorithmic steps: (a) EIT and (b) tensorial kernel approaches.

electrical (i.e., voltages and/or currents) measurements made only at the boundary. In tactile systems, a conductive patch is used as sensitive skin; when a contact on the sensitive skin happens, the conductivity distribution changes as a consequence and the EIT reconstruction problem consists in finding the distribution of conductivity due to the contact; the result is an image displaying the distribution of conductivity and consequently the contact force/pression distribution. The EIT reconstruction problem is a nonlinear inverse problem where a unique solution does not exist (ill-conditioned problem). The starting point for EIT is Maxwell's equations for electromagnetics. A commonly used approach to solve the mathematical model represented by the Maxwell differential equations is the finite element method (FEM) [112]. At each iteration, the multiplication of a sparse $M \times M$ matrix by $1 \times M$ vector (where M is the number of potential measurements on the boundary) is implemented. In Ref. [5], the ill-conditioned problem has been solved by using the generalized Tikhonov regularization [113–115]. The solution figures out a matrix $\delta\sigma_{K \times L}$ containing the difference between the conductivity of the input contact and a conductivity σ_0 taken as reference. K is the number of elements in the FEM mesh and L is the number of injection current patterns [25].

The classification of touch modalities is done using "LogitBoost" classifier. Table 23.4 shows the operations needed to implement the EIT method.

23.4.2 Tensorial Kernel

Machine learning approaches have not been developed to handle data that are represented in the form of tensors, that is, n-dimension input vectors; hence, the implementation of some feature extraction process is needed to map tensor signals into multidimensional vectors, with the risk of compromising the original structural and topological information [116,117]. The ML-based approach in Ref. [101] proposes a tensor-based morphology of tactile signals, in a marked analogy with image tensor constructed from face images for face recognition application [36].

Tensorial kernel approach consists of first computing the singular value decomposition (SVD) [37] of the input tactile $N' \times N'$ matrix. The study of the computational requirements for the SVD is based on the *one-sided Jacobi*

Table 23.4 The computational requirements [5]; N_t is the number of training data.

Operation	FEM	Tikhonov regularization	Classification
Addition/subtraction	$M^2(M-1) + \frac{M}{6}(20M^2 - 3M - 5)$	$\frac{K}{6}(20K^2 - 5 + 6KM - 3K) + KM(L + K - 1) - KL$	$M \times (2M+3) \times 3N_t$
Multiplication	M^3	$KM(2K+L)$	$M \times (M+6) \times 3N_t$
Division	0	MK	$2 \times M \times 3N_t$
Square root	0	K^2	0

Table 23.5 The computational requirements [101]; N is the dimension of the matrix, K^* is the number of SVD iterations, and N_{sv} is the number of support vectors.

Operation	SVD	Kernel function	Classification
Addition/subtraction	$\frac{3}{2}N'(N'-1)\left[2N'^2(N'-1)+4\right]K^*$	$3N'(2N'^2-1)$	$(1+3Nt) \times N_{sv}$
Multiplication	$\frac{3}{2}N'(N'-1)(2N'^3+3)K^*$	$3N'^2(2N'+1)+5$	$3Nt \times N_{sv}$
Division	$\frac{9}{2}N'(N'-1)K^*$	0	0
Square-root	$3N'(N'-1)K^*$	0	0

algorithm that provides high accuracy and convergence in about $K^* = 5$–10 iterations [38]. Following step is the computation of the kernel function for a couple of SVDs, the first corresponds to the tensor input and the second to the tensor representing a predefined class extracted from the training data. Table 23.5 shows the operations needed to implement the tensorial kernel approach.

In order to assess the computational burden, a case study with the setups used in Refs [5,101] has been considered. The task is to classify an input touch with duration of $\tau = 1$ s into three touch modalities; the number of training data is set to $N_t = 100$. In Ref. [5], a total of $M = 224$ V measurements were needed at each acquisition step. A total of $K = 2240$ elements in the FEM mesh were used for the forward solution. The number of iterations of the LogitBoost classifier needed to converge is $M' = 25$. In Ref. [101], the input matrix has size 8×8 sensors; for the duration of touch $\tau = 1$ s, three different matrix sizes ($N' = 8$, $N' = 8$, and $N' = 64$) have to be decomposed by the SVD. The number of SVD iterations using the one-sided Jacobi algorithm is $K^* = 8$; the number of support vectors has been set to $N_{sv} = 50$. Table 23.6 shows the results of the case study in terms of number of operations.

Following Table 23.6 results, about 42 Giga-operations/s [5] and about 31 Giga-operations/s [101] are needed for a real-time single-touch classification (three classes). The requirements for the processing unit are very challenging; for instance, let us consider the very well-known ARM Cortex processor family [39]. The ARM Cortex-R are high-performance processors, meeting challenging real-time constraints: they offer high-performance computing solutions for real-time

Table 23.6 Number of operations needed for a single-input touch classification [5,101].

Operation	EIT	Tensorial kernel
Addition/subtraction	3.97×10^{10}	1.56×10^{10}
Multiplication	2.28×10^9	1.58×10^{10}
Division	6.36×10^5	6.48×10^5
Square root	5.01×10^6	4.32×10^5

embedded systems. The Cortex-R7 processor architecture reduces latency and enables symmetric multiprocessing in a dual-core configuration. Cortex-R7 can achieve 3 GIPS/core: The dual-core architecture is not able to achieve the target requirements as highlighted in Table 23.6.

One possible approach to tackle this issue could be to design a multiprocessor Cortex-R7 architecture on FPGA. Alternative solution is to design a dedicated ASIC either on FPGA or on a standard cell technology; although this solution is expensive, it allows the optimization of the parallel implementation and to fulfill target requirements.

23.5 Conclusions

The chapter reviews and assesses current trends in the development of an electronic skin. We addressed such an issue from a system perspective, starting from basic building components (materials, electronics) and approaching toward the complete skin system, taking into the due course embedded tactile data processing. We first focused our discussion and literature review on capacitive, piezoelectric, optical, and resistive/piezoresistive sensors, which are well-established and most relevant and promising tactile transduction techniques. As a next step, while the most demanding applications require capabilities of both perceiving the environment, interpreting it, and of basing decisions and actions on perception, we focused on the step before, consisting in analyzing data processing methods that provide useful tactile information while leaving the human brain the responsibility of deciding on how to proceed.

The embedded electronic system raises a number of issues: It must be compliant with the e-skin structure, that is, flexible, it must consume low power (heat dissipation is another cumbersome issue to be tackled), it must be implemented in the real-time complex processing tasks with a huge amount of tactile data to manage, it must be robust against electrical noise and mechanical damages, and must be resilient. All such demanding requirements are very difficult to achieve with currently available approaches; relevant research and engineering efforts must be devoted to this scope.

References

1 Mittendorfer, P. and Cheng, G. (2012) Integrating discrete force cells into multi-modal artificial skin. 12th IEEE-RAS International Conference on Humanoid Robots, Nov. 29–Dec. 1, Business Innovation Center Osaka, Japan.
2 Bossuyt, F., Vervust, T., and Vanfleteren, J. (2013) Stretchable electronics technology for large area applications: fabrication and mechanical characterization. *IEEE Transactions on Components, Packaging and Manufacturing Technology*, **3** (2), 229–235.
3 van den Brand, J., de Kok, M., Sridhar, A., Cauwe, M., Verplancke, R., Bossuyt, F., de Baets, J., and Vanfleteren, J. (2014) Flexible and stretchable electronics for wearable healthcare. The 44th European Solid State Device

Research Conference (ESSDERC), pp. 206–209. doi: 10.1109/ESSDERC.2014.6948796

4 Rogers, J.A., Someya, T., and Huang, Y. (2010) Materials and mechanics for stretchable electronics. *Science*, **327** (5973), 1603–1607.

5 Silvera Tawil, D., Rye, D., and Velonaki, M. (2014) Interpretation of social touch on an artificial arm covered with an EIT-based sensitive skin. *Journal of Social Robotics*, **6** (4), 489–505.

6 Gastaldo, P., Pinna, L., Seminara, L., Valle, M., and Zunino, R. (2015) A tensor-based approach to touch modality classification by using machine learning. *Robotics and Autonomous Systems*, **63**, 268–278.

7 Ramadan, K.S., Sameoto, D., and Evoy, S. (2014) A review of piezoelectric polymers as functional materials for electromechanical transducers. *Smart Materials and Structures*, **23** (3), 033001.

8 Ma, C.-W., Lin, T.-H., and Yang, Y.-J. (2015) Tunneling piezoresistive tactile sensing array fabricated by a novel fabrication process with membrane filters. 28th IEEE International Conference on Micro Electro Mechanical Systems (MEMS), Jan. 18–22, 2015, pp. 249–252. doi: 10.1109/MEMSYS.2015.7050935

9 Massaro, A., Spano, F., Lay-Ekuakille, A., Cazzato, P., Cingolani, R., and Athanassiou, A. (2011) Design and characterization of a nanocomposite pressure sensor implemented in a tactile robotic system, instrumentation and measurement. *IEEE Transactions on Instrumentation and Measurement*, **60** (8), 2967–2975.

10 Saraf, R. and Maheshwari, V. (2008) Tactile devices to sense touch on a par with a human finger. *Angewandte Chemie, International Edition*, **47**, 7808–7826.

11 Kim, J., Lee, M., Shim, H.J., Ghaffari, R., Cho, H.R., Son, D., Jung, Y.H., Soh, M., Choi, C., Jung, S., Chu, K., Jeon, D., Lee, S.-T., Kim, J.H., Choi, S.H., Hyeon, T., and Kim, D.-H. (2014) Stretchable silicon nanoribbon electronics for skin prosthesis. *Nature Communications*, **5** (5747). doi: 10.1038/ncomms6747

12 Lumelsky, V.J., Shur, M.S., and Wagner, S. (2001) Sensitive skin. *IEEE Sensors Journal*, **1** (1). doi: 10.1109/JSEN.2001.923586

13 Hammock, M.L., Chortos, A., Tee, B.C.-K., Tok, J.B.-H., and Bao, Z. (2013) 25th anniversary article: the evolution of electronic skin (e-skin): a brief history, design considerations, and recent progress. *Advanced Materials*, **25**, 5997–6038.

14 Yousef, H., Boukallel, M., and Althoefer, K. (2011) Tactile sensing for dexterous in-hand manipulation in robotics: a review. *Sensors and Actuators A*, **167**, 171–187.

15 Mehrizi, A., Dargahi, J., and Najarian, S. (2009) *Artificial Tactile Sensing in Biomedical Engineering*, McGraw-Hill.

16 Dahiya, R.S., Metta, G., Valle, M., and Sandini, G. (2010) Tactile sensing: from humans to humanoids. *IEEE Transactions on Robotics*, **26** (1), 1–20.

17 Stassi, S., Cauda, V., Canavese, G., and Fabrizio Pirri, C. (2014) Flexible tactile sensing based on piezoresistive composites: a review. *Sensors*, **14** (3), 5296–5332.

18 Tiwana, M.I., Redmond, S.J., and Lovell, N.H. (2012) A review of tactile sensing technologies with applications in biomedical engineering. *Sensors and Actuators A*, **179**, 17–31.

19 Gerratt, A.P., Michaud, H.O., and Lacour, S.P. (2015) Elastomeric electronic skin for prosthetic tactile sensation. *Advanced Functional Materials*, **25** (15), 2287–2295.

20 Park, S., Kim, H., Vosgueritchian, M., Cheon, S., Kim, H., Koo, J.H., Kim, T.R., Lee, S., Schwartz, G., Chang, H., and Bao, Z. (2014) Stretchable energy-harvesting tactile electronic skin capable of differentiating multiple mechanical stimuli modes. *Advanced Materials*, **26**, 7324–7332.

21 Tsai, T.-H., Tsai, H.-C., and Wu, T.-K. (2014) A CMOS micromachined capacitive tactile sensor with integrated readout circuits and compensation of process variations. *IEEE Transactions on Biomedical Circuits and Systems*, (99), 1–9.

22 Maiolino, P., Maggiali, M., Cannata, G., Metta, G., and Natale, L. (2013) A flexible and robust large scale capacitive tactile system for robots. *IEEE Sensors Journal*, **13** (10), 3910–3917.

23 Dobrzynska, J.A. and Gijs, M.A.M. (2020) Polymer-based flexible capacitive sensor for three-axial force measurements. *Journal of Micromechanics and Microengineering*, **23** (1), 015009

24 Cheng, M.-Y., Tsao, C.-M., Lai, Y.-Z., and Yang, Y.-J. (2011) The development of a highly twistable tactile sensing array with stretchable helical electrodes. *Sensors and Actuators A*, **166** (2), 226–233.

25 Tawil, D.S., Rye, D., and Velonaki, M. (2011) Improved image reconstruction for an EIT-based sensitive skin with multiple internal electrodes. *IEEE Transactions on Robotics*, **27** (3), 425–435.

26 Howe, R.D. and Cutkosky, M.R. (1993) Dynamic tactile sensing: perception of fine surface features with stress rate sensing. *IEEE Transactions on Robotics and Automation*, **9** (2), 140–151.

27 Dahiya, R.S., Adami, A., Pinna, L., Collini, C., Valle, M., and Lorenzelli, L. (2014) Tactile sensing chips with POSFET array and integrated interface electronics. *IEEE Sensors Journal*, **14** (10), 3448–3457.

28 Seminara, L., Pinna, L., Valle, L.M., Basirico, L., Loi, A., Cosseddu, P., Bonfiglio, A., Ascia, A., Biso, M., Ansaldo, A., Ricci, D., and Metta, G. (2013) Piezoelectric polymer transducer arrays for flexible tactile sensors. *IEEE Sensors Journal*, **13** (10), 4022–4029.

29 Faramarzi, S., Ghanbari, A., Chen, X.Q., and Wang, W.H. (2009) A PVDF based 3D force sensor for micro and nano manipulation IEEE International Conference on Control and Automation (ICCA'09), Dec. 9–11, 2009. pp. 867–871. doi: 10.1109/ICCA.2009.5410237

30 Cotton, D.P.J., Chappell, P.H., Cranny, A., White, N.M., and Beeby, S.P. (2007) A novel thick-film piezoelectric slip sensor for a prosthetic hand. *IEEE Sensors Journal*, **7** (5), 752–761.

31 Yun, S., Park, S., Park, B., Kim, Y., Park, S.K., Nam, S., and Kyung, K.-U. (2014) Polymer-waveguide-based flexible tactile sensor array for dynamic response. *Advanced Materials*, **26**, 4474–4480.

32 Ramuz, M., Tee, B.C.-K., Tok, J.B.-H., and Bao, Z. (2012) Transparent, optical, pressure-sensitive artificial skin for large-area stretchable electronics. *Advanced Materials*, **24**, 3223–3227.

33 Dahiya, R.S., Mittenderfer, P., Valle, M., Cheng, G., and Lumelsky, V.J. (2013) Directions toward effective utilization of tactile skin: a review. *IEEE Sensors Journal*, **13** (11), 4121–4138.

34 Chortos, A. and Bao, Z. (2014) Skin-inspired electronic devices. *Materials Today*, **17** (7), 321–331.

35 Sokhanvar, S., Dargahi, J., Najarian, S., and Arbatani, S. (2012) Tactile sensing technologies, in *Tactile Sensing and Displays: Haptic Feedback for Minimally Invasive Surgery and Robotics*, John Wiley & Sons, Ltd, Chichester, UK.

36 Jinseok, L. and Daijin, K. (2009) Tensor-based AAM with continuous variation estimation: application to variation-robust face recognition. *IEEE Transactions on Pattern Analysis and Machine Intelligence*, **31** (6), 1102–1116.

37 Ledesma-Carrillo, L.M., Cabal-Yepez, E., Romero-Troncoso, R.de J., Garcia-Perez, A., Osornio-Rios, R.A., and Carozzi, T.D. (2011) Reconfigurable FPGA based unit for singular value decomposition of large $m \times n$ matrices. International Conference on Reconfigurable Computing and FPGAs.

38 Wang, X. and Zambreno., J. (2014) An FPGA implementation of the Hestenes–Jacobi algorithm for singular value decomposition. IEEE 28th International Parallel & Distributed Processing Symposium Workshops.

39 infocenter.arm.com/help/index.jsp?topic=/com.arm.doc.set.cortexr/index.html.

40 Tiwana, M.I., Shashank, A., Redmond, S.J., and Lovell, N.H. (2011) Characterization of a capacitive tactile shear sensor for application in robotic and upper limb prostheses. *Sensors and Actuators A*, **165** (2), 164–172.

41 Chen, C.-C., Chang, P.-Z., and Shih, W.-P. (2012) Flexible tactile sensor with high sensitivity utilizing botanical epidermal cell natural micro-structures. *IEEE Sensors Journal*, 1–4.

42 Lei, K.F., Lee, K.-F., and Lee, M.-Y. (2014) A flexible PDMS capacitive tactile sensor with adjustable measurement range for plantar pressure measurement. *Microsystem Technologies*, **20** (7), 1351–1358.

43 Wang, Y.R., Zheng, J.M., Ren, G.Y., Zhang, P.H., and Xu, C. (2011) A flexible piezoelectric force sensor based on PVDF fabrics. *Smart Materials and Structures*, **20** (4), 045009.

44 Kim, M.S., Jo, S.E., Kang, D.H., Ahn, H.R., and Kim, Y.J. (2013) A dome shaped piezoelectric tactile sensor array using controlled inflation technique. The 17th International Conference on Solid-State Sensors, Actuators and Microsystems (Transducers & Eurosensors XXVII), June 16–20, 2013, pp. 1891–1894. doi: 10.1109/Transducers.2013.6627161

45 Beccai, L., Roccella, S., Ascari, L., Valdastri, P., Sieber, A., Carrozza, M., and Dario, P. (2008) Development and experimental analysis of a soft compliant tactile microsensor for anthropomorphic artificial hand. *IEEE/ASME Transactions on Mechatronics*, **13**, 158–168.

46 Cosseddu, P., Viola, F., Lai, S., Raffo, L., Seminara, L., Pinna, L., Valle, M., Dahiya, R., and Bonfiglio, A. (2014) Tactile sensors with integrated piezoelectric polymer and low voltage organic thin-film transistors. *IEEE Sensors Journal*, pp. 1734–1736. doi: 10.1109/ICSENS.2014.6985358

47 Maita, F., Maiolo, L., Minotti, A., Pecora, A., Ricci, D., Metta, G., Scandurra, G., Giusi, G., Ciofi, C., and Fortunato, G. (2015) Ultra-flexible tactile piezoelectric sensor based on low-temperature polycrystalline silicon thin film transistor technology. *IEEE Sensors Journal*. doi: 10.1109/JSEN.2015.2399531.

48 Fearing, R.S. (1990) Tactile sensing mechanisms. *International Journal of Robotics Research*, **9** (3), 3–23.

49 Schmitz, A., Maiolino, P., Maggiali, M., Natale, L., Cannata, G., and Metta, G. (2011) Methods and technologies for the implementation of large-scale robot tactile sensors. *IEEE Transactions on Robotics*, **27** (3), 389–400.

50 Metta, G., Vernon, D., Natale, L., Nori, F., and Sandini, G. (2008) The iCub humanoid robot: an open platform for research in embodied cognition. Presented at IEEE Workshop on Performance Metrics Intelligent Systems, San Jose, CA.

51 Dautenhahn, K., Nehaniva, C.L., Waltersa, M.L., Robins, B., Kose-Bagcia, H., Mirzaa, N.A., and Blow, M. (2009) KASPAR: a minimally expressive humanoid robot for human–robot interaction research. *Applied Bionics and Biomechanics*, **6** (3), 369–397.

52 Dahl, T.S. and Palmer, A. (2010) Touch-triggered protective reflexes for safer robots. Proceedings of the International Symposium on New Frontiers in Human–Robot Interactions, p. 7.

53 Argall, B.D. and Billard, A.G. (2010) A survey of tactile human–robot interactions. *Robotics and Autonomous Systems*, **58**, 1159–1176.

54 Sokhanvar, S., Packrisamy, M., and Dargahi, J. (2007) A multifunctional PVDF-based tactile sensor for minimally invasive surgery. *Smart Materials and Structures*, **16**, 989–998.

55 Aucouturier, J.-J., Ikeuchi, K., Hirukawa, H., Nakaoka, S., Shiratori, T., Kudoh, S., Kanehiro, F., Ogata, T., Kozima, H., Okuno, H.G., Michalowski, M.P., Ogai, Y., Ikegami, T., Kosuge, K., Takeda, T., and Hirata, Y. (2008) Cheek to chip: dancing robots and AI's future. *IEEE Intelligent Systems*, **23** (2), 1541–1672.

56 Miyashita, T., Tajika, T., Ishiguro, H., Kogure, K., and Hagita, N. (2007) Haptic communication between humans and robots, in *Robotics Research*, vol. 28, Springer, Berlin, Germany, pp. 525–536.

57 Mukai, T., Onishi, M., Odashima, T., Hirano, S., and Luo, Z. (2008) Development of the tactile sensor system of a human-interactive robot RI-MAN. *IEEE Transactions on Robotics*, **24** (2), 505–512.

58 Nishio, S., Ishiguro, H., and Hagita, N. (2007) Geminoid: teleoperated android of an existing person, in *Humanoid Robots: New Developments* (ed. A.C. de Pina Filho), I-Tech, Vienna, Austria.

59 Wettels, N., Santos, V., Johansson, R., and Loeb, G. (2008) Biomimetic tactile sensor array. *Advanced Robotics*, **22**, 829–849.

60 Yamada, Y., Morizono, T., Umetani, Y., and Takahashi, H. (2005) Highly soft viscoelastic robot skin with a contact object-location-sensing capability. *IEEE Transactions on Industrial Electronics*, **52** (4), 960–968.

61 Lee, H.-K., Chung, J., Chang, S.-I., and Yoon, E. (2008) Normal and shear force measurement using a flexible polymer tactile sensor with embedded multiple capacitors. *Journal of Microelectromechanical Systems*, **17** (4), 934–942.

62 Lee, H.K., Chung, J., Chang, S.I., and Yoon, E. (2011) Real-time measurement of the three-axis contact force distribution using a flexible capacitive polymer tactile sensor. *Journal of Micromechanics and Microengineering*, **21** (3), 035010.

63 Barbaro, M., Caboni, A., Cosseddu, P., Mattana, G., and Bonfiglio, A. (2010) Active devices based on organic semiconductors for wearable applications. *IEEE Transactions on Information Technology in Biomedicine*, **14** (3), 758–766.

64 Dahiya, R.S., Adami, A., Collini, C., and Lorenzelli, L. (2013) POSFET tactile sensing arrays using CMOS technology. *Sensors and Actuators A*, **202**, 226–232.

65 Mannsfeld, S.C.B., Tee, B.C., Stoltenberg, R.M., Chen, C.V.H., Barman, S., Muir, B.V.O., Sokolov, A.N., Reese, C., and Bao, Z. (2010) Highly sensitive flexible pressure sensors with microstructured rubber dielectric layers. *Nature Materials*, **9** (10), 859–864.

66 Iwata, H. and Sugano, S. (2005) Human–robot-contact-state identification based on tactile recognition. *IEEE Transactions on Industrial Electronics*, **52** (6), 1468–1477.

67 Takahashi, H., Nakai, A., Thanh-Vinh, N., Matsumoto, K., and Shimoyama, I. (2013) A triaxial tactile sensor without crosstalk using pairs of piezoresistive beams with sidewall doping. *Sensors and Actuators A*, **199**, 43–48.

68 Noda, K., Onoe, H., Iwase, E., Matsumoto, K., and Shimoyama, I. (2012) Flexible tactile sensor for shear stress measurement using transferred sub-μm-thick Si piezoresistive cantilevers. *Journal of Micromechanics and Microengineering*, **22** (11). doi: 10.1088/0960-1317/22/11/115025

69 Dargahi, J., Rao, N., and Sokhanvar, S. (2006) Design and microfabrication of a hybrid piezoelectric-capacitive tactile sensor. *Sensor Review*, **26** (3), 186–192.

70 da Rocha, J.G.V. and Lanceros-Mendez, S. (2009) Capacitive sensor for three-axis force measurements and its readout electronics. *IEEE Transactions on Instrumentation and Measurement*, **58**, 2830–2836.

71 Mittendorfer, P. and Cheng, G. (2011) Humanoid multimodal tactile-sensing modules. *IEEE Transaction on Robotics*, **27** (3), 401–410.

72 Schmidt, P.A., Maël, E., and Würtz, R.P. (2006) A sensor for dynamic tactile information with applications in human–robot interaction and object exploration. *Robotics and Autonomous Systems*, **54** (12), 1005–1014.

73 Qasaimeh, M., Sokhanvar, S., Dargahi, J., and Kahrizi, M. (2009) PVDF-based microfabricated tactile sensor for minimally invasive surgery. *Journal of Microelectromechanical Systems*, **18** (1), 195–207.

74 Choi, W.-C. (2010) Polymer micromachined flexible tactile sensor for three-axial loads detection. *Transactions on Electrical and Electronic Materials*, **11** (3), 130–133.

75 Huang, Y., Sohgawa, M., Yamashita, K., Kanashima, T., Okuyama, M., Noda, M., and Noma, H. (2008) Fabrication and normal/shear stress responses of tactile sensors of polymer/Si cantilevers embedded in PDMS and urethane gel elastomers. *IEEE Transactions on Sensors and Micromachines*, **128**, 193–197.

76 Kim, K., Lee, K.R., Lee, D.S., Cho, N.-K., Kim, W.H., Park, K.-B., Park, H.-D., Kim, Y.K., Park, Y.-K., and Kim, J.-H. (2006) A silicon-based flexible tactile sensor for ubiquitous robot companion applications. *Journal of Physics*, **34**, 399–403.

77 Valdastri, P., Roccella, S., Beccai, L., Cattin, E., Menciassi, A., Carrozza, M.C., and Dario, P. (2005) Characterization of a novel hybrid silicon three-axial force sensor. *Sensors and Actuators A*, **123–124**, 249–257.

78 Yussof, H., Ohka, M., Kobayashi, H., Takata, J., Yamano, M., and Nasu, Y. (2007) Development of an optical three-axis tactile sensor for object handing tasks in humanoid robot navigation system. *Studies in Computational Intelligence*, **76**, 43–51.

79 Liu, H., Nguyen, K.C., Perdereau, V., Bimbo, J., Back, J., Godden, M., Seneviratne, L.D., and Althoefer, K. (2015) Finger contact sensing and the application in dexterous hand manipulation. *Autonomous Robots*, **39** (1), 25–41.

80 Fumagalli, M., Ivaldi, S., Randazzo, M., Natale, L., Metta, G., Sandini, G., and Nori, F. (2012) Force feedback exploiting tactile and proximal force/torque sensing. *Autonomous Robots*, **33**, 381–398.

81 De Maria, G., Natale, C., and Pirozzi, S. (2012) Force/tactile sensor for robotic applications. *Sensors and Actuators. A*, **175**, 60–72.

82 Kawaguchi, H., Someya, T., Sekitani, T., and Sakurai, T. (2005) Cut-and-paste customization of organic FET integrated circuit and its application to electronic artificial skin. *IEEE Journal of Solid-State Circuits*, **40** (1), 177–185.

83 Yang, Y.-J., Cheng, M.-Y., Chang, W.-Y., Tsao, L.-C., Yang, S.-A., Shih, W.-P., Chang, F.-Y., Chang, S.-H., and Fan, K.-C. (2008) An integrated flexible temperature and tactile sensing array using PI–copper films. *Sensors and Actuators A*, **143**, 143–153.

84 Yang, Y.-J., Cheng, M.-Y., Shih, S.-C., Huang, X.-H., Tsao, C.-M., Chang, F.-Y., and Fan, K.-C. (2010) A 32×32 temperature and tactile sensing array using PI–copper films. *International Journal of Advanced Manufacturing Technology*, **46**, 945–956.

85 Cirillo, A., Cirillo, P., De Maria, G., Natale, C., and Pirozzi, S. (2014) An artificial skin based on optoelectronic technology. *Sensors and Actuators A*, **212**, 110–122.

86 Seminara, L., Capurro, M., and Valle, M. (2015) Tactile data processing method for the reconstruction of contact force distributions. *Mechatronics*, **27**, 28–37

87 Arabshahi, S.A. and Jiang, Z. (2005) Development of a tactile sensor for Braille pattern recognition: sensor design and simulation. *Smart Materials and Structures*, **14**, 1569–1578.

88 Decherchi, S., Gastaldo, P., Dahiya, R.S., Valle, M., and Zunino, R. (2011) Tactile-data classification of contact materials using computational intelligence. *IEEE Transactions on Robotics*, **37** (3), 635–639.

89 Kim, S.H., Engel, J., Liu, C., and Jones, D.L. (2005) Texture classification using a polymer-based MEMS tactile sensor. *Journal of Micromechanics and Microengineering*, **15** (5), 912–920.

90 Hosoda, K., Tada, Y., and Asada, M. (2006) Anthropomorphic robotic soft fingertip with randomly distributed receptors. *Robotics and Autonomous Systems*, **54**, 104–109.

91 Jamali, N. and Sammut, S. (2011) Majority voting: material classification by tactile sensing using surface texture. *IEEE Transactions on Robotics*, **27**, 508–521.

92 Maheshwari, V. and Saraf, R.F. (2006) High-resolution thin-film device to sense texture by touch. *Science*, **312** (5779), 501–1504.

93 Sinapov, J., Sukhoy, V., Sahai, R., and Stoytchev, A. (2011) Vibrotactile recognition and categorization of surfaces by a humanoid robot. *IEEE Transactions on Robotics*, **27** (3), 488–497.

94 Oddo, C.M., Controzzi, M., Beccai, L., Cipriani, C., and Carrozza, M.C. (2011) Roughness encoding for discrimination of surfaces in artificial active-touch. *IEEE Transactions on Robotics*, **27** (3), 522–533.

95 Francomano, M.T., Accoto, D., and Guglielmelli, E. (2013) Artificial sense of slip: a review. *IEEE Sensors Journal*, **13** (7), 2489–2498.

96 Najarian, S., Dargahi, J., and Mehrizi, A. (2009) *Artificial Tactile Sensing in Biomedical Engineering*, McGraw Hill Professional.

97 Ho, A. and Hirai, S. (2014) *Mechanics of Localized Slippage in Tactile Sensing: And Application to Soft Sensing Systems*, Springer Tracts in Advanced Robotics, vol. 99, Springer International Publishing, Cham, Switzerland.

98 Romano, J.M., Hsiao, K., Niemayer, G., Chitta, S., and Kuchenbecker, K.J. (2011) Human-inspired robotic grasp control with tactile sensing. *IEEE Transactions on Robotics*, **99**, 1–13.

99 Cooney, M.D., Nishio, S., and Ishiguro, H. (2014) Importance of touch for conveying affection in a multimodal interaction with a small humanoid robot. *International Journal of Humanoid Robotics*, **12**. doi: 10.1142/S0219843615500024

100 Tawil, D.S., Rye, D., and Velonaki, M. (2012) Interpretation of the modality of touch on an artificial arm covered with an EIT-based sensitive skin. *International Journal of Robotics Research*, **31**, 1627–1641.

101 Gastaldo, P., Pinna, L., Seminara, L., Valle, M., and Zunino, R. (2014) A tensor-based pattern recognition framework for the interpretation of touch modality in artificial skin systems. *IEEE Sensors Journal*, **14** (7), 2216–2225.

102 Stiehl, W.D. and Breazeal, C. (2005) Affective touch for robotic companions, affective computing and intelligent interaction. *Lecture Notes in Computer Science*, **3784**, 747–754.

103 Yohanan, S. and MacLean, K. (2012) The role of affective touch in human–robot interaction: human intent and expectations in touching the haptic creature. *International Journal of Social Robotics*, **4**, 163–180.

104 Cutkosky, M.R., Howe, R.D., and Provancher, W. (2007) Force and tactile sensors, in *Springer Handbook of Robotics*, Springer, Berlin, pp. 455–476.

105 De Rossi, D., Caiti, A., Bianchi, R., and Canepa, G. (1991) Fine-form tactile discrimination through inversion of data from a skin-like sensor. Proceedings of the 1991 IEEE International Conference on Robotics and Automation, Sacramento, CA, pp. 398–403.

106 Cranny, A., Cotton, D.P.J., Chappell, P.H., Beeby, S.P., and White, N.M. (2005) Thick-film force, slip and temperature sensors for a prosthetic hand. *Measurement Science and Technology*, **16**, 1–11.

107 Cranny, A., Cotton, D.P.J., Chappell, P.H., Beeby, S.P., and White, N.M. (2005) Thick-film force and slip sensors for a prosthetic hand. *Sensors and Actuators A*, **123–124**, 162–171.

108 Howe, R.D. and Cutkosky, M.R. (1989) Sensing skin acceleration for slip and texture perception. Proceedings on the IEEE International Conference on Robotics and Automation, pp. 145–150.

109 Mingrino, A., Bucci, A., Magni, R., and Dario, P. (1994) Slippage control in hand prostheses by sensing grasping forces and sliding motion. *Proceedings of the IEEE/RSJ International Conference on Intelligent Robots and Systems*, **3** (3), 1803–1809.

110 Scholkopf, B. and Smola, A.J. (2001) *Learning with Kernels*, MIT Press, Cambridge, MA.

111 Goger, D., Gorges, N., and Worn, H. (2009) Tactile sensing for an anthropomorphic robotic hand: hardware and signal processing. Proceedings of the IEEE International Conference on Robotics and Automation, Kobe, Japan, May 12–17, pp. 895–901.

112 Silvester, P. and Ferrary, R. (1996) *Finite Elements for Electrical Engineers*, 3rd edn, Cambridge University Press, Cambridge, UK.

113 Lionheart, W., Polydorides, N., and Borsic, A. (2005) The reconstruction problem, in *Electrical Impedance Tomography: Methods, History and Applications* (ed. D.S. Holder), Institute of Physics, Bristol, UK, pp. 3–63.

114 Adler, A. and Guardo, R. (1996) Electrical impedance tomography: regularized imaging and contrast detection. *IEEE Transactions on Medical Imaging*, **15** (2), 170–179.

115 Adler, A. and Lionheart, W.R.B. (2006) Uses and abuses of EIDORS: an extensible software base for EIT. *Physiological Measurement*, **27**, S25–S42.

116 Signoretto, M., De Lathauwerb, L., and Suykens, J.A.K. (2011) A kernel based framework to tensorial data analysis. *Neural Network*, **24** (8), 861–874.

117 Zhao, Q., Zhou, G., Adali, T., Zhang, L., and Cichocki, A. (2013) Kernelization of tensor-based models for multiway data analysis: processing of multidimensional structured data. *IEEE Signal Processing Magazine*, **30** (4), 137–148.

24

Intelligent Materials in Machine Tool Applications: A Review

Hans-Christian Möhring

Otto-von-Guericke-University Magdeburg, Institute of Manufacturing Technology and Quality Management (IFQ), Universitätsplatz 2, Geb. 12, 39106 Magdeburg, Germany

In this chapter, an overview of the use of intelligent materials in manufacturing technology applications is given. Examples of realized systems and prototypes are introduced and experiences and main results are summarized. Thus, a variety of approaches and potentials are shown and discussed. The chapter is structured and focused with respect to the different types of intelligent materials which are predominantly exploited in machine tools and machine tool components. However, similarities to other fields of applications can be observed and technology transfer possibilities are obvious. A general overview of the use of construction materials in machine tool applications is given in Ref. [1]. Park *et al.* provide a review of the use of active materials for machining processes in Ref. [2]. Hurlebaus and Gaul summarize the use of intelligent or "smart" materials with respect to structural dynamics [3].

In principle, intelligent materials are applied in order to implement either sensory or actuator functionality and to integrate this inherently into a machine component or tool structure. Sensory and actuator functionality in this context particularly pertains to deformations, strains and stresses, forces and pressures, as well as temperatures. Based on this, also accelerations and vibrations are captured. This information is utilized for monitoring of process conditions or machine states and for control purposes such as semipassive, semiactive, or active damping of machines and processes [4]. Furthermore, energy harvesting and structural health monitoring can be implemented.

By intelligent materials, in contrast to conventional mechatronic system design, due to the material inherent capability, extremely high degrees of functional integration can be achieved. Consequently, intelligent or smart materials show high potential with respect to the realization of mechatronic and adaptronic components and subsystems and were intensively applied in related research work [5–8].

24.1 Applications of Shape Memory Alloys (SMA)

Shape memory alloys (SMA) have the ability to reassume their initial shape after deformation when temperature is increased [9]. Jani *et al.* and Humbeeck give an overview of SMA research and applications [10,11]. Based on the change of shape or, if preloaded, the change of force due to heating, SMAs can be used in grippers [12,13], clamping mechanisms [14–16], active compensators to maintain certain force conditions [17], or positioning systems [18–21].

Meier *et al.* introduced an actuator concept based on SMA wires for applications in small machine tools for small workpieces [18]. In each actuator, a central bolt is moved by the elongation and contraction of SMA wires. Individual strokes in a range of 5–5000 μm can be achieved. By combination of up to seven actuators, total displacements of up to 15 mm of a driven axis are realized. Based on a change of the electric resistance of the SMA actuators during phase transformation, controlled systems can be created without mechanical sensors such as stroke sensors.

Bucht *et al.* presented an approach of "self-sufficient" SMA actuators for compensating losses in the prestress of ball screw drives due to heating of the system [17]. Instead of SMA wires, compact body actuators are applied here. Besides an externally controlled solution in which the SMA actuator is heated by an active element, a self-adaptive concept is possible in which the actuator is driven by heat losses due to friction of the ball screw drive itself.

Because of the superelasticity of shape memory alloys, these materials can favorably be used also for flexure hinges allowing extreme deformations [22,23]. Furthermore, concepts for thermal control based on SMA actuation and phase change material were researched [24]. In addition, SMAs were investigated regarding their use in adaptive tuned vibration absorbers. By heating, the elastic modulus of the SMA changes and, thus, the natural frequency of the absorber can be adjusted. The heat treatment should avoid hysteresis in the martensitic transformation. For this, an R-phase transformation is discussed in Ref. [25].

24.2 Applications of Piezoelectric Ceramics

Piezoelectric ceramics are the mostly used smart materials in machine tool applications [5,26–28]. Piezoelectric ceramics are often integrated as sensors for monitoring of process conditions and machine tool states [29]. Furthermore, piezoactuators are used for precision positioning and compensation of deflections [30–37], as an additional high dynamic and high-precision compensator in hybrid drives [38,39], for semiactive and active damping [40–43], and for ultrahigh speed motion and targeted vibration excitation [44]. An exemplary "fast tool servo" based on piezoactuation is introduced in Ref. [30].

Piezoceramics are integrated into mechatronic and adaptronic systems and mechanical structures in the form of stack actuators [41], patch transducers [45], or thin film elements with piezoresistivity [46]. Piezoceramics possess a low elongation-to-size ratio (approximately 1: 1000), but high power density and stiffness. Furthermore, piezoceramics can be applied in a broad dynamic bandwidth. In order to overcome limits regarding positioning paths, mechanical or hydraulic transmission systems can be used [47–49].

A piezofluidic actuator for precision movement of large masses is presented in Refs [50–52]. In this case, the piezoactuators are used as a drive for a micropump and for the necessary valves to control the hydraulic flow that moves the piston of the actuator system.

In Ref. [46] an investigation of piezoresistive thin layer force sensors (produced by Fraunhofer IST) based on amorphous diamond-like carbon (DLC) is presented (Figure 24.1). The aim of this sensor application was to observe a reduction of prestress due to wear in a ball-screw drive by a highly integrated system. The analyses show that a temperature compensation of the sensor system is necessary, which can also be implemented by use of the thin-layer technology. The stiffness of the mechanical components is not affected significantly by this type of sensor integration. Furthermore, only low energy consumption emerges when applying these sensors.

An integration of piezopatch transducers into sensory CFRP elements is presented in Ref. [4] (Figure 24.2). These elements are used inside an intelligent clamping system for the identification and active reduction of vibrations during milling of thin-walled workpieces (Figure 24.3).

The displacement behavior of piezoceramic actuators possesses significant hysteresis characteristics. In principle, this can be avoided or compensated by the following strategies [53]:

- Low-level signal operation
- Usage of charge amplifiers instead of voltage as activation parameter
- Implementation of model-based precontrol strategies [54–56]
- Application of closed-loop control [57].

In Ref. [58], a piezo-based device for reducing axial vibrations in ball-screw drives is presented. The applied controller is based on a model of the feed axis including the actuator sensor unit. By use of the piezo-based system, a significantly improved step response could be achieved.

Some research work focuses on the application of piezoactuators for compensating tool deflections and vibrations in milling by modified spindle systems [59–63]. In these applications, the piezoactuators are used for moving an elastically supported milling spindle (Figure 24.4). Changes of the length of the actuators move the tool tip within a certain workspace, and thus allow a high dynamic correction of tool deviations due to quasi-static and dynamic process forces.

Latest applications utilize piezoactuators in tooling systems for microstructuring of tribological surfaces and for writing of information at workpieces [64,65].

As can be seen, the quantity and variety of piezoapplications is huge and the number of exemplary prototypes is still increasing. Of course, the field of application is not limited to production equipment. A review about vibration control of civil structures by piezoceramics is given, for example, in Ref. [66]. Intensive research deals with structural health monitoring by integration of piezoceramic transducers [67,68].

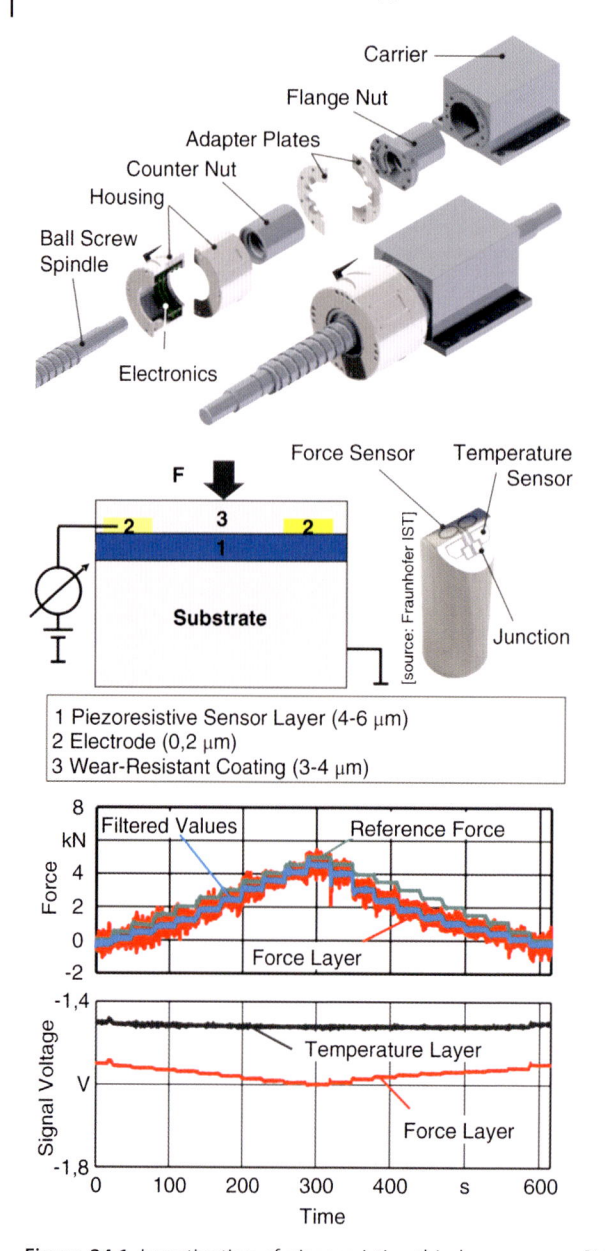

Figure 24.1 Investigation of piezoresistive thin layer sensors [46].

24.3 Applications of Magnetostrictive Materials

An overview of the function and use of magnetostrictive materials is given in Ref. [69]. As shown by Eda and Ohmura, giant magnetostrictive actuators can achieve several times the power and displacement of piezoactuators [70].

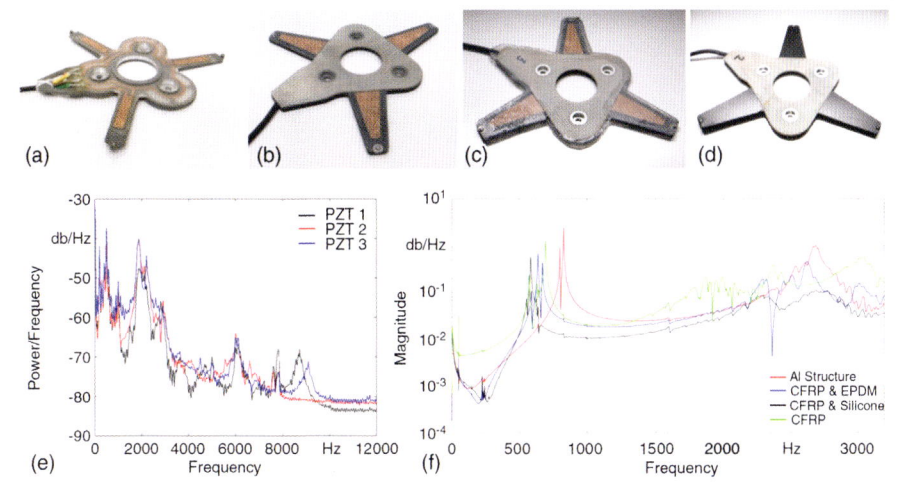

Figure 24.2 Sensory elements based on piezopatch transducers. (a) 3D printed aluminum structure with attached piezopatches. (b) CFRP structure with integrated piezopatches. (c) Sensory CFRP structure with silicone cover. (d) Sensory CFRP structure with EPDM cover. (e) Comparison of three sensors in one CFRP structure. (f) Frequency response characteristics of different sensor systems [4].

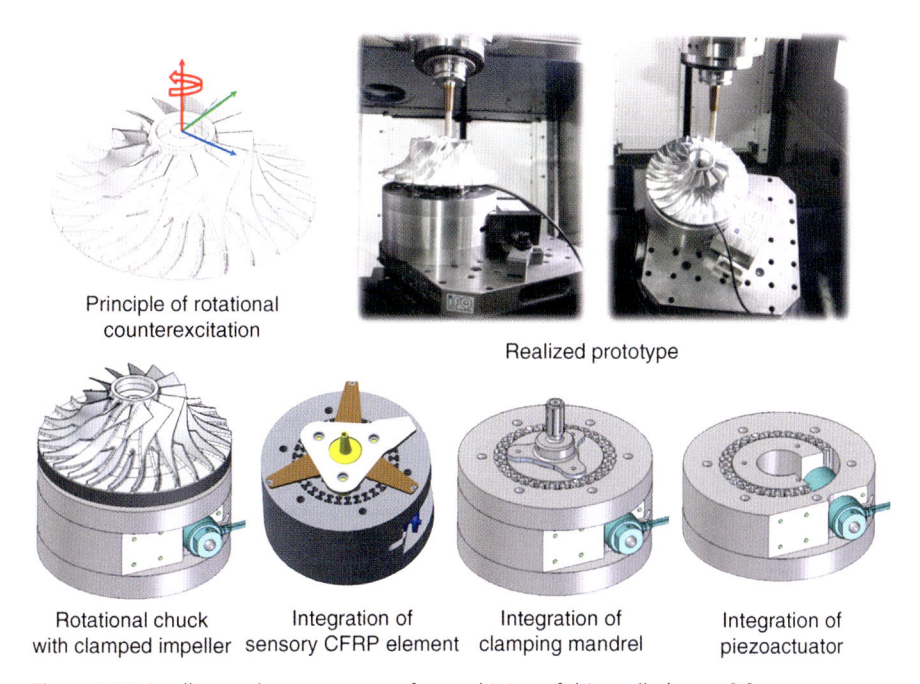

Figure 24.3 Intelligent clamping system for machining of thin-walled parts [4].

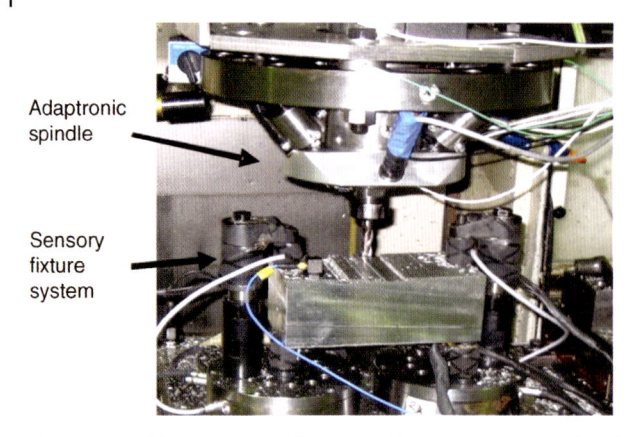

Figure 24.4 Adaptronic spindle system for compensation of tool deflections [63].

Furthermore, magnetostrictive actuators provide high Curie temperature and noncontact driving without electrodes. Yoshioka *et al.* investigated the application of a giant magnetostrictive actuator in a rotary–linear motion platform [71]. The actuator is placed at the rotary part of a spindle. It is controlled by the magnetic flux that is generated by a coil and magnetic core in the stationary outer component of the system. Application of a magnetic field forces the actuator to change its length and, thus, to move the rotating shaft at the spindle in axial direction. The stroke of a realized prototype achieves nearly 25 μm. Linear actuation is possible without wiring at the rotating part of the system. Experiments revealed that a position resolution of 10 nm at 0 rpm and of 1 μm at 4000 rpm is possible.

Guo *et al.* presented a magnetostrictive vibrating polisher for ultraprecision finishing of microaspheric molds [72]. The tool prototype was tested successfully in finishing experiments. The experimental results verified the capability of the realized approach. The form deviations of a test workpiece with a diameter of 2 mm and a maximum tangential angle of 45° could be reduced from 0.18 μm after grinding to 0.09 μm after polishing. A major benefit of the analyzed solution is its miniaturization potential.

Other applications of magnetostrictive materials include attenuation and isolation of vibrations [73,74]. In order to overcome the problem of nonlinearities in high-frequency control of magnetostrictive actuators, model-based approaches can be applied [75].

24.4 Applications of Electro- and Magnetorheological Fluids

By applying an electric or magnetic field, the viscosity of electrorheologic fluids (ERFs) and magnetorheologic fluids (MRFs) can be manipulated, respectively. This effect was used in many systems with the aim to allow a controlled and improved damping. An overview about magnetic field-driven actuators is given in Ref. [76].

Aoyama and Inasaki [77–78] and Shinno and Hashizume [79] have presented ERF applications for machine tools. Aoyama presented gel structured ERF in order to overcome the problems of sealing and sedimentation of ER particles [78]. The ERF was used in a clamping system for an aerostatic slider also providing variable damping functionality.

Ramkumar and Ganesan used an ERF core inside a composite sandwich box column and obtained an increased stiffness and natural frequency by applying an electric field [80]. In Ref. [81], an ERF damper for semiactive vibration control of rotor systems is presented. Dynamic models of the damper and its amplifier are derived for control purposes. An artificial intelligent (AI) feedback controller is synthesized based on the system model. Wang and Fei presented a boring bar with tunable stiffness containing ERF [82]. Chatter suppression could be achieved under a certain range of electric field and cutting conditions.

Magnetorheological fluids behave similar to ERF but react on magnetic instead of electric fields. In Ref. [83], MRF-based shock absorbers are introduced that mitigate chatter occurrence in milling of thin floor parts in aerospace industry. Ghiotti *et al.* presented MRF dampers for reducing the shock response in sheet metal blanking [84]. A significant vibration reduction could be achieved by means of the damping solution.

Uhlmann and Bayat used magnetorheological ferrofluids for precision positioning devices [85]. Weinert *et al.* developed a damping system for deep-hole drilling using MRF [86,87]. An elastic and infinitely variable torsional clamping of the drilling tool was realized in order to dissipate the vibration energy within the damping device. Two coils controlled the magnetic field and hence the mechanical transmission behavior of the damping system. Further MRF damping systems can be found in Refs [88–93].

Besides damping, other applications of ERF and MRF can be found in literature with respect to niche solutions. As an example, Lanzetta and Iagnemma presented an MRF-based gripper utilizing controllable wet adhesion in Ref. [94]. It was shown that the realized end effector prototype could hold for 1 min with 5 kPa.

24.5 Intelligent Structures and Components

Intelligent or smart materials basically enable the implementation of functional integrated structures. As can be seen in the sections above, until now, the applications of intelligent materials in machine tools predominantly concern the exploitation of the material capabilities for sensory purposes or for the realization of actuators. Besides the sensing or actuating elements, additional subsystems are necessary in order to realize full measurement chains or (controlled) drive systems. A power supply including energy transmission, an information processing unit together with a data communication system, a controller, and a user interface are also required. The entire integration of these coupled subsystems leads to intelligent machine components that provide an autonomous enhanced functionality.

Besides the use of smart materials, an inherent combination of mechanical structures and sensing or actuating elements can be utilized in order to realize intelligent machine components. Exemplary implementations involve the integration of fiber optic Bragg gratings in hybrid material structures [95,96], the creation of microstructured strain, temperature, and wear sensors at tool and machine component surfaces [97–100], or the emplacement of sensory microelectromechanical systems (MEMS) [101,102]. An analysis of micromachined surface accelerometer structures is presented in Ref. [103]. Spray-coated sensors for cutting force measuring based on polymeric nanocomposites (consisting of polyvinylidene difluoride polymer reinforced with multiwalled carbon nanotubes) are introduced by Park et al. in Ref. [104].

Structures with sensing capabilities can also be implemented and embedded into mechanical systems by printing processes that deposit conductive materials [105–110].

Intensive research work is focused on the integration and application of sensor elements in tools for machining operations. Sensor-equipped grinding tools and systems are presented and discussed in Refs [111–113]. An integration of sensory elements in cutting tools with defined cutting edges can be seen in Refs [102,114,115]. Figure 24.5 shows the integration principle of a MEMS acceleration sensor close to the tip of long and slender milling cutters.

Further exemplary sensor-integrated intelligent components are presented in Ref. [116] regarding microrolling and in Ref. [117] regarding injection molding. In Ref. [118], the use of a fiber Bragg grating thermal sensor that was attached to a tool for monitoring and optimization of micro–nano scratching processes is introduced. The integration of fiber Bragg gratings during 3D printing of strain sensory structures is shown in Ref. [119]. Denkena et al. investigate the approach of sensory workpieces that gather information about their processing and application by themselves [120].

A huge field of application of intelligent structures and components in production and manufacturing environments is constituted by assembly systems, manipulators, grasping, and gripping devices [12,121–123].

In general, a trend toward even smaller and miniaturized microsensors can be recognized. An exemplary overview of miniaturized force sensors can be found in

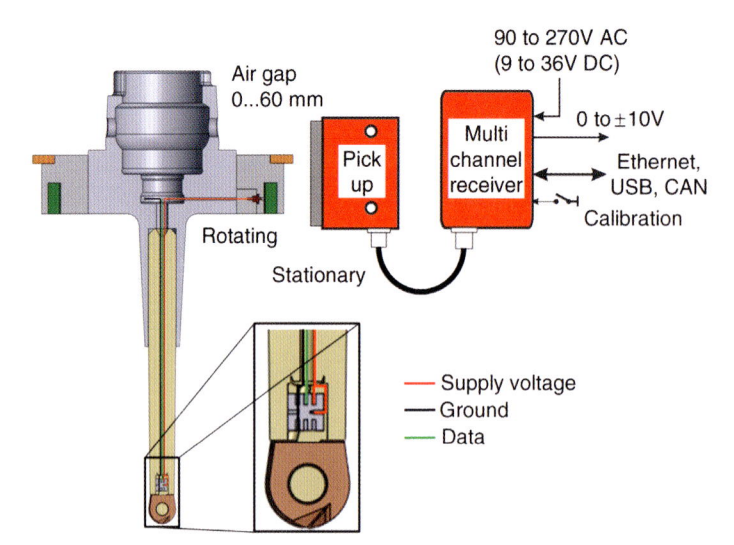

Figure 24.5 Integration of MEMS acceleration sensor in milling tool [102].

Ref. [124]. The small dimensions of thin layer and microsensors allow the (distributed) integration of multiple sensing elements into the same structure leading to sensor arrays and sensory networks [125–127]. Different types of sensors can be combined in such networks in order to allow the measurement and observation of multicriteria monitoring information (e.g., strains, forces, temperatures, and accelerations). A future vision of distributed multicriteria sensor networks is to achieve an information gathering capability that can be compared with entities that can be found in Nature. With respect to tactile, force-, strain-, and temperature-sensitive devices, the human skin provides a benchmark for such pioneering approaches [128]. Inherent integration of multiple sensory and also active elements can be recognized especially together with the use of polymer and composite materials [129–131].

When implementing distributed sensor and actuator elements, a challenging task concerns the communication of measurement signals and control data as well as the power supply of the individual units. Wired solutions necessitate a high mounting effort. Wireless communication solutions, on the other hand, must be robust and stable within the harsh manufacturing environment and even in shielded conditions [132]. A principal approach for distributed energy supply involves energy harvesting technologies [133–137].

24.6 Summary and Conclusion

As a conclusion, the following main findings can be summarized:

- Functional and smart materials possess a very high potential with respect to the creation and implementation of sensory and active "intelligent components" in manufacturing applications.
- The number and variety of realized solutions and prototypes are already quite high and still increasing; exemplary devices for many different machining operations and production processes can be recognized.
- The fundamental characteristics of the various intelligent materials are widely understood and some methods for the design and layout of related components are available.
- However, each individual implementation reveals specific technical challenges and requires adapted approaches and procedures in order to gain the full potential of the material exploitation and to achieve optimum performance.
- Although numerous prototypes, layouts, modeling approaches, and simulations exist, a universal and common design methodology is still missing.
- Besides the integration and utilization of sensory and active elements in mechanical structures and machinery components, the energy supply and the communication of measurement signals and control data constitutes a challenging task, especially with respect to distributed systems.

Even though a huge amount of information about investigations and exemplary implementations of intelligent materials and structures can be found in literature, the application in industrial manufacturing environments is marginal until now. The research systems have to be evaluated under harsh conditions and their robustness and stability have to be proven. Furthermore, the costs of implemented prototypes are hardly acceptable in most manufacturing scenarios. Especially with respect to the electronics, low-cost solutions are desired. The economic benefits of intelligent machine components have to be made visible. The further research and development of intelligent materials will also lead to more efficient and affordable (mass) production and generation technologies. In addition, the need for comprehensive gathering of process and machine state information in real time will increase. This constitutes a precondition for the realization of "Industry 4.0" at the process and machinery level.

References

1 Möhring, H.-C., Brecher, C., Abele, E., Fleischer, J., and Bleicher, F. (2015) Materials in machine tool structures. *CIRP Annals: Manufacturing Technology*, **64/2**, 725–748.
2 Park, G., Bement, M.T., Hartman, D.A., Smith, R.E., and Farrar, C.R. (2007) The use of active materials for machining processes: a review. *International Journals of Machine Tools & Manufacture*, **47**, 2189–2206.
3 Hurlebaus, S. and Gaul, L. (2006) Smart structure dynamics. *Mechanical Systems and Signal Processing*, **20**, 255–281.
4 Möhring, H.-C., Wiederkehr, P., Lerez, C., Schmitz, H., Goldau, H., and Czichy, C. (2016) Sensor integrated CFRP structures for intelligent fixtures. *Procedia Technology*, **26**, 120–128.
5 Neugebauer, R., Denkena, B., and Wegener, K. (2007) Mechatronic systems for machine tools. *Annals of the CIRP*, **56/2**, 657–686.
6 Isermann, R. (2008) Mechatronic systems: innovative products with embedded control. *Control Engineering Practice*, **16**, 14–29.
7 Korkmaz, S. (2011) A review of active structural control: challenges for engineering informatics. *Computers and Structures*, **89**, 2113–2132.
8 Smith, S.T. and Seugling, R.M. (2006) Sensor and actuator considerations for precision, small machines. *Precision Engineering*, **30**, 245–264.
9 Lagoudas, D.C. (ed.) (2008) *Shape Memory Alloys: Modeling and Engineering Applications*, Springer Science+Business Media. ISBN 978-0-387-47684-1.
10 Jani, J.M., Leary, M., Subic, A., and Gibson, M.A. (2013) A review of shape memory alloy research, applications and opportunities. *Materials & Design*, **56**, 1078–1113.
11 Humbeeck, J.V. (1999) Non-medical applications of shape memory alloys. *Materials Science and Engineering*, **A273–275**, 134–148.
12 Fantoni, G., Santochi, M., Dini, G., Tracht, K., Scholz-Reiter, B., Fleischer, J., Lien, T.K., Seliger, G., Reinhart, G., Franke, J., Hansen, H.N., and Verl, A. (2014) Grasping devices and methods in automated production processes. *CIRP Annals: Manufacturing Technology*, **63/2**, 679–701.
13 Bellouard, Y., Clavel, R., Gotthardt, R., Bidaux, J.E., and Sidler, T. (1998) A new concept of monolithic shape memory alloy micro-devices used in microrobotics. *Proceedings of Actuator*, **98**, 499–502.

14 Malukhin, K., Sung, H., and Ehmann, K. (2012) A shape memory alloy based tool clamping device. *Journal of Materials Processing Technology*, **212**, 735–744.

15 Shin, W.C., Ro, S.-K., Park, H.-W., and Park, J.-K. (2009) Development of a micro/meso-tool clamp using a shape memory alloy for applications in microspindle units. *International Journal of Machine Tools and Manufacture*, **49**, 579–585.

16 Qin, Y. (2006) Forming-tool design innovation and intelligent tool-structure/system concepts. *International Journal of Machine Tools & Manufacture*, **46**, 1253–1260.

17 Bucht, A., Junker, T., Pagel, K., Drossel, W.-G., and Neugebauer, R. (2011) Design and modeling of a self-sufficient shape-memory-actuator. Proceedings of SPIE Conference Active and Passive Smart Structures and Integrated Systems, March 7–10, 2011, San Diego, CA. ISBN 978-0-8194-8539-7.

18 Meier, H., Pollmann, J., and Czechowicz, A. (2013) Design and control strategies for SMA actuators in a feed axis for precision machine tools. *Production Engineering: Research and Development*, **7**, 547–553.

19 Pagel, K., Drossel, W.-G., and Zorn, W. (2013) Multi-functional shape-memory-actuator with guidance function. *Production Engineering: Research and Development*, **7**, 491–496.

20 Sadeghzadeh, A., Asua, E., Feuchtwanger, J., Etxebarria, V., and Garcia-Arribas, A. (2012) Ferromagnetic shape memory alloy actuator enabled for nanometric position control using hysteresis compensation. *Sensors and Actuators A*, **182**, 122–129.

21 Lan, C.-C., Wang, J.-H., and Fan, C.-H. (2009) Optimal design of rotary manipulators using shape memory alloy wire actuated flexures. *Sensors and Actuators A*, **153/2**, 258–266.

22 Hesselbach, J. and Raatz, A. (2004) Performance of pseudo-elastic flexure hinges in parallel robots for micro-assembly tasks. *Annals of the CIRP*, **53/1**, 329–332.

23 Rivin, E.I. (2007) Precision compensators using giant superelasticity effect. *Annals of the CIRP*, **56/1**, 391–394.

24 Neugebauer, R., Drossel, W.-G., Bucht, A., and Ohsenbrügge, C. (2012) Control of thermal flow in machine tools using shape memory alloys, world academy of science, engineering and technology. *International Scholarly and Scientific Research & Innovation*, **6/11**, 490–494.

25 Williams, K.A., Chiu, G.T.-C., and Bernhard, R.J. (2005) Dynamic modelling of a shape memory alloy adaptive tuned vibration absorber. *Journal of Sound and Vibration*, **280**, 211–234.

26 Hesselbach, J. (ed.) (2011) *Adaptronik für Werkzeugmaschinen*, Shaker Verlag, Aachen, Germany. ISBN 978-3-8322-9809-8.

27 Crawley, E.F. and DeLuis, J. (1987) Use of piezoelectric actuators as elements of intelligent structures. *AIAA Journal*, **25/10**, 1373–1385.

28 Janocha, H. (2004) *Actuators: Basics and Applications*, Springer, Berlin. ISBN 3-540-61564-4.

29 Teti, R., Jemielniak, K., O'Donnell, G., and Dornfeld, D. (2010) Advanced monitoring of machining operations. *CIRP Annals: Manufacturing Technology*, **59/2**, 717–739.

30 Altintas, Y. and Woronko, A. (2002) A piezo tool actuator for precision turning of hardened shafts. *Annals of the CIRP*, **51/1**, 303–306.

31 van Brussel, H., Reynaerts, D., Vanherck, P., Versteyhe, M., and Devos, S. (2003) A nanometer-precision, ultra-stiff piezostepper stage for ELID-grinding. *CIRP Annals: Manufacturing Technology*, **52/1**, 317–322.

32 Matsumoto, K., Hatamura, Y., and Nakao, M. (2000) Actively controlled compliance device for machining error reduction. *Annals of the CIRP*, **49/1**, 313–316.

33 Chiu, W.M., Lam, F.W., and Gao, D. (2002) An overhung boring bar servo system for on-line correction of machining errors. *Journal of Materials Processing Technology*, **122**, 286–292.

34 Mulders, P.C., van der Wolf, A.C.H., Heuvelman, C.J., and Snijder van Wissekerke, M.P. (1986) A robot-arm with compensation for bending. *Annals of the CIRP*, **35/1**, 305–308.

35 Schneider, U., Drust, M., Puzik, A., and Verl, A. (2013) Compensation of errors in robot machining with a parallel 3D-piezo compensation mechanism. *Procedia CIRP*, **7**, 305–310.

36 Möhring, H.-C., Gümmer, O., and Fischer, R. (2011) Active error compensation in contour-controlled grinding. *CIRP Annals: Manufacturing Technology*, **60/1**, 429–432.

37 Denkena, B. and Gümmer, O. (2013) Active tailstock for precise alignment of precision forged crankshafts during grinding. *Procedia CIRP*, **12**, 121–126.

38 Chang, S.F. and Lee, B.J. (2001) A lead screw stepper equipped with piezo driven ultra-high resolution linear positioner. 2nd EUSPEN International Conference, Turin, Italy.

39 Elfizy, A.T., Bone, G.M., and Elbestawi, M.A. (2005) Design and control of a dual-stage feed drive. *International Journal of Machine Tools & Manufacture*, **45**, 153–165.

40 Aggogeri, F., Al-Bender, F., Brunner, B., Elsaid, M., Mazzola, M., Merlo, A., Ricciardi, D., de la Rodriguez, M.O., and Salvi, E. (2013) Design of piezo-based AVC system for machine tool applications. *Mechanical Systems and Signal Processing*, **36**, 53–65.

41 Denkena, B. and Harms, A. (2004) A tool adaptor for vibration damping in CNC lathes. 37th CIRP International Seminar on Manufacturing Systems, Budapest, Hungary.

42 Liu, J. (1993) An application of active dynamic absorber to the milling process and its experimental verification: vibration and control of mechanical systems. ASME Design Technical Conference, Albuquerque, NM.

43 Rose, M., Keimer, R., Breitbach, E.J., and Campanile, L.F. (2004) Parallel robots with adaptronic components. *Journal of Intelligent Material Systems and Structures*, **15**, 763–769.

44 Weck, M. and Özmeral, H. (1996) A fast tool servo system based on electrodynamic and piezoelectric actuators. *Production Engineering*, **III** (2), 57–60.

45 Gall, M., Thielicke, B., and Schmidt, I. (2009) Integrity of piezoceramic patch transducers under cyclic loading at different temperatures. *Smart Materials and Structures*, **18**, 1–10.

46 Möhring, H.-C. and Betram, O. (2012) Integrated autonomous monitoring of ball screw drives. *CIRP Annals: Manufacturing Technology*, **61**/1, 355–358.

47 Denkena, B. and Götz, T. (2004) Effects of a preload control unit on machine tool spindles. *Production Engineering*, **2**, 175–178.

48 Kim, H.S. and Cho, Y.M. (2009) Design and modeling of a novel 3-DOF precision micro-stage. *Mechatronics*, **19**, 598–608.

49 Li, Y. and Xu, Q. (2009) Modeling and performance evaluation of a flexure-based XY parallel micromanipulator. *Mechanism and Machine Theory*, **44**, 2127–2152.

50 Pluemer, S., Voges-Schwieger, K., Denkena, B., and Behrens, B.-A. (2012) Development of an advanced micro-positioning system for increasing operation accuracy: consideration of high pressure sealing. *Production Engineering Research and Development*, **6**, 213–218.

51 Pluemer, S. (2013) Piezo-hydraulisches Mikro-Stellsystem zur Feinpositionierung großer Lasten. Dr.-Ing. Dissertation, Leibniz Universität Hannover, Germany.

52 Denkena, B., Möhring, H.-C., Hesse, P., and Simon, S. (2007) Präzision im Kleinformat. Konzeption eines piezohydraulischen Servo-Mikrostellsystems. *wt Werkstattstechnik*, **97**/3, 115–119.

53 Gnad, G. (2005) Ansteuerungskonzept für piezoelektrische Aktoren. Dr.-Ing. Dissertation, Otto-von-Guerickce-Universität Magdeburg, Germany.

54 Ge, P. (1995) Modelling of hysteresis in piezoceramic actuator. *Precision Engineering*, **17**/3, 211–221.

55 Janocha, H. and Kuhnen, K. (2002) Inverse Steuerung für den Großsignalbetrieb von Piezoaktoren. *Automatisierungstechnik*, **50**/9, 439–450.

56 Last, B. (1998) Analyse und Modellierung von Hystereseeigenschaften in piezoelektrischen Aktoren zum Zweck der Reglersynthese. Dr.-Ing. Dissertation, Fortschrittsberichte VDI, Reihe 8, Nr. 715, Düsseldorf, Germany.

57 Schitter, G., Stark, R.W., and Stemmer, A. (2002) Fast feedback control of piezo-electric actuators. 8th International Conference on New Actuators, pp. 430–433.

58 Neugebauer, R., Pagel, K., Bucht, A., Wittstock, V., and Pappe, A. (2010) Control concept for piezo-based actuator–sensor-units for uniaxial vibration damping in machine tools. *Production Engineering Research and Development*, **4**, 413–419.

59 Denkena, B., Möhring, H.-C., and Will, J.C. (2007) Tool deflection compensation with an adaptronics milling spindle. International Conference on Smart Machining Systems (ICSMS'07), Gaithersburg, MD.

60 Will, J.C. (2008) Adaptronische Spindeleinheit zur Abdrängungs- und Schwingungskompensation in Fräsprozessen. Dr.-Ing. Dissertation, Leibniz Universität Hannover, Germany.

61 Neugebauer, R., Drossel, W.-G., Bucht, A., Kranz, B., and Pagel, K. (2010) Control design and experimental validation of an adaptive spindle support for enhanced cutting processes. *CIRP Annals: Manufacturing Technology*, **59**/1, 373–376.

62 Denkena, B., Möhring, H.-C., and Will, J.C. (2005) Design and layout of an adaptronic spindle system. *CIRP Journal of Manufacturing Systems*, **34/3**, 247–257.

63 Möhring, H.-C., Litwinski, K.M., Gümmer, Olaf (2010) Process monitoring with sensory machine tool components. *CIRP Annals: Manufacturing Technology*, **59/1**, 383–386.

64 Denkena, B., Kästner, J., Knoll, G., Brandt, S., Bach, F.-W., Drößler, B., Reithmeier, E., and Bretschneider, M. (2008) Mikrostrukturierung funktionaler Oberflächen. *wt Werkstattstechnik*, **98/6**, 486–494.

65 Denkena, B., Köhler, J., Mörke, T., and Gümmer, O. (2012) High-performance cutting of micro patterns. *Procedia CIRP*, **1**, 144–149.

66 Song, G., Sethi, V., and Li, H.-N. (2006) Vibration control of civil structures using piezoceramic smart materials: a review. *Engineering Structures*, **28**, 1513–1524.

67 Sun, M., Staszewski, W.J., Swamy, R.N., and Li, Z. (2008) Application of low-profile piezoceramic transducers for health monitoring of concrete structures. *NDT & E International*, **41**, 589–595.

68 Kehlenbach, M. and Hanselka, H. (2003) Automated structural integrity monitoring based on broadband lamb wave excitation and matched filtering. 44th AIAA/ASME/ASCE/AHS/ASC Structures, Structural Dynamics, and Materials Conference, Norfolk, VA, April 7–10, 2003. ISBN: 1-563-47625-8.

69 Olabi, A.G. and Grunwald, A. (2008) Design and application of magnetostrictive materials. *Materials & Design*, **29**, 469–483.

70 Eda, H. and Ohmura, E. (1992) Ultraprecise machine tool equipped with a giant magnetostriction actuator: development of new materials and their application. *Annals of the CIRP*, **41/1**, 421–424.

71 Yoshioka, H., Shinno, H., and Sawano, H. (2013) A newly developed rotary–linear motion platform with a giant magnetostrictive actuator. *CIRP Annals: Manufacturing Technology*, **62/1**, 371–374.

72 Guo, J., Morita, S.-y., Hara, M., Yamagata, Y., and Higuchi, T. (2012) Ultra-precision finishing of micro-aspheric mold using a magnetostrictive vibrating polisher. *CIRP Annals: Manufacturing Technology*, **61/1**, 371–374.

73 Hiller, M.W., Bryant, M.D., and Umegaki, J. (1989) Attenuation and transformation of vibration through active control of magnetostrictive terfenol. *Journal of Sound and Vibration*, **134/3**, 507–519.

74 Morales, A.L., Nieto, A.J., Chicharro, J.M., and Pintado, P. (2012) Vibration isolation of unbalanced machinery using an adaptive–passive magnetoelastic suspension. *Journal of Sound and Vibration*, **331**, 263–275.

75 Braghin, F., Cinquemani, S., and Resta, F. (2011) A model of magnetostrictive actuators for active vibration control. *Sensors and Actuators A*, **165**, 342–350.

76 Janocha, H. (2001) Application potential of magnetic field driven new actuators. *Sensors and Actuators A*, **91**, 126–132.

77 Aoyama, T. and Inasaki, I. (1997) Application of electrorheological fluid dampers to machine tool elements. *Annals of the CIRP*, **46/1**, 309–313.

78 Aoyama, T. (2004) Development of gel structured electrorheological fluids and their application for the precision clamping mechanism of aerostatic sliders. *CIRP Annals: Manufacturing Technology*, **53/1**, 325–329.

79 Shinno, H. and Hashizume, H. (1998) Nanometer positioning of a linear motor-driven ultraprecision aerostatic table system with electrorheological fluid dampers. *Annals of the CIRP*, **48/1**, 289–292.

80 Ramkumar, K. and Ganesan, N. (2009) Vibration and damping of composite sandwich box column with viscoelastic/electrorheological fluid core and performance comparison. *Materials & Design*, **30**, 2981–2994.

81 Lim, S., Park, S.-M., and Kim, K.-I. (2005) AI vibration control of high-speed rotor systems using electrorheological fluid. *Journal of Sound and Vibration*, **284**, 685–703.

82 Wang, M. and Fei, R. (1999) Chatter suppression based on nonlinear vibration characteristic of electrorheological fluids. *International Journal of Machine Tools & Manufacture*, **39**, 1925–1934.

83 Diaz-Tena, E., Lopez de Lacalle Marcaide, L.N., CampaGomez, F.J., and ChairesBocanegra, D.L. (2013) Use of magnetorheological fluids for vibration reduction on the milling of thin floor parts. *Procedia Engineering*, **63**, 835–842.

84 Ghiotti, A., Regazzo, P., Bruschi, S., and Bariani, P.F. (2010) Reduction of vibrations in blanking by MR dampers. *CIRP Annals: Manufacturing Technology*, **59/1**, 275–278.

85 Uhlmann, E. and Bayat, N. (2006) High precision positioning with ferrofluids as an active medium. *Annals of the CIRP*, **55/1**, 415–418.

86 Weinert, K. and Kersting, M. (2007) Adaptronic chatter damping system for deep hole drilling. CIRP International Conference on Smart Machining Systems, NIST National Institute of Standards and Technologies (NIST), March 13–15, 2007, Gaithersburg, MD.

87 Weinert, K., Biermann, D., Iovkov, I., and Kersting, M. (2011) Aktive Dämpfung von Ratterschwingungen beim Einlippentiefbohren, in *Adaptronik für Werkzeugmaschinen: Forschung in Deutschland* (ed. J. Hesselbach), Shaker Verlag, Aachen, Germany, pp. 244–262. ISBN 978-3-8322-9809-8.

88 Milecki, A. (2001) Investigation and control of magneto-rheological fluid dampers. *International Journal of Machine Tools & Manufacture*, **41**, 379–391.

89 Paul, P.S., Varadarajan, A.S., Lawrance, G., and Sunil, K.S. (2012) Analysis of turning tool holder with MR fluid damper: International Conference on Modeling, Optimization and Computing (ICMOC'12). *Procedia Engineering*, **38**, 2572–2578.

90 Mei, D., Kong, T., Shih, A.J., and Chen, Z. (2009) Magnetorheological fluid-controlled boring bar for chatter suppression. *Journal of Materials Processing Technology*, **209**, 1861–1870.

91 Hong, S.R., Wereley, N.M., Choi, Y.T., and Choi, S.B. (2008) Analytical and experimental validation of a nondimensional Bingham model for mixed-mode magnetorheological dampers. *Journal of Sound and Vibration*, **312**, 399–417.

92 Paul, P.S., Varadarajan, A.S., Vasanth, X.A., and Lawrance, G. (2013) Effect of magnetic field on damping ability of magnetorheological damper during hard turning. *Archives of Civil and Mechanical Engineering*, **14**, 433–443.

93 Ahmed, G.M.S., Reddy, P.R., and Seetharamaiah, N. (2013) Experimental evaluation of metal cutting coefficients under the influence of magneto-rheological damping in end milling process. *Procedia Engineering*, **64**, 435–445.

94 Lanzetta, M. and Iagnemma, K. (2013) Gripping by controllable wet adhesion using a magnetorheological fluid. *CIRP Annals: Manufacturing Technology*, **62/1**, 21–25.

95 Meo, F., Merlo, A., Rodriguez, M., Brunner, B., Fleck, N.A., Lu, T.J., Mai, S.P., Srikantha Phani, A., and Woodhouse, J. (2008) Advanced hybrid mechatronic materials for ultra precise and high performance machining systems design, in *Innovative Production Machines and Systems: Fourth I*PROMS Virtual International Conference* (eds D.T. Pham, E.E. Eldukhri, and A.J. Soroka), Whittles Publishing.

96 Möhring, H.-C., Wiederkehr, P., Baumann, J., König, A., Spieker, C., and Müller, M. (2017) Intelligent hybrid material slide component for machine tools. *Journal of Machine Engineering*, **17** (1), 17–30.

97 Denkena, B., Litwinski, K.M., Brouwer, D., and Boujnah, H. (2014) Design and analysis of a prototypical sensory Z-slide for machine tools. *Production Engineering: Research and Development*, **7**, 9–14.

98 Overmeyer, L., Duesing, J.F., Suttmann, O., and Stute, U. (2012) Laser patterning of thin film sensors on 3D surfaces. *CIRP Annals: Manufacturing Technology*, **61/1**, 215–218.

99 Klocke, F. and Schmitz, R. (2001) Entwicklungsschritte zum "Intelligenten Werkzeug". *wt Werkstattstechnik*, **91/5**, 248–254.

100 Sugita, N., Ishii, K., Furusho, T., Harada, K., and Mitsuishi, M. (2015) Cutting temperature measurement by a micro-sensor array integrated on the rake face of a cutting tool. *CIRP Annals: Manufacturing Technology*, **64/1**, 77–80.

101 Ko, W.H. (2007) Trends and frontiers of MEMS. *Sensors and Actuators A*, **136**, 62–67.

102 Möhring, H.-C., Nguyen, Q.P., Kuhlmann, A., Lerez, C., Nguyen, L.T., and Misch, S. (2016) Intelligent tools for predictive process control. *Procedia CIRP*, **57**, 539–544.

103 Aoyagi, S., Kumagai, S., Yoshikawa, D., and Isono, Y. (2007) Surface micromachined accelerometer using ferroelectric substrate. *Sensors and Actuators A*, **139**, 88–94.

104 Park, S.S., Parmar, K., Shajari, S., and Sanati, M. (2016) Polymeric carbon nanotube nanocomposite-based force sensors. *CIRP Annals: Manufacturing Technology*, **65/1**, 361–364.

105 Borghetti, M., Serpelloni, M., Sardini, E., and Pandini, S. (2016) Mechanical behavior of strain sensors based on PEDOT:PSS and silver nanoparticles inks deposited on polymer substrate by inkjet printing. *Sensors and Actuators A*, **243**, 71–80.

106 Cho, C. and Ryuh, Y. (2016) Fabrication of flexible tactile force sensor using conductive ink and silicon elastomer. *Sensors and Actuators A*, **237**, 72–80.

107 Debéda, H., Lucat, C., and Pommier-Budinger, V. (2016) Printed piezoelectric materials for vibration-based damage detection. *Procedia Engineering*, **168**, 708–712.

108 Enser, H., Kulha, P., Sell, J.K., Jakoby, B., Hilber, W., Strauß, B., and Schatzl-Linder, M. (2016) Printed strain gauges embedded in organic coatings. *Procedia Engineering*, **168**, 822–825.

109 Hay, G.I., Southee, D.J., Evans, P.S.A., Harrison, D.J., Simpson, G., and Ramsey, B.J. (2007) Examination of silver–graphite lithographically printed resistive strain sensors. *Sensors and Actuators A*, **135**, 534–546.

110 Zhang, Y., Anderson, N., Bland, S., Nutt, S., and Jursich, G. (2017) All-printed strain sensors: building blocks of the aircraft structural health monitoring system. *Sensors and Actuators A*, **253**, 165–172.

111 Varghese, B., Pathare, S., Gao, R., Guo, C., and Malkin, S. (2000) Development of a sensor-integrated "Intelligent" grinding wheel for in-process monitoring. *Annals of the CIRP*, **49/1**, 231–234.

112 Brinksmeier, E., Heinzel, C., and Meyer, L. (2005) Development and application of a wheel based process monitoring system in grinding. *Annals of the CIRP*, **54/1**, 301–304.

113 Hatamura, Y., Nagao, T., Mitsuishi, M., and Tanaka, H. (1989) Development of a force controlled automatic grinding system for actual NC machining centers. *Annals of the CIRP*, **38/1**, 343–346.

114 Spiewak, S. (1992) Instrumented milling cutter for in-process measurement of spindle error motion. *Annals of the CIRP*, **41/1**, 429–432.

115 Barthelmä, F. (2016) Improved machining processes by sensor and actuator integration. Proceedings of the 3rd Wiener Produktionstechnik Kongress, Adaptive and Smart Manufacturing, September 28–29, 2016, Vienna, Austria.

116 Fan, Z., Ng, M.-K., Gao, R.X., Cao, J., and Smith, E.F. (2012) Real-time monitoring of pressure distribution in microrolling through embedded capacitive sensing. *CIRP Annals: Manufacturing Technology*, **61/1**, 367–370.

117 Gao, R.X., Fan, Z., and Kazmer, D.O. (2008) Injection molding process monitoring using a self-energized dual-parameter sensor. *CIRP Annals: Manufacturing Technology*, **57/1**, 389–393.

118 Mandal, S., Roy, S., Chatterjee, K., Haldar, S., Vijay, V., and Hanumaiah, N. (2015) Fiber Bragg grating sensor for cutting speed optimization and burr reduction in micro-nano scratching. *Procedia Technology*, **19**, 327–332.

119 Zeleny, R. and Vcelak, J. (2016) Strain measuring 3D printed structure with embedded fibre Bragg grating. *Procedia Engineering*, **168**, 1338–1341.

120 Denkena, B., Dahlmann, D., and Mücke, M. (2016) Sensorische Werkstücke. *wt Werkstattstechnik*, **106-11/12**, 815–820.

121 Santochi, M. and Dini, G. (1998) Sensor technology in assembly systems. *Annals of the CIRP*, **47/2**, 503–524.

122 Yousef, H., Boukallel, M., and Althoefer, K. (2011) Tactile sensing for dexterous in-hand manipulation in robotics: a review. *Sensors and Actuators A*, **167**, 171–187.

123 Kappassov, Z., Corrales, J.-A., and Perdereau, V. (2015) Tactile sensing in dexterous robot hands: review. *Robotics and Automation Systems*, **74**, 195–220.

124 Baki, P., Szekely, G., and Kosa, G. (2013) Design and characterization of a novel, robust, tri-axial force sensor. *Sensors and Actuators A*, **192**, 101–110.

125 Borenstein, J., Koren, Y., and Weil, R. (1987) Hierarchically structured multisensor system for an intelligent mobile robot. *Annals of the CIRP*, **36/1**, 331–334.

126 Akyildiz, I.F. and Kasimoglu, I.H. (2004) Wireless sensor and actor networks: research challenges. *Ad Hoc Networks*, **2/4**, 351–367.

127 Song, G., Song, A., and Huang, W. (2005) Distributed measurement system based on networked smart sensors with standardized interfaces. *Sensors and Actuators A*, **120**, 147–153.

128 Lang, W., Jakobs, F., Tolstosheeva, E., Sturm, H., Ibragimov, A., Kesel, A., Lehmhus, D., and Dicke, U. (2011) From embedded sensors to sensorial materials: the road to function scale integration. *Sensors and Actuators A*, **171**, 3–11.

129 McEvoy, M.A. and Correll, N. (2015) Thermoplastic variable stiffness composites with embedded, networked sensing, actuation, and control. *Journal of Composite Materials*, **49/15**, 1799–1808.

130 Canavese, G., Stassi, S., Fallauto, C., Corbellini, S., Cauda, V., Camarchia, V., Pirola, M., and Pirri, C.F. (2014) Piezoresistive flexible composite for robotic tactile applications. *Sensors and Actuators A*, **208**, 1–9.

131 Paskarbeit, J., Annunziata, S., Basa, D., and Schneider, A. (2013) A self-contained, elastic joint drive for robotics applications based on a sensorized elastomer coupling: design and identification. *Sensors and Actuators A*, **199**, 56–66.

132 Gao, R.X. and Kazmer, D.O. (2012) Multivariate sensing and wireless data communication for process monitoring in RF-shielded environment. *CIRP Annals: Manufacturing Technology*, **61/1**, 523–526.

133 Anton, S.R. and Sodano, H.A. (2007) A review of power harvesting using piezoelectric materials (2003–2006). *Smart Materials and Structures*, **16**, R1–R21.

134 Asai, T., Araki, Y., and Ikago, K. (2017) Energy harvesting potential of tuned inertial mass electromagnetic transducers. *Mechanical Systems and Signal Processing*, **84**, 659–672.

135 Shaikh, F.K. and Zeadally, S. (2016) Energy harvesting in wireless sensor networks: a comprehensive review. *Renewable and Sustainable Energy Reviews*, **55**, 1041–1054.

136 Ferdous, R.Md., Reza, A.W., and Siddiqui, M.F. (2016) Renewable energy harvesting for wireless sensors using passive RFID tag technology: a review. *Renewable and Sustainable Energy Reviews*, **58**, 1114–1128.

137 Selvan, K.V. and Ali, M.S.M. (2016) Micro-scale energy harvesting devices: review of methodological performances in the last decade. *Renewable and Sustainable Energy Reviews*, **54**, 1035–1047.

25

New Markets/Opportunities through Availability of Product Life Cycle Data

Thorsten Wuest,[1] Karl Hribernik,[2] and Klaus-Dieter Thoben[2,3]

[1]West Virginia University, Industrial and Management Systems Engineering, P.O. Box 6070, Morgantown, WV 26506-6070, USA
[2]BIBA – Bremer Institut für Produktion und Logistik GmbH at the University of Bremen, ICT Applications for Production, Hochschulring 20, 28359 Bremen, Germany
[3]University of Bremen, ISIS Sensorial Materials Scientific Centre, Wiener Str. 12, 28359 Bremen, Germany

In this chapter, an outlook on the future potential of sensorial materials within the context of product life cycle data and information management is presented. The outlook character of this chapter is based on the circumstances that sensorial materials are somewhat still under development or at least in a prototypical state [1]. In order to connect to potential application scenarios of sensorial materials, different case studies with a focus on product life cycle application are illustrated. These case studies employ no sensorial materials at this point but regular sensors or Product Embedded Information Devices (PEIDs). Based on this, opportunities and limitations of using sensorial materials in such applications are discussed for each use case.

The chapter is structured as follows: First, a short introduction to closed-loop, item-level product life cycle management (PLM) is presented as a basis for the following application case studies. Following, case studies utilizing product life cycle information and/or data are derived and discussed. Based on these case studies, the potential of sensorial materials in this context is elaborated, with a focus on the scenario-specific opportunities and limitations of such an application. Concluding, a brief discussion of the overall potential sensorial materials present within the context of PLM is discussed and major findings are highlighted.

25.1 Product Life Cycle Management

In this section, the product life cycle and the corresponding PLM concept is introduced. The following case studies are based on the product life cycle concept and the understanding of data and information handling.

In general, there are two basic views on product life cycles and PLM, depending on the discipline. In business and related subjects, the product life cycle is

Material-Integrated Intelligent Systems: Technology and Applications, First Edition.
Edited by Stefan Bosse, Dirk Lehmhus, Walter Lang, and Matthias Busse.
© 2018 Wiley-VCH Verlag GmbH & Co. KGaA. Published 2018 by Wiley-VCH Verlag GmbH & Co. KGaA.

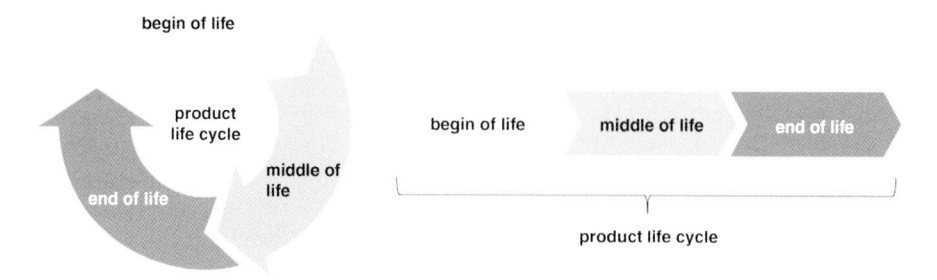

Figure 25.1 Variations of the representation of product life cycles in engineering [5].

understood more from an economic perspective, looking, for example, at the revenue of a certain product family over time [2].

In engineering, the product life cycle is understood differently, emphasizing more on physical product properties and product development tasks among others. Basically, engineering view on PLM expands on the concept of product data management (PDM) and focuses to include data and information generated and used beyond design and manufacturing [3,4]. Besides merely handling product- and process-related data, PLM also has to take into account the interdependencies of information and communication between all of the stakeholders involved in the product life cycle [5].

Common graphical representations of the product life cycle encompass three phases – begin-of-life (BOL), middle-of-life (MOL), and end-of-life (EOL) – arranged either in a circle or a straight line (see Figure 25.1). The linear form represents the product life cycle "from the cradle to the grave," while the circular form stresses PLM "from cradle to cradle" with a focus on recycling, refurbishing, and reuse: The final phase feeds back into the first [5]. In the following section focusing on closed-look, item-level PLM, this "cradle to cradle" perspective is discussed in more detail.

Manufacturers and other stakeholders along the product life cycle are increasingly becoming aware of the benefits inherent in managing the associated data and information. Especially, today's products are becoming increasingly complicated. For example, the amount of component parts is increasing. Simultaneously, development, manufacturing, and usage cycles are accelerating [6] and production is being distributed globally. These trends highlight the need for innovative concepts for structuring and handling product-related data and information efficiently throughout the entire life cycle. Above all, customer demand for more customization and variation stresses the need for a PLM at item and not merely type level.

This emphasis on increased information and data needs throughout the life cycle highlights the demand for solutions that allow gathering and managing such data and information throughout the life cycle. This stands especially true for the phases out of the control of the manufacturer, the usage phase. In this case, the developments in the area of sensor technology come into play. As of today, mostly regular sensors are employed where considered useful. However, as this is still under development and the full potential of information and data usage has yet to be discovered, new developments like sensorial materials have to be considered for their future potential in this domain. Before this is conducted in the following use cases, the specific domain of closed-loop, item-level PLM is introduced as this domain is considered to be the one with the most urgent needs for innovative sensor solutions by the authors.

25.1.1 Closed-Loop and Item-Level PLM

As presented in the previous section, conventional views of PLM tend to often stress the first phase of the product life cycle, due to its beginnings in PDM and CAD. Processes highlighted here are product design, development, production (logistics), and sales. Emerging approaches such as closed-loop PLM [7,8] take a holistic view upon the entire product life cycle, from product ideation to end-of-life processes, ideally also the end of one life cycle into the beginning of the next. It thus puts forward a paradigm shift from "cradle to grave" to "cradle to cradle" [9]. An example is the refurbishment of components from decommissioned products for use in new ones. The aim of closed-loop PLM is to close information gaps between the different phases and processes of the product life cycle of individual products. This can be backward, for example, providing usage data to design processes, or forward, for example, providing production and assembly information to recycling processes. It deals with products not as classes or variants, but as individual items (item-level). Additionally, item-level PLM allows focusing more on the individual product and therefore creates the basis to, for example, monitor product quality on an individual level rather than on a batch level.

Being in the focus of researchers and practitioners for over a decade, data management in PLM still has to face the challenge of a gap between the existing reality and the expected features in data management [4]. Especially in the area of item-level product data management along complex manufacturing chains are still many issues to be solved, for example, a generally accepted and interchangeable format that contains all relevant information. The developed concept contributes to this development in the abstract way of systemizing product information around the product state. This remaining challenges and their impact on the application and utilization of sensorial materials in a closed-loop, item-level PLM context are elaborated in the following sections.

25.1.2 Data and Information in PLM

Overall, it can be stated that information management is considered more technical than, for example, knowledge management and that knowledge is backed by information and data. An important differentiation is that data and information can be stored relatively easily compared to knowledge, which is always connected to a person [10]. The differentiation in explicit and implicit knowledge is crucial for the transferability and applicability. In this regard, information is relatively easy to store and transfer, data even easier. However, information itself does foster realizations or new findings; it has to be connected to a context in order to become knowledge [11]. This contextual supplement may be derived from the PLM perspective.

Information management and the more technical term information management systems are strongly linked to ICT and related technical solution. In this area, a lot of progress has been made over the last years. As stated above, the understanding of information management and information management systems does not distinguish itself sharply from, for example, PLM systems, also with strong links to ICT. However, IM is closely connected to the quality of the data and information to be managed [12]. Garetti and Terzi [13] highlight that the "product" information and data management is representing a key aspect of product-centric and product-driven approaches, also emphasizing a strong overlap between the two. In practical application, the information management system can be a part of the overall PLM-centric information management approach. Looking at sensorial materials as a source for PLM data, the emphasis on data quality is rather important. For further reading, Refs [14–16] are recommended.

Nevertheless, the research focus of existing information management systems is mostly focused on how already existing information has to be managed [17,18] or what existing information management system should be chosen [19,20]. The general principles of information management [21–23], the right information at the right time in the right granularity at the right place in the right quality, can be seen as the general vision of the information-centric closed-loop, item-level PLM approach presented here.

In order to emphasize the data and information perspective in PLM, the product has to be understood to be more than "just" the physical product – It has to include the information infrastructure to some extent. One possible aspect of such a product understanding is described in the next section, where intelligent or smart products are briefly introduced together with related concepts like the Product Avatar and Semantic Mediator.

25.1.3 Supporting Concepts for Data and Information Integration in PLM

In this section, selected information and data-driven concepts supporting the practical application of closed-loop, item-level PLM applications are briefly introduced. In this chapter, the illustration of Intelligent Products [24], the Product Avatar [25], and the Semantic Mediator [26] is only an introduction; for more detailed explanations of the concepts, further reading is advised.

Intelligent Products (also known as smart products [27]) are often considered a prerequisite for successful closed-loop, item-level PLM application. McFarlane *et al.* define the Intelligent Product as "a physical and information based representation of an item [. . .] which possesses a unique identification, is capable of communicating effectively with its environment, can retain or store data about itself, deploys a language to display its features, production requirements, etc., and is capable of participating in or making decisions *relevant* to its own destiny" [28]. The degree of intelligence inherited by an Intelligent Product may vary to some degree. It may range from simple data processing to complex proactive behavior [29]. Intelligent Products can make use of, for example, sensors and embedded computing throughout their life cycle s in order to collect data.

One concept to representing and managing the complex data and information flows connected to an individual product, for example, captured by sensorial materials and communicated by an Intelligent Product, is the Product Avatar. This concept describes a digital representation of a physical product, which exposes functionality and information to stakeholders of the product's life cycle via a user interface(s) [30]. One distinguishing factor toward other concepts is that it is not a simple data base, but actively selects, combines, and communicates relevant information targeted specifically to individual stakeholders' needs. At the same time, the Product Avatar also collects information and thus supports the life cycle approach especially during usage (MOL).

The concept presented next, the Semantic Mediator, is gaining increasing importance as with the increase of data sources along the life cycle also the variety of, for example, (data) formats, collaborators, and repositories make (automated) data exchange more complex. The Semantic Mediator targets the challenge for data integration in order to avoid semantic interoperability conflicts using semantic approaches [26]. The applicability of the concept has been evaluated in multiple real-life examples, for example, in complex data semantic integration problems in logistics, closed-loop PLM problems, and other domains.

In the next section, the three case studies provide a practical perspective on sensor usage in closed-loop, item-level PLM applications. Furthermore, the benefits and challenges faced by potential utilization of sensorial materials are elaborated as a basis for the discussion.

25.2 Case Studies

In this section, three case studies are presented. For each case study, first the background is illustrated briefly before the sensors used are described. Next, the potential application scenarios of sensorial materials within the case studies context are presented before the limitations and opportunities of such an utilization are discussed.

25.2.1 Case Study 1: Life Cycle of Leisure Boats

The first case study focuses on the development of a variety of services addressing intelligent maintenance, sustainability, upgrades, and the used boat market. These services actively take the whole product life cycle into account and are mostly based on data and information. The areas where the services are supposed to have an impact on are rather wide, from enhancing design processes with knowledge from the entire life cycle to the usage phase applying intelligent maintenance concepts and technologies to improve safety, reliability, and quality to the end-of-life phase by empowering sustainability issues such as increased reusability, refurbishing, and reuse of used boats and component parts as well as their ultimate retirement [31]. In this context, the application of a variety of sensors is essential in the leisure boat environment, especially during the MOL phase.

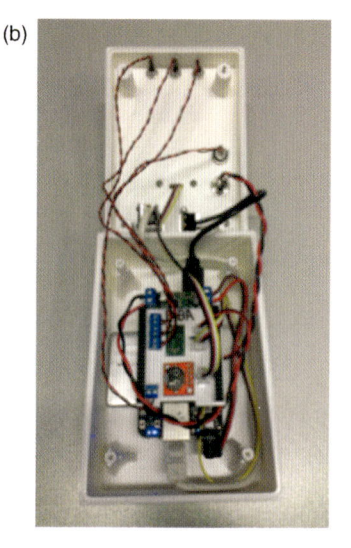

Figure 25.2 (a) UMG closed – outer view on hardware. (b) UMG open – internal view on hardware.

25.2.1.1 Sensors Used

In the project, the required sensors were combined in a so-called Universal Marine Gateway (UMG) (see Figure 25.2a). This sensor box entails a multi-purpose sensor, the "MinIMU-9 v2," which combines a gyro, accelerometer, and compass. Furthermore, two more sensor units are implemented in the UMG, a pressure sensor (MPL3115A2) and a combined temperature/humidity sensor (HIH6130). Besides the sensing units, the UMG contained a BeagleBone as the computing basis, a Wireless LAN unit, and a real-time clock that ensured the correct time even when the power supply was not given (Figure 25.2b).

The focus of the UMG's sensor deployment was to capture MOL data of a leisure boat. The final stage of the UMG, Prototype MK.III (see Figure 25.3), consists of a fully functional and live size boat where a set of sensors, based on the findings of the earlier stage tests, is implemented. The UMG sensor package can be connected to the "NMEA 2000" standardized bus system for boats. With this

Figure 25.3 Prototype boat for sensor integration and testing [32].

transmission protocol, all established sensors such as engine monitor, GPS, Wind transducer, and so on can be connected.

This UMG was tested under realistic settings and different scenarios capturing a variety of usage data. Today, it is employed in operational use by a major boat OEM and a commercial version based on the developed UMG will be offered by a project partner.

25.2.1.2 Potential Application of Sensorial Materials

Looking at the implemented sensor technology in this use case, to this point, no sensorial materials were used. For most data capturing points for PLM data during the usage phase of a boat, the applied regular sensors are a good fit. However, for selected applications, sensorial materials may offer an advantage over "regular" sensors in this specific scenario.

An example for a possible application of sensorial materials in the leisure boat use case may be an implementation within the structure monitoring of the hull. Boat designers are in need of information regarding experienced stress (e.g., from waves) affecting the rigidity of the hull in order to optimize the parameters (e.g., thickness, material) in the next generation. Whereas regular sensors can provide data for specific areas of the hull, the hull with/made of sensorial material may provide a better foundation from the captured data.

Another application, probably more likely to be useful in sailing boats, may be a sensorial material used in the mast. Especially in rough weather, a reliable indication of critical strain indicated by, for example, change of color may help to avoid damage to the boat and sailors. Furthermore, when the mast suffers from internal damage, an indication by the sensorial material may help authorities in the judgment of a boat's seaworthiness.

25.2.1.3 Limitations and Opportunities of Sensorial Materials

As mentioned before, no sensorial materials were actually applied during this case study. The limitations and opportunities discussed in the following are somewhat theoretical at this point.

At this point, regular sensors are able to collect most of the desired data and information for PLM-based improvement. In this context, sensorial materials may present an opportunity in the area of security critical parts. In this area sensorial materials (or transducer materials) may have advantages as they are directly integrated in the material and hard/impossible to be tempered with (if designed that way).

The example of the hull and mast as potential application cases for sensorial materials on boats can be extended to other critical parts such as anchor, ropes and sails, and so on. The information and data collected by such sensorial materials can not only help owners and authorities to judge the boats seagoing state but also a potential buyer. Thinking this further, sensorial material in the hull may also help EOL providers to learn if the boat was used in hazardous environment when it is being disassembled, so appropriate countermeasures may be taken.

Of course, if sensorial materials become cheaper than and similarly reliable as regular sensors, the application in all the above scenarios is possible.

Looking at the limitations, the main question is that are sensorial materials capable of providing the same data and information in the right quality and are they as reliable as regular sensors yet.

25.2.2 Case Study 2: PROMISE – Product Life Cycle Management and Information Using Smart Embedded Systems

The second case study is based on an application scenario investigated within the project PROMISE,[1] which studied the whole information flow from design, production and use-service-maintenance or MOL to retirement (EOL), and deals with predictive maintenance in the construction vehicle sector.

Manufacturers of construction vehicles offer a variety of services throughout the product life cycle, from maintenance and fleet management to rental and the certification of rebuilt machines. The ownership and operating costs of these vehicles are invariably linked to machine reliability and durability in the construction vehicle sector. Consequently, increasing the reliability of the main structures of the vehicles is key to reducing life cycle costs and thus making the products more attractive to the customers.

In this case study, approaches to structural health monitoring that support the product life cycle management of construction vehicles were investigated. A technical system was developed and evaluated in a real-life industrial environment.

The objective of the technical system was to optimize the life cycles of the structures of the vehicles in question. This was achieved by implementing a system for predictive maintenance and improving the design of structural parts, thanks to fatigue monitoring damage occurring to major structural elements such as the boom of a hydraulic excavator. In addition, the predictive maintenance system further made use of field data on machine use and knowledge of application severity.

25.2.2.1 Sensors Used

The project developed an overall concept and IT architecture for closed-loop PLM based on data gathered in the field throughout the product life cycle. The architecture specifies PEIDs (see Table 25.1) with associated firmware and software components and tools for decision-making based on data gathered through a product life cycle. The overall goal was to enable and exploit the seamless flow and traceability of life cycle data for different closed-loop PLM applications, such as predictive maintenance.

The predictive maintenance system implemented in this construction vehicle case study described above made use of sensors embedded into the vehicles (type 2 PEIDs). The sensors collected data on fatigue status and application severity parameters, such as the machine configuration and historical payload. "Crack first" sensors were located around weak areas of the structure to monitor the fatigue damage near welding seams. The machine configuration described the current equipment on the vehicle. The application severity was described by

---

1 Product Lifecycle Management and Information using Smart Embedded Systems, 15.11.2004–14.05.2008, EU IST-NMP-IMS.

Table 25.1 PEID classification according to the PROMISE Consortium, 2008.

Type	Identification	Data storage	Sensors	Data processing	Connectivity
Type 0	✓				Passive
Type 1	✓	✓			Passive
Type 2	✓	✓	(✓)	+	Wireless
Type 3	✓	✓	✓	++	Wireless
Type 4	✓	✓	✓	+++	Always

the total payload on the cylinders as well as by the number of times the lifter and ripper had to be moved. By reasoning on the field data, the machine configuration, and the application severity knowledge, inspection intervals could be determined that took into account the actual usage and condition of the vehicle.

25.2.2.2 Potential Application of Sensorial Materials

Even though no sensorial materials were applied, the objectives and applications of the case study indicate that sensorial materials may be very beneficial in this area. A sensorial metal could be used in place of the current main structure material, in this case of the excavator's boom, to measure load and strain and detect damage. Additionally, a sensorial material could be directly introduced into the welding seam. This could help detect weak points in the seam during manufacture or repair, and additionally be used to monitor the load and straining on the seams in the vehicle's MOL and predict damage by detecting cracks at an early stage.

Additional applications for sensorial materials within the project context could be in the area of traceability of individual products (or even parts of a product) as this is of major concern for many applications in a PLM context. This can be hazardous materials, dangerous goods (e.g., guns/ammunition), counterfeiting activities (e.g., perfumes), or drug monitoring. All these applications would profit from a build traceability and even more when the sensorial material may provide specific case-relevant information. In case of drugs, this could be if the drugs were stored in the right environment concerning humidity and/or temperature. Wrong storage of such sensitive products might prove dangerous to future users and reliable warning systems are essential to prevent accidents or misuse.

25.2.2.3 Limitations and Opportunities of Sensorial Materials

Sensorial materials such as intelligent coatings are already being applied to make strain and damage to structures visible to the human eye. This goes a long way to making inspection and repair processes more cost- and time-effective. However, currently available intelligent coatings are not able to directly communicate structure strain and damage to an IT system such as a closed-loop PLM or a predictive maintenance system. Further research is required in order to facilitate a reliable communication between intelligent coatings and IT systems without secondary, for example, optical, sensors.

Fundamental research has already shown that in principle, materials such as metals can be made sensorial in order to measure and communicate load, strain, and damage [33]. With the maturation and widespread availability of these technologies, sensorial materials could have a significant impact on the construction vehicle manufacturers ability to offer predictive maintenance and related services, increasing the reliability of the main structures, reducing life cycle costs, and increasing their vehicles' market attraction.

A major advantage of sensorial materials in the above-sketched context and beyond is the high threshold of manipulation. As the sensorial material is embedded/merged in the material, it is hard to almost impossible to temper with the measurements without leaving a trace. This is crucial for the above illustrated drug storage alarm system.

Today, a limiting factor is the limited variety of applications (e.g., GPS, acceleration, etc.) of sensorial materials compared to regular sensors. However, that may change in the near future.

25.2.3 Case Study 3: Composite Bridge

The third case study is based on an EU funded project iREMO – Intelligent Reactive Polymer Composites Molding. In this project, the goal was to enhance and improve the production processes of composite material by sensor usage to reduce existing uncertainties to reduce the scrap rate. In addition, the design of products is affected by safety margins. The detailed objective of iREMO is the development of a sensor-based environment for the monitoring and control of an important step in the production: the curing of the resin that is brought into moldings. The aim is to derive corrective actions (e.g., cooling of certain areas of a molding) through the measurement of the temperature and viscosity of the resin [34].

The application scenarios within the project were composite parts of cars, boats, and infrastructure (bridge). Especially the bridge production scenario is interesting in the context of sensorial materials. In this bridge scenario, the sensors (Figure 25.4), monitoring the resin arrival, viscosity rise, and gelation as

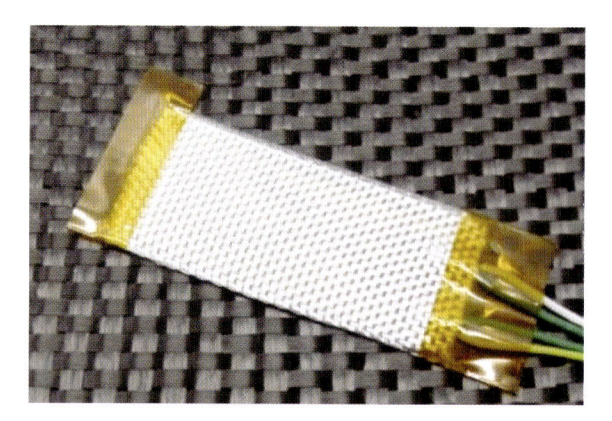

Figure 25.4 Flexible sensor used in composite bridge production [35].

well as the end-of-cure are placed within the component and left there after the product is finished. In the other scenarios, the sensors are placed in the cast or the molding machine.

25.2.3.1 Sensors Used

In the whole project, different sensors were employed. However, in this use case, the focus is on the flexible sensor used in the monitoring of the bridge part production (Figure 25.4). This sensor is put in the component and later left there, just cutting the connecting wires. In this application, the data are only used in the BOL phase, during manufacturing, and the sensor is useless during the further life cycle phases MOL and EOL.

25.2.3.2 Potential Application of Sensorial Materials

In this scenario, the potential of sensorial materials is considered rather high. In this case, the physical sensor is left in the component without having a purpose further down the life cycle. A sensorial material may provide the needed information during the manufacturing phase (resin arrival, viscosity rise, gelation, end-of-cure) and thus satisfy the objective of the project.

25.2.3.3 Limitations and Opportunities of Sensorial Materials

In case of employment of sensorial materials instead of a regular sensor, which is put in the component itself, the advantages are the following: The parts structure would not be interrupted, which may lead to a different stress allocation upon deployment as originally planned. With sensorial materials used, the structure of the material would be consistent.

Furthermore, instead of having a useless sensor in the part, the sensorial material could provide essential data for structural health monitoring of the bridge. This information is crucial for critical infrastructure. Additionally, it provides the chance to replace individual components where it is indicated that a failure is possible by the monitoring data instead of replacing the whole bridge. From another perspective, this could mean during the end of life, individual parts, which have experienced a lower level of stress, can be resold and put back in the MOL phase – thus reducing waste.

25.3 Potential of Sensorial Materials in PLM Application

In this chapter, first the basics of item-level, closed-loop PLM was presented. In this domain, data and information from sensors are key to the development of new products and services. Based on three use cases, the potential usage of sensorial materials in PLM applications is illustrated and limitations and opportunities are discussed.

Overall, it can be observed that there are several application cases that might profit from the usage of sensorial materials over regular sensors. One major advantage is that the installation of sensorial materials is most likely easier than a regular, external, or embedded sensor as it is already in the respective part. A very important benefit is the integration in security-sensitive applications or crucial infrastructure. The reason for this is that it is nearly impossible (at least very complicated) to temper with the sensorial material as it is (theoretically) possible with external sensors.

Looking at the utilization from a PLM perspective, the continuous monitoring of the products environment offers lot of potential. With regular sensors, the life cycle is often just partly covered. This stands especially true if the sensorial material secures its own power supply. As sensorial materials are actually part of the product, it offers the potential for a complete (seamless) observation over all phases of the life cycle (for that specific product). This corresponds perfectly with the intention of closed-loop, item-level PLM. In future application, this will most likely play an increasingly important role. Another point is the fact that just in time observation of processes (parameters) and usage are possible as the sensorial material is directly involved and no external infrastructure has to be taken into account. A visionary outlook discussing the potential of PLM, sensor integration, additive and cloud manufacturing, and automated design is available here [36].

However, these advantages are to a certain point expectations. As of today, a limiting factor is the small amount of available variety of applications (e.g., GPS and acceleration) of sensorial materials compared to regular sensors. However, this may change in the near future and then sensorial materials may find their way in day-to-day operations on a large scale.

Acknowledgment

This work has partly been funded by the European Commission through the BOMA – Boat Management project in FP7 SME-2011-1 "Research for SMEs," the IREMO – Intelligent reactive polymer composite molding project in FP7 NMP-2008-3.2.2 Self-Learning Production Systems, and the PROMISE – Product Lifecycle Management and Information using Smart Embedded Systems project in FP6 IP-507100. The authors gratefully acknowledge the support of the Commission and all BOMA, IREMO, and PROMISE project partners.

References

1 Lehmhus, D., Lang, W., and Bosse, S. (2013) Material-integrated sensing and intelligent introduction overview. Proceedings of the EPoSS General Assembly and Annual Forum 2013, Tyndall National Institute, Cork, Ireland.

2 Meffert, H., Burmann, C., and Kirchgeorg, M. (2008) *Marketing: Grundlagen marktorientierter Unternehmensführung*, Gabler, p. 924.

3 Paul, R. and Paul, G. (2008) Engineering data management and product data management: roles and prospects. Proceedings of the Fourth International Bulgarian-Greek Conference Computer Science, September 2008, Kavala, Greece, pp. 614–619.

4 Fasoli, T., Terzi, S., Jantunen, E., Kortelainen, J., Sääski, J., and Salonen, T. (2011) Challenges in data management in product life cycle engineering, in *Glocalized*

Solutions for Sustainability in Manufacturing (eds J. Hesselbach and C. Herrmann), Springer, Heidelberg, pp. 525–530.

5 Wuest, T., Hribernik, K., and Thoben, K. (2012) Can a product have a facebook? A new perspective on product avatars in product lifecycle management, in *Product Lifecycle Management: Towards Knowledge-Rich Enterprises. Proceedings of the 9th International Conference on Product Lifecycle Management. Montréal, Canada* (eds L. Rivest, A. Bouraz, and B. Louhichi), Springer.

6 Sendler, U. (2009) *Das PLM-Kompendium. Referenzbuch des Produkt-Lebenszyklus-Managements*, Springer, Berlin.

7 Jun, H.-B., Kiritsis, D., and Xirouchakis, P. (2007) Research issues on closed-loop PLM. *Computers in Industry*, **58** (8–9), 855–868.

8 Kiritsis, D., Bufardi, A., and Xirouchakis, P. (2003) Research issues on product lifecycle management and information tracking using smart embedded systems. *Advanced Engineering Informatics*, **17** (3–4), 189–202.

9 Pokharel, S. and Mutha, A. (2009) Perspectives in reverse logistics: a review. *Resources, Conservation and Recycling*, **53** (4), 175–182.

10 Probst, G.J.B., Raub, S., and Romhardt, K. (2006) *Wissen managen*, Gabler, Wiesbaden, Germany, p. 307. doi: 10.1007/978-3-8349-9343-4

11 Haun, M. (2002) *Handbuch Wissensmanagement. Grundlagen und Umsetzung, Systeme und Praxisbeispiele*, Springer, Berlin.

12 Storey, V.C., Dewan, R.M., and Freimer, M. (2012) Data quality: setting organizational policies. *Decision Support Systems*, **54** (1), 434–442.

13 Garetti, M. and Terzi, S. (2004) Product lifecycle management: definition, trends and open issues. Proceedings of the 3rd international Conference on Advances in Production Engineering, June 17–19, 2004, Warsaw, Poland.

14 Wellsandt, S., Wuest, T., Hribernik, K., and Thoben, K.-D. (2015) Information quality in PLM: a product design perspective, in *Advances in Production Management Systems: Innovative Production Management Towards Sustainable Growth*, IFIP Advances in Information and Communication Technology (eds S. Umeda *et al.*), vol. 459, Springer, pp. 515–523. doi: 10.1007/978-3-319-22756-6_63

15 Wuest, T., Tinscher, R., Porzel, R., and Thoben, K.-D. (2014) Experimental research data quality in materials science. *International Journal of Advanced Information Technology*, **6** (4), 1–16.

16 Wuest, T., Wellsandt, S., and Thoben, K.-D. (2015) Information quality in PLM: a production process perspective. 12th International Conference on Product Lifecycle Management (PLM), October 2015, Doha, Qatar, pp. 19–21.

17 Choe, J. (2004) The relationships among management accounting information, organizational learning and production performance. *Journal of Strategic Information Systems*, **13** (1), 61–85.

18 Hicks, B.J. (2007) Lean information management: understanding and eliminating waste. *International Journal of Information Management*, **27** (4), 233–249.

19 Beach, R., Muhlemann, A.P., Price, D.H.R., Paterson, A., and Sharp, J.A. (2000) The selection of information systems for production management: an evolving problem. *International Journal of Production Economics*, **64** (1–3), 319–329.

20 Gunasekaran, A. and Ngai, E.W. (2004) Information systems in supply chain integration and management. *European Journal of Operational Research*, **159** (2), 269–295.

21 Augustin, S. (1990) *Information als Wettbewerbsfaktor: Informationslogistik – Herausforderung an das Management*, TÜV Media GmbH, Köln, Germany.

22 Jehle, E. (1999) *Produktionswirtschaft*, Verlag Recht und Wirtschaft, Heidelberg, Germany.

23 Hoke, G.E.J. (2011) Shoring up information governance with GARP. *Information Management Journal*, **45** (1), 26–31.

24 Meyer, G.G., Främling, K., and Holmström, J. (2009) Intelligent products: a survey. *Computers in Industry*, **60**, 137–148.

25 Wuest, T., Hribernik, K., and Thoben, K.-D. (2015) Accessing servitisation potential of PLM data by applying the product avatar concept. *Production Planning and Control*, **26** (14–15), 1198–1218.

26 Hribernik, K., Kramer, C., Hans, C., and Thoben, K.-D. (2010) A semantic mediator for data integration in autonomous logistics processes, in *Enterprise Interoperability IV: Making the Internet of the Future for the Future of Enterprise* (eds K. Poppelwell, J. Harding, R. Poler, and R. Chalmeta), Springer, London, pp. 157–167.

27 Anderl, R., Picard, A., and Albrecht, K. (2013) Smart product engineering, in *23rd CIRP Design Conference* (eds M. Abramovici and R. Stark), Springer, Berlin, pp. 1–10. doi: 10.1007/978-3-642-30817-8

28 McFarlane, D., Sarma, S., Chirn, J.L., Wong, C.Y., and Ashton, K. (2003) Auto ID systems and intelligent manufacturing control. *Engineering Applications of Artificial Intelligence*, **16** (4), 365–376.

29 Kärkkäinen, M., Holmström, J., Främling, K., and Artto, K. (2003) Intelligent products: a step towards a more effective project delivery chain. *Computers in Industry*, **50** (2), 141–151.

30 Wuest, T., Hribernik, K., and Thoben, K. (2013) Digital representations of intelligent products: product avatar 2.0, in *Smart Product Engineering*, LNPE (eds M. Abramovici, R. Stark, M. Abramovici, and R. Stark), Springer, Berlin, pp. 675–684.

31 BOMA – Boat Management (2014) http://www.2020-horizon.com/BOMA-Boat-Management(BOMA)-s2779.html (retrieved August 2014).

32 Hribernik, K., Wuest, T., and Thoben, K.-D. (2013) Towards product avatars representing middle-of-life information for improving design, development and manufacturing processes, in *NEW PROLAMAT 2013*, IFIP Advances in Information and Communication Technology (eds G.L. Kovacs and D. Kochan), vol. 411, IFIP International Federation for Information Processing, pp. 85–96.

33 Behrens, B.A., Voges-Schwieger, K., Schrödter, J., and Jocker, J. (2012) Load sensitive control arm based on martensitic phase transformation. Proceedings of the 1st Joint International Symposium (SysInt), pp. 52–55.

34 Ghrairi, Z., Hribernik, K.A., Hans, C., and Thoben, K. (2012) Intelligent wireless communication devices for efficient data transfer and machine control. 2nd International Conference on Communications, Computing and Control Applications (CCCA'12), pp. 1–6. doi: 10.1109/CCCA.2012.6417891

35 Pantelelis, N., Bansal, A., Harismendy, I., Mezzacasa, R., and Bistekos, E. (2013) Intelligent monitoring of the infusion of large carbon-fibre reinforced structural parts. First International Symposium on Automated Composites Manufacturing, April 11–12, 2013, Concordia University, Montreal, Canada. Available at http://www.synthesites.com/resources/files/pantelelis-Canada.pdf (retrieved August 14, 2014).

36 Lehmhus, D., Wuest, T., Wellsandt, S., Bosse, S., Kaihara, T., Thoben, K.-D., and Busse, M. (2015) Cloud-based automated design and additive manufacturing: a usage data-enabled paradigm shift. *Sensors*, **15**, 32079–32122.

26

Human–Computer Interaction with Novel and Advanced Materials

Tanja Döring, Robert Porzel, and Rainer Malaka

University of Bremen, Technologiezentrum Informatik und Informationstechnik, Postfach 330 440, 28334 Bremen, Germany

26.1 Introduction

Computing has become ubiquitous and increasingly gets embedded into our surrounding [1]. This means, computing power gets more and more interweaved with physical materials, taking on diverse forms and appearances, for example, articulated in application fields such as the *Internet of Things* [2] and *cyber-physical systems* [3]. The integration of novel, unconventional, and advanced materials into user interfaces enables new ways of human–computer interaction (HCI), for example, multimodal, tangible, and embodied interaction. Novel and advanced materials can significantly change form factors and interactivity of user interfaces, and recent research results have exemplarily demonstrated how this development will potentially shape interactions in the future. On the one hand, novel materials extend the material canon of HCI, for example, toward textiles, liquids, substrates, or deposable materials; on the other hand, especially the developments regarding new materials and processes for functional components allow the replacement of classical rigid hardware parts. Here, one of the innovative fields is printed electronics [4], for instance. Printing electronics on different materials such as paper, polymer films, or wood allows for a range of novel form factors such as bendable or foldable devices and large surfaces. Due to the huge potential for interaction design that derives from novel material solutions, recently an increasing number of HCI research contributions focus on novel materials for interaction and their fabrication processes [5–8]. Especially smart materials with their properties that can be controlled by external stimuli (e.g., magnetic or electric fields, temperature, or moisture) are interesting for interactive applications embedded into physical materials [9–11].

As human–computer interaction moves toward interactive environments with interactive everyday objects, advances in materials will increasingly shape the appearance of and interaction with computational technology. In this chapter, we present and discuss approaches, potentials, and challenges of user interfaces with

Material-Integrated Intelligent Systems: Technology and Applications, First Edition.
Edited by Stefan Bosse, Dirk Lehmhus, Walter Lang, and Matthias Busse.
© 2018 Wiley-VCH Verlag GmbH & Co. KGaA. Published 2018 by Wiley-VCH Verlag GmbH & Co. KGaA.

novel and advanced materials. This is done from the perspective of HCI, the scientific discipline that is concerned with the design, evaluation, and implementation of interactive computing systems for human use, and with the study of the phenomena surrounding them. As an interdisciplinary field, HCI integrates perspectives from computer science, psychology, design, and social sciences, among others.

This chapter focuses on three aspects of human–computer interaction with novel and advanced materials. First, we introduce guiding interaction paradigms, user interfaces, and interaction visions within HCI research that benefit from or demand for the integration of novel materials into user interfaces. Second, we present applications and scenarios as explored in recent research that either can be seen as precursors of advanced material-based interfaces or directly explore the potentials of novel and advanced materials for interaction. Third, we discuss opportunities and challenges of the current development toward novel user interfaces with advanced materials.

26.2 New Forms of Human–Computer Interaction

For the last decades, the personal computer represented the dominant setting for interacting with digital data. Today, advances in technology have made a range of very diverse interaction techniques possible. In consumer products, gestural interaction on mobile touch screen devices, speech interaction on smartphones, or full-body interaction in camera-based computer games are already established interaction techniques, for example. Within HCI research, especially the idea of *natural interaction* [12] has evolved into a guiding theme that argues for using peoples' everyday experiences for designing interaction techniques. This means the human bodies with all their senses and their environment, including surfaces, objects, and the space around, are increasingly integrated into interaction styles. These new forms of human–computer interactions offer huge potentials for novel and advanced materials. In this section, we introduce some of the dominating underlying concepts that lay the foundations for advanced material-based interfaces.

An important concept for current interaction design is *embodied interaction*. While human–computer interaction used to mainly focus on hardware and software as artifacts that are separated from the environment of their users, the concept of embodied interaction strongly relies on an integration of interaction between humans and computers into the material and social environment. This new quality of situatedness of interaction techniques with computers into our daily environment makes a "new perspective on the relationship between people and systems necessary" (13], p. 192). Dourish called this new perspective *embodied interaction* and described it as follows: "Embodied Interaction is the creation, manipulation, and sharing of meaning through engaged interaction with artifacts" [13]. Embodiment is an important constituent of human intelligence. It is not only abstract cognition that allows for intelligent acting in the world but rather the interplay with artifacts in the world [14]. Our actions are embedded in material, cultural, and social conditions. Human culture and our cognition developed in long tradition, spanning thousands of years, of using haptic real objects that represent abstract concepts and ideas. Thus, intelligent interaction in and with the environment can benefit from how the artifacts in the environment are shaped and the affordances they provide. Even though computers have raised our human potential for problem solving significantly, their embodiment was for decades restricted to mice, keyboards, and monitors and their understanding of their external context was minimal. The digital and the real world were distinctly separated. Drawing from theories of embodiment in psychology, sociology, and philosophy, Klemmer et al. [15] discuss five themes of embodiment that matter within interaction design: "thinking through doing," "performance," "visibility," "risk," and "thick practice." With "thinking through doing," the authors address research from embodied cognition [16] that "regard bodily activity as being essential to understanding human cognition" ([15], p. 141). Consequently, applying the ideas of embodied interaction implicates that the humans' bodily skills are leveraged within interaction and that the natural surroundings, including spaces, objects, and surfaces, become interactive.

Tangible user interfaces (TUIs) are among the central concepts that put the material into focus to computing technology. The central idea of tangible user interfaces is to connect real objects with digital data and thus to enable the interaction with the data via interaction with the objects. The word "tangible" comes from the Latin word *tangere*, which means "to touch, to grasp." Users can literally grasp data and manipulate it by manipulating objects with their hands. Tangible user interfaces became popular in the research community with the work conducted by Ishii and Ullmer et al. at MIT Media Lab in the 1990s and 2000s. In their "Tangible Bits" paper in 1997, they introduce the overall concept and part of ubiquitous computing where the whole world around us potentially becomes an interface [17]. Ishii and Ullmer describe TUIs as follows: "The key idea of tangible interfaces is giving physical form to digital information. These physical forms serve both as representations and controls for their interwoven digital bindings and associations. TUIs make digital information directly manipulable with our hands and perceptible through our foreground and peripheral senses" ([18], p. 466). Thus, the materials used for physical devices and the shapes they come in directly embody digital information. Hemmert ([19], p. 36) defines this aspect as *representational embodiment* as opposed to *experiential embodiment*, which refers one's experience of the world through having a living body ([19], p. 44). To shape the representational embodiment of a device, interaction designers have to face material decisions as a very central design issue of the interaction.

Among the limitations of using classical physical objects as tangible user interfaces is that they normally are rigid and static, whereas digital data is malleable: "digital objects are easy to create, modify, replicate, and distribute" ([20], p. 205). To compensate these limitations, actuation and self-actuation of physical components have evolved to a major research strand. Poupyrev et al. defined actuated interfaces as interfaces "in which physical components move in a way that can be detected by the user. There are many types of actuation, for example: change in spatial position of objects or their parts, e.g. their position, orientation; change in speed of motion of objects or their parts, e.g. speed of rotation, speed of linear motion, direction of motion; change in surface texture of objects or their parts, e.g. visible or perceived by touch; change in force applied to the user, e.g. change in force amplitude, direction, or torque." [20], p. 206). All these forms of actuation would allow more powerful forms of tangible interaction. Recently, efforts regarding shape-change in user interfaces have been made under the term *shape-changing interfaces*, which use "physical change of shape as input or output" [21], p. 735). Analyzing the design space for shape-changing interfaces, Rasmussen *et al.* discussed four aspects of shape change: the types of shape change, the dynamics of change, approaches to interact with shape-changing interfaces, and purposes of using shape change [21]. They found, for example, different functional aims ranging from using shape-change to communicate information, for dynamic affordances, or to provide haptic feedback as well as hedonic aims where shape-change is applied for aesthetical, emotional, or stimulative reasons.

A literally radical idea of interactive material that transcends these ideas for tangible and shape-changing interfaces is the vision of *radical atoms* by Ishii *et al.* [22]. This long-term vision hypothesizes a new generation of materials that can dynamically change their form and appearance as well as can be transformed by users: "Radical Atoms is based on a hypothetical, extremely malleable, and dynamic physical material that is bidirectionally coupled with an underlying digital model (bits), so that dynamic changes of the physical form can be reflected in the digital states in real time, and vice versa" [22], p. 45). In this sense, radical atoms would be as malleable and flexible as pixels on a monitor screen, but contain due to their physicality many more properties and address more senses that can be used for interaction. In more detail, Ishii *et al.* postulate three requirements for radical atoms [22]. First, they need to be able to *transform* according to user input or underlying computational state. Second, they should *conform* to constraints. Third, radical atoms should be able to *inform* the user by communicating their transformational capabilities (i.e., they need to offer dynamic affordances). While this vision strongly relies on advances in disciplines such as materials or chemical engineering, it is already possible to explore dedicated aspects of this general idea at larger scales. For example, the "ZeroN" prototype [23] presents an antigravity interaction element enabled by computer-controlled magnetic leviation; the "materiable" interaction technique [24] allows the rendering of dynamic material properties such as flexibility, elasticity, and viscosity with a shape-changing interface; "Biologic" [25] presents a material that expands when exposed to moisture, which can be used for shape-change. Especially the last example demonstrates that the potential material diversity for human–computer interaction is tremendous. In Biologic, the shape-change is based on a specific moisture-sensitive bacterium that is applied to textiles or other flexible materials. Organic and natural materials have recently found their way into the material spectrum for user interfaces, for example, as part of ephemeral user interfaces [26] that integrate plants, water, food, bubbles, fog, or ice for input and output. Among other materials that transcend the classical repertoire of electronic parts and plastics and that have been explored within HCI research for a while are interactive fabrics [5], textile sensors [27], interactive paper [28], or interactive clay [29,30]. Especially advances in the fabrication of interactive materials with novel properties and qualities make many of these interactive objects and surfaces possible, for instance, printable sensors [31] and displays [32] or 3D printing of deformable materials [6–8]. In future, the efforts to integrate "functional as well as structural materials, in order to print complete working systems" (33), p. 101) might lead to a situation, where a typical user interface design task rather focuses on the design and invention of a new material instead of selecting an existing one.

26.3 Applications and Scenarios

While in the 1990s computer systems dominantly were designed to support tasks within work contexts, with the so-called "third wave of HCI" [34] computing nowadays gets integrated into all aspects of our lives also supporting humans in domestic, personal, and leisure domains. In the following, we present interactive prototypes that explore the ideas of embodied interaction, tangible user interfaces, shape-changing interfaces, and radical atoms in different application scenarios, encompassing personal devices, everyday domestic contexts, learning, collaboration, and entertainment. The chosen examples present precursors or early explorations of user interfaces with advanced materials and, overall, demonstrate potentials how diverse physical, advanced, and novel materials can shape human–computer interaction in the future.

26.3.1 Domestic and Personal Devices

Examples for user interfaces with novel and advanced materials can be found as part of domestic and personal devices. These systems put a focus on material qualities that fit the home context and the personal demands, for example, in aesthetical or practical regard. In the following, we present *the Marble Answering Machine*, a tangible answering machine concept that uses physical marbles for interaction, *Living Wall*, an interactive wallpaper, *Sprout I/O* and *Shutters*, ambient textile information displays, and *FlexCase*, a flexible sensing and display cover for mobile devices.

26.3.1.1 The Marble Answering Machine

The Marble Answering Machine [35], which was conceptually designed as part of a master thesis project at the Royal College of Art in London, is an answering machine, where incoming messages are presented in the form of real marbles. If a message has been left on the machine, a marble rolls down a track, where it stays together with other marbles, each of them presenting messages. At a glance, a user can see the number of messages. By manipulating them, she can either play a message (by putting the associated marble into a specific track in the machine), erase it (by putting it into a hole), save it for other family members (by putting it into a dedicated bowl), annotate it (by putting them onto a "todo" shelf, adding further notes simply by writing with a pen onto the shelf), or call back the person that left the message (by directly putting the marble into the telephone). This concept of a tangible user interface nicely shows how digital data, in this case the voice messages, can be represented by graspable objects and how functions with this data can be triggered by manipulations of the real objects. This potentially leads to more meaningful ways to interact than just by pressing on arrays of buttons that are arbitrarily mapped to certain functions.

26.3.1.2 Living Wall: An Interactive Wallpaper

One example how everyday domestic environments can become interactive in an unobtrusive way by integrating novel materials has been presented by Buechley *et al.* [36]. They presented "Living Wall," a wallpaper with ambient input and output facilities that can be used for applications such as interactive lighting, environmental sensing, appliance control, or as ambient information display. The Living Wall approach, like other works from the high–low tech group at MIT Media Lab, is characterized by a beautiful integration of high and low tech materials. This integration is not based on hiding technical and functional components but on inventing novel functional components from diverse materials and with new form factors. Living Wall contains a layer in which the whole circuitry needed to connect sensors, actuators, and communication elements is painted with a conductive paint that at the same time is part of the decorative pattern of the wall. Beneath the wallpaper there exists a layer of magnetic paint so that electronic modules can be freely and flexibly attached by magnets to dedicated areas on the wall. Among these components are, for example, a microcontroller, LEDs, motors, shape-changing paper flowers, and several sensors measuring light, temperature, and motion. With this setup, a broad range of domestic applications are possible without the need for classical computer-based devices.

26.3.1.3 Sprout I/O and Shutters: Ambient Textile Information Displays

Other works have explored the opportunities of ambient information displays in domestic and everyday environments with shape-changing interfaces. Coelho and Zigelbaum [37], for example, explored the potential of combining shape memory alloys (SMA) with fabrics and built a number of design probes. Among these was "Sprout I/O," a grid of 6 × 6 stretchy textile strands with integrated shape memory alloys that can be bent forward and backward. Sprout I/O presents an early prototype for an interactive fabric, in which angle, speed, and direction of the single strands can be computationally controlled. At the same time, the fabric senses touch and provide rich visual and tactile qualities. While it presents only an exploration, potential application contexts for systems such as Sprout I/O could be interactive carpets or clothes that are used as ambient displays to present information (e.g., for indoor navigation). Similarly, the "Follow the Grass" prototype sketched out scenarios for interactive artificial shape-changing grass [38]. Recently, other approaches used the natural shape-changing abilities of 3D printed hair structures for sensing and actuation [7]. In a further prototype, "Shutters," Coelho and Zigelbaum [37] designed dynamic permeability within fabrics through creating a grid of louvers cut into a felt sheet and embedding shape memory alloy strands. Structures like these can be envisioned to be used as architectural elements controlling light and airflow or as kinetic and shadow displays in the future.

26.3.1.4 FlexCase: A Flexible Sensing and Display Cover

Extending the form factors and interaction opportunities of rigid multitouch devices toward bendable devices that allow for richer manipulations and more diverse haptic feedback has evolved to an important research topic within human–computer interaction. Recently, several approaches for bendable mobile devices [39] as well as more general interaction concepts for flexible devices [40,41] have been presented. Additionally, rich sensing capabilities that go beyond simple touch sensing allow novel, more fine-grained interaction techniques. Based on their previous work on a flexible piezoelectric sensor that combines touch, pressure, and bend sensing [42,43], Rendl et al. built FlexCase, an interactive smartphone cover with input and output facilities and the capability to communicate with the smartphone [44]. It allows bend and touch input and also contains an e-paper display that can be used as secondary display together with the display of the mobile phone. Due to this combination of materials, a range of novel interaction techniques and application scenarios is possible. For instance, having the flip cover opened with the sensor and e-paper display on the left and the mobile phone on the right, the device could be used like a book with two pages and page flipping mechanisms by bending a corner of the left page forward or backward. Other scenarios the authors describe include bend sensing for map interaction or zooming pictures, bimanual interaction for content transfer between pages, using continuous bend input for gaming, or back-of-device interaction for navigating content.

26.3.2 Learning, Collaboration, and Entertainment

Application areas such as learning, collaboration, and entertainment benefit from the potentials of tangible and shape-changing interfaces as well as full-body interaction. Especially within learning, a number of concepts suggest the use of physical materials. If these become interactive and get flexible properties through smart materials, a large variety of future application options evolve. In the following, we present novel approaches to support learning with interactive objects, introduce the *inForm* system, a tool that supports remote collaboration, and describe *the soap bubble interface*, an entertainment installation that explores unusual interaction materials.

26.3.2.1 Tangibles for Learning and Creativity

Using physical representations and three-dimensional space to support learning and creativity has a long history [45]. Many studies have shown that "objects-to-think-with" [46] help understand abstract processes and problem solving tasks (e.g., Refs [47,48] or, more recent Ref. [49]). For example, Bruner's model of three different representations – enactive, iconic, and symbolic [48] – that highlights the value of tactile learning for novices was very influential. Among other important benefits of using physical objects are *externalized representations for epistemic actions* [50,51] or the support of collaboration. Based on these and other concepts, theories, and experiences about physical objects for learning and creativity, a number of interactive systems have been built that further explore these potentials. Early examples are Lego Mindstormes™, a robotic construction kit that evolved from research conducted at the MIT Media Lab [52,53], or *AlgoBlocks*, a prototype using tangible blocks to program a graphical representation of a submarine [54]. Recently, prototypes have been built for diverse learning contexts such as math learning [55], storytelling [56], or supporting children with special needs [57]. Döring and Beckhaus [58] presented an interactive tabletop system for art historians that combined interactions with paper cards with digital systems such as image databases and Web pages in order to support creative work processes in a spatial and collaborative setting. Given these many novel opportunities that arise for the integration of physical and digital learning and creativity support systems, novel and smart materials could even leverage these potentials further. As Marshall [59] analyzed, dedicated material properties, for example, "mass, texture, temperature or malleability" ([59], p. 165), could be utilized to support learning. Especially the integration of the representation and manipulation of three-dimensional models in physical and digital settings through computationally controlled malleable materials could further enhance many learning scenarios in future.

26.3.2.2 inFORM: Supporting Remote Collaboration through Shape Capture and Actuation

Current shape-changing interfaces are regarded as precursors of radical atoms: While they already realize transformations triggered by users, systems, or the environments, their scale is much larger than envisioned, their resolution is still low, and they generally are composed of many different materials that still need to be assembled and programmed in extensive manner. Nevertheless, recent shape-changing interfaces show in compelling ways how powerful user interfaces of the future could be, given that further material innovations support a smaller size, a higher resolution, and to higher degrees integrated interactive materials that unify rich physical and computational capabilities. The "inFORM" system, built by Follmer *et al.* [60,61], explores these potentials by providing a grid of 30×30 pins in cuboid shapes arranged in a plane of approximately 40 cm^2. Each of the pins can be separately moved up and down by actuator motors. Due to a connection via rods that transmit bidirectional force, the stiffness of a pin can also be varied, which allows different haptic experiences when pins are pushed down manually. Together with depth camera sensing above the plane that tracks the user and surface objects, this setting allows for a broad variety of applications. It is able to realize dynamic affordances (e.g., the system presents user interface elements that change their shape according to the user's actions or the system state), dynamic constraints (i.e., interaction can be guided by limiting interaction possibilities physically), object actuation (i.e., objects like marbles can be moved on the surface by the system itself by rising and lowering dedicated pins), or physical rendering of content (e.g., by physically presenting the upper surface of 3D models or bar charts). For example, with the inFORM platform, it is possible to built a working prototype of the above-mentioned "Marble Answering Machine" that transports marbles over the surface and collects them in dedicated holes. A further compelling application scenario is remote collaboration with physical telepresence. While current systems that support remote collaboration basically provide video and sound channels as well as the exchange of digital data, inFORM allows remote manipulation, for example. A remote person's arm movements can be physically rendered by the system, such that remote manipulations of objects are (in restricted ways) possible. Given a symmetric setup with inFORM systems at both remote locations, rendered physical objects could be manipulated concurrently.

26.3.2.3 The Soap Bubble Interface

Especially in the domain of entertainment interfaces, aspects such as "joy of use" [62] and "user experience" [63,64] play important roles. To leverage and explore how unusual interaction materials influence these, Döring *et al.* built a soap bubble interface [65,66], an installation in which room lighting and ambient sound could collaboratively be controlled by generating and moving soap bubbles that float on a liquid surface. Due to the nature of soap bubbles as playful, cheerful, and unconventional material for interaction, an engaging and playful atmosphere was created that motivated potential users to interact with the installation. The soap bubble interface consists of a soap bubble basin with a transparent acrylic surface and a top thin layer of colored liquid as well as a housing with self-constructed soap bubble generator and integrated fog machine behind. When the soap bubble machine is used via a small control panel to generate single bubbles, either filled with fog or plain, in desired sizes, these fall onto the basin where they float as half spheres on the liquid layer until they burst. Below the basin is a camera to track the size and location of the bubbles as well as a projector for visual output on the fog-filled bubbles. Thus, the amount, location, and sizes of the bubbles on the surface can be potentially used to control an application. Users can generate soap bubbles and once they float on the surface, they can try to move them around, often by (a variation of) one of the three dominating interaction techniques: by gently touching a bubble with a wet finger and by moving it around, by bending forward and blowing a bubble into a desired direction, or by creating airflow with the hands or an object. The techniques vary in their degree of preciseness, but also in the degree of needed body movement and skillfulness. An interesting aspect here is that the interaction techniques are not specified. Users became creative in trying out and inventing diverse gestural and full body techniques. The resulting techniques are purely inspired and guided by the specific material properties of soap bubbles. This prototype demonstrates that it will be beneficial to explore novel and motivational materials for interaction to inspire interesting, surprising and engaging interaction techniques. Furthermore, it shows how interactive organic shapes that can be generated, destroyed, and deformed and come along with input and output capabilities could be used for interactions in the future. Novel and advanced materials could provide a wealth of new opportunities for user interfaces with more diverse interaction techniques.

26.4 Opportunities and Challenges

The examples presented above demonstrate that the integration of novel and advanced materials has potentials to shape the future of human–computer interaction in many regards: Everyday objects can seamlessly be extended by additional interactive functions, embodied interaction with advanced materials will allow novel interaction techniques and application scenarios, and our surrounding will fuse with computation. Computational technology does not necessarily come in rigid and cuboid-like shapes anymore, it could be organic, soft, or even shape-changing. Advanced materials will increasingly influence *explicit interaction,* which offers new interaction techniques like bending gestures, and *implicit interaction* in the form of sensing components. Overall, there are a number of developments that will give novel materials increasing impacts on shaping the interaction with interactive devices. Among these are interweaving of the digital and physical in ubiquitous computing, increased attention to hedonic aspects and user experience of user interfaces beyond usability, opportunities for novel kinds of multisensory and full-body interaction, the rise of personal fabrication and novel fabrication techniques, and sustainability as topic in interaction design (see also Ref. [67])

Considering these diverse opportunities for novel forms of human–computer interaction in the age of ubiquitous computing enabled by advanced materials, challenges will lie in establishing an inter- and transdisciplinary field for human–computer interaction with advanced materials. Here, areas such as *material fabrication, material application,* and *material appreciation* (see also Ref. [68] for a product design perspective on these areas) will build important pillars that encompass different aspects of advanced materials for interaction. Thus, disciplines such as materials engineering, computer science, interaction design, product design, art, architecture, psychology, or anthropology as well as social science could all contribute to this transdisciplinary field. Within HCI, it has been accepted that in order to design successful user experiences, it needs a holistic approach covering knowledge, skills, and views from different disciplines. With the rise of user interfaces, novel and advanced materials demands for even more disciplines to participate, as physicality and computational behavior further fuse and interactive artifacts increasingly get integrated into everyday environments. Here, appropriate terminology, processes, methods, and tools are needed to support the collaboration of different stakeholders. An area that currently pushes interdisciplinary innovations that combine computation and advanced materials is the research field of *computational fabrication,* compare the newly founded *symposium on computational fabrication* (scf.acm.org).

As the integration of advanced materials into user interfaces allows for novel forms of human–computer interaction that are highly integrated into physical materials in everyday contexts, new models are needed to support both humans in understanding the interaction and the interactive artifacts in developing intelligent behavior. If everyday objects are provided with interactive functionality, it could be difficult for a user to detect which surfaces are interactive and what interaction techniques are implemented, to understand how the new functionality works, what effects it has, and how it is linked to the traditional use of the object. Concepts such as *mapping, affordances,* and *constraints* have already been applied within HCI [69] and especially within tangible user interfaces in order to support the users' understanding of interaction opportunities with objects. While these approaches offer potentials, they also come along with limitations as they do not provide sufficient cues to explain the novel and magical functionality of digital artifacts that go beyond mimicking reality. Hornecker [70] discussed this using the example of affordances for tangible user interfaces.

While we need new models for the interaction with smart and advanced materials that are seamlessly integrated into everyday materials, there is also a need for new models that support a context-aware and adaptive behavior of interactive artifacts. As intelligent and adaptive behavior is grounded on simulation-based interaction of a model of oneself within a model of the given situational context, any enhancement of self- and context-awareness – provided by smart sensing materials – can lead to a corresponding increase in adaptability and intelligent behavior. Only entities that know what they are and that are aware of the situation they are in can react intelligently to a given stimuli. The embodiment view of a live artificial intelligence [71] has already brought about a paradigm shift in the way adaptive and resilient robotic agents are constructed and programmed [72]. Nevertheless, other sensing objects, such as smartphones or houses, still employ conventional techniques. Therefore, an additional challenge and opportunity lies in the question of what kind of "mental models" sensing objects can develop in order to become context-aware and context-adaptive.

26.5 Conclusions

This chapter gives an introduction to human–computer interaction with novel and advanced materials. We provided an overview of current developments within human–computer interaction toward interactive environments with interactive everyday objects in the era of ubiquitous computing. Nowadays, physical materials and computation increasingly fuse, leading to novel forms of interactions between humans and technology that are guided and inspired by concepts such as *embodied interaction*, *tangible user interfaces*, *shape-changing interfaces*, and *radical atoms*. Novel, advanced, and smart materials have the potential to radically influence or even shape interaction design in the future as they provide new opportunities to put these guiding concepts into practice. In this chapter, we presented a number of example applications that already use smart materials or can be regarded as precursors for advanced-material-based interfaces. As we show, these application contexts also lie beyond the conventional Internet of Things and industry-centered scenarios that often focus on the integration of novel sensing materials and sensor networks. As example areas, we presented *domestic and personal devices* as well as *learning, collaboration, and entertainment*. The presented prototype systems in these areas demonstrate diverse interaction techniques that are enabled by the integration of physical and advanced materials. Finally, we discussed opportunities and challenges of this emerging field, which will benefit from interdisciplinary research on materials for interaction.

References

1 Weiser, M. (1991) The computer for the 21st century. *Scientific American*, **265** (3), 94–104.

2 Atzori, L., Iera, A., and Morabito, G. (2010) The Internet of Things: a survey. *Computer Networks*, **54** (15), 2787–2805.
3 Broy, M. and Schmidt, A. (2014) Challenges in engineering cyber-physical systems. *Computer*, **47** (2), 70–72.
4 Steimle, J. (2015) Printed electronics for human–computer interaction. *Interactions*, **22** (3), 72–75.
5 Poupyrev, I., Gong, N.-W., Fukuhara, S., Karago-Zler, M.E., Schwesig, C., and Robinson, K.E. (2016) Project Jacquard: interactive digital textiles at scale. Proceedings of the 2016 CHI Conference on Human Factors in Computing Systems (CHI'16), New York, NY, ACM, pp. 4216–4227.
6 Wang, G., Yao, L., Wang, W., Ou, J., Cheng, C.-Y., and Ishii, H. (2016) xPrint: a modularized liquid printer for smart materials deposition. Proceedings of the 2016 CHI Conference on Human Factors in Computing Systems (CHI'16), New York, NY, ACM, pp. 5743–5752.
7 Ou, J., Dublon, G., Cheng, C.-Y., Heibeck, F., Willis, K., and Ishii, H. (2016) Cilllia: 3D printed micro-pillar structures for surface texture, actuation and sensing. Proceedings of the 2016 CHI Conference on Human Factors in Computing Systems (CHI'16), New York, NY, ACM, pp. 5753–5764.
8 Groeger, D., Chong Loo, E., and Steimle, J. (2016) HotFlex: post-print customization of 3D prints using embedded state change. Proceedings of the 2016 CHI Conference on Human Factors in Computing Systems (CHI'16), New York, NY ACM, pp. 420–432.
9 Gandhi, M.V. and Thompson, B.S. (1992) *Smart Materials and Structures*, Chapman & Hall.
10 Ritter, A. (2007) *Smart Materials in Architektur, Innenarchitektur und Design*, Birkhäuser.
11 Minuto, A. and Nijholt, A. (2013) Smart material interfaces as a methodology for interaction: a survey of SMIs' state of the art and development, in Proceedings of the Second International Workshop on Smart Material Interfaces: Another Step to a Material Future (SMI'13), ACM, New York, NY, pp. 1–6.
12 Hinckley, K. and Wigdor, D. (2012) Input technologies and techniques, in *The Human–Computer Interaction Handbook, Fundamentals, Evolving Technologies, and Emerging Applications* (ed. J.A. Jacko), 3rd edn, CRC Press, pp. 95–132.
13 Dourish, P. (2004) *Where the Action Is: The Foundations of Embodied Interaction*, MIT Press.
14 Malaka, R. (2008) Intelligent user interfaces for ubiquitous computing, in *Handbook of Research on Ubiquitous Computing Technology for Real Time Enterprises*, Information Science Reference, pp. 470–486.
15 Klemmer, S.R., Hartmann, B., and Takayama, L. (2006) How bodies matter: five themes for interaction design. Proceedings of the 6th Conference on Designing Interactive Systems (DIS'06), New York, NY, ACM, pp. 140–149.
16 Johnson, M. (1987) The body in the mind, in *The Bodily Basis of Meaning, Imagination, and Reason*, The University of Chicago Press, Chicago.
17 Ishii, H. and Ullmer, B. (1997) Tangible bits: towards seamless interfaces between people, bits and atoms. Proceedings of the SIGCHI Conference on

Human Factors in Computing Systems (CHI'97), New York, NY, ACM, pp. 234–241.

18 Ishii, H. and Ullmer, B. (2012) Tangible user interfaces, in *The Human–Computer Interaction Handbook: Fundamentals, Evolving Technologies, and Emerging Applications* (ed. J. Jacko), CRC Press, pp. 465–490.

19 Hemmert, F. (2014) Encountering the digital: representational and experiential embodiment in tangible user interfaces. Ph.D. thesis, Universität der Künste Berlin, Germany.

20 Poupyrev, I., Nashida, T., and Okabe, M. (2007) Actuation and tangible user interfaces: the Vaucanson duck, robots, and shape displays. Proceedings of the 1st International Conference on Tangible and Embedded Interaction (TEI'07), New York, NY, ACM, pp. 205–212.

21 Rasmussen, M.K., Pedersen, E.W., Petersen, M.G., and Hornbæk, K. (2012) Shape-changing interfaces: a review of the design space and open research questions. Proceedings of the SIGCHI Conference on Human Factors in Computing Systems (CHI'12), New York, NY, ACM, pp. 735–744.

22 Ishii, H., Lakatos, D., Bonanni, L., and Labrune, J.-B. (2012) Radical atoms: beyond tangible bits, toward transformable materials. *Interactions*, **19** (1), 38–51.

23 Lee, J., Post, R., and Ishii, H. (2011) ZeroN: mid-air tangible interaction enabled by computer controlled magnetic levitation. Proceedings of the 24th Annual ACM Symposium on User Interface Software and Technology, pp. 327–336.

24 Nakagaki, K., Vink, L., Counts, J., Windham, D., Leithinger, D., Follmer, S., and Ishii, H. (2016) Materiable: rendering dynamic material properties in response to direct physical touch with shape changing interfaces. Proceedings of the 2016 CHI Conference on Human Factors in Computing Systems (CHI'16), New York, NY, ACM, pp. 2764–2772.

25 Yao, L., Ou, J., Cheng, C.-Y., Steiner, H., Wang, W., Wang, G., and Ishii, H. (2015) bioLogic: Natto cells as nanoactuators for shape changing interfaces. Proceedings of the 33rd Annual ACM Conference on Human Factors in Computing Systems (CHI'15), New York, NY, ACM, pp. 1–10.

26 Döring, T., Sylvester, A., and Schmidt, A. (2013) A design space for ephemeral user interfaces. Proceedings of the 7th International Conference on Tangible, Embedded and Embodied Interaction (TEI'13), New York, NY, ACM, pp. 75–82.

27 Perner-Wilson, H., Buechley, L., and Satomi, M. (2011) Handcrafting textile interfaces from a kit-of-no-parts. Proceedings of the Fifth International Conference on Tangible, Embedded, and Embodied Interaction (TEI'11), New York, NY, ACM, pp. 61–68.

28 Qi, J. and Buechley, L. (2014) Sketching in circuits: designing and building electronics on paper. Proceedings of the SIGCHI Conference on Human Factors in Computing Systems (CHI'14), New York, NY, ACM, pp. 1713–1722.

29 Reed, M. (2009) Prototyping digital clay as an active material. Proceedings of the 3rd International Conference on Tangible and Embedded Interaction (TEI'09), New York, NY, ACM, pp. 339–342.

30 Herrlich, M., Braun, A., and Malaka, R. (2012) Towards bimanual control for virtual sculpting, in *Mensch & Computer 2012: interaktiv informiert –*

allgegenwärtig und allumfassend!? (Hrsg. H. Reiterer and O. Deussen), Oldenbourg Verlag, München, Germany, pp. S.173–182.

31 Weigel, M., Lu, T., Bailly, G., Oulasvirta, A., Majidi, C., and Steimle, J. (2015) iSkin: flexible, stretchable and visually customizable on-body touch sensors for mobile computing. Proceedings of the SIGCHI Conference on Human Factors in Computing Systems (CHI'15).

32 Olberding, S., Wessely, M., and Steimle, J. (2014) PrintScreen: fabricating highly customizablethin-film touch-displays. Proceedings of the 27th Annual ACM Symposium on User Interface Software and Technology.

33 Gershenfeld, N. (2005) *Fab: The Coming Revolution on Your Desktop: From Personal Computers to Personal Fabrication*, Basic Books.

34 Bødker, S. (2015) Third-wave HCI, 10 years later: participation and sharing. *Interactions*, **22** (5), 24–31.

35 Bishop, D. (2009) Visualising and physicalising the intangible product: "What happened to that bloke who designed the marble answer machine?" Proceedings of the 3rd International Conference on Tangible and Embedded Interaction, 2. doi: 10.1145/1517664.1517667.

36 Buechley, L., Mellis, D., Perner-Wilson, H., Lovell, E., and Kaufmann, B. (2010) Living wall: programmable wallpaper for interactive spaces. Proceedings of the 18th ACM International Conference on Multimedia, New York, NY, ACM, pp. 1401–1402.

37 Coelho, M. and Zigelbaum, J. (2011) Shape-changing interfaces. *Personal and Ubiquitous Computing*, **15** (2), 161–173.

38 Minuto, A., Huisman, G., and Nijholt, A. (2012) Follow the grass: a smart material interactive pervasive display. Proceedings of the 11th International Conference on Entertainment Computing (ICEC'12), Berlin, Springer, pp. 144–157.

39 Strohmeier, P., Burstyn, J., Carrascal, J.P., Lévesque, V., and Vertegaal, R. (2016) ReFlex: a flexible smartphone with active haptic feedback for bend input. Proceedings of the 10th International Conference on Tangible and Embedded Interaction, pp. 185–192. doi: http://dx.doi.org/10.1145/2839462.2839494

40 Khalilbeigi, M., Lissermann, R., Mühlhäuser, M., and Steimle, J. (2011) Xpaaand: interaction techniques for rollable displays. Proceedings of the SIGCHI International Conference on Human Factors in Computing Systems, pp. 2729–2732. doi: 10.1145/1978942.1979344

41 Khalilbeigi, M., Lissermann, R., Kleine, W., and Steimle, J. (2012) FoldMe: interacting with double-sided foldable displays. Proceedings of the Sixth International Conference on Tangible and Embedded Interaction, pp. 33–40. doi: 10.1145/2148131.2148142

42 Rendl, C., Greindl, P., Haller, M., Zirkl, M., Stadlober, B., and Hartmann, P. (2012) PyzoFlex: printed piezoelectric pressure sensing foil. Proceedings of the 25th Annual ACM Symposium on User Interface Software and Technology (UIST'12), New York, NY, ACM, pp. 509–518. doi: 10.1145/2148131.2148142

43 Rendl, C., Kim, D., Fanello, S., Parzer, P., Rhemann, C., Taylor, J., Zirkl, M., Scheipl, G., Rothländer, T., Haller, M., and Izadi, S. (2014) FlexSense: a transparent self-sensing deformable surface. Proceedings of the 27th Annual ACM Symposium on User Interface Software and Technology (UIST'14), New York, NY, ACM, pp. 129–138. doi: 10.1145/2642918.2647405

44 Rendl, C., Kim, D., Parzer, P., Fanello, S., Zirkl, M., Scheipl, G., Haller, M., and Izadi, S. (2016) FlexCase: enhancing mobile interaction with a flexible sensing and display cover. Proceedings of the 2016 CHI Conference on Human Factors in Computing Systems (CHI'16), New York, NY, pp. 5138–5150. doi: 10.1145/2858036.2858314

45 O'Malley, C. and Fraser, D.S. (2004) Literature Review in Learning with Tangible Technologies. A NESTA Futurelab Research Report – Report 12.

46 Papert, S. and Harel, I. (1991) Situating constructionism, in *Constructionism*, (Hg. P. Seymour and H. Idit), Ablex Publishing Corporation, Norwood NJ, pp. 1–12.

47 Piaget, J. (1953) How children form mathematical concepts. *Scientific American*, **189** (5), 74–77.

48 Bruner, J. (1966) *Toward a Theory of Instruction*, WW Norton, New York.

49 Antle, A.N. and Wang, S. (2013) Comparing motor-cognitive strategies for spatial problem solving with tangible and multi-touch interfaces. Proceedings of the 7th International Conference on Tangible and Embedded Interaction, pp. 65–72. doi: 10.1145/2460625.2460635.

50 Kirsh, D. (1995) The intelligent use of space. *Artificial Intelligence*, **73** (1–2), 31–68.

51 Manches, A. and O'Malley, C. (2012) Tangibles for learning: a representational analysis of physical manipulation. *Personal and Ubiquitous Computing*, **16** (4), 405–419.

52 Papert, S. (1980) *Mindstorms: Children, Computers and Powerful Ideas*, Harvester, Brighton, UK.

53 Resnick, M. (1993) Behavior construction kits. *Communications of the ACM*, **36** (7), 64–71.

54 Suzuki, H. and Kato, H. (1995) Interaction-level support for collaborative learning: AlgoBlock – an open programming language. The First International Conference on Computer Support for Collaborative Learning (CSCL'95), pp. 349–355.

55 Girouard, A., Solovey, E.T., Hirshfield, L.M., Ecott, S., Shaer, O., and Jacob, R.J.K. (2007) Smart Blocks: a tangible mathematical manipulative. First International Conference on Tangible and Embedded Interaction, pp. 183–186.

56 Ryokai, K. and Cassell, J. (1999) StoryMat: a play space for collaborative storytelling. Proceedings of CHI'99, ACM, pp. 272–273.

57 Hengeveld, B., Voort, R., Hummels, C., de Moor, J., van Balkom, H., Overbeeke, K., and van der Helm, A. (2008) The development of linguabytes: an interactive tangible play and learning system to stimulate the language development of toddlers with multiple disabilities. *Advances in Human–Computer Interaction*, **13**, 381086:1–381086.

58 Döring, T. and Beckhaus, S. (2007) The Card Box at Hand: exploring the potentials of a paper-based tangible interface for education and research in art history. Proceedings of Tangible and Embedded Interaction 2007 (TEI'07), ACM, pp. 87–90.

59 Marshall, P. (2007) Do tangible interfaces enhance learning? Proceedings of the 1st International Conference on Tangible and Embedded Interaction (TEI'07), ACM, New York, NY, pp. 163–170.

60 Follmer, S., Leithinger, D., Olwal, A., Hogge, A., and Ishii, H. (2013) inFORM: dynamic physical affordances and constraints through shape and object actuation. Proceedings of the 26th Annual ACM Symposium on User Interface Software and Technology (UIST '13), New York, NY, ACM, pp. 417–426.

61 Leithinger, D., Follmer, S., Olwal, A., and Ishii, H. (2014) Physical telepresence: shape capture and display for embodied, computer-mediated remote collaboration. Proceedings of the 27th Annual ACM Symposium on User Interface Software and Technology (UIST'14), New York, NY, ACM, pp. 461–470.

62 Hatscher, M. (2001) Joy of use: determinanten der Freude bei der Software-Nutzung, in *Mensch & Computer 2001: 1. Fachübergreifende Konferenz* (eds H. Oberquelle, R. Oppermann, and J. Krause) B. G. Teubner, Stuttgart, pp. 445–446.

63 Norman, D. (2002) Emotion & design: attractive things work better. *Interactions*, **9** (4), 36–42.

64 Hassenzahl, M. (2014) User experience and experience design, in *The Encyclopedia of Human–Computer Interaction*, 2nd edn (eds M. Soe-Gaard and R.F. DAM), The Interaction Design Foundation.

65 Sylvester, A., Döring, T., and Schmidt, A. (2010) Liquids, smoke, and soap bubbles: reflections on materials for ephemeral user interfaces. Proceedings of the Fourth International Conference on Tangible, Embedded, and Embodied Interaction (TEI'10), New York, NY, ACM, pp. 269–270.

66 Döring, T., Sylvester, A., and Schmidt, A. (2012) Exploring material-centered design concepts for tangible interaction. CHI'12 Extended Abstracts on Human Factors in Computing Systems (CHI EA '12), New York, NY, ACM, pp. 1523–1528.

67 Döring, T. (2017) A materials perspective on human–computer interaction: case studies on tangible, gestural, and ephemeral user interfaces. Doctoral dissertation, University of Bremen.

68 Doordan, D.P. (2003) On materials. *Design*, **19** (4), 3–8.

69 Norman, D. (1988) *The Design of Everyday Things*, Perseus Books.

70 Hornecker, E. (2012) Beyond affordance: tangibles' hybrid nature. Proceedings of the Sixth International Conference on Tangible, Embedded and Embodied Interaction (TEI'12), New York, NY, ACM, pp. 175–182.

71 Steels, L. (2006) Fifty years of AI: from symbols to embodiment – and back, in *50 Years of Artificial Intelligence*, Lecture Notes in Computer Science, Springer, pp. 18–28.

72 Bongard, J., Zykov, V., and Lipson, H. (2006) Resilient machines through continuous self-modeling. *Science*, **314** (5802), 1118–1121.

Index

a

accelerometer model   101
  for one axis   72
acoustic data transfer   356
acoustic emission (AE) techniques   546
ACROSOMA process   171
acrylate coating   177
acrylic sheet adhesive (Dupont LF0100)   142
activated carbon   454
active fiber composite (AFC)   201
active messaging   381
activity-transition graph (ATG)   394
  dynamic   395
actuators   34, 58
  concept   596
  functionality   204, 595
adapting control strategies   36
adapting system behavior   38
adaptive learning   332
adaptive path finding (routing)   416
adaptronic spindle system   600
ADA software language   344
A/D converter   84
additional subsystems   601
additive manufacturing (AM)   222, 243
  business cases   223
  controlled multimaterial capabilities   228
  generalized process sequence   218
  processes   217, 220, 236, 239
  sensors/functional structures   237
  rapid prototyping to rapid tooling   219
  sensor/electronics integration   217, 228, 233
    in additive manufacturing   224
    advent of   224
    case studies   230
    cavity-based sensor/electronic system integration   236
    demand for customization   224
    direct write (DW) techniques   224
    importance of smart products   224
    multiprocess hybrid manufacturing systems   239
    process overview   224
    single AM platform for structural electronics fabrication   243
  techniques   233
    cloud-based design and manufacturing scenario   231
adhesive films   177
ad hoc network   6
Aerosol Jet^{TM} printing   178, 240, 243
aerospace   57
Aerospace Recommended Practice (ARP)   6461 535
  guidelines   167
agent-based systems   388
  learning   415, 416

Material-Integrated Intelligent Systems: Technology and Applications, First Edition.
Edited by Stefan Bosse, Dirk Lehmhus, Walter Lang, and Matthias Busse.
© 2018 Wiley-VCH Verlag GmbH & Co. KGaA. Published 2018 by Wiley-VCH Verlag GmbH & Co. KGaA.

agent behaviors   394
  class   395
  data processing, features
    autonomy   387
    creation   388
    inheritance/replication   388
    interaction   388
    mobility   387
  models   391
    agents and objects   393
    communication   393
    computation   393
    mobility   393
    reconfiguration   393
    replication   393
  used to manage/distribute
    energy   444
agent characteristics   394
agent implementation, with pipelined
    multiprocess architecture   409
agent models   392
  computation models   389, 391
agent processing platform (APP)   20
  heterogeneous networks support   406
  JAM platform   411
  PAVM platform   410
  PAVM virtual machine, and code
      architecture   410
  PCSP platform   407
  program code frame   411
  programmable vs. application-specific
      platforms   406
  SEJAM/JAM simulator world   414
  technologies   404
agent programming models   397
Airbus A320   539
Airbus A350   22, 532
aircraft applications   202
AlgoBlocks   636
algorithmic design errors   343
algorithms/computational models   283
Alias frequency
  caused by undersampling   262
alkaline battery (Zn/MnO_2)   459
alkaline fuel cell (AFC)   468
aluminum tube
  piezoceramic ring integrated   202

ambient and pervasive networks
    (APN)   380
  heterogeneous network
      environment   381
ambient energy, types   480
  light energy   502
  mechanical   481
  radio-frequency electromagnetic
      energy   506
  thermal energy   496
ambient intelligence   32
Ambient Textile Information
    Displays   634
ambient vibration energy   494
Ambimax   450
Amdahl's law   292
AMF format   217
amplitude   110
analog data processing   264, 440
  quantizes   263
  signal to noise ratio (SNR)   273
analog signal processing (ASP)   16, 259,
    426
  sensor   259
analog-to-digital converters (ADC)   16,
    71, 259, 262, 337
  architectures   268
  basic architecture of   274
  circuits   426
  coarse and fine conversion   276
  dynamic range   267
  first-order sigma-delta   271
  ideal, transfer curve   264
  linear errors   265
  nonlinear errors   265
  normalized power dissipation   277
  power dissipation   267
  quantization   263
  range and resolution   264
  sampling   262
  second-order sigma-delta   272
  transfer curve   266
  zooming and moving window   278
analytic hierarchy process (AHP)   49
analyzing current situation   36
angular velocity   73
ANSI EIA/TIA-232-F   357

ANSI EIA/TIA-232 standard   357
antenna, monopole   356
aperture uncertainty   263
application severity   620
ARALL beam   201
ARM Cortex-R   584
Armijo rule   315
artificial intelligence (AI)   20, 553, 601
  concepts   20
ASTM, definition   217
ASTM F2792-12a   224
  defined   217
ASTM F2792, additive manufacturing
    processes   225
A2T abstraction tool   72
ATG-based agent programming
    language (AAPL)
  simulation of   410
  statements   398
atmospheric pressure CVD   94
Au electrode   95
autoclave   176
automated prepreg layup
    processes   173
automobile technology   31
automotive ethernet   358
autonomous agent systems   347

b

ball-screw drives   597
Banach/Hilbert spaces, infinite-
    dimensional   308
barely visible impact damage
    (BVID)   167
batteries   458
  based electronics   450
  capacity   463
  chemistries   459
battery manufacturing, three-
    dimensional
  based on template deposition   465
  integrated all-solid-state battery to be
    applied as energy supply unit
    to   466
BeagleBone   618
beginning-of-life (BoL)   9, 11
belief–desire–intention (BDI)   392

agent model   389
bendability   13
  bender geometries   122
  bending deformations   203
  bend loss coefficient   124
benzocyclobutene (BCB)   86
biaxial magnetoelectric generator   494
bidirectional communication   363
Big Data   385
binary coding   295
binary-weighted capacitor array   274
binder jetting processes   227
Biologic   632
bit error rate (BER)   359
3-bit parallel flash converter   275
black box   42
block diagram simulation   58
Bluetooth low energy (BLE)   359
boat's seaworthiness   619
body area network (BAN)   379
Boeing   787 532
  Dreamliner   22
Bosch processes   92
boundary element method (BEM)   58
boundary-value problem   305
broadcasting strategies   376
bulk micromachining process   96
bulk piezopolymers   488
bus request-acknowledge protocol   369

c

cables   109
camera-based computer games   630
cantilever-based gas sensor   100
cantilever structure   485
capacitive detection   101
capacitive e-skin, applied to different
    robots   579
capacitive technology   574
  sensing   99
  tactile sensors   574
capacitors   452, 453, 575
  classification of   453
  electrodes   101
  working principle   452
carbon-based materials   454
carbon fiber filaments   177

carbon fiber-reinforced polymers (CFRP) 167, 203
  laminates 181
  material 181
  online measured pressure values 186
  resonance frequency, Influence of 182
  structure 599
carbon monoxide 469
carbon nanotube (CNT) 501, 574
Cartesian direction 372
case studies 617
  composite bridge 622
    limitations and opportunities of sensorial materials 623
    potential application of sensorial materials 623
    sensors used 623
  leisure boats, life cycle of 617
    limitations and opportunities of sensorial materials 619
    potential application of sensorial materials 619
    sensors used 618
  model development and system-level simulation 71
  PROMISE-product life cycle management/information using smart embedded systems 620
  sensorial materials
    limitations/opportunities of 621
    potential application of 621
    uses 620
  sensor node for drift-free limb tracking 69
  system architecture 71
casting/molding processes 9
Cauchy stress tensor 302, 319
C4.5 decision trees 337
cellular/lattice gas automata 6
cellular networks 506
cellular polypropylene (PP) 489
centralized processing
  architectures 365
  sensor fusion 344
central neutral plane 178
central processing unit (CPU) 434

ceramic 462
channels 354
charge redistribution architecture using binary-weighted capacitor array 274
chatter suppression 601
chemical–electrical converter 459
chemical interaction 100
chemically amplified resist (CAR) 90
chemical vapor deposition (CVD) 87
chip package miniaturization technologies 140
  applications 143
  circuit integration
    embedding 142
    stacking 143
  ultrathin chip package technology (UTCP) 140
chip-to-chip communication 357
Cholesky factor, covariance matrix 308
circuit design 180
circular radio transmission ranges 365
circumferential strain 120
civil structures 116
cladding 109, 111, 123, 124, 125
clamping mechanisms 596
CLIP processes 236
closed-loop operation 87
  PLM applications 620
cloud-based design and manufacturing (CBDM) 22
cloud computing 18, 363, 385
  large-scale sensor network applications 385
  manufacturing 229
  multiple goals 385
  storage and computation 385
coarse conversion 276
code navigator 414
coefficient of thermal expansion (CTE) 12
cognitive functions 36, 430
  information processing 34
  processing 34
  regulation 34
  science 34
collaboration 387, 635

Collaborative Research Centre (CRC) 614 31, 38
combinatorial logic 294
common fault 347
common-mode rejection ratio (CMRR) 261, 262
common-mode voltage gain 262
communication 179, 294, 343, 387, 388, 389, 393, 426, 430, 614
  agents, interaction of 396
  channels 370
  failures 343, 344, 377
  hardware 351
    applications 351
    data throughput/latency 353
    efficiency 352
    electrical wires 354
    embedded, requirements for 352
    error protection 353
    maintainability 353
    material-integrated intelligent systems 351
    physical communication classes 354, 356
    polling *vs.* event-triggered data transfer 353
    robustness regarding environmental conditions 353
    security 353
    wireless electromagnetic transfer 355
  link failures 343
  networks 363, 377
  reliability 377
  robustness 377
  structures 347
  system 31, 34, 427
  yarns 156
communication unit (CMU) 259
comparative vacuum monitoring (CVM) 168, 546
  sensors, surface-attached 562
compensation circuit 87
complementary metal-oxide-semiconductor (CMOS) 13
  circuit designs 267
  compatible materials 486
  compatible microfabrication techniques 495
  compatible silicon-based microfabrication 491
  compatible technique 14, 17, 180, 426, 485
complex models, 3D visualization of 58
complex pattern recognition 287
composite manufacturing processes 171
composite materials 112, 113, 127, 161, 162, 168
  aerospace structures 107
  embedded FOS 114
  modern automobiles 112
  SHM, challenges 128
  surface-mounted FOS 115
computational complexity 572
computational efficiency 58
computational fabrication 638
computational requirements 582, 583, 584
  electrical impedance tomography 582
  tensorial kernel 583
computational technology 638
computation time 293
  bounds 293
computer-integrated manufacturing (CIM) 22
computer numerical control (CNC) 22
conceptual design 39
  subsystem level 42, 47
  system level 42
CONceptual design Specification technique for ENgineering of complex Systems (CONSENS) 39
conditioning 34
conductivity 86
  fillers 574
  polymer 574
ConPro language 344
contaminants
  organic and inorganic 89
continuous fiber-reinforced polymers 170

continuous liquid interface production (CLP) 227
core–cladding interface 109
corning SMF28e fiber 125
corrosion 457
Cortex-R7 processor 585
cosimulation 60
cost-benefit ratio 31, 44
CPU communication hardware 351
crack first sensors 620
crack growth rate 114
crack initiation 114
creativity
  tangibles 635
crystalline rare earth-iron alloy TbDyFe (Terfenol-D) 493
CSP-packaged ZL70102 144
cube networks 365
Cu conductors, polyimide flex 139
Curie temperatures 10, 484, 600
curing process 164
cut-through (Ct) 372
  message routing, store-and-forward 371
cyber-physical system (CPS) 19, 33, 390, 629
Czochralski process 88

d

damage detection system 545
damage estimation, of a technical structure 331
damage monitoring 535, 553
damage-tolerant design
  context 537
  strategy 561
Darcy's law 166
data acquisition
  physical model 302
  system 260
data analysis 110, 288
database management 58
data-centric approach 368
  directed diffusion algorithms 377
data communication 440
data converter architectures 268
  flash 275
  integrating 269
  low-power designs 276
  sigma-delta converters 270
  successive approximation 273
data delivery 387
data dependency classes of algorithms 294
data fusion 287, 344, 389
data latency 294
data mining process 287, 288
datapath selectors 295
data processing 179, 287, 387
  algorithm classes
    used with single and multiple sensor nodes 286
  architecture 293
  classes 285
  decentralized architectures 425
  errors 347
  factors impact on design/performance 292
  failures 343
  nodes 345
  parallel 287, 291
  pipeline 293
  in RTL architectures 295
data processing unit (DPU) 259
data quality 616
data reduction 289
data sinks 368
data storage 287
data throughput rate 352
data transmission latency 73
3D batteries, 3D architectures for 464
decimation 270
decision-level fusion 344
decision-making processes 344
decision rule (DR) learners 332
decision tree learner (DTL) 332, 334
decomposing computation 291
dedicated short-range communications (DSRC) 358
deep reactive ion etching (DRIE) 92
deformations 571
  deformable materials, 3D printing 633
  of FOLS 124

mechanical measures/CTE
    mismatch   14
  model   418
deposition techniques   93
  lift-off process   94
  wafer bonding   95
designs, moving window approach   277
diamond-like carbon (DLC)   597
dielectric analysis (DEA)   165, 167
dielectric capacitors   453
dielectric elastomers   489
dielectric materials   86, 452
dielectric systems   165
differential nonlinearity (DNL)   265
differential signal or differential pair (DS/DP)   354
digital analog converter (DAC)   270
  capacitors   274
digital circuit   426
  optimization strategies   426
digital data processing   426, 430, 440
digital filtering   270
digital front-end (DFE)   71
digital hardware   58
digital information   631
digital light processing (DLP)   227
  process   236, 244
digital logic   17
digital micromirror device   83
digital processing units   337
digital signal processing (DSP)   15
  circuits   84
digital storage (DS)   15, 259
direct component (DC)
  error   263
directed diffusion routing   376
direct print/cure (DPC)   244
  process   244, 245
Dirichlet boundary condition   319
discrepancy principle   310
disposable sensors   167
distributed learning   332
  event-based learning   333
distributed material-embedded sensor networks   19

distributed sensor network (DSN)   20, 378
  embedded sensor   19
domain-spanning conceptual design   39
  concept integration   47
  conceptual design, on (sub-)system level   46
  dependability, analysis of   48
  dynamical behavior, analysis of   48
  economic efficiency, analysis of   47
  evaluate the solution   50
  phases   41
  planning/clarifying task   42
domain-specific design
  development   39
  tools   41
domestic/personal devices   633
dopant diffusion   10
doping   109
3D pressure/force sensing   574
3D printers   111
2.5D rigid thermoplastic circuits   152
dry etching   92
dual-core architecture   585
Dual Logo   357
dual process material jetting platforms   236
dual-slope integrating architecture   269
DuPont™ Kapton   182
DuraAct™ modules   560
dynamic activity-transition graphs   394
dynamic ad hoc network topologies   370
dynamic current   426
dynamic incremental learning   332
dynamic power management (DPM)   434
  CPU-centric DPM   434, 435
    EDF algorithm   435
    LEDF algorithm   435
  EDS algorithm   437
  I/O-centric DPM   434, 437
    LEDES algorithm   439
dynamic redundancy   347

e

ebeam lithography   89
ECAD tool   41
EC FP7 TERASEL project   153
eco-friendly hybrid composites   480
economic efficiency model   31, 32, 43, 44
EDA tools   56, 59
EEG monitors   380
EEPROM system   143
  chips   144
  memory device   143
  in single UTCP   145
effective number of bit (ENOB)   267, 270
elastic circuits   145
  printed circuit board   145
  stretchable displays   149
  thin film metal   148
  wearable light therapy   148
elastomer composites   574
elastomer membrane   489
electrical communication   356
electrical connections   137
electrical energy   451, 452
electrical impedance tomography (EIT)   572, 582
  algorithmic steps   582
electrical interconnections   154
electrical resistivity   165
electric cars   113
electric charge, from direct piezoelectric effect   209
electric conductivity   454
electric insulator   452
electricity generating performance   201
electroactive polymer (EAP)   489
  films   489
electrochemical capacitors   453
electrochemical stability   457
electrochemical storage
  devices   459
  units   454
electrode layer thickness   455
electrode materials   454
electrolyte materials   457
electrolytic capacitors   453
electromagnetic (EM)   506
  based communication methods   356
  energy   449, 450
  induction   451, 493
  noise   110
  sensors   550
  vibration energy harvester, schematic representation   492
electromagnetic compatibility (EMC)   552
electromagnetic interference (EMI)   353, 574, 575
electromechanical conversion   489
electromechanical coupling   485
electromechanical transduction   481, 490
electronic double-layer capacitor (EDLC)   454
electronics
  circuits   146, 571
    design level   59
  conductivities   463
  cosimulation of   60
  devices   452, 506
  failures   343
  functionality   76
  integration   179
  materials integration of   180
  modules, conductive yarns   155
  printing   629
  skin (e-skin)   571
electroplating process   87
electrorheologic fluid (ERF)   600
  applications of   600
electrostatic sensing   58
electrostatic vibration energy harvesters, comparison of   491
electrostrictive polymers   489
ELTRAN   88
embedded-AMS (E-AMS)   60
embedded systems   59
  design   58
  electronics   180
  embodied interaction   630
  networks   413

schematic overview   142
sensor   623
energy-aware systems
  address minimization of energy consumption   425
  aims   425
Energy-aware Algorithmic Scaling   17
energy cells   461
energy consumers   432
energy consumption   36, 425, 428, 449
  by transmitting data, using wireless communication   440
energy converters   432
energy density   454, 468
energy distribution   445
energy efficiency   376
energy generation performance   481
energy harvesting (EH)   431, 449, 451, 479, 480, 595
  batteries, technologies   479
  device   431
  mechanical   451
  thermal   451
energy management (EM) units   259, 425, 426, 430, 431
energy scavenging   450, 451, 479
energy sources   431
energy storage (ES)   16, 259, 432, 450, 452
  devices   454
  historical background   452
energy supply
  decentralized architectures   425
  and management   21
energy transmission   551, 601
entertainment   635
ε-entropy   334, 335
  computation   335
environmental temperature   74
epoxy-based structural adhesive   205
equivalent series resistance (ESR)   454
Erichsen testing machine   206
error detection or error correction codes (EDC/ECC)   353
e-skin system   572
etching   91
ethernet   357

Open Alliance BroadR-Reach™ (OABR)   357
Euclidean space   302
Euler angles   72, 74, 75
  estimation   75, 76
event and distributed agent-based learning of noisy sensor data   416
event-based sensor data distribution strategies   375
event-based sensor distribution behavior   416
Everlast   450
experiential embodiment   631
explicit interaction   638
explicit minimizers, for linear models   311
extraordinary cycle life   454
extravehicular activity (EVA)   359
extrinsic Fabry–Pérot interferometer (EFPI)   108, 111, 118
  operation   118
  sensors, sensitivities of   119

f

fabrication
  of sheet metal compounds   204
Fabry–Pérot sensors   118
failure mode and criticality analysis (FMCA)   48
failure mode and effect analysis (FMEA)   48
failure mode, effect, and failure cost analysis (FMEFA)   48
fault tolerance   430
fault tree analysis (FTA)   48
feature extraction   344
feature-level fusion   344
feedback controller   601
ferroelectric nanowires   486
ferroelectric piezoceramics   484
ferroelectric polymers   574
fetch–execute cycle   291
fiber Bragg grating (FBG)   108, 119, 176, 542
  hydrostatic pressure sensor   114
  in optical fibers   548

fiber Bragg grating (FBG) (continued)
  sensors 119, 177
    optical, sensing principle of 117
    strain 110
    thermal 602
fiber–metal laminates (FML) 8, 201
fiber microbend sensors 122
fiber-optic loop sensor (FOLS) 108, 123
fiber-optic sensor (FOS) 107, 110, 166, 178, 179, 187, 549
  for structural monitoring 112, 115
    aerospace structures 115
    technology for SHM applications 107
fiber orientation 171
fiber permeability 166
fiber placement 175
fiber-reinforced composites 112, 201
fiber-reinforced polymer (FRP) 161, 164, 192
  composite production 166, 170, 171
fiber-reinforced structures 203
fiber volume content 162
field-driven actuators 600
field-effect transistors (FET) 17
Field Programmable Gate Arrays (FPGA) 17
filament winding process 171, 175
finite-dimensional vector space 308
  linear vector space 311
finite element analysis 120
finite element method (FEM) 57, 583
  simulations 12, 59, 337, 417
finite element model-based simulation (FEM) 285
finite state machine (FSM) 287, 295
fixed-point iteration relies, on scalar function 312
Flame Retardant 4 (FR4) 137
Flash ADCs 275
FlexCase 635
flex foil manufacturing technologies 145
flexibility 14
  circuits 138
  electronics 148
  printed circuits 180
  PVDF-based piezoelectric-pyroelectric-photovoltaic hybrid cell 502
  schematic representation 498
flexible circuit board (FCB) 141
flexible printed circuit board (FPCB) 575
flexoelectric effect 487
flexographic printing 111
flits 372
fluoropolymers 489
foglets 6
foil-based sensor 110
foil technology 180
Follow the Grass prototype 634
force–time graphs 127
FORTH programming 296
FPGA implementations 367
Fraunhofer Institute for Manufacturing Technology and Advanced Materials (IFAM) 238
Fréchet differentiability 314
free-space optical communication (FSO) 360
frequency modulation sensors 117
  Bragg grating sensors 117
  Fabry–Pérot interferometer sensor 118
fresnel reflection 187
fuel cells 467, 468
fully interpenetrating nanoscale battery, steps needed to assemble 467
functionality
  characterization of 208
  integrated structures 601
  nets 169
functionally graded materials (FGM) 222
fundamental concepts, schematic representation 13
fused deposition modeling (FDM) systems
  polymer material 242
fusion architectures 344, 345
fusion process, levels 344

g

Galfenol 493
galvanic cells 458
gap-closing converters 491
gap-closing electrostatic generator, schematic representation 490
gas sensors 100
gate-level simulation 428
Gaussian elimination 311
Gauss–Newton method 318
generalized minimal residual (GMRES) method 311
generation-to-generation product optimization 229
generic hypercube networks 365
GigaDock 360
GLARE® 201
glass fibers 128
  composite structures 110
glass transition temperature 176
glass-type lithium phosphorous oxynitride (LiPON) 462
global monitoring 168
GPS-based solutions 556
graded-index fibers 109
gradient descent schemes 314
graphene 454
grid intelligence 430
GSM900 base station 506
Guidelines for the Implementation of Structural Health Monitoring on Fixed Wing Aircraft 535
Gustafson's law 292
gyroscope 73

h

half-Heusler compounds 500
Hanke–Raus rule 316
hardware description language (HDL) 60
harvesters 451
hearing aids 144
heart rate monitoring 144
heat conduction 305
  inverse 307
heat flux reversal 11
heating coil 88
heat losses 596
helicopter fuselage 548
Heliomote 450
hertz contact theory 120
heterogeneous systems 389
  components/subsystems 57
hierarchical networks 365
HIFSuite tools 72
high frequency (HF) 181
high-k fillers 574
high-level synthesis (HLS) 20
high-quality standard 165
high-temperature electronic (HTE) 10
high-temperature fuel cells 469
host architectures 389
hot embossing process 111
human-computer interaction (HCI) 629, 630, 638
  applications/scenarios 633
  challenges 637
  design methodology 59
  integrates 630
  new forms 630
  with novel and advanced materials 629
  research 629
  soap bubble interface 637
  third wave of 633
human culture 630
human intelligence 630
human-machine interaction 24
human-machine interface 34, 46
human nervous system 3
human–robot interaction (HRI) 579
humidity sensors 83, 84
HW/SW systems 60
hybrid capacitors (HCs) 454
hybrid data processing approach 287
hybrid drives 596
hybrid gap-closing/overlap configurations 491
hybrid integration approach 6
hybrid manufacturing system
  building, functionalization, and secondary processes 243
Hybrid mechanical energy harvesting 493

hybrid solar/vibration energy harvester
  schematic representation of   503
hydrated ionomeric membrane   489
hydraulic transmission systems   597
hydrofluoric acid   86
hysteresis   573, 575

i

iCub robot   575
  humanoid robot   575
ideal S/H circuit, output of   262
IDTexEx report   228
IEEE-488   356
IEEE 1609   359
IEEE 488/GPIB protocol   369
IEEE 802.11p   358
IEEE standardized Bluetooth   359
IEEE 802.11 standards   359
image analysis   288
image processing, parallelized   388
image sensors   83
IMEC, thermoformed lighting module   153
implicit interaction   638
incremental learning   332
inductive coupling   179
industrial agents, behavior model classes
  deliberative agents   391
  hybrid agents   392
  reactive agents   391
industrial exploitation   56
industrialized society, development stages   23
industrial manufacturing environments   603
industrial, scientific, and medical (ISM) bands   181
Industry 4.0   604
inelastic material behavior   207
iNEMO Engine Lite   72
iNEMO M1 System-on-Board   70
infinite-dimensional theory   308
information communication technology (ICT)   16
information, levels of   281
  load monitoring   282

structural control, adaptive and morphing structures   282
structural monitoring
  control   281
  health   282
  tactile sensing   283
information processing   34
  methodologies   17
  unit   34
inFORM platform   636
Infrared Data Association (IrDA)   360
Infrared Link Management Protocol (IrLMP)   360
Infrastructure-as-a-Service (IaaS)   229
inkjet printing   111
  microstructures   456
  nozzles   83
inkjet printing module (IPM)   240
IN–OUT   272
in-process sensor integration   230
Institute for Microsensors, -Actuators and -Systems (IMSAS)   190
  embedded microscale capacitive sensor   190
instrumentation amplifier   261
  for high input impedance   261
integral nonlinearity   267
integrated circuit (IC)   83, 180, 357
  design   56, 60
  devices   85
  packaging technologies   56
Integrated Project (IP)   56
integrating capacitor
  vs. time, charge of   269
intelligent materials   595
  clamping system   599
  in manufacturing technology applications   595
  mechatronic systems   34
intelligent networked systems   34
Intelligent Product   617
intelligent sensing systems
  modules contributing   20
intelligent/smart materials   595
intelligent technical systems   31, 32, 34, 601, 602
  characteristics   33, 34

adaptive   33
anticipative   33
robust   33
user-friendly   33
design methodology for   38
  domain-spanning conceptual design   41
  domain-specific conceptual design   50
  miniaturization of electronics   32
  networking of information systems   32
  software technology, as driver for innovations   32
intensity modulation sensors   110, 122
  fiber microbend sensors   122
  fiber-optic loop sensor   123
interaction model   389
interdigital planar capacitive sensors   188
  complex relative permittivity of resin   188
  in'situ measuring of the resistivity ρ of resin by   191
  relative permittivity and the loss factor for a capacitor with parallel plate   188
interdigitated array   188
interdigitated electrode (IDE)   177
  limitation   484
interferometric sensors   547
internal objectives   36
international technology roadmap   450
Internet   57, 413
  clouds, material-integrated intelligent systems   18
  Internet-of-Everything   20, 24
  Internet-of-Thing (IoT)   16, 140, 386, 387, 629
interprocess communication (IPC)   291
InteRRap architecture   392
intrinsic/extrinsic fiber optic sensor principles   108
intrinsic sensors   108
inverse heatconduction   305
inverse numerical approaches   287, 301

inversion problem   305
ion beam lithography   89
ion conductivity   165, 457
ionic polymer metal composite (IPMC)   481
  mechanical-to-electrical energy conversion   489
ion viscosity   189
  vs. degree of resin cure   189
iPeM-Modell   39
iPhone 6   449
ISIS Sensorial Materials Scientific Centre   24
isotropic/anisotropic wet etching   91
iterative regularization schemes   313

j

JAM AgentJS processing platform   413
JavaScript (JS)   389
  software language   344
Java software language   344
jetting testbed module (JTM)   240
JSON-like tree presentation   414
junction-based routing   371

k

Kalman filter   75
  algorithm   72
Kalman noise covariance matrices   74
Kapton PCB   185
KASPAR humanoid robot   578
Keysight SystemVue environment   72
k-Nearest-neighbor   337
  classificator   337
Kozeny–Carman equation   166

l

Lagrange equation   311
Lamb waves   168, 547, 560
Lamé parameters   319
laminar piezomodules   203
laminates   8, 162
Landweber iteration   315, 316, 320, 321, 322, 323, 325, 326
  discrepancy principle   326
  linear and nonlinear models   316
  for load monitoring   319

Landweber regularization 316, 326
laser beam melting 239
latter matrix–vector equation 311
lead zirconate titanate (PZT) 201
  nanofibers, nanogenerator 487
  sensor 10
    single-patch 550
learning 635
  algorithm 332
  of data streams 331
  with noise 333
  rewarded behavior 390
  tangibles 635
least significant bit (LSB) 263, 264
Leclanché cell 459
LEDs 634
  engine 149, 150
  islands 151
Lego Mindstormes™ 635
Lepskii's balancing principle 316, 318
Levenberg–Marquardt iteration 317, 318
LF0100 adhesive 142
life cycle costs 44
Li-Fi Consortium 360
lift-off process 94
light energy harvester (LEH) 502
  operation principle, schematic representation of 503
light modulations 108, 110
light transmitter 356
Li-ion technology 460
  battery manufacturing 461
  microbatteries 462
limb movement recordings 73
limb position estimation algorithms 73
linear actuation 600
linear errors 265
linear models, explicit minimizers 311
liquid composite manufacturing (LCM) 163
lithium-ion technology 462
lithography 88
  development 90
  exposure 90
  fabrications technique 456
  photoresist coating, and baking 89
  substrate preparation 89
Living Wall 633
  interactive wallpaper 634
load-bearing structures 5
load–execute–store model 291
loading–unloading cycles 123
load monitoring (LM) 20
  algorithmic and information level hierarchy 282
  Landweber iteration 319
load reconstruction
  numerical examples 318
load scenarios R250
  normalized capacitance and computed residual performance 211
load strain, fiber 209
local information processing 389
LogitBoost classifier 583, 584
long fiber-reinforced polymers 170
lookup table (LUT) 336
loop sensors 127
low-energy device schedule (LEDES) algorithm 439
low-power processors 427
low-power smart sensor systems, design 426
low-pressure CVD 94
low-temperature fuel cells 469
low-voltage differential signaling (LVDS) 355

m

MAC address 380
machine learning (ML) 21, 287, 329, 331, 336, 415, 572
  algorithms 332
  association learning 330
  classifier systems 330
  clustering 331
  regression 331
  reinforcement learning 330
  semisupervised learning 330
  software language 344
  supervised learning 329
  techniques 288
  unsupervised learning 329

machine tool
  applications   595
  components   595
Mach-Zehnder interferometer   547
macrofiber composite (MFC)   201
  integrated rectangular deep drawn cup   213
macro/mesoscale generators   485
macroscopic systems   83
  mechanical properties   178
magnet/coil assembly   482
magnetic field forces   600
magnetic shape memory alloy (MSMA)   493
magnetization   451
magnetometer   83
magnetorheologic fluid (MRF)   600, 601
  damping systems   601
  gripper utilizing controllable wet adhesion   601
  shock absorbers   601
magnetostrictive materials   451
  applications of   598
Maintenance on Demand (MoD)   532
mandrel   177
mapping function   329, 391, 638
Marble Answering Machine   633, 636
Marcuse models   126
mass flow sensor   97
mass-spring system   101
material appreciation   638
material extrusion transfers   227
material-integrated systems   56, 76, 171
  intelligent systems   168, 228
    fundamental elements   7
  material-integrated sensing system (MISS)   386
    clouds   386
material jetting   227
MATLAB framework   76, 430
MATLAB-SIMULINK approach   58
MATLAB tool   41
matrix materials, used in fiber-reinforced polymers   173
max. comfort   42, 43

max. dependability   42
max. driving speed   42
max. energy efficiency   36
max. safety   36, 42
max. user satisfaction   42
MCAD tool   41
mean time to failure (MTTF)   346
mechanical compliance   13
mechanical components   36
mechanical energy   450, 451
mechanical energy harvester (MEH)   480–496, 481, 487, 492
  classification of   481
  devices   482
  electromagnetic micropower generators   491
  electrostatic micropower generators   490
  hybrid micropower generators   493
  main transduction mechanisms, comparison of   483
  micropower generators based on electroactive polymers (EAP)   489
  piezoelectric micropower generators   482
  triboelectric nanogenerators   492
  wideband/nonlinear micropower generators   494
mechanical stability   112
mechanical-to-electrical energy conversion   481, 486
mechatronic system   31, 32, 33, 34, 36
mega sample per second (MSPS)   273
MEMS gyroscope, one-axis   103
mental models   639
mesh-like networks   375
  redundancy   366
  two-dimensional   365
mesh sensor network
  two-dimensional   376
message typical format   370
message-based communication   378
message formats   370
message passing   363, 370
message passing network   368
metallic AM part   238

metalorganic chemical vapor
    deposition   94
methanol   469
MFC
  actuator function   209
  applied
  foam core   202
  elastic and inelastic parameters   207
  Gauss curvature   207
  load regime with loading/unloading
    phase   207, 210
  location detection   203
M8528P1
  bending specimen   204
  performance degradation   210
microbend deformer geometries   122
microbending optical fiber sensors   123
microbend sensors   123
microbial fuel cells   469
microchannel plate (MCP)   464
microchip-level sensor network
    designs   367
microchip resources   367
microcombustors   470
microelectromechanical system
    (MEMS)   55, 450, 602
  acceleration sensor   602
  accelerometer   83
  capacitive humidity sensor   100
  gyroscopes   83, 102
  micro/MEMS-scale generators   485
  microphones   83
  power supplies for   450
  scale piezoelectric VEH   486
    with clamped-clamped multibeam
      configuration   484
  sensor   8, 84, 95
    accelerometer   101
    cantilever-based gas sensor   99
    capacitive sensors   99
    gyroscope   102
    humidity sensor   99
    piezoresistive sensor   97
    resistive sensors   95
    sensing quantity   85
    thermoresistive sensor   96
    working principle   84

  specific physics   58
  subsystem   77
  systems   112
  technology   110
  thermoresistive sensor   97
  three-axis MEMS accelerometer   102
  transducers   74, 84
  validated MEMS IP design   60
microelectronics   449, 453
microfabrication technologies   84, 87
  silicon wafers   87
microheater   97
micromachining   84
  surface accelerometer
    structures   602
  surface and bulk   95
  technologies   83, 99
micro-nano scratching processes   602
micropower generator   479
microscale devices
  construction of   455
  interdigital sensor   190
  planar capacitive sensor   190
  sensors   112
  supercapacitors   454
microsupercapacitors   454, 457
  materials in   456
microTEGs   497
middle-of-life (MoL)   9, 11
miniaturization   55, 107, 110, 111,
    112
  sensor   112
Miniaturized Smart Systems   57
MinIMU-9 v2 618
MIT Media Lab   631, 634
mobile ad hoc networks   365
mobile agents
  autonomous behavior model   388
  based computation   386
  layers occupied by   389
modal analysis   168
model-based computing   301
modeling errors   304
model order reduction (MOR)   58
modern automobiles, large-scale
    composite parts in   113
modified vacuum infusion process   173

modular manufacturing platform
    (MMP)   239
  functionalization processes   241
M-of-N independent modules   347
M-of-N system designs   347
molded interconnect device (MID)
    technique   229
molten carbonate fuel cell (MCFC)   468
monocrystalline silicon seed   88
Monte Carlo simulation   420
Moore's law   4, 18
Morozov's discrepancy principle   316
MOSFET   261
most significant bit (MSB)   276
  capacitor   275
motoric skills   34
MPPT design   450
multiagent system (MAS)   20, 387
  mobile autonomous agents   388
multiband/wideband RF energy
    harvesters   507
multicriteria sensor networks   603
multidomain cosimulation
    approach   59
multidomain modeling approaches   60
multidomain simulation and
    optimization   57
multijet modeling (MJM)   244
multilevel design methodology   60
multimode
  optical fiber   109
  single-mode fibers   109
multiphysics models   75
multiple bolted joint, health
    monitoring   212
multiple input/multiple output
    (MIMO)   359
multiple optical fibers   109
multiprocess hybrid manufacturing
    systems   236
multiprogram system   389
multipurpose sensor   618
multi-RTL architectures, parallel   287
multisensor   344
  computations   286
  data fusion system   344
  preprocessing chains   344

sensory, inherent integration   603
multiwall carbon nanotubes
    (MWCNTs)   454, 602
MUNED WiCkeD RSM model   72

n

NanoCleave   88
nanoflexoelectricity   487
nanoimprint lithography   111
nanorods, template deposition   465
nanoscale
  generators   486
  sensors   112
NAO humanoid robot   578
natural interaction   630
navigation system   101
nearest-neighborhood decision tree
    learning   334
network configuration   364
network connectivity graph (NCG)   363
network domains   381
networking   34, 575
  and communication   17
  protocol
    defines   366
    redundancy   365
  switch   368
  topologies   364, 365
    common   364
    logical   365
Netzsch/Lambient Technologies
    LLC   191
neural networks   331, 337
  algorithms   331
  approaches   288
neuromusculoskeletal system   73
Newton methods   58, 312, 318
  type regularization schemes   317
Ni/Au finish, printed circuit board   138
nickel–metal hydride (NiMH)   459
noise
  absolute level   304
  data   304
    coping with   306
    simulated   323
  interference parameters   352
  shaping   270

nonadditive manufacturing processes 229
noncognitive regulation 34
nondestructive evaluation and testing (NDE/NDT) methods 170, 532
nondestructive testing 170
  material's behavior 301
nondigital domain 56
nonfunctional signals 74
nongenerative energy 450
  supply 450
nonhardware services 57
nonlinear error 266
nonlinear Landweber iteration 314
nonphotosensitive polyimide PI2611 141
nonplanar surface functionalization
  modular manufacturing platform 240
nonpolynomial (NP)
  complexity class 329
  problem 425
n sensor values 283
N-version system 347
NXP PHY devices 358
Nyquist criteria 271

o

Ohmic contact 87
Ohm's law 98
OLED lighting devices 228
Olympus Stylus camera 139
omission faults 343
one-pair ethernet (OPEN) alliance 357
online process monitoring 165
OpenDSS simulator 59
OpenSim 75
  environment 73
  online database 73
  open-source platform 73
open systems interconnection (OSI) 367
  layer model 367
operating system (OS) 296, 389, 434
operational amplifier (OPAMP) 260
  circuits 261
operational monitoring systems 545

OPNET simulator 59
optical-based tactile sensors 574
optical communication hardware 360
optical fibers 109, 119, 176, 177
  methods 167
  microbend sensor 122
  refractometers 167
  sensors 111
  types of 108
optical measurements 165
optical sensors 110, 112, 128, 180
  tactile 574
  WGM 119
organic printed/conventional silicon microelectronics
  qualitative comparison 14
organic surface protection (OSP) 137
oversampling 270
oxidation rate 469
oxygen supply 227

p

packaging materials 140
panoscopic approach 500
parallelization rules 293
parallel system 346
parameter identification, physical models 302
partial differential equation (PDE) 57
passive messaging 387
passive monitoring 168
passive wireless power transfer 181
path finding 363, 371
  algorithms 364
path redundancy 365
path selectors 295
pattern-level fusion 344
pattern recognition 344
Pavlov's dogs 34
PCSP platform architecture 412
  PAVM agent processing platforms 407
peer-to-peer networks 375
performance attributes 44
permeability 165
perovskite-type solar cells 451

personal area networks (PAN)   379
  heterogeneous network
    environment   381
Petri net   428
phase change material   596
phone conversation   83
phosphoric acid   86
phosphoric acid fuel cell (PAFC)   468
photodetector   110, 112
photodiode   110
photolithography technique   89, 111
photon flux   112
photoresistive sensor   95
photoresist mask   142
photovoltaic (PV) cell   504
  dye-sensitized solar cell (DSSC)   504
  first generation   504
  hybrid organic-inorganic perovskite
    cells   504
  organic (or polymer) solar cells
    (OSCs)   505
  second generation   504
  third generation   504
photovoltaic effect   502
physical node structures, mapping of
    logical on   406
physical vapor deposition (PVD)   87
pick-and-place system   242
piezoactuators   597
piezo-based device   597
piezoceramics   596, 597
  actuators possesses
    displacement behavior   597
  elongation-to-size ratio   596
  fibers   202
  foils, prepackaged   201
  materials   211
  ring, with steel end caps   202
piezocomposite generating element
    (PCGE)   201
piezoelectric (PZT)
  ceramics   596
    applications   596
  effects   58, 168
  energy harvesters   481
  generators   482
  materials   10, 451
    methods   166
  nanocomposites   488, 496
  nanogenerators   486
  nanoparticle-polymer composite
    foam   488
  nanostructures   486
  patches   556
  polymers   487
  sensors   166, 180, 574
    arrays   168
    tactile for 3D force sensing   574
  transducers   202
    elements   202
  transduction   482
  triboelectric generators   493
  VEHs, fabrication methods   485
  vibration energy harvesters
    operating modes of   483
piezoelectric vibration energy harvesters
    (P-VEH)   482
  cantilever-type   485
  ferroelectric piezoceramics   482
  nonferroelectric crystalline
    materials   482
piezoelectric wafer active sensor
    (PWAS)   168, 181, 547
piezofluidic actuator   597
piezometal compounds
  fabrication of   205
  forming of   205
piezomodules   204
piezopatch transducers   597
  sensory elements   599
piezoresistive effect   98
piezoresistive pressure sensor   98
piezoresistive strain gauges   98, 547
  strain sensors   548
piezoresistive thin layer sensors,
    investigation   598
piezoresistors   98, 101
pipelined multi-VM architecture
    411
Piranha solution   89
pitch-catch mode   168
planning   34
  clarifying the task   42
plasma-enhanced CVD   94

plastic optical fibers (POF) 574
Poisson's ration 319
polarization 110
polyacetylene (PA) 501
polyamide (PA) 162, 488
polyaniline (PANI) 501
  material 457
polycarbonate (PC) 153
polydimethyl siloxane (PDMS) 146
polyetheretherketone (PEEK) 162
polyethylene naphthalate (PEN) 139
polyethylene terephthalate (PET) 139
polyimide (PI) 139, 141, 177
  membrane (HD4110) 141
PolyJet 243
  standard materials 244
  technology 244
polymer-based sensor foils 110
polymer fibers 547
  fiber-Bragg grating-type sensors 176
  optical fibers 176
polymer waveguides 180
  based sensors 574
polymethyl methacrylate (PMMA) 120
polyolefin foil, inner-bonded 203
polyphenylene sulfide (PPS) 162
polypropylene (PP) 162
polypyrrole (PPy) 501
  electrode material 456
  material 457
polysilicon 88
  ingot 88
polystyrene (PS) 153
polythiophene (PTH) 501
  material 457
porous material 166
positioning systems 596
postexposure bake (PEB) 90
powder bed fusion 227
power analysis workflow 428, 429
power cells 461
power conditioning circuit 479
power consumption 426, 427, 430, 575
power conversion efficiency (PCE) 504
power dissipation 277, 427
powering intelligent systems, routes
    of 450

power management 427
power minimization techniques 359
power model of sensor node, using
    toolbox components 433
power optimization 432
power saving
  benefit for 436
  on device level at runtime
    address 429
predicted load positions 338, 339
prediction (analysis and classification)
    algorithm 336
preforming processes (dry fibers) 171
prepreg processes 173, 175
pressure-sensitive silicone adhesive
    (PSA) 152
pressure sensors 184
preventive maintenance 540
  diagnostic (fault detection) 540
  prognostic (fault prevision) 540
principle data processing, with register-
    transfer architectures 294
principle energy harvesting
    architecture 431
principle learning and classification
    algorithms 335
principle sensor node architecture, with
    energy harvesting 442
printed circuit board (PCB)
  based stretchable circuit
    technology 148
  components assembly 137
  fabrication 138, 140
  manufacturing 143, 147
  structuring to thin film-based
    processing 148
printing
  components 111
  electronics 629
probability of detection (PoD) 170, 552
process-dependent size 233
processing element (PE) 379
processing layers, of distributed sensing
    application 389
  aggregation 389
  application 389
  sensing 389

Product Avatar   616, 617
product data management (PDM)   614
  common graphical representations   614
    begin-of-life (BOL)   614
    end-of-life (EOL)   614
    middle-of-life (MOL)   614
product embedded information devices (PEIDs)   613
  classification   621
product life cycle management (PLM)   613, 614
  based improvement   619
  closed-loop and item-level   615
  data and information in   615
  engineering   614
  supporting concepts for data and information integration in   616
programmable matter stress, definitions of   6
projection-based methods   58
project management   39
Prometheus   450
proportional-integral-differential (PID)   21
protocol-based routing, in changing networks   370
proton-exchange membrane fuel cell (PEMFC)   468
  advantages   469
  component   469
prototype boat, for sensor integration and testing   618
pseudocapacitive material   458
pseudocapacitors (PC)   454
pulse-echo mode   168
pultrusion processes   171, 174
punch force *vs.* displacement   208
PVDF-based polymers   502
pyroelectric energy generation (PEG)   501
pyroelectric materials   501
PZT. *see* lead zirconate titanate (PZT)

q
Q-learning   390
quad-core Intel Core i7 - 2670QM   75
quad flatpack (QFP)   143

quality control data   169
quality-of-service (QoS)   21, 416
quantization noise approaches   272
quartz capillary   119
quasi-isotropic (QI)   162

r
race conditions   343
radiation energy harvesters   451
radiation harvesters
  light energy harvesters (LEH)   502
  radio frequency   506
radical atoms   632
radio frequency (RF)   83, 352, 355, 506
  combining configuration   507
  energy harvester   506
  energy harvesting   506
  identification (RFID)   179
  RF-DC conversion efficiency   507
  waves   451
radio frequency identification (RFID)   179
  integration   236
  systems   239
    material-integrated   238
  technology   179
  transponder   238
radio waves   181
RailCab   49
raw sensor   572
  level fusion   344
reactive ion etching (RIE)   92
real microbatteries   463
real-time capability   353
  data processing platforms   287
real-time monitoring, of resin cure   191
real-time signal processing systems   291
reconfigurable (dynamic) activity-transition graphs (DATG)   392
rectangular cup, with double curvature deep-drawing   206
reduced-order model (ROM)   58, 60
redundancy, in networks   365
reflection coefficient   181
refractive index (RI)   109, 111
region of interest (ROI)   417
register-based machines   296

registers, and addressable cell memories (RAM)   295
register-transfer level architectures   294
regression algorithms   331
regularization parameter
  rules for choice   309
reinforcing fibers, used in fiber-reinforced polymers   172
relaxation parameter   308
reliability   17, 343, 344, 416
  in distributed systems   378
  reliability R 346
remote procedure call (RPC)   367
Renner model   126
representational embodiment   631
representative volume element (RVE)
  MFC homogenization approach   207
representative volume element (RVF)   206
residual stresses   14
resin cure
  measurement   184
  monitoring   166
resin flow, monitoring of   166
resin front by means of simulation, analytical modeling   166
resin properties   165
resin transfer infusion (RTI)   163
  process   164
resin transfer molding (RTM)   163, 174
  classical process   164
  processes   163, 164, 174
resistance-based systems   165
resistive sensors   277
resistivity drop   156
resolution   344
  corresponding values   265
resonance frequency   84
reusable sensors   167
reusable tool mount sensor   189
rigid multitouch devices   635
ROBOSKIN e-skin system
  capacitive e-skin system   575
  integration into iCub   578
robotic platform   575
robotic tactile sensing system   579

robust system design, on system level   343, 344, 345
  distributed neighborhood detection   347
  dynamic self-organizing systems   348
  M-of-N systems   346
  resource failures   379
  version systems   347
routing   370
  adaptive Δ-Distance with Backtracking   374
  agent-based   376
  communication failures   377
  cut-through and wormhole   372
  data-centric and event-based   375
  decisions   376
  Δ-distance   372
  geographic   371
  in large-scale wireless sensor networks   371
  store and forward   371
  strategy   375
  tables   364
R-phase transformation   596
R2R printing   507
RS-232   357
RSI wrap tear-down   150
RTL architectures
  with datapath pipelining   294
  data processing   287, 295
runtime reconfiguration   379

S

SaaS. see sensors as a service (SaaS); Software-as-a-Service (SaaS)
SAE ARP6461   535, 536
sample-and-hold (S/H)   262
  circuit, typical errors   263
SARISTU project   560
sata acquisition systems   259
scanning electron microscope (SEM) image   162, 163
scanning probe lithography   89
scientific data analysis used in data mining processes   288
scientific data mining   287
  correlation techniques   289

autocorrelation function   289
   cross-correlation analysis   290
   discrete correlation function   289
feature selection   289
feature transformation   290
screen printing module (SPM)   240
Seebeck effect   451, 497
Seiko AGS Kinetic quartz watch   491
SEJAM MAS simulator   417
SEJAM simulation   415
selective laser melting (SLM)   238, 456
  SLS from polymer   222
self-aligning mask   141
self-optimization process   31, 32, 36, 38
  functionality   46
self-organizing agent systems   416
self-organizing capabilities   19
self-organizing system (SoS)   20
  MAS methods   555
self-resonance   182
semantic mediator   616, 617
semiconductors   450
  materials   98
  piezoresistive materials   98
  semiconducting stabilizers   501
semifinished fabrics   175
SEM imaging. *see* scanning electron microscope (SEM) image
sensing   380
  capabilities   602
  heterogeneous   363
  instrumentation   119
  materials properties   5
  principles   108
sensorial materials   7, 619, 624
  embedding a sensor node   431
  potential, in PLM application   623
sensorization, modular manufacturing platform   240
sensors   75, 110, 169, 178, 304, 344, 417, 430, 572
  applications   211
  axis   74
  box   618
  calibration   74
  computation/data processing   285, 288, 319, 331, 428, 572

communication model   286
corruption   378
fusion algorithms   293, 345
information extracted from   580
number of sensors   285
spatial computation model   285
spatial network model   286
temporal computation model   285
temporal processing model   285
for cure monitoring of composites   114
disorder   378
distributed computing   378, 379, 572
energy-aware communication   440
  caching-based communication   441
  clustering   441
  data-centric and event-based communication   441
errors   304
fabrication, trends in   110
failures   343
fair-loss and finite duplication   377
fiber   177
flexible   622
footprint   169
fusion   17, 72, 344
  algorithm   74, 75
  library   72
information   344
  on material behavior   304
integration   169
  in AM   230
  in fiber-reinforced polymers   175, 183
  intelligent components   602
  material integration   9
  processes   170
  scenarios   233
  using additive manufacturing techniques   234
interfaces   378
measurements   302, 304
  of surface strain/surface deformation   303
monitoring   288

sensors (*continued*)
  networks   15, 345, 347, 363, 364, 387, 417, 419, 425, 551
    applications   379
    characterization   381
    distributed energy management   443
    energy distribution in   442
    no creation/causality   377
    no delivery/live-lock   378
    nodes   19, 425, 430, 432
      activity   426
      density   425
      processes   16, 259
    noise   420
    pitch   572
    placement   170
    preprocessing   344, 345
    self-organizing   319
    and sensing principles employed in SHM of composite materials   169
    in sensor clouds   363
    signal processing system   8, 344
    size   572
    Stubborn delivery   377
    systems   233
      for functionality reason   10
      integration   222
    technologies, in material-integrated sensing   8
    transfer functions   73
sensors as a service (SaaS)   375
sensory   595. *see also* sensors
  CFRP elements   597
  CFRP structure   599
  data   533
  networks   603
  yarns   155
Separation by IMplantation of Oxygen (SIMOX)   88
sequential data processing   287
serial communication interface (SCI)   356, 357
serial infrared (SIR)   360
serial, parallel, and mixed system designs   346
serial peripheral interface (SPI)   357
serial wired communication   357
service announcement messages (SAM)   359
service loop, functions in   391
SeSAm simulator   410
S/H amplifier (SHA)   273
shape-changing interfaces   632, 633, 636
shape memory alloy (SMA)   481, 596, 634
  actuation, thermal control   596
  actuators   596
  applications 596
super-elasticity 596
  wires 596
shear force sensing   573
sheet lamination   227
short fiber-reinforced polymers   170
shrinkage operator   313
shutters   633
sigma-delta ADCs   268, 270
sigma-delta modulator, frequency domain model   272
sigma-delta waveforms, analog input voltage   271
signal characteristics   260
signal communication yarns   155
signal conversion   275
signal processing algorithms   287
signal to noise ratio (SNR)
  vs. oversampling ratio   273
signal-to-noise ratio (SNR)   112, 267, 344
silicon   87
  makers   57
  material   85
  MEMS   85
  silicon dioxide   86, 91
  silicon IC   85
  silicon nitride   86, 91
  springs   87
  wafer   84, 87, 88
silicon-on-insulator (SOI)   88
silver nanoparticle-based inks   245
simulation   419, 430, 444, 445
  environment   74
  levels   68
  methods   165

order reduction 58
platform 75
results in static conditions 75
Simulink framework 430
Simulink tool 60
single-input touch classification, operations needed 584
single-mode fibers 110
single-point-of-failure (SPoF) 18
single-process sensor integration solutions 245
single-sensor-based computations 286
single sensor processing architecture 344
single wall carbon nanotubes (SWCNTs) 454
singular value decomposition (SVD) 583
skin integration, into iCub forearm 578
skin mechanical structure 573
materials 573
transducers 573
slip detection 581
SMAC platform 56, 60, 75
ADS and thermal simulation 63
Automated EM - Circuit Cosimulation in ADS 64
(Co)Simulation Levels and the Design-Domains Matrix 67
EMPro Extension and ADS Integration 64
HIF Suite Toolsuite 65
MEMS+ platform 66
platform overview 61, 62
simulation and cosimulation in 68
System C-SystemVue Cosimulation 61
smart algorithmic energy management (SAEM) 21
smart bed linen capable, of detecting moisture 156
smart communication 440
Smart Cut 88
smart dust 6
smart electronic systems, design challenges 56
smart fabrics 352

smart fabrics and interactive textile (SFIT) 352, 480
SmartLayer™ approach 177
SmartLayer™ patches 560
SmartLayer™ sensors 550
Smart-Material Corp. 204
smart materials 601
smart phones 83, 639
smart products 616
smart sensor network 17
smart sensor nodes 17, 259
architecture 260
material-integrated network 6
passive sensors to networks 19
smart sensors 19
smart system 55, 57
codesign 57
design 56
SnAgCu (SAC) 138
soap bubble interface 637
Society of Automotive Engineers (SAE) 535
soft-shrinkage iteration 312
software agent technologies 58
autonomy 390
properties 389
reactivity proactiveness 390
social ability 390
Software-as-a-Service (SaaS) 229
software support 39
software technology 31
solar energy harvesting 451
solid oxide fuel cell (SOFC) 468
solid-state electrolyte 463
sparse matrix techniques 58
spatial parallelism 294
spatial temperature gradient 451
Special Interest Group (SIG) 357, 359
spectral modulation 110
spin-on polyimide allows 148
sponge architecture 463
spray-coated sensors 602
Sprout I/O 633, 634
stacked resin-soaked plies 173
starting yield stress 207
state-transition (SN) Petri net 428
static current 426

Index

static learning 332
static redundancy 347
statistical pattern recognition 331
stepped-index fibers 109
stereolithography 218
stiffness tensor 302, 308
STMicroelectronics 72
store and forward (SF) algorithm 371
stored energy 452
strain 110, 451
  energy harvesters 481
  gauges 319
  measurement 419
  space, performance degradation model 210
stray capacitances 574
stream-based data communication 375
stress 101
  boundary conditions 303
stretchability 13, 571, 574
  circuits
    two-step molding process 147
  cycles 147
  electrical interconnections fabrication process 148
  flexible networks 180
  interposers 156
  LED 151
    microdisplay 151
  MEH devices 486
  modules 154
  polymer 146
  substrate fabrication, on rigid carrier 146
  system 145
structural health monitoring (SHM) 5, 107, 167, 179, 531, 595
  advantages and drawbacks 545
  aerospace fatigue design principles 538
  aerospace industry 556
  algorithmic and information level hierarchy 282
  application areas and exemplary case studies 557
  areas and application and case studies 555
  categories 534
  CBM-skip 539
  challenge 542
  civil engineering 556
  classes and types of sensors 543
  classification 540
  communication 550
  component-wise breakdown 541
  concepts 168
  condition-based maintenance (CBM) 539
  cyclic and nontypical 531
  damage tolerance 537
  data evaluation approaches 553
    and principles 552
  development trends from simple 563
  direct operating cost factors in commercial aircraft operation 539
  distributed and parallel 389
  full size representative commercial aircraft door 560
  fundamental axioms 532
  global and local 546
  helicopter fuselage panel with stiffeners 549, 550
  implications of material integration 561
  information expected 534
  Lamb waves 548
  listening sensor system 545
  maintenance scheduling 538
  methodologies 107
  motivations for implementation 536, 537
  organization of 536
  piezoelectric sensors 545
  railway industry 556
  sensor and actuator elements 542
  sensor networks 534
  sensors uses 548
  solutions 168
  supported scheduled maintenance 539
  system levels of capability 533
  wind energy plants 556

successive approximation register
    (SAR)   273
  conversion method   277
  SAR-ADC architecture   279
supercapacitors (SC)   453
supervised machine learning   331
supporting remote collaboration   636
support vector machines (SVM)   332
surface deformation
    measurements   303
surface integration
  on nonplanar   239
surface micromachining process   96
surface mount devices (SMD)   145
surface mount technology (SMT)   138
  component assembly   139
surface strain sensor measurements   320
switched circuit network   368
switched networks   368, 369
  vs. message passing   368
switching frequency   426
symposium, on computational
    fabrication   638
synchronous serial communication
    interface standards I²C (I²S)   357
synergy   416
syntactic foam, reinforced with glass
    fibers   121
SystemC-AMS   58
  conversions   72
  modules   72
SystemC module   72, 75, 76
  SystemVue cosimulation   72
system design flow   59
system development methodologies   7
system development tools   7
system flexibility   571
system-in-foil packaging
    approaches   14
system in package (SiP)   4, 56
system on chip (SoC)   4, 20
  design language   344
  hardware implementations   373
System-on-Package (SiP)   77
system operation adaptation   49
  adaptation of systems objectives   49
  analysis of current situation   49

of systems behavior   50
system partitioning, of sensor node   70
system reliability   347

t

tactile information processing   579
tactile sensing system
  design requirements   573
  technologies, comparison study   576
tactile sensors   575
  based on piezoelectric materials   574
tailored fiber placement (TFP)   174, 182
Tangible Bits paper   631
tangible user interface (TUI)   631, 639
TCP/IP   367
temperature compensation   74
temperature electronics material,
    exemplary overview   11
temperature insensitivity   119
temperature-sensitive devices   603
temperature sensors   83, 96
temporal parallelization rule   293
tensile strain monitoring   573
tensile strength (MPa)
  von Mises stress distribution   208
tensorial kernel approaches   582
Terfenol-D   493
testing machine
  stretch-drawing of rotationally
      symmetric cups   206
tetramethylammonium hydroxide
    (TMAH)   90
textile-based circuits   153
  applications   155
  electronic module integration
      technology   154
theoretic entropy   334
thermal compression bonding
    techniques   95
thermal diffusivity   305
thermal energy harvesters   450, 451
  TEGs, implementation of   499
  thermal-to-electrical energy
      conversion techniques   501
  thermoelectric generators   496
  thermoelectric materials/
      efficiency   499

thermal expansion coefficients 119
thermal loads 10
  production processes, qualitative comparison of 10
thermal power dissipation 426
thermal production processes 12
thermal simulation 55
thermal stability 10
thermal system 450
thermal-to-electrical energy conversion 499
thermocouples 186
thermoelectric generator (TEG) 451, 496
  device 499, 500
    thermocouple (TC) 497
    operation principle 498
thermoelectric (TE) materials 497
  historical progression 500
thermoelectrics community 499
thermophotovoltaics 502
thermoplastic elastomer (TPE) 152
  film 153
thermoplastic matrix 173
thermoplastic polyurethane (TPU) 152
thermoset polymer 218
thermoset resin 164
thin film Cu deposition techniques for electrical interconnectivity 141
thin film ("2D") microbatteries 462
thin film stretchable interconnections fabrication process 149
thin-walled workpieces 597
three-dimensional microbattery concepts 463
  future prospects 466
through-silicon vias (TSVs) 56
Tikhonov functional 309, 312, 313, 314, 315, 317, 318
Tikhonov minimization problem 311
Tikhonov regularization 306, 308, 309, 315, 316, 317, 318, 319, 320, 321, 322, 324, 326
Timed Data Flow (TDF) 72
toolbox 430
  for energy analysis 430
    energy consumer block 434
    energy converter blocks 433
    energy source block 432
    energy storage block 433
  for energy, simulation 430
tool prototype 600
total internal reflection 125
touch sensors 83
TP blanks, automated stamping of 175
TPM-modules 177
track-and-hold circuit 262
traditional machine learning 415
transduction technologies 573, 574
transformations (FFT) analysis 212, 287
transistor technologies 426
transition metal oxide 457
transmission power 356
transversal fiber orientation 179
triboelectric nanogenerators (TENGs) 488, 492, 496
  implementation, material selection 493
tuple in database 420
two-dimensional mesh-grid topology 417
two-dimensional sensor network 340

u

UDP/IP protocol stack 367
ultralow power circuit techniques 180
ultrasonic probing approach 547
ultrasonic transducers 167
ultrasonic waves 168, 203
ultrasound reflection sensing 183
ultrasound signal 167
ultrasound transducer 184
ultrathin chip package (UTCP) 140, 141
  carrier 143
  packages 143
  packaging technology 141
ultraviolet (UV) light 89
UML's sensor deployment 618
Universal Marine Gateway (UMG) 618
  closed/ UMG open view on hardware 618
Universal Serial Bus (USB) 357

unsupervised learning 331, 415
user interfaces 632, 638
UV stability 146

V

vacuum-assisted resin transfer molding (VARTM) 165
 sequence 174
vacuum-based thin film deposition techniques 500
vacuum forming process 152
vacuum infusion processes 173
validation 56, 75
vat photopolymerization 227
VDI guideline 2206 38
very fast decision tree (VFDT) 333
vibration-based analyses 287, 548, 550
vibration detection 573
vibration energy 493
vibration (kinetic) energy harvester (VEH) 481
 devices 495
 electromagnetic 489, 491
 electrostatic 490, 491
 piezoelectric 489
 technology 496
vibration tests 127
Villari effect 451, 493
virtual machine (VM) 18, 296
 architectures 295
 monitors 297
viscoelasticity 87
viscosity 165, 457, 623
vision hypothesizes 632
visual sensor network 389
VLSI design 480
V-Model 39
voided charged polymers 489
volume integration 240
von Mises stress distribution 207
Voxel8 243, 244

W

Wafer bonding 95
Wafer-scale implementations 492
waveguide 110
Wearable ECG patch containing 144
WEB data taxonomy 385
Web pages 636
wedge geometries 121
wet etching 91
Wh algorithms 372
Wheatstone bridge circuit 98
whispering gallery mode (WGM) 119
 sensor instrumentation 119
wide area networks (WAN) 380
WiFi stations 506
wind energy plants 534
 rotor blades 555
wire-bonded piezoresistive pressure sensor chip 185
wired communication 356, 427
 electrical communication
  single-ended (SE) 354
 hardware 356
wired solutions 603
wireless access in vehicular environment (WAVE) 358
wireless body area network (WBAN) 380, 480
wireless (RΓ) communication 358, 427, 551
 hardware 358
  application examples 359
  standards 358
 solutions 603
wireless networks 377
wireless power transfer 480
wireless sensing, electronics for 181
wireless sensor module 507
 powered by RF energy harvester 503
wireless sensor network (WSN) 352, 479
wireless sensors
 low-power VLSI design 480
 nodes 479
wireless transmission 440
WLAN (Wi-Fi) standard IEEE 802.11 358
World Wide Web 33
wormhole (Wh) routing 372

x

x-ray
 analysis 208
 lithography 89

y

Young's modulus 12, 13

z

ZeroN prototype 632
ZigBee Alliance 359
zinc–carbon battery 459
ZnO nanorods, flexible
  nanogenerator 487
ZnO nanowires 486